CALCULUS IN CONTEXT

CALCULUS IN CONTEXT

THE FIVE COLLEGE CALCULUS PROJECT

JAMES CALLAHAN
Smith College

DAVID COX
Amherst College

KENNETH HOFFMAN
Hampshire College

DONAL O'SHEA
Mount Holyoke College

HARRIET POLLATSEK
Mount Holyoke College

LESTER SENECHAL
Mount Holyoke College

W. H. Freeman and Company
New York

Library of Congress Cataloging-in-Publication Data

Calculus in context : the Five College Calculus Project / James
Callahan . . . [et al.].
    p. cm.
    Includes index.
    ISBN 0-7167-2630-0
    1. Calculus.   II. Callahan, James.
    QA303.C1757 1995   94-45886
    515'.35--dc20   CIP

Printed in the United States of America

    1 2 3 4 5 6 7 8 9 0   H C   9 9 8 7 6 5

Contents

Preface

We tell our "creation story" and then describe how it led to this text, spelling out our starting points, our curricular goals, our functional goals, and our view of the impact of technology.

Origins. The story of the Five College Calculus Project began almost thirty years ago, when the Five Colleges were only Four: Amherst, Mount Holyoke, Smith, and the large Amherst campus of the University of Massachusetts. These four resolved to create a new institution that would be a site for educational innovation at the undergraduate level; by 1970, Hampshire College was enrolling students and enlisting faculty.

Early in their academic careers, Hampshire students grapple with primary sources in all fields—in economics and ecology, as well as in history and literature. And journal articles don't shelter their readers from home truths: if a mathematical argument is needed, it is used. In this way, students in the life and social sciences found, sometimes to their surprise and dismay, that they needed to know calculus if they were to master their chosen fields. Yet the calculus they needed was not, by and large, the calculus that was actually being taught. The journal articles dealt directly with the relation between quantities and their rates of change—in other words, with differential equations.

Confronted with a clear need, these students asked for help. By the mid-1970s, Michael Sutherland and Kenneth Hoffman were teaching a course for them. The core of the course was calculus, but calculus as it is *used* in contemporary science. Mathematical ideas and techniques grew out of scientific questions. Given a process, students had to recast it as a model; most often, the model was a set of differential equations. To solve the differential equations, they used numerical methods implemented on a computer.

The course evolved and prospered quietly at Hampshire. More than a decade passed before several of us at the other four institutions paid some attention to it. We liked its fundamental premise, that differential equations belong at the center of calculus. What astounded us, though, was the revelation that differential equations could really *be* at the center—thanks to the use of computers.

This book is the result of our efforts to translate the Hampshire course for a wider audience. The typical student in calculus has not been driven to study calculus to come to grips with his or her own scientific questions—as those pioneering students had. If calculus is to emerge organically in the minds of the larger student population, a way must be found to involve that population in a spectrum of scientific and mathematical questions. Hence, calculus *in context*. Moreover, those contexts must be understandable to students with no special scientific training, and the mathematical issues they raise must lead to the central ideas of the calculus—to differential equations, in fact.

Coincidentally, the country turned its attention to the undergraduate science curriculum, and it focused on the calculus course. The National Science Foundation created a program to support calculus curriculum development. To carry out our plans, we requested funds for a five-year project; we were fortunate to receive the only multi-year curriculum development grant awarded in the first year of the NSF program. This text is the outcome of our effort.

Designing the curriculum. We believe that calculus can be for students what it was for Euler and the Bernoullis: a language and a tool for exploring the whole fabric of science. We also believe that much of the mathematical depth and vitality of calculus lies in connections to other sciences. The mathematical questions that arise are compelling in part because the answers matter to other disciplines. We began our work with a "clean slate," *not* by asking what parts of the traditional course to include or discard. Our starting points are thus our summary of what calculus is really about. Our curricular goals are what we aim to convey about the subject in the course. Our functional goals describe the attitudes and behaviors we hope our students will adopt in using calculus to approach scientific and mathematical questions.

Starting Points

- Calculus is fundamentally a way of dealing with functional relationships that occur in scientific and mathematical contexts. The techniques of calculus must be subordinate to an overall view of the questions that give rise to these relationships.

- Technology radically enlarges the range of questions we can explore and the ways we can answer them. Computers and graphing calculators are much more than tools for teaching the traditional calculus.
- The concept of a dynamical system is central to science. Therefore, differential equations belong at the center of calculus, and technology makes this possible *at the introductory level.*
- The process of successive approximation is a key tool of calculus, even when the outcome of the process—the limit—cannot be explicitly given in closed form.

Curricular Goals

- Develop calculus in the context of scientific and mathematical questions.
- Treat systems of differential equations as fundamental objects of study.
- Construct and analyze mathematical models.
- Use the method of successive approximations to define and solve problems.
- Develop geometric visualization with hand-drawn and computer graphics.
- Give numerical methods a more central role.

Functional Goals

- Encourage collaborative work.
- Enable students to use calculus as a language and a tool.
- Make students comfortable tackling large, messy, ill-defined problems.
- Foster an experimental attitude towards mathematics.
- Help students appreciate the value of approximate solutions.
- Teach students that understanding grows out of working on problems.

Impact of Technology

- Differential equations can now be solved numerically, so they can take their rightful place in the introductory calculus course.
- The ability to handle data and perform many computations makes exploring messy, real-world problems possible.
- Since we can now deal with credible models, the role of modeling becomes much more central to the subject.

By studying the text you can see how we have pursued the curricular goals. Each goal is addressed within the first chapter, which begins with questions about describing and analyzing the spread of a contagious disease. A model is built: a model that is actually a system of coupled nonlinear differential equations. We then begin a numerical assault on those equations, and the door is opened to a solution by successive approximations.

Our implementation of the functional goals is also evident. The text has many more words than the traditional calculus book—it is a book to be read. The exercises make unusual demands on students. Most are not just variants of examples that have been worked in the text. In fact, the text has rather few "template" examples.

Shifts in emphasis. It will also become apparent to you that the text reflects substantial shifts in emphasis in comparison to the traditional course. Here are some of the most striking:

How the Emphasis Shifts:

Increase	Decrease
concepts	techniques
geometry	algebra
graphs	formulas
brute force	elegance
numerical	closed-form
solutions	solutions

Because we all value elegance, let us explain what we mean by "brute force." Euler's method is a good example. It is a general method of wide applicability. Of course when we use it to solve a differential equation like $y'(t) = t$, we are using a sledgehammer to crack a peanut. But at least the sledgehammer *works.* Moreover, it works with coconuts (like $y' = y(1 - y/10)$), and it will even knock down a house (like $y' = \cos^2(t)$). Students also see the elegant special methods that can be invoked to solve $y' = t$ and $y' = y(1 - y/10)$ (separation of variables and partial fractions are discussed in Chapter 11), but they understand that they are fortunate indeed when a real problem will succumb to such methods.

Audience. Our curriculum is not aimed at a special readership. On the contrary, we think that calculus is one of the great bonds that unifies science. All students should have an opportunity to see how the language and tools of calculus help forge that bond. We emphasize that this is not a "service" course of calculus "with applications," but rather a course rich in mathematical ideas that will serve all students well, including mathematics majors.

The student population in the first semester course is especially diverse. In fact, because many students take only one semester, the first six chapters stand alone as a reasonably complete course. We have also tried to present the contexts of broadest interest first. The emphasis on the physical sciences increases in the second half of the book.

Handbook and supplements. We have prepared an *Instructor's Handbook* based on our experiences and those of colleagues at other schools with specific suggestions for how to use the text. We urge prospective instructors to consult it, because this course differs substantially from the calculus courses most of us have learned from and taught in the past. The *Handbook* also includes solutions to all text exercises, sample syllabi, and examples of tests and quizzes. Some software programs are available at no charge for use with this text. These include graphing and differential equations utilities. The *Handbook* explains how to obtain them.

Acknowledgments

Certainly this book would not have been possible without the support of the National Science Foundation and of Five Colleges, Inc. We particularly want to thank Louise Raphael who, as the first director of the calculus program at the National Science Foundation, had faith in us and recognized the value of what had already been accomplished at Hampshire College when we began our work. Five College coordinators Conn Nugent and Lorna Peterson supported and encouraged our efforts, and Five College treasurer and business manager Jean Stabell has assisted us in countless ways throughout the project.

We are very grateful to the members of our Advisory Board: to Peter Lax, for his faith in us and his early help in organizing and chairing the Board; to Solomon Garfunkel, for his advice on politics and publishing; to Barry Simon, for using our text and giving us his thoughtful and imaginative suggestions for improving it; to John Neuberger, for his passionate convictions; to Gilbert Strang, for his support of a radical venture; to John Truxal, for his detailed commentaries and insights into the world of engineering.

Among our colleagues, James Henle of Smith College deserves special thanks. Besides his many contributions to our discussions of curriculum and pedagogy, he developed the computer programs that have been so valuable for our teaching: Graph, Slinky, Superslinky, and Tint. Jeff Gelbard and Fred Henle extended Jim's programs to the Macintosh and to DOS Windows and X Windows. All this software is available on anonymous ftp at emmy.smith.edu. Mark Peterson, Robert Weaver, and David Cox also developed software that has been used by our students.

Several of our colleagues made substantial contributions to our frequent editorial conferences and helped with the writing of early drafts. We offer thanks to David Cohen, Robert Currier, and James Henle at Smith; David Kelly at Hampshire; and Frank Wattenberg at the University of

Massachusetts. Mary Beck, who is now at the University of Virginia, gave heaps of encouragement and good advice as a co-teacher of the earliest version of the course at Smith. Anne Kaufmann, an Ada Comstock Scholar at Smith, assisted us with extensive editorial reviews from the student perspective.

Two of the most significant new contributions to this edition are the appendix for graphing calculators and a complete set of solutions to all the exercises. From the time he first became aware of our project, Benjamin Levy has been telling us how easy and natural it would be to adapt our Basic programs for graphing calculators. He has always used them when he taught *Calculus in Context*, and he created the appendix that contains translations of our programs for most of the graphing calculators in common use today. Lisa Hodsdon, Diane Jamrog, and Marcia Lazo have worked long hours over an entire summer to solve all the exercises and to prepare the results as LATEX documents for inclusion in the *Instructor's Handbook*. We think both these contributions do much to make the course more useful to a wider audience.

We appreciate the contributions of our colleagues who participated in numerous debriefing sessions at semester's end and gave us comments on the evolving text. We thank George Cobb, Giuliana Davidoff, Alan Durfee, Janice Gifford, Mark Peterson, Margaret Robinson, and Robert Weaver at Mount Holyoke; Michael Albertson, Ruth Haas, Mary Murphy, Marjorie Senechal, Patricia Sipe, and Gerard Vinel at Smith. We also learned from the reactions of our colleagues in other disciplines who participated in faculty workshops on Calculus in Context.

We profited a great deal from the comments and reactions of early users of the text. We extend our thanks to Marian Barry at Aquinas College, Peter Dolan and Mark Halsey at Bard College, Donald Goldberg and his colleagues at Occidental College, Benjamin Levy at Beverly High School, Joan Reinthaler at Sidwell Friends School, Keith Stroyan at the University of Iowa, and Paul Zorn at St. Olaf College. Later users who have helped us are Judith Grabiner and Jim Hoste at Pitzer College; Allen Killpatrick, Mary Scherer, and Janet Beery at the University of Redlands; and Barry Simon at Cal Tech.

Dissemination grants from the NSF have funded regional workshops for faculty planning to adopt *Calculus in Context*. We are grateful to Donald Goldberg, Marian Barry, Janet Beery, and to Henry Warchall of the University of North Texas for coordinating workshops.

We owe a special debt to our students over the years, especially those who assisted us in teaching, but also those who gave us the benefit of their thoughtful reactions to the course and the text. Seeing what they were learning encouraged us at every step.

We continue to find it remarkable that our text is to be published the way we want it, not softened or ground down under the pressure of anonymous reviewers seeking a return to the mean. All this is due to the bold and generous stance of W. H. Freeman and Company. Robert Biewen, its president, understands —more than we could ever hope—what we are trying to do, and he has given us his unstinting support. Our acquisitions editors, Jeremiah Lyons and Holly Hodder, have inspired us with their passionate belief that our book has something new and valuable to offer science education. Christine Hastings, our project editor, has shown heroic patience and grace in shaping the book itself against our often contrary views. We thank them all.

To the Student

This book is different. In a typical high school math text, each section has a "technique" that you practice in a series of exercises very like the examples in the text. In this course you will be learning to use calculus both as a tool and as a language. As with any other language, a certain amount of time will need to be spent learning and practicing the formal rules. For instance, the conjugation of *être* must be second nature to you if you are to be able to read French. In calculus, too, there are a number of manipulations that must become automatic so that you can focus clearly on the content of what is being said. It is important to realize that becoming good at these techniques is not the goal of learning calculus any more than becoming good at conjugations is the goal of learning French.

Up to now, most of the problems you have met in math classes have had definite answers such as "17," or "the circle with radius 1.75 and center at (2,3)." Such definite answers are satisfying and even comforting. Yet many interesting and important questions, like "How far is it to the planet Pluto?" or "How many people are there with sickle-cell anemia?" or "What are the solutions to the equation $x^5 + x + 1 = 0$?" can't be answered exactly. Instead, we have ways to *approximate* the answers, and the more time or money or both we are willing to spend, the better our approximations may be. Although many calculus problems have exact answers, such problems often tend to be special or atypical in some way. Therefore, as you learn how to deal with these "nice" problems, you will also be developing ways of making good approximations to the solutions of the less well-behaved (and more common!) problems.

The computer or the graphing calculator is a tool that you will need for this course. We don't assume that you know anything about this technology ahead of time. Everything necessary is covered as we go along.

You can't learn mathematics simply by reading or watching others. The only way to really learn the material is to work on problems yourself. Grappling with problems yields understanding in much the same way that writing and discussion promote clarity of thought. It allows you to identify strengths and weaknesses in your grasp of the material.

One of the most important intellectual skills you can develop is that of exploring questions on your own. Don't simply shut your mind down when you come to the end of an assigned problem. These problems have been designed to prod your thinking and to get you wondering about the concepts being explored. See if you can invent and answer variations on the problem. Does the problem suggest other questions? The ability to ask good questions of your own devising is as important as being able to answer questions posed by others.

We encourage you to work with others on the exercises. Two or three of you working on a problem will often accomplish much more than any of you would working alone. You will stimulate one another's imaginations, combine differing insights into a greater whole, and keep up one another's spirits in the frustrating times. This is particularly effective if you first spend time working on the material alone. Many students find it helpful to schedule a regular time together to work on problems.

Above all, take time to pause and admire the beauty and power of what you are learning. Aside from its utility, calculus is one of the most elegant and richly structured creations of the human mind. It deserves to be profoundly admired on those grounds alone. Enjoy!

CHAPTER 1

A CONTEXT FOR CALCULUS

Calculus gives us a language to describe how quantities are related to one another, and it gives us a set of computational and visual tools for exploring those relationships. Usually, we want to understand how quantities are related in the context of a particular problem—it might be in chemistry, or public policy, or mathematics itself. In this chapter we take a single context—an infectious disease spreading through a population—to see how calculus emerges and how it is used.

➢ 1.1 THE SPREAD OF DISEASE

Making a Model

Many human diseases are contagious: you "catch" them from someone who is already infected. Contagious diseases are of many kinds. Smallpox, polio, and plague are severe and even fatal, while the common cold and the childhood illnesses of measles, mumps, and rubella are usually relatively mild. Moreover, you can catch a cold over and over again, but you get measles only once. A disease like measles is said to "confer immunity" on someone who recovers from it. Some diseases have the potential to affect large segments of a population; they are called *epidemics* (from the Greek words *epi*, upon + *demos*, the people). *Epidemiology* is the scientific study of these diseases.

Some properties of contagious diseases

An epidemic is a complicated matter, but the dangers posed by contagion—and especially by the appearance of new and uncontrollable diseases—compel

The idea of a
mathematical model

us to learn as much as we can about the nature of epidemics. Mathematics offers a very special kind of help. First, we can try to draw out of the situation its essential features and describe them mathematically. This is calculus as *language*. We substitute an "ideal" mathematical world for the real one. This mathematical world is called a **model**. Second, we can use mathematical insights and methods to analyze the model. This is calculus as *tool*. Any conclusion we reach about the model can then be interpreted to tell us something about the reality.

To give you an idea how this process works, we'll build a model of an epidemic. Its basic purpose is to help us understand the way a contagious disease spreads through a population—to the point where we can even predict what fraction falls ill, and when. Let's suppose the disease we want to model is like measles. In particular,

- it is mild, so anyone who falls ill eventually recovers;
- it confers permanent immunity on every recovered victim.

In addition, we will assume that the affected population is large but fixed in size and confined to a geographically well-defined region. To have a concrete image, you can imagine the elementary school population of a big city.

At any time, that population can be divided into three distinct classes:

Susceptible: those who have never had the illness and can catch it;
Infected: those who currently have the illness and are contagious;
Recovered: those who have already had the illness and are immune.

The quantities that our
model analyzes

Suppose we let S, I, and R denote the number of people in each of these three classes, respectively. Of course, the classes are all mixed together throughout the population: on a given day, we may find persons who are susceptible, infected, and recovered in the same family. For the purpose of organizing our thinking, though, we'll represent the whole population as separated into three "compartments" as in the following diagram:

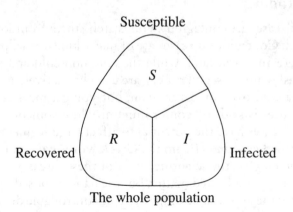

The whole population

The goal of our model is to determine what happens to the numbers S, I, and R over the course of time. Let's first see what our knowledge and experience of childhood diseases might lead us to expect. When we say there is a "measles outbreak," we mean that there is a relatively sudden increase in the number of cases, and then a gradual decline. After some weeks or months, the illness virtually disappears. In other words, the number I is a **variable**; its value changes over time. One possible way that I might vary is shown in the following graph.

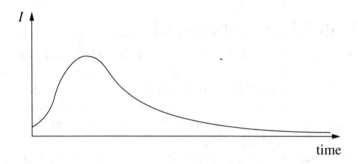

During the course of the epidemic, susceptibles are constantly falling ill. Thus we would expect the number S to show a steady decline. Unless we know more about the illness, we cannot decide whether everyone eventually catches it. In graphical terms, this means we don't know whether the graph of S levels off at zero or at a value above zero. Finally, we would expect more and more people in the recovered group as time passes. The graph of R should therefore climb from left to right. The graphs of S and R might take the following forms:

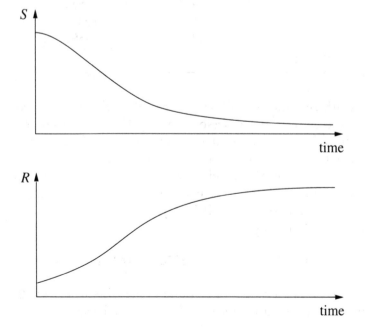

Some quantitative
questions that the
graphs raise

While these graphs give us an idea about what might happen, they raise some new questions, too. For example, because there are no scales marked along the axes, the first graph does not tell us how large I becomes when the infection reaches its peak, nor when that peak occurs. Likewise, the second and third graphs do not say how rapidly the population either falls ill or recovers. A good model of the epidemic should give us graphs like these and it should also answer the quantitative questions we have already raised—for example: When does the infection hit its peak? How many susceptibles eventually fall ill?

A Simple Model for Predicting Change

Suppose we know the values of S, I, and R today; can we figure out what they will be tomorrow, or the next day, or a week or a month from now? Basically, this is a problem of predicting the future. One way to deal with it is to get an idea how S, I, and R are *changing*. To start with a very simple example, suppose the city's Board of Health reports that the measles infection has been spreading at the rate of 470 new cases per day for the last several days. If that rate continues to hold steady and we start with 20,000 susceptible children, then we can expect 470 fewer susceptibles with each passing day. The immediate future would then look like this:

Days after today	Accumulated number of new infections	Remaining number of susceptibles
0	0	20000
1	470	19530
2	940	19060
3	1410	18590
⋮	⋮	⋮

Knowing rates, we can
predict future values

Of course, these numbers will be correct only if the infection continues to spread at its present rate of 470 persons per day. If we want to follow S, I, and R into the future, our example suggests that we should pay attention to the **rates** at which these quantities change. To make it easier to refer to them, let's denote the three rates by S', I', and R'. For example, in the illustration above, S is changing at the rate $S' = -470$ persons per day. We use a minus sign here because S is *decreasing* over time. If S' stays fixed we can express the value of S after t days by the following formula:

$$S = 20000 + S' \cdot t = 20000 - 470\,t \text{ persons.}$$

Check that this gives the values of S found in the table when $t = 0, 1, 2,$ or 3. How many susceptibles does it say are left after 10 days?

Our assumption that $S' = -470$ persons per day amounts to a mathematical characterization of the susceptible population—in other words, a model! Of course it is quite simple, but it led to a formula that told us what value we could expect S to have at any time t.

The equation $S' = -470$ is a model

The model will even take us backwards in time. For example, two days ago the value of t was -2; according to the model, there were

$$S = 20000 - 470 \times -2 = 20940$$

susceptible children then. There is an obvious difference between going backwards in time and going forwards: we already know the past. Therefore, by letting t be negative we can generate values for S that can be checked against health records. If the model gives good agreement with *known* values of S, we become more confident in using it to predict *future* values.

To predict the value of S using the rate S' we clearly need to have a starting point—a known value of S from which we can measure changes. In our case that starting point is $S = 20000$. This is called the **initial value** of S, because it is given to us at the "initial time" $t = 0$. To construct the formula $S = 20000 - 470t$, we needed to have an initial value as well as a rate of change for S.

Predictions depend on the initial value, too

In the following pages we will develop a more complex model for all three population groups that has the same general design as this simple one. Specifically, the model will give us information about the rates S', I', and R', and with that information we will be able to predict the values of S, I, and R at any time t.

The Rate of Recovery

Our first task will be to model the recovery rate R'. We look at the process of recovering first, because it's simpler to analyze. An individual caught in the epidemic first falls ill and then recovers—recovery is just a matter of time. In particular, someone who catches measles has the infection for about fourteen days. So if we look at the entire infected population today, we can expect to find some who have been infected less than one day, some who have been infected between one and two days, and so on, up to fourteen days. Those in the last group will recover today. In the absence of any definite information about the fourteen groups, let's assume they are the same size. Then 1/14-th of the infected population will recover today:

$$\text{today the change in the recovered population} = \frac{I \text{ persons}}{14 \text{ days}}.$$

There is nothing special about today, though; I has a value at any time. Thus we can make the same argument about any other day:

$$\text{every day the change in the recovered population} = \frac{I \text{ persons}}{14 \text{ days}}.$$

This equation is telling us about R', the rate at which R is changing. We can write it more simply in the form

$$R' = \frac{1}{14}I \text{ persons per day.}$$

We call this a **rate equation**. Like any equation, it links different quantities together. In this case, it links R' to I. The rate equation for R is the first part of our model of the measles epidemic.

Are you uneasy about our claim that one-fourteenth of the infected population recovers every day? You have good reason to be. After all, during the first few days of the epidemic almost no one has had measles the full fourteen days, so the recovery rate will be much less than 1/14 persons per day. About a week before the infection disappears altogether there will be no one in the early stages of the illness. The recovery rate will then be much greater than 1/14 persons per day. Evidently our model is not a perfect mirror of reality!

Don't be particularly surprised or dismayed by this. Our goal is to gain insight into the workings of an epidemic and to suggest how we might intervene to reduce its effects. So we start off with a model which, while imperfect, still captures some of the workings. The simplifications in the model will be justified if we are led to inferences which help us understand how an epidemic works and how we can deal with it. If we wish, we can then refine the model, replacing the simple expressions with others that mirror the reality more fully.

Notice that the rate equation for R' does indeed give us a tool to predict future values of R. For suppose today 2100 people are infected and 2500 have already recovered. Can we say how large the recovered population will be tomorrow or the next day? Since $I = 2100$,

$$R' = \frac{1}{14} \times 2100 = 150 \text{ persons per day}.$$

Thus 150 people will recover in a single day, and twice as many, or 300, will recover in two. At this rate the recovered population will number 2650 tomorrow and 2800 the next day.

These calculations assume that the rate R' holds steady at 150 persons per day for the entire two days. Since $R' = I/14$, this is the same as assuming that I holds steady at 2100 persons. If instead I varies during the two days we would have to adjust the value of R' and, ultimately, the future values of R as well. In fact, I *does* vary over time. We shall see this when we analyze how the infection is transmitted. Then, in chapter 2, we'll see how to make the adjustments in the values of R' that will permit us to predict the value of R in the model with as much accuracy as we wish.

Other Diseases

What can we say about the recovery rate for a contagious disease other than measles? If the period of infection of the new illness is k days, instead of 14, and if we assume that $1/k$ of the infected people recover each day, then the new recovery rate is

$$R' = \frac{I \text{ persons}}{k \text{ days}} = \frac{1}{k} I \text{ persons per day.}$$

If we set $b = 1/k$, we can express the recovery rate equation in the form

$$R' = bI \text{ persons per day.}$$

The constant b is called the **recovery coefficient** in this context.

Let's incorporate our understanding of recovery into the compartment diagram. For the sake of illustration, we'll separate the three compartments. As time passes, people "flow" from the infected compartment to the recovered. We represent this flow by an arrow from I to R. We label the arrow with the recovery coefficient b to indicate that the flow is governed by the rate equation $R' = bI$.

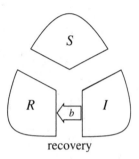

recovery

The Rate of Transmission

Since susceptibles become infected, the compartment diagram above should also have an arrow that goes from S to I and a rate equation for S' to show how S changes as the infection spreads. While R' depends only on I, because recovery involves only waiting for people to leave the infected population, S' will depend on both S and I, because transmission involves contact between susceptible and infected persons.

Here's a way to model the transmission rate. First, consider a single susceptible person on a single day. On average, this person will contact only a small fraction, p, of the infected population. For example, suppose there are 5000 infected children, so $I = 5000$. We might expect only a couple of them—let's say 2—will be in the same classroom with our "average" susceptible. So the fraction of contacts is $p = 2/I = 2/5000 = .0004$. The 2 contacts themselves can be expressed as $2 = (2/I) \cdot I = pI$ contacts per day per susceptible.

To find out how many daily contacts the *whole* susceptible population will have, we can just multiply the average number of contacts per susceptible person by the number of susceptibles: this is $pI \cdot S = pSI$.

Contacts are proportional to both S and I

Not all contacts lead to new infections; only a certain fraction q do. The more contagious the disease, the larger q is. Since the number of daily contacts is pSI, we can expect $q \cdot pSI$ new infections per day (i.e., to convert contacts to infections, multiply by q). This becomes aSI if we define a to be the product qp.

Recall, the value of the recovery coefficient b depends only on the illness involved. It is the same for all populations. By contrast, the value of a depends on the general health of a population and the level of social interaction between its members. Thus, when two different populations experience the same illness, the values of a could be different. One strategy for dealing with an epidemic is to alter the value of a. Quarantine does this, for instance; see the exercises.

Since each new infection decreases the number of susceptibles, we have the rate equation for S:

Here is the second piece of the *S-I-R* model

$$S' = -aSI \text{ persons per day}.$$

The minus sign here tells us that S is decreasing (since S and I are positive). We call a the **transmission coefficient**.

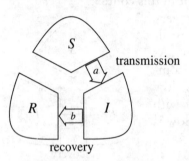

Just as people flow from the infected to the recovered compartment when they recover, they flow from the susceptible to the infected when they fall ill. To indicate the second flow let's add another arrow to the compartment diagram. Because this flow is due to the transmission of the illness, we will label the arrow with the transmission coefficient a. The compartment diagram now reflects all aspects of our model.

We haven't talked about the units in which to measure a and b. They must be chosen so that any equation in which a or b appears will balance. Thus, in $R' = bI$ the units on the left are persons/day; since the units for I are persons, the units for b must be 1/(days). The units in $S' = -aSI$ will balance only if a is measured in 1/(person-day).

The reciprocals have more natural interpretations. First of all, $1/b$ is the number of days a person needs to recover. Next, note that $1/a$ is measured in person-days (i.e., persons × days), which are the natural units in which to measure exposure. Here is why. Suppose you contact 3 infected persons for each of 4 days. That gives you the same exposure to the illness that you get from 6 infected persons in 2 days—both give 12 "person-days" of exposure. Thus, we can interpret $1/a$ as the level of exposure of a typical susceptible person.

Completing the Model

The final rate equation we need—the one for I'—reflects what is already clear from the compartment diagram: every loss in I is due to a gain in R, while every gain in I is due to a loss in S.

Here is the complete *S-I-R* model

$$S' = -aSI$$
$$I' = aSI - bI$$
$$R' = bI$$

If you add up these three rates you should get the overall rate of change of the whole population. The sum is zero. Do you see why?

You should not draw the conclusion that the only use of rate equations is to model an epidemic. Rate equations have a long history, and they have been put to many uses.

Isaac Newton (1642–1727) introduced them to model the motion of a planet around the sun. He showed that the same rate equations could model the motion of the moon around the earth and the motion of an object falling to the ground. Newton created calculus as a tool to analyze these equations. He did the work while he was still an undergraduate—on an extended vacation, curiously enough, because a plague epidemic was raging at Cambridge!

Today we use Newton's rate equations to control the motion of earth satellites and the spacecraft that have visited the moon and the planets. We use other rate equations to model radioactive decay, chemical reactions, the growth and decline of populations, the flow of electricity in a circuit, the change in air pressure with altitude—just to give a few examples. You will have an opportunity in the following chapters to see how they arise in many different contexts, and how they can be analyzed using the tools of calculus.

The following diagram summarizes, in a schematic way, the relation between our model and the reality it seeks to portray.

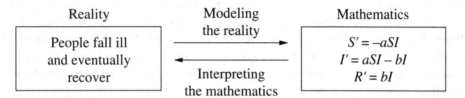

The diagram calls attention to several facts. First, the model is a part of mathematics. It is distinct from the reality being modeled. Second, the model is based on a simplified interpretation of the epidemic. As such, it will not match the reality exactly; it will be only an **approximation**. Thus, we cannot expect the values of S, I, and R that we calculate from the rate equations to give us the exact sizes of the susceptible, infected, and recovered populations. Third, the connection between reality and mathematics is a two-way street. We have already traveled one way by constructing a mathematical object that reflects some aspects of the epidemic. This is model-building. Presently we will travel the other way. First we need to get mathematical answers to mathematical questions; then we will see what those answers tell us about the epidemic. This is interpretation of the model. Before we begin the interpretation, we must do some mathematics.

The model is part of mathematics; it only approximates reality

Analyzing the Model

Now that we have a model we shall analyze it as a mathematical object. We will set aside, at least for the moment, the connection between the mathematics and the reality. Thus, for example, you should not be concerned when our calculations produce a value for S of 44,446.6 persons at a certain time—a value that will never be attained in reality. In the following analysis S is just a numerical quantity that varies with t, another numerical quantity. Using only mathematical tools we must now extract from the rate equations the information that will tell us just how S and I and R vary with t.

We already took the first steps in that direction when we used the rate equation $R' = I / 14$ to predict the value of R two days into the future (see page 6).

Rates are continually
changing; this affects
the calculations

We assumed that I remained fixed at 2100 during those two days, so the rate $R' = 2100/14 = 150$ was also fixed. We concluded that if $R = 2500$ today, it will be 2650 tomorrow and 2800 the next day.

A glance at the full S-I-R model tells us those first steps have to be modified. The assumption we made—that I remains fixed—is not justified, because I (like S and R) is continually changing. As we shall see, I actually increases over those two days. Hence, over the same two days, R' is not fixed at 150, but is continually increasing also. That means that R' becomes larger than 150 during the first day, so R will be larger than 2650 tomorrow.

The fact that the rates are continually changing complicates the mathematical work we need to do to find S, I, and R. In chapter 2 we will develop tools and concepts that will overcome this problem. For the present we'll assume that the rates S', I', and R' stay fixed for the course of an entire day. This will still allow us to produce reasonable estimates for the values of S, I, and R. With these estimates we will get our first glimpse of the predictive power of the S-I-R model. We will also use the estimates as the starting point for the work in chapter 2 that will give us precise values. Let's look at the details of a specific problem.

The Problem

Consider a measles epidemic in a school population of 50,000 children. The recovery coefficient is $b = 1/14$. For the transmission coefficient we choose $a = .00001$, a number within the range used in epidemic studies. We suppose that 2100 people are currently infected and 2500 have already recovered. Since the total population is 50,000, there must be 45,400 susceptibles. Here is a summary of the problem in mathematical terms:

Rate equations:

$$S' = -.00001SI$$
$$I' = .00001SI - I/14$$
$$R' = I/14$$

Initial values: when $t = 0$,

$$S = 45400 \qquad I = 2100 \qquad R = 2500$$

Tomorrow.   From our earlier discussion, $R' = 2100/14 = 150$ persons per day, giving us an estimated value of $R = 2650$ persons for tomorrow. To estimate S we use

$$S' = -.00001\, SI = -.00001 \times 45400 \times 2100 = -953.4 \text{ persons/day}.$$

Hence we estimate that tomorrow

$$S = 45400 - 953.4 = 44446.6 \text{ persons.}$$

Since $S + I + R = 50000$ and we have $S + R = 47096.6$ tomorrow, a final subtraction gives us $I = 2903.4$ persons. (Alternatively, we could have used the rate equation for I' to estimate I.)

The fractional values in the estimates for S and I remind us that the S-I-R model describes the behavior of the epidemic only approximately.

Several days hence.   According to the model, we estimate that tomorrow $S = 44446.6$, $I = 2903.4$, and $R = 2650$. Therefore, from the new I we get a new approximation for the value of R' tomorrow; it is

$$R' = \frac{1}{14} I = \frac{1}{14} \times 2903.4 = 207.4 \text{ persons/day.}$$

Hence, two days from now we estimate that R will have the new value $2650 + 207.4 = 2857.4$. Now follow this pattern to get new approximations for S' and I', and then use those to estimate the values of S and I two days from now.

The pattern of steps that just carried you from the first day to the second will work just as well to carry you from the second to the third. Pause now and do all these calculations yourself. See exercises 15 and 16 on page 20. If you round your calculated values of S, I, and R to the nearest tenth, they should agree with those in the following table.

Stop and do the calculations

Estimates for the first three days

t	S	I	R	S'	I'	R'
0	45400.0	2100.0	2500.0	-953.4	803.4	150.0
1	44446.6	2903.4	2650.0	-1290.5	1083.1	207.4
2	43156.1	3986.5	2857.4	-1720.4	1435.7	284.7
3	41435.7	5422.1	3142.1			

Yesterday.   We already pointed out, on page 5, that we can use our models to go *backwards* in time, too. This is a valuable way to see how well the model fits reality, because we can compare estimates that the model generates with health records for the days in the recent past.

To find how S, I, and R change when we go one day into the future we multiplied the rates S', I', and R' by a time step of $+1$. To find how they change when we go one day into the past we do the same thing, except that we must now use a time step of -1. According to the table above, the rates at time $t = 0$ (i.e., today) are

$$S' = -953.4 \qquad I' = 803.4 \qquad R' = 150.0.$$

Therefore we estimate that, one day ago,

$$S = 45400 + (-953.4 \times -1) = 45400 + 953.4 = 46353.4$$
$$I = 2100 + (803.4 \times -1) = 2100 - 803.4 = 1296.6$$
$$R = 2500 + (150.0 \times -1) = 2500 - 150.0 = 2350.0$$

Just as we would expect with a spreading infection, there are more susceptibles yesterday than today, but fewer infected or recovered. In the exercises for section 1.3 you will have an opportunity to continue this process, tracing the epidemic many days into the past. For example, you will be able to go back and see when the infection started—that is, find the day when the value of I was only about 1.

Go forward a day and then back again

There and back again. What happens when we start with tomorrow's values and use tomorrow's rates to go back one day—back to today? We should get $S = 45400$, $I = 2100$, and $R = 2500$ once again, shouldn't we? Tomorrow's values are

$$S = 44446.6 \qquad I = 2903.4 \qquad R = 2650.0$$
$$S' = -1290.5 \qquad I' = 1083.1 \qquad R' = 207.4$$

To go backwards one day we must use a time step of -1. The predicted values are thus

$$S = 44446.6 + (-1290.6 \times -1) = 45737.2$$
$$I = 2903.4 + (1083.1 \times -1) = 1820.3$$
$$R = 2650.0 + (207.4 \times -1) = 2442.6$$

These are *not* the values that we had at the start, when $t = 0$. In fact, it's worth noting the difference between the original values and those produced by "going there and back again."

	Original value	There and back again	Difference
S	45400	45737.1	337.1
I	2100	1820.3	-279.7
R	2500	2442.6	-57.4

Do you see why there are differences? We went forward in time using the rates that were current at $t = 0$, but when we returned we used the rates that were current at $t = 1$. Because these rates were different, we didn't get back where we started. These differences do not point to a flaw in the model; the problem lies with the way we are trying to extract information from the model. As we have been making estimates, we have assumed that the rates don't change over the course of a whole day. We already know that's not true, so the values that we have been getting are not exact. What this test adds to our knowledge is a way to measure just *how* inexact those values are—as we do in the table above.

The differences measure how rough an estimate is

In chapter 2 we will solve the problem of rough estimates by recalculating all the quantities ten times a day, a hundred times a day, or even more. When we do the computations with shorter and shorter time steps, we will be able to see how the estimates improve. We will even be able to see how to get values that are mathematically exact!

Delta Notation

This work has given us some insights about the way our model predicts future values of S, I, and R. The basic idea is very simple: *determine how S, I, and R change*. Because these changes play such an important role in what we do, it is worth having a simple way to refer to them. Here is the notation that we will use:

$$\Delta x \text{ stands for a \textbf{change} in the quantity } x$$

The symbol "Δ" is the Greek capital letter *delta*; it corresponds to the Roman letter "D" and stands for **difference**.

Delta notation gives us a way to refer to changes of all sorts. For example, in the table on page 11, between day 1 and day 3 the quantities t and S change by

$$\Delta t = \quad 2 \quad \text{days}$$
$$\Delta S = -3010.9 \quad \text{persons.}$$

We sometimes refer to a change as a **step**. For instance, in this example we can say there is a "t step" of 2 days, and an "S step" of -3010.9 persons. In the calculations that produced the table on page 11 we "stepped into" the future, a day at a time. Finally, delta notation gives us a concise and vivid way to describe the relation between rates and changes. For example, if S changes at the constant rate S', then under a t step of Δt, the value of S changes by

Δ stands for a change, a difference, or a step

$$\Delta S = S' \cdot \Delta t.$$

Using the computer as a tool. Suppose we wanted to find out what happens to S, I, and R after a month, or even a year. We need only repeat—30 times, or 365 times—the three rounds of calculations we used to go three days into the future. The computations would take time, though. The same is true if we wanted to do ten or one hundred rounds of calculations per day—which is the approach we'll take in chapter 2 to get more accurate values. To save our time and effort we will soon begin to use a computer to do the repetitive calculations.

A computer does calculations by following a set of instructions called a **program**. Of course, if we had to give a million instructions to make the computer carry out a million steps, there would be no savings in labor. The trick is to write a program with just a few instructions that can be repeated over and over again to do all the calculations we want. The usual way to do this is to arrange the instructions in a **loop**. To give you an idea what a loop is, we'll look at the S-I-R calculations. They form a loop. We can see the loop by making a flow chart.

The Flow Chart

We'll start by writing down the three steps that take us from one day to the next:

Step I.   Given the current values of S, I, and R, we get current S', I', and R' by using the rate equations.

$$S' = -aSI$$
$$I' = aSI - bI$$
$$R' = bI$$

Step II.   Given the current values of S', I', and R', we find the changes ΔS, ΔI, and ΔR over the course of a day by using the equations

$$\Delta S \text{ persons} = S' \frac{\text{persons}}{\text{day}} \times 1 \text{ day}$$

$$\Delta I \text{ persons} = I' \frac{\text{persons}}{\text{day}} \times 1 \text{ day}$$

$$\Delta R \text{ persons} = R' \frac{\text{persons}}{\text{day}} \times 1 \text{ day}$$

Step III.   Given the current values of ΔS, ΔI, and ΔR, we find the new values of S, I, and R a day later by using the equations

$$\text{new } S = \text{current } S + \Delta S$$
$$\text{new } I = \text{current } I + \Delta I$$
$$\text{new } R = \text{current } R + \Delta R.$$

Each step takes three numbers as **input**, and produces three new numbers as **output**. Note that the output of each step is the input of the next, so the steps flow together. The diagram below, called a **flow chart**, shows us how the steps are connected.

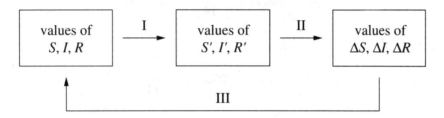

The flow chart forms a loop

The calculations form a **loop**, because the output of step III is the input for step I. If we go once around the loop, the output of step III gives us the values of S, I, and R *on the following day.* The steps do indeed carry us into the future.

Each step involves calculating three numbers. If we count each calculation as a single instruction, then it takes nine instructions to carry the values of S, I, and R one day into the future. To go a million days into the future, we need add only one more instruction: "Go around the loop a million times." In this way, a computer program with only ten instructions can carry out a million rounds of calculations!

Later in this chapter (section 1.3) you will find a real computer program that lists these instructions (for three days instead of a million, though). Study the program to see which instructions accomplish which steps. In particular, see how it makes a loop. Then run the program to check that the computer reproduces the values you already computed by hand. Once you see how the program works, you can modify it to get further information—for example, you can find out what happens to S, I, and R thirty days into the future. You will even be able to plot the graphs of S, I, and R.

A computer program will carry out the three steps

> Rate equations have always been at the heart of calculus, and they have been analyzed using mechanical and electronic computers for as long as those tools have been available. Now that small powerful computers have begun to appear in the classroom, it is possible for beginning calculus students to explore interesting and complex problems that are modeled by rate equations. Computers are changing how mathematics is done and how it is learned.

Analysis without a computer.   A computer is a powerful tool for exploring the S-I-R model, but there are many things we can learn about the model without using a computer. Here is an example.

According to the model, the rate at which the infected population grows is given by the equation

$$I' = .00001\, SI - I/14 \quad \text{persons/day.}$$

In our example, $I' = 803.4$ at the outset. This is a positive number, so I increases initially. In fact, I will continue to increase as long as I' is positive. If

I' ever becomes negative, then I decreases. So let's ask the question: when is I' positive, when is it negative, and when is it zero? By factoring out I in the last equation we obtain

$$I' = I\left(.00001\,S - \frac{1}{14}\right) \text{persons/day}.$$

Consequently $I' = 0$ if either

$$I = 0 \qquad \text{or} \qquad .00001\,S - \frac{1}{14} = 0.$$

The first possibility $I = 0$ has a simple interpretation: there is no infection within the population. The second possibility is more interesting; it says that I' will be zero when

$$.00001\,S - \frac{1}{14} = 0 \qquad \text{or} \qquad S = \frac{100000}{14} \approx 7142.9.$$

If S is *greater* than $100000/14$ and I is positive, then you can check that the formula

$$I' = I\left(.00001\,S - \frac{1}{14}\right) \text{persons/day}$$

tells us I' is positive—so I is increasing. If, on the other hand, S is *less than* $100000/14$, then I' is negative and I is decreasing. So $S = 100000/14$ represents a **threshold**. If S falls below the threshold, I decreases. If S exceeds the threshold, I increases. Finally, I reaches its peak when S equals the threshold.

The presence of a threshold value for S is purely a mathematical result. However, it has an interesting interpretation for the epidemic. As long as there are at least 7143 susceptibles, the infection will spread, in the sense that there will be more people falling ill than recovering each day. As new people fall ill, the number of susceptibles declines. Finally, when there are fewer than 7143 susceptibles, the pattern reverses: each day more people will recover than will fall ill.

If there were fewer than 7143 susceptibles in the population *at the outset*, then the number of infected would only decline with each passing day. The infection would simply never catch hold. The clear implication is that the noticeable surge in the number of cases that we associate with an "epidemic" disease is due to the presence of a large susceptible population. If the susceptible population lies below a certain threshold value, a surge just isn't possible. This is a valuable insight—and we got it with little effort. We didn't need to make lengthy calculations or call on the resources of a computer; a bit of algebra was enough.

The threshold determines whether there will be an epidemic

Exercises

Reading a Graph

The graphs on page 3 have no scales marked along their axes, so they provide only *qualitative* information. The graphs below do have scales, so you can now answer *quantitative* questions about them. For example, on day 20 there are about 18,000 susceptible people. Read the graphs to answer the following questions. (Note: $S + I + R$ is *not* constant in this example, so these graphs cannot be solutions to our model.)

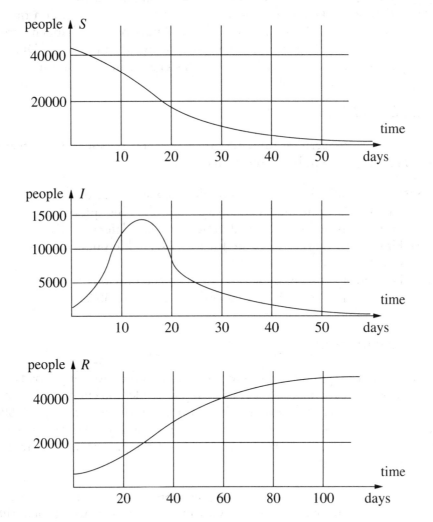

1. When does the infection hit its peak? How many people are infected at that time?
2. Initially, how many people are susceptible? How many days does it take for the susceptible population to be cut in half?

3. How many days does it take for the recovered population to reach 25,000? How many people *eventually* recover? Where did you look to get this information?

4. On what day is the size of the infected population increasing most rapidly? When is it decreasing most rapidly? How do you know?

5. How many people caught the illness at some time during the first 20 days? (Note that this is not the same as the number of people who are infected on day 20.) Explain where you found this information.

6. Copy the graph of R as accurately as you can, and then superimpose a sketch of S on it. Notice that the time scales on the *original* graphs of S and R are different. Describe what happened to the graph of S when you superimposed it on the graph of R. Did it get compressed or stretched? Was this change in the horizontal direction or the vertical?

A Simple Model

These questions concern the rate equation $S' = -470$ persons per day that we used to model a susceptible population on pages 4–5.

7. Suppose the initial susceptible population was 20,000 on Wednesday. Use the model to answer the following questions.
   a) How many susceptibles will be left ten days later?
   b) How many days will it take for the susceptible population to vanish entirely?
   c) How many susceptibles were there on the previous Sunday?
   d) How many days before Wednesday were there 30,000 susceptibles?

Mark Twain's Mississippi

The Lower Mississippi River meanders over its flat valley, forming broad loops called ox-bows. In a flood, the river can jump its banks and cut off one of these loops, getting shorter in the process. In his book *Life on the Mississippi* (1884), Mark Twain suggests, with tongue in cheek, that some day the river might even vanish! Here is a passage that shows us some of the pitfalls in using rates to predict the future and the past.

> In the space of one hundred and seventy six years the Lower Mississippi has shortened itself two hundred and forty-two miles. That is an average of a trifle over a mile and a third per year. Therefore, any calm person, who is not blind or idiotic, can see that in the Old Oölitic Silurian Period, just a million years ago next November, the Lower Mississippi was upwards of one million three hundred thousand miles long, and stuck out over the Gulf of Mexico like a fishing-pole. And by the same token any person can see that seven hundred and forty-two years from now the Lower Mississippi will be only a mile and

three-quarters long, and Cairo [Illinois] and New Orleans will have joined their streets together and be plodding comfortably along under a single mayor and a mutual board of aldermen. There is something fascinating about science. One gets such wholesale returns of conjecture out of such a trifling investment of fact.

Let L be the length of the Lower Mississippi River. Then L is a variable quantity we shall analyze.

8. According to Twain's data, what is the exact **rate** at which L is changing, in miles per year? What approximation does he use for this rate? Is this a reasonable approximation? Is this rate *positive* or *negative*? Explain. In what follows, use Twain's approximation.

* 9. Twain wrote his book in 1884. Suppose the Mississippi that Twain wrote about had been 1100 miles long; how long would it have become in 1990?

10. Twain does not tell us how long the Lower Mississippi was in 1884 when he wrote the book, but he does say that 742 years later it will be only $1\frac{3}{4}$ miles long. How long must the river have been when he wrote the book?

11. Suppose t is the number of years since 1884. Write a formula that describes how much L has changed in t years. Your formula should complete the equation

    the change in L in t years $= \ldots$.

*12. From your answer to question 10, you know how long the river was in 1884. From question 11, you know how much the length has changed t years after 1884. Now write a formula that describes how long the river is t years later.

13. Use your formula to find what L was a million years ago. Does your answer confirm Twain's assertion that the river was "upwards of 1,300,000 miles long" then?

14. Was the river ever 1,300,000 miles long; will it ever be $1\frac{3}{4}$ miles long? (This is called a **reality check**.) What, if anything, is wrong with the "trifling investment of fact" which led to such "wholesale returns of conjecture" that Twain has given us?

The Measles Epidemic

We consider once again the specific rate equations

$$S' = -.00001\,SI$$
$$I' = .00001\,SI - I/14$$
$$R' = I/14$$

---

*Exercises with an asterisk have answers in Appendix C.

discussed in the text on pages 10–13. We saw that at time $t = 1$,

$$S = 44446.6 \qquad I = 2903.4 \qquad R = 2650.0.$$

15. Calculate the current rates of change S', I', and R' when $t = 1$, and then use these values to determine S, I, and R one day later.

16. In the previous question you found S, I, and R when $t = 2$. Using these values, calculate the rates S', I', and R' and then determine the new values of S, I, and R when $t = 3$. See the table on page 11.

∗17. *Double the time step.* Go back to the starting time $t = 0$ and to the initial values

$$S = 45400 \quad I = 2100 \quad R = 2500.$$

Recalculate the values of S, I, and R at time $t = 2$ by using a time step of $\Delta t = 2$. You should perform only a single round of calculations, and use the rates S', I', and R' that are current at time $t = 0$.

18. *There and back again.* In the text we went one day into the future and then back again to the present. Here you'll go forward two days from $t = 0$ and then back again. There are two ways to do this: with a time step of $\Delta t = \pm 2$ (as in the previous question), and with a pair of time steps of $\Delta t = \pm 1$.

   a) $(\Delta t = \pm 2)$. Using the values of S, I, and R at time $t = 2$ that you just got in the previous question, calculate the rates S', I', and R'. Then using a time step of $\Delta t = -2$, estimate new values of S, I, and R at time $t = 0$. How much do these new values differ from the original values 45,400, 2100, 2500?

   b) $(\Delta t = \pm 1)$. Now make a new start, using the values

$$S = 43156.1 \qquad I = 3986.5 \qquad R = 2857.4$$
$$S' = -1720.4 \qquad I' = 1435.7 \qquad R' = 284.7$$

   that occur when $t = 2$ if we make estimates with a time step $\Delta t = 1$. (These values come from the table on page 11.) Using two rounds of calculations with a time step of $\Delta t = -1$, estimate another set of new values for S, I, and R at time $t = 0$. How much do these new values differ from the original values 45,400, 2100, 2500?

   c) Which process leads to a *smaller* set of differences: a single round of calculations with $\Delta t = \pm 2$, or two rounds of calculations with $\Delta t = \pm 1$? Consequently, which process produces better estimates—in the sense in which we used to measure estimates on page 12?

19. *Quarantine.* One of the ways to treat an epidemic is to keep the infected away from the susceptible; this is called *quarantine*. The intention is to reduce the chance that the illness will be transmitted to a susceptible person. Thus, quarantine alters the *transmission coefficient*.

a) Suppose a quarantine is put into effect that cuts in half the chance that a susceptible will fall ill. What is the new transmission coefficient?

*b) On page 16 it was determined that whenever there were fewer than 7143 susceptibles, the number of infected would decline instead of grow. We called 7143 a *threshold* level for S. Changing the transmission coefficient, as in part (a), changes the threshold level for S. What is the new threshold?

*c) Suppose we start with $S = 45,400$. Does quarantine eliminate the epidemic, in the sense that the number of infected immediately goes down from 2100, without ever showing an increase in the number of cases?

d) Since the new transmission coefficient is not small enough to guarantee that I never goes up, can you find a smaller value that *does* guarantee I never goes up? Continue to assume we start with $S = 45400$.

e) Suppose the initial susceptible population is 45,400. What is the *largest* value that the transmission coefficient can have and still guarantee that I never goes up? What level of quarantine does this represent? That is, do you have to reduce the chance that a susceptible will fall ill to one-third of what it was with no quarantine at all, to one-fourth, or what?

Other Diseases

20. Suppose the spread of an illness similar to measles is modeled by the following rate equations:

$$S' = -.00002\,SI$$
$$I' = .00002\,SI - .08\,I$$
$$R' = .08\,I$$

Note: The initial values $S = 45400$, and so forth, that we used in the text do not apply here.

a) Roughly how long does someone who catches this illness remain infected? Explain your reasoning.

b) How large does the susceptible population have to be in order for the illness to take hold—that is, for the number of cases to increase? Explain your reasoning.

c) Suppose 100 people in the population are currently ill. According to the model, how many (of the 100 infected) will recover during the next 24 hours?

*d) Suppose 30 *new* cases appear during the same 24 hours. What does that tell us about S'?

*e) Using the information in parts (c) and (d), can you determine how large the current susceptible population is?

21. a) Construct the appropriate S-I-R model for a measleslike illness that lasts for 4 days. It is also known that a typical susceptible person meets only about 0.3% of infected population each day, and the infection is transmitted in only one contact out of six.

*b) How small does the susceptible population have to be for this illness to fade away without becoming an epidemic?

22. Consider the general S-I-R model for a measleslike illness:

$$S' = -aSI$$
$$I' = aSI - bI$$
$$R' = bI$$

a) The threshold level for S—below which the number of infected will only decline—can be expressed in terms of the transmission coefficient a and the recovery coefficient b. What is that expression?

b) Consider two illnesses with the same transmission coefficient a; assume they differ only in the length of time someone stays ill. Which one has the lower threshold level for S? Explain your reasoning.

What Goes Around Comes Around

Some relatively mild illnesses, like the common cold, return to infect you again and again. For a while, right after you recover from a cold, you are immune. But that doesn't last; after some weeks or months, depending on the illness, you become susceptible again. This means there is now a flow from the recovered population to the susceptible. These exercises ask you to modify the basic S-I-R model to describe an illness where immunity is temporary.

23. Draw a compartment diagram for such an illness. Besides having all the ingredients of the diagram on page 8, it should depict a flow from R to S. Call this **immunity loss**, and use c to denote the coefficient of immunity loss.

24. Suppose immunity is lost after about 6 weeks. Show that you can set $c = 1/42$ per day, and explain your reasoning carefully. A suggestion: adapt the discussion of recovery in the text.

25. Suppose this illness lasts 5 days and it has a transmission coefficient of .00004 in the population we are considering. Suppose furthermore that the total population is fixed in size (as was the case in the text). Write down rate equations for S, I, and R.

*26. We saw in the text that the model for an illness that confers permanent immunity has a threshold value for S in the sense that when S is above the threshold, I increases, but when it is below, I decreases. Does *this* model have the same feature? If so, what is the threshold value?

27. For a mild illness that confers permanent immunity, the size of the recovered population can only grow. This question explores what happens when immunity is only temporary.

   *a) Will R increase or decrease if

$$S = 45400 \qquad I = 2100 \qquad R = 2500\,?$$

   *b) Suppose we shift 20000 susceptibles to the recovered population (so $S = 25400$ and $R = 22500$), leaving I unchanged. Will R increase or decrease?

   c) Using a total population of 50,000, give two other sets of values for S, I, and R that lead to a decreasing R.

   d) In fact, the relative sizes of I and R determine whether R will increase or decrease. Show that

$$\text{if } I > \tfrac{5}{42}R, \quad \text{then } R \text{ will increase;}$$

$$\text{if } I < \tfrac{5}{42}R, \quad \text{then } R \text{ will decrease.}$$

   Explain your argument clearly. A suggestion: consider the rate equation for R'.

28. *The steady state.* Any illness that confers only temporary immunity can appear to linger in a population forever. You may not always have a cold, but someone does, and eventually you catch another one. ("What goes around comes around.") Individuals gradually move from one compartment to the next. When they return to where they started, they begin another cycle.

   Each compartment (in the diagram you drew in question 23) has an *inflow* and an *outflow*. It is conceivable that the two exactly balance, so that the *size* of the compartment doesn't change (even though its individual occupants do). When this happens for all three compartments simultaneously, the illness is said to be in a **steady state**. In this question you explore the steady state of the model we are considering. Recall that the total population is 50,000.

   *a) What must be true if the inflow and outflow to the I compartment are to balance?

   *b) What must be true if the inflow and outflow to the R compartment are to balance?

   c) If neither I nor R is changing, then the model must be at the steady state. Why?

   d) What is the value of S at the steady state?

   *e) What is the value of R at the steady state? A suggestion: you know $R + I = 50000 -$ (the steady-state value of S). You also have a connection between I and R at the steady state.

➢   ## 1.2   THE MATHEMATICAL IDEAS

A number of important mathematical ideas have already emerged in our study of an epidemic. In this section we pause to consider them, because they have a "universal" character. Our aim is to get a fuller understanding of what we have done so we can use the ideas in other contexts.

> *We often draw out of a few particular experiences a lesson that can be put to good use in new settings. This process is the essence of mathematics, and it has been given a name— abstraction—which means literally "drawing from." Of course abstraction is not unique to mathematics; it is a basic part of the human psyche.*

Functions

A **function** describes how one quantity depends on another. In our study of a measles epidemic, the relation between the number of susceptibles S and the time t is a function. We write $S(t)$ to denote that S is a function of t. We can also write $I(t)$ and $R(t)$ because I and R are functions of t, too. We can even write $S'(t)$ to indicate that the rate S' at which S changes over time is a function of t. In speaking, we express $S(t)$ as "S of t" and $S'(t)$ as "S prime of t."

Functions and their notation

You can find functions everywhere. The amount of postage you pay for a letter is a function of the weight of the letter. The time of sunrise is a function of what day of the year it is. The crop yield from an acre of land is a function of the amount of fertilizer used. The position of a car's gasoline gauge (measured in centimeters from the left edge of the gauge) is a function of the amount of gasoline in the fuel tank. On a polygraph ("lie detector") there is a pen that records breathing; its position is a function of the amount of expansion of the lungs. The volume of a cubical box is a function of the length of a side. The last is a rather special kind of function because it can be described by an algebraic formula: if V is the volume of the box and s is the length of a side, then $V(s) = s^3$.

Most functions are *not* described by algebraic formulas, however. For instance, the postage function is given by a set of verbal instructions and the time of sunrise is given by a table in an almanac. The relation between a gas gauge and the amount of fuel in the tank is determined simply by making measurements. There is no algebraic formula that tells us how the number of susceptibles, S, depends upon t, either. Instead, we find $S(t)$ by carrying out the steps in the flow chart on page 15 until we reach t days into the future.

A function has input and output

In the function $S(t)$ the variable t is called the **input** and the variable S is called the **output**. In the sunrise function, the day of the year is the input and the time of sunrise is the output. In the function $S(t)$ we think of S as *depending on t*, so t is also called the independent variable and S the dependent variable. The set of values that the input takes is called the **domain** of the function. The set of values that the output takes is called the **range**.

The idea of a function is one of the central notions of mathematics. It is worth highlighting:

> A function is a rule that specifies how the value of one variable, the input, determines the value of a second variable, the output.

Notice that we say *rule* here, and not *formula*. This is deliberate. We want the study of functions to be as broad as possible, to include all the ways one quantity is likely to be related to another in a scientific question.

Some Technical Details

It is important not to confuse an expression like $S(t)$ with a product; $S(t)$ does *not* mean $S \times t$. On the contrary, the expression $S(1.4)$, for example, stands for the output of the function S when 1.4 is the input. In the epidemic model we then interpret it as the number of susceptibles that remain 1.4 days after today.

We have followed the standard practice in science by letting the single letter S designate both the *function*—that is, the *rule*—and the *output* of that function. Sometimes, though, we will want to make the distinction. In that case we will use two different symbols. For instance, we might write $S = f(t)$. Then we are still using S to denote the output, but the new symbol f stands for the function rule.

The symbols we use to denote the input and the output of a function are just names; if we change them, we don't change the function. For example, here are three ways to describe the same function g:

$$g : \text{multiply the input by 5, then subtract 3}$$
$$g(x) = 5x - 3$$
$$g(u) = 5u - 3.$$

It is important to realize that the *formulas* we just wrote in the last two lines are merely shorthand for the instructions stated in the first line. If you keep this in mind, then absurd-looking combinations like $g(g(2))$ can be decoded easily by remembering g of *anything* is just 5 times that thing, minus 3. We could thus evaluate $g(g(2))$ from the inside out (which is usually easier) as

$$g(g(2)) = g(5 \cdot 2 - 3) = g(7) = 5 \cdot 7 - 3 = 32,$$

or we could evaluate it from the outside in as

$$g(g(2)) = 5g(2) - 3 = 5(5 \cdot 2 - 3) - 3 = 5 \cdot 7 - 3 = 32$$

as before.

Suppose f is some other rule, say $f(t) = t^2 + 4t - 1$. Remember that this is just shorthand for "Take the input (whatever it is), square it, add four times the input, and subtract 1." We could then evaluate

$$f(g(3)) = f(5 \cdot 3 - 3) = f(12) = 12^2 + 4 \cdot 12 - 1 = 144 + 48 - 1 = 191,$$

while

$$g(f(3)) = g(20) = 97.$$

Part of math is simply learning the language

This process of **chaining** functions together by using the output of one function as the input for another turns out to be very important later in this course, and will be taken up again in chapter 3. For now, though, you should treat it simply as part of the formal language of mathematics, requiring a knowledge of the rules but no cleverness. It is analogous to learning how to conjugate verbs in French class—it's not very exciting for its own sake, but it allows us to read the interesting stuff later on!

A particularly important class of functions is composed of the **constant functions** which give the same output for every input. If h is the constant function that always gives back 17, then in formula form we would express this as $h(x) = 17$. Constant functions are so simple you might feel you are missing the point, but that's all there is to it!

Graphs

A graph is a function rule given visually

A graph describes a function in a visual form. Sometimes—as with a seismograph or a lie detector, for instance—this is the *only* description we have of a particular function. The usual arrangement is to put the input variable on the horizontal axis and the output on the vertical—but it is a good idea when you are looking at a particular graph to take a moment to check; sometimes, the opposite convention is used! This is often the case in geology and economics, for instance.

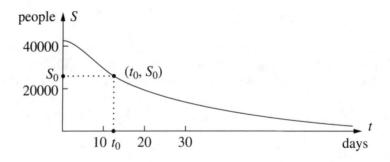

Sketched above is the graph of a function $S(t)$ that tells how many susceptibles there are after t days. Given any t_0, we "read" the graph to find $S(t_0)$, as follows: from the point t_0 on the t-axis, go vertically until you reach the graph; then go horizontally until you reach the S-axis. The value S_0 at that point is the output $S(t_0)$. Here t_0 is about 13 and S_0 is about 27,000; thus, the graph says that $S(13) \approx 27000$, or about 27,000 susceptibles are left after 13 days.

Linear Functions

Changes in Input and Output

Suppose y is a function of x. Then there is some rule that answers the question: what is the value of y for any given x? Often, however, we start by knowing the value of y for a particular x, and the question we really want to ask is: how does y respond to *changes* in x? We are still dealing with the same function—just looking at it from a different point of view. This point of view is important; we use it to analyze functions [like $S(t)$, $I(t)$, and $R(t)$] that are defined by rate equations.

> If y depends on x, then Δy depends on Δx

The way Δy depends on Δx can be simple or it can be complex, depending on the function involved. The simplest possibility is that Δy and Δx are **proportional**:

$$\Delta y = m \cdot \Delta x, \quad \text{for some constant } m.$$

> The defining property of a linear function

Thus, if Δx is doubled, so is Δy; if Δx is tripled, so is Δy. A function whose input and output are related in this simple way is called a **linear function**, because the graph is a straight line. Let's take a moment to see why this is so.

The Graph of a Linear Function

The graph consists of certain points (x, y) in the x, y-plane. Our job is to see how those points are arranged. Fix one of them, and call it (x_0, y_0). Let (x, y) be any other point on the graph. Draw the line that connects this point to (x_0, y_0), as we have done in the figure at the right. Now set

$$\Delta x = x - x_0 \qquad \Delta y = y - y_0.$$

By definition of a linear function, $\Delta y = m \cdot \Delta x$, as the figure shows, so the slope of this line is $\Delta y / \Delta x = m$. Recall that m is a constant; thus, if we pick a new point (x, y), the slope of the connecting line won't change.

Since (x, y) is an arbitrary point on the graph, what we have shown is that **every point on the graph lies on a line of slope m through the point** (x_0, y_0). But there is only one such line—and all the points lie on it! That line must be the graph of the linear function.

> A linear function is one that satisfies $\Delta y = m \cdot \Delta x$; its graph is a straight line whose slope is m.

Rates, Slopes, and Multipliers

The interpretation of m as a slope is just one possibility; there are two other interpretations that are equally important. To illustrate them we'll use Mark Twain's vivid description of the shortening of the Lower Mississippi River (see page 18). This will also give us the chance to see how a linear function emerges in context.

Twain says "the Lower Mississippi has shortened itself ... an average of a trifle over a mile and a third per year." Suppose we let L denote the length of the river, in miles, and t the time, in years. Then L depends on t, and Twain's statement implies that L is a *linear* function of t, in the sense in which we have just defined a linear function. Here is why. According to our definition, there must be some number m which makes $\Delta L = m \cdot \Delta t$. But notice that Twain's statement has exactly this form if we translate it into mathematical language. Convince yourself that it says

Stop and do the translation

$$\Delta L \text{ miles} = -1\tfrac{1}{3}\,\frac{\text{miles}}{\text{year}} \times \Delta t \text{ years}.$$

Thus we should take m to be $-1\tfrac{1}{3}$ miles per year.

The role of m here is to convert one quantity (Δt years) into another (ΔL miles) by multiplication. All linear functions work this way. In the defining equation $\Delta y = m \cdot \Delta x$, multiplication by m converts Δx into Δy. Any change in x produces a change in y that is m times as large. For this reason we give m its second interpretation as a **multiplier**.

> It is easier to understand why the usual symbol for slope is m—instead of s—when you see that a slope can be interpreted as a multiplier.

It is important to note that, in our example, m is not simply $-1\tfrac{1}{3}$; it is $-1\tfrac{1}{3}$ *miles per year*. In other words, m is the **rate** at which the river is getting shorter. All linear functions work this way, too. We can rewrite the equation $\Delta y = m \cdot \Delta x$ as a ratio

$$m = \frac{\Delta y}{\Delta x} = \text{ the rate of change of } y \text{ with respect to } x.$$

For these reasons we give m its third interpretation as a **rate of change**.

> For a linear function satisfying $\Delta y = m \cdot \Delta x$,
> the coefficient m is
> rate of change, slope, and multiplier.

We already use y' to denote the rate of change of y, so we can now write $m = y'$ when y is a linear function of x. In that case we can also write

$$\Delta y = y' \cdot \Delta x.$$

This expression should recall a pattern very familiar to you. (If not, change y to S and x to t!) It is the fundamental formula we have been using to calculate future values of S, I, and R. We can approach the relation between y and x the same way. That is, if y_0 is an "initial value" of y, when $x = x_0$, then *any* value of y can be calculated from

$$y = y_0 + y' \cdot \Delta x \quad \text{or} \quad y = y_0 + m \cdot \Delta x.$$

Units

Suppose x and y are quantities that are measured in specific units. If y is a linear function of x, with $\Delta y = m \cdot \Delta x$, then m must have units too. Since m is the multiplier that "converts" x into y, the units for m must be chosen so they will convert x's units into y's units. In other words,

If x and y have units, so does m

$$\text{units for } y = \text{units for } m \times \text{units for } x.$$

This implies

$$\text{units for } m = \frac{\text{units for } y}{\text{units for } x}.$$

For example, the multiplier in the Mississippi River problem converts years to miles, so it must have units of miles per year. The rate equation $R' = bI$ in the S-I-R model is a more subtle example. It says that R' is a linear function of I. Since R' is measured in persons per day and I is measured in persons, we must have

$$\text{units for } b = \frac{\text{units for } R'}{\text{units for } I} = \frac{\dfrac{\text{persons}}{\text{day}}}{\text{persons}}$$

Formulas for Linear Functions

The expression $\Delta y = m \cdot \Delta x$ declares that y is a linear function of x, but it doesn't quite tell us what y itself *looks like* directly in terms of x. In fact, there are several equivalent ways to write the relation $y = f(x)$ in a formula, depending on what information we are given about the function.

The initial-value form. Here is a very common situation: we know the value of y at an "initial" point—let's say $y_0 = f(x_0)$—and we know the rate of change—let's say it is m. Then the graph is the straight line of slope m that passes through the point (x_0, y_0). The formula for f is

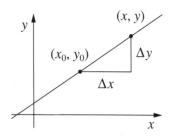

$$y = y_0 + \Delta y = y_0 + m \cdot \Delta x = y_0 + m(x - x_0) = f(x).$$

What you should note particularly about this formula is that it expresses y in terms of the initial data x_0, y_0, and m, as well as x. Since those data consists of a point (x_0, y_0) and a slope m, the initial-value formula is also referred to as the **point-slope form** of the equation of a line. It may be more familiar to you with that name.

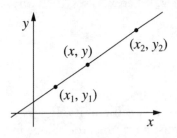

The interpolation form. This time we are given the value of y at *two* points—let's say $y_1 = f(x_1)$ and $y_2 = f(x_2)$. The graph is the line that passes through (x_1, y_1) and (x_2, y_2), and its slope is therefore

$$m = \frac{y_2 - y_1}{x_2 - x_1}.$$

Now that we know the slope of the graph we can use the point-slope form (taking (x_1, y_1) as the "point," for example) to get the equation. We have

$$y = y_1 + m(x - x_1) = y_1 + \frac{y_2 - y_1}{x_2 - x_1}(x - x_1) = f(x).$$

Notice how, once again, y is expressed in terms of the initial data, which consists of the two points (x_1, y_1) and (x_2, y_2).

The process of finding values of a quantity between two given values is called **interpolation**. Since our new expression does precisely that, it is called the interpolation formula. (Of course, it also finds values outside the given interval.) Since the initial data are two points, the interpolation formula is also called the **two-point formula** for the equation of a line.

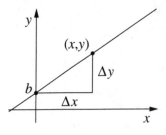

The slope-intercept form. This is a special case of the initial-value form that occurs when $x_0 = 0$. Then the point (x_0, y_0) lies on the y-axis, and it is frequently written in the alternate form $(0, b)$. The number b is called the **y-intercept**. The equation is

$$y = mx + b = f(x).$$

In the past you may have thought of this as *the* formula for a linear function, but for us it is only one of several. You will find that we will use the other forms more often.

Functions of Several Variables

Language and Notation

A function can have
several input variables

Many functions depend on more than one variable. For example, sunrise depends on the day of the year, but it also depends on the latitude (position north or south of the equator) of the observer. Likewise, the crop yield from an acre of land depends on the amount of fertilizer used, but it also depends on the amount of rainfall, on the composition of the soil, on the amount of weeding done, to mention just a few of the other variables that a farmer has to contend with.

The rate equations in the S-I-R model also provide examples of functions with more than one input variable. The equation

$$I' = .00001\, SI - I/14$$

says that we need to specify both S and I to find I'. We can say that

$$F(S, I) = .00001\, SI - I/14$$

is a function whose input is the **ordered pair** of variables (S, I). In this case F is given by an algebraic formula. While many other functions of several variables also have formulas—and they are extremely useful—not all functions do. The sunrise function, for example, is given by a two-way table (see page 147) that shows the time of sunrise for different days of the year and different latitudes.

As a technical matter it is important to note that the input variables S and I of the function $F(S, I)$ above appear in a particular *order*, and that order is part of the definition of the function. For example, $F(1, 0) = 0$, but $F(0, 1) = -1/14$. (Do you see why? Work out the calculations yourself.)

Parameters

Suppose we rewrite the rate equation for I', replacing .00001 and 1/14 with the general values a and b:

$$I' = aSI - bI.$$

This makes it clear that I' depends on a and b, too. But note that a and b are not variables in quite the same way that S and I are. For example, a and b will vary if we switch from one disease to another or from one population to another. However, they will stay fixed while we consider a particular disease in a particular population. By contrast, S and I will *always* be treated as variables. We call a quantity like a or b a **parameter**.

To emphasize that I' depends on the parameters as well as S and I, we can write I' as the output of a new function

$$I' = I'(S, I, a, b) = aSI - bI$$

whose input is the set of *four* variables (S, I, a, b), in that order. The variables $S, I,$ and R must also depend on the parameters, too, and not just on t. Thus, we should write $S(t, a, b)$, for example, instead of simply $S(t)$. We implicitly used the fact that $S, I,$ and R depend on a and b when we discovered there

Some functions depend on parameters

was a threshold for an epidemic (page 16). In exercise 22 of section 1.1 (page 22), you made the relation explicit. In that problem you show I will simply decrease over time (i.e., there will be no "burst" of infection) if

$$S < \frac{b}{a}.$$

There are even more parameters lurking in the S-I-R problem. To uncover them, recall that we needed *two* pieces of information to estimate S, I, and R over time:

1. The rate equations;
2. The initial values S_0, I_0, and R_0.

We used $S_0 = 45400$, $I_0 = 2100$, and $R_0 = 2500$ in the text, but if we had started with other values, then S, I, and R would have ended up being different functions of t. Thus, we should really write

$$S = S(t, a, b, S_0, I_0, R_0)$$

to tell a more complete story about the inputs that determine the output S. Most of the time, though, we do *not* want to draw attention to the parameters; we usually write just $S(t)$.

Further Possibilities

Steps I, II, and III on page 14 are also functions, because they have well-defined input and output. They are unlike the other examples we have discussed up to this point because they have more than one output variable. You should see, though, that there is nothing more difficult going on here.

In our study of the S-I-R model it was natural not to separate functions that have one input variable from those that have several. This is the pattern we shall follow in the rest of the book. In particular, we will want to deal with parameters, and we will want to understand how the quantities we are studying depend on those parameters.

The Beginnings of Calculus

While functions, graphs, and computers are part of the general fabric of mathematics, we can also abstract from the S-I-R model some important aspects of the calculus itself. The first of these is the idea of a **rate of change**. In this chapter we just assumed the idea was intuitively clear. However, there are some important questions not yet answered; for example, how do you deal with a quantity whose rate of change is itself always changing? These questions, which lead to the fundamental idea of a **derivative**, are taken up in chapter 3.

Rate equations—more commonly called **differential equations**—lie at the very heart of calculus. We will have much more to say about them, because many processes in the physical, biological, and social realms can be modeled by rate equations. In our analysis of the S-I-R model, we used rate equations to estimate future values by assuming that rates stay fixed for a whole day at a time. The discussion called "there and back again" on pages 12–13 points up the shortcomings of this assumption. In chapter 2 we will develop a procedure, called Euler's method, to address this problem. In chapter 4 we will return to differential equations in a general way, equipped with Euler's method and the concept of the derivative.

How the next three chapters are connected

Exercises

Functions and Graphs

1. Sketch the graph of each of the following functions. Label each axis and mark a scale of units on it. For each line that you draw, indicate
   i. its slope;
   ii. its y-intercept;
   iii. its x-intercept (where it crosses the x-axis).
   a) $y = -\frac{1}{2}x + 3$
   b) $y = (2x - 7)/3$
   c) $5x + 3y = 12$

2. Graph the following functions. Put labels and scales on the axes.
   a) $V = .3Z - 1$ b) $W = 600 - P^2$.

3. Sketch the graph of each of the following functions. Put labels and scales on the axes. For each graph that you draw, indicate
   i. its y-intercept;
   ii. its x-intercept(s).
   For part (d) you will need the **quadratic formula**

   $$x = \frac{-b \pm \sqrt{b^2 - 4ac}}{2a}$$

   for the roots of the **quadratic equation** $ax^2 + bx + c = 0$.
   a) $y = x^2$ c) $y = (x + 1)^2$
   b) $y = x^2 + 1$ d) $y = 3x^2 + x - 1$

The next four questions refer to these functions:

$$
\begin{aligned}
c(x, y) &= 17 & &\text{a constant function} \\
j(z) &= z & &\text{the identity function} \\
r(u) &= 1/u & &\text{the reciprocal function} \\
D(p, q) &= p - q & &\text{the difference function}
\end{aligned}
$$

$$s(y) = y^2 \qquad \text{the squaring function}$$

$$Q(v) = \frac{2v + 1}{3v - 6} \qquad \text{a rational function}$$

$$H(x) = \begin{cases} 5 & \text{if } x < 0 \\ x^2 + 2 & \text{if } 0 \le x < 6 \\ 29 - x & \text{if } 6 \le x \end{cases}$$

$$T(x, y) = r(x) + Q(y)$$

4. Determine the following values:

$$c(5, -3) \qquad j(17) \qquad c(a, b) \qquad j(u^2 + 1)$$

$$j(c(3, -5)) \qquad s(1.1) \qquad r(1/17) \qquad Q(0)$$

$$Q(2) \qquad Q(3/7) \qquad D(5, -3) \qquad D(-3, 5)$$

$$H(1) \qquad H(7) \qquad H(4) \qquad H(H(H(-3)))$$

$$r(s(-4)) \qquad r(Q(3)) \qquad Q(r(3)) \qquad T(3, 7)$$

*5. True or false. Give reasons for your answers: if you say true, explain why; if you say false, give an example that shows why it is false.
   a) For every nonzero number x, $r(r(x)) = j(x)$.
   b) If $a > 1$, then $s(a) > 1$.
   c) If $a > b$, then $s(a) > s(b)$.
   d) For all real numbers a and b, $s(a + b) = s(a) + s(b)$.
   e) For all real numbers a, b, and c, $D(D(a, b), c) = D(a, D(b, c))$.

6. Find all numbers x for which $Q(x) = r(Q(x))$.

7. The **natural domain** of a function f is the largest possible set of real numbers x for which $f(x)$ is defined. For example, the natural domain of $r(x) = 1/x$ is the set of all nonzero real numbers.
   a) Find the natural domains of Q and H.
   *b) Find the natural domains of $P(z) = Q(r(z))$; $R(v) = r(Q(v))$.

   *c) What is the natural domain of the function $W(t) = \sqrt{\dfrac{1 - t^2}{t^2 - 4}}$?

Computer Graphing

The purpose of these exercises is to give you some experience using a "graphing package" on a computer. This is a program that will draw the graph of a function $y = f(x)$ whose formula you know. You must type in the formula, using the following symbols to represent the basic arithmetic operations:

To indicate	Type
Addition	+
Subtraction	-
Multiplication	*
Division	/
Exponentiation	^

The caret "^" appears above the "6" on a keyboard (Shift-6). Here is an example:

to enter :

$$\frac{7x^5 - 9x^2}{x^3 + 1}$$

type :

`(7*x^5 - 9*x^2)/(x^3 + 1)`

The parentheses you see here are important. If you do not include them, the computer will interpret your entry as

$$7x^5 - \frac{9x^2}{x^3} + 1 = 7x^5 - \frac{9}{x} + 1 \neq \frac{7x^5 - 9x^2}{x^3 + 1}.$$

In some graphing packages, you do not need to use * to indicate a multiplication. If this is true for the package you use, then you can enter the fractional expression above in the somewhat simpler form

`(7x^5 - 9x^2)/(x^3 + 1).`

To do the following exercises, follow the specific instructions for the graphing package you are using.

8. Graph the function $f(x) = .6x + 2$ on the interval $-4 \leq x \leq 4$.
  a) What is the y-intercept of this graph? What is the x-intercept?
  b) Read from the graph the value of $f(x)$ when $x = -1$ and when $x = 2$. What is the difference between these y values? What is the difference between the x values? According to these differences, what is the slope of the graph? According to the *formula*, what is the slope?

*9. Graph the function $f(x) = 1 - 2x^2$ on the interval $-1 \leq x \leq 1$.
  a) What is the y-intercept of this graph? The graph has two x-intercepts; use algebra to find them.
     You can also find an x-intercept using the computer. The idea is to **magnify** the graph near the intercept until you can determine as many decimal places in the x coordinate as you want. For a start, graph the function on the interval $0 \leq x \leq 1$. You should be able to see that the graph on your computer monitor crosses the x-axis somewhere around .7. Regraph $f(x)$ on the interval $.6 \leq x \leq .8$. You should then be able to determine that the

x-intercept lies between .7 and .8. This means $x = .7\ldots$; that is, you know the value of the first decimal place of the x-intercept.

b) Regraph $f(x)$ on the interval $.70 \leq x \leq .71$ to get two decimal places of accuracy in the location of the x-intercept. Continue this process until you have at least 7 places of accuracy. What is the x-intercept?

The circular functions.   Graphing packages "know" the familiar functions of trigonometry. Trigonometric functions are qualitatively different from the functions in the preceding problems. Those functions are defined by algebraic formulas, so they are called **algebraic functions**. The trigonometric functions are defined by explicit "recipes," but *not* by algebraic formulas; they are called **transcendental functions**. For calculus, we usually use the definition of the trigonometric functions as **circular functions**. This definition begins with a unit circle centered at the origin. Given the input number t, locate a point P on the circle by tracing an arc of length t along the circle from the point $(1, 0)$. If t is positive, trace the arc counterclockwise; if t is negative, trace it clockwise. Because the circle has radius 1, the arc of length t subtends a central angle of **radian** measure t.

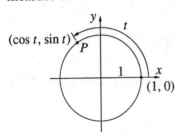

The circular (or trigonometric) functions $\cos t$ and $\sin t$ are defined as the coordinates of the point P,

$$P = (\cos t, \sin t).$$

The other trigonometric functions are defined in terms of the sine and cosine:

$$\tan t = \sin t / \cos t \qquad \sec t = 1 / \cos t$$
$$\cot t = \cos t / \sin t \qquad \csc t = 1 / \sin t.$$

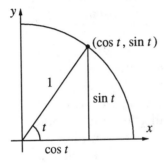

Notice that when t is a positive acute angle, the circle definition agrees with the right triangle definitions of the sine and cosine:

$$\sin t = \frac{\text{opposite}}{\text{hypotenuse}} \quad \text{and} \quad \cos t = \frac{\text{adjacent}}{\text{hypotenuse}}.$$

However, the circle definitions of the sine and cosine have the important advantage that they produce functions whose domains are the set of *all* real numbers. (What are the domains of the tangent, secant, cotangent, and cosecant functions?)

In calculus, angles are always measured in radians. To convert between radians and degrees, notice that the circumference of a unit circle is 2π, so the radian measure of a semicircular arc is half of this, and thus we have

$$\pi \text{ radians } = 180 \text{ degrees.}$$

As the course progresses, you will see why radians are used rather than degrees (or mils or any other unit for measuring angles)—it turns out that the formulas important to calculus take their simplest form when angles are expressed in radians.

Simplicity determines the choice of radians for measuring angles

Graphing packages "know" the trigonometric functions in exactly this form: circular functions with the input variable given in radians. You might wonder, though, how a computer or calculator "knows" that $\sin(1) = .017452406\ldots$. It certainly isn't drawing a very accurate circle somewhere and measuring the y coordinate of some point. While the circular function approach is a useful way to think about the trigonometric functions conceptually, it isn't very helpful if we actually want values of the functions. One of the achievements of calculus, as you will see later in this text, is that it provides effective methods for computing values of functions like the circular functions that aren't given by algebraic formulas.

How do we get values for the circular functions?

The following exercises let you review the trigonometric functions and explore some of the possibilities using computer graphing.

10. Graph the function $f(x) = \sin(x)$ on the interval $-2 \le x \le 10$.
   a) What are the x-intercepts of $\sin(x)$ on the interval $-2 \le x \le 10$? Determine them to two decimal places accuracy.
   ***b)** What is the largest value of $f(x)$ on the interval $-2 \le x \le 10$? Which value of x makes $f(x)$ largest? Determine x to two decimal places accuracy.
   c) Regraph $f(x)$ on the very small interval $-.001 \le x \le .001$. Describe what you see. Can you determine the slope of this graph?

11. Graph the function $f(x) = \cos(x)$ on the interval $0 \le x \le 14$. On the same screen graph the *second* function $g(x) = \cos(2x)$.
   a) How far apart are the x-intercepts of $f(x)$? How far apart are the x-intercepts of $g(x)$?
   b) The graph of $g(x)$ has a pattern that repeats. How wide is this pattern? The graph of $f(x)$ also has a repeating pattern; how wide is *it*?
   c) Compare the graphs of $f(x)$ and $g(x)$ to one another. In particular, can you say that one of them is a stretched or compressed version of the other? Is the compression (or stretching) in the vertical or the horizontal direction?
   d) Construct a *new* function $f(x)$ whose graph is the same shape as the graph of $g(x) = \cos(2x)$, but make the graph of $f(x)$ twice as tall as the graph of $g(x)$. [A suggestion: Either deduce what $f(x)$

should be, or make a guess. Then test your choice on the computer. If your choice doesn't work, think how you might modify it, and then test your modifications the same way.]

12. The aim here is to find a solution to the equation $\sin x = \cos(3x)$. There is no purely *algebraic* procedure to solve this equation. Because the sine and cosine are not defined by *algebraic* formulas, this should not be particularly surprising. (Even for algebraic equations, there are only a few very special cases for which there are formulas like the quadratic formula. In chapter 5 we will look at a method for solving equations when formulas can't help us.)
   a) Graph the two functions $f(x) = \sin(x)$ and $g(x) = \cos(3x)$ on the interval $0 \le x \le 1$.
   *b) Find a solution of the equation $\sin(x) = \cos(3x)$ that is accurate to six decimal places.
   c) Find *another* solution of the equation $\sin(x) = \cos(3x)$, accurate to four decimal places. Explain how you found it.

13. Use a graphing program to make a sketch of the graph of each of the following functions. In each case, make clear the domain and the range of the function, where the graph crosses the axes, and where the function has a maximum or a minimum.
   a) $F(w) = (w - 1)(w - 2)(w - 3)$ d) $e(x) = x - \dfrac{1}{x}$

   b) $Q(a) = \dfrac{1}{a^2 + 5}$ e) $g(u) = \sqrt{\dfrac{u - 1}{u + 1}}$

   c) $E(x) = x + \dfrac{1}{x}$ f) $M(u) = \dfrac{u^2 - 2}{u^2 + 2}$

14. Graph on the same screen the following three functions:

$$f(x) = 2^x \qquad g(x) = 3^x \qquad h(x) = 10^x.$$

   Use the interval $-2 \le x \le 1.2$.
   a) Which function has the largest value when $x = -2$?
   b) Which is climbing most rapidly when $x = 0$?
   *c) Magnify the picture at $x = 0$ by resetting the size of the interval to $-.0001 \le x \le .0001$. Describe what you see. Estimate the slopes of the three graphs.

Proportions, Linear Functions, and Models

15. Go back to the three functions given in problem 1. For each function, choose an initial value x_0 for x, find the corresponding value y_0 for y, and express the function in the form $y - y_0 = m \cdot (x - x_0)$.

16. You should be able to answer all parts of this problem without ever finding the equations of the functions involved.

a) Suppose $y = f(x)$ is a linear function with multiplier $m = 3$. If $f(2) = -5$, what is $f(2.1)$? $f(2.0013)$? $f(1.87)$? $f(922)$?

b) Suppose $y = G(x)$ is a linear function with multiplier $m = -2$. If $G(-1) = 6$, for what value of x is $G(x) = 8$? $G(x) = 0$? $G(x) = 5$? $G(x) = 491$?

c) Suppose $y = h(x)$ is a linear function with $h(2) = 7$ and $h(6) = 9$. What is $h(2.046)$? $h(2 + a)$?

17. In Massachusetts there is a sales tax of 5%. The tax T, in dollars, is proportional to the price P of an object, also in dollars. The constant of proportionality is $k = 5\% = .05$. Write a formula that expresses the sales tax as a linear function of the price, and use your formula to compute the tax on a television set that costs $289.00 and a toaster that costs $37.50.

18. Suppose $W = 213 - 17Z$. How does W change when Z changes from 3 to 7; from 3 to 3.4; from 3 to 3.02? Let ΔZ denote a change in Z and ΔW the change thereby produced in W. Is $\Delta W = m\,\Delta Z$ for some constant m? If so, what is m?

19. a) In the following table, q is a linear function of p. Fill in the blanks in the table.

p	-3	0		7	13		π
q	7		4	1		0	

b) Find a formula to express Δq as a function of Δp, and another to express q as a function of p.

20. *Thermometers.* There are two common scales in use to measure the temperature, called **Fahrenheit degrees** and the **Celsius degrees**. Let F and C, respectively, be the temperature on each of these scales. Each of these quantities is a linear function of the other; the relation between them in determined by the following table:

Physical measurement	C	F
Freezing point of water	0	32
Boiling point of water	100	212

a) Which represents a larger change in temperature, a Celsius degree or a Fahrenheit degree?

b) How many Fahrenheit degrees does it take to make the temperature go up one Celsius degree? How many Celsius degrees does it take to make it go up one Fahrenheit degree?

*c) What is the multiplier m in the equation $\Delta F = m \cdot \Delta C$? What is the multiplier μ in the equation $\Delta C = \mu \cdot \Delta F$? (The symbol μ is the Greek letter *mu*.) What is the relation between μ and m?

d) Express F as a linear function of C. Graph this function. Put scales and labels on the axes. Indicate clearly the slope of the graph and its vertical intercept.

e) Express C as a linear function of F and graph this function. How are the graphs in parts (d) and (e) related? Give a clear and detailed explanation.

f) Is there any temperature that has the same reading on the two temperature scales? What is it? Does the temperature of the air ever reach this value? Where?

21. *The greenhouse effect.* The concentration of carbon dioxide (CO_2) in the atmosphere is increasing. The concentration is measured in parts per million (PPM). Records kept at the South Pole show an increase of .8 PPM per year during the 1960s.

a) At that rate, how many years does it take for the concentration to increase by 5 PPM; by 15 PPM?

b) At the beginning of 1960 the concentration was about 316 PPM. What would it be at the beginning of 1970; at the beginning of 1980?

c) Draw a graph that shows CO_2 concentration as a function of the time since 1960. Put scales on the axes and label everything clearly.

*d) The *actual* CO_2 concentration at the South Pole was 324 PPM at the beginning of 1970 and 338 PPM at the beginning of 1980. Plot these values on your graph, and compare them to your calculated values.

e) Using the actual concentrations in 1970 and 1980, calculate a new rate of increase in concentration. Using that rate, estimate what the increase in CO_2 concentration was between 1970 and 1990. Estimate the CO_2 concentration at the beginning of 1990.

f) Using the rate of .8 PPM per year that held during the 1960s, determine how many years before 1960 there would have been *no* carbon dioxide at all in the atmosphere.

22. *Thermal expansion.* Measurements show that the length of a metal bar increases in proportion to the increase in temperature. An aluminum bar that is exactly 100 inches long when the temperature is 40°F becomes 100.0052 inches long when the temperature increases to 80°F.

a) How long is the bar when the temperature is 60°F? 100°F?

*b) What is the multiplier that connects an increase in length ΔL to an increase in temperature ΔT?

c) Express ΔL as a linear function of ΔT.

d) How long will the bar be when $T = 0°F$?

e) Express L as a linear function of T.

f) What temperature change would make $L = 100.01$ inches?

g) For a *steel* bar that is also 100 inches long when the temperature is 40°F, the relation between ΔL and ΔT is $\Delta L = .00067\,\Delta T$. Which expands more when the temperature is increased; aluminum or steel?

h) How long will this steel bar be when $T = 80°F$?

23. *Falling bodies.* In the simplest model of the motion of a falling body, the velocity increases in proportion to the increase in the time that the body has been falling. If the velocity is given in feet per second, measurements show the constant of proportionality is approximately 32.

a) A ball is falling at a velocity of 40 feet/sec after 1 second. How fast is it falling after 3 seconds?

b) Express the change in the ball's velocity Δv as a linear function of the change in time Δt.

c) Express v as a linear function of t.

The model can be expanded to keep track of the *distance* that the body has fallen. If the distance d is measured in feet, the units of d' are feet per second; in fact, $d' = v$. So the model describing the motion of the body is given by the rate equations

$$d' = v \quad \text{feet per second;}$$

$$v' = 32 \quad \text{feet per second per second.}$$

d) At what rate is the distance increasing after 1 second? After 2 seconds? After 3 seconds?

e) Is d a linear function of t? Explain your answer.

   In many cases, the rate of change of a variable quantity is proportional to the quantity itself. Consider a human population as an example. If a city of 100,000 is increasing at the rate of 1500 persons per year, we would expect a similar city of 200,000 to be increasing at the rate of 3000 persons per year. That is, if P is the population at time t, then the **net growth rate** P' is proportional to P:

$$P' = kP.$$

In the case of the two cities, we have

$$P' = 1500 = kP = k \times 100000 \quad \text{so} \quad k = \frac{1500}{100000} = .015.$$

24. In the equation $P' = kP$, above, explain why the units for k are

$$\frac{\text{persons per year}}{\text{person}}.$$

The number k is called the **per capita growth rate**. ("Per capita" means "per person"—"per *head*," literally.)

25. *Poland and Afghanistan.* In 1985 the per capita growth rate in Poland was 9 persons per year per thousand persons. (That is, $k = 9/1000 = .009$.) In Afghanistan it was 21.6 persons per year per thousand.

   a) Let P denote the population of Poland and A the population of Afghanistan. Write the equations that govern the growth rates of these populations.

   b) In 1985 the population of Poland was estimated to be 37.5 million persons, that of Afghanistan 15 million. What are the net growth rates P' and A' (as distinct from the *per capita* growth rates)? Comment on the following assertion: When comparing two countries, the one with the larger per capita growth rate will have the larger net growth rate.

   c) On the average, how long did it take the population to increase by one person in Poland in 1985? What was the corresponding time interval in Afghanistan?

26.*a) *Bacterial growth.* A colony of bacteria on a culture medium grows at a rate proportional to the present size of the colony. When the colony weighed 32 grams it was growing at the rate of 0.79 gram per hour. Write an equation that links the growth rate to the size of the population.

   b) What is ΔP if $\Delta t = 1$ minute? Estimate how long it would take to make $\Delta P = .5$ gram.

27. *Radioactivity.* In radioactive decay, radium slowly changes into lead. If one sample of radium is twice the size of a second lump, then the larger sample will produce twice as much lead as the second in any given time. In other words, the rate of decay is proportional to the amount of radium present. Measurements show that 1 gram of radium decays into lead at the rate of 1/2337 gram per year. Write an equation that links the decay rate to the size of the radium sample. How does your equation indicate that the process involves *decay* rather than *growth*?

28. *Cooling.* Suppose a cup of hot coffee is brought into a room at 70°F. It will cool off, and it will cool off *faster* when the temperature difference between the coffee and the room is greater. The simplest assumption we can make is that the rate of cooling is proportional to this temperature difference (this is called Newton's law of cooling). Let C denote the temperature of the coffee, in °F, and C' the rate at which it is cooling, in °F per minute. The new element here is that C' is proportional, not to C, but to the *difference* between C and the room temperature of 70°F.

   a) Write an equation that relates C' and C. It will contain a proportionality constant k. How did you indicate that the coffee is *cooling* and not *heating up*?

   b) When the coffee is at 180°F, it is cooling at the rate of 9°F per minute. What is k?

c) At what rate is the coffee cooling when its temperature is 120°F?

d) Estimate how long it takes the temperature to fall from 180°F to 120°F. Then make a better estimate, and explain why it is better.

➤ 1.3   USING A PROGRAM

Computers

A computer changes the way we can use calculus as a tool, and it vastly enlarges the range of questions that we can tackle. No longer need we back away from a problem that involves a lot of computations. There are two aspects to the power of a computer. First, it is fast. It can do a million additions in the time it takes us to do one. Second, it can be programmed. By arranging computations into a loop (page 15), we can construct a program with only a few instructions that will carry out millions of repetitive calculations.

The purpose of this section is to give you practice using a computer program that estimates values of S, I, and R in the epidemic model. As you will see, it carries out the three rounds of calculations you have already done by hand. It also contains a loop that will allow you to do a hundred, or a million, rounds of calculations with no extra effort.

Nearly all that we say about computers applies equally well to graphing calculators. In Appendix A you will find translations of the following program— and all the others that appear later in text—for a number of popular calculators.

You can use a graphing calculator too.

The Program SIR

The program on the following page calculates values of S, I, and R. It is a set of instructions—sometimes called **code**—that is designed to be read by you and by a computer. These instructions mirror the operations we performed by hand to generate the table on page 11. The code here is similar to what it would be in most programming languages. The line numbers, however, are not part of the program; they are there to help us refer to the lines. A computer reads the code one line at a time, starting at the top. Each line is a complete instruction which causes the computer to do something. The purpose of nearly every instruction in this program is to assign a numerical value to a symbol. Watch for this as we review the program.

Read a program line by line from the top

The first line, $t = 0$, is the instruction "Give t the value 0." The next four lines are similar. Notice in the fifth line how Δt is typed out as `deltat`. It is a common practice for the name of a variable to be several letters long. A few lines later S' is typed out as `Sprime`, for instance. The instruction on the sixth line is the first that does not assign a value to a symbol. Instead, it causes the computer to print the following on the screen:

Lines 1–5

Line 6

0   45400   2100   2500

Program: SIR

```
 1   t = 0
 2   S = 45400
 3   I = 2100
 4   R = 2500
 5   deltat = 1
 6   PRINT t, S, I, R
 7   FOR k = 1 TO 3
 8           Sprime = -.00001 * S * I
 9           Iprime = .00001 * S * I - I / 14      } Step I
10           Rprime = I / 14
11           deltaS = Sprime * deltat
12           deltaI = Iprime * deltat
13           deltaR = Rprime * deltat
14           t = t + deltat
15           S = S + deltaS
16           I = I + deltaI
17           R = R + deltaR
18           PRINT t, S, I, R
19   NEXT k
```

Line 7

Lines 8–10

Skip over the line that says FOR k = 1 TO 3. It will be easier to understand after we've read the rest of the program.

Look at the first three indented lines. You should recognize them as coded versions of the rate equations

$$S' = -.00001\, SI$$

$$I' = .00001\, SI - I/14$$

$$R' = I/14$$

for the measles epidemic. (The program uses $*$ to denote multiplication.) They are instructions to assign numerical values to the symbols S', I', and R'. For instance, Sprime = -.00001 * S * I (line 8) says

Give S' the value $-.00001\, SI$;

use the current values of S and I to get $-.00001\, SI$.

Now the computer knows that the current values of S and I are 45400 and 2100, respectively. (Can you see why?) So it calculates the product $-.00001 \times 45400 \times 2100 = -953.4$ and then gives S' the value -953.4. There is an extra step to calculate the product.

Notice that the first three indented lines are bracketed together and labeled "Step I," because they carry out Step I in the flow chart. The next three in- Lines 11–13 dented lines carry out Step II in the flow chart. They assign values to three more symbols—namely ΔS, ΔI, and ΔR—using the current values of S', I', R' and Δt.

The next four indented lines present a puzzle. They don't make sense if we Lines 14–17 read them as ordinary mathematics. In an expression like `t = t + deltat`, we would cancel the t's and conclude `deltat` = 0. The lines *do* make sense when we read them as computer instructions, however. As a computer instruction, `t = t + deltat` says

Make the new value of t equal to the current value of $t + \Delta t$.

(To make this clear, some computer languages express this instruction in the form `let t = t + deltat`.) Once again we have an instruction that assigns a numerical value to a symbol, but this time the symbol (t, in this case) already has a value before the instruction is carried out. The instruction gives it a *new* value. (Here the value of t is changed from 0 to 1.) Likewise, the instruction `S = S + deltaS` gives S a new value. What was the old value, and what is the new?

Compare the three lines of code that produce new values of S, I, and R with How a program computes new values the original equations that we used to define Step III back on page 14:

$$\text{S = S + deltaS} \qquad \text{new } S = \text{current } S + \Delta S$$
$$\text{I = I + deltaI} \qquad \text{new } I = \text{current } I + \Delta I$$
$$\text{R = R + deltaR} \qquad \text{new } R = \text{current } R + \Delta R$$

The words "new" and "current" aren't needed in the computer code because they are automatically understood to be there. Why? First of all, a symbol (like S) always has a *current* value, but an instruction can give it a *new* value. Second, a computer instruction of the form A = B is always understood to mean "new A = current B."

> Notice that the instructions A = B and B = A mean different things. The second says "new B = current A." Thus, in A = B, A is altered to equal B, while in B = A, B is altered to equal A. To emphasize that the symbol on the left is always the one affected, some programming languages use a modified equals sign, as in A := B. We sometimes read this as "A gets B."

The next line is another PRINT statement, exactly like the one on line 6. Line 18 It causes the current values of t, S, I, and R to be printed on the computer monitor screen. But this time what appears is

$$1 \qquad 44446.6 \qquad 2903.4 \qquad 2650$$

The values were changed by the previous four instructions. It is important to remember that the computer carries out instructions in the order they are written. Had the second PRINT statement appeared right after line 13, say, the old values of t, S, I, and R would have appeared on the monitor screen a second time.

Lines 7 and 19—the loop

We will take the last line and line 7 together. They are the instructions for the loop. Consider the situation when we reach the last line. The variables t, S, I, and R now have their "day 1" values. To continue, we need an instruction that will get us back to line 8, because the instructions on lines 8–17 will convert the current (day 1) values of t, S, I, and R into their "day 2" values. That's what lines 7 and 19 do.

Here is the meaning of the instruction FOR k = 1 TO 3 on line 7:

FOR k = 1 TO 3

Give k the value 1, and be prepared later to
give it the value 2 and then the value 3.

The variable k plays the role of a **counter**, telling us how many times we have gone around the loop. Notice that k did not appear in our hand calculations. However, when we said we had done three rounds of calculations, for example, we were really saying $k = 3$.

After the computer reads and executes line 7, it carries out all the instructions from lines 8 to 18, arriving finally at the last line. The computer then interprets the instruction NEXT k on the last line as follows:

NEXT k

Give k the next value that the FOR command allows, and
move back to the line immediately after the FOR command.

How the program stops

After the computer carries out this instruction, k has the value 2 and the computer is set to carry out the instruction on line 8. It then executes that instruction, and continues down the program, line by line, until it reaches line 19 once again. This sets the value of k to 3 and moves the computer back to line 8. Once again it continues down the program to line 19. This time there is no allowable value that k can be given, so the program stops.

The NEXT k command is different from all the others in the program. It is the only one that directs the computer to go to a different line. That action causes the program to **loop**. Because the loop involves all the indented instructions between the FOR statement and the NEXT statement, it is called a **FOR–NEXT loop**. This is just one kind of loop. Computer programs can contain other sorts that carry out different tasks. In the next chapter we will see how a **DO–WHILE loop** is used.

Exercises

The Program SIR

The object of these exercises is to verify that the program SIR works the way the text says it does. Follow the instructions for running a program on the computer you are using.

1. Run the program to confirm that it reproduces what you have already calculated by hand (table, page 11).

2. On a copy of the program, mark the instructions that carry out the following tasks:
   a) Give the input values of S, I, and R;
   b) Say that the calculations take us 1 day into the future;
   c) Carry out step II (see page 14);
   d) Carry out step III;
   e) Give us the output values of S, I, and R;
   f) Take us once around the whole loop;
   g) Say how many times we go around the loop.

3. Delete all the lines of the program from line 7 onward (or else type in the first 6 lines). Will this program run? What will it do? Run it and report what you see. Is this what you expected?

4. Starting with the original SIR program on page 44, delete lines 7 and 19. These are the ones that declare the FOR–NEXT loop. Will this program run? What will it do? Run it and report what you see. Is this what you expected?

5. Using the 17-line program you constructed in the previous question, remove the PRINT statement from the last line and insert it between what appear as lines 13 and 14 on page 44. Will this program run? What will it do? Run it and report what you see. Is this what you expected?

6. Starting with the original SIR program on page 44, change line 7 so it reads FOR k = 26 TO 28. Thus, the counter k takes the values 26, 27, and 28. Will this program run? What will it do? Run it and report what you see. Is this what you expected?

Programs to Practice On

In this section there are a number of short programs for you to analyze and run.

Program 1	Program 2	Program 3
A = 2	A = 2	A = 2
B = 3	B = 3	B = 3
A = B	B = A	A = A + B
PRINT A, B	PRINT A, B	B = A + B
		PRINT A, B

7. When Program 1 runs it will print the values of A and B that are current when the program stops. What values will it print? Type in this program and run it to verify your answers.

8. What will Program 2 do when it runs? Type in this program and run it to verify your answers.

9. After each line in Program 3 write the values that A and B have *after* that line has been carried out. What values of A and B will it print? Type in this program and run it to verify your answers.

The next three programs have an element not found in the program SIR. In each of them, there is a FOR–NEXT loop, and the counter k actually appears in the statements within the loop.

Program 4
```
FOR k = 1 TO 5
    A = k ^ 3
    PRINT A
NEXT k
```

Program 5
```
FOR k = 1 TO 5
    A = k ^ 3
NEXT k
PRINT A
```

Program 6
```
x = 0
FOR k = 1 TO 5
    x = x + k
    PRINT k, x
NEXT k
```

10. What output does Program 4 produce? Type in the code and run the program to confirm your answer.

11. What is the difference between the code in Program 5 and the code in Program 4? What is the output of Program 5? Does it differ from the output of Program 4? If so, why?

*12. What output does Program 6 produce? Type in the code and run the program to confirm your answer.

Program 7
```
A = 0
B = 0
FOR k = 1 TO 5
    A = A + 1
    B = A + B
    PRINT A, B
NEXT k
```

Program 8
```
A = 0
B = 1
FOR k = 1 TO 5
    A = A + B
    B = A + B
    PRINT A, B
NEXT k
```

Program 9
```
A = 0
B = 1
FOR k = 1 TO 5
    A = A + B
    B = A + B
NEXT k
PRINT A, B
```

13. Program 7 prints five lines of output. What are they? Type in the program and run it to confirm your answers.

14. What is the output of Program 8? Type in the program and run it to confirm your answers.

15. Describe exactly how the *codes* for Programs 8 and 9 differ. How do the *outputs* differ?

Analyzing the Measles Epidemic

16. Alter the program SIR to have it calculate estimates for S, I, and R over the first *six* days. Construct a table that shows those values.

17. Alter the program to have it estimate the values of S, I, and R for the first *thirty* days.
    *a) What are the values of S, I, and R when $t = 30$?
    *b) According to these figures, on what day does the infection peak? What values do you get for S, I, and R?
    c) How can you reconcile the value you just got for S with the value we obtained algebraically on page 16?

18. By adding an appropriate PRINT statement after line 10 you can also get the program to print values for S', I', and R'. Do this, and check that you get the values shown in the table on page 11.

*19. According to these estimates, on what day do the largest number of new infections occur? How many are there? Explain where you got your information.

20. On what day do you estimate that the largest number of recoveries occurs? Do you see a connection between this question and 17b?

*21. On what day do you estimate the infected population grows most rapidly? Declines most rapidly? What value does I' have on those days?

22. a) Alter the original SIR program so that it will go *backward* in time, with time steps of 1 day. Specify the changes you made in the program. Use this altered program to obtain estimates for the values of S, I, and R yesterday. Compare your estimates with those in the text (page 12).
    b) Estimate the values of S, I, and R *three* days before today.

23. According to the S-I-R model, when did the infection begin? That is, how many days before today was the estimated value of I approximately 1?

24. *There and back again.* Use the SIR program, modified as necessary, to carry out the calculations described in exercise 18 on page 20. Do your computer results agree with those you obtained earlier?

25. In problems 22–26 at the end of section 1.2 you set up rate equations to model some other systems. Choose a couple of these and think of some interesting questions that could be answered using a suitably modified SIR program. Make the modifications and report your results.

➤ 1.4 CHAPTER SUMMARY

The Main Ideas

- Natural processes like the spread of disease can often be described by **mathematical models**. Initially, this involves identifying **numerical quantities** and relations between them.

- A relation between quantities often takes the form of a **function**. A function can be described in many different ways: **graphs**, **tables**, and **formulas** are among the most common.
- **Linear functions** make up a special but important class. If y is a linear function of x, then $\Delta y = m \cdot \Delta x$, for some constant m. The constant m is a **multiplier**, **slope**, and **rate of change**.
- If $y = f(x)$, then we can consider the **rate of change** y' of y with respect to x. A mathematical model whose variables are connected by **rate equations** can be analyzed to predict how those variables will change.
- Predicted changes are **estimates** of the form $\Delta y = y' \cdot \Delta x$.
- The computations that produce estimates from rate equations can be put into a **loop**, and they are readily carried out on a **computer**.
- A computer increases the **scope** and **complexity** of the problems we can consider.

Expectations

- You should be able to work with functions given in various forms, to find the output for any given input.
- You should be able to read a graph. You should also be able to construct the graph of a linear function directly, and the graph of a more complicated function using a computer graphing package.
- You should be able to determine the natural domain of a function given by a formula.
- You should be able to express proportional quantities by a linear function, and interpret the constant of proportionality as a multiplier.
- Given any two of these quantities for a linear function—multiplier, change in input, change in output—you should be able to determine the third.
- You should be able to model a situation in which one variable is proportional to its rate of change.
- Given the value of a quantity that depends on time and given its rate of change, you should be able to estimate values of the quantity at other times.
- For a set of quantities determined by rate equations and initial conditions, you should be able to estimate how the quantities change.
- Given a set of rate equations, you should be able to determine what happens when one of the quantities reaches a maximum or minimum, or remains unchanged over time.
- You should be able to understand how a computer program with a FOR–NEXT loop works.

Chapter Exercises
A Model of an Orchard

If an apple orchard occupies one acre of land, how many trees should it contain so as to produce the largest apple crop? This is an example of an **optimization** problem. The word *optimum* means "best possible"—especially, the best under a given set of conditions. These exercises seek an optimum by analyzing a simple mathematical model of the orchard. The model is the function that describes how the total yield depends on the number of trees.

An immediate impulse is just to plant a lot of trees, on the principle "more trees, more apples." But there is a catch: if there are too many trees in a single acre, they crowd together. Each tree then gets less sunlight and nutrients, so it produces fewer apples. For example, the relation between the *yield per tree*, Y, and the *number of trees*, N, may be like that shown in the graph on the left, below.

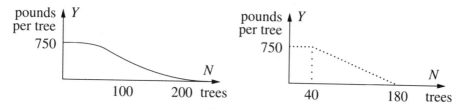

When there are only a few trees, they don't get in each other's way, and they produce at the maximum level—say, 750 pounds per tree. Hence the graph starts off level. At some point, the trees become too crowded to produce *anything*! In between, the yield per tree drops off as shown by the curved middle part of the graph.

We want to choose N so that the *total yield*, T, will be as large as possible. We have $T(N) = Y(N) \cdot N$, but since we don't know $Y(N)$ very precisely, it is difficult to analyze $T(N)$. To help, let's replace $Y(N)$ by the approximation shown in the graph on the right. Now carry out an analysis using this graph to represent $Y(N)$.

1. Find a formula for the straight segment of the new graph of $Y(N)$ on the interval $40 \le N \le 180$. What is the formula for $T(N)$ on the same interval?

2. What are the formulas for $T(N)$ when $0 \le N \le 40$ and when $180 \le N$? Graph T as a function of N. Describe the graph in words.

*3. What is the maximum possible total yield T? For which N is this maximum attained?

4. Suppose the endpoints of the sloping segment were P and Q, instead of 40 and 180, respectively. Now what is the formula for $T(N)$? (Note

that P and Q are *parameters* here. Different values of P and Q will give different models for the behavior of the total output.) How many trees would then produce the maximum total output? Expect the maximum to depend on the parameters P and Q.

Rate Equations

Do the following exercises by hand. You may wish to check your answers by using suitable modifications of the program SIR.

5. *Radioactivity.* From exercise 27 on page 42 we know a sample of R grams of radium decays into lead at the rate

$$R' = -\frac{1}{2337} R \quad \text{gram per year.}$$

Using a step size of 10 years, estimate how much radium remains in a .072 gram sample after 40 years.

6. *Poland and Afghanistan.* If P and A denote the populations of Poland and Afghanistan, respectively, then their net per capita growth rates imply the following equations:

$$P' = .009\,P \quad \text{persons per year;}$$
$$A' = .0216\,A \quad \text{persons per year.}$$

(See exercise 25 on page 42.) In 1985, $P = 37.5$ million, $A = 15$ million. Using a step size of 1 year, estimate P and A in 1990.

7. *Falling bodies.* If d and v denote the distance fallen (in feet) and the velocity (in feet per second) of a falling body, then the motion can be described by the following equations:

$$d' = v \quad \text{feet per second;}$$
$$v' = 32 \quad \text{feet per second per second.}$$

(See exercise 23 on page 41.) Assume that when $t = 0$, $d = 0$ feet and $v = 10$ feet/sec. Using a step size of 1 second, estimate d and v after 3 seconds have passed.

CHAPTER 2

SUCCESSIVE APPROXIMATIONS

In this chapter we continue exploring the mathematical implications of the S-I-R model. In the last chapter we calculated future values of S, I, and R by assuming that the rates S', I', and R' stayed fixed for a whole day. Since the rates are *not* fixed—they change with S, I, and R—the values of S, I, and R we obtained have to be considered as estimates only. In this chapter we will see how to build a succession of better and better estimates that get us as close as we wish to the true values implied by the model.

This method of **successive approximation** is a basic tool of calculus. It is the one fundamentally new process you will encounter, the ingredient that sets calculus apart from the mathematics you have already studied. With it you will be able to solve a vast array of problems that other methods can't handle.

➤ 2.1 MAKING APPROXIMATIONS

In chapter 1 we looked at the specific S-I-R model:

$$S' = -.00001\,SI$$
$$I' = .00001\,SI - I/14$$
$$R' = I/14$$

with initial values at time $t = 0$:

$$S = 45400 \qquad I = 2100 \qquad R = 2500.$$

Rate equations tell us
where to go next

We originally developed this model as a *description* of the relations among the different components of an epidemic. Almost immediately, though, we began using the rate equations in the model as a *recipe* for predicting what happens over the course of the epidemic: If we know at some time t the values of $S(t), I(t)$, and $R(t)$, then the equations tell us how to estimate values of the functions at other times. We used this approach in the last chapter to move backwards and forwards in time, calculating the values of S, I, and R as we went.

While we got numbers, there were some questions about how accurate these numbers were—that is, how exactly they represented the values implied by the model. In the process we called "there and back again" we used current values of S, I, and R to find the rates, used these rates to go forward one day, recalculated the rates, and came back to the present—and we got different values from the ones we started with! Resolving this discrepancy will be an important feature of the technique developed in this section.

The Longest March Begins with a Single Step

So far, in generating numbers from the S-I-R rate equations, we have assumed that the rates remained constant over an entire day, or longer. Since the rates aren't constant—they depend on the values of S, I, and R, which are always changing—the values we calculated for the variables at times other than the given initial time are, at best, estimates. These estimates, while incorrect, are not useless. Let's see how they behave in the "there and back again" process of chapter 1 as we recalculate the rates more and more frequently, producing a sequence of approximations to the values we are looking for.

There and Back Again Again

On pages 11–12 in chapter 1 we used the rate equations to go forward a day and come back again. We started with the initial values

$$S(0) = 45400 \qquad I(0) = 2100 \qquad R(0) = 2500,$$

calculated the rates, went forward a day to $t = 1$, recalculated the rates, and came back a day to $t = 0$. We ended up with the estimates

$$S(0) = 45737.1 \qquad I(0) = 1820.3 \qquad R(0) = 2442.6,$$

which are rather far from the values of $S(0), I(0)$, and $R(0)$ we started with.

A clue to the resolution of this discrepancy appeared in problem 18, page 20. There you were asked to go forward two days and come back again in two different ways, using $\Delta t = 2$ (a total of 2 steps) in the first case and $\Delta t = 1$ (a total of 4 steps) in the second. Here are the resulting values calculated for $S(0)$ in each case and the discrepancy between this value and the original value $S(0) = 45400$:

Step size	New $S(0)$	Discrepancy
$\Delta t = 2$	46717.6	1317.6
$\Delta t = 1$	46021.3	621.3

While the discrepancy is fairly large in either case, $\Delta t = 1$ clearly does better than $\Delta t = 2$. But if smaller is better, why stop at $\Delta t = 1$? What happens if we take even smaller time steps, get the corresponding new values of S, I, and R, and use these values to recalculate the rates each time?

Smaller steps generate a smaller discrepancy

Recall that the rate S' (or I' or R') is simply the multiplier which gives ΔS—the (estimated) change in S—for a given change Δt in t

$$\Delta S = S' \cdot \Delta t.$$

This relation holds for any value of Δt, integer or not. Once we have this value for ΔS, we can then calculate

$$\text{new (estimated) } S = \text{current (estimated) } S + \Delta S$$

in the usual way. Note that we have written "(estimated)" throughout to emphasize the fact that if S' is not constant over the entire time Δt, then the value we get for ΔS will typically be only an approximation to the real change in S.

Let's try going forward one day and coming back, using different values for Δt. As we reduce Δt the number of calculations will increase. The program SIR we used in the last chapter can still be used to do the tedious calculations. Thus if we decide to use 10 steps of size $\Delta t = .1$, we would just change two lines in that program:

```
deltat = .1
FOR k = 1 TO 10
```

If we now run SIR with these modifications we can verify the following sequence of values (the values have been rounded off, and the PRINT statement has been modified to show the new values of the rates at each step as well):

Estimated values of S, I, and R for step sizes $\Delta t = .1$

t	$S(t)$	$I(t)$	$R(t)$	$S'(t)$	$I'(t)$	$R'(t)$
0.0	45 400.0	2100.0	2500.0	−953.4	803.4	150.0
0.1	45 304.7	2180.3	2515.0	−987.8	832.1	155.7
0.2	45 205.9	2263.6	2530.6	−1023.3	861.6	161.7
0.3	45 103.6	2349.7	2546.7	−1059.8	892.0	167.8
⋮	⋮	⋮	⋮	⋮	⋮	⋮
1.0	44 278.7	3042.9	2678.4	−1347.7	1130.0	217.4

Having arrived at $t = 1$, we can now use SIR to turn around and go back to $t = 0$. Here's how:

- Change the initial line of the program to $t = 1$ to reflect our new starting time.
- Change the next three lines to use the values we just calculated for $S(1)$, $I(1)$, and $R(1)$ as our *starting* values in SIR.
- Change the value of deltat to be $-.1$ (so each time the program executes the command $t = t + \texttt{deltat}$ it reduces the value of t by .1).

The sign of deltat *determines whether we move forward or backward in time*

With these changes SIR will yield the desired estimates for $S(0)$, $I(0)$, and $R(0)$, and we get

$$S(0) = 45433.5 \quad I(0) = 2072.3 \quad R(0) = 2494.3.$$

This is clearly a considerable improvement over the values obtained with $\Delta t = 1$.

With this promising result, the obvious thing to do is to try even smaller values of Δt, perhaps $\Delta t = .01$. We could continue using SIR, making the needed modifications each time. Instead, though, let's rewrite SIR slightly to make it better suited to our current needs. Look at the program SIRVALUE below and compare it with SIR. (You will find translations of SIRVALUE for various graphing calculators in Appendix A.)

Program: SIRVALUE

```
tinitial = 0
tfinal = 1
t = tinitial
S = 45400
I = 2100
R = 2500
numberofsteps = 10
deltat = (tfinal - tinitial)/numberofsteps
FOR k = 1 TO numberofsteps
    Sprime = -.00001 * S * I
    Iprime = .00001 * S * I - I / 14
    Rprime = I / 14
    deltaS = Sprime * deltat
    deltaI = Iprime * deltat
    deltaR = Rprime * deltat
    t = t + deltat
    S = S + deltaS
    I = I + deltaI
    R = R + deltaR
NEXT k
PRINT t, S, I, R
```

Program: SIR

```
t = 0
S = 45400
I = 2100
R = 2500
deltat = .1
FOR k = 1 TO 10
    Sprime = -.00001 * S * I
    Iprime = .00001 * S * I - I / 14
    Rprime = I / 14
    deltaS = Sprime * deltat
    deltaI = Iprime * deltat
    deltaR = Rprime * deltat
    t = t + deltat
    S = S + deltaS
    I = I + deltaI
    R = R + deltaR
    PRINT t, S, I, R
NEXT k
```

You will see that the major change is to place the PRINT statement outside the loop, so only the final values of S, I, and R get printed. This speeds up the work, since otherwise, with $\Delta t = .001$, for instance, we would be asking the computer to print out 1000 lines—about 30 screens of text! Another change is that the value of deltat no longer needs to be specified—it is automatically determined by the values of tinitial, tfinal, and numberofsteps.

As written above, SIR and SIRVALUE both run for 10 steps of size .1. By changing the value of numberofsteps in the program we can quickly get estimates for $S(1)$, $I(1)$, and $R(1)$ for a wide range of values for Δt. Moreover, once we have these estimates we can use SIRVALUE again to go backwards in time to $t = 0$, by making changes similar to those we made in SIR earlier. First, we need to change the value of tinitial to 1 and the value of tfinal to 0. Notice that this automatically will make deltat a negative quantity, so that each time we run through the loop we step back in time. Second, we need to set the starting values of S, I, and R to the values we just obtained for $S(1)$, $I(1)$, and $R(1)$. With these changes, SIRVALUE will give us the corresponding estimated values for $S(0)$, $I(0)$, and $R(0)$.

With a computer we can generate lots of data and look for patterns

If we use SIRVALUE with Δt ranging from 1 to .00001 (which means letting numberofsteps range from 1 to 100,000) we produce the table below. This table lists the computed values of $S(1)$, $I(1)$, and $R(1)$ for each Δt, followed by the estimated value of $S(0)$ obtained by running SIRVALUE backwards in time from these new values, and, finally, the discrepancy between this estimated value of $S(0)$ and the original value $S(0) = 45400$.

Estimated values of S, I, and R when $t = 1$,
for step sizes $\Delta t = 10^{-N}$, $N = 0, \dots, 5$,
together with the corresponding backwards estimate for $S(0)$.

Δt	$S(1)$	$I(1)$	$R(1)$	New $S(0)$	Discrepancy
1.0	44 446.6	2903.4	2650.0	45 737.0626	337.0626
0.1	44 278.6648	3042.9241	2678.4111	45 433.4741	33.4741
0.01	44 257.8301	3060.1948	2681.9751	45 403.3615	3.3615
0.001	44 255.6960	3061.9633	2682.3406	45 400.3363	.3363
0.000 1	44 255.4821	3062.1406	2682.3773	45 400.0336	.0336
0.000 01	44 255.4607	3062.1584	2682.3809	45 400.0034	.0034

There are several striking features of this table. The first is that if we go forward one day and come back again, we can get back as close as we want to our initial value of $S(0)$ *provided we recalculate the rates frequently enough*. After 200,000 rounds of calculations ($\Delta t = .00001$) we ended up only .0034 away from our starting value. In fact, there is a clear pattern to the values of the errors as we decrease the step size. In the exercises it is left for you to explore this pattern and show that similar results hold for I and for R.

Smaller steps generate a discrepancy which can be made as small as we like

A second feature is that as we read down the column under $S(1)$, we find each digit **stabilizes**—that is, after changing for a while, it eventually becomes fixed at a particular value. The initial digits 44 are the first to stabilize, and that happens by the time $\Delta t = .1$. Then the third digit 2 stabilizes, when $\Delta t = .01$. Roughly speaking, one more digit stabilizes at each successive level. The table is revealing to us, digit by digit, the true value of $S(1)$. By the fifth stage we learn that the integer part of $S(1)$ is 44255. By the sixth stage we can say that the true value of $S(1)$ is 44255.4 . . . .

When we write $S(1) = 44255.4$ . . . we are expressing $S(1)$ to **one decimal place accuracy**. This says, first, that the decimal expansion of $S(1)$ begins with exactly the six digits shown and, second, that there are further digits after the 4 (represented by the three dots, "..."). In this case, we can identify further digits simply by continuing the table. Since our step sizes have the form $\Delta t = 10^{-N}$, we just need to increase N. For example, to express $S(1)$ accurately to six decimal places, we need to stabilize the first eleven digits in our estimates of $S(1)$. The table suggests that Δt should probably be about 10^{-10}—that is, $N = 10$.

The true value of $S(1)$ emerges through a process that generates a sequence of successive approximations. We say $S(1) = 44255.4$ . . . is the **limit** of this sequence as Δt is made smaller and smaller or, equivalently, as N is made larger and larger. We also say that the sequence of successive approximations **converges** to the limit $S(1)$. Here is a mathematical notation that expresses these statements more compactly:

$$S(1) = \lim_{\Delta t \to 0} \{\text{the estimate of } S(1)\}$$

or, equivalently,

$$S(1) = \lim_{N \to \infty} \{\text{the estimate of } S(1) \text{ with } \Delta t = 10^{-N}\}.$$

The symbol ∞ stands for "infinity," and the expression $N \to \infty$ is often pronounced "as N goes to infinity." However, it is often more instructive to say "as N gets larger and larger, without bound."

You should check that similar patterns are occurring in the $I(1)$ and $R(1)$ columns as well.

The limit concept lies at the heart of calculus. Later on we'll give a precise definition, but you should first see limits at work in a number of contexts and begin to develop some intuitions about what they are. This approach mirrors the historical development of calculus—mathematicians freely used limits for well over a century before a careful, rigorous definition was developed.

One Picture Is Worth a Hundred Tables

As we noted, the program SIRVALUE prints out only the final values of S, I, and R because it would typically take too much space to print out the in-

Approximations lead to exact values

termediate values. However, if instead of printing these values we plot them graphically, we can convey all this intermediate information in a compact and comprehensible form.

Suppose, for instance, that we wanted to record the calculations leading up to $S(3)$ by plotting all the points. The graphs at the right plot all the pairs of values (t, S) that are calculated along the way for the cases $\Delta t = 1$ (4 points), $\Delta t = .1$ (31 points), and $\Delta t = .01$ (301 points).

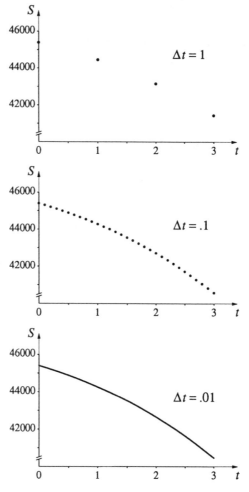

By the time we get to steps of size .01, the resulting graph begins to look like a continuous curve. This suggests that instead of simply plotting the points we might want to draw lines connecting the points as they're calculated.

We can easily modify SIRVALUE to do this—the only changes will be to replace the PRINT command with a command to draw a line and to move this command inside the loop (so that it is executed every time new values are computed). We will also need to add a line or two at the beginning to tell the computer to set up the screen to plot points. This usually involves opening a **window**— that is, specifying the horizontal and vertical ranges the screen should depict. Since programming languages vary slightly in the way this is done, we use italicized text "*Set up GRAPH-ICS*" to make clear that this statement is **not** part of the program—you will have to express this in the form your programming language specifies. Similarly, the command

"*Plot the line from* (t, S)

to (t + deltat, S + deltaS)"

will have to be stated in the correct format for your language. The computational core of SIRVALUE is unchanged. Here is what the new program looks like if we want to use $\Delta t = .1$ and connect the points with straight lines:

Program: SIRPLOT

```
Set up GRAPHICS
tinitial = 0
tfinal = 3
t = tinitial
S = 45400
I = 2100
R = 2500
numberofsteps = 30
deltat = (tfinal - tinitial)/numberofsteps
FOR k = 1 TO numberofsteps
      Sprime = -.00001 * S * I
      Iprime = .00001 * S * I - I / 14
      Rprime = I / 14
      deltaS = Sprime * deltat
      deltaI = Iprime * deltat
      deltaR = Rprime * deltat
      Plot the line from (t, S)
         to (t + deltat, S + deltaS)
      t = t + deltat
      S = S + deltaS
      I = I + deltaI
      R = R + deltaR
NEXT k
```

If we had wanted just to plot the points, we could have used a command of the form *Plot the point* (t, S) in place of the command to plot the line, moving this command down two lines so it came after we had computed the new values of t and S. We would also need to place that command before the loop so that the initial point corresponding to $t = 0$ is plotted.

When we "connect the dots" like this we emphasize graphically the underlying assumption we have been making in all our estimates: that the function $S(t)$ is linear (i.e., it is changing at a constant rate) over each interval Δt. Let's see what the graphs look like when we do this for the three values of Δt we used above. To compare the results more readily we'll plot the graphs on the same set of axes. (We will look at a program for doing this in the next section.) We get the following picture:

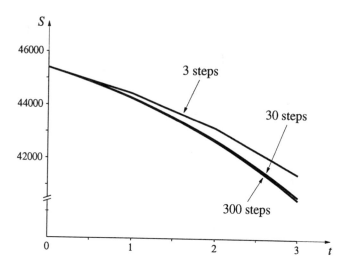

The graphs become indistinguishable from each other and increasingly look like smooth curves as the number of segments increases. If we plotted the 3000-step graph as well, it would be indistinguishable from the 300-step graph at this scale. If we now shift our focus from the end value $S(3)$ and look at all the intermediate values as well, we find that each graph gives an approximate value for $S(t)$ for *every* value of t between 0 and 3. We are seeing the entire function $S(t)$ over this interval.

Graphs made up of line segments look like smooth curves if the segments are short enough

Just as we wrote

$$S(3) = \lim_{N \to \infty} \{\text{the estimate of } S(3) \text{ with } \Delta t = 10^{-N}\},$$

we can also write

$$\text{graph of } S(t) = \lim_{N \to \infty} \{\text{line-segment approximations with } \Delta t = 10^{-N}\}.$$

The way we see the graph of $S(t)$ emerging from successive approximations is our first example of a fundamental result. It has wide-ranging implications which will occupy much of our attention for the rest of the course.

Piecewise Linear Functions

Let's examine the implications of this approach more closely by considering the "one-step" ($\Delta t = 3$) approximation to $S(t)$ and the "three-step" ($\Delta t = 1$) approximation over the time interval $0 \le t \le 3$. In the first case we are making the simplifying assumption that S decreases at the rate $S' = -953.4$ persons per day for the entire three days. In the second case we use three shorter steps of length $\Delta t = 1$, with the slopes of the corresponding segments given by the table on page 11 in chapter 1, summarized below (note that since $\Delta t = 1$ day we have that the magnitude of $\Delta S = S' \cdot \Delta t$ is the same as the magnitude of S'):

t	S	S'
0	45 400.0	-953.4
1	44 446.6	$-1\,290.5$
2	43 156.1	$-1\,720.4$
3	41 435.7	

Here are the corresponding graphs we produce:

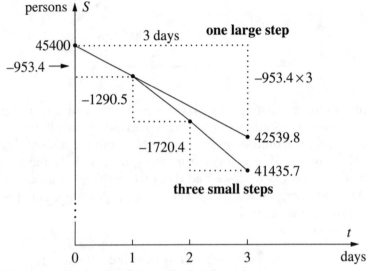

Two approximations to S during the first three days

The "One-Step" Estimate

Assuming that S decreases at the rate $S' = -953.4$ persons per day for the entire three days is equivalent to assuming that S follows the upper graph—a straight line with slope -953.4 persons/day. In other words, the one-step approach approximates S by a **linear function** of t. If we use the notation $S_1(t)$ to denote this (one-step) linear approximation, we have

One-step estimate: $S(t) \approx S_1(t) = 45400 - 953.4t.$

Because the one-step estimate is actually a function, we can find the value of $S_1(t)$ for *all* t in the interval $0 \leq t \leq 3$, not just $t = 3$, and thereby get corresponding estimates for $S(t)$ as well. For example,

$$S_1(2) = 45400 - 953.4 \times 2 = 43493.2$$

$$S_1(1.7) = 45400 - 953.4 \times 1.7 = 43779.22.$$

The "Three-Step" Estimate

With three smaller steps of size $\Delta t = 1$, we get a function whose graph is composed of three line segments, each starting at the (t, S) point at the beginning of each day and with a slope equal to the corresponding rate of new infections at the beginning of the day. Let's call this function $S_3(t)$. The three-step estimate $S_3(t)$ is hence not a linear function, strictly speaking. However, since its graph is made up of several straight pieces, it is called a **piecewise linear function**. Recalling that the equation of a line through the point (x_0, y_0) with slope m can be written in the form $y = m(x - x_0) + y_0$, we can use the values for S (which correspond to the y-values) and S' (which give us the slopes of the segments) at times $t = 0$, $t = 1$, and $t = 2$ calculated above to get an explicit formula for $S_3(t)$:

A single function may not be specified by a single equation

$$S_3(t) = \begin{cases} y = -953.4(t - 0) + 45400 & \text{if } 0 \le t \le 1 \\ y = -1290.5(t - 1) + 44446.6 & \text{if } 1 \le t \le 2 \\ y = -1720.4(t - 2) + 43156.1 & \text{if } 2 \le t \le 3 \end{cases}$$

Note that we have used $\le$ in the defining formulas since at the values $t = 1$ and $t = 2$ it doesn't matter which equation we use. This is equivalent to saying that the straight line segments are connected to each other. Since the slopes of the three segments of $S_3(t)$ are progressively more negative, the piecewise linear graph gets progressively steeper as t increases. This explains why the value $S_3(3)$ is *lower* than the value of $S_1(3)$. While it is rare that we would actually need to write down an explicit formula like this for the piecewise-linear approximation—it is easier, and usually more informative, just to define $S_3(t)$ by its graph—it is nevertheless important to realize that there really is an approximating function defined for all values of t in the interval $[0, 3]$, not just at the finite set of t values where we make the recalculations.

By the time we are dealing with the 300-step function $S_{300}(t)$ we can't even tell by its graph that it is piecewise linear unless we zoom in very close. In principle, though, we could still write down a simple linear formula for each of its segments (see the exercises).

An Appraisal

The graph of $S_1(t)$ gives us a rough idea of what is happening to the true function $S(t)$ during the first three days. It starts off at the same rate as S', but subsequently the rates move apart. The value of S_1' never changes, while S' changes with the (ever-changing) values of S and I.

The graph of $S_3(t)$ is a distinct improvement because it changes its direction twice, modifying its slope at the beginning of each day to come back into agreement with the rate equation. But since the three-step graph is still piecewise linear, it continues to suffer from the same shortcoming as the one-step: once we restrict our attention to a single straight segment (for example, where $1 \leq t \leq 2$), the three-step graph *also* has a constant slope, while S' is always changing. Nevertheless, $S_3(t)$ does satisfy the rate equation in our original model three times—at the beginning of each segment—and isn't too far off at other times. When we get to $S_{300}(t)$ we have a function which satisfies the rate equation at 300 times and is very close in between.

Each of these graphs gives us an idea of the behavior of the true function $S(t)$ during the time interval $0 \leq t \leq 3$. None is strictly correct, but none is hopelessly wrong, either. All are **approximations** to the truth. Moreover, $S_3(t)$ is a **better** approximation than $S_1(t)$—because it reflects at least some of the variability in S'—and $S_{300}(t)$ is better still. Thus, even before we have a clear picture of the shape of the true function $S(t)$, we would expect it to be closer to $S_{300}(t)$ than to $S_3(t)$. As we saw above, when we take piecewise linear approximations with smaller and smaller step sizes, it is reasonable to think that they will approach the true function S in the limit. Expressing this in the notation we have used before,

$$\text{the function } S(t) = \lim_{N \to \infty} \{\text{the chain of linear functions with } \Delta t = 10^{-N}\}.$$

Approximate versus Exact

You may find it unsettling that our efforts give us only a sequence of approximations to $S(3)$, and not the exact value, or only a sequence of piecewise linear approximations to $S(t)$, not the "real" function itself. In what sense can we say we "know" the number $S(3)$ or the function $S(t)$? The answer: in the same sense that we "know" a number like $\sqrt{2}$ or π. There are two distinct aspects to the way we know a number. On the one hand, we can **characterize** a number precisely and completely:

What does it mean to "know" a number like π?

π: the ratio of the circumference of a circle to its diameter;

$\sqrt{2}$: the positive number whose square is 2.

On the other hand, when we try to **construct** the decimal expansion of a number, we usually get only approximate and incomplete results. For example, when we do calculations by hand we might use the rough estimates $\sqrt{2} \approx 1.414$ and $\pi \approx 3.1416$. With a desktop computer we might have $\sqrt{2} \approx 1.414\,213\,562\,373\,095$ and $\pi \approx 3.141\,592\,653\,589\,793$, but these are still approximations, and we are really saying

$$\sqrt{2} = 1.414\,213\,562\,373\,09\ldots$$

$$\pi = 3.141\,592\,653\,589\,79\ldots.$$

The *complete* decimal expansions for $\sqrt{2}$ and π are unknown! The exact values exist as limits of approximations that involve successively longer strings of digits, but we never see the limits—only approximations. In the final section of this chapter we will see ways of generating these approximations for $\sqrt{2}$ and for π.

What we say about π and $\sqrt{2}$ is true for $S(3)$ in exactly the same way. We can *characterize* it quite precisely, and we can *construct* approximations to its numerical value to any desired degree of accuracy. Here, for example, is a characterization of $S(3)$:

> The *S-I-R* problem for which $a = .00001$ and $b = 1/14$ and for which $S = 45400$, $I = 2100$, $R = 2500$ when $t = 0$ determines three functions $S(t)$, $I(t)$, and $R(t)$. The number $S(3)$ is the value that the function $S(t)$ has when $t = 3$.

You should try to extend this argument to describe the sense in which we "know" the function $S(t)$ by knowing its piecewise-linear approximations. Try to convince yourself that this is operationally no different from the way we "know" functions like $f(x) = \sqrt{x}$. In each instance we can **characterize** the function completely, but we can only **construct** an approximation to most values of the function or to its graph.

All this discussion of approximations may strike you as an unfortunate departure from the accuracy and precision you may have been led to expect in mathematics up until now. In fact, it is precisely this ability to make quick and accurate approximations to problems that is one of the most powerful features of mathematics. This is what goes on every time you use your calculator to evaluate log 3 or sin 37. Your calculator doesn't really know what these numbers are—but it does know how to approximate them quickly to 12 decimal places. Similar kinds of approximations are also at the heart of how bridges are built and spaceships are sent to the moon.

A Caution: The fact that computers and calculators are really only dealing with approximations when we think they are being exact occasionally leads to problems, the most common of which involves **roundoff errors**. You can probably generate a relatively harmless manifestation of this on your computer with the SIRVALUE program. Modify the PRINT line so it prints out the final value of t to 10 or 12 digits, and try running it with a high value for numberofsteps, say 1 million or 10 million. You would expect the final value of t to be exactly 1 in every case, since you are adding deltat = 1/numberofsteps to itself numberofsteps times. The catch is that the computer doesn't store the exact value 1/numberofsteps unless numberofsteps is a power of 2. In all other cases it will only be using an approximation, and if you add up enough quantities that are slightly off, their cumulative error will begin to show. We will encounter a somewhat less benign manifestation of roundoff error in the next chapter.

Being able to approximate a number to 12 decimal places is usually as good as knowing its value precisely

However ...

Exercises

There and Back Again

1. a) Look at the table on page 57. What is your best guess of the *exact* value of $I(1)$? (Use the "..." notation introduced on page 58.)
    b) What is the exact value of $R(1)$?

2. We noted that the discrepancy (the difference between the new estimate for $S(0)$ and the original value) seemed to decrease as Δt decreased.
    a) What is your best estimate (using only the information in the table) for the value of Δt needed to produce a discrepancy of .001?
    b) More generally, express as precisely as you can the apparent relation between the size of the discrepancy and the size of Δt.

3. a) Suppose you wanted to try going three days forward and then coming back, using $\Delta t = .01$. What changes would you have to make in SIRVALUE to do this?
    b) Make a table similar to the one on page 55 for going three days forward and coming back for $\Delta t = 1, .1, .01,$ and .001.
    c) In this new table how does the size of the discrepancy for a given value of Δt compare with the value in the original table?
    d) What value of Δt do you think you would need to determine the integer parts of $S(3)$ and $R(3)$ exactly?

Piecewise Linear Functions

*4. Using the three-step approximation on page 63, what is $S_3(1.7)$? What is $S_3(2.5)$?

5. How would you modify SIRVALUE to get $S_{3000}(3)$? Do it; what do you get?

6. What additional changes would you make to get the values of t, S, and S' at the beginning of the 193rd segment of $S_{300}(t)$? [HINT: You only need to alter the FOR k = 1 TO numberofsteps line (since you don't want to go all the way to the end) and the PRINT line. (Note that after running the loop for, say, 20 times, the values of t and S are the values for the beginning of the twenty-first segment, while the value of Sprime will still be the slope of the twentieth segment.)]

7. Suppose we wanted to determine the value of $S_{300}(2.84135)$.
    a) In which of the 300 segments of the graph of $S_{300}(t)$ would we look to find this information?
    b) What are the values of t, S, and S' at the beginning of this segment?
    c) What is the equation of this segment?
    d) What is $S_{300}(2.84135)$?

8. How would you modify SIRVALUE to calculate estimates for S, I, and R when $t = -6$, using $\Delta t = .05$? Do it; what do you get?

9. We want to use SIRPLOT to look at the graph of $S(t)$ over the first 20 days, using $\Delta t = .01$.
   a) What changes would we have to make in the program?
   b) Sketch the graph you get when you make these changes.
   c) If you wanted to plot the graph of $I(t)$ over this same time interval, what additional modifications to SIRPLOT would be needed? Make them, and sketch the result. When does the infection appear to hit its peak?
   d) Modify SIRPLOT to sketch on the same graph all three functions over the first 70 days. Sketch the result.

The DO-WHILE Loop

A difficulty in giving a precise answer to the last question was that we had to get all the values for 20 days, then go back to estimate by eye when the peak occurred. It would be helpful if we could write a program that ran until it reached the point we were looking for, and then stopped. To do this, we need a different kind of loop—a *conditional* loop that keeps looping only while some specified condition is true. A DO-WHILE loop is one useful way to do this. Here's how the modified SIRPLOT program would look:

```
Set up GRAPHICS
tinitial = 0
t = tinitial
S = 45400
I = 2100
R = 2500
Iprime = .00001 * S * I - I / 14
deltat = .01
DO WHILE Iprime > 0
        Sprime = -.00001 * S * I
        Iprime = .00001 * S * I - I / 14
        Rprime = I / 14
        deltaS = Sprime * deltat
        deltaI = Iprime * deltat
        deltaR = Rprime * deltat
        Plot the line from (t, S)
           to (t + deltat, S + deltaS)
        t = t + deltat
        S = S + deltaS
        I = I + deltaI
        R = R + deltaR
   LOOP
PRINT t - deltat
```

The changes we have made are:

- Since we don't know what the final time will be, eliminate the `tfinal = 20` and the `numberofsteps = 2000` commands.
- Since the condition in our loop is keyed to the value of I', we have to calculate the initial value of I' before the loop starts.
- Instead of `deltat = (tfinal - tinitial)/numberofsteps` use the statement `deltat = .01`.
- The key change is to replace the `FOR k = 1 TO numberofsteps` line with the line `DO WHILE Iprime > 0`.
- To denote the end of the loop we replace the `NEXT k` command with the command `LOOP`.
- After the `LOOP` command add the line `PRINT t - deltat`. (The reason we had to back up one step at the end is because the way the program was written, the computer takes one final step with a negative value for `Iprime` before it stops.)

The net effect of all this is that the program will continue working as before, calculating values and plotting points (`t, S`), but *only as long as the condition in the* `DO WHILE` *statement is true.* The condition we used here was that I' had to be positive—this is the condition that ensures that values of I are still getting bigger. While this condition is true, we can always get a larger value for I by going forward another increment Δt. As soon as the condition is false—that is, as soon as I' is negative—the values for I will be decreasing, which means we have passed the peak and so want to stop.

10. Make these modifications; what value for t do you get?

11. You could modify the `PRINT t` command to also print out other quantities.
    a) What is the value of I at its peak?
    b) What is the value of S when I is at its peak? Does this agree with the threshold value we predicted in chapter 1?

12. Suppose you change the initial value of S to be 5400 and run the previous program. Now what happens? Why?

13. Suppose we wanted to know how long the epidemic lasts. We could use DO-WHILE and keep stepping forward using, say $\Delta t = .1$, so long as $I \geq 1$. As soon as I was less than 1 we would want to stop and see what the value of t was.
    *a) What modifications would you make in SIRVALUE to get this information?
    *b) Run your modified program. What value do you get for t?
    c) Run the program using $\Delta t = .01$. Now what is your estimate for the duration of the epidemic? What can you say about the actual time required for I to drop below 1?

14. a) If we think the epidemic started with a single individual, we can go backwards in time until I is no longer greater than 1, and see what the corresponding time is. Do this for $\Delta t = -1, -.1,$ and $-.01$. What is your best estimate for the time that the infection arrived?

b) How many Recovereds were there at the start of the epidemic? This would be the people who had presumably been infected in a previous epidemic and now had immunity.

➤ 2.2 THE MATHEMATICAL IMPLICATIONS— EULER'S METHOD

Approximate Solutions

In the last section we approximated the function $S(t)$ by piecewise-linear approximations using steps of size $\Delta t = 1, .1,$ and $.01$. This process can clearly be extended to produce approximations with an *arbitrary* number of steps. For any given step size Δt, the result is a piecewise linear graph whose segments are Δt days wide. This graph then provides us with an estimate for $S(t)$ for every value of t in the interval $0 \le t \le 3$. We call this process of obtaining a function by constructing a sequence of increasingly better approximations **Euler's method**, after the Swiss mathematician Leonhard Euler (1707–1783). Euler was interested in the general problem of finding the functions determined by a set of rate equations, and in 1768 he proposed this method to approximate them. The method is conceptually simple and can indeed be used to get solutions for an enormous range of rate equations. For this reason we will make it a basic tool.

To begin to get a sense of the general utility of Euler's method, let's use it in a new setting. Here is a simple problem that involves just a single variable y that depends on t.

$$\text{rate equation:}\quad y' = .1y\left(1 - \frac{y}{1000}\right),$$

$$\text{initial condition:}\quad y = 100 \text{ when } t = 0.$$

To solve the problem we must find the function $y(t)$ determined by this rate equation and initial condition. We'll work without a context, in order to emphasize the purely mathematical nature of Euler's method. However, this rate equation is one member of a family called **logistic equations** that are frequently used in population models. We will explore this context in the exercises.

Let's construct the function approximating $y(t)$ on the interval $0 \le t \le 75$, using $\Delta t = .5$. We can make the suitable modifications in SIRPLOT or SIR-VALUE to get this approximation. Suppose we want to view the approximation graphically. Here's what the modified SIRPLOT would look like:

Program: modified SIRPLOT

```
Set up GRAPHICS
tinitial = 0
tfinal = 75
t = tinitial
y = 100
numberofsteps = 150
deltat = (tfinal - tinitial)/numberofsteps
FOR k = 1 TO numberofsteps
     yprime = .1 * y * (1 - y / 1000)
     deltay = yprime * deltat
     Plot the line from (t, y)
        to (t + deltat, y + deltay)
     t = t + deltat
     y = y + deltay
NEXT k
```

As before, words in italics, like "*Plot the line from,*" need to be translated into the specific formulation required by the computer language you are using.

When we run this program, we get the following graph (axes and scales have been added):

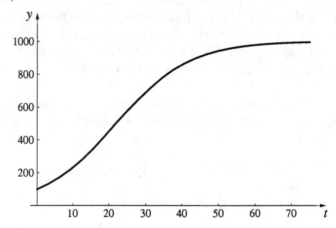

As before, what we see here is only an approximation to the true solution $y(t)$. How can we get some idea of how good this approximation is?

Exact Solutions

For any chosen step size, we can produce an *approximate* solution to a rate equation problem. We will call such an approximation an **Euler approximation**. In the last section we saw that we can improve the accuracy of the approximation by making the steps smaller and using more of them.

For example, we have already found an approximate solution to the problem

$$y' = .1y\left(1 - \frac{y}{1000}\right); \qquad y(0) = 100$$

on the interval $0 \le t \le 75$, using 150 steps of size $\Delta t = 1/2$. Consider a **sequence** of Euler approximations to this problem that are obtained by increasing the number of steps from one stage to the next. To be systematic, let the first approximation have 1 step, the next 2, the next 4, and so on. (The important feature is that the number of steps increases from one approximation to the next, not necessarily that they double—going up by powers of 10 would be just as good. A slight advantage in using powers of 2 is to maximize computer accuracy.) The number of steps thus has the form 2^{j-1}, where $j = 1, 2, 3, \ldots$. If we use $y_j(t)$ to denote the approximating function with 2^{j-1} steps, then we have an unending list:

$$y_1(t): \quad \text{Euler's approximation with 1 step}$$
$$y_2(t): \quad \text{Euler's approximation with 2 steps}$$
$$y_3(t): \quad \text{Euler's approximation with 4 steps}$$
$$\vdots \qquad\qquad \vdots$$
$$y_j(t): \quad \text{Euler's approximation with } 2^{j-1} \text{ steps}$$
$$\vdots \qquad\qquad \vdots$$

Here are the graphs of $y_j(t)$ for $j = 1, 2, \ldots, 7$ plus the graph of $y_{11}(t)$:

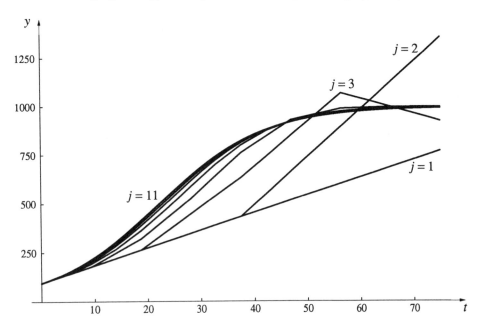

These functions form a sequence of **successive approximations** to the true solution $y(t)$, which is obtained by taking the **limit**, as we did in the last section:

$$y(t) = \lim_{j \to \infty} y_j(t).$$

Functions and graphs can be limits, too

Earlier we noticed how the digits in the estimates for $S(3)$ stabilized. If we plot the approximations $y_j(t)$ together we'll find that they stabilize, too. Each graph in the sequence is different from the preceding one, but the differences diminish the larger j becomes. Eventually, when j is large enough, the graph of y_{j+1} does not differ noticeably from the graph of y_j. That is, the position of the graph **stabilizes** in the coordinate plane. In this example, at the scale in the graph above, this happens around $j = 11$. If we had drawn the graph of y_{15} or y_{20}, it would not have been distinguishable from the graph of y_{11}. It is this entire process of calculating a sequence of successive approximations using increasingly many steps as far as is needed to get the desired level of stabilization that is meant when we talk about Euler's method.

Euler's method is the process of finding solutions through a sequence of successive approximations

The program SEQUENCE shown below plots 14 Euler approximations to $y(t)$, increasing the number of steps by a factor of 2 each time. It demonstrates how the graphs of $y_j(t)$ stabilize to define $y(t)$ as their limit.

Program: SEQUENCE
A sequence of graphs for $y' = .1y(1 - y/1000); y(0) = 100$

```
Set up GRAPHICS
FOR j = 1 TO 14
    tinitial = 0
    tfinal = 75
    t = tinitial
    y = 100
    numberofsteps = 2 ^ (j - 1)
    deltat = (tfinal - tinitial) / numberofsteps
    FOR k = 1 TO numberofsteps
        yprime = .1 * y * (1 - y / 1000)
        deltay = yprime * deltat
        Plot the line from (t, y)
          to (t + deltat, y + deltay)
        Color the line with color  j
        t = t + deltat
        y = y + deltay
    NEXT k
NEXT j
```

Program:
modified SIRPLOT

Notice that SEQUENCE contains the program SIRPLOT embedded in a loop that executes SIRPLOT 14 times. In this way SEQUENCE plots 14 different graphs. The only new element that has been added to SIRPLOT is *"Color the line with color* j*"*. When you express this in your programming language it instructs the computer to draw the j th graph using color number j in the computer's "palette." In the exercises you are asked to use the program SEQUENCE to explore the solutions to a number of rate equation problems.

Approximate Solutions versus Exact

By constructing successive approximations to the solution of a rate equation problem, using a sequence of step sizes `deltat` = Δt that shrink to 0, we obtain the *exact* solution in the limit.

In practice, though, all we can ever get are particular approximations. However, we can control the level of precision in our approximations by adjusting the step size. If we are dealing with a model of some real process, then this is typically all we need. For example, when it comes to interpreting the S-I-R model, we might be satisfied to predict that there will be about 40500 susceptibles remaining in the population after three days. The table on page 57 indicates we would get that level of precision using a step size of about $\Delta t = 10^{-2}$. Greater precision than this may be pointless, because the modeling process—which converts reality to mathematics—is itself only an approximation.

The question—In what sense do we know a number?—that we asked in the last section applies equally to the functions we obtain using Euler's method. That is, even if we can characterize a function quite precisely as the solution of a particular rate equation, we may be able to evaluate it only approximately.

A Caution

We have now seen how to take a set of rate equations and find approximations to the solution of these equations to any degree of accuracy desired. It is important to remember that all these mathematical manipulations are only drawing inferences about the model. We are essentially saying that *if* the original equations capture the internal dynamics of the situation being modeled, *then* here is what we would expect to see. It is still essential at some point to go back to the reality being modeled and check these predictions to see whether our original assumptions were in fact reasonable, or need to be modified. As Alfred North Whitehead has said:

> There is no more common error than to assume that, because prolonged and accurate mathematical calculations have been made, the application of the result to some fact of nature is absolutely certain.

Exercises

Approximate Solutions

1. Modify SIRVALUE and SIRPLOT to analyze the population of Poland (see exercise 25 of chapter 1, page 42). We assume the population $P(t)$ satisfies the conditions

$$P' = .009\,P \quad \text{and} \quad P(0) = 37,500,000,$$

where t is years since 1985. We want to know P 100 years into the future; you can assume that P does not exceed 100,000,000.
a) Estimate the population in 2085.
b) Sketch the graph that describes this population growth.

The Logistic Equation

Suppose we were studying a population of rabbits. If we turn 100 rabbits loose in a field and let $y(t)$ be the number of rabbits at time t measured in months, we would like to know how this function behaves. The next several exercises are designed to explore the behavior of the rate equation

$$y' = .1y\left(1 - \frac{y}{1000}\right); \qquad y(0) = 100$$

and see why it might be a reasonable model for this system.

*2. By modifying SIRVALUE in the way we modified SIRPLOT to get SEQUENCE, obtain a sequence of estimates for $y(37)$ that allows you to specify the exact value of $y(37)$ to two decimal places accuracy.

3. a) Referring to the graph of $y(t)$ obtained in the text on page 70, what can you say about the behavior of y as t gets large?
   b) Suppose we had started with $y(0) = 1000$. How would the population have changed over time? Why?
   c) Suppose we had started with $y(0) = 1500$. How would the population have changed over time? Why?
   d) Suppose we had started with $y(0) = 0$. How would the population have changed over time? Why?
   e) The number 1000 in the denominator of the rate equation is called the **carrying capacity** of the system. Can you give a physical interpretation for this number?

4. Obtain graphical solutions for the rate equation for different values of the carrying capacity. What seems to be happening as the carrying capacity is increased? (Don't restrict yourself to $t = 37$ here.) In this problem and the next you should sketch the different solutions on the same set of axes.

5. Keep everything in the original problem unchanged except for the constant .1 out front. Obtain graphical solutions with the value of this constant = .05, .2, .3, and .6. How does the behavior of the solution change as this constant changes?

6. Returning to the original logistic equation, modify SIRVALUE or DO-WHILE to find the value for t such that $y(t) = 900$.

7. Suppose we wanted to fit a logistic rate equation to a population, starting with $y(0) = 100$. Suppose further that we were comfortable with the 1000 in the denominator of the equation, but weren't sure about the .1 out front. If we knew that $y(20) = 900$, what should the value for this constant be?

Using SEQUENCE

8. Each Euler approximation is made up of a certain number of straight line segments. What instruction in the program SEQUENCE determines the number of segments in a particular approximation? The first graph drawn has only a single segment. How many does the fifth have? How many does the fourteenth have?

9. What is the slope of the first graph? What are the slopes of the two parts of the second graph? (You should be able to answer these questions without resorting to a computer.)

10. Modify the line in the program SEQUENCE which determines the number of steps by having it read `numberofsteps = j`, and run the modified program. Again, we are getting a sequence of approximations, with the number of steps increasing each time, but the approximations don't seem to be getting all that close to anything. Explain why this modified program isn't as effective for our purposes as the original.

11. Modify SEQUENCE to produce a sequence of Euler approximations to the function $y(t)$ that satisfies the conditions

$$y' = .2y(5 - y) \quad \text{and} \quad y(0) = 1.$$

on the interval $0 \le t \le 10$. (You need to change the final t value in the program, and you also need to ensure that the graphs will fit on your screen.)

   a) What is $y(10)$? (If you add the line PRINT j, y just before the line NEXT j, a sequence of 14 estimates for $y(10)$ will appear on the screen with the graphs.)

   *b) Make a rough sketch of the graph that is the limit of these approximations. The right half of the limit graph has a distinctive feature; what is it?

   c) Without doing any calculations, can you estimate the value of $y(50)$? How did you arrive at this value?

   d) Change the **initial condition** from $y(0) = 1$ to $y(0) = 9$. Construct the sequence of Euler approximations beginning with `numberofsteps` equal to 1, and make a rough sketch of the limit graph. What is $y(10)$ now? Explain why the first several approximations look so strange.

12. Modify SEQUENCE to construct a sequence of Euler approximations for the population of Poland (from exercise 1, above). Sketch the limit graph $P(t)$, and mark the values of $P(0)$ and $P(100)$ at the two ends.

13. Construct a sequence of Euler approximations to the function $y(t)$ that satisfies the conditions

$$y' = 2t \quad \text{and} \quad y(0) = 0$$

over the interval $0 \le t \le 2$. Note that this time the rate y' is given in terms of t, not y. Euler's method works equally well. Using your sequence of approximations, estimate $y(2)$. How accurate is your estimate?

14. Construct a sequence of Euler approximations to the function $y(t)$ that satisfies the conditions

$$y' = \frac{4}{1 + t^2} \quad \text{and} \quad y(0) = 0$$

over the interval $0 \le t \le 1$. Estimate $y(1)$. How accurate is your estimate? [Note: The exact value of $y(1)$ is π, which your estimates may have led you to expect. By using special methods we shall develop much later, we can prove that $y(1) = \pi$.]

▷ 2.3 APPROXIMATE SOLUTIONS

Our efforts to find the functions that were determined by the rate equations for the S-I-R model have brought to light several important issues:

- We often have to deal with a question that does not have a simple, straightforward answer; perhaps we are trying to determine a quantity [like the square root of 2, or $S(3)$ in the S-I-R model], to find some function [like $S(t)$], or to understand a process (like an epidemic, or buying and selling in a market). An **approximation** can get us started.

- In many instances, we can make repeated improvements in the approximation. If these **successive approximations** get arbitrarily close to the unknown, and they do it quickly enough, that may answer the question for all practical purposes. In many cases, there is no alternative.

- The information that successive approximations give us is conveyed in the form of a **limit**.

- The method of successive approximations can be used to evaluate many kinds of mathematical objects, including numbers, graphs, and functions.

- **Limit processes** give us a valuable tool to probe difficult questions. They lie at the heart of calculus.

> *Even the process of building a mathematical model for a physical system can be seen as an instance of successive approximations. We typically start with a simple model (such as the S-I-R model) and then add more and more features to it (e.g., in the case of the S-I-R model we might divide the population into different subgroups, have the parameters in the model depend on the season of the year, make immunity of limited duration, etc.). Is it always possible, at least in theory, to get a sequence of approximating mathematical models that approaches reality in the limit?*

In the following chapters we will apply the process of successive approximation to many different kinds of problems. For example, in chapter 3 the problem will be to get a better understanding of the notion of a rate of change of one quantity with respect to another. Then, in chapter 4, we will return to the task of solving rate equations using Euler's method. Chapter 6 introduces the integral, defining it through a sequence of successive approximations. As you study each chapter, pause to identify the places where the method of successive approximations is being used. This can give you insight into the special role that calculus plays within the broader subject of mathematics.

To illustrate the general utility of the method, we end this chapter by returning to the problem raised in section 2.1 of constructing the values of $\sqrt{2}$ and π to an arbitrary number of decimal places.

Calculating π—The Length of a Curve

Humans were grappling with the problem of calculating π at least 3000 years ago. In his work *Measurement of the Circle*, Archimedes (287–212 B.C.) used the method of successive approximations to calculate $\pi = 3.14\ldots$. He did this by starting with a circle of diameter 1, constructing an inscribed and a circumscribed hexagon, and calculating the lengths of their perimeters. The perimeter of the circumscribed hexagon was clearly an overestimate for π, while the perimeter of the inscribed hexagon was an underestimate. He then improved these estimates by going from hexagons to inscribed and circumscribed 12-sided polygons and again calculating the perimeters. He repeated this process of doubling the number of sides until he had inscribed and circumscribed polygons with 96 sides. These left him with his final estimate

$$3.1409\ldots = 3\frac{284\frac{1}{4}}{2017\frac{1}{4}} < \pi < 3\frac{667\frac{1}{2}}{4673\frac{1}{2}} = 3.1428\ldots.$$

In grade school we learned a nice simple formula for the length of a circle, but that was about it. We were never taught formulas for the lengths of other simple curves like elliptic or parabolic arcs, for a very good reason—there are no such formulas. There are various physical approaches we might take. For example, we could get a rough approximation by laying a piece of string along the curve, then picking up the string and measuring it with a ruler. Instead of a physical solution, we can use the essence of Archimedes' insight of approximating a circle by an inscribed "polygon"—what we have earlier called a piecewise linear graph—to determine the length of any curve. The basic idea is reminiscent of the way we made successive approximations to the functions $S(t)$, $I(t)$, and $R(t)$ in the first section of this chapter. Here is how we will approach the problem:

- Approximate the curve by a chain of straight line segments;
- Measure the lengths of the segments;
- Use the sum of the lengths as an approximation to the true length of the curve.

Repeat this process over and over, each time using a chain that has shorter segments (and therefore more of them) than the last one. The length of the curve emerges as the limit of the sums of the lengths of the successive chains.

Distance Formula If we are given two points $P_1(x_1, y_1)$ and $P_2(x_2, y_2)$ in the plane, then the distance between them is just

$$d = \sqrt{\Delta x^2 + \Delta y^2}$$
$$= \sqrt{(x_2 - x_1)^2 + (y_2 - y_1)^2}$$

That this follows directly from the Pythagorean theorem can be seen from the picture below:

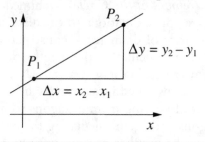

We'll demonstrate how this process works on a parabola. Specifically, consider the graph of $y = x^2$ on the interval $0 \le x \le 1$. At the right we have sketched the graph and our initial approximation. It is a piecewise linear approximation with two segments whose endpoints have equally spaced x-coordinates.

We can use the distance formula to find the lengths of the two segments.

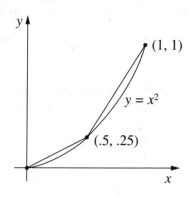

$$\text{first segment}: \sqrt{(.5 - 0)^2 + (.25 - 0)^2} = .559016994$$

$$\text{second segment}: \sqrt{(1 - .5)^2 + (1 - .25)^2} = .901387819$$

Their total length is the sum

$$.559016994 + .901387819 = 1.460404813.$$

The following program prints out the lengths of the two **segments** and their **total** length.

Program: LENGTH
Estimating the length of $y = x^2$ over $0 \le x \le 1$

```
DEF fnf (x) = x ^ 2
xinitial = 0
xfinal = 1
numberofsteps = 2
deltax = (xfinal - xinitial) / numberofsteps
total = 0
FOR k = 1 TO numberofsteps
        xl = xinitial + (k - 1) * deltax
        xr = xinitial + k * deltax
        yl = fnf(xl)
        yr = fnf(xr)
        segment = SQR((xr - xl) ^ 2 + ( yr - yl) ^ 2)
        total = total + segment
        PRINT k, segment
NEXT k
PRINT numberofsteps, total
```

Finding Roots with a Computer

When we casually turn to our calculator and ask it for the value of $\sqrt{2}$, what does it really do? Like us, the calculator can only add, subtract, multiply, and divide. Anything else we ask it to do must be reducible to these operations. In particular, the calculator doesn't really "know" the value of $\sqrt{2}$. What it does know is how to approximate $\sqrt{2}$ to, say, 12 significant figures using only elementary arithmetic. In this section we will look at two ways we might do this. Apart from the fact that both approaches use successive approximations, they are remarkably different in flavor. One works graphically, using a computer graphing package, and the other is a numerical algorithm that is about 4000 years old.

Calculators and computers really work by making approximations

A Geometric Approach

Exercise 9 on page 35 considered the problem of finding the roots of $f(x) = 1 - 2x^2$. A bit of algebra confirms that $\sqrt{2}/2$ is a root—that is, $f(\sqrt{2}/2) = 0$. The question is: what is the *numerical value* of $\sqrt{2}/2$?

We'll answer this question by constructing a sequence of approximations that add digits, one at a time, to an estimate for $\sqrt{2}/2$. Since the root lies at the point where the graph of f crosses the x-axis, we just magnify the graph at this point over and over again, "trapping" the point between x values that can be made arbitrarily close together.

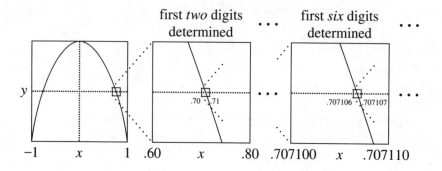

If we make each stage a tenfold magnification over the previous one, then, as we zoom in on the next smaller interval that contains the root, one more digit in our estimate will be stabilized. The first six stages are described in the table below. They tell us $\sqrt{2}/2 = .707106\ldots$ to six decimal places accuracy.

Since this method of finding roots requires only that we be able to plot successive magnifications of the graph of f on a computer screen, the method can be applied to any function that can be entered into a computer.

The positive root of $1 - 2x^2$

When the root lies between:		The decimal expansion
lower value	upper value	of the root begins with
.70	.80	.7
.700	.710	.70
.7070	.7080	.707
.70710	.70720	.7071
.707100	.707110	.70710
.7071060	.7071070	.707106
⋮	⋮	⋮

An Algebraic Approach—The Babylonian Algorithm

About 4000 years ago Babylonian builders had a method for constructing the square root of a number from a sequence of successive approximations. To demonstrate the method, we'll construct $\sqrt{5}$. We want to find x so that

$$x^2 = 5 \quad \text{or} \quad x = \frac{5}{x}.$$

The second expression may seem a peculiar way to characterize x, but it is at least equivalent to the first. The advantage of the second expression is that it gives us *two* numbers to consider: x and $5/x$.

For example, suppose we guess that $x = 2$. Of course this is incorrect, because $2^2 = 4$, not 5. The two numbers we get from the second expression are 2 and $5/2 = 2.5$. One is smaller than $\sqrt{5}$, the other is larger (because $2.5^2 = 6.25$). Perhaps their *average* is a better estimate. The average is 2.25, and $2.25^2 = 5.0625$.

Although 2.25 is not $\sqrt{5}$, it is a better estimate than either 2.5 or 2. If we change x to 2.25, then

$$x = 2.25, \quad \frac{5}{x} = 2.222, \quad \text{and their average is } 2.236.$$

Is 2.236 a better estimate than 2.5 or 2.222? Indeed it is: $2.236^2 = 4.999696$. If we now change x to 2.236, a remarkable thing happens:

$$x = 2.236, \quad \frac{5}{x} = 2.236, \quad \text{and their average is } 2.236!$$

In other words, if we want accuracy to three decimal places, we have already found $\sqrt{5}$. All the digits have stabilized.

Suppose we want *greater* accuracy? Our routine readily obliges. If we set $x = 2.236000$, then

$$x = 2.236000, \quad \frac{5}{x} = 2.236136, \quad \text{and their average is } 2.236068.$$

In fact, $\sqrt{5} = 2.236068\ldots$ is accurate to six decimal places.

Here is a summary of the argument we have just developed, expressed in terms of $\sqrt{a}$ for an arbitrary positive number a.

> If x is an estimate for $\sqrt{a}$,
> then the average of x and a/x
> is a better estimate.

Once we choose an *initial* estimate, this argument constructs a sequence of successive approximations to $\sqrt{a}$. The process of constructing the sequence is called the **Babylonian algorithm** for square roots.

A procedure that tells us how to carry out a sequence of steps, one at a time, to reach a specific goal is called an **algorithm**. Many algebraic processes are algorithms. The word is a Latinization of the name of the Arab astronomer Muhammad al-Khwārizmī (c. A.D. 780 –850). The title of his seminal book *Hisāb al-jabr wal-muqā bala* (A.D. 830)—usually referred to simply as *al-Jabr*—has a Latin form that is even more familiar to mathematics students.

The Babylonian algorithm takes the current estimate x for $\sqrt{a}$ that we have at each stage and says "replace x by the average of x and a/x." This kind of instruction is ideally suited to a computer, because A = B in a computer program means "replace the current value of A by the current value of B." In the program BABYLON printed below, the algorithm is realized by a FOR-NEXT loop with a single line that reads "x = (x + a / x) / 2".

The three-step procedure that we used in chapter 1 to obtain values of S, I, and R in the epidemic problem is also an algorithm, and for that reason it was a straightforward matter to express it as the computer program SIRVALUE.

Program: BABYLON
An algorithm to find $\sqrt{a}$

```
a = 5
x = 2
n = 6
FOR k = 1 TO n
    x = (x + a / x) / 2
    PRINT x
NEXT k
```

Output:
2.25
2.236111
2.236068
2.236068
2.236068
2.236068

Exercises

In many of the questions below you are asked to do things like divide an arc into 2,000,000 segments or find a certain number to 8 decimal places. This assumes you have fast computers available. If you are using slower machines or programmable calculators, you should certainly feel free to scale back what is called for—perhaps to 200,000 segments or 6 decimal places. Use your own common sense; there's no value in sitting in front of a screen for an hour waiting for an answer to emerge. To approximate a length by 200,000 segments will take 10 times as long as to approximate it by 20,000 segments, which in turn will take 10 times as long as the 2,000 segment approximation. If you start with the cruder approximations, you should be able to get a good sense of what is reasonable to attempt with the facilities you have available, and modify what the problems call for accordingly.

Use common sense in deciding how close an approximation to make

*1. By using a computer to graph $y = x^2 - 2^x$, find the solutions of the equation $x^2 = 2^x$ to four decimal places accuracy.

The Program LENGTH

2. Run the program LENGTH to verify that it gives the lengths of the individual segments and their total length.

3. What line in the program gives the instruction to work with the function $f(x) = x^2$? What line indicates the number of segments to be measured?

4. Each segment has a left and a right endpoint. What lines in the program designate the x- and y-coordinates of the left endpoint; the right endpoint?

5. Where in the program is the length of the kth segment calculated? The segment is treated as the hypotenuse of a triangle whose length is measured by the Pythagorean theorem. How is the base of that triangle denoted in the program? How is the altitude of that triangle denoted?

*6. Modify the program so that it uses 20 segments to estimate the length of the parabola. What is the estimated value?

7. Modify the program so that it estimates the length of the parabola using 200, 2,000, 20,000, 200,000, and 2,000,000 segments. Compare your results with those tabulated below. (To speed the process up, you will certainly want to delete the PRINT k, segment statement that appears inside the loop. Do you see why?)

Number of line segments	Sum of their lengths
2	1.460 404 813
20	1.478 756 512
200	1.478 940 994
2 000	1.478 942 839
20 000	1.478 942 857
200 000	1.478 942 857
2 000 000	1.478 942 857

8. What is the length of the parabola $y = x^2$ over the interval $0 \le x \le 1$, correct to 8 decimal places? What is the length, correct to 12 decimal places?

9. Starting at the origin, and moving along the parabola $y = x^2$, where are you when you've gone a total distance of 10?

10. Modify the program to find the length of the curve $y = x^3$ over the interval $0 \le x \le 1$. Find a value that is correct to 8 decimal places.

11. *Back to the circle.* Consider the unit circle centered at the origin. The Pythagorean theorem shows that a point (x, y) is on the circle if and only if $x^2 + y^2 = 1$. If we solve this for y in terms of x, we get $y = \pm\sqrt{1 - x^2}$, where the plus sign gives us the upper half of the circle and the minus sign gives the lower half. This suggests that we look at the function $g(x) = \sqrt{1 - x^2}$. The arclength of $g(x)$ over the interval $-1 \leq x \leq 1$ should then be exactly π.

 a) Divide the interval into 100 pieces—what is the corresponding length?

 b) How many pieces do you have to divide the interval into to get an accuracy equal to that of Archimedes?

 c) Find the length of the curve $y = g(x)$ over the interval $-1 \leq x \leq 1$, correct to eight decimal places accuracy.

12. This question concerns the function $h(x) = \dfrac{4}{1 + x^2}$.

 a) Sketch the graph of $y = h(x)$ over the interval $-2 \leq x \leq 2$.

 b) Find the length of the curve $y = h(x)$ over the interval $-2 \leq x \leq 2$.

13. Find the length of the curve $y = \sin x$ over the interval $0 \leq x \leq \pi$.

The Program BABYLON

14. Run the program BABYLON on a computer to verify the tabulated estimates for $\sqrt{5}$.

*15. In whatever computer language you are using, it should be possible to tell the computer to print out more decimals. Your teacher can tell you how this is handled on your computers. Modify BABYLON to run with at least 14-digit precision in this and the following problems. Also modify the program so it prints out the square of the estimate each time as well. What is the estimated value of $\sqrt{5}$ in this circumstance? How many steps were needed to reach this value? Use the square of this estimate as a measure of its accuracy. What is the square?

16. Use the Babylonian algorithm to find $\sqrt{80}$.

 a) First use 2 as your initial estimate. How many steps are needed for the calculations to stabilize—that is, to reach a value that doesn't change from one step to the next?

 b) Since $9^2 = 81$, a good first estimate for $\sqrt{80}$ is 9. How many steps are needed this time for the calculations to stabilize? Are the final values in (a) and (b) the same?

17. Use the Babylonian algorithm to find $\sqrt{250}$ and $\sqrt{1990}$. If you use 2 as the initial estimate in each case, how many steps are needed for the calculations to stabilize? If you use the integer nearest to the final

answer as your initial estimate, then how many steps are needed? Square your answers to measure their accuracy.

18. The Babylonian algorithm is considered to be **very fast,** in the sense that each stage roughly doubles the number of digits that stabilize. Does your work on the preceding exercises confirm this observation? By comparison, is the routine that got the estimates for $S(3)$ (computed with the program SIRVALUE) faster or slower than the Babylonian algorithm?

> ## 2.4   CHAPTER SUMMARY

The Main Ideas

- The exact numerical value of a quantity may not be known; the value is often given by an **approximation.**

- A numerical quantity is often given as the **limit** of a sequence of **successive approximations.**

- When a particular digit in a sequence of successive approximations **stabilizes,** that digit is assumed to appear in the limit.

- **Euler's method** is a procedure to construct a sequence of increasingly better **approximations** of a function defined by a set of rate equations and initial conditions. Each approximation is a piecewise linear function.

- The exact function defined by a set of rate equations and initial conditions can be expressed as the **limit** of a sequence of **successive Euler approximations** with smaller and smaller step sizes.

Expectations

- You should be able to use a program to **construct a sequence** of estimates for S, I, and R, given a specific S-I-R model with initial conditions.

- You should be able to **modify** the SIRVALUE and SIRPLOT programs to construct a sequence of estimates for the values of functions defined by other rate equations and initial conditions.

- You should be able to use programs that construct a sequence of **Euler approximations** for the function defined by a rate equation with an initial condition. The programs should provide both tabular and graphical output.

- You should be able to estimate the values of the roots of an equation $f(x) = 0$ using a **computer graphing package.**

- You should be able to find a square root using the **Babylonian algorithm.**

- You should be able to find the length of any piece of any curve.

Chapter Exercises

1. a) We have considered the logistic equation

$$y' = .1y\left(1 - \frac{y}{1000}\right)$$

On page 70 we looked at the resulting graph of y versus t, using the starting value $y(0) = 100$. There is another graphical interpretation, though, that is also instructive. Note that the logistic equation specifies y' as a function of y. Sketch the graph of this function. That is, plot y values on the horizontal axis and y' values on the vertical axis; points on the graph will thus be of the form (y, y'), where the y'-coordinate is the value given by the logistic equation for the given y value. What shape does the graph have?

b) For what value of y does this graph take on its largest value? Where does this y value appear in the graph of $y(t)$ versus t?

c) For what values of y does the graph of y' versus y cross the y-axis? Where do these y values appear in the graph of $y(t)$ versus t?

Grids on Graphs

When writing a graphing program it is often useful to have the computer draw a grid on the screen. This makes it easier to estimate numerical values, for instance. We can use a simple FOR–NEXT loop inside a program to do this. For instance, suppose we had written a program (e.g., SIRPLOT or SEQUENCE) with the graphics already in place. Suppose the screen window covered values from 0 to 100 horizontally, and 0 to 50,000 vertically. If we wanted to draw 21 vertical lines (including both ends) spaced 5 units apart and 11 horizontal lines spaced 5,000 units apart, the following two loops inserted in the program would work:

```
FOR k = 0 TO 20
        Plot the line from (5 * k, 0)
        to (5 * k, 50000)
NEXT k
FOR k = 0 TO 10
        Plot the line from (0, 5000 * k)
        to (100, 5000 * k)
NEXT k
```

You should make sure you see how these loops work and that you can modify them as needed.

2. If the screen window runs from −20 to 120 horizontally, and 250 to 750 vertically, how would you modify the loops above to create a vertical grid spaced 10 units apart and a horizontal grid spaced 25 units apart? Answer:

```
FOR k = 0 TO 15
        Plot the line from (-20 + 10 * k, 250)
        to (-20 + 10 * k, 750)
NEXT k
FOR k = 0 TO 20
        Plot the line from (-20, 250 + 25 * k)
        to (120, 250 + 25 * k)
NEXT k
```

3. Go back to our basic S-I-R model. Modify SIRPLOT to calculate and plot on the same graph the values of $S(t)$, $I(t)$, and $R(t)$ for t going from 0 to 120, using a step size of $\Delta t = .1$. Include a grid with a horizontal spacing of 5 days, and a vertical spacing of 2000 people. You should get something that looks like the following figure:

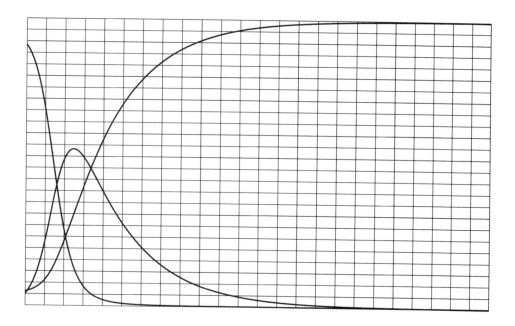

CHAPTER **3**

THE DERIVATIVE

In developing the S-I-R model in chapter 1 we took the idea of the rate of change of a population as intuitively clear. The rate at which one quantity changes with respect to another is a central concept of calculus and leads to a broad range of insights. The chief purpose of this chapter is to develop a fuller understanding—both analytic and geometric—of the connection between a function and its rate of change. To do this we will introduce the concept of the **derivative** of a function.

➤ 3.1 RATES OF CHANGE

The Changing Time of Sunrise

The sun rises at different times, depending on the date and location. At 40°N latitude (New York, Beijing, and Madrid are about at this latitude) in the year 1990, for instance, the sun rose at

The time of sunrise is a function

$$
\begin{aligned}
&7:16 \quad \text{on January 23} \\
&5:58 \quad \text{on March 24} \\
&4:52 \quad \text{on July 25}
\end{aligned}
$$

Clearly the time of sunrise is a function of the date. If we represent the time of sunrise by T (in hours and minutes) and the date by d (the day of the year), we can express this functional relation in the form $T = T(d)$. For example, from the table above we find $T(23) = 7{:}16$. It is not obvious from the table, but it is also true that the rate at which the time of sunrise is changing is different at different times of the year—T' varies as d varies. We can see how the rate varies by looking at some further data for sunrise at the same latitude, taken from *The Nautical Almanac for the Year 1990*:

Date	Time	Date	Time	Date	Time
January 20	7:18	March 21	6:02	July 22	4:49
23	7:16	24	5:58	25	4:52
26	7:14	27	5:53	28	4:54

Calculate the rate using earlier and later dates

Let's use this table to estimate the rate at which the time of sunrise is changing on January 23. We'll use the times three days earlier and three days later, and compare them. On January 26 the sun rose 4 minutes earlier than on January 20. This is a change of -4 minutes in 6 days, so the rate of change is

$$\frac{-4 \text{ minutes}}{6 \text{ days}} \approx -.67 \text{ minute per day}.$$

We say this is the **rate** at which sunrise is changing on January 23, and we write

$$T'(23) \approx -.67 \, \frac{\text{minute}}{\text{day}}.$$

The rate is negative because the time of sunrise is decreasing—the sun is rising earlier each day.

Similarly, we find that around March 24 the time of sunrise is changing approximately $-9/6 = -1.5$ minutes per day, and around July 25 the rate is $5/6 \approx +.8$ minute per day. The last value is positive, since the time of sunrise is increasing—the sun is rising later each day in July. Since March 24 is the 83rd day of the year and July 25 is the 206th, using our notation for rate of change we can write

$$T'(83) \approx -1.5 \, \frac{\text{minutes}}{\text{day}}; \qquad T'(206) \approx .8 \, \frac{\text{minute}}{\text{day}}.$$

Notice that, in each case, we have calculated the rate on a given day by using times shortly before and shortly after that day. We will continue this pattern wherever possible. In particular, you should follow it when you do the exercises about a falling object, at the end of the section.

Once we have the rates, we can estimate the time of sunrise for dates not given in the table. For instance, January 28 is five days after January 23, so the total change in the time of sunrise from January 23 to January 28 should be approximately

$$\Delta T \approx -.67 \frac{\text{minute}}{\text{day}} \times 5 \text{ days} = -3.35 \text{ minutes.}$$

In whole numbers, then, the sun rose 3 minutes earlier on January 28 than on January 23. Since sunrise was at 7:16 on the 23rd, it was at 7:13 on the 28th.

By letting the change in the number of days be negative, we can use this same reasoning to tell us the time of sunrise on days shortly *before* the given dates. For example, March 18 is −6 days away from March 24, so the change in the time of sunrise should be

$$\Delta T \approx -1.5 \frac{\text{minutes}}{\text{day}} \times -6 \text{ days} = +9 \text{ minutes.}$$

Therefore, we can estimate that sunrise occurred at $5:58 + 0:09 = 6:07$ on March 18.

Changing Rates

Suppose instead of using the tabulated values for March we tried to use our January data to *predict* the time of sunrise in March. Now March 24 is 60 days after January 23, so the change in the time of sunrise should be approximately

$$\Delta T \approx -.67 \frac{\text{minute}}{\text{day}} \times 60 \text{ days} = -40.2 \text{ minutes,}$$

and we would conclude that sunrise on March 24 should be at about $7:16 - 0:40 = 6:37$, which is more than half an hour later than the actual time! This is a problem we met often in estimating future values in the *S-I-R* model. When we use the formula above to estimate ΔT, we implicitly assume that the time of sunrise changes at the fixed rate of −.67 minute per day over the entire 60-day time-span. But this turns out not to be true: the rate actually varies, and the variation is too great for us to get a useful estimate. Only with a much smaller time-span does the rate not vary too much.

Predictions over long time spans are less reliable

Here is the same lesson in another context. Suppose you are traveling in a car along a busy road at rush hour and notice that you are going 50 miles per hour. You would be fairly confident that in the next 30 seconds (1/120 of an hour) you will travel about

$$\Delta \, \text{position} \approx 50 \, \frac{\text{miles}}{\text{hour}} \times \frac{1}{120} \, \text{hour} = \frac{5}{12} \, \text{mile} = 2200 \, \text{feet}.$$

The actual value ought to be within 50 feet of this, making the estimate accurate to within about 2% or 3%. On the other hand, if you wanted to estimate how far you would go in the next 30 minutes, your speed would probably fluctuate too much for the calculation

$$\Delta \, \text{position} \approx 50 \, \frac{\text{miles}}{\text{hour}} \times \frac{1}{2} \, \text{hour} = 25 \, \text{miles}$$

to have the same level of reliability.

Other Rates, Other Units

In the S-I-R model the rates we analyzed were **population growth rates**. They told us how the three populations changed over time, in units of persons per day. If we were studying the growth of a colony of mold, measuring its size by its weight (in grams), we could describe *its* population growth rate in units of grams per hour. In discussing the motion of an automobile, the rate we consider is the **velocity** (in miles per hour), which tells us how the distance from some starting point changes over time. We also pay attention to the rate at which *velocity* changes over time. This is called **acceleration,** and can be measured in miles per hour per hour.

While many rates do involve changes with respect to time, other rates do not. Two examples are the survival rate for a disease (survivors per thousand infected persons) and the dose rate for a medicine (milligrams per pound of body weight). Other common rates are the annual birth rate and the annual death rate, which might have values like 19.3 live births per 1,000 population and 12.4 deaths per 1,000 population. Any quantity expressed as a percentage, such as an interest rate or an unemployment rate, is a rate of a similar sort. An unemployment rate of 5%, for instance, means 5 unemployed workers per 100 workers. There are many other examples of rates in the economic world that make use of a variety of units—exchange rates (e.g., francs per dollar), marginal return (e.g., dollars of profit per dollar of change in price).

Sometimes we even want to know the rate of change of one rate with respect to another rate. For example, automobile fuel economy (in miles per gallon—the first rate) changes with speed (in miles per hour—the second rate), and we can measure the rate of change of fuel economy with speed. Take a car that goes 22 miles per gallon of fuel at 50 miles per hour, but only 19 miles per gallon at 60 miles per hour. Then its fuel economy is changing approximately at the rate

Examples of rates

The rate of change of a rate

$$\frac{\Delta \text{ fuel economy}}{\Delta \text{ speed}} = \frac{19 - 22 \text{ miles per gallon}}{60 - 50 \text{ miles per hour}}$$

$$= -.3 \text{ mile per gallon } per \text{ mile per hour.}$$

Exercises

A Falling Object

These questions deal with an object that is held motionless 10,000 feet above the surface of the ocean and then dropped. Start a clock ticking the moment it is dropped, and let D be the number of feet it has fallen after the clock has run t seconds. The following table shows some of the values of t and D:

Time (seconds)	Distance (feet)
0	0.00
1	15.07
2	56.90
3	121.03
4	203.76
5	302.00
6	413.16
7	535.10

1. What units do you use to measure velocity—that is, the rate of change of distance with respect to time—in this problem?

2. a) Make a careful graph that shows these eight data points. Put *time* on the horizontal axis. Label the axes and indicate the units you are using on each.
   b) The slope of any line drawn on this time–distance graph has the units of a *velocity*. Explain why.

*3. Make three estimates of the velocity of the falling object at the 2 second mark using the distances fallen between these times:
   i) from 1 second to 2 seconds;
   ii) from 2 seconds to 3 seconds;
   iii) from 1 second to 3 seconds.

4. a) Each of the estimates in the previous question corresponds to the slope of a particular line you can draw in your graph. Draw those lines and label each with the corresponding velocity.

b) Which of the three estimates in question 3 do you think is best? Explain your choice.

5. Using your best method, estimate the velocity of the falling object after 4 seconds have passed.

6. Is the object speeding up or slowing down as it falls? How can you tell?

7. Approximate the velocity of the falling object after 7 seconds have passed. Use your answer to estimate the number of feet the object has fallen after 8 seconds have passed. Do you think your estimate is too high or too low? Why?

8. For those of you attending a school at which you pay tuition, find out what the tuition has been for each of the last four years at your school.

   a) At what rate has the tuition changed in each of the last three years? What are the units? In which year was the rate the greatest?

   b) A more informative rate is often the **inflation rate** which doesn't look at the dollar change per year, but at the percentage change per year—the dollar change in tuition in a year's time expressed as a percentage of the tuition at the beginning of the year. What is the tuition inflation rate at your school for each of the last three years? How do these rates compare with the rates you found in part (a)?

   c) If you were interested in seeing how the inflation rate was changing over time, you would be looking at the rate of change of the inflation rate. What would the units of this rate be? What is the rate of change of the inflation rate at your school for the last two years?

   d) Using all this information, what would be your estimate for next year's tuition?

9. Your library should have several reference books giving annual statistics of various sorts. A good one is the *Statistical Yearbook* put out by the United Nations with detailed data from all over the world on manufacturing, transportation, energy, agriculture, tourism, and culture. Another is *Historical Statistics of the United States, Colonial Times—1970*. Select an interesting quantity and compare its growth rate at different times or for different countries. Calculate this growth rate over a stretch of four or five years and report whether there are any apparent patterns. Calculate the rate of change of this growth rate and interpret its values.

10. Oceanographers are very interested in the **temperature profile** of the part of the ocean they are studying. That is, how does the temperature T (in degrees Celsius) vary with the depth d (measured in meters). A typical temperature profile might look something like the following:

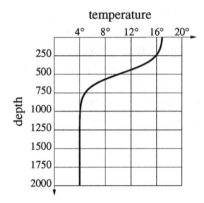

(Note that since water is densest at 4°C, water at that temperature settles to the bottom.)

a) What are the units for the rate of change of temperature with depth?

b) In this graph will the rate be positive or negative? Justify your answer.

c) At what rate is the temperature changing at a depth of 0 meters? 500 meters? 1000 meters?

d) Sketch a possible temperature profile for this location if the surface is iced over.

e) This graph is not oriented the way our graphs have been up until now. Why do you suppose oceanographers (and geologists and atmospheric scientists) often draw graphs with axes positioned like this?

➤ 3.2 MICROSCOPES AND LOCAL LINEARITY

The Graph of Data

This section is about seeing rates geometrically. We know from chapter 1 that we can visualize the rate of change of a *linear* function as the slope of its graph. Can we say the same thing about the sunrise function? The graph of this function appears at the right; it

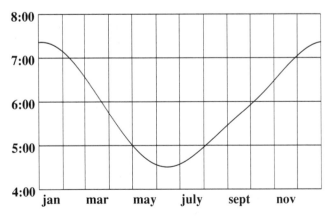

plots the time of sunrise (over the course of a year at 40°N latitude), as a function of the date. The graph is curved, so the sunrise function is not linear. There is no immediately obvious connection between rate and slope. In fact, it isn't even clear what we might mean by the *slope* of this graph! We can make it clear by using a **microscope**.

Zoom in on the graph with a microscope

Imagine we have a microscope that allows us to "zoom in" on the graph near each of the three dates we considered in section 3.1. If we put each magnified image in a window, then we get the following:

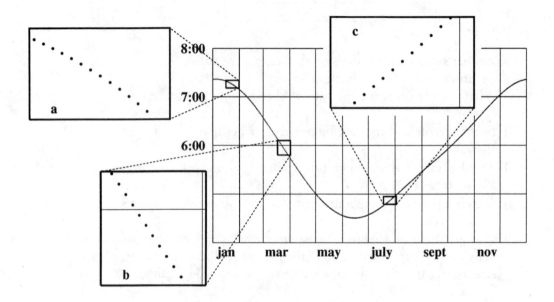

Notice how different the graph looks under the microscope. First of all, it now shows up clearly as a collection of separate points—one for each day of the year. Second, the points in a particular window lie on a line that is essentially straight. The straight lines in the three windows have very different slopes, but that is only to be expected.

The graph looks straight under a microscope

What is the connection between these slopes and the rates of change we calculated in the last section? To decide, we should calculate the slope in each window. This involves choosing a pair of points (d_1, T_1) and (d_2, T_2) on the graph and calculating the ratio

$$\frac{\Delta T}{\Delta d} = \frac{T_2 - T_1}{d_2 - d_1}.$$

In window **a** we'll take the two points that lie three days on either side of the central date, January 23. These points have coordinates (20, 7:18) and (26, 7:14) (table, page 90). The slope is thus

$$\frac{\Delta T}{\Delta d} = \frac{7{:}14 - 7{:}18}{26 - 20} = \frac{-4 \text{ minutes}}{6 \text{ days}} = -.67 \frac{\text{minute}}{\text{day}}.$$

If we use the same approach in the other two windows we find that the line in window **b** has slope -1.5 min/day, while the line in window **c** has slope $+.8$ min/day. These are exactly the same calculations we did in section 3.1 to determine the rate of change of the time of sunrise around January 23, March 24, and July 25, and they produce the same values we obtained there:

Slope and rate calculations are the same

$$T'(23) \approx -.67 \frac{\text{min}}{\text{day}} \qquad T'(83) \approx -1.5 \frac{\text{min}}{\text{day}} \qquad T'(206) \approx .8 \frac{\text{min}}{\text{day}}.$$

This is a crucial observation which we use repeatedly in other contexts; let's pause and state it in general terms:

> The rate of change of a function at a point is equal to the slope of its graph at that point, if the graph looks straight when we view it under a microscope.

The Graph of a Formula

Rates and slopes are really the same thing—that's what we learn by using a microscope to view the graph of the sunrise function. But the sunrise graph consists of a finite number of disconnected points—a very common situation when we deal with data. In such cases it doesn't make sense to magnify the graph too much. For instance, we would get no useful information from a window that was narrower than the space between the data points. There is no such limitation if we use a microscope to look at the graph of a function given by a formula, though. We can zoom in as close as we wish and still see a continuous curve or line. By using a high-power microscope, we can learn even more about rates and slopes.

High-power magnification is possible with a formula

Consider this rather complicated-looking function:

$$f(x) = \frac{2 + x^3 \cos x + 1.5^x}{2 + x^2}.$$

Let's find $f'(27)$, the rate of change of f when $x = 27$. We need to zoom in on the graph of f at the point $(27, f(27)) = (27, 69.859043)$. We do this in stages, producing a succession of windows that run clockwise from the upper left. Notice how the graph gets straighter with each magnification.

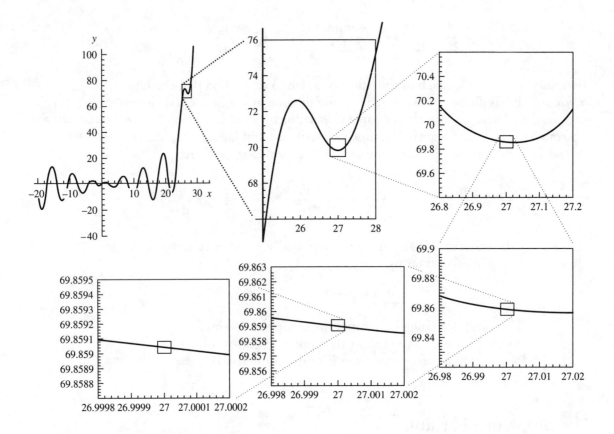

We will need a way to describe the part of the graph that we see in a window. Let's call it the **field of view**. The field of view of each window is only one-tenth as wide as the previous one, and the field of view of the last window is only one-millionth of the first! The last window shows what we would see if we looked at the original graph with a million-power microscope.

The field of view of a window

> *The microscope used to study functions is real, but it is different from the one a biologist uses to study microorganisms. Our microscope is a computer graphing program that can "zoom in" on any point on a graph. The computer screen is the window you look through, and you determine the field of view when you set the size of the interval over which the graph is plotted.*

Here is our point of departure: **the rate $f'(27)$ is the slope of the graph of f when we magnify the graph enough to make it look straight.** But how much is enough? Which window should we use? The following table gives the slope $\Delta y / \Delta x$ of the line that appears in each of the last four windows in the sequence. For Δx we take the difference between the x coordinates of the points at the ends of the line, and for Δy we take the difference between the y coordinates. In particular, the width of the field of view in each case is Δx.

Δx	Δy	$\Delta y / \Delta x$
.04	$-1.081\,508\,24 \times 10^{-2}$	$-.270\,377\,066$
.004	$-1.089\,338\,27 \times 10^{-3}$	$-.272\,334\,556$
.0004	$-1.089\,416\,49 \times 10^{-4}$	$-.272\,354\,131$
.00004	$-1.089\,417\,28 \times 10^{-5}$	$-.272\,354\,327$

As you can see, it *does* matter how much we magnify. The slopes $\Delta y / \Delta x$ in the table are not quite the same, so we don't yet have a definite value for $f'(27)$. The table gives us an idea how we *can* get a definite value, though. Notice that the slopes become more and more alike the more we magnify. In fact, under successive magnifications the first five digits of $\Delta y / \Delta x$ have **stabilized**. We saw in chapter 2 how to think about a sequence of numbers whose digits stabilize one by one. We should treat the values of $\Delta y / \Delta x$ as **successive approximations** to the slope of the graph. The exact value of the slope is then the **limit** of these approximations as the width of the field of view shrinks to zero:

The slope is a limit

$$f'(27) = \text{the slope of the graph} = \lim_{\Delta x \to 0} \frac{\Delta y}{\Delta x}.$$

In the limit process we take $\Delta x \to 0$ because Δx is the width of the field of view. Since five digits of $\Delta y / \Delta x$ have stabilized, we can write

$$f'(27) = -.27235\ldots.$$

To find $f'(x)$ at some other point x, proceed the same way. Magnify the graph at that point repeatedly, until the value of the slope stabilizes. The method is very powerful. In the exercises you will have an opportunity to use it with other functions.

By using a microscope of arbitrarily high power, we have obtained further insights about rates and slopes. In fact, with these insights we can now state definitively what we mean by the slope of a curved graph and the rate of change of a function.

> The **slope** of a graph at a point is the *limit* of the slopes seen in a microscope at that point, as the field of view shrinks to zero.
>
> The **rate of change** of a function at a point is the slope of its graph at that point. Thus the rate of change is also a limit.

To calculate the value of the slope of the graph of $f(x)$ when $x = a$, we have to carry out a limit process. We can break down the process into these four steps:

1. Magnify the graph at the point $(a, f(a))$ until it appears straight.

2. Calculate the slope of the magnified segment.

3. Repeat steps 1 and 2 with successively higher magnifications.

4. Take the limit of the succession of slopes produced in step 3.

Local Linearity

A microscope gives a local view

The crucial property of a microscope is that it allows us to look at a graph **locally**, that is, in a small neighborhood of a particular point. The two functions we have been studying in this section have curved graphs—like most functions. But *locally*, their graphs are straight—or nearly so. This is a remarkable property, and we give it a name. We say these functions are **locally linear**. In other words, a locally linear function looks like a linear function, locally.

The graph of a linear function has a definite slope at every point, and so does a *locally* linear function. For a linear function, the slope is easy to calculate, and it has the same value at every point. For a locally linear function, the slope is harder to calculate; it involves a limit process. The slope also varies from point to point.

All the standard functions are locally linear at almost all points

How common is local linearity? All the standard functions you already deal with are locally linear almost everywhere. To see why we use the qualifying phrase "almost everywhere," look at what happens when we view the graph of $y = f(x) = x^{2/3}$ with a microscope. At any point other than the origin, the graph is locally linear. For instance, if we view this graph over the interval from 0 to 2 and then zoom in on the point $(1, 1)$ by two successive powers of 10, here's what we see:

As the field of view shrinks, the graph looks more and more like a straight line. Using the highest magnification given, we estimate the slope of the graph—and hence the rate of change of the function—to be

$$f'(1) \approx \frac{\Delta y}{\Delta x} = \frac{1.006656 - .993322}{1.01 - .99} = \frac{.013334}{.02} = .6667.$$

Similarly, if we zoom in on the point $(.001, .01)$ we get:

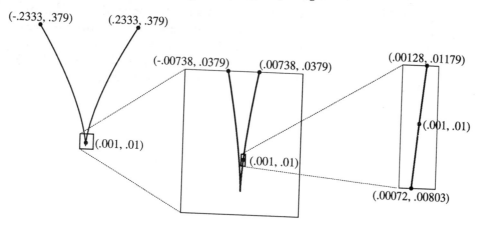

In the last window the graph looks like a line of slope

$$f'(.001) \approx \frac{.0118 - .0082}{.00128181 - .00074254} = \frac{.0036}{.00053937} = 6.674.$$

At the origin, though, something quite different happens:

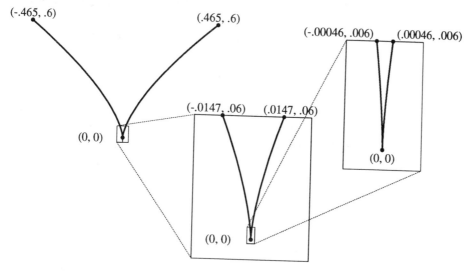

The graph simply looks more and more sharply pointed the closer we zoom in to the origin—it never looks like a straight line. However, the origin turns out to be the only point where the graph does not eventually look like a straight line.

Fractals are locally *nonlinear* objects

In spite of these examples, it is important to realize that local linearity is a very special property. There are some functions that fail to be locally linear anywhere! Such functions are called **fractals**. No matter how much you magnify the graph of a fractal at any point, it continues to look nonlinear—bent and "pointy" in various ways. In recent years fractals have been used in problems where the more common (locally linear) functions are inadequate. For instance, they describe irregular shapes like coastlines and clouds, and they model the way molecules are knocked about in a fluid (this is called *Brownian motion*). However, calculus does not deal with such functions. On the contrary,

> Calculus studies functions that are locally linear almost everywhere.

Exercises

Using a Microscope

1. Use a computer microscope to do the following. (A suggestion: First look at each graph over a fairly large interval.)
   *a) With a window of size $\Delta x = .002$, estimate the rate $f'(1)$ where $f(x) = x^4 - 8x$.
   b) With a window of size $\Delta x = .0002$, estimate the rate $g'(0)$ where $g(x) = 10^x$.
   c) With a window of size $\Delta t = .05$, estimate the slope of the graph of $y = t + 2^{-t}$ at $t = 7$.
   d) With a window of size $\Delta z = .0004$, estimate the slope of the graph of $w = \sin z$ at $z = 0$.

2. Use a computer microscope to determine the following values, correct to one decimal place. Obtain estimates using a *sequence* of windows, and shrink the field of view until the first two decimal places stabilize. Show *all* the estimates you constructed in each sequence.
   *a) $f'(1)$ where $f(x) = x^4 - 8x$.
   b) $h'(0)$ where $h(s) = 3^s$.
   c) The slope of the graph of $w = \sin z$ at $z = \pi/4$.
   d) The slope of the graph of $y = t + 2^{-t}$ at $t = 7$.
   e) The slope of the graph of $y = x^{2/3}$ at $x = -5$.

3. For each of the following functions, magnify its graph at the indicated point until the graph appears straight. Determine the equation of that straight line. Then verify that your equation is correct by plotting it as a second function in the same window in which you are viewing the given function. (The two graphs should "share phosphor"!)

a) $f(x) = \sin x$ at $x = 0$;
b) $\varphi(t) = t + 2^{-t}$ at $t = 7$;
c) $H(x) = x^{2/3}$ at $x = -5$.

4. Consider the function that we investigated in the text:

$$f(x) = \frac{2 + x^3 \cos x + 1.5^x}{2 + x^2}.$$

a) Determine $f(0)$.
b) Make a sketch of the graph of f on the interval $-1 \le x \le 1$. Use the same scale on the horizontal and vertical axes so your graph shows slopes accurately.
c) Sketch what happens when you magnify the previous graph so the field of view is only $-.001 \le x \le .001$.
d) Estimate the slope of the line you drew in the part (c).
e) Estimate $f'(0)$. How many decimal places of accuracy does your estimate have?
f) What is the equation of the line in part (c)?

5. A function that occurs in several different contexts in physical problems is

$$g(x) = \frac{\sin x}{x}.$$

Use a graphing program to answer the following questions.
a) Estimate the rate of change of g at the following points to two-decimal place accuracy:

$$g'(1), \qquad g'(2.79), \qquad g'(\pi), \qquad g'(3.1).$$

b) Find three values of x where $g'(x) = 0$.
c) In the interval from 0 to 2π, where is g decreasing the most rapidly? At what rate is it decreasing there?
d) Find a value of x for which $g'(x) = -.25$.
e) Although $g(0)$ is not defined, the function $g(x)$ seems to behave nicely in a neighborhood of 0. What seems to be true about $g(x)$ and $g'(x)$ when x is near 0?
f) According to your graphs, what value does $g(x)$ approach as $x \to 0$? What value does $g'(x)$ approach as $x \to 0$?

Rates from Graphs; Graphs from Rates

6. a) Sketch the graph of a function f that has $f(1) = 1$ and $f'(1) = 2$.
   b) Sketch the graph of a function f that has $f(1) = 1$, $f'(1) = 2$, and $f(1.1) = -5$.

7. A and B start off at the same time, run to a point 50 feet away, and return, all in 10 seconds. A graph of distance from the starting point as a function of time for each runner appears below. It tells where each runner is during this time interval.

a) Who is in the lead during the race?

b) At what time(s) is A farthest ahead of B? At what time(s) is B farthest ahead of A?

c) Estimate how fast A and B are going after one second.

d) Estimate the velocities of A and B during each of the 10 seconds. Be sure to assign *negative* velocities to times when the distance to the starting point is *shrinking*. Use these estimates to sketch graphs of the velocities of A and B versus time. (Although the velocity of B changes rapidly around $t = 5$, assume that the graph of B's distance *is* locally linear at $t = 5$.)

e) Use your graphs in (d) to answer the following questions. When is A going faster than B? When is B going faster than A? Around what time is A running at -5 feet/second (i.e., running 5 feet/second *toward* the starting point)?

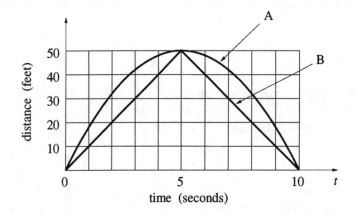

8. For each of the following functions draw a graph that reflects the given information. Restate the given information in the language and notation of rate of change, paying particular attention to the units in which any rate of change is expressed.

a) A woman's height h (in inches) depends on her age t (in years). Babies grow very rapidly for the first two years, then more slowly until the adolescent growth spurt; much later, many women actually become shorter because of loss of cartilage and bone mass in the spinal column.

b) The number R of rabbits in a meadow varies with time t (in years). In the early years food is abundant and the rabbit population grows rapidly. However, as the population of rabbits approaches the "carrying capacity" of the meadow environment, the growth rate slows, and the population never exceeds the carrying capacity. Each year, during the harsh conditions of winter, the population dies back slightly, although it never gets quite as low as its value the previous year.

c) In a fixed population of couples who use a contraceptive, the average number N of children per couple depends on the effectiveness E (in percent) of the contraceptive. If the couples are using a contraceptive of low effectiveness, a small increase in effectiveness has a small effect on the value of N. As we look at contraceptives of greater and greater effectiveness, small additional increases in effectiveness have larger and larger effects on N.

9. If we graph the distance traveled by a parachutist in freefall as a function of the length of time spent falling, we would get a picture something like the following:

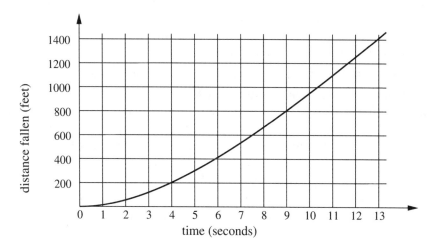

a) Use this graph to make estimates of the parachutist's velocity at the end of each second.

b) Describe what happens to the velocity as time passes.

c) How far do you think the parachutist would have fallen by the end of 15 seconds?

10. True or false. If you think a statement is true, give your reason; if you think a statement is false, give a counterexample—that is, an example that shows why it must be false.

a) If $g'(t)$ is positive for all t, we can conclude that $g(214)$ is positive.

b) If $g'(t)$ is positive for all t, we can conclude that $g(214) > g(17)$.

c) Bill and Samantha are driving separate cars in the same direction along the same road. At the start Samantha is 1 mile in front of

Bill. If their speeds are the same at every moment thereafter, at the end of 20 minutes Samantha will be 1 mile in front of Bill.

d) Bill and Samantha are driving separate cars in the same direction along the same road. They start from the same point at 10 A.M. and arrive at the same destination at 2 P.M. the same afternoon. At some time during the four hours their speeds must have been exactly the same.

When Local Linearity Fails

11. The absolute value function $f(x) = |x|$ is *not* locally linear at $x = 0$. Explore this fact by zooming in on the graph at $(0, 0)$. Describe what you see in successively smaller windows. Is there any change?

12. Find three points where the function $f(x) = |\cos x|$ fails to be locally linear. Sketch the graph of f to demonstrate what is happening.

13. Zoom in on the graph of $y = x^{4/5}$ at $(0, 0)$. In order to get an accurate picture, be sure that you use the same scales on the horizontal and vertical axes. Sketch what you see happening in successive windows. Is the function $x^{4/5}$ locally linear at $x = 0$?

14. Is the function $x^{4/5}$ locally linear at $x = 1$? Explain your answer.

15. This question concerns the function $K(x) = x^{10/9}$.
    a) Sketch the graph of K on the interval $-1 \le x \le 1$. Compare K to the absolute value function $|x|$. Are they similar or dissimilar? In what ways? Would you say K is locally linear at the origin, or not?
    b) Magnify the graph of K at the origin repeatedly, until the field of view is no bigger than $\Delta x \le 10^{-10}$. As you magnify, be sure the scales on the horizontal and vertical axes remain the same, so you get a true picture of the slopes. Sketch what you see in the final window.
    c) After using the microscope, do you change your opinion about the local linearity of K at the origin? Explain your response.

➤ 3.3 THE DERIVATIVE

Definition

One of our main goals in this chapter is to make precise the notion of the rate of change of a function. In fact, we have already done that in the last section. We defined the rate of change of a function at a point to be the slope of its graph at that point; we defined the slope, in turn, by a four-step limit process. Thus, the precise definition of a rate of change involves a limit, and it involves geometric visualization—we think of a rate as a slope. We introduce a new word—*derivative*—to embrace both of these concepts as we now understand them.

> The **derivative** of the function $f(x)$ at $x = a$ is its rate of change at $x = a$, which is the same as the slope of its graph at $(a, f(a))$. The derivative of f at a is denoted $f'(a)$.

Later in this section we will extend our interpretation of the derivative to include the idea of a *multiplier*, as well as a rate and a slope. Besides providing us with a single word to describe rates, slopes, and multipliers, the term "derivative" also reminds us that the quantity $f'(a)$ is *derived* from information about the function f in a particular way. It is worth repeating here the four steps by which we derive $f'(a)$:

1. Magnify the graph at the point $(a, f(a))$ until it appears straight.
2. Calculate the slope of the magnified segment.
3. Repeat steps 1 and 2 with successively higher magnifications.
4. Take the limit of the succession of slopes produced in step 3.

We can express this limit in analytic form in the following way:

$$f'(a) = \lim_{\Delta x \to 0} \frac{\Delta y}{\Delta x} = \lim_{h \to 0} \frac{f(a + h) - f(a - h)}{2h}.$$

The **difference quotient**

$$\frac{\Delta y}{\Delta x} = \frac{f(a + h) - f(a - h)}{2h}$$

is the usual way we estimate the slope of the magnified graph of f at the point $(a, f(a))$. As the following figure shows, the calculation involves two points equally spaced on either side of $(a, f(a))$.

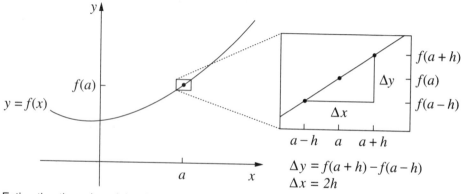

$$\Delta y = f(a + h) - f(a - h)$$
$$\Delta x = 2h$$

Estimating the value of the derivative $f'(a)$

Choosing Points in a Window

To estimate the slope in the window above, we chose two particular points, $(a - h, f(a - h))$ and $(a + h, f(a + h))$. However, *any* two points in the window would give us a valid estimate. Our choice depends on the situation. For example, if we are working with formulas, we want simple expressions. In that case we would probably replace $(a - h, f(a - h))$ by $(a, f(a))$. We do that in the window on the left. The resulting slope is

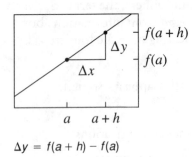

$$\frac{\Delta y}{\Delta x} = \frac{f(a + h) - f(a)}{h}.$$

$\Delta y = f(a + h) - f(a)$

While the limiting value of $\Delta y / \Delta x$ doesn't depend on the choices you make, the *estimates* you produce with a fixed Δx can be closer to or farther from the true value. The exercises will explore this.

Data Versus Formulas

The derivative is a limit. To find that limit we have to be able to zoom in arbitrarily close, to make Δx arbitrarily small. For functions given by data, that is usually impossible; we can't use any Δx smaller than the spacing between the data points. Thus, a data function of this sort does not have a derivative, strictly speaking. However, by zooming in as much as the data allow, we get the most precise description possible for the rate of change of the function. In these circumstances it makes a difference which points we choose in a window to calculate $\Delta y / \Delta x$. In the exercises you will have a chance to see how the precision of your estimate depends on which points you choose to calculate the slope.

A data function might not have a derivative

For a function given by a formula, it is possible to find the value of the derivative exactly. In fact, the derivative of a function given by a formula is itself given by a formula. Later in this chapter we will describe some general rules which will allow us to produce the formula for the derivative without going through the successive approximation process each time. In chapter 5 we will discuss these rules more fully.

There are rules for finding derivatives

Practical Considerations

The derivative is a limit, and there are always practical considerations to raise when we discuss limits. As we saw in chapter 2, we cannot expect to construct the entire decimal expansion of a limit. In most cases all we can get are a specified number of digits. For example, in section 3.2 we found that

$$\text{if}\quad f(x) = \frac{2 + x^3 \cos x + 1.5^x}{2 + x^2}, \quad \text{then}\quad f'(27) = -.27235\ldots.$$

The same digits without the "..." give us **approximations**. Thus we can write $f'(27) \approx -.27235$; we also have $f'(27) \approx -.2723$ and $f'(27) \approx -.272$. Which approximation is the right one to use depends on the context. For example, if f appears in a problem in which all the other quantities are known only to one or two decimal places, we probably don't need a very precise value for $f'(27)$. In that case we don't have to carry the sequence of slopes $\Delta y / \Delta x$ very far. For instance, if we want to know $f'(27)$ to three decimal places, and so justify writing $f'(27) \approx -.272$, we only need to continue the zooming process until the slopes $\Delta y / \Delta x$ all have values that begin $-.2723\ldots$. By the table on page 99, $\Delta x = .0004$ is sufficient.

Language and Notation

- If f has a derivative at a, we also say f **is differentiable at** a. If f is differentiable at every point a in its domain, we say f **is differentiable**.

- Do *locally linear* and *differentiable* mean the same thing? The awkward case is a function whose graph is vertical at a point (for example, $y = \sqrt[3]{x}$ at the origin). On the one hand, it makes sense to say that the function is locally linear at such a point, because the graph looks straight under a microscope. On the other hand, the derivative itself is undefined, because the line is vertical. So the function is locally linear, but not differentiable, at that point.

  There is another way to view the matter. We can say, instead, that a vertical line *does* have a slope, and its value is *infinity* (∞). From this point of view, if the graph of f is vertical at $x = a$, then $f'(a) = \infty$. In other words, f *does* have a derivative at $x = a$; its value just happens to be ∞.

  Which view is *right*? Neither; we can choose either. Our choice is a matter of convention. (In some countries cars travel on the left; in others, on the right. That's a convention, too.) However, we will follow the second alternative. One advantage is that we will be able to use the derivative to indicate where the graph of a function is vertical. Another is that *locally linear* and *differentiable* then mean exactly the same thing.

  > Convention: if the graph of f is vertical at a, write $f'(a) = \infty$

- Suppose $y = f(x)$ and the quantities x and y appear in a context in which they have units. Then the derivative of $f'(x)$ *also* has units, because it is the rate of change of y with respect to x. The units for the derivative must be

$$\text{units for } f' = \frac{\text{units for output } y}{\text{units for input } x}.$$

We have already seen several examples—persons per day, miles per hour, milligrams per pound, dollars of profit per dollar change in price—and we will see many more.

Leibniz's notation

- There are several notations for the derivative. You should be aware of them because they are all in common use and because they reflect different ways of viewing the derivative. We have been writing the derivative of $y = f(x)$ as $f'(x)$. Leibniz wrote it as dy/dx. This notation has several advantages. It resembles the quotient $\Delta y/\Delta x$ that we use to approximate the derivative. Also, because dy/dx looks like a rate, it helps remind us that a derivative is a rate. Later on, when we consider the chain rule to find derivatives, you'll see that it can be stated very vividly using Leibniz's notation.

> *The German philosopher Gottfried Wilhelm Leibniz (1646–1716) developed calculus about the same time Newton did. While Newton dealt with derivatives in more or less the way we do, Leibniz introduced a related idea which he called a differential—'infinitesimally small' numbers which he would write as dx and dy.*

Newton's notation

The other notation still encountered is due to Newton. It occurs primarily in physics and is used to denote rates with respect to time. If a quantity y is changing over time, then the Newton notation expresses the derivative of y as $\dot{y}$ (that's the variable y with a dot over it).

The Microscope Equation

A Context: Driving Time

If you make a 400-mile trip at an average speed of 50 miles per hour, then the trip takes 8 hours. Suppose you increase the average speed by 2 miles per hour. How much time does that cut off the trip?

One way to approach this question is to start with the basic formula

$$\text{speed} \times \text{time} = \text{distance}.$$

Travel time depends on speed

The distance is known to be 400 miles, and we really want to understand how *time T* depends on *speed s*. We get T as a function of s by rewriting the last equation:

$$T \text{ hours} = \frac{400 \text{ miles}}{s \text{ miles per hour}}.$$

To answer the question, just set $s = 52$ miles per hour in this equation. Then $T = 7.6923$ hours, or about 7 hours, 42 minutes. Thus, compared to the original 8 hours, the higher speed cuts 18 minutes off your driving time.

What happens to the driving time if you increase your speed by 4 miles per hour, or 5, instead of 2? What happens if you go slower, say 2 or 3 miles per hour slower? We could make a fresh start with each of these questions and answer them, one by one, the same way we did the first. But taking the questions one at a time misses the point. What we really want to know is the general pattern:

If I'm traveling at 50 miles per hour, how much does any given increase in speed decrease my travel time?

How does travel time respond to changes in speed?

We already know how T and s are related: $T = 400/s$ hours. This question, however, is about the connection between a *change* in speed of

$$\Delta s = s - 50 \text{ miles per hour}$$

and a *change* in arrival time of

$$\Delta T = T - 8 \text{ hours.}$$

To answer it we should change our point of view slightly. It is not the relation between s and T, but between Δs and ΔT, that we want to understand.

Since we are considering speeds s that are only slightly above or below 50 miles per hour, Δs will be small. Consequently, the arrival time T will be only slightly different from 8 hours, so ΔT will also be small. Thus we want to study small changes in the function $T = 400/s$ near $(s, T) = (50, 8)$. The natural tool to use is a microscope.

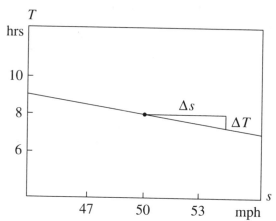

How travel time changes with speed around 50 miles per hour

In the microscope window above we see the graph of $T = 400/s$, magnified at the point $(s, T) = (50, 8)$. The field of view was chosen so that s can take values about 6 mph above or below 50 mph. The graph looks straight, and its slope is $T'(50)$. In the exercises you are asked to determine the value of $T'(50)$; you should find $T'(50) \approx -.16$. [Later on, when we have rules for finding the derivative of a formula, you will see that $T'(50) = -.16$ *exactly*.] Since the quotient $\Delta T / \Delta s$ is also an estimate for the slope of the line in the window, we can write

The slope of the graph in the microscope window

$$\frac{\Delta T}{\Delta s} \approx -.16 \text{ hour } per \text{ mile per hour.}$$

If we multiply both sides of this approximate equation by Δs miles per hour, we get

$$\Delta T \approx -.16 \, \Delta s \text{ hours.}$$

How travel time changes with speed

This equation answers our question about the general pattern relating changes in travel time to changes in speed. It says that the changes are *proportional*. For every mile per hour increase in speed, travel time decreases by about .16 hour, or about $9\frac{1}{2}$ minutes. Thus, if the speed is 1 mph over 50 mph, travel time is cut by about $9\frac{1}{2}$ minutes. If we double the increase in speed, that doubles the savings in time: if the speed is 2 mph over 50 mph, travel time is cut by about 19 minutes. Compare this with a value of about 18 minutes that we got with the exact equation $T = 400/s$.

Δs and ΔT now have a special meaning

Notice that we are using Δs and ΔT in a slightly more restricted way than we have previously. Up to now, Δs measured the horizontal distance between *any* two points on a graph. Now, however, Δs just measures the horizontal distance from the fixed point $(s, T) = (50, 8)$ (marked with a large dot) that sits at the center of the window. Likewise, ΔT just measures the vertical distance from this point. The central point therefore plays the role of an origin, and Δs and ΔT are the *coordinates* of a point measured from that origin. To underscore the fact that Δs and ΔT are really coordinates, we have added a Δs-axis and a ΔT-axis in the window below. Notice that these coordinate axes have their own labels and scales.

Δs and ΔT are coordinates in the window

Every point in the window can therefore be described in two different coordinate systems. The two different sets of coordinates of the point labeled P, for instance, are $(s, T) = (53, 7.52)$ and $(\Delta s, \Delta T) = (3, -.48)$. The first pair says "When your speed is 53 mph, the trip will take 7.52 hrs." The second pair says "When you increase your speed by 3 mph, you will decrease travel time by .48 hr." Each statement can be translated into the other, but each statement has its own point of reference.

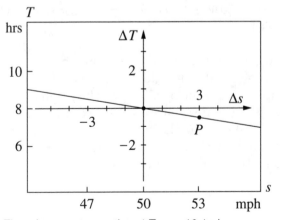

The microscope equation: $\Delta T \approx -.16 \, \Delta s$ hours

We call $\Delta T \approx -.16\,\Delta s$ the **microscope equation** because it tells us how the microscope coordinates Δs and ΔT are related.

In fact, we now have two different ways to describe how travel time is related to speed. They can be compared in the following table.

	Global	Local
Coordinates:	s, T	$\Delta s, \Delta T$
Equation:	$T = 400/s$	$\Delta T \approx -.16\,\Delta s$
Properties:	Exact nonlinear	Approximate linear

We say the microscope equation is *local* because it is intended to deal only with speeds near 50 miles per hour. There is a different microscope equation for speeds near 40 miles per hour, for instance. By contrast, the original equation is *global*, because it works for all speeds. While the global equation is exact, it is also nonlinear; this can make it more difficult to compute. The microscope equation is approximate but linear; it is easy to compute. It is also easy to put into words:

Global vs. local descriptions

At 50 miles per hour, the travel time of a 400-mile journey decreases $9\frac{1}{2}$ minutes for each mile per hour increase in speed.

The connection between the global equation and the microscope equation is shown in the following illustration.

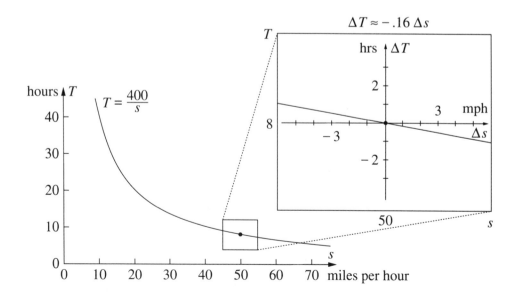

Local Linearity and Multipliers

The reasoning that led us to a microscope equation for travel time can be applied to any locally linear function. If $y = f(x)$ is locally linear, then at $x = a$ we can write

$$\boxed{\text{The microscope equation:} \quad \Delta y \approx f'(a) \cdot \Delta x}$$

We know an equation of the form $\Delta y = m \cdot \Delta x$ tells us that y is a linear function of x, in which m plays the role of slope, rate, and multiplier. The microscope equation therefore tells us that **y is a linear function of x when x is near a**— at least approximately. In this almost-linear relation, the derivative $f'(a)$ plays the role of slope, rate, and multiplier.

The microscope equation is the analytic form of local linearity

The microscope equation is just the idea of local linearity expressed analytically rather than geometrically—that is, by a formula rather than by a picture. Here is a chart that shows how the two descriptions of local linearity fit together.

$y = f(x)$ is locally linear at $x = a$:

Geometrical	Analytical
When magnified at $(a, f(a))$, the graph of f is almost straight, and the slope of the line is $f'(a)$.	When x is near a, y is almost a linear function of x, and the multiplier is $f'(a)$.

Microscope window

Microscope equation

$$\boxed{\Delta y \approx f'(a) \cdot \Delta x}$$

Of course, the graph in a microscope window is not *quite* straight. The analytic counterpart of this statement is that the microscope equation is not *quite* exact—the two sides of the equation are only approximately equal. We write "$\approx$" instead of "$=$". However, we can make the graph look even straighter by increasing the magnification—or, what is the same thing, by decreasing the field of view. Analytically, this increases the exactness of the microscope equation. Like a laboratory microscope, our microscope is most accurate at the center of the field of view, with increasing aberration toward the periphery!

In the microscope equation $\Delta y \approx f'(a) \cdot \Delta x$, **the derivative is the multiplier that tells how y responds to changes in x.** In particular, a small increase in x produces a change in y that depends on the sign and magnitude of $f'(a)$ in the following way:

- $f'(a)$ is large and positive $\Rightarrow$ large increase in y,
- $f'(a)$ is small and positive $\Rightarrow$ small increase in y,
- $f'(a)$ is large and negative $\Rightarrow$ large decrease in y,
- $f'(a)$ is small and negative $\Rightarrow$ small decrease in y.

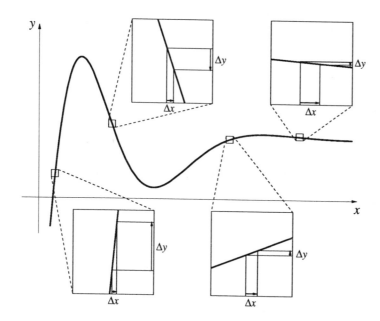

For example, suppose we are told the value of the derivative is 2. Then any small change in x induces a change in y approximately twice as large. If, instead, the derivative is $-1/5$, then a small change in x produces a change in y only one-fifth as large, and in the opposite direction. That is, if x increases, then y decreases, and vice versa.

The microscope equation should look familiar to you. It has been with us from the beginning of the course. Our "recipe" $\Delta S = S' \cdot \Delta t$ for predicting future values of S in the S-I-R model is just the microscope equation for the function $S(t)$. (Although we wrote it with an "$=$" instead of an "$\approx$" in the first chapter, we noted that ΔS would provide us only an *estimate* for the new value of S.) The success of Euler's method in producing solutions to rate equations depends fundamentally on the fact that the functions we are trying to find are locally linear.

The microscope equation is the recipe for building solutions to rate equations

The derivative is one of the fundamental concepts of the calculus, and one of its most important roles is in the microscope equation. Besides giving us a tool for building solutions to rate equations, the microscope equation helps us do estimation and error analysis, the subject of the next section.

We conclude with a summary that compares linear and locally linear functions. Note there are two differences, but only two: 1) the equation for local linearity is only an approximation; 2) it holds only locally—that is, near a given point.

If $y = f(x)$ is linear, then $\Delta y = m \cdot \Delta x$; the constant m is rate, slope, and multiplier for all x.

If $y = f(x)$ is locally linear, then $\Delta y \approx f'(a) \cdot \Delta x$; the derivative $f'(a)$ is rate, slope, and multiplier for x near a.

Exercises

Computing Derivatives

1. Sketch graphs of the following functions and use these graphs to determine which function has a derivative that is always positive (except at $x = 0$, where neither the function nor its derivative is defined).

$$y = \frac{1}{x} \qquad y = \frac{-1}{x} \qquad y = \frac{1}{x^2} \qquad y = \frac{-1}{x^2}$$

What feature of the graph told you whether the derivative was positive?

2. For each of the functions f below, approximate its derivative at the given value $x = a$ in two different ways. First, use a computer microscope (i.e., a graphing program) to view the graph of f near $x = a$. Zoom in until the graph looks straight and find its slope. Second, use a calculator to find the value of the quotient

$$\frac{f(a + h) - f(a - h)}{2h}$$

for $h = .1, .01, .001, \ldots, .000001$. Based on these values of the quotients, give your best estimate for $f'(a)$, and say how many decimal places of accuracy it has.

*a) $f(x) = 1/x$ at $x = 2$.   c) $f(x) = x^3$ at $x = 200$.

b) $f(x) = \sin(7x)$ at $x = 3$.   d) $f(x) = 2^x$ at $x = 5$.

3. In a later section we will establish that the derivative of $f(x) = x^3$ at $x = 1$ is exactly 3: $f'(1) = 3$. This question concerns the freedom we have to choose points in a window to estimate $f'(1)$ (see page 108). Its

purpose is to compare two quotients, to see which gets closer to the exact value of $f'(1)$ for a fixed "field of view" Δx. The two quotients are

$$Q_1 = \frac{\Delta y}{\Delta x} = \frac{f(a + h) - f(a - h)}{2h} \quad \text{and} \quad Q_2 = \frac{\Delta y}{\Delta x} = \frac{f(a + h) - f(a)}{h}.$$

In this problem $a = 1$.

a) Construct a table that shows the values of Q_1 and Q_2 for each $h = 1/2^k$, where $k = 0, 1, 2, \ldots, 8$. If you wish, you can use this program to compute the values:

```
a = 1
FOR k = 0 TO 8
        h = 1 / 2 ^ k
        q1 = ((a + h) ^ 3 - (a - h) ^ 3) / (2 * h)
        q2 = ((a + h) ^ 3 - a ^ 3) / h
        PRINT h, q1, q2
NEXT k
```

b) How many digits of Q_1 stabilize in this table? How many digits of Q_2?

*c) Which is a better estimator, Q_1 or Q_2? To indicate how much better, give the value of h for which the *better* estimator provides an estimate that is as close as the best estimate provided by the *poorer* estimator.

4. Repeat all the steps of the last question for the function $f(x) = \sqrt{x}$ at $x = 9$. The exact value of $f'(9)$ is $1/6$.

Comment: Note that, in section 3.1, we estimated the rate of change of the sunrise function using an expression like Q_1 rather than one like Q_2. The previous exercises should persuade you this was deliberate. We were trying to get the most representative estimates, given the fact that we could not reduce the size of Δx arbitrarily.

5. At this point you will find it convenient to write a more general derivative-finding program. You can modify the program in problem 3 to do this by having a DEF command at the beginning of the program to specify the function you are currently interested in. For instance, if you insert the command DEF fnf (x) = x ^ 3 at the beginning of the program, how could you simplify the lines specifying q1 and q2? If you then wanted to calculate the derivative of another function at a different x-value, you would only need to change the DEF specification and, depending on the point you were interested in, the a = 1 line.

6. Use one of the methods of problem 2 to estimate the value of the derivative of each of the following functions at $x = 0$:

$$y = 2^x, \quad y = 3^x, \quad y = 10^x, \quad \text{and} \quad y = (1/2)^x.$$

These are called **exponential** functions, because the input variable x appears in the exponent. How many decimal places accuracy do your approximations to the derivatives have?

7. In this problem we look again at the exponential function $f(x) = 2^x$ from the previous problem.
    a) Use the rules for exponents to put the quotient

$$\frac{f(a + h) - f(a)}{h}$$

   in the simplest form you can.
   b) We know that

$$f'(0) = \lim_{h \to 0} \frac{f(h) - f(0)}{h}.$$

   Use this fact, along with the algebraic result of part (a), to explain why $f'(a) = f'(0) \cdot 2^a$.

8. Apply all the steps of the previous question to the exponential function $f(x) = b^x$ with an arbitrary base b. Show that $f'(x) = f'(0) \cdot b^x$.

9. a) For which values of x is the absolute value function $y = |x|$ differentiable?
    b) At each point where $y = |x|$ is differentiable, find the value of the derivative.

The Microscope Equation

10. Write the microscope equation for each of the following functions at the indicated point. (To find the necessary derivative, consult problem 2.)
    *a) $f(x) = 1/x$ at $x = 2$.
    b) $f(x) = \sin(7x)$ at $x = 3$.
    c) $f(x) = x^3$ at $x = 200$.
    d) $f(x) = 2^x$ at $x = 5$.

11. This question uses the microscope equation for $f(x) = 1/x$ at $x = 2$ that you constructed in the previous question.
    a) Draw the graph of what you would see in the microscope if the field of view is .2 unit wide.

*b) If we take $x = 2.05$, what is Δx in the microscope equation? What estimate does the microscope equation give for Δy? What estimate does the microscope equation then give for $f(2.05) = 1/2.05$? Calculate the true value of $1/2.05$ and compare the two values; how far is the microscope estimate from the true value?

c) What estimate does the microscope equation give for $1/2.02$? How far is this from the true value?

d) What estimate does the microscope equation give for $1/1.995$? How far is this from the true value?

12. This question concerns the travel time function $T = 400/s$ hours, discussed in the text.

a) How many hours does a 400-mile trip take at an average speed of 40 miles per hour?

b) Find the microscope equation for T when $s = 40$ miles per hour.

c) At what rate does the travel time decrease as speed increases around 40 mph—in hours *per* mile per hour?

*d) According to the microscope equation, how much travel time is saved by increasing the speed from 40 to 45 mph?

e) According to the microscope equation based at 50 mph (as done in the text), how much time is *lost* by decreasing the speed from 50 to 45 mph?

*f) The last two parts both predict the travel time when the speed is 45 mph. Do they give the same result?

13. a) Suppose $y = f(x)$ is a function for which $f(5) = 12$ and $f'(5) = .4$. Write the microscope equation for f at $x = 5$.

b) Draw the graph of what you would see in the microscope. Do you need a formula for f itself in order to do this?

*c) If $x = 5.3$, what is Δx in the microscope equation? What estimate does the microscope equation give for Δy? What estimate does the microscope equation then give for $f(5.3)$?

d) What estimates does the microscope equation give for the following: $f(5.23), f(4.9), f(4.82), f(9)$? Do you consider these estimates to be equally reliable?

14. a) Suppose $z = g(t)$ is a function for which $g(-4) = 7$ and $g'(-4) = 3.5$. Write the microscope equation for g at $t = -4$.

b) Draw the graph of what you see in the microscope.

c) Estimate $g(-4.2)$ and $g(-3.75)$.

d) For what value of t near -4 would you estimate that $g(t) = 6$? For what value of t would you estimate $g(t) = 8.5$?

15. If $f(a) = b, f'(a) = -3$, and if k is small, which of the following is the best estimate for $f(a + k)$?

$$a + 3k, \ b + 3k, \ a + 3b, \ b - 3k, \ a - 3k, \ 3a - b, \ a^2 - 3b, \ f'(a + k)$$

16. If f is differentiable at a, which of the following, for small values of h, are reasonable estimates of $f'(a)$?

$$\frac{f(a + h) - f(a - h)}{h} \qquad \frac{f(a + h) - f(a - h)}{2h}$$

$$\frac{f(a + h) - f(b)}{h} \qquad \frac{f(a + 2h) - f(a - h)}{3h}$$

17. Suppose a person has traveled D feet in t seconds. Then $D'(t)$ is the person's velocity at time t; $D'(t)$ has units of feet per second.

a) Suppose $D(5) = 30$ feet and $D'(5) = 5$ feet/second. Estimate the following:

$$D(5.1) \qquad D(5.8) \qquad D(4.7)$$

b) If $D(2.8) = 22$ feet, while $D(3.1) = 26$ feet, what would you estimate $D'(3)$ to be?

18. Fill in the blanks.

*a) If $f(3) = 2$ and $f'(3) = 4$, a reasonable estimate of $f(3.2)$ is _____.

b) If $g(7) = 6$ and $g'(7) = .3$, a reasonable estimate of $g(6.6)$ is

_____.

c) If $h(1.6) = 1$, $h'(1.6) = -5$, a reasonable estimate of $h(\underline{\hspace{1cm}})$ is 0.

*d) If $F(2) = 0$, $F'(2) = .4$, a reasonable estimate of $F(\underline{\hspace{1cm}})$ is .15.

e) If $G(0) = 2$ and $G'(0) = $ _____, a reasonable estimate of $G(.4)$ is 1.6.

f) If $H(3) = -3$ and $H'(3) = $ _____, a reasonable estimate of $H(2.9)$ is -1.

19. In manufacturing processes the profit is usually a function of the number of units being produced, among other things. Suppose we are studying some small industrial company that produces n units in a week and makes a corresponding weekly profit of P. Assume $P = P(n)$.

a) If $P(1000) = \$500$ and $P'(1000) = \$2/$unit, then

$$P(1002) \approx \underline{\hspace{1cm}} \qquad P(995) \approx \underline{\hspace{1cm}} \qquad P(\underline{\hspace{1cm}}) \approx \$512$$

b) If $P(2000) = \$3000$ and $P'(2000) = -\$5/$unit, then

$$P(2010) \approx \underline{\hspace{1cm}} \qquad P(1992) \approx \underline{\hspace{1cm}} \qquad P(\underline{\hspace{1cm}}) \approx \$3100$$

c) If $P(1234) = \$625$ and $P(1238) = \$634$, what is an estimate for $P'(1236)$?

➤ 3.4 ESTIMATION AND ERROR ANALYSIS

Making Estimates

The Expanding House

In the book *The Secret House—24 hours in the strange and unexpected world in which we spend our nights and days* (Simon and Schuster, 1986), David Bodanis describes many remarkable events that occur at the microscopic level in an ordinary house. At one point he explains how sunlight heats up the structure, stretching it imperceptibly in every direction through the day until it has become several cubic inches larger than it was the night before. Is it plausible that a house can become several cubic inches larger as it expands in the heat of the day? In particular, how much longer, wider, and taller would it have to become if it were to grow in volume by, let us say, 3 cubic inches?

How much does a house expand in the heat?

For simplicity, assume the house is a cube 200 inches on a side. (This is about 17 feet, so the house is the size of a small, two-story cottage.) If s is the length of a side of *any* cube, in inches, then its volume is

$$V = s^3 \text{ cubic inches.}$$

Our question is about how V changes with s when s is about 200 inches. In particular, we want to know which Δs would yield a ΔV of 3 cubic inches. This is a natural question for the microscope equation

$$\Delta V \approx V'(200) \cdot \Delta s.$$

According to exercise 2c in the previous section, we can estimate the value of $V'(200)$ to be about 120,000, and the appropriate units for V' are cubic inches per inch. Thus

$$\Delta V \approx 120000 \, \Delta s$$

$$3 \text{ cubic inches} \approx 120000 \, \frac{\text{cubic inches}}{\text{inches}} \times \Delta s \text{ inches,}$$

so $\Delta s \approx 3/120000 = .000025$ inch—many times thinner than a human hair!

This value is much too small. To get a more realistic value, let's suppose the house is made of wood and the temperature increases about 30°F from night to day. Then measurements show that a 200-inch length of wood will actually become about $\Delta s = .01$ inch longer. Consequently the volume will actually expand by about

$$\Delta V \approx 120000 \, \frac{\text{cubic inches}}{\text{inches}} \times .01 \text{ inch} = 1200 \text{ cubic inches.}$$

This increase is 400 times as much as Bodanis claimed; it is about the size of a small computer monitor. So even as he opens our eyes to the effects of thermal expansion, Bodanis dramatically understates his point.

Estimates Versus Exact Values

Don't lose sight of the fact that the values we derived for the expanding house are *estimates*. In some cases we can get the exact values. Why don't we, whenever we can?

For example, we can calculate exactly how much the volume increases when we add $\Delta s = .01$ inch to $s = 200$ inches. The increase is from $V = 200^3 = 8{,}000{,}000$ cubic inches to

$$V = (200.01)^3 = 8001200.060001 \text{ cubic inches.}$$

Thus, the *exact* value of ΔV is 1200.060001 cubic inches. The estimate is off by only about .06 cubic inch. This isn't very much, and it is even less significant when you think of it as a percentage of the volume (namely 1200 cubic inches) being calculated. The percentage is

$$\frac{.06 \text{ cubic inch}}{1200 \text{ cubic inches}} = .00005 = .005\%.$$

That is, the difference is only 1/200 of 1% of the calculated volume.

To get the exact value we had to cube two numbers and take their difference. To get the estimate we only had to do a single multiplication. Estimates made with the microscope equation are always easy to calculate—they involve only linear functions. Exact values are usually harder to calculate. As you can see in the example, the extra effort may not gain us extra information. That's one reason why we don't always calculate exact values when we can.

Here's another reason. Go back to the question: How large must Δs be if $\Delta V = 3$ cubic inches? To get the exact answer, we must solve for Δs in the equation

$$3 = \Delta V = (200 + \Delta s)^3 - 200^3$$
$$= 200^3 + 3(200)^2 \Delta s + 3(200)(\Delta s)^2 + (\Delta s)^3 - 200^3.$$

Simplifying, we get

$$3 = 120000\, \Delta s + 600(\Delta s)^2 + (\Delta s)^3.$$

This is a cubic equation for Δs; it *can* be solved, but the steps are complicated. Compare this with solving the microscope equation:

$$3 = 120000\, \Delta s.$$

Thus, another reason we don't calculate exact values at every opportunity is that the calculations can be daunting. The microscope estimates are always straightforward.

Perhaps the most important reason, though, is the insight that calculating V' gave us. Let's translate into English what we have really been talking about:

Exact values can be harder to calculate than estimates.

In dealing with a cube 200 inches on a side, any small change (measured in inches) in the length of the sides produces a change (measured in cubic inches) in the volume approximately 120,000 times as great.

This is, of course, simply another instance of the point we have made before, that small changes in the input and the output are related in an (almost) linear way, even when the underlying function is complex. Let's continue this useful perspective by looking at error analysis.

Propagation of Error

From Measurements to Calculations

We can view all the estimates we made for the expanding house from another perspective—the lack of precision in measurement. To begin with, just think of the house as a cubical box that measures 200 inches on a side. Then the volume must be 8,000,000 cubic inches. But measurements are never exact, and any uncertainty in measuring the length of the side will lead to an uncertainty in calculating the volume. Let's say your measurement of length is accurate to within .5 inch. In other words, you believe the true length lies between 199.5 inches and 200.5 inches, but you are uncertain precisely where it lies within that interval. How uncertain does that make your calculation of the volume?

There is a direct approach to this question: we can simply say that the volume must lie between $199.5^3 = 7,940,149.875$ cubic inches and $200.5^3 = 8,060,150.125$ cubic inches. In a sense, these values are almost too precise. They don't reveal a general pattern. We would like to know how an uncertainty—or error—in measuring the length of the side of a cube propagates to an error in calculating its volume.

How uncertain is the calculated value of the volume?

Let's take another approach. If we measure s as 200 inches, and the true value differs from this by Δs inches, then Δs is the error in measurement. That produces an error ΔV in the calculated value of V. The microscope equation for the expanding house (page 121) tells us how ΔV depends on Δs when $s = 200$:

$$\Delta V \approx 120000 \, \Delta s.$$

Since we now interpret Δs and ΔV as errors, the microscope equation becomes the **error propagation** equation:

$$\text{error in } V \text{ (cu. in.)} \approx 120000 \left(\frac{\text{cu. in.}}{\text{inch}}\right) \times \text{error in } s \text{ (inches)}.$$

Thus, for example, an error of 1/2 inch in measuring s propagates to an error of about 60,000 cubic inches in calculating V. This is about 35 cubic feet, the size of a large refrigerator! Putting it another way:

The microscope equation describes how errors propagate

$$\text{if } s = 200 \pm .5 \text{ inch, then } V \approx 8,000,000 \pm 60,000 \text{ cubic inches.}$$

If we keep in mind the error propagation equation $\Delta V \approx 120000\,\Delta s$, we can quickly answer other questions about measuring the same cube. For instance, suppose we wanted to determine the volume of the cube to within 5,000 cubic inches. How accurately would we have to measure the side? Thus we are given $\Delta V = 5000$, and we conclude $\Delta s \approx 5000/120000 \approx .04$ inch. This is just a little more than 1/32 inch.

Relative Error

Suppose we have a second cube whose side is twice as large ($s = 400$ inches), and once again we measure its length with an error of 1/2 inch. Then the error in the calculated value of the volume is

$$\Delta V \approx V'(400) \cdot \Delta s = 480000 \times .5 = 240{,}000 \text{ cubic inches.}$$

[In the exercises you will be asked to show $V'(400) = 480{,}000$.] The error in our calculation for the bigger cube is four times what it was for the smaller cube, even though the length was measured to the same accuracy in both cases! There is no mistake here. In fact, the volume of the second cube is eight times the volume of the first, so the numbers we are dealing with in the second case are roughly eight times as large. We should not be surprised if the error is larger, too.

> **Bigger numbers have bigger errors**

In general, we must expect that the size of an error will depend on the size of the numbers we are working with. We expect big numbers to have big errors and small numbers to have small errors. In a sense, though, an error of 1 inch in a measurement of 50 inches is no worse than an error of 1/10-th of an inch in a measurement of 5 inches: both errors are 1/50-th the size of the quantity being measured.

A watchmaker who measures the tiny objects that go into a watch only as accurately as a carpenter measures lumber would never make a watch that worked; likewise, a carpenter who takes the pains to measure things as accurately as a watchmaker does would take forever to build a house. The scale of allowable errors is dictated by the scale of the objects they work on.

> **Absolute and relative error**

The errors Δx we have been considering are called **absolute errors**; their values depend on the size of the quantities x we are working with. To reduce the effect of differences due to the size of x, we can look instead at the error as a *fraction* of the number being measured or calculated. This fraction $\Delta x / x$ is called **relative error**. Consider two measurements: one is 50 inches with an error of ± 1 inch; the other is 2 inches with an error of $\pm.1$ inch. The *absolute* error in the second measurement is much smaller than in the first, but the *relative* error is $2\frac{1}{2}$ times larger. (The first relative error is .02 inch per inch, the second is .05 inch per inch.) To judge how good or bad a measurement really is, we usually take the relative error instead of the absolute error.

Let's compare the propagation of relative and absolute errors. For example, the absolute error in calculating the volume of a cube whose side measures s is

$$\Delta V \approx V'(s) \cdot \Delta s.$$

The absolute errors are proportional, but the multiplier $V'(s)$ depends on the size of s. (We saw above that the multiplier is 120,000 cubic inches per inch when $s = 200$ inches, but it grows to 480,000 cubic inches per inch when $s = 400$ inches.)

In section 3.5, which deals with formulas for derivatives, we will see that $V'(s) = 3s^2$. If we substitute $3s^2$ for $V'(s)$ in the propagation equation for absolute error, we get

$$\Delta V \approx 3s^2 \cdot \Delta s.$$

To see how relative error propagates, we should divide this equation by $V = s^3$:

$$\frac{\Delta V}{V} \approx \frac{3s^2 \cdot \Delta s}{s^3} = 3\frac{\Delta s}{s}.$$

The relative errors are proportional, but the multiplier is always 3; it doesn't depend on the size of the cube, as it did for absolute errors.

Return to the case where $\Delta s = .5$ inch and $s = 200$ inches. Since Δs and s have the same units, the relative error $\Delta s / s$ is "dimensionless"—it has no units. We can, however, describe $\Delta s / s$ as a *percentage*: $\Delta s / s = .5/200 = .25\%$, or $1/4$ of 1%. For this reason, relative error is sometimes called **percentage error**. It tells us the error in measuring a quantity as a *percentage* of the value of that quantity. Since the percentage error in volume is

Percentage error is relative error

$$\frac{\Delta V}{V} = \frac{60,000 \text{ cu. in.}}{8,000,000 \text{ cu. in.}} = .0075 = .75\%$$

we see that the percentage error in volume is 3 times the percentage error in length—*and this is independent of the length and volumes involved*. This is what the propagation equation for relative error says: A 1% error in measuring s, whether $s = .0002$ inch or $s = 2000$ inches, will produce a 3% error in the calculated value of the volume.

Exercises

Estimation

1. **a)** Suppose you are going on a 110-mile trip. Then the time T it takes to make the trip is a function of how fast you drive:

$$T(v) = \frac{110 \text{ miles}}{v \text{ miles per hour}} = 110\,v^{-1} \text{ hours}.$$

If you drive at $v = 55$ miles per hour, T will be 2 hours. Use a computer microscope to calculate $T'(55)$ and write an English sentence interpreting this number.

b) More generally, if you and a friend are driving separate cars on a 110-mile trip, and you are traveling at some velocity v, while her speed is 1% greater than yours, then her travel time is less. How much less, as a percentage of yours? Use the formula $T'(v) = -110v^{-2}$, which can be obtained using rules given in the next section.

2. **a)** Suppose you have 600 square feet of plywood which you are going to use to construct a cubical box. Assuming there is no waste, what will its volume be?

   b) Find a general formula which expresses the volume V of the box as a function of the area A of plywood available.

   c) Use a microscope to determine $V'(600)$, and express its significance in an English sentence.

   d) Use this multiplier to estimate the additional amount of plywood you would need to increase the volume of the box by 10 cubic feet.

   e) In the original problem, if you had to allow for wasting 10 square feet of plywood in the construction process, by how much would this decrease the volume of the box?

   f) In the original problem, if you had to allow for wasting 2% of the plywood in the construction process, by what percentage would this decrease the volume of the box?

3. **a)** Let $R(s) = 1/s$. You can use the fact that $R'(s) = -1/s^2$, to be established in section 3.5.

   b) Since $R(100) = $ _____ and $R'(100) = $ _____, we can make the following approximations:

   $$1/97 \approx \text{_____} \qquad 1/104 \approx \text{_____} \qquad R(\text{_____}) \approx .0106 \,.$$

4. Using the fact that the derivative of $f(x) = \sqrt{x}$ is $f'(x) = 1/(2\sqrt{x})$, you can estimate the square roots of numbers that are close to perfect squares.

   a) For instance $f(4) = $ _____ and $f'(4) = $ _____, so $\sqrt{4.3} \approx$ _____.

   b) Use the values of $f(4)$ and $f'(4)$ to approximate $\sqrt{5}$ and $\sqrt{3.6}$.

   c) Use the values of $f(100)$ and $f'(100)$ to approximate $\sqrt{101}$ and $\sqrt{99.73}$.

Error Analysis

5. **a)** If you measure the side of a square to be 12.3 inches, with an uncertainty of $\pm.05$ inch, what is your relative error?

   b) What is the area of the square? Write an error propagation equation that will tell you how uncertain you should be about this value.

   c) What is the relative error in your calculation of the area?

   d) If you wanted to calculate the area with an error of less than 1 square inch, how accurately would you have to measure the length of the side? If you wanted the error to be less than .1 square inch, how accurately would you have to measure the side?

6. a) Suppose the side of a square measures x meters, with a possible error of Δx meters. Write the equation that describes how the error in length propagates to an error in the area. (The derivative of $f(x) = x^2$ is $f'(x) = 2x$; see section 3.5.)

   b) Write an equation that describes how the *relative* error in length propagates to a *relative* error in area.

7. You are trying to measure the height of a building by dropping a stone off the top and seeing how long it takes to hit the ground, knowing that the distance d (in feet) an object falls is related to the time of fall, t (in seconds), by the formula $d = 16t^2$. You find that the time of fall is 2.5 seconds, and you estimate that you are accurate to within a quarter of a second. What do you calculate the height of the building to be, and how much uncertainty do you consider your calculation to have?

8. You see a flash of lightning in the distance and note that the sound of thunder arrives 5 seconds later. You know that at 20°C sound travels at 343.4 m/sec. This gives you an estimate of

$$5 \text{ sec} \times 343.4 \, \frac{\text{meters}}{\text{sec}} = 1717 \text{ meters}$$

for the distance between you and the spot where the lightning struck. You also know that the velocity v of sound varies as the square root of the temperature T measured in degrees Kelvin (the Kelvin temperature = Celsius temperature + 273), so

$$v(T) = k\sqrt{T}$$

for some constant k.

   a) Use the information given here to determine the value of k.

   b) If your estimate of the temperature is off by 5 degrees, how far off is your estimate of the distance to the lightning strike? How significant is this source of error likely to be in comparison with the imprecision with which you measured the 5-second time lapse? (Suppose your uncertainty about the time is .25 second.) Give a clear analysis justifying your answer.

9. We can measure the distance to the moon by bouncing a laser beam off a reflector placed on the moon's surface and seeing how long it takes the beam to make the round trip. If the moon is roughly 400,000 km away, and if light travels at 300,000 km/sec, how accurately do we have to be able to measure the length of the time interval to be able to determine the distance to the moon to the nearest tenth of a meter?

▷ 3.5 A GLOBAL VIEW

Derivative as Function

Up to now we have looked upon the derivative as a *number*. It gives us information about a function at a *point*—the rate at which the function is changing, the slope of its graph, the value of the multiplier in the microscope equation. But the numerical value of the derivative varies from point to point, and these values can also be considered as the values of a new function—the derivative function—with its own graph. Viewed this way the derivative is a *global* object.

The derivative is a function in its own right

The connection between a function and its derivative can be seen very clearly if we look at their graphs. To illustrate, we'll use the function $I(t)$ that describes how the size of an infected population varies over time, from the S-I-R problem we analyzed in chapter 1. The graph of I appears below, and directly beneath it is the graph of I', the derivative of I. The graphs are lined up vertically: the values of $I(a)$ and $I'(a)$ are recorded on the same vertical line that passes through the point $t = a$ on the t-axis.

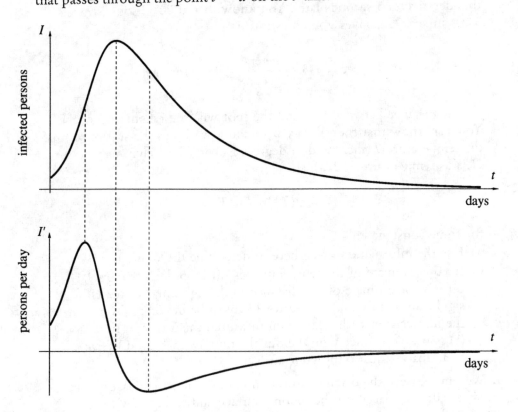

The height of I is the slope of I

To understand the connection between the graphs, keep in mind that the derivative represents a slope. Thus, at any point t, the *height* of the lower graph (I') tells us the *slope* of the upper graph (I). At the points where I is increasing, I' is positive—that is, I' lies *above* its t-axis. At the point where I is increasing most rapidly, I' reaches its highest value. In other words, where the graph of I is steepest, the graph of I' is highest. At the point where I is decreasing most rapidly, I' has its lowest value.

Next, consider what happens when I itself reaches its maximum value. Since I is about to switch from increasing to decreasing, the derivative must be about to switch from positive to negative. Thus, at the moment when I is largest, I' must be zero. Note that the highest point on the graph of I lines up with the point where I' crosses the t-axis. Furthermore, if we zoomed in on the graph of I at its highest point, we would find a horizontal line—in other words, one whose slope is zero.

All functions and their derivatives are related the same way that I and I' are. In the following table we list the various features of the graph of a function; alongside each is the corresponding feature of the graph of the derivative.

Function	Derivative
Increasing	Positive
Decreasing	Negative
Horizontal	Zero
Steep (rising or falling)	Large (positive or negative)
Gradual (rising or falling)	Small (positive or negative)
Straight	Horizontal

By using this table, you should be able to make a rough sketch of the graph of the derivative, when you are given the graph of a function. You can also read the table from right to left, to see how the graph of a function is influenced by the graph of its derivative.

For instance, suppose the graph of the function $L(x)$ is

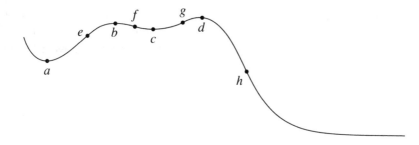

Then we know that its derivative L' must be 0 at points a, b, c, and d; that the derivative must be positive between a and b and between c and d, negative otherwise; that the derivative takes on relatively large values at e and g (positive) and at f and h (negative); that the derivative must approach 0 at the right endpoint and be large and negative at the left endpoint. Putting all this together we conclude that the graph of the derivative L' must look something like the following:

Finding a derivative "by eye"

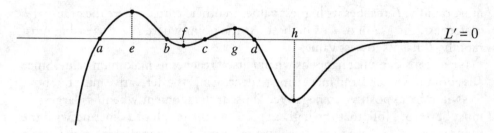

Conversely, suppose all we are told about a certain function G is that the graph of its derivative G' looks like this:

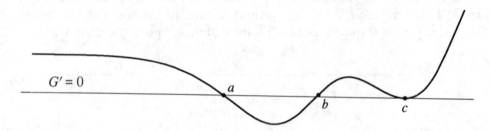

Then we can infer that the function G itself is decreasing between a and b and is increasing everywhere else; that the graph of G is horizontal at a, b, and c; that both ends of the graph of G slope upward from left to right—the left end more or less straight, the right getting steeper and steeper.

Formulas for Derivatives

Basic Functions

If a function is given by a formula, then its derivative also has a formula, defined for the points where the function is locally linear. The formula is produced by a definite process, called **differentiation**. In this section we look at some of the basic aspects, and in the next we will take up the chain rule, which is the key to the whole process. Then in chapter 5 we will review differentiation systematically.

Most formulas are constructed by combining only a few basic functions in various ways. For instance, the formula

$$3x^7 - \frac{\sin x}{8\sqrt{x}},$$

uses the basic functions x^7, $\sin x$, and $\sqrt{x}$. In fact, since $\sqrt{x} = x^{1/2}$, we can think of x^7 and $\sqrt{x}$ as two different instances of a single basic "power of x"— which we can write as x^p.

The following table lists some of the more common basic functions with their derivatives. The number c is an arbitrary constant, and so is the power p. The last function in the table is the exponential function with base b. Its

Formulas are combinations of basic functions

derivative involves a parameter k_b that varies with the base b. For instance, exercise 6 in section 3.3 established that, when $b = 2$, then $k_2 \approx .69$. Exercise 7 established, for any base b, that k_b is the value of the derivative of b^x when $x = 0$. We will have more to say about the parameter k_b in the next chapter.

Function	Derivative
c	0
x^p	px^{p-1}
$\sin x$	$\cos x$
$\cos x$	$-\sin x$
$\tan x$	$\sec^2 x$
b^x	$k_b \cdot b^x$

Remember that the input to the trigonometric functions is always measured in radians; the above formulas are not correct if x is measured in degrees. There are similar formulas if you insist on using degrees, but they are more complicated. This is the principal reason we work in radians—the formulas are nice!

For example,

- The derivative of $1/x = x^{-1}$ is $-x^{-2} = -1/x^2$;
- The derivative of $\sqrt{w} = w^{1/2}$ is $\frac{1}{2}w^{-1/2} = \dfrac{1}{2\sqrt{w}}$;
- The derivative of x^π is $\pi x^{\pi-1}$; and
- The derivative of π^x is $k_\pi \cdot \pi^x \approx 1.4\pi^x$.

Compare the last two functions. The first, x^π, is a **power function**—it is a power of the input x. The second, π^x, is an **exponential function**—the input x appears in the exponent. When you differentiate a power function, the exponent drops by 1; when you differentiate an exponential function, the exponent doesn't change.

Basic Rules

Since basic functions are combined in various ways to make formulas, we need to know how to differentiate *combinations*. For example, suppose we add the basic functions $g(x)$ and $h(x)$ to get $f(x) = g(x) + h(x)$. Then f is differentiable, and $f'(x) = g'(x) + h'(x)$. Actually, this is true for *all* differentiable functions g and h, not just basic functions. It says: "The rate at which f changes is the sum of the separate rates at which g and h change." Here are some examples that illustrate the point.

The addition rule

$$\text{If} \quad f(x) = \tan x + x^{-6}, \quad \text{then} \quad f'(x) = \sec^2 x - 6x^{-7}.$$

$$\text{If} \quad f(w) = 2^w + \sqrt{w}, \quad \text{then} \quad f'(w) = k_2 2^w + \frac{1}{2\sqrt{w}} \quad (\text{and } k_2 \approx .69).$$

Likewise, if we multiply any differentiable function g by a constant c, then the product $f(x) = cg(x)$ is also differentiable and $f'(x) = cg'(x)$. This says: "If f is c times as large as g, then f changes at c times the rate of g." Thus the derivative of $5 \sin x$ is $5 \cos x$. Likewise, the derivative of $(5x)^2$ is $50x$. (This took an extra calculation.) However, the rule does *not* tell us how to find the derivative of $\sin(5x)$, because $\sin(5x) \neq 5\sin(x)$. We will need the chain rule to work this one out. (See section 3.6.)

The rules about sums and constant multiples of functions are just the first of several basic rules for differentiating combinations of functions. We will describe how to handle products and quotients of functions in chapter 5. For the moment we summarize in the following table the rules we have already covered.

Function	Derivative
$f(x) + g(x)$	$f'(x) + g'(x)$
$c \cdot f(x)$	$c \cdot f'(x)$

With just the few facts already laid out we can differentiate a variety of functions given by formulas. In particular, we can differentiate any **polynomial** function:

$$P(x) = a_n x^n + a_{n-1} x^{n-1} + \cdots + a_2 x^2 + a_1 x + a_0.$$

Here $a_n, a_{n-1}, \ldots, a_2, a_1, a_0$ are various constants, and n is a positive integer, called the **degree** of the polynomial. A polynomial is a sum of constant multiples of integer powers of the input variable. A polynomial of degree 1 is just a linear function.

All the rules presented up to this point are illustrated in the following examples; note that the first three involve polynomials.

Function	Derivative
$7x + 2$	7
$5x^4 - 2x^3$	$20x^3 - 6x^2$
$5x^4 - 2x^3 + 17$	$20x^3 - 6x^2$
$3u^{15} + .5u^8 - \pi u^3 + u - \sqrt{2}$	$45u^{14} + 4u^7 - 3\pi u^2 + 1$
$6 \cdot 10^z + 17/z^5$	$6 \cdot k_{10}\, 10^z - 85/z^6$
$3 \sin t - 2t^3$	$3 \cos t - 6t^2$
$\pi \cos x - \sqrt{3} \tan x + \pi^2$	$-\pi \sin x - \sqrt{3} \sec^2 x$

The first two functions have the same derivative because they differ only by a constant, and the derivative of a constant is zero. The constant k_{10} that appears in the fourth example is approximately 2.30.

Exercises

Sketching the Graph of the Derivative

1. Sketch the graphs of two different functions that have the same derivative. (For example, can you find two *linear* functions that have the same derivative?)

2. Here are the graphs of four related functions: s, its derivative s', another function $c(t) = s(2t)$, and *its* derivative $c'(t)$. The graphs are out of order. Label them with the correct names s, s', c, and c'.

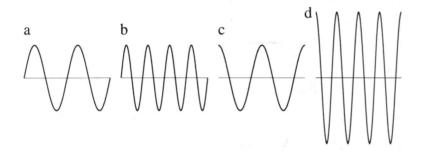

3. a) Suppose a function $y = g(x)$ satisfies $g(0) = 0$ and $0 \leq g'(x) \leq 1$ for all values of x in the interval $0 \leq x \leq 3$. Explain carefully why the graph of g must lie entirely in the triangular region shaded below:

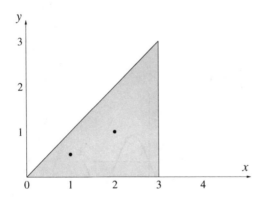

   b) Suppose you learn that $g(1) = .5$ and $g(2) = 1$. Draw the smallest shaded region in which you can guarantee that the graph of g must lie.

4. Suppose h is differentiable over the interval $0 \leq x \leq 3$. Suppose $h(0) = 0$, and that

$$
\begin{aligned}
.5 \leq h'(x) \leq 1 &\quad \text{for} \quad 0 \leq x \leq 1 \\
0 \leq h'(x) \leq .5 &\quad \text{for} \quad 1 \leq x \leq 2 \\
-1 \leq h'(x) \leq 0 &\quad \text{for} \quad 2 \leq x \leq 3
\end{aligned}
$$

Draw the smallest shaded region in the x, y-plane in which you can guarantee that the graph of $y = h(x)$ must lie.

5. For each of the functions graphed below, sketch the graph of its derivative.

i.

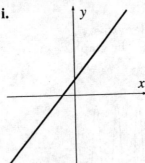

ii.

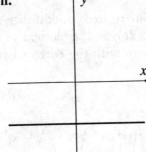

iii.

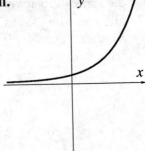

iv.

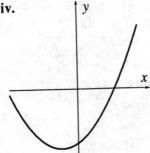

v.

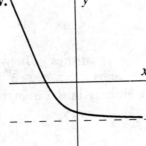

vi.

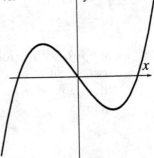

vii.

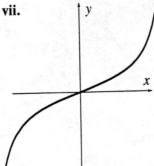

viii.

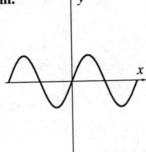

ix.

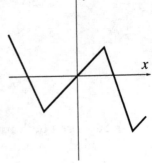

Differentiation

6. Find formulas for the derivatives of the following functions; that is, *differentiate* them.

   a) $f(x) = 3x^7 - .3x^4 + \pi x^3 - 17$.

*b) $g(x) = \sqrt{3}\,\sqrt{x} + \dfrac{7}{x^5}$.

c) $h(w) = 2w^8 - \sin w + \dfrac{1}{3w^2}$.

d) $R(u) = 4\cos u - 3\tan u + \sqrt[3]{u}$.

e) $V(s) = \sqrt[4]{16} - \sqrt[4]{s}$.

f) $F(z) = \sqrt{7}\cdot 2^z + (1/2)^z$.

g) $P(t) = -\dfrac{a}{2}t^2 + v_0 t + d_0$ (a, v_0, and d_0 are constants).

7. Use a computer graphing utility for this exercise. Graph on the same screen the following three functions:
   i) the function f given below, on the indicated interval;
   ii) the function $g(x) = (f(x + .01) - f(x - .01))/.02$ that estimates the slope of the graph of f at x;
   iii) the function $h(x) = f'(x)$, where you use the differentiation rules to find f'.
   a) $f(x) = x^4$ on $-1 \le x \le 1$.
   b) $f(x) = x^{-1}$ on $1 \le x \le 8$.
   c) $f(x) = \sqrt{x}$ on $.25 \le x \le 9$.
   d) $f(x) = \sin x$ on $0 \le x \le 2\pi$.
   The graphs of g and h should coincide—or "share phosphor"—in each case. Do they?

8. In each case below, find a function $f(x)$ whose derivative $f'(x)$ is
   a) $f'(x) = 12x^{11}$.
   *b) $f'(x) = 5x^7$.
   c) $f'(x) = \cos x + \sin x$.
   d) $f'(x) = ax^2 + bx + c$.
   e) $f'(x) = 0$.
   f) $f'(x) = \dfrac{5}{\sqrt{x}}$.

*9. What is the slope of the graph of $y = x - \sqrt{x}$ at $x = 4$? At $x = 100$? At $x = 10000$?

10. a) For which values of x is the function $x - x^3$ increasing?
    b) Where is the graph of $y = x - x^3$ rising most steeply?
    c) At what points is the graph of $y = x - x^3$ horizontal?
    d) Make a sketch of the graph of $y = x - x^3$ that reflects all these results.

11. a) Sketch the graph of the function $y = 2x + \dfrac{5}{x}$ on the interval $.2 \le x \le 4$.
    *b) Where is the lowest point on that graph? Give the value of the x-coordinate *exactly*.

12. What is the slope of the graph of $y = \sin x + \cos x$ at $x = \pi/4$?

13. a) Write the microscope equation for $y = \sin x$ at $x = 0$.
    *b) Using the microscope equation, estimate the following values: $\sin.3$, $\sin.007$, $\sin(-.02)$. Check these values with a calculator. (Remember to set your calculator to radian mode!)

14. a) Write the microscope equation for $y = \tan x$ at $x = 0$.
    b) Estimate the following values: $\tan.007$, $\tan.3$, $\tan(-.02)$. Check these values with a calculator.

15. a) Write the microscope equation for $y = \sqrt{x}$ at $x = 3600$.
    *b) Use the microscope equation to estimate $\sqrt{3628}$ and $\sqrt{3592}$. How far are these estimates from the values given by a calculator?

16. If the radius of a spherical balloon is r inches, its volume is $\frac{4}{3}\pi r^3$ cubic inches.
    a) At what rate does the volume increase, in cubic inches per inch, when the radius is 4 inches?
    b) Write the microscope equation for the volume when $r = 4$ inches.
    c) When the radius is 4 inches, approximately how much does it increase if the volume is increased by 50 cubic inches?
    d) Suppose someone is inflating the balloon at the rate of 10 cubic inches of air per second. If the radius is 4 inches, at what rate is it increasing, in inches per second?

17. A ball is held motionless and then dropped from the top of a 200-foot-tall building. After t seconds have passed, the distance from the ground to the ball is $d = f(t) = -16t^2 + 200$ feet.
    a) Find a formula for the velocity $v = f'(t)$ of the ball after t seconds. Check that your formula agrees with the given information that the initial velocity of the ball is 0 feet/second.
    b) Draw graphs of both the velocity and the distance as functions of time. What time interval makes physical sense in this situation? (For example, does $t < 0$ make sense? Does the distance formula make sense after the ball hits the ground?)
    c) At what time does the ball hit the ground? What is its velocity then?

18. A second ball is tossed straight up from the top of the same building with a velocity of 10 feet per second. After t seconds have passed, the distance from the ground to the ball is $d = f(t) = -16t^2 + 10t + 200$ feet.
    a) Find a formula for the velocity of the second ball. Does the formula agree with given information that the initial velocity is +10 feet per second? Compare the velocity formulas for the two balls; how are they similar, and how are they different?
    b) Draw graphs of both the velocity and the distance as functions of time. What time interval makes physical sense in this situation?

c) Use your graph to answer the following questions. During what period of time is the ball rising? During what period of time is it falling? When does it reach the highest point of its flight?

*d) How high does the ball rise?

19. a) What is the velocity formula for a third ball that is thrown *downward* from the top of the building with a velocity of 40 feet per second? Check that your formula gives the correct initial velocity.

   b) What is the distance formula for the third ball? Check that it satisfies the initial condition (namely, that the ball starts at the top of the building).

   c) When does this ball hit the ground? How fast is it going then?

20. A steel ball is rolling along a 20-inch-long straight track so that its distance from the midpoint of the track (which is 10 inches from either end) is $d = 3 \sin t$ inches after t seconds have passed. (Think of the track as aligned from left to right. Positive distances mean the ball is to the right of the center; negative distances mean it is to the left.)

   a) Find a formula for the velocity of the ball after t seconds. What is happening when the velocity is positive; when it is negative; when it equals zero? Write a sentence or two describing the motion of the ball.

  *b) How far from the midpoint of the track does the ball get? How can you tell?

   c) How fast is the ball going when it is at the midpoint of the track? Does it ever go faster than this? How can you tell?

21. A forester who wants to know the height of a tree walks 100 feet from its base, sights to the top of the tree, and finds the resulting angle to be 57 degrees.

   a) What height does this give for the tree?

   b) If the measurement of the angle is certain only to 5 degrees, what can you say about the uncertainty of the height found in part (a)? (Note: You need to express angles in *radians* to use the formulas from calculus: π radians = 180 degrees.)

22. a) In the preceding problem, what percentage error in the height of the tree is produced by a 1-degree error in measuring the angle?

   b) What would the percentage error have been if the angle had been 75 degrees instead of 57 degrees? 40 degrees?

   c) If you can measure angles to within 1-degree accuracy and you want to measure the height of a tree that's roughly 150 feet tall by means of the technique in the preceding problem, how far away from the tree should you stand to get your best estimate of the tree's height? How accurate would your answer be?

➤ 3.6   THE CHAIN RULE

Combining Rates of Change

Let's return to the expanding house that we studied in section 3.4. When the temperature T increased, every side s of the house got longer; when s got longer, the volume V got larger. We already discussed how V responds to changes in s, but that's only part of the story. What we'd really like to know is this: Exactly how does the volume V respond to changes in temperature T? We can work this out in stages: first we see how V responds to changes in s, and then how s responds to changes in T.

Stage 1.   Recall that our "house" is a cube that measures 200 inches on a side. The microscope equation from section 3.4 describes how V responds to changes in s:

How volume responds
to changes in length

$$\Delta V \approx 120000 \, \frac{\text{cubic inches of volume}}{\text{inch of length}} \cdot \Delta s \text{ inches.}$$

Stage 2.   Physical experiments with wood show that a 200-inch length of wood increases about .0004 inch in length per degree Fahrenheit. This is a *rate*, and we can build a second microscope equation with it:

How length responds
to changes in
temperature

$$\Delta s \approx .0004 \, \frac{\text{inch of length}}{\text{degree F}} \cdot \Delta T \text{ degrees F,}$$

where ΔT measures the change in temperature, in degrees Fahrenheit.

We can combine the two stages because Δs appears in both. Replace Δs in the first equation by the right-hand side of the second equation. The result is

$$\Delta V \approx 120000 \, \frac{\text{cubic inches}}{\text{inch}} \times .0004 \, \frac{\text{inch}}{\text{degree F}} \cdot \Delta T \text{ degrees F.}$$

We can condense this to

How volume responds
to changes in
temperature

$$\Delta V \approx 48 \, \frac{\text{cubic inches}}{\text{degree F}} \cdot \Delta T \text{ degrees F.}$$

This is a *third* microscope equation, and it shows directly how the volume of the house responds to changes in temperature. It is the answer to our question.

As always, the multiplier in a microscope equation is a rate. The multiplier in the third microscope equation, 48 cubic inches/degree F, tells us the rate at which *volume changes with respect to temperature*. Thus, if the temperature increases by 10 degrees between night and day, the house will become about 480 cubic inches larger. Recall that Bodanis (see section 3.4) said that the house might become only a few cubic inches larger—say, $\Delta V = 3$ cubic inches. If we solve the microscope equation

$$3 \approx 48 \cdot \Delta T$$

for ΔT, we see that the temperature would have risen only one-sixteenth of a degree Fahrenheit!

The rate that appears as the multiplier in the third microscope equation is the product of the other two:

$$48 \, \frac{\text{cubic inches}}{\text{degree F}} = 120000 \, \frac{\text{cubic inches}}{\text{inch}} \times .0004 \, \frac{\text{inch}}{\text{degree F}}.$$

How the rates combine

Each of these rates is a derivative:

$$\underbrace{48 \, \frac{\text{cubic inches}}{\text{degree F}}}_{dV/dT} = \underbrace{120000 \, \frac{\text{cubic inches}}{\text{inch}}}_{dV/ds} \times \underbrace{.0004 \, \frac{\text{inch}}{\text{degree F}}}_{ds/dT}.$$

We wrote the derivatives in Leibniz's notation because it's particularly helpful in keeping straight what is going on. For instance, dV/dT indicates very clearly the rate at which volume is changing with respect to *temperature*, and dV/ds the rate at which it is changing with respect to *length*. These rates are quite different—they even have different units—but the notation V' does not distinguish between them. In Leibniz's notation, the relation between the three rates takes this striking form:

$$\frac{dV}{dT} = \frac{dV}{ds} \cdot \frac{ds}{dT}.$$

This relation is called the **chain rule** for the variables T, s, and V. (We'll see in a moment what this has to do with chains.)

The chain rule is a consequence of the way the three microscope equations are related to each other. We can see how it emerges directly from the microscope equations if we replace the numbers that appear as multipliers in those equations by the three derivatives. To begin, we write

$$\Delta V \approx \frac{dV}{ds} \cdot \Delta s \qquad \text{and} \qquad \Delta s \approx \frac{ds}{dT} \cdot \Delta T.$$

Then, combining these equations, we get

$$\Delta V \approx \frac{dV}{ds} \cdot \frac{ds}{dT} \cdot \Delta T.$$

In fact, this is the microscope equation for V in terms of T, which can be written more directly as

$$\Delta V \approx \frac{dV}{dT} \cdot \Delta T.$$

In these two expressions we have the same microscope equation, so the multipliers must be equal. Thus, we recover the chain rule:

$$\frac{dV}{ds} \cdot \frac{ds}{dT} = \frac{dV}{dT}.$$

> *Recall that Leibniz worked directly with differentials, like dV and ds, so a derivative was a genuine fraction. For him, the chain rule is true simply because we can cancel the two appearances of "ds" in the derivatives. For us, though, a derivative is not really a fraction, so we need an argument like the one in the text to establish the rule.*

Chains and the Chain Rule

Let's analyze the relationships between the three variables in the expanding house problem in more detail. There are three functions involved: volume is a function of length: $V = V(s)$; length is a function of temperature: $s = s(T)$; and finally, volume is a function of temperature, too: $V = V(s(T))$. To visualize these relationships better, we introduce the notion of an **input-output diagram**. The input-output diagram for the function $s = s(T)$ is just $T \to s$. It indicates that T is the input of a function whose output is s. Likewise $s \to V$ says that volume V is a function of length s. Since the output of $T \to s$ is the input of $s \to V$, we can make a *chain* of these two diagrams:

An input-output chain

$$T \longrightarrow s \longrightarrow V.$$

The result describes a function that has input T and output V. It is thus an input-output diagram for the third function $V = V(s(T))$.

We could also write the input-output diagram for the third function simply as $T \to V$; in other words,

$$T \longrightarrow V \qquad \text{equals} \qquad T \longrightarrow s \longrightarrow V.$$

A chain and its links

We say that $T \to s \to V$ is a **chain** that is made up of the two **links** $T \to s$ and $s \to V$. Since each input-output diagram represents a function, we can attach a derivative that describes the rate of change of the output with respect to the input:

$$T \xrightarrow{\frac{ds}{dT}} s \qquad s \xrightarrow{\frac{dV}{ds}} V \qquad T \xrightarrow{\frac{dV}{dT}} V$$

Here is a single picture that shows all the relationships:

$$\frac{dV}{dT}$$

$$T \xrightarrow{} s \xrightarrow{} V$$
$$\frac{ds}{dT} \qquad \frac{dV}{ds}$$

We can thus relate the derivative dV/dT of the whole chain to the derivatives dV/ds and ds/dT of the individual links by

$$\frac{dV}{dT} = \frac{dV}{ds} \cdot \frac{ds}{dT}.$$

The same argument holds for any chain of functions. If u is a function of x, and if y is some function of u, then a small change in x produces a small change in u and hence in y. The total multiplier for the chain is simply the product of the multipliers of the individual links:

$$\boxed{\textbf{The chain rule:} \quad \frac{dy}{dx} = \frac{dy}{du} \cdot \frac{du}{dx}}$$

Moreover, an obvious generalization extends this result to a chain containing more than two links.

A Simple Example

We can sometimes use the chain rule without giving it much thought. For instance, suppose a bookstore makes an average profit of $3 per book, and its sales are increasing at the rate of 40 books per month. At what rate is its monthly profit increasing, in dollars per month? Does it seem clear to you that the rate is $120 per month?

Let's analyze the question in more detail. There are three variables here:

$$\textbf{time} \quad t \quad \text{measured in months;}$$
$$\textbf{sales} \quad s \quad \text{measured in books;}$$
$$\textbf{profit} \quad p \quad \text{measured in dollars.}$$

The two known rates are

$$\frac{dp}{ds} = 3 \, \frac{\text{dollars}}{\text{book}} \qquad \text{and} \qquad \frac{ds}{dt} = 40 \, \frac{\text{books}}{\text{month}}.$$

The rate we seek is dp/dt, and we find it by the chain rule:

$$\frac{dp}{dt} = \frac{dp}{ds} \cdot \frac{ds}{dt}$$

$$= 3 \frac{\text{dollars}}{\text{book}} \times 40 \frac{\text{books}}{\text{month}}$$

$$= 120 \frac{\text{dollars}}{\text{month}}.$$

Chains, in General

The chain rule applies whenever the output of one function is the input of another. For example, suppose $u = f(x)$ and $y = g(u)$. Then $y = g(f(x))$, and we have:

$$x \xrightarrow{\quad u \quad} y \qquad \frac{dy}{dx} = \frac{dy}{du} \cdot \frac{du}{dx}$$

Let's take

$$u = x^2 \qquad \text{and} \qquad y = \sin(u);$$

then $y = \sin(x^2)$, and it is not at all obvious what the derivative dy/dx ought to be. None of the basic rules in section 3.5 covers this function. However, those rules *do* cover $u = x^2$ and $y = \sin(u)$:

$$\frac{du}{dx} = 2x \qquad \text{and} \qquad \frac{dy}{du} = \cos(u).$$

We can now get dy/dx by the chain rule:

$$\frac{dy}{dx} = \frac{dy}{du} \cdot \frac{du}{dx} = \cos(u) \cdot 2x.$$

Since we are interested in y as a function of x—rather than u—we should rewrite dy/dx so that it is expressed entirely in terms of x:

$$\text{If} \quad y = \sin(x^2), \quad \text{then} \quad \frac{dy}{dx} = 2x \cos(u) = 2x \cos(x^2).$$

Let's start over, using the function names f and g we introduced at the outset:

$$u = f(x) \quad \text{and} \quad y = g(u), \quad \text{so} \quad y = g(f(x)).$$

The third function, $y = g(f(x))$, needs a name of its own; let's call it h. Thus

$$y = h(x) = g(f(x)).$$

Composition of functions

We say that h is **composed** of g and f, and h is called the **composite**, or the **composition**, of g and f.

The problem is to find the derivative h' of the composite function, knowing g' and f'. Let's translate all the derivatives into Leibniz's notation.

$$h'(x) = \frac{dy}{dx} \qquad g'(u) = \frac{dy}{du} \qquad f'(x) = \frac{du}{dx}.$$

We can now invoke the chain rule:

$$h'(x) = \frac{dy}{dx} = \frac{dy}{du} \cdot \frac{du}{dx} = g'(u) \cdot f'(x).$$

Although h' is now expressed in terms of g' and f', we are not yet done. The variable u that appears in $g'(u)$ is out of place—because h is a function of x, not u. [We got to the same point in the example; the original form of the derivative of $\sin(x^2)$ was $2x \cos(u)$.] The remedy is to replace u by $f(x)$; we can do this because $u = f(x)$ is given.

The chain rule: $h'(x) = g'(f(x)) \cdot f'(x)$ when $h(x) = g(f(x))$

There is a certain danger in a formula as terse and compact as this that it loses all conceptual meaning and becomes simply a formal string of symbols to be manipulated blindly. You should always remember that the expression in the box is just a mathematical statement of the intuitively clear idea that when two functions are chained together, with the output of one serving as the input of the other, then the combined function has a multiplier which is simply the product of the multipliers of the two constituent functions.

Using the Chain Rule

The chain rule will allow us to differentiate nearly any formula. The key is to recognize when a given formula can be written as a chain—and then, how to write it.

EXAMPLE 1.   Here is a problem first mentioned on page 132: What is the derivative of $y = \sin(5x)$? If we set

$$y = \sin(u) \qquad \text{where} \qquad u = 5x,$$

then we find immediately

$$\frac{dy}{du} = \cos(u) \qquad \text{and} \qquad \frac{du}{dx} = 5.$$

Thus, by the chain rule we see

$$\frac{dy}{dx} = \frac{dy}{du} \cdot \frac{du}{dx} = \cos(u) \cdot 5 = 5\cos(5x).$$

EXAMPLE 2. $w = 2^{\cos z}$. Set

$$w = 2^u \qquad \text{and} \qquad u = \cos z.$$

Then, once again, the basic rules from section 3.5 are sufficient to differentiate the individual links:

$$\frac{dw}{du} = k_2\, 2^u \qquad \text{and} \qquad \frac{du}{dz} = -\sin z.$$

The chain rule does the rest:

$$\frac{dw}{dz} = \frac{dw}{du} \cdot \frac{du}{dz} = k_2\, 2^u \cdot (-\sin z) = -k_2 2^{\cos z} \sin z.$$

EXAMPLE 3. $p = \sqrt{7t^3 + \sin^2 t}$. This presents several challenges. First let's make a chain:

$$p = \sqrt{u} \qquad \text{where} \qquad u = 7t^3 + \sin^2 t.$$

The basic rules give us $dp/du = 1/2\sqrt{u}$, but it is more difficult to deal with u. Let's at least introduce separate labels for the two terms in u:

$$q = 7t^3 \qquad \text{and} \qquad r = \sin^2 t.$$

Then

$$\frac{du}{dt} = \frac{dq}{dt} + \frac{dr}{dt} \qquad \text{and} \qquad \frac{dq}{dt} = 21t^2.$$

The remaining term $r = \sin^2 t = (\sin t)^2$ can itself be differentiated by the chain rule. Set

$$r = v^2 \qquad \text{where} \qquad v = \sin t.$$

Then

$$\frac{dr}{dv} = 2v \qquad \text{and} \qquad \frac{dv}{dt} = \cos(t),$$

so

$$\frac{dr}{dt} = \frac{dr}{dv} \cdot \frac{dv}{dt} = 2v \cos t = 2 \sin t \cos t.$$

The final step is to assemble all the pieces:

$$\frac{dp}{dt} = \frac{dp}{du} \cdot \frac{du}{dt} = \frac{1}{2 \sqrt{u}} \cdot \left(21t^2 + 2 \sin t \cos t\right) = \frac{21t^2 + 2 \sin t \cos t}{2 \sqrt{7t^3 + \sin^2 t}}$$

By breaking down a complicated expression into simple pieces, and applying the appropriate differentiation rule to each piece, it is possible to differentiate a vast array of formulas. You may meet two sorts of difficulties: you may not see how to break down the expression into simpler parts; and you may overlook a step. Practice helps overcome the first, and vigilance the second.

Here is an example of the second problem: find the derivative of $y = -3 \cos(2x)$. The derivative is *not* $3 \sin(2x)$; it is $6 \sin(2x)$. Besides remembering to deal with the constant multiplier -3, and with the fact that there is a minus sign in the derivative of $\cos u$, you must not overlook the link $u = 2x$ in the chain that connects y to x.

Exercises

1. Use the chain rule to find dy/dx, when y is given as a function of x in the following way.
   a) $y = 5u - 3$, where $u = 4 - 7x$.
   *b) $y = \sin u$, where $u = 4 - 7x$.
   c) $y = \tan u$, where $u = x^3$.
   d) $y = 10^u$, where $u = x^2$.
   e) $y = u^4$, where $u = x^3 + 5$.

2. Find the derivatives of the following functions.
   a) $F(x) = (9x + 6x^3)^5$.

   *b) $G(w) = \sqrt{4w^2 + 1}$.

   c) $S(w) = \sqrt{(4w^2 + 1)^3}$.

d) $R(x) = \dfrac{1}{1-x}.$

$$\left[\text{Hint: Think of } \frac{1}{1-x} \text{ as } (1-x)^{-1}. \right]$$

e) $D(z) = 3\tan\left(\dfrac{1}{z}\right).$

f) $\text{dog}(w) = \sin^2(w^3 + 1).$

g) $\text{pig}(t) = \cos(2^t).$

h) $\text{wombat}(x) = 5^{1/x}.$

3. If $h(x) = (f(x))^6$, where f is some function satisfying $f(93) = 2$ and $f'(93) = -4$, what is $h'(93)$?

4. If $H(x) = F(x^2 - 4x + 2)$, where F is some function satisfying $F'(2) = 3$, what is $H'(4)$?

* 5. If $f(x) = (1 + x^2)^5$, what are the numerical values of $f'(0)$ and $f'(1)$?

6. If $h(t) = \cos(\sin t)$, what are the numerical values of $h'(0)$ and $h'(\pi)$?

7. If $f'(x) = g(x)$, which of the following defines a function which also must have g as its derivative?

$$f(x + 17) \qquad f(17x) \qquad 17f(x) \qquad 17 + f(x) \qquad f(17)$$

8. Let $f(t) = t^2 + 2t$ and $g(t) = 5t^3 - 3$. Determine all of the following:
$f'(t)$, $g'(t)$, $g(f(t))$, $f(g(t))$, $g'(f(t))$, $f'(g(t))$, $(f(g(t)))'$, $(g(f(t)))'$.

9. a) What is the derivative of $f(x) = 2^{-x^2}$?

b) Sketch the graphs of f and its derivative on the interval $-2 \le x \le 2$.

* c) For what values(s) of x is $f'(x) = 0$? What is true about the graph of f at the corresponding points?

d) Where does the graph of f have positive slope, and where does it have negative slope?

10. a) With a graphing utility, find the point x where the function $y = 1/(3x^2 - 5x + 7)$ takes its maximum value. Obtain the numerical value of x accurately to two decimal places.

* b) Find the derivative of $y = 1/(3x^2 - 5x + 7)$, and determine where it takes the value 0.

c) Using part (b), find the *exact* value of x where $y = 1/(3x^2 - 5x + 7)$ takes its maximum value.

d) At what point is the graph of $y = 1/(3x^2 - 5x + 7)$ rising most steeply? Describe how you determined the location of this point.

11. a) Write the microscope equation for the function $y = \sin \sqrt{x}$ at $x = 1$.

b) Using the microscope equation, estimate the following values:
$\sin \sqrt{1.05}$, $\sin \sqrt{.9}$.

12. **a)** Write the microscope equation for $w = \sqrt{1 + x}$ at $x = 0$.
    b) Use the microscope equation to estimate the values of $\sqrt{1.1056}$ and $\sqrt{.9788}$. Compare your estimates with the values provided by a calculator.

13. When the sides of a cube are 5 inches, its surface area is changing at the rate of 60 square inches per inch increase in the side. If, at that moment, the sides are increasing at a rate of 3 inches per hour, at what rate is the area increasing: Is it 60, 3, 63, 20, 180, 5, or 15 square inches per hour?

14. Find a function $f(x)$ for which $f'(x) = 3x^2(5 + x^3)^{10}$. Find a function $p(x)$ for which $p'(x) = x^2(5 + x^3)^{10}$. A useful way to proceed is to guess. For instance, you might guess $f(x) = (5 + x^3)^{11}$. While this guess isn't correct, it suggests what modification you might make to get the answer.

15. Find a function $g(t)$ for which $g'(t) = t / \sqrt{1 + t^2}$.

▷ 3.7 PARTIAL DERIVATIVES

Let's return to the sunrise function once again. The time of sunrise depends not only on the date, but on our latitude. In fact, if we are far enough north or south, there are days when the sun never rises at all. We give in the table below the time of sunrise at eight different latitudes on March 15, 1990.

Latitude	36°N	38°N	40°N	42°N	44°N	46°N	48°N	50°N
Mar 15	6:10	6:11	6:12	6:13	6:13	6:13	6:14	6:14

Thus on March 15, the time of sunrise increases as latitude increases.

Clearly what this shows is that the time of sunrise is actually a function of two independent inputs: the date and the latitude. If T denotes the time of sunrise, then we will write $T = T(d, \lambda)$ to make explicit the dependence of T on both the date d and the latitude λ. To capture this double dependence, we need information like the following table:

The time of sunrise depends on latitude as well as on the date

Latitude	36°N	38°N	40°N	**42°N**	44°N	46°N	48°N	50°N
Mar 3	6:24	6:27	6:31	**6:33**	6:34	6:36	6:38	6:40
7	6:20	6:22	6:25	**6:26**	6:27	6:29	6:30	6:32
11	6:15	6:17	6:19	**6:19**	6:20	6:21	6:22	6:23
15	**6:10**	**6:11**	**6:12**	**6:13**	**6:13**	**6:13**	**6:14**	**6:14**
19	6:06	6:06	6:06	**6:06**	6:06	6:06	6:06	6:06
23	6:01	6:00	5:59	**5:59**	5:58	5:58	5:58	5:57
27	5:56	5:54	5:53	**5:52**	5:51	5:50	5:49	5:48

Thus we can say $T(74, 42°N) = 6:13$ (March 15 is the seventy-fourth day of the year). Note, though, that at this date and place the time of sunrise is changing in two very different senses:

First: At 42°N, during the eight days between March 11 and March 19, the time of sunrise gets 13 minutes earlier. We thus would say that on March 15 at 42°N, sunrise is changing at -1.63 minutes/day.

Second: On the other hand, on March 15 we see that the time of sunrise varies by 1 minute as we go from 40°N to 44°N. We would thus say that at 42°N the rate of change of sunrise as the latitude varies is approximately 1 minute/4° = +.25 min/deg of latitude.

Two quite different rates are at work here, one with respect to time, the other with respect to latitude.

We need a notation which allows us to talk about the different rates at which a function can change, when that function depends on more than one variable. A rate of change is, of course, a derivative. But since a change in one input produces only part of the change that a function of several variables can experience, we call the rate of change with respect to any one of the inputs a **partial derivative**. If the value of z depends on the variables x and y according to the rule $z = F(x, y)$, then we denote the rate at which z is changing with respect to x when $x = a$ and $y = b$ by

A function of several variables has several rates of change

$$F_x(a, b) \qquad \text{or by} \qquad \frac{\partial z}{\partial x}(a, b).$$

Partial derivatives

We call this rate the **partial derivative of F with respect to x**. Similarly, we define the partial derivative of F with respect to y to be the rate at which z is changing when y is varied. It is denoted

$$F_y(a, b) \qquad \text{or} \qquad \frac{\partial z}{\partial y}(a, b).$$

There is nothing conceptually new involved here; to calculate either of these partial derivatives you simply hold one variable constant and go through the same limiting process as before for the input variable of interest. Note that, to call attention to the fact that there is more than one input variable present, we write

$$\frac{\partial z}{\partial x} \qquad \text{rather than} \qquad \frac{dz}{dx},$$

as we did when x was the only input variable.

To calculate the partial derivative of F with respect to x at the point (a, b), we can use

$$F_x(a, b) = \frac{\partial z}{\partial x}(a, b) = \lim_{\Delta x \to 0} \frac{F(a + \Delta x, b) - F(a, b)}{\Delta x}.$$

Similarly,

$$F_y(a, b) = \frac{\partial z}{\partial y}(a, b) = \lim_{\Delta y \to 0} \frac{F(a, b + \Delta y) - F(a, b)}{\Delta y}.$$

By using this notation for partial derivatives, we can cast some of our earlier statements about the sunrise function $T = T(d, \lambda)$ in the following form:

$$T_d(74, 42°N) \approx -1.63 \text{ min per day};$$

$$T_\lambda(74, 42°N) \approx +.25 \text{ min per degree.}$$

Partial Derivatives as Multipliers

For any given date d and latitude λ we can write down two microscope equations for the sunrise function $T(d, \lambda)$. One describes how the time of sunrise responds to changes in the date, the other to changes in the latitude. Let's consider variations in the time of sunrise in the vicinity of March 15 and 42°N.

The partial derivative $T_d(74, 42°N)$ of T with respect to d is the multiplier in the first of these microscope equations:

$$\Delta T \approx T_d(74, 42°N) \cdot \Delta d.$$

The microscope equation for dates

For example, from March 15 to March 17 ($\Delta d = 2$ days), we would expect the time of sunrise to change by

$$\Delta T \approx -1.63 \frac{\text{min}}{\text{day}} \times 2 \text{ days} = -3.3 \text{ minutes.}$$

Thus, we would expect the time of sunrise on March 17 at 42°N to be approximately

$$T(76, 42°N) \approx 6{:}09.7.$$

The partial derivative $T_\lambda(74, 42°N)$ of T with respect to λ is the multiplier in the second microscope equation:

$$\Delta T \approx T_\lambda(74, 42°N) \cdot \Delta \lambda.$$

The microscope equation for latitudes

If, say, we moved 1° north, to 43°N, we would expect the time of sunrise on March 7 to change by

$$\Delta T \approx .25 \frac{\text{min}}{\text{deg}} \times 1 \text{ degree} = .25 \text{ minute.}$$

The time of sunrise on March 15 at 43°N would therefore be

$$T(74, 43°N) \approx 6:13.25.$$

We have seen what happens to the time of sunrise from March 15 to March 17 if we stay at 42°N, and we have seen what happens to the time on March 15 if we move from 42°N to 43°N. Can we put these two pieces of information together? That is, can we determine the time of sunrise on March 17 at 43°N? This involves changing *both* the date and the latitude.

The total change   To determine the total change we shall just combine the two changes ΔT we have already calculated. Making the date two days later moves the time of sunrise 3.3 minutes *earlier*, and traveling one degree north makes the time of sunrise .25 minute *later*, so the net effect would be to change the time of sunrise by

$$\Delta T \approx -3.3 \text{ min} + .25 \text{ min} \approx -3 \text{ minutes}.$$

This puts the time of sunrise at $T(76, 43°N) \approx 6:10$.

We can formulate this idea more generally in the following way: partial derivatives are not only multipliers for gauging the separate effects that changes in each input have on the output, but they also serve as multipliers for gauging the *cumulative* effect that changes in all inputs have on the output. In general, if $z = F(x, y)$ is a function of two variables, then near the point (a, b), the combined change in z caused by small changes in x and y can be stated by the *full* microscope equation:

> **The full microscope equation: $\Delta z \approx F_x(a, b) \cdot \Delta x + F_y(a, b) \cdot \Delta y$**

As was the case for the functions of one variable, there is an important class of functions for which we may write "=" instead of "≈" in this relation, the linear functions. The most general form of a **linear function of two variables** is $z = F(x, y) = mx + ny + c$, for constants m, n, and c.

In the exercises you will have an opportunity to verify that for a linear function $z = F(x, y) = mx + ny + c$ and for all (a, b), we know that $F_x(a, b) = m$ and $F_y(a, b) = n$, and the full microscope equation $\Delta z = F_x(a, b) \cdot \Delta x + F_y(a, b) \cdot \Delta y$ is true for all values of Δx and Δy.

Formulas for Partial Derivatives

To find a partial derivative, treat the other variable as a constant   No new rules are needed to find the formulas for the partial derivatives of a function of two variables that is given by a formula. To find the partial derivative with respect to one of the variables, just treat the other variable as a constant and follow the rules for functions of a single variable. (The basic rules are described in section 3.5, and the chain rule in section 3.6.) We give two examples to illustrate the method.

EXAMPLE 1. For $z = F(x, y) = 3x^2y + 5y^2 \sqrt{x}$, we have

$$F_x(x, y) = 3y(2x) + 5y^2 \frac{1}{2\sqrt{x}} = 6xy + \frac{5y^2}{2\sqrt{x}}$$

$$F_y(x, y) = 3x^2 + 10y \sqrt{x}.$$

EXAMPLE 2. For $w = G(u, v) = 3u^5 \sin v - \cos v + u$, we have

$$\frac{\partial w}{\partial u} = 15u^4 \sin v + 1$$

$$\frac{\partial w}{\partial v} = 3u^5 \cos v + \sin v.$$

The formulas for derivatives and the combined multiplier effect of partial derivatives allow us to determine the effect of changes in length and in width on the area of a rectangle. The area A of the rectangle is a simple function of its dimensions l and w, $A = F(l, w) = lw$. The partial derivatives of the area are then

$$F_l(l, w) = w \quad \text{and} \quad F_w(l, w) = l.$$

Changes Δl and Δw in the dimensions produce a change

$$\Delta A \approx w \cdot \Delta l + l \cdot \Delta w$$

The full microscope equation for rectangular area

in the area. The picture below shows that the *exact* value of ΔA includes an additional term—namely, $\Delta l \cdot \Delta w$—that is not in the approximation $w \cdot \Delta l + l \cdot \Delta w$. The difference, $\Delta l \cdot \Delta w$, is *very* small when the changes Δl and Δw are small. In chapter 5 we will have a further look at the nature of this approximation.

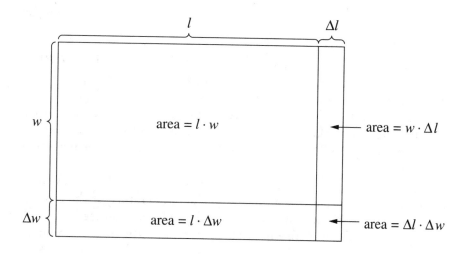

Exercises

1. Use differentiation formulas to find the partial derivatives of the following functions.

   a) $x^2 y$.

   b) $\sqrt{x + y}$.

   c) $x^2 y + 5x^3 - \sqrt{x + y}$.

   d) 10^{xy}.

   e) $\dfrac{y}{x}$.

   f) $\sin \dfrac{y}{x}$.

   g) $17 \dfrac{x^2}{y^3} - x^2 \sin y + \pi$.

   h) $\dfrac{uv}{5} + \dfrac{5}{uv}$.

   i) $2 \sqrt{x} \sqrt[3]{y} - 7 \cos x$.

   *j) $x \tan y$.

2. On March 7 in the Northern Hemisphere, the farther south you are the earlier the sun rises. The sun rises at 6:25 on this date at 40°N. If we had been far enough south, we could have experienced a 6:25 sunrise on March 5. Near what latitude did this happen?

3. The volume V of a given quantity of gas is a function of the temperature T (in degrees Kelvin) and the pressure P. In a so-called ideal gas the functional relationship between volume and pressure is given by a particularly simple rule called the **ideal gas law**:

$$V(T, P) = R \frac{T}{P},$$

   where R is a constant.

   *a) Find formulas for the partial derivatives $V_T(T, P)$ and $V_P(T, P)$.

   b) For a particular quantity of an ideal gas called a *mole*, the value of R can be expressed as 8.3×10^3 newton-meters per degree Kelvin. (The *newton* is the unit of force in the *MKS* unit system.) Check that the units in the ideal gas law are consistent if V is measured in cubic meters, T in degrees Kelvin, and P in newtons per square meter.

   c) Suppose a mole of gas at 350 degrees Kelvin is under a pressure of 20 newtons per square meter. If the temperature of the gas increases by 10 degrees Kelvin and the volume increases by 1 cubic meter, will the pressure increase or decrease? By about how much?

4. Write the formula for a linear function $F(x, y)$ with the following properties:

$$F_x(x, y) = .15 \quad \text{for all } x \text{ and } y$$
$$F_y(x, y) = 2.31 \quad \text{for all } x \text{ and } y$$
$$F(4,1) = 8$$

5. The purpose of this exercise is to verify the claims made in the text for the linear function $z = F(x, y) = mx + ny + c$, where m, n, and c are constants.
   a) Use the differentiation rules to find the partial derivatives of F.
   b) Use the *definition* of the partial derivative $F_x(a, b)$ to show that $F_x(a, b) = m$ for any input (a, b). That is, show that the value of

$$\frac{F(a + \Delta x, b) - F(a, b)}{\Delta x}$$

   exactly equals m, no matter what a and b are.
   c) Compute the exact value of the change

$$\Delta z = F(a + \Delta x, b + \Delta y) - F(a, b)$$

   corresponding to changing a by Δx and b by Δy.

6. Suppose $w = G(u, v) = \dfrac{uv}{3 + v}$.
   a) Approximate the value of the partial derivative $G_u(1, 2)$ by computing $\Delta w / \Delta u$ for $\Delta u = \pm.1, \pm.01, \ldots, \pm.00001$.
   b) Approximate the value of $G_v(1, 2)$ by computing $\Delta w / \Delta v$ for $\Delta v = \pm.1, \pm.01, \ldots, \pm.00001$.
   c) Write the full microscope equation for $G(u, v)$ at $(u, v) = (1, 2)$.
   d) Use the full microscope equation to approximate $G(.8, 2.1)$. How close is your approximation to the true value of $G(.8, 2.1)$?

7. a) A rectangular piece of land has been measured to be 51 feet by 2034 feet. What is its area?
   b) The narrow dimension has been measured with an accuracy of 4 inches, but the long dimension is accurate only to 10 feet. What is the error, or uncertainty, in the calculated area? What is the percentage error?

8. Suppose $z = f(x, y)$ and

$$f(3, 12) = 240 \quad f_x(3, 12) = 7 \quad f_y(3, 12) = 4.$$

   a) Estimate $f(4, 12)$, $f(3, 13)$, $f(4, 13)$, $f(4, 10)$.
   b) When $x = 3$ and $y = 12$, how much does a 1% increase in x cause z to change? How much does a 1% increase in y cause z

to change? Which has the larger effect: a 1% increase in x or a 1% increase in y?

9. Let $P(K, L)$ represent the monthly profit, in thousands of dollars, of a company that produces a product using capital whose monthly cost is K thousand dollars and labor whose monthly cost is L thousand dollars. The current levels of expense for capital and labor are $K = 23.5$ and $L = 39.0$. Suppose now that company managers have determined

$$\frac{\partial P}{\partial K} (23.5, 39.0) = -.12 \qquad \frac{\partial P}{\partial L} (23.5, 39.0) = -.20.$$

a) Estimate what happens to the monthly profit if monthly capital expenses increase to $24,000.

b) Each typical person added to the work force increases the monthly labor expense by $1,500. Estimate what happens to the monthly profit if one more person is added to the work force. What, therefore, is the rate of change of profit, in thousands of dollars per person? Is the rate positive or negative?

c) Suppose managers respond to increased demand for the product by adding three workers to the labor force. What does that do to monthly profit? If the managers want to keep the profit level unchanged, they could try to alter capital expenses. What change in K would leave profit unchanged after the three workers are added? (This is called a **trade-off**.)

10. A forester who wants to know the height of a tree walks 100 feet from its base, sights to the top of the tree, and finds the resulting angle to be 57 degrees.

a) What height does this give for the tree?

b) If the 100-foot measurement is certain only to 1 foot and the angle measurement is certain only to 5 degrees, what can you say about the uncertainty of the height measured in part (a)? (Note: You need to express angles in *radians* to use calculus: π radians = 180 degrees.)

c) Which would be more effective: improving the accuracy of the angle measurement, or improving the accuracy of the distance measurement? Explain.

➤ 3.8 CHAPTER SUMMARY

The Main Ideas

- The functions we study with the calculus have graphs that are **locally linear**; that is, they look approximately straight when magnified under a microscope.

- The **slope of the graph** at any point is the **limit** of the slopes seen under a microscope at that point.

- The **rate of change** of a function at a point is the slope of its graph at that point, and thus is also a **limit**. Its dimensional units are (units of output)/(unit of input).

- The **derivative** of $f(x)$ at $x = a$ is the name given to both the rate of change of f at a and the slope of the graph of f at $(a, f(a))$.

- The derivative of $y = f(x)$ at $x = a$ is written $f'(a)$. The **Leibniz notation** for the derivative is dy/dx.

- To calculate the derivative $f'(a)$, make **successive approximations** using $\Delta y / \Delta x$:

$$f'(a) = \lim_{\Delta x \to 0} \frac{\Delta y}{\Delta x} = \lim_{h \to 0} \frac{f(a + h) - f(a - h)}{2h} = \lim_{h \to 0} \frac{f(a + h) - f(a)}{h}.$$

- The **microscope equation** $\Delta y \approx f'(a) \cdot \Delta x$ describes the relation between x and $y = f(x)$ as seen under a microscope at $(a, f(a))$; Δx and Δy are the **microscope coordinates**.

- The microscope equation describes how the output changes in response to small changes in the input. The response is proportional, and the derivative $f'(a)$ plays the role of **multiplier**, or scaling factor.

- The microscope equation expresses the **local linearity** of a function in analytic form. The microscope equation is exact for **linear** functions.

- The microscope equation describes **error propagation** when one quantity, known only approximately, is used to calculate another.

- The **derivative function** is the rule that assigns to any x the number $f'(x)$.

- The derivative of a function gives information about the shape of the graph of the function, and conversely.

- If a function is given by a formula, its derivative also has a formula. There are formulas for the derivatives of the **basic functions**, and there are **rules** for the derivatives of combinations of basic functions.

- The **chain rule** gives the formula for the derivative of a **chain**, or **composite** of functions.

- Functions that have more than one input variable have **partial derivatives**. A partial derivative is the rate at which the output changes with respect to one variable when we hold all the others constant.

- If a multi-input function is given by a formula, its partial derivatives also have formulas that can be found using the same rules that apply to single-input functions.

- A function $z = F(x, y)$ of two variables also has a **microscope equation**:

$$\Delta z \approx F_x(a, b) \cdot \Delta x + F_y(a, b) \cdot \Delta y.$$

The partial derivatives are the **multipliers** in the microscope equation.

Expectations

- You should be able to approximate $f'(a)$ by zooming in on the graph of f near a and calculating the slope of the graph on an interval on which the graph appears straight.
- You should be able to approximate $f'(a)$ using a table of values of f near a.
- From the microscope equation $\Delta y \approx f'(a) \cdot \Delta x$, you should be able to estimate any one of Δx, Δy, and $f'(a)$ if given the other two.
- If $y = f(x)$ and there is an error in the measured value of x, you should be able to determine the absolute and relative error in y.
- You should be able to sketch the graph of f' if you are given the graph of f.
- You should be able to use the basic differentiation rules to find the derivative of a function given by a formula that involves sums of constant multiples of x^p, $\sin x$, $\cos x$, $\tan x$, or b^x.
- You should be able to break down a complicated formula into a chain of simple pieces.
- You should be able to use the chain rule to find the derivative of a chain of functions. This could involve several independent steps.
- For $z = F(x, y)$, you should be able to approximate any one of Δz, Δx, Δy, $F_x(a, b)$, and $F_y(a, b)$, if given the other four.
- You should be able to find formulas for partial derivatives using the basic rules and the chain rule for finding formulas for derivatives.

CHAPTER 4

DIFFERENTIAL EQUATIONS

The rate equations with which we began our study of calculus are called **differential equations** when we identify the rates of change that appear within them as derivatives of functions. Differential equations are essential tools in many areas of mathematics and the sciences. In this chapter we explore three of their important uses:

- **Modeling** problems using differential equations;
- **Solving** differential equations, both through numerical techniques like Euler's method and, where possible, through finding formulas which satisfy the equations;
- **Defining** new functions by differential equations.

We also introduce two important functions—the **exponential function** and the **logarithmic function**—which play central roles in the theory of solving differential equations. Finally, we introduce the operation of **antidifferentiation** as an important tool for solving some special kinds of differential equations.

➤ 4.1 MODELING WITH DIFFERENTIAL EQUATIONS

To analyze the way an infectious disease spreads through a population, we asked how three quantities S, I, and R would vary over time. This was difficult

157

to answer; we found no simple, direct relation between S (or I or R) and t. What we *did* find, though, was a relation between the variables S, I, and R and their rates S', I', and R'. We expressed the relation as a set of rate equations. Then, given the rate equations and initial values for S, I, and R, we used Euler's method to estimate the values at any time in the future. By constructing a sequence of successive approximations, we were able to make these estimates as accurate as we wished.

There are two ideas here. The first is that we can write down equations for the rates of change that reflect important features of the process we seek to model. The second is that these equations *determine* the variables as functions of time, so we can make predictions about the real process we are modeling. Can we apply these ideas to other processes?

Differential equations and initial value problems

To answer this question, it will be helpful to introduce some new terms. What we have been calling rate equations are more commonly called **differential equations**. (The name is something of a historical accident. Since the equations involve functions and their derivatives, we might better call them *derivative* equations.) Euler's method treats the differential equations for a set of variables as a prescription for finding future values of those variables. However, in order to get started, we must always specify the initial values of the variables—their values at some given time. We call this specification an **initial condition**. The differential equations together with an initial condition is called an **initial value problem**. Each initial value problem determines a set of functions which we find by using Euler's method.

> If we use Leibniz's notation for derivatives, a differential equation like $S' = -aSI$ takes the form $dS/dt = -aSI$. If we then treat dS/dt as a quotient of the individual differentials dS and dt (see page 110), we can even write the equation as $dS = -aSI\,dt$. Since this expresses the differential dS in terms of the differential dt, it was natural to call it a differential equation. Our approach is similar to Leibniz's, except that we don't need to introduce infinitesimally small quantities, which differentials were for Leibniz. Instead, we write $\Delta S \approx -aSI\,\Delta t$ and rely on the fact that the accumulated error of the resulting approximations can be made as small as we like.

To illustrate how differential equations can be used to describe a wide range of processes in the physical, biological, and social sciences, we'll devote this section to a number of ways to model and analyze the long-term behavior of animal populations. To be specific, we will talk about rabbits and foxes, but the ideas can be adapted to the population dynamics of virtually all living things (and many nonliving systems such as chemical reactions).

In each model, we will begin by identifying the relevant variables. Then, we will try to establish how those variables change over time. Of course, no model can hope to capture every feature of the process we seek to describe, so we begin simply. We choose just one or two elements that seem particularly important. After examining the predictions of our simple model and checking how well they correspond to reality, we make modifications. We might include more features of the population dynamics, or we might describe the same fea-

Models can provide successive approximations to reality

tures in different ways. Gradually, through a succession of refinements of our original simple model, we hope for descriptions that come closer and closer to the real situation we are studying.

Single-Species Models: Rabbits

The Problem

If we turn 2000 rabbits loose on a large, unpopulated island that has plenty of food for the rabbits, how might the number of rabbits vary over time? If we let $R = R(t)$ be the number of rabbits at time t (measured in months, let us say), we would like to be able to make some predictions about the function $R(t)$. It would be ideal to have a formula for $R(t)$—but this is not usually possible. Nevertheless, there may still be a great deal we can say about the behavior of R. To begin our explorations we will construct a model of the rabbit population that is obviously too simple. After we analyze the predictions it makes, we'll look at various ways to modify the model so that it approximates reality more closely.

The First Model

Let's assume that, at any time t, the rate at which the rabbit population changes is proportional to the number of rabbits present at that time. For instance, if there were twice as many rabbits, then the *rate* at which new rabbits appear will also double. In mathematical terms, our assumption takes the form of the differential equation

$$\frac{dR}{dt} = kR \frac{\text{rabbits}}{\text{month}}.$$

Constant per capita growth

The multiplier k is called the **per capita growth rate** (or the **reproductive rate**), and its units are rabbits per month per rabbit. Per capita growth is discussed in problem 22 in chapter 1, section 1.2.

For the sake of discussion, let's suppose that $k = .1$ rabbit per month per rabbit. This assumption means that, on the average, one rabbit will produce .1 new rabbit every month. In the S-I-R model of chapter 1, the reciprocals of the coefficients in the differential equations had natural interpretations. The same is true here for the per capita growth rate. Specifically, we can say that $1/k = 10$ months is the average length of time required for a rabbit to produce one new rabbit.

Since there are 2000 rabbits at the start, we can now state a clearly defined initial value problem for the function $R(t)$:

$$\frac{dR}{dt} = .1R \qquad R(0) = 2000.$$

Use Euler's method to find $R(t)$

By modifying the program SIRPLOT, we can readily produce the graph of the function $R(t)$ determined by this problem. Before we do that, though, let's first consider some of the implications that we can draw out of the problem without the graph.

Since $R'(t) = .1R(t)$ rabbits per month and $R(0) = 2000$ rabbits, we see that the initial rate of growth is $R'(0) = 200$ rabbits per month. If this rate were to persist for 20 years (= 240 months), R would have increased by

$$\Delta R = 240 \text{ months} \times 200 \frac{\text{rabbits}}{\text{month}} = 48000 \text{ rabbits,}$$

yielding altogether

$$R(240) = R(0) + \Delta R = 2000 + 48000 = 50000 \text{ rabbits}$$

at the end of the 20 years. However, since the population R is always increasing, the differential equation tells us that the growth rate R' will *also* always be increasing. Consequently, 50,000 is actually an underestimate of the number of rabbits predicted by this model.

Let's restate our conclusions in a graphical form. If R' were always 200 rabbits per month, the graph of R plotted against t would just be a straight line whose slope is 200 rabbits/month. But R' is always increasing, so the slope of the graph should increase from left to right. This will make the graph curve upward. In fact, SIRPLOT will produce the following graph of $R(t)$:

The graph of R curves up

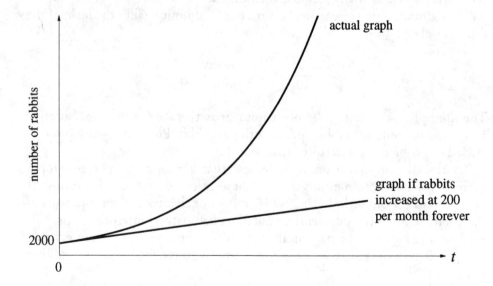

Later, we will see that the function $R(t)$ determined by this initial value problem is actually an exponential function of t, and we will even be able to write down a formula for $R(t)$, namely,

$$R(t) = 2000\,(1.10517)^t.$$

This model is too simple to be able to describe what happens to a rabbit population very well. One of the obvious difficulties is that it predicts that the rabbit population just keeps growing—forever. For example, if we used the formula for $R(t)$ given above, our model would predict that after 20 years ($t = 240$) there will be more than 50 *trillion* rabbits! While rabbit populations can, under good conditions, grow at a nearly constant per capita rate for a surprisingly long time (this happened in Australia during the nineteenth century), our model is ultimately unrealistic.

> It is a good idea to think qualitatively about the functions determined by a differential equation and make some rough estimates before doing extensive calculations. Your sketches may help you see ways in which the model doesn't correspond to reality. Or, you may be able to catch errors in your computations if they differ noticeably from what your estimates led you to expect.

The Second Model

One way out of the problem of unlimited growth is to modify the equation in the first model to reflect the fact that any given ecological system can support only some finite number of creatures over the long term. This number is called the **carrying capacity** of the system. We expect that when a population has reached the carrying capacity of the system, the population should neither grow nor shrink. At carrying capacity, a population should hold steady—its rate of change should be zero. To be specific, let's suppose that in our example the carrying capacity of the island is 25,000 rabbits.

The carrying capacity of the environment

What we would like to do, then, is to find an expression for R' which is similar to the equation in the first model when the number of rabbits R is near 2000, but which approaches 0 as R approaches 25,000. One model which captures these features is the **logistic equation**, first proposed by the Belgian mathematician Otto Verhulst in 1845:

$$R' = kR \left(1 - \frac{R}{b}\right) \frac{\text{rabbits}}{\text{month}}.$$

Logistic growth

In this equation, the coefficient k is called the **natural growth rate**. It plays the same role as the per capita growth rate in the first equation , and it has the same units—rabbits per month per rabbit. The number b is the **carrying capacity**; it is measured in rabbits. (We first saw the logistic equation on pages 69–73.) Notice also that we have written the derivative of R in the simpler form R', a practice we will continue for the rest of the section.

If the carrying capacity of the island is 25,000 rabbits, and if we keep the natural growth rate at .1 rabbit per month per rabbit, then the logistic equation for the rabbit population is

$$R' = .1R \left(1 - \frac{R}{25000}\right) \frac{\text{rabbits}}{\text{month}}.$$

Check to see that this equation really does have the behavior claimed for it—namely, that a population of 25,000 rabbits neither grows nor declines. Notice also that R' is positive as long as R is less than 25,000, so the population increases. However, as R approaches 25,000, R' will get closer and closer to 0, so the graph will become nearly horizontal. (What would happen if the island ever had more than 25,000 rabbits?)

The graph of $R(t)$ levels off near $R = 25000$

These observations about the qualitative behavior of $R(t)$ are consistent with the following graph, produced by a modified version of the program SIRPLOT. For comparison, we have also graphed the exponential function produced by the first model. Notice that the two graphs "share ink" when R is near 2000, but diverge later on.

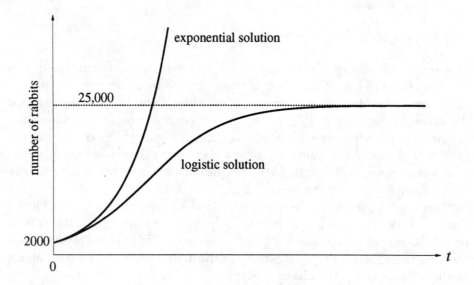

By modifying the program SIRVALUE, we can even obtain numerical answers to specific questions about the two models. For example, after 30 months under constant per capita growth, the rabbit population will be more than 40,000—well beyond the carrying capacity of the island. Under logistic growth, though, the population will be only about 16,000.

In the following figure we display several functions that are determined by the logistic equation

$$R' = .1R \left(1 - \frac{R}{25000} \right)$$

when different initial conditions are given. Each graph therefore predicts the future for a different initial population $R(0)$. One of the graphs is just the t-axis itself. What does this graph predict about the rabbit population? What other graph is just a straight line, and what initial population will lead to this line?

> *While the logistic equation above was developed to model a physical problem in which only values of R with R ≥ 0 have any meaning, the mathematical problem of finding solutions for the resulting differential equation makes sense for all values of R. We have drawn three graphs resulting from initial values R(0) < 0. While this growth behavior of "antirabbits" is of little practical interest in this case, there may well be other physical problems of an entirely different sort which lead to the same mathematical model, and in which the solutions below the t-axis are crucial.*

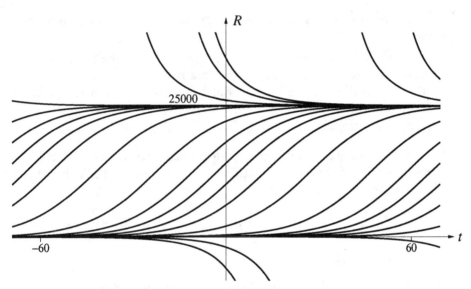

Solutions to the logistic equation $R' = .1R(1 - R/25000)$

Two-Species Models: Rabbits and Foxes

No species lives alone in an environment, and the same is true of the rabbits on our island. The rabbits will probably have to deal with predators of various sorts. Some are microscopic—disease organisms, for example—while others loom as obvious threats. We will enrich our population model by adding a second species—foxes—that will prey on the rabbits. We will continue to suppose that the rabbits live on abundant native vegetation, and we will now assume that the rabbits are the sole food supply of the foxes. Can we say what will happen? Will the number of foxes and rabbits level off and reach a "steady state" where their numbers don't vary? Or will one species perhaps become extinct?

Introduce predators

Let F denote the number of foxes, and R the number of rabbits. As before, measure the time t in months. Then F and R are functions of t: $F(t)$ and $R(t)$. We seek differential equations that describe how the growth rates F' and R' are related to the population sizes F and R. We make the following assumptions.

- In the absence of foxes, the rabbit population grows logistically.
- The population of rabbits declines at a rate proportional to the product $R \cdot F$. This is reasonable if we assume rabbits never die of old age— they just get a little too slow. Their death rate, which depends on the number of fatal encounters between rabbits and foxes, will then be approximately proportional to both R and F—and thus to their product. (This is the same kind of interaction effect we used in our epidemic model to predict the rate at which susceptibles become infected.)
- In the absence of rabbits, the foxes die off at a rate proportional to the number of foxes present.
- The fox population increases at a rate proportional to the number of encounters between rabbits and foxes. To a first approximation, this says that the birth rate in the fox population depends on maternal fox nutrition, and this depends on the number of rabbit-fox encounters, which is proportional to $R \cdot F$.

Our assumptions are about birth and death rates, so we can convert them quite naturally into differential equations. Pause here and check that the assumptions translate into these differential equations:

Lotka–Volterra equations
with bounded growth

$$R' = aR\left(1 - \frac{R}{b}\right) - cRF \;=\; aR - \frac{a}{b}R^2 - cRF$$
$$F' = dRF - eF$$

These are the **Lotka–Volterra equations with bounded growth.** The coefficients a, b, c, d, and e are **parameters**—constants that have to be determined through field observations in particular circumstances.

An Example

To see what kind of predictions the Lotka–Volterra equations make, we'll work through an example with specific values for the parameters. Let

$$a = .1 \text{ rabbit per month per rabbit}$$
$$b = 10000 \text{ rabbits}$$
$$c = .005 \text{ rabbit per month per rabbit-fox}$$
$$d = .00004 \text{ fox per month per rabbit-fox}$$
$$e = .04 \text{ fox per month per fox}$$

(Check that these five parameters have the right units.) These choices give us the specific differential equations

$$R' = .1R - .00001R^2 - .005RF$$
$$F' = .00004RF - .04F$$

To use this model to follow R and F into the future, we need to know the initial sizes of the two populations. Let's suppose that there are 2000 rabbits and 10 foxes at time $t = 0$. Then the two populations will vary in the following way over the next 250 months.

The graphs of R and F

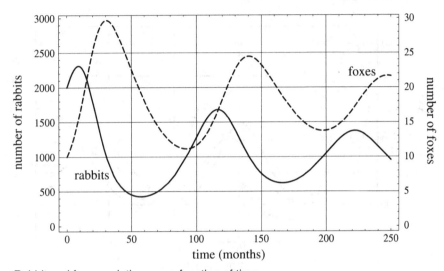

Rabbit and fox populations as a function of time

A variant of the program SIRPLOT was used to produce these graphs. Notice that it plots $100F$ rather than F itself. This is because the number of foxes is about 100 times smaller than the number of rabbits. Consequently, $100F$ and R are about the same size, so their graphs fit nicely together on the same screen.

The graphs have several interesting features. There are different scales for the R and the F values, because the program plots $100F$ instead of F. The peak fox population is about 30, while the peak rabbit population is about 2300. The rabbit and fox populations rise and fall in a regular manner. They rise and fall less with each repeat, though, and if the graphs were continued far enough into the future, we would see R and F level off to nearly constant values.

The illustration below shows what happens to an initial rabbit population of 2000 in the presence of three different initial fox populations $F(0)$. Note that the peak rabbit populations are different, and they occur at different times. The size of the intervals between peaks also depends on $F(0)$.

How rabbits respond to changes in the initial fox population

We have looked at three models, each a refinement of the preceding one. The first was the simplest. It accounted only for the rabbits, and it assumed the rabbit population grew at a constant per capita rate. The second was also restricted to rabbits, but it assumed logistic growth to take into account the carrying capacity of the environment. The third introduced the complexity of a second species preying on the rabbits. In the exercises you will have an opportunity to explore these and other models. Remember that when you use Euler's method to find the functions determined by an initial value problem, you must construct a sequence of successive approximations, until you obtain the level of accuracy desired.

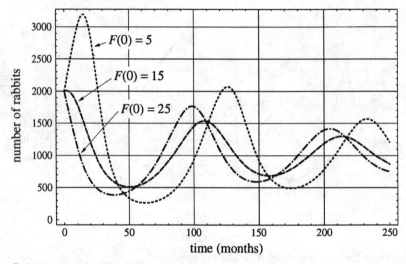

Rabbit populations for different initial fox populations

Exercises

Single-Species Models

1. *Constant per capita growth.* This question considers the initial value problem given in the text:

    $$R' = .1R \text{ rabbits per month}; \qquad R(0) = 2000 \text{ rabbits.}$$

    a) Use Euler's method to determine how many rabbits there are after 6 months. Present a table of successive approximations from which you can read the exact value to whole-number accuracy.

    b) Determine, to whole-number accuracy, how many rabbits there are after 24 months.

    *c) How many months does it take for the rabbit population to reach 25,000?

2. *Logistic growth.* The following questions concern a rabbit population described by the logistic model

    $$R' = .1R \left(1 - \frac{R}{25000}\right) \text{ rabbits per month.}$$

a) What happens to a population of 2000 rabbits after 6 months, after 24 months, and after 5 years? To answer each question, present a table of successive approximations that allows you to give the exact value to the nearest whole number.

b) Sketch the functions determined by the logistic equation if you start with either 2000 or 40,000 rabbits. (Suggestion: You can modify the program SIRPLOT to answer this question.) Compare the two functions. How do they differ? In what ways are they similar?

3. *Seasonal factors.* Living conditions for most wild populations are not constant throughout the year—due to factors like drought or cold, the environment is less supportive during some parts of the year than at others. Partially in response to this, most animals don't reproduce uniformly throughout the year. This problem explores ways of modifying the logistic model to reflect these facts.

a) For the eastern cottontail rabbit, most young are born during the months of March–May, with reduced reproduction during June–August, and virtually no reproduction during the other 6 months of the year. Write a program to generate the solution to the differential equation $R' = kR(1 - R/25000)$, where $k = .2$ during March, April, and May; $k = .05$ during June, July, and August; and $k = 0$ the rest of the year. Start with an initial population of 2000 rabbits on January 1. You may find that using the IF . . . THEN construction in your program is a convenient way to incorporate the varying reproductive rate.

b) How would you modify the model to take into account the fact that rabbits don't reproduce during their first season?

4. *World population.* Here is a model for world population that assumes constant per capita growth. The world's population in 1990 was about 5 billion, and data show that birth rates range from 35 to 40 per thousand people per year and death rates from 15 to 20 per thousand per year. Take this to imply a net constant growth rate of 20 per thousand per year or, what is the same, a constant per capita growth rate of $20/1000 = .02$ per year.

a) Write a differential equation for P that expresses this assumption. Use P to denote the world population, measured in billions.

b) According to the differential equation in (a), at what rate (in billions of persons per year) was the world population growing in 1990?

∗c) By applying Euler's method to this model, using the initial value of 5 billion in 1990, estimate the world population in the years 1980, 2000, 2040, and 2230. Present a table of successive approximations that stabilize with one decimal place of accuracy (in billions). What step size did you have to use to obtain this accuracy?

5. *Supergrowth.* Another model for the world population, one that actually seems to fit recent population data fairly well, assumes "supergrowth"—the rate P' is proportional to a *higher power* of P, rather than to P itself. The model is

$$P' = .015P^{1.2}.$$

As in the previous exercise, assume that P is measured in billions, and the population in 1990 was about 5 billion.

a) According to this model, at what rate (in billions of persons per year) was the population growing in 1990?

b) Using Euler's method, estimate the world population in the years 1980, 2000, and 2040. Use successive approximations until you have one decimal place of accuracy (in billions). What step size did you have to use to obtain this accuracy?

c) Use an Euler approximation with a step size of .1 to estimate the world population in the year 2230. What happens if you repeat your calculation with a step size of .01? [Comment: Something strange is going on here. We will look again at this model in the next section.]

Two-Species Models

Here are some other differential equations that model a predator–prey interaction between two species.

6. *The May model.* This model has been proposed by the contemporary ecologist, R. M. May, to incorporate more realistic assumptions about the encounters between predators (foxes) and their prey (rabbits). So that you can work with quantities that are about the same size (and therefore plot them on the same graph), let y be the number of foxes and let x be the number of rabbits *divided by* 100. We are thus measuring rabbits in units of what we could call "centirabbits."

While a term like "centirabbits" is deliberately whimsical, it echoes the common and sensible practice of choosing units that allow us to measure things with numbers that are neither too small nor too large. For example, we wouldn't describe the distance from the earth to the moon in millimeters, and we wouldn't describe the mass of a raindrop in kilograms.

In his model, May makes the following assumptions.

- In the absence of foxes, the rabbits grow logistically.
- The number of rabbits a single fox eats in a given time period is a function $D(x)$ of the number of rabbits available. $D(x)$ varies from 0 if there are no rabbits available to some value c (the **saturation**

value) if there is an unlimited supply of rabbits. The total number of rabbits consumed in the given time period will thus be $D(x) \cdot y$.

- The fox population is governed by the logistic equation, and the carrying capacity is proportional to the number of rabbits.

a) Explain why

$$D(x) = \frac{c\,x}{x + d}$$

(where d is some constant) might be a reasonable model for the function $D(x)$. Include a sketch of the graph of D in your discussion. What is the role of the parameter d? That is, what feature of rabbit-fox interactions is reflected by making d smaller or larger?

b) Explain how the following system of equations incorporates May's assumptions.

$$x' = ax\left(1 - \frac{x}{b}\right) - \frac{cxy}{x + d}$$

$$y' = ey\left(1 - \frac{y}{fx}\right)$$

The parameters a, b, c, d, e, and f are all positive.

c) Assume you begin with 2000 rabbits and 10 foxes. (Be careful: $x(0) \ne 2000$.) What does May's model predict will happen to the rabbits and foxes over time if the values of the parameters are $a = .6$, $b = 10$, $c = .5$, $d = 1$, $e = .1$, and $f = 2$? Use a suitable modification of the program SIRPLOT.

d) Using the same parameters, describe what happens if you begin with 2000 rabbits and 20 foxes; with 1000 rabbits and 10 foxes; with 1000 rabbits and 20 foxes. Does the eventual long-term behavior depend on the initial condition? How does the long-term behavior here compare with the long-term behavior of the two populations in the Lotka–Volterra model of the text?

e) Using 2000 rabbits and 20 foxes as the initial values, let's see how the behavior of the solutions is affected by changing the values of the parameter c, the saturation value for the number of rabbits (measured in centirabbits, remember) a single fox can eat in a month. Keeping all the other parameters (a, b, d, ...) fixed at the values given above, get solution curves for $c = .5$, $c = .45$, $c = .4$, ..., $c = .15$, and $c = .1$. The solutions undergo a qualitative change somewhere between $c = .3$ and $c = .25$. Describe this change. Can you pinpoint the crucial value of c more closely? This phenomenon is an example of **Hopf bifurcation**, which we will look at more closely in chapter 8. The May model undergoes a Hopf bifurcation as you vary each of the other parameters as well. Choose a couple of them and locate approximately the associated bifurcation values.

7. *The Lotka–Volterra equations.* This model for predator and prey interactions is slightly simpler than the "bounded growth" version we consider in the text. It is important historically, though, because it was one of the first mathematical population models, proposed as a way of understanding why the harvests of certain species of fish in the Adriatic Sea exhibited cyclical behavior over the years. For the sake of variety, let's take the prey to be hares and the predators to be lynx.

   Let $H(t)$ denote the number of hares at time t, and $L(t)$ the number of lynx. This model, the basic Lotka–Volterra model, differs from the bounded growth model in only one respect: it assumes the hares would experience constant per capita growth if there were no lynx.

   a) Explain why the following system of equations incorporates the assumptions of the basic model. (The parameters a, b, c, and d are all positive.)

   $$H' = aH - bHL$$
   $$L' = cHL - dL.$$

   (These are called the **Lotka–Volterra equations**. They were developed independently by the Italian mathematical physicist Vito Volterra in 1925–1926, and by the mathematical ecologist and demographer Alfred James Lotka a few years earlier. Though simplistic, they form one of the principal starting points in ecological modeling.)

   b) Explain why a and b have the units hares per month per hare and hares per month per hare-lynx, respectively. What are the units of c and d? Explain why.

   Suppose time t is measured in months, and suppose the parameters have values

   $$
   \begin{aligned}
   a &= .1 && \text{hare per month per hare} \\
   b &= .005 && \text{hare per month per hare-lynx} \\
   c &= .00004 && \text{lynx per month per hare-lynx} \\
   d &= .04 && \text{lynx per month per lynx.}
   \end{aligned}
   $$

   This leads to the system of differential equations

   $$H' = .1H - .005HL$$
   $$L' = .00004HL - .04L.$$

c) Suppose that you start with 2000 hares and 10 lynx—that is, $H(0) = 2000$ and $L(0) = 10$. Describe what happens to the two populations. A good way to do this is to draw graphs of the functions $H(t)$ and $L(t)$. It will be convenient to have the H scale run from 0 to 3000, and the L scale from 0 to 50. If you modify the program SIRPLOT, have it plot H and $60L$.

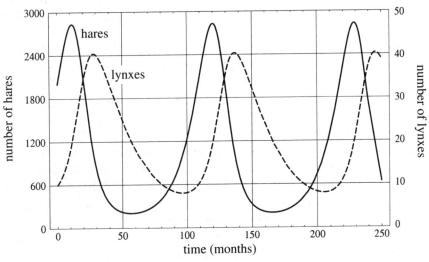

Hare and lynx populations as a function of time

You should get graphs like those above. Notice that the hare and lynx populations rise and fall in a fashion similar to the rabbits and foxes, but here they oscillate—returning *periodically* to their original values.

d) What happens if you keep the same initial hare population of 2000, but use different initial lynx populations? Try $L(0) = 20$ and $L(0) = 50$. (In each case, use a step size of .1 month.)

e) Start with 2000 hares and 10 lynx. From part (c), you know the solutions are periodic. The goal of this part is to analyze this periodic behavior. You can do this with your program in part (c), but you may prefer to replace the FOR-NEXT loop in your program by a variety of DO-WHILE loops (see page 67). First find the maximum number of hares. What is the length of one **period** for the hare population? That is, how long does it take the hare population to complete one cycle (e.g., to go from one maximum to the next)? Find the length of one period for the lynx. Do the hare and lynx populations have the same periods?

f) Plot the hare populations over time when you start with 2000 hares and, successively, 10, 20, and 50 lynx. Is the hare population periodic in each case? What is the period? Does it vary with the size of the initial lynx population?

Fermentation

Wine is made by yeast; yeast digests the sugars in grape juice and produces alcohol as a waste product. This process is called fermentation. The alcohol is toxic to the yeast, though, and the yeast is eventually killed by the alcohol. This stops fermentation, and the liquid has become wine, with about 8–12% alcohol.

Although alcohol isn't a "species," it acts like a predator on yeast. Unlike the other predator–prey problems we have considered, though, the yeast does not have an unlimited food supply. The following exercises develop a sequence of models to take into account the interactions between sugar, yeast, and alcohol.

8. a) In the first model assume that the sugar supply is not depleted, that no alcohol appears, and that the yeast simply grows *logistically*. Begin by adding .5 lb of yeast to a large vat of grape juice whose carrying capacity is 10 lbs of yeast. Assume that the natural growth rate of the yeast is .2 lb of yeast per hour, per pound of yeast. Let $Y(t)$ be the number of pounds of live yeast present after t hours; what differential equation describes the growth of Y?

b) Graph the solution $Y(t)$, for example, by using a suitable modification of the program SIRPLOT. Indicate on your graph approximately when the yeast reaches one-half the carrying capacity of the vat, and when it gets to within 1% of the carrying capacity.

c) Suppose you use a second strain of yeast whose natural growth rate is only half that of the first strain of yeast. If you put .5 lb of *this* yeast into the vat of grape juice, when will it reach one-half the carrying capacity of the vat, and when will it get to within 1% of the carrying capacity? Compare these values to the values produced by the first strain of yeast: are they larger, or smaller? Sketch, on the same graph as in part (b), the way this yeast grows over time.

9. a) Now consider how the yeast produces alcohol. Suppose that waste products are generated at a rate proportional to the amount of yeast present; specifically, suppose each pound of yeast produces .05 lb of alcohol per hour. (The other major waste product is carbon dioxide gas, which bubbles out of the liquid.) Let $A(t)$ denote the amount of alcohol generated after t hours. Write a differential equation that describes the growth of A.

b) Consider the toxic effect of the alcohol on the yeast. Assume that yeast cells die at a rate proportional to the amount of alcohol present, and also to the amount of yeast present. Specifically, assume that, in each pound of yeast, a pound of alcohol will kill

.1 lb of yeast per hour. Then, if there are Y lbs of yeast and A lbs of alcohol, how many pounds of yeast will die in one hour? Modify the original logistic equation for Y (strain 1) to take this effect into account. The modification involves subtracting off a new term that describes the rate at which alcohol kills yeast. What is the new differential equation?

c) You should now have two differential equations describing the rates of growth of yeast and alcohol. The equations are **coupled**, in the sense that the yeast equation involves alcohol, and the alcohol equation involves yeast. Assuming that the vat contains, initially, .5 lb of yeast and no alcohol, describe by means of a graph what happens to the yeast. How close does the yeast get to carrying capacity, and when does this happen? Does the fermentation end? If so, when; and how much alcohol has been produced by that time? (Note that since Y will never get all the way to 0, you will need to adopt some convention like $Y \leq .01$ to specify the end of fermentation.)

10. What happens if the rates of toxicity and alcohol production are different? Specifically, increase the rate of alcohol production by a factor of 5—from .05 to .25 lb of alcohol per hour, per pound of yeast—and at the same time reduce the toxicity rate by the same factor—from .10 to .02 lb of yeast per hour, per pound of alcohol and pound of yeast. How do these changes affect the time it takes for fermentation to end? How do they affect the amount of alcohol produced? What happens if only the rate of alcohol production is changed? What happens if only the toxicity rate is reduced?

11. a) The third model will take into account that the sugar in the grape juice is consumed. Suppose the yeast consumes .15 lb of sugar per hour, per lb of yeast. Let $S(t)$ be the amount of sugar in the vat after t hours. Write a differential equation that describes what happens to S over time.

b) Since the carrying capacity of the vat depends on the amount of sugar in it, the carrying capacity must now vary. Assume that the carrying capacity of S lb of sugar is $.4S$ lb of yeast. How much sugar is needed to maintain a carrying capacity of 10 lb of yeast? How much is needed to maintain a carrying capacity of 1 lb of yeast? Rewrite the logistic equation for yeast so that the carrying capacity is $.4S$ lb, instead of 10 lb, of yeast. Retain the term you developed in 9(b) to reflect the toxic impact of alcohol on the yeast.

c) There are now three differential equations. Using them, describe what happens to .5 lb of yeast that is put into a vat of grape juice that contains 25 lb of sugar at the start. Does all the sugar disappear? Does all the yeast disappear? How long does it take before there is only .01 lb of yeast? How much sugar is left then? How much alcohol has been produced by that time?

Newton's Law of Cooling

Suppose that we start off with a freshly brewed cup of coffee at 90°C and set it down in a room where the temperature is 20°C. What will the temperature of the coffee be in 20 minutes? How long will it take the coffee to cool to 30°C?

If we let the temperature of the coffee be Q (in °C), then Q is a function of the time t, measured in minutes. We have $Q(0) = 90$°C, and we would like to find the value t_1 for which $Q(t_1) = 30$°C.

It is not immediately apparent how to give Q as a function of t. However, we can describe the *rate* at which a liquid cools off, using **Newton's law of cooling**: the rate at which an object cools (or warms up, if it's cooler than its surroundings) is proportional to the *difference* between its temperature and that of its surroundings.

12. In our example, the temperature of the room is 20°C, so Newton's law of cooling states that $Q'(t)$ is proportional to $Q - 20$, the difference between the temperature of the liquid and the room. In symbols, we have

$$Q' = -k\,(Q - 20)$$

where k is some positive constant.

a) Why is there a minus sign in the equation? The particular value of k would need to be determined experimentally. It will depend on things like the size and shape of the cup, how much sugar and cream you use, and whether you stir the liquid. Suppose that k has the value of .1°C per minute per °C of temperature difference. Then the differential equation becomes:

$$Q' = -.1(Q - 20) \quad \text{°C per minute.}$$

*b) Use Euler's method to determine the temperature Q after 20 minutes. Write a table of successive approximations with smaller and smaller step sizes. The values in your table should stabilize to the second decimal place.

c) How long does it take for the temperature Q to drop to 30°C? Use a DO-WHILE loop to construct a table of successive approximations that stabilize to the second decimal place.

13. On a hot day, a cold drink warms up at a rate approximately proportional to the difference in temperature between the drink and its surroundings. Suppose the air temperature is 90°F and the drink is

initially at 36°F. If Q is the temperature of the drink at any time, we shall suppose that it warms up at the rate

$$Q' = -.2(Q - 90) \quad °F \text{ per minute.}$$

According to this model, what will the temperature of the drink be after 5 minutes, and after 10 minutes? In both cases, produce values that are accurate to two decimal places.

14. In our discussion of cooling coffee, we assumed that the coffee did not heat up the room. This is reasonable because the room is large, compared to the cup of coffee. Suppose, in an effort to keep it warmer, we put the coffee into a small insulated container—such as a microwave oven (which is turned off). We must assume that the coffee *does* heat up the air inside the container. Let A be the air temperature in the container, and Q the temperature of the coffee. Then both A and Q change over time, and Newton's law of cooling tells us the *rates* at which they change. In fact, the law says that both Q' and A' are proportional to $Q - A$. Thus,

$$Q' = -k_1(Q - A)$$
$$A' = k_2(Q - A),$$

where k_1 and k_2 are positive constants.
a) Explain the signs that appear in these differential equations.
b) Suppose $k_1 = .3$ and $k_2 = .1$. If $Q(0) = 90°C$ and $A(0) = 20°C$, when will the temperature of the coffee be 40°C? What is the temperature of the air at this time? Your answers should be accurate to one decimal place.
c) What does the temperature of the coffee become eventually? How long does it take to reach that temperature?

S-I-R Revisited

Consider the spread of an infectious disease that is modeled by the S-I-R differential equations

$$S' = -.00001\, SI$$
$$I' = .00001\, SI - .08\, I$$
$$R' = .08\, I.$$

Take the initial condition of the three populations to be

$$S(0) = 35,400 \text{ persons}$$
$$I(0) = 13,500 \text{ persons}$$
$$R(0) = 22,100 \text{ persons.}$$

*15. How many susceptibles are left after 40 days? When is the largest number of people infected? How many susceptibles are there at that time? Explain how you could determine the last number *without* using Euler's method.

16. What happens as the epidemic "runs its course"? That is, as more and more time goes by, what happens to the numbers of infecteds and susceptibles?

17. One of the principal uses of a mathematical model is to get a qualitative idea of how a system will behave with different initial conditions. For instance, suppose we introduce 100 infected individuals into a population. How will the spread of the infection depend on the size of the population? Assume the same S-I-R differential equations that were used in the previous exercise, and draw the graphs of $S(t)$ for initial susceptible population sizes $S(0)$ ranging from 0 to 45,000 in increments of 5000 [that is, take $S(0) = 0, 5000, 10000, \ldots, 45000$]. In each case assume that $R(0) = 0$ and $I(0) = 100$. Use these graphs to argue that the larger the initial susceptible population, the more rapidly the epidemic runs its course.

18. Draw the graphs of $I(t)$ for the same initial conditions as in the previous problem. Using these graphs you can demonstrate that the larger the susceptible population, the larger will be the fraction of the population that is infected during the worst stages of the epidemic. Do this by constructing a table displaying I_{max}, t_{max}, and P_{max}, where I_{max} is the maximum value of $I(t)$, t_{max} is the time at which this maximum occurs [that is, $I_{max} = I(t_{max})$], and P_{max} is the ratio of I_{max} to the initial susceptible population: $P_{max} = I_{max}/S(0)$. The table below gives the first three sets of values.

$S(0)$	I_{max}	P_{max}	t_{max}
5 000	100	0.02	0
10 000	315	0.03	> 100
15 000	2071	0.14	66
⋮	⋮	⋮	⋮

Your table should show that there is a time when over half the population is infected if $S(0) = 45,000$, while there is never a time when more than one-fourth of the population is infected if $S(0) = 20000$.

Constructing Models

Systems in which we know a number of quantities at a given time and would like to know their values at a future time (or know at what future time they will attain given values) occur in many different contexts. The following are some systems for discussion. Can any of these be modeled as initial value problems? What information would you need to resolve the question? Make some reasonable assumptions about the missing information and write down an initial value problem which would model the system.

19. We deposit a fixed sum of money in a bank, and we'd like to know how much will be there in ten years.

20. We know the diameter of the mold spot growing on a cheese sandwich is 1/4 inch, and we'd like to know when its diameter will be one inch.

21. We know the fecal bacterial and coliform concentrations in a local swimming hole, and we'd like to know when they fall below certain prescribed levels (which the Board of Health deems safe).

22. We know what the temperature and rainfall are today, and we'd like to know what both will be one week from today.

23. We know what the winning lottery number was yesterday, and we'd like to know what the winning number will be the day after tomorrow.

24. We know where the earth, sun, and moon are in relation to each other now, and how fast and in what direction they are moving. We would like to be able to predict where they are going to be at any time in the future. We know the gravity of each affects the motions of the others by determining the way their velocities are changing.

➤ 4.2   SOLUTIONS OF DIFFERENTIAL EQUATIONS

Differential Equations Are Equations

Until now, we have viewed a system of differential equations as a set of *instructions* for "stepping into the future" (or the past). Put another way, an initial value problem was treated as a prescription for using Euler's method to determine a set of functions which were then given either graphically or in tabular form.

Differential equations give instructions for Euler's method

In this section we take a new point of view: we will think of differential equations as *equations* for which we would like to find *solutions* in terms of functions which can be given by explicit formulas. While it is unfortunately the case that most differential equations do not have solutions which can be given by formulas, there are enough important classes of equations where such solutions do exist to make them worth studying. When such solutions can be found, we have a very powerful tool for examining the behavior of the phenomenon being modeled.

To see what this means, let's look first at equations in algebra. Consider the equation $x^2 = x + 6$. As it stands, this is neither true nor false. We make it true or false, though, when we substitute a particular number for x. For example, $x = 3$ makes the equation true, because $3^2 = 3 + 6$. On the other hand, $x = 1$ makes the equation false, because $1^2 \neq 1 + 6$. Any number that makes an equation true is called a *solution* to that equation. In fact, $x^2 = x + 6$ has exactly two solutions: $x = 3$ and $x = -2$.

Equations have solutions

We can view differential equations the same way. Consider, for example, the differential equation

$$\frac{dy}{dt} = \frac{1}{2y}.$$

Because it involves the expression dy/dt, we understand that y is a function of t. As it stands, the differential equation is neither true nor false. We make it true or false, though, when we substitute a particular *function* for y. For example, $y = \sqrt{t} = t^{1/2}$ makes the differential equation true. To see this, first look at the left-hand side of the equation:

Substitute $y = \sqrt{t}$ into the differential equation

$$\frac{dy}{dt} = \tfrac{1}{2}t^{-1/2} = \frac{1}{2\sqrt{t}}.$$

Now look at the right-hand side:

$$\frac{1}{2y} = \frac{1}{2\sqrt{t}}.$$

The two sides of the equation are equal, so the substitution $y = \sqrt{t}$ makes the equation true.

The function $y = t^2$, however, makes the differential equation *false*. The left-hand side is

$$\frac{dy}{dt} = 2t,$$

but the right-hand side is

$$\frac{1}{2y} = \frac{1}{2t^2}.$$

Since $2t$ and $1/2t^2$ are different functions, the two sides are unequal and the equation is therefore false.

A solution makes the equation true

We say that $y = \sqrt{t}$ is a **solution** to this differential equation. The function $y = t^2$ is *not* a solution. To decide whether a particular function is a solution when the function is given by a formula, notice that we need to be able to differentiate the formula.

If we view differential equations simply as instructions for carrying out Euler's method, we need only the microscope equation $\Delta y \approx y' \cdot \Delta t$ in order to find functions. However, if we want to find functions that are solutions to differential equations from our new point of view, we first need to introduce the idea of the derivative and the rules for differentiating functions.

Just as an algebraic equation can have more than one solution, so can a differential equation. In fact, we can show that $y = \sqrt{t + C}$ is a solution to the differential equation

$$\frac{dy}{dt} = \frac{1}{2y},$$

for any value of the constant C. To evaluate the left-hand side dy/dt, we need the chain rule (section 3.6). Let's write

$$y = \sqrt{u} \quad \text{where} \quad u = t + C.$$

Then the left-hand side is the function

$$\frac{dy}{dt} = \frac{dy}{du} \cdot \frac{du}{dt} = \frac{1}{2\sqrt{u}} \cdot 1 = \frac{1}{2\sqrt{t + C}}.$$

Since the right-hand side of the differential equation is

$$\frac{1}{2y} = \frac{1}{2\sqrt{t + C}},$$

the two sides are equal—no matter what value C happens to have. This proves that every function of the form $y = \sqrt{t + C}$ is a solution to the differential equation. Since there are infinitely many values that C can take, the differential equation has infinitely many different solutions!

A differential equation can have infinitely many solutions

If a differential equation arises in modeling a physical or biological process, the variables involved must also satisfy an initial condition. Suppose we add an initial condition to our differential equation:

$$\frac{dy}{dt} = \frac{1}{2y} \quad \text{and} \quad y(0) = 5.$$

Does *this* problem have a solution—that is, can we find a function $y(t)$ that is a solution to the differential equation and also satisfies the condition $y(0) = 5$?

Notice $y = \sqrt{t}$ is not a solution to this new problem. Although it satisfies the differential equation, it fails to satisfy the initial condition:

$$y(0) = \sqrt{0} = 0 \neq 5.$$

Perhaps one of the other solutions to the differential equation will work. When we evaluate the solution $y = \sqrt{t + C}$ at $t = 0$ we get

$$y(0) = \sqrt{0 + C} = \sqrt{C}.$$

An initial value problem
has only one solution

We want this to equal 5, and it will if $C = 25$. Thus, $y = \sqrt{t + 25}$ is a solution to the initial value problem. Furthermore, the only value of C which will make $y(0) = 5$ is $C = 25$, so the initial value problem has only one solution of the form $\sqrt{t + C}$. Here is the graph of this solution:

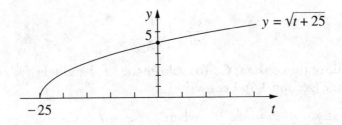

As always, you can use Euler's method to find the function determined by an initial value problem, and you can graph that function using the program SIRPLOT, for example. How will *that* graph compare with this one? In the exercises you can explore this question.

Checking Solutions Versus Finding Solutions

Notice that we have only *checked* whether a given function solves an initial value problem; we have not *constructed* a formula to solve the problem. By this point you are probably wondering where the given solutions came from.

It is helpful to continue exploring the parallels with solutions of algebraic equations. In the case of the equation $x^2 = x + 6$, there are, of course, methods to find solutions. One possibility is to rewrite $x^2 = x + 6$ in the form $x^2 - x - 6 = 0$. By factoring $x^2 - x - 6$ as

$$x^2 - x - 6 = (x - 3)(x + 2)$$

we can see that either $x - 3 = 0$ (so $x = 3$), or $x + 2 = 0$ (so $x = -2$). Another method is to use the **quadratic formula**

$$x = \frac{-b \pm \sqrt{b^2 - 4ac}}{2a}$$

for the roots of the quadratic function $ax^2 + bx + c$. In our case, the quadratic formula yields

$$x = \frac{-(-1) \pm \sqrt{1 - 4 \cdot 1 \cdot (-6)}}{2 \cdot 1} = \frac{1 \pm \sqrt{1 + 24}}{2} = \frac{1 \pm 5}{2},$$

so again we find that x must be either 3 or -2.

Thus we have at least two different methods for finding solutions to this particular equation. The methods we use to solve an algebraic equation depend very much on the equation we face. For example, there is no way to find a solution to $\sin x = 2^x$ by factoring, or by using a "magic formula" like the quadratic formula. Nevertheless, there are methods that *do* work. In chapters 1 and 2 we dealt with similar problems by using a computer graphing utility that could zoom in on the point of intersection of two graphs. In chapter 5 we will introduce another tool, the Newton–Raphson method, for finding roots. These are both powerful methods, because they will work with nearly all algebraic equations. It is important to recognize that these numerical methods really do solve the problem, even though they do not give solutions in closed form the way the quadratic formula does.

There are special methods to solve particular equations, but other methods work generally

The situation is entirely analogous in dealing with differential equations. The methods we use to solve a differential equation depend on the equation we face. A course in differential equations provides methods for finding formulas that solve many different kinds of differential equations. The methods are like the quadratic formula in algebra, though—they give a complete solution, but they work only with differential equations that have a very specific form. This course will not attempt to survey the methods that find such formulas, although in the next sections we will see effective methods for dealing with some special subcases.

It is important to realize, though, that Euler's method is always there if we can't think of anything more clever, and it really does provide solutions. In fact, most initial value problems have one, *and only one*, solution, and Euler's method can be used to determine this unique solution. If we can also find a formula for the solution, then it must be the same solution as that produced by Euler's method. In more advanced courses you will see a proof that this is true in general, provided some mild conditions are satisfied. To emphasize the importance of this idea, we give it a name:

Most initial value problems have one, and only one, solution. Euler's method will find it.

> **Existence and Uniqueness Principle:**
> Under most conditions, an initial value problem has one and only one solution.

The existence and uniqueness principle is one of the most important mathematical results in the theory of differential equations.

We will continue to rely primarily on Euler's method, which generates solutions for nearly all differential equations.

However, there are clear benefits to having a formula for the solution to a differential equation, allowing us to investigate questions that we can't answer very well if we only have solutions given by Euler's method. In this section, we will look at some of those benefits.

World Population Growth
Two Models

In the exercises in the last section, we looked at two different models that seek to describe how the world population will grow. One model assumed constant per capita growth—rate of change proportional to population size. The other assumed "supergrowth"—rate of change proportional to a *higher* power of the population size. Let's write P for the population size in the constant per capita growth model and Q for the population size in the supergrowth model. In both cases, the population is expressed in billions of persons and time is measured in years, with $t = 0$ in 1990. In this notation, the two models are

Models for world
population growth

$$\text{constant per capita:} \quad \frac{dP}{dt} = .02\,P \qquad P(0) = 5$$

$$\text{supergrowth:} \quad \frac{dQ}{dt} = .015\,Q^{1.2} \quad Q(0) = 5$$

By using Euler's method, we discover that the two models predict fairly similar results over sixty years, although the supergrowth model lives up to its name by predicting larger populations than the constant per capita growth model as time passes:

t	P	Q
-10	4.09	4.08
0	5.00	5.00
10	6.11	6.18
50	13.59	15.94

These estimates are accurate to one decimal place, and that level of accuracy was obtained with the step size $\Delta t = .1$.

However, the predictions made by the models differ widely over longer time spans. If we use Euler's method to estimate the populations after 240 years, we get

Δt	$P(240)$	$Q(240)$
.1	6.046×10^2	1.979×10^{10}
.01	6.073×10^2	2.573×10^{11}
.001	6.075×10^2	3.825×10^{11}

As the step size decreases from .1 to .01 to .001, the estimates of the constant per capita growth model $P(240)$ behave as we have come to expect: already three digits have stabilized. But in the estimates of the supergrowth model, not even one digit of $Q(240)$ has stabilized.

In this section we will see that there are actually *formulas* for the functions $P(t)$ and $Q(t)$. These formulas will illuminate the reason behind the differences in speed of stabilization in the estimates.

A Formula for the Supergrowth Model

Without asking how the following formula might have been derived, let's check that it is indeed a solution to the supergrowth initial value problem.

$$Q(t) = \left(\frac{1}{\sqrt[5]{5}} - .003\, t \right)^{-5}$$

First of all, the formula satisfies the initial condition $Q(0) = 5$: *Checking the initial condition*

$$Q(0) = \left(\frac{1}{\sqrt[5]{5}} \right)^{-5} = (\sqrt[5]{5})^5 = 5.$$

To check that it also satisfies the differential equation, we must evaluate the two sides of the differential equation *Checking the differential equation*

$$\frac{dQ}{dt} = .015\, Q^{1.2}.$$

Let's begin by evaluating the left-hand side. To differentiate $Q(t)$, we will write *Left-hand side* Q as a *chain* of functions:

$$Q = u^{-5} \qquad \text{where} \qquad u = \frac{1}{\sqrt[5]{5}} - .003\, t.$$

Since $Q = u^{-5}$, $dQ/du = (-5)u^{-6}$. Also, since u is just a linear function of t in which the multiplier is $-.003$, we have $du/dt = -.003$. Consequently,

$$\begin{aligned}
\frac{dQ}{dt} &= \frac{dQ}{du} \cdot \frac{du}{dt} \\
&= (-5)\, u^{-6} \cdot (-.003) \\
&= .015\, u^{-6}.
\end{aligned}$$

Ordinarily, we would "finish the job" by substituting for u its formula in terms of t. However, in this case it is clearer to just leave the left-hand side in this form.

To evaluate the right-hand side of the differential equation (which is the *Right-hand side* expression $.015\, Q^{1.2}$), we would expect to substitute for Q its formula in terms of t. But since in evaluating the left-hand side, we expressed things in terms of u, let's do the same thing here. Since $Q = u^{-5}$,

$$Q^{1.2} = Q^{6/5} = (u^{-5})^{6/5} = u^{-5 \cdot 6/5} = u^{-6}.$$

Therefore, the right-hand side is equal to $.015\,u^{-6}$. But so is the left-hand side, so $Q(t)$ is indeed a solution to the differential equation

$$\frac{dQ}{dt} = .015Q^{1.2}.$$

Notice two things about this result. First, when we work with formulas we have greater need for algebra to manipulate them. For example, we needed one of the laws of exponents, $(a^b)^c = a^{bc}$, to evaluate the right-hand side. Second, we found it simpler to express Q in terms of the intermediate variable u, instead of the original input variable t. In another computation, it might be preferable to replace u by its formula in terms of t. You need to choose your algebraic strategy to fit the circumstances.

Behavior of the Supergrowth Solution

It was convenient to use a negative exponent in the formula for $Q(t)$ when we wanted to differentiate Q. However, to understand what the formula tells us about supergrowth, it will be more useful to write Q as

$$Q(t) = \left(\frac{1}{1/\sqrt[5]{5} - .003\,t} \right)^5.$$

This way makes it clear that Q is a fraction, and we can see its denominator. In particular, this fraction is not defined when the denominator is zero—that is, when

$$\frac{1}{\sqrt[5]{5}} - .003\,t = 0, \quad \text{or} \quad t = \frac{1/\sqrt[5]{5}}{.003} = 241.6\ldots \quad \text{years after 1990.}$$

Q "blows up" as $t \to 241.6\ldots$

Consider what happens, though, as t approaches this special value $241.6\ldots$. The denominator isn't *yet* zero, but it is *approaching* zero, so the fraction Q is becoming infinite. This means that the supergrowth model predicts the world population will become infinite in about 240 years!

Let's see what the predicted population size is when $t = 240$ (which is the year 2230 A.D.), shortly before Q becomes infinite. We have

$$Q(240) = \left(\frac{1}{\sqrt[5]{5}} - (.003)(240) \right)^{-5} \approx 4.0088 \times 10^{11}.$$

Remember that Q expresses the population in *billions* of people, so the supergrowth model predicts about 4×10^{20} people (i.e., 400 quintillion!) in the year 2230. Refer back to our estimates of $Q(240)$ using Euler's method (page 182). Although not even one digit of the estimates had stabilized, at least the final one (with a step size of .001) had reached the right power of ten. In fact, estimates made with still smaller step sizes do eventually approach the value given by the formula for Q:

Step size	Q(240)
.1	0.1979×10^{11}
.01	2.5727×10^{11}
.001	2.8249×10^{11}
.0001	3.9999×10^{11}
.00001	4.0069×10^{11}

Let's look at the relationship between the Euler approximations of Q and the formula for Q graphically. Here are graphs produced by a modification of the program SEQUENCE.

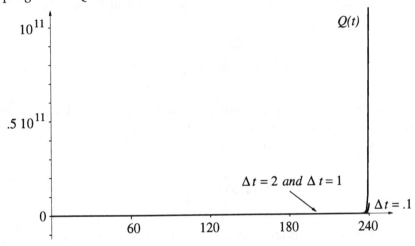

The range of values of Q for $0 \leq t \leq 240$ is so immense that these graphs are useless. In a case like this, it is helpful to rescale the vertical axis so that the space between one power of 10 and the next is the same. In other words, instead of seeing 1, 2, 3, ..., we see 10^1, 10^2, 10^3, ... This is called a **logarithmic** scale. Here's what happens to the graphs if we put a logarithmic scale on the vertical axis:

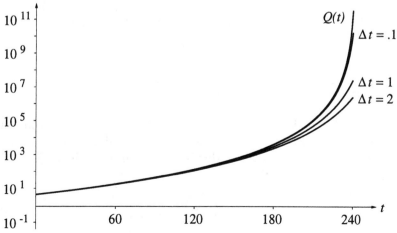

Euler approximations and the formula for Q

The second graph makes it clearer that the Euler approximations do indeed approach the graph of the function given by our formula, but they approach more and more slowly, the closer t approaches 241.6....

> Graphs made with logarithmic scales are particularly useful when the numbers being plotted cover a wide range of values. When just one axis is logarithmic, the result is called a semi-log plot; when both axes are logarithmic, the result is called a log–log plot.

The initial value problem has a solution only for $t < 241.6...$

Since $Q(t)$ becomes infinite when $t = 241.6...$, we must conclude that the solution to the original initial value problem is meaningful only for $t < 241.6...$. Of course, the formula for Q works quite well when $t > 241.6...$. It just has no meaning as the size of a population. For instance, when $t = 260$ we get

$$Q(260) = \left(\frac{1}{\sqrt[5]{5}} - (.003)(260) \right)^{-5} \approx (-.05522)^{-5} \approx -1.948 \times 10^6.$$

In other words, the function determined by the initial value problem is defined only on intervals around $t = 0$ that do not contain $t = 241.6...$.

The formula for $Q'(t)$ is informative too:

$$Q'(t) = .015 \left(\frac{1}{1/\sqrt[5]{5} - .003t} \right)^6.$$

Q' blows up as $t \to 241.6...$

Since Q' has the same denominator as Q, it becomes infinite the same way Q does: $Q'(t) \to \infty$ as $t \to 241.6...$. Because Euler's method uses the microscope equation $\Delta Q \approx Q' \cdot \Delta t$ to predict the next value of Q, we can now understand why the estimates of $Q(240)$ were so slow to stabilize: as $Q' \to \infty$, $\Delta Q \to \infty$, too.

A Formula for Constant Per Capita Growth

The constant per capita growth model for the world population that we are considering is

$$\frac{dP}{dt} = .02\,P \qquad P(0) = 5.$$

This differential equation has a very simple form; if $P(t)$ is a solution, then the derivative of P is just a multiple of P. We have already seen in chapter 3 that *exponential* functions behave this way (exercises 5–7 in section 3.3). For example, if $P(t) = 2^t$, then

$$\frac{dP}{dt} = .69 \cdot 2^t = .69\,P.$$

Of course, the multiplier that appears here is .69, not .02, so $P(t) = 2^t$ is not a solution to our problem.

However, the multiplier that appears when we differentiate an exponential function changes when we change the base. That is, if $P(t) = b^t$, then $P'(t) = k_b \cdot b^t$, where k_b depends on b. Here is a sample of values of k_b for different bases b:

Exponential functions satisfy $dy/dt = ky$

b	k_b
.5	$-.693147\ldots$
2	$.693147\ldots$
3	$1.098612\ldots$
10	$2.302585\ldots$

Notice that k_b gets larger as b does. Since .02 lies between $-.693147$ and $+.693147$, the table suggests that the value of b we want lies somewhere between .5 and 2.

We can say even more about the multiplier. Since $P'(t) = k_b \cdot P(t)$ and $P(t) = b^t$, we find

$$P'(0) = k_b \cdot P(0) = k_b \cdot b^0 = k_b \cdot 1 = k_b.$$

In other words, k_b is the *slope of the graph of* $P(t) = b^t$ *at the origin.*

Thus, we will be able to solve the differential equation $dP/dt = .02\,P$ if we can find an exponential function $P(t) = b^t$ whose graph has slope .02 at the origin. This is a problem that we can solve with a computer microscope. Pick a value of b and graph b^t. Zoom in on the graph at the origin and measure the slope. If the slope is more than .02, choose a smaller value for b; if the slope is less than .02, choose a larger value for b. Repeat this process, narrowing down the possibilities for b until the slope is as close to .02 as you wish. Eventually, we get

The correct exponential function has slope .02 at the origin

$$P(t) = (1.0202)^t.$$

You should check that $P'(0) = .02000\ldots$; see the exercises.

Thus $P(t) = (1.0202)^t$ solves the differential equation $P' = .02\,P$. However, it does not satisfy the initial condition, because

$$P(0) = (1.0202)^0 = 1 \neq 5.$$

This is easy to fix: $P(t) = 5 \cdot (1.0202)^t$ satisfies both conditions. More generally, $P(t) = C\,(1.0202)^t$ satisfies the initial condition $P(0) = C$ as well as the differential equation $P'(t) = .02\,P(t)$. To check the initial condition, we compute

$$P(0) = C\,(1.0202)^0 = C \cdot 1 = C.$$

The differential equation is also satisfied:

$$P'(t) = (C\,(1.0202)^t)' = C \cdot ((1.0202)^t)' = C \cdot (.02\,(1.0202)^t) = .02\,P(t).$$

So we have verified that the solution to our problem is

The formula for P

$$P(t) = 5\,(1.0202)^t.$$

Because exponential functions are involved, constant per capita growth is commonly called **exponential growth**. In the figure below we compare exponential growth $P(t)$ to "supergrowth" $Q(t)$. The two graphs agree quite well when $t < 50$. Notice that population is plotted on a logarithmic scale (a *semi–log* plot). This makes the graph of P a straight line!

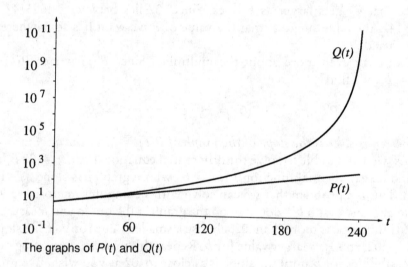

The graphs of $P(t)$ and $Q(t)$

Differential Equations Involving Parameters

The S-I-R model contained two parameters—the transmission and recovery coefficients a and b. When we used Euler's method to analyze S, I, and R, we were working numerically. To do the computations, we had to give the parameters definite numerical values. That made it more difficult to deal with our questions about the effects of *changing* the parameters. As a result, we took other approaches to explore those questions. For example, we used algebra to see that there was a threshold for the spread of the disease: if there were fewer than b/a people in the susceptible population, the infection would fade away.

How do parameters affect solutions?

This is the situation generally. Euler's method can be used to produce solutions to a very broad range of initial value problems. However, if the model includes parameters, then we usually want to know how the solutions are affected when the parameters change. Euler's method is a rather clumsy tool for investigating this question. Other methods—ones that don't require the values of the parameters to be fixed—sometimes work better. One possibility is to start with a formula.

The supergrowth problem illustrates both how questions about parameters can arise and how a formula for the solution can be useful in answering the questions. One of the most striking features of the supergrowth model is that it predicts the population becomes infinite in 241.6 . . . years. That prediction was based on an initial population of 5 billion and a growth constant of .015. Suppose those values turn out to be incorrect, and we need to start with different values. Will that change the prediction? If so, how?

Supergrowth parameters

We should treat the initial population and the growth constant as parameters—that is, as quantities that *can* vary, although they will have fixed values in any specific situation that we consider. Suppose we let A denote the size of the initial population, and k the growth constant. If we incorporate these parameters into the supergrowth model, the initial value problem takes this form:

$$\frac{dQ}{dt} = k\, Q^{1.2} \qquad Q(0) = A$$

Here is the formula for a function that solves this problem:

The supergrowth solution with parameters

$$Q(t) = \left(\frac{1}{\sqrt[5]{A}} - .2kt \right)^{-5}$$

Notice that, when $A = 5$ and $k = .015$, this formula reduces to the one we considered earlier.

Let's check that the formula does indeed solve the initial value problem. First, the initial condition:

Checking the formula

$$Q(0) = \left(\frac{1}{\sqrt[5]{A}} - .2k \cdot 0 \right)^{-5} = \left(\frac{1}{\sqrt[5]{A}} \right)^{-5} = (\sqrt[5]{A})^5 = A.$$

Next, the differential equation: To differentiate $Q(t)$ we introduce the chain

$$Q = u^{-5} \qquad \text{where} \qquad u = \frac{1}{\sqrt[5]{A}} - .2kt.$$

We see that $dQ/du = -5u^{-6}$. Since u is a linear function of t in which the multiplier is $-.2k$, we also have $du/dt = -.2k$. Thus, by the chain rule,

$$\frac{dQ}{dt} = \frac{dQ}{du} \cdot \frac{du}{dt} = -5u^{-6} \cdot (-.2k) = ku^{-6}.$$

That is the left-hand side of the differential equation. To evaluate the right-hand side, we use the fact that $Q = u^{-5}$. Thus

$$kQ^{1.2} = kQ^{6/5} = k(u^{-5})^{6/5} = ku^{-5 \cdot 6/5} = ku^{-6}.$$

Since both sides equal ku^{-6}, they equal each other, proving that $Q(t)$ is a solution to the differential equation.

Next, we ask when the population becomes infinite. Exactly as before, this will happen when the denominator of the formula for $Q(t)$ becomes zero:

The time to infinity

$$\frac{1}{\sqrt[5]{A}} - .2kt = 0, \qquad \text{or} \qquad t = \frac{1}{.2k\sqrt[5]{A}}.$$

Here, in fact, is a *formula* that tells us how each of the parameters A and k affects the time it takes for the population to become infinite.

Let's use τ (the Greek letter "tau") to denote the "time to infinity." For example, if we double the initial population, so $A = 10$ billion people, while keeping the original growth constant $k = .015$, then the time to infinity is

$$\tau = \frac{1}{.003 \times \sqrt[5]{10}} \approx 210.3 \text{ years.}$$

By contrast, if we double the growth rate, to $k = .030$, while keeping the original $A = 5$, then the time to infinity is only

$$\tau = \frac{1}{.006 \times \sqrt[5]{5}} \approx 120.8 \text{ years.}$$

Conclusion: Doubling the growth rate has a much greater impact than doubling the initial population.

Uncertainty in the size of τ

For any specific growth rate and initial population, we can always calculate the time to infinity. But we can actually do more; the formula for τ allows us to do an *error analysis* along the patterns described in section 3.4. For example, suppose we are uncertain of our value of the growth rate k; there may be an error of size Δk. How uncertain does that make us about the calculated value of τ? Likewise, if the current world population A is known only with an error of ΔA, how uncertain does *that* make τ? Also, how are the *relative* errors related? Let's do this analysis, assuming that $k = .015$ and $A = 5$.

How an error in k propagates

Our tool is the error propagation equation—which is the microscope equation. If we deal with k first, then

$$\Delta \tau \approx \frac{\partial \tau}{\partial k} \cdot \Delta k.$$

We have used partial derivatives because τ is a function of *two* variables, A as well as k. If we write

$$\tau = \frac{1}{.2\sqrt[5]{A}}k^{-1},$$

then the differentiation rules yield

$$\frac{\partial \tau}{\partial k} = -1 \cdot \frac{1}{.2\sqrt[5]{A}}k^{-2} = \frac{-1}{.2\sqrt[5]{5}} \times (.015)^{-2} \approx -16106.$$

Thus $\Delta\tau \approx -16106 \cdot \Delta k$. For example, if the uncertainty in the value of $k = .015$ is $\Delta k = \pm.001$, then the uncertainty in τ is about ∓ 16 years.

To determine how an error in A propagates to τ, we first write

How an error in A
propagates

$$\tau = \frac{1}{.2k}A^{-1/5}.$$

Then

$$\frac{\partial \tau}{\partial A} = -\frac{1}{5} \cdot \frac{1}{.2k}A^{-6/5} = \frac{-1}{5 \times .2 \times .015} \times 5^{-6/5} \approx -9.7.$$

The error propagation equation is thus $\Delta\tau \approx -9.7 \cdot \Delta A$. If the uncertainty in the world population is about 100 million persons, so $\Delta A = \pm.1$, then the uncertainty in τ is less than 1 year.

To complete the analysis, let's compare relative errors. This involves a lot of algebra. To see how an error in k propagates, we have

Relative errors

$$\Delta\tau \approx -\frac{\Delta k}{.2k^2\sqrt[5]{A}} \qquad \text{and} \qquad \tau = \frac{1}{.2k\sqrt[5]{A}}.$$

We can therefore compute that a given relative error in k propagates as

$$\frac{\Delta\tau}{\tau} \approx -\frac{\Delta k}{.2k^2\sqrt[5]{A}} \cdot \frac{.2k\sqrt[5]{A}}{1} = -\frac{\Delta k}{k}.$$

Thus, a 1% error in k leads to a 1% error in τ, although the sign is reversed.

To analyze how a given relative error in A propagates, we start with

$$\Delta\tau \approx -\frac{1}{5} \cdot \frac{\Delta A}{.2kA^{6/5}}.$$

Then

$$\frac{\Delta\tau}{\tau} \approx -\frac{1}{5} \cdot \frac{\Delta A}{.2kA^{6/5}} \cdot \frac{.2k\sqrt[5]{A}}{1} = -\frac{1}{5} \cdot \frac{\Delta A}{A}.$$

How sensitive τ is to errors in k and A

This says that it takes a 5% error in A to produce a 1% error in τ. Consequently, the time to infinity τ *is 5 times more sensitive to errors in k than to errors in A.*

The exercises in this section will give you an opportunity to check that a particular formula is a solution to an initial value problem that arises in a variety of contexts. Later in this chapter, we will make a modest beginning on the much harder task of *finding* solutions given by formulas for special initial value problems. There are more sophisticated methods for finding formulas, when the formulas exist, and they provide powerful tools for some important problems, especially in physics. However, most initial value problems we encounter cannot be solved by formulas. This is particularly true when two or more variables are needed to describe the process being modeled. The tool of widest applicability is Euler's method. This isn't so different from the situation in algebra, where exact solutions given by formulas (e.g., the quadratic formula) are also relatively rare, and numerical methods play an important role. (Section 5.5 presents the Newton–Raphson method for solving algebraic equations by successive approximation.) In most cases that will interest us, there are simply no formulas to be found—the limitation lies in the mathematics, not the mathematicians.

Special methods give formulas; general methods are numerical

Exercises

In exercises 1–4, verify that the given formula is a *solution* to the initial value problem.

1. *Powers of y.*
   a) $y' = y^2$, $y(0) = 5$: $y(t) = 1/(\frac{1}{5} - t)$.

   b) $y' = y^3$, $y(0) = 5$: $y(t) = 1/\sqrt{\frac{1}{25} - 2t}$.

   c) $y' = y^4$, $y(0) = 5$: $y(t) = 1/\sqrt[3]{\frac{1}{125} - 3t}$.

   d) Write a general formula for the solution of the initial value problem $y' = y^n$, $y(0) = 5$, for any integer $n > 1$.

   e) Write a general formula for the solution of the initial value problem $y' = y^n$, $y(0) = C$, for any integer $n > 1$ and any constant $C \geq 0$.

2. *Powers of t.*
   a) $y' = t^2$, $y(0) = 5$: $y(t) = \frac{1}{3}t^3 + 5$.
   b) $y' = t^3$, $y(0) = 5$: $y(t) = \frac{1}{4}t^4 + 5$.
   c) $y' = t^4$, $y(0) = 5$: $y(t) = \frac{1}{5}t^5 + 5$.
   d) Write a general formula for the solution of the initial value problem $y' = t^n$, $y(0) = 5$ for any integer $n > 1$.
   e) Write a general formula for the solution of the initial value problem $y' = t^n$, $y(0) = C$ for any integer $n > 1$ and any constant C.

3. *Sines and cosines.*
   a) $x' = -y$, $y' = x$, $x(0) = 1$, $y(0) = 0$: $x(t) = \cos t$, $y(t) = \sin t$

b) $x' = -y$, $y' = x$, $x(0) = 0$, $y(0) = 1$: $x(t) = \cos(t + \pi/2)$,
$y(t) = \sin(t + \pi/2)$

4. *Exponential functions.*
   a) $y' = 2.3\,y$, $y(0) = 5$: $y(t) = 5 \cdot 10^t$.
   b) $y' = 2.3\,y$, $y(0) = C$: $y(t) = C \cdot 10^t$.
   c) $y' = -2.3\,y$, $y(0) = 5$: $y(t) = 5 \cdot 10^{-t}$.
   d) $y' = 4.6\,ty$, $y(0) = 5$: $y(t) = 5 \cdot 10^{t^2}$.

5. *Initial conditions.*
   ∗a) Choose C so that $y(t) = \sqrt{t + C}$ is a solution to the initial value
   problem

$$y' = \frac{1}{2y} \quad y(3) = 17.$$

   b) Choose C so that $y(t) = -1/(t + C)$ is a solution to the initial
   value problem

$$y' = y^2 \quad y(0) = -5.$$

   c) Choose C so that $y(t) = -1/(t + C)$ is a solution to the initial
   value problem

$$y' = y^2 \quad y(2) = 3.$$

World Population Growth with Parameters

6. a) Using a graphing utility or a calculator, show that the derivative
   of $P(t) = (1.0202)^t$ at the origin is approximately .02: $P'(0) \approx .02$.
   Since quick convergence is desirable, use

$$\frac{\Delta P}{\Delta t} = \frac{P(0 + h) - P(0 - h)}{2h} = \frac{(1.0202)^h - (1.02020)^{-h}}{2h}$$

   b) By using more decimal places to get higher precision, show that
   $P(t) = (1.0202013)^t$ satisfies $P'(0) = .02$ even more exactly.

7. a) Show that the function $y = 2^{t/.69}$ satisfies the differential equation
   $dy/dt = y$. Use the chain rule: $y = 2^u$, $u = t/.69$. (Recall that
   $k_2 = .69\ldots$.)
   b) Show that the function $y = 2^{kt/.69}$ satisfies the differential equation
   $dy/dt = k\,y$.
   c) Show that the function $P(t) = A \cdot 2^{kt/.69}$ is a solution to the initial
   value problem

$$\frac{dP}{dt} = kP \qquad P(0) = A.$$

Note that this describes a population that grows at the constant per capita rate k from an initial size of A.

8. a) Show that the function $y = 10^{t/2.3}$ satisfies the differential equation $dy/dt = y$. Use the chain rule: $y = 10^u$, $u = t/2.3$. (Recall that $k_{10} = 2.3\ldots$.)

b) Show that the function $y = 10^{kt/2.3}$ satisfies the differential equation $dy/dt = ky$.

c) Show that the function $P(t) = A \cdot 10^{kt/2.3}$ is a solution to the initial value problem

$$\frac{dP}{dt} = kP \qquad P(0) = A.$$

This formula provides an alternative way to describe a population that grows at the constant per capita rate k from an initial size of A.

9.*a) The formula $P(t) = 5 \cdot 2^{kt/.69}$ describes how an initial population of 5 billion will grow at a constant per capita rate of k persons per year per person. Use this formula to determine how many years t it will take for the population to double, to 10 billion persons.

b) Suppose the initial population is A billion, instead of 5 billion. What is the doubling time then?

c) Suppose the initial population is 5 billion, and the per capita growth rate is .02, but that value is certain only with an error of Δk. How much uncertainty is there in the doubling time that you found in part (a)?

Newton's Law of Cooling

There are formulas that describe how a body cools, or heats up, to match the temperature of its surroundings. See the exercises on Newton's law of cooling in section 4.1. Consider first the model

$$\frac{dT}{dt} = -.1(T - 20) \qquad T(0) = 90,$$

introduced on page 174 to describe how a cup of coffee cools.

10. Show that the function $y = 2^{-.1t/.69}$ is a solution to the differential equation $dy/dt = -.1y$. (Use the chain rule: $y = 2^u$, $u = -.1t/.69$.)

11. a) Show that the function

$$T = 70 \cdot 2^{-.1t/.69} + 20$$

is a solution to the initial value problem $dT/dt = -.1(T - 20)$, $T(0) = 90$. This is the temperature T of a cup of coffee, initially at

90°C, after t minutes have passed in a room whose temperature is 20°C.

b) Use the formula in part (a) to find the temperature of the coffee after 20 minutes. Compare this result with the value you found in exercise 12b, page 174.

c) Use the formula in part (a) to determine how many minutes it takes for the coffee to cool to 30°C. In doing the calculations you will find it helpful to know that $1/7 = 2^{-2.8}$. Compare this result with the value you found in exercise 12c, page 174.

12. a) A cold drink is initially at $Q = 36°F$ when the air temperature is 90°F. If the temperature changes according to the differential equation

$$\frac{dQ}{dt} = -.2(Q - 90)°F \text{ per minute,}$$

show that the function $Q(t) = 90 - 54 \cdot 2^{-.2t/.69}$ describes the temperature after t minutes.

b) Use the formula to find the temperature of the drink after 5 minutes and after 10 minutes. Compare your results with the values you found in exercise 13, page 174.

13. Find a formula for a function that solves the initial value problem

$$\frac{dQ}{dt} = -k(Q - A) \qquad Q(0) = B.$$

A Leaking Tank

The rate at which water leaks from a small hole at the bottom of a tank is proportional to the square root of the height of the water surface above the bottom of the tank. Consider a cylindrical tank that is 10 feet tall and stands on one of its circular ends, which is 3 feet in diameter. Suppose the tank is currently half full, and is leaking at a rate of 2 cubic feet per hour.

14. a) Let $V(t)$ be the volume of water in the tank t hours from now. Explain why the leakage rate can be written as the differential equation

$$V'(t) = -k \sqrt{V(t)},$$

for some positive constant k. (The issue to deal with is this: Why is it permissible to use the square root of the *volume* here, when the rate is known to depend on the square root of the *height?*)

*b)** Determine the value of k and explain *why* k has this value.

15. a) How much water leaks out of the tank in 12 hours? In 24 hours? Use Euler's method, and compute successive approximations until your results stabilize.
   b) How many hours does it take for the tank to empty?

16. a) Use the differentiation rules to show that any function of the form

$$V(t) = \begin{cases} \dfrac{k^2}{4}(C - t)^2 & \text{if } 0 \le t \le C \\ 0 & \text{if } C < t \end{cases}$$

satisfies the differential equation.
   b) For the situation we are considering, what is the value of C? According to this solution, how long does it take for the tank to empty? Compare this result with your answer using Euler's method.
   c) Sketch the graph of $V(t)$ for $0 \le t \le 2C$, taking particular care to display the value $V(0)$ in terms of k and C.

Motion

Newton created the calculus to study the motion of the planets. He said that all motion obeys certain basic laws. One law says that the velocity of an object changes only if a force acts on the body. Furthermore, the *rate* at which the velocity changes is proportional to the force. By knowing the forces that act on a body we can construct—and then solve—a differential equation for the velocity.

Falling bodies—with gravity. A body falling through the air starts up slowly but picks up speed as it falls. Its velocity is thus changing, so there must be a force acting. We call the force that pulls objects to the earth **gravity**. Near the earth, the strength of gravity is essentially constant.

Suppose an object is x meters above the surface of the earth after t seconds have passed. Then, by definition, its velocity is

$$v = \frac{dx}{dt} \text{ meters/second.}$$

According to Newton's laws of motion, the force of gravity causes the velocity to change, and we can write

$$\frac{dv}{dt} = -g.$$

Here g is a constant whose numerical value is about 9.8 meters/second per second. Since x and v are positive when measured *upwards*, but gravity acts *downwards*, a minus sign is needed in the equation for dv/dt. (The derivative of velocity is commonly called **acceleration**, and g is called the **acceleration due to gravity**.)

17. Verify that $v(t) = -gt + v_0$ is a solution to the differential equation $dv/dt = -g$ with initial velocity v_0.

18. Since $dx/dt = v$, and since $v(t) = -gt + v_0$, the *position* x of the body satisfies the differential equation

$$\frac{dx}{dt} = -gt + v_0 \text{ meters/second.}$$

Find a formula for $x(t)$ that solves this differential equation. This function describes how a body moves under the force of gravity.

19. Suppose the initial position of the body is x_0, so that the position x is a solution to the initial value problem

$$\frac{dx}{dt} = -gt + v_0 \qquad x(0) = x_0 \text{ meters.}$$

Find a formula for $x(t)$.

20. a) Suppose a body is held motionless 200 meters above the ground, and then released. What values do x_0 and v_0 have? What is the formula for the motion of this body as it falls to the ground?
b) How far has the body fallen in 1 second? In 2 seconds?
c) How long does it take for the body to reach the ground?

Falling bodies—with gravity and air resistance. As a body falls, air pushes against it. Air resistance is thus another force acting on a falling body. Since air resistance is slight when an object moves slowly but increases as the object speeds up, the simplest model we can make is that the force of air resistance is proportional to the velocity: force $= -bv$ (reality turns out to be somewhat more complicated than this). The multiplier b is positive, and the minus sign tells us that the direction of the force is always opposite the velocity.

The forces of gravity and air resistance combine to change the velocity:

$$\frac{dv}{dt} = -g - bv \text{ meters/second per second.}$$

21. Show that

$$v(t) = \frac{g}{b}(2^{-bt/.69} - 1) \text{ meters/second}$$

is a solution to this differential equation that also satisfies the initial condition $v(0) = 0$ meters/second.

22. **a)** Show that the position $x(t)$ of a body that falls against air resistance from an initial height of x_0 meters is given by the formula

$$x(t) = x_0 - \frac{g}{b}t - \frac{g}{b^2}(2^{-bt/.69} - 1) \text{ meters.}$$

b) Suppose the coefficient of air resistance is $b = .2$ per second. If a body is held motionless 200 meters above the ground, and then released, how far will it fall in 1 second? In 2 seconds? Compare these values with those you obtained assuming there was no air resistance.

c) How long does it take for the body to reach the ground? (Use a computer graphing package to get this answer.) Compare this value with the one you obtained assuming there was no air resistance. How much does air resistance add to the time?

23. **a)** According to the equation $dv/dt = -g - bv$, there is a velocity v_T at which the force of air resistance exactly balances the force of gravity, and the velocity doesn't change. What is v_T, expressed as a function of g and b? Note: v_T is called the **terminal velocity** of the body. Once the body reaches its terminal velocity, it continues to fall at that velocity.

b) What is the terminal velocity of the body in the previous exercise?

The oscillating spring. Springs can smooth out life's little irregularities (as in the suspension of a car) or amplify and measure them (as in earthquake detection devices). Suppose a spring that hangs from a hook has a weight at its end. Let the weight come to rest. Then, when the weight moves, let x denote the position of the weight above the rest position. (If x is negative, this means the weight is below the rest position.) If you pull down on the weight, the spring pulls it back up. If you push up on the weight, the spring (and gravity) push it back down. This push is the **spring force**.

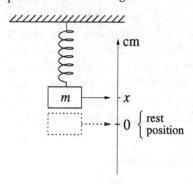

The simplest assumption is that the spring force is proportional to the amount x that the spring has been stretched: force $= -c^2x$. The constant c^2 is customarily written as a square to emphasize that it is positive. The minus sign tells us the force pushes down if $x > 0$ (so the weight is above the rest position), but it pushes up if $x < 0$.

If $v = dx/dt$ is the velocity of the weight, then Newton's law of motion says

$$\frac{dv}{dt} = -c^2x.$$

Suppose we move the weight to the point $x = a$ on the scale, hold it motionless momentarily, and then release it at time $t = 0$. This determines the initial value problem

$$x' = v \qquad x(0) = a$$
$$v' = -c^2 x \qquad v(0) = 0.$$

24. a) Show that

$$x(t) = a\cos(ct) \qquad v(t) = -ac\sin(ct)$$

is a solution to the initial value problem.

    b) What range of values does x take on? That is, how far does the weight move from its rest position?

25. a) Use a graphing utility to compare the graphs of $y = \cos(x)$, $y = \cos(2x)$, $y = \cos(3x)$, and $y = \cos(.5x)$. Based on your observations, explain how the value of c affects the nature of the motion $x(t) = a\cos(ct)$ for a fixed value of a.

    b) How long does it take the weight to complete one cycle (from $x = a$ back to $x = a$) when $c = 1$? The motion of the weight is said to be **periodic**, and the time it takes to complete one cycle is called its **period**.

    c) What is the period of the motion when $c = 2$? When $c = 3$? Does the period depend on the initial position a?

    d) Write a formula that expresses the period of the motion in terms of the parameters a and c.

26. a) The parameter c depends on two things, the mass m of the weight and the stiffness k of the spring:

$$c = \sqrt{\frac{k}{m}}.$$

Write a formula that expresses the period of the motion of the weight in terms of m and k.

    b) Suppose you double the weight on the spring. Does that *increase* or *decrease* the period of the motion? Does your answer agree with your intuitions?

    c) Suppose you put the first weight on a second spring that is twice as stiff as the first (i.e., double the value of k). Does that *increase* or *decrease* the period of the motion? Does your answer agree with your intuitions?

    d) When you calculate the period of the motion using your formula from part (a), suppose you know the actual value of the mass only to within 5%. How accurately do you know the period—as a percentage of the calculated value?

➤ 4.3 THE EXPONENTIAL FUNCTION

The Equation $y' = ky$

As we have seen, initial value problems *define* functions—as their solutions. They therefore provide us with a vast, if somewhat bewildering, array of new functions. Fortunately, a few differential equations—in fact, the very simplest—arise over and over again in an astonishing variety of contexts. The functions they define are among the most important in mathematics.

A simple and natural model of growth and decay

One of the simplest differential equations is $dy/dt = ky$, where k is a constant. It is also one of the most useful. We used it in chapter 1 to model the populations of Poland and Afghanistan, as well as bacterial growth and radioactive decay. In this chapter, it was our initial model of a rabbit population and one of our models of the world population. Later, we will use it to describe how money accrues interest in a bank and how radiation penetrates solid objects.

In this section we will look at the solutions to differential equations of this form from two different vantage points. On the one hand, we already have named functions which solve such equations—the exponential functions. On the other hand, the fact that Euler's method produces the same functions will allow us to prove properties of such functions and to compute their values effectively.

Exponential solutions— variable base . . .

In chapter 3 we established that the solutions to $dy/dt = ky$ are exponential functions. Specifically, for each base b, the exponential function $y = b^t$ was a solution to $dy/dt = k_b \cdot b^t = k_b \cdot y$, where k_b was the slope of the graph of $y = b^t$ at the origin. In this approach, if the constant k changes, we must change the base b so that $k_b = k$.

. . . and fixed base

Exercise 7 in the last section (page 193) opened up a new possibility: for the *fixed* base 2, the function

$$y = 2^{kt/.6931...}$$

was a solution to the differential equation $dy/dt = kt$, no matter what value k took. There was nothing special about the base 2, of course. In the next exercise, we saw that the functions

$$y = 10^{kt/2.3025...}$$

would serve equally well as solutions.

In fact, we can show that, for any base b, the functions

$$y = b^{kt/k_b}$$

are also solutions to $dy/dt = ky$. Construct the chain

$$y = b^u \qquad \text{where} \qquad u = kt/k_b.$$

Then $dy/du = k_b \cdot b^u = k_b \cdot y$, while $du/dt = k/k_b$. Thus, by the chain rule we have

$$\frac{dy}{dt} = \frac{dy}{du} \cdot \frac{du}{dt} = k_b y \cdot \frac{k}{k_b} = ky.$$

If we express solutions to $dy/dt = ky$ by exponential functions with a *fixed* base, it is easy to alter the solution if the growth constant k changes. We just change the value of k in the exponent of b^{kt/k_b}. Let's see how this works when $b = 2$ and $b = 10$:

Advantages of a fixed base

Differential equation	Solution base 2	Solution base 10
$\dfrac{dy}{dt} = .16y$	$2^{.231t}$	$10^{.069t}$
$\dfrac{dy}{dt} = .18y$	$2^{.260t}$	$10^{.078t}$

Notice that the growth constant k gets "swallowed up" in the exponent of the solution when k has a specific numerical value. The number that appears in the exponent is k divided by $k_2 = .6931\ldots$ (when the base is 2) and by $k_{10} = 2.3025\ldots$ (when the base is 10).

The most vivid solution to $dy/dt = ky$ would use the base b for which $k_b = 1$ *exactly*. There is such a base, and it is always denoted e. (We will determine the value of e in a moment.) Since $k_e = 1$, k would stand out in the exponent:

The base e

Differential equation	Solution base e
$\dfrac{dy}{dt} = .16y$	$e^{.16t}$
$\dfrac{dy}{dt} = .18y$	$e^{.18t}$

The simplicity and clarity of this expression have led to the universal adoption of the base e for describing exponential growth and decay—that is, for describing solutions to $dy/dt = ky$.

The use of the symbol e to denote the base dates back to a paper that Euler wrote at age 21, entitled Meditatio in experimenta explosione tormentorum nuper instituta *(Meditation upon recent experiments on the firing of cannons), where the symbol e was used sixteen times. It is now in universal use. The number e is, like π, one of the most important and ubiquitous numbers in mathematics.*

By design, $y = e^t$ is a solution to the differential equation $dy/dt = y$. In particular, the slope of the graph of $y = e^t$ at the origin is exactly 1. As we have just seen, the function $y = e^{kt}$ is a solution to the differential equations $dy/dt = ky$ whose growth constant is k. Finally:

$y = C \cdot e^{kt}$ is the solution to the initial value problem

$$\frac{dy}{dt} = ky \qquad y(0) = C.$$

We can check this quickly. The initial condition is satisfied because $e^0 = 1$, so $y(0) = C \cdot e^{k \cdot 0} = C \cdot 1 = C$. The differential equation is satisfied because

$$(C \cdot e^{kt})' = C \cdot (e^{kt})' = C \cdot k e^{kt} = ky.$$

We used the differentiation rule for a constant multiple of a function, and we used the fact that the derivative of e^{kt} was already established to be $k e^{kt}$.

The Number e

The number e is determined by the property that $k_e = 1$. Since this number is the slope of the graph of $y = e^t$ at the origin, one way to find e is with a computer microscope. Pick an approximation E for e and graph $y = E^t$. Zoom in the graph at the origin and measure the slope. If the slope is more than 1, choose a smaller approximation; if the slope is less than 1, choose a larger value. Repeat this process, narrowing down the value of e until you know its value to as many decimal places as you wish.

*Finding e with a
computer microscope*

We already know $E = 2$ is too small, because the slope of $y = 2^t$ at the origin is .69. Likewise, $E = 3$ is too large, because the slope of $y = 3^t$ at the origin is 1.09. Thus $2 < e < 3$, and is closer to 3 than to 2. At the next stage we learn that 2.7 is too small (slope = .9933) but 2.8 is too large (slope = 1.0296). Thus, at least we know $e = 2.7\ldots$. Several stages later we would learn $e = 2.71828\ldots$.

*Under successive
magnifications:
$e = 2.71828\ldots$*

While the method just described for finding the value of e works, it is somewhat ponderous. We can take a very different approach to finding the numerical value of e by using the fact that e is defined by an initial value problem. Here is the idea: e is the value of the function e^t when $t = 1$, and $y(t) = e^t$ is the solution to the initial value problem

*Finding e by Euler's
method*

$$y' = y \qquad y(0) = 1.$$

We can then find $e = y(1)$ in the usual way by solving this initial value problem using Euler's method. Due to some convenient algebraic simplifications, this approach yields powerful insights about the nature of e.

Suppose we take n steps to go from $t = 0$ to $t = 1$. Then the step size is $\Delta t = 1/n$. The following table shows the calculations:

Finding $y(1)$ by Euler's method when $y' = y$ and $y(0) = 1$

t	y	$y' = y$	$\Delta y = y' \cdot \Delta t$
0	1	1	$1 \cdot 1/n$
$1/n$	$1 + 1/n$	$1 + 1/n$	$(1 + 1/n) \cdot 1/n$
$2/n$	$(1 + 1/n)^2$	$(1 + 1/n)^2$	$(1 + 1/n)^2 \cdot 1/n$
$3/n$	$(1 + 1/n)^3$	$(1 + 1/n)^3$	$(1 + 1/n)^3 \cdot 1/n$
$\vdots$	$\vdots$	$\vdots$	$\vdots$
n/n	$(1 + 1/n)^n$		

The entries in the y column need to be explained. The first two should be clear: $y(0)$ is the initial value 1, and $y(1/n) = y(0) + \Delta y = 1 + 1/n$. To get from any entry to the next we must do the following:

$$\begin{aligned} \text{new } y &= \text{current } y + \Delta y \\ &= \text{current } y + y' \cdot \Delta t \\ &= \text{current } y + \text{current } y \cdot \Delta t \\ &= \text{current } y \cdot (1 + \Delta t) \\ &= \text{current } y \cdot (1 + 1/n) \end{aligned}$$

The new y is the current y multiplied by $(1 + 1/n)$. Since the second y is itself $(1 + 1/n)$, the third will be $(1 + 1/n)^2$, the fourth will be $(1 + 1/n)^3$, and so on.

Euler's method with n steps therefore gives us the following estimate for $e = y(1) = y(n/n)$:

$$e \approx (1 + 1/n)^n.$$

We can calculate these numbers on a computer. In the following table we give values of $(1 + 1/n)^n$ for increasing values of n. By the time $n = 2^{40}$ (about 10^{12}), eleven digits of e have stabilized.

n	$(1 + 1/n)^n$
2^0	2.0
2^4	2.638
2^8	2.712 992
2^{12}	2.717 950 081
2^{16}	2.718 261 089 905
2^{20}	2.718 280 532 282
2^{24}	2.718 281 747 448
2^{28}	2.718 281 823 396
2^{32}	2.718 281 828 142
2^{36}	2.718 281 828 439
2^{40}	2.718 281 828 458

Expressing *e* as a limit

The *true* value of *e* is the limit of these approximations as we take *n* arbitrarily large:

$$e = \lim_{n \to \infty} (1 + 1/n)^n = 2.71828182845904\ldots$$

We can generalize the preceding to get an expression for e^T for any value of T. In exactly the same way as we did above, divide the interval from 0 to T into n pieces, each of width $\Delta t = T/n$. Starting from $t = 0$ and $y(0) = 1$ and applying Euler's method, we find, using the same algebraic simplifications, that after n steps the value for t will be T and y will be $(1 + T/n)^n$. Since these approximations approach the true value of the function as $n \to \infty$, we have that

$$e^T = \lim_{n \to \infty} (1 + T/n)^n \quad \text{for any value of } T.$$

Differential Equations Define Functions

There is an important point underlying the operations we just performed having to do with the question of **computability**. While it may be appalling to think about doing it by hand, there is nothing conceptually difficult about evaluating an expression like $(1 + 1/1000)^{1000}$—all we need are ordinary addition, division, and multiplication. In fact, for any differential equation, Euler's method generates a solution using only ordinary arithmetic.

Euler's method uses only arithmetic

By contrast, think for a moment about the earlier method for evaluating *e* by evaluating expressions like $(2.718^{.0001} - 2.718^{-.0001})/.0002$ and seeing whether we get a value bigger than or less than 1. While a calculator or a computer will readily give us a value, how does it "know" what $2.718^{.0001}$ is? The fact is, it doesn't have a built-in exponentiator which lets it know immediately what the value of this expression is any more than we do. A computer—like humans—can essentially only add, subtract, and multiply. Any other operation has to be reduced to these operations somehow. Thus when we use a computer to evaluate something like $2.718^{.0001}$, we actually trigger a fairly elaborate program having little directly to do with raising numbers to powers, which produces an approximation to the desired number. It turns out that if you use the x^y key on your calculator to evaluate 2^5, your calculator doesn't come up with the answer by multiplying 2 by itself 5 times, but uses this more complicated program.

Defining a function is not always the same as being able to evaluate it

There is often a large gap between naming and defining a function, and being able to compute values for it to four or five decimals. Think about the trigonometric functions for a moment. You have probably seen several definitions of the cosine function by now—as the ratio of the adjacent side over the hypotenuse of a right triangle, or as the x-coordinate of a point moving around a circle of radius 1. Yet neither of these definitions would help you calculate $\cos(2)$ to five decimals. It turns out that most methods for evaluating functions are based on the way the derivatives of the functions behave. While we will have much more to say about this in chapter 10, Euler's method is a good first example of this.

Returning to exponents, think what would be involved in evaluating $2^{\sqrt{3}}$ using the precalculus concept of exponents. We might first get a series of rational approximations to $\sqrt{3} = 1.73205081\ldots$: $17/10 = 1.7$, $173/100 = 1.73$, $433/250 = 1.732$, and so on. We would then calculate

$$2^{17/10} = (\sqrt[10]{2})^{17}$$

$$2^{173/100} = (\sqrt[100]{2})^{173}$$

$$2^{433/250} = (\sqrt[250]{2})^{433}$$

$$\vdots \qquad \vdots$$

Even evaluating the first of these approximations would involve finding the 10th root of 2 and raising it to the 17th power, which would not be easy. We would continue with these approximations until the desired number of digits remained fixed.

By contrast, evaluating $e^{\sqrt{3}}$ by Euler's method is very straightforward. As we saw above, it reduces to evaluating $(1 + 1.73205\ldots/n)^n$ for increasing values of n until the desired number of digits remains fixed. Moreover, this same process works just as well for any kind of exponent—positive or negative, rational or irrational.

In fact, *all* the properties of the exponential function follow from the fact that it is the solution to its initial value problem, so we could have made this the definition in the first place. This would have given us the benefit of coherence (not having to distinguish among different kinds of exponents) and direct computability. It would also directly reflect the primary reason the exponential function is important, namely, that its rate of change is proportional to its value. Since the process of deducing the properties of a function from its defining equation will be important later on, and since it is a good exercise in some of the theoretical ideas we've been developing, let's see how this works.

We will assume *nothing* about the function $y = e^t$. Instead, we begin simply with the observation that each initial value problem defines a function—its solution. Therefore, the specific problem

$$y' = y \qquad y(0) = 1$$

defines a function; we call it $y = \exp(t)$. At the outset, all we know about the function $\exp(t)$ is that

Defining exp(t)

$$\exp'(t) = \exp(t) \qquad \exp(0) = 1.$$

As before, we can use Euler's method to evaluate $\exp(1)$, which we will call e. From these facts alone we want to *deduce* that $\exp(t) = e^t$ for all values of

t. We will actually show this only for all *rational* values of t, since there is, as we've seen, a bit of hand-waving about what it means to raise a number to an irrational power. The following theorem is the key to establishing this result.

THEOREM 1.   *For any real numbers r and s,*

$$\exp(r + s) = \exp(r) \cdot \exp(s).$$

We will prove this result shortly, but let's see what we can deduce from it. First off, note that

$$\exp(2) = \exp(1 + 1) = \exp(1) \cdot \exp(1) = (\exp(1))^2 = e^2.$$

Notice that we invoked Theorem 1 to equate $\exp(1 + 1)$ with $\exp(1) \cdot \exp(1)$. In a similar way,

$$\exp(3) = \exp(2 + 1) = \exp(2) \cdot \exp(1) = (\exp(1))^2 \cdot \exp(1) = (\exp(1))^3 = e^3.$$

Repeating this argument for any positive integer m, we obtain:

COROLLARY 1.   *For any positive integer m,*

$$\exp(m) = (\exp(1))^m = e^m.$$

We can also express $\exp(t)$ in terms of e when t is a *negative* integer.

Negative integers

We begin with another consequence of Theorem 1:

$$1 = \exp(0) = \exp(-1 + 1) = \exp(-1) \cdot \exp(1).$$

This says $e = \exp(1)$ is the reciprocal of $\exp(-1)$:

$$\exp(-1) = (\exp(1))^{-1} = e^{-1}.$$

Since $-2 = -1 - 1$, $-3 = -2 - 1$, and so forth, we can eventually show that:

COROLLARY 2.   *For any negative integer $-m$,*

$$\exp(-m) = \exp(-1 - 1 - \cdots - 1) = (\exp(-1))^m = (\exp(1))^{-m} = e^{-m}.$$

Rational numbers

We can even do the same thing with fractions. Here's how to deal with $\exp(1/3)$, for example:

$$\begin{aligned}
\exp(1) &= \exp(1/3 + 1/3 + 1/3) \\
&= \exp(1/3) \cdot \exp(1/3) \cdot \exp(1/3) \\
&= (\exp(1/3))^3,
\end{aligned}$$

so $\exp(1/3)$ is the cube root of $\exp(1)$:

$$\exp(1/3) = (\exp(1))^{1/3} = e^{1/3}.$$

A similar argument will show that:

COROLLARY 3.   *For any positive integer n,*

$$\exp(1/n) = e^{1/n}.$$

Finally, we can deal with any rational number m/n:

$$\exp(m/n) = (\exp(1/n))^{m} = (e^{1/n})^{m} = e^{m/n}.$$

This leads to:

THEOREM 2.   *For any rational number r, $\exp(r) = (\exp(1))^{r} = e^{r}$.*

In other words, Theorem 1 implies that the function $\exp(t)$ is the same function as the exponential function e^{t}—at least when t is a rational number m/n, as claimed. We could now prove that $\exp(t) = e^{t}$ when t is an irrational number, which would require being clearer about what it means to raise a number to an irrational power than most high school texts are.

We adopt a more attractive option. Since $\exp(t)$ and e^{t} agree at rational values of t, and since $\exp(t)$ is well-defined for *all* values of t—including irrational numbers—we *define* e^{t} for irrational values of t by setting it equal to $\exp(t)$.

Proof of Theorem 1:   The proof uses the *Existence and Uniqueness Principle* for differential equations we articulated earlier (page 181): If two functions satisfy the same differential equation and satisfy the same initial conditions, then they have to be the same function.

The proof of Theorem 1

Theorem 1 involves two fixed real numbers, r and s. We fix one of them, say r, to define two new functions of t:

$$P(t) = \exp(r + t) \qquad Q(t) = \exp(r) \cdot \exp(t).$$

We shall show that *both* of these functions are solutions to the same initial value problem:

$$\frac{dy}{dt} = y \qquad y(0) = \exp(r).$$

[Remember, $\exp(r)$ is a constant, because r is fixed.]

If we show this, it will then follow that $P(t)$ and $Q(t)$ must be the same function.

Since

$$P(0) = \exp(r + 0) = \exp(r)$$
$$Q(0) = \exp(r) \cdot \exp(0) = \exp(r) \cdot 1 = \exp(r),$$

$P(t)$ and $Q(t)$ both satisfy the initial condition $y(0) = \exp(r)$.

Next we show that they both satisfy the differential equation $y' = y$:

$$Q'(t) = (\exp(r) \cdot \exp(t))' = \exp(r) \cdot (\exp(t))' = \exp(r) \cdot \exp(t) = Q(t),$$

so $Q(t)$ is a solution.

To differentiate $P(t)$ we construct a chain:

$$P = \exp(u) \qquad \text{where} \qquad u = r + t.$$

Then $dP/du = \exp(u)$ and $du/dt = 1$, so

$$P'(t) = \frac{dP}{du} \cdot \frac{du}{dt} = \exp(u) \cdot 1 = P(t) \cdot 1 = P(t),$$

so $P(t)$ is also a solution.

Therefore $P(t)$ and $Q(t)$ must be the same function. It follows then that $P(t) = Q(t)$ for all values of t, in particular for $t = s$. But this means that

$$\exp(r + s) = P(s) = Q(s) = \exp(r) \cdot \exp(s),$$

which is exactly the statement of Theorem 1, and so completes the proof.

Now that we have established $\exp(x) = e^x$, we will call $\exp(x)$ **the exponential function** and we will use the forms e^x and $\exp(x)$ interchangeably. The following theorem summarizes several more properties of the exponential function.

THEOREM 3.   *For any real numbers r and s,*

$$\exp(s) > 0$$

$$\exp(-s) = \frac{1}{\exp(s)}$$

$$\exp(r - s) = \frac{\exp(r)}{\exp(s)}$$

$$\exp(rs) = (\exp(r))^s$$

$$= (\exp(s))^r.$$

To make the statements in this theorem seem more natural, you should stop and translate them from $\exp(x)$ to e^x. Proofs will be covered in the exercises.

Exponential Growth

The function $\exp(x) = e^x$, like polynomials and the sine and cosine functions, is defined for all real numbers. Nevertheless, it behaves in a way that is quite different from any of those functions.

One difference occurs when x is large, either positive or negative. The sine function and the cosine function stay bounded between $+1$ and -1 over their entire domain. By contrast, every polynomial "blows up" as $x \to \pm\infty$. In this regard, the exponential function is a hybrid. As $x \to -\infty$, $\exp(x) \to 0$. As $x \to +\infty$, however, $\exp(x) \to +\infty$.

Let's look more closely at what happens to power functions x^n and the exponential function e^x as $x \to \infty$. Both kinds of functions "blow up" but they do so at quite different rates, as we shall see. Before we compare power and exponential functions directly, let's compare one power of x with another— say x^2 with x^5. As $x \to \infty$, both x^2 and x^5 get very large. However, x^2 is only a small fraction of the size of x^5, and that fraction becomes smaller the larger x becomes. The following table demonstrates this. Even though x^2 grows enormous, we interpret the fact that $x^2/x^5 \to 0$ to mean that x^2 *grows more slowly than* x^5.

How fast do x^n and e^x become infinite?

x	x^2	x^5	x^2/x^5
10	10^2	10^5	10^{-3}
100	10^4	10^{10}	10^{-6}
1000	10^6	10^{15}	10^{-9}
$\downarrow$	$\downarrow$	$\downarrow$	$\downarrow$
∞	∞	∞	0

It should be clear to you that we can compare *any* two powers of x this way. We will find that x^p grows more slowly than x^q if, and only if, $p < q$. To prove this, we must see what happens to the ratio x^p/x^q, as $x \to +\infty$. We can write $x^p/x^q = 1/x^{q-p}$, and the exponent $q - p$ that appears here is positive, because $q > p$. Consequently, as $x \to \infty$, $x^{q-p} \to \infty$ as well, and therefore $1/x^{q-p} \to 0$. This completes the proof.

x^p grows more slowly than x^q if $p < q$

x	x^{50}	e^x	x^{50}/e^x
100	$\sim 10^{100}$	$\sim 10^{43}$	$\sim 10^{56}$
200	10^{115}	10^{86}	10^{28}
300	10^{123}	10^{130}	10^{-7}
400	10^{130}	10^{173}	10^{-44}
500	10^{134}	10^{217}	10^{-83}
$\downarrow$	$\downarrow$	$\downarrow$	$\downarrow$
∞	∞	∞	0

How does e^x compare to x^p? To make it tough on e^x, let's compare it to x^{50}. We know already that x^{50} grows faster than any lower power of x. The table above compares x^{50} to e^x. However, the numbers involved are so large that the table shows only their *order of magnitude*—that is, the number of digits they contain. At the start, x^{50} is *much* larger than e^x. However, by the time $x = 500$, the ratio x^{50}/e^x is so small its first 82 decimal places are zero!

e^x grows more rapidly than any power of x

So x^{50} grows more slowly than e^x, and so does any lower power of x. Perhaps a higher power of x would do better. It does, but ultimately the ratio $x^p/e^x \to 0$, no matter how large the power p is. We don't yet have all the tools needed to prove this, but we will after we introduce the logarithm function in the next section.

The speed of exponential growth has had an impact in computer science. In many cases, the number of operations needed to calculate a particular quantity is a power of the number of digits of precision required in the answer. Sometimes, though, the number of operations is an *exponential* function of the number of digits. When that happens, the number of operations can quickly exceed the capacity of the computer. In this way, some problems that can be solved by an algorithm that is straightforward in theoretical terms are intractable in practical terms.

Exercises

The Exponential Functions b^t

1. Use a graphing utility or a calculator to approximate the slopes of the following functions at the origin and show:
   a) If $f(t) = (2.71)^t$, then $f'(0) < 1$.
   b) If $g(t) = (2.72)^t$, then $g'(0) > 1$.
   c) Use parts (a) and (b) to explain why $2.71 < e < 2.72$.

2. a) In the same way find the value of the parameter k_b for the bases $b = .5, .75$, and $.9$ accurate to three decimal places.
   b) What is the shape of the graph of $y = b^t$ when $0 < b < 1$? What does that imply about the *sign* of k_b for $0 < b < 1$? Explain your reasoning.

Differentiating Exponential Functions

3. Differentiate the following functions.
   a) $7e^{3x}$
   b) Ce^{kx}, where C and k are constants.
   c) $1.5e^t$
   *d) $1.5e^{2t}$
   e) $2e^{3x} - 3e^{2x}$
   f) $e^{\cos t}$

4. Find partial derivatives of the following functions.
   *a) e^{xy}
   b) $3x^2e^{2y}$
   c) $e^u \sin v$
   d) $e^{u \sin(v)}$

Powers of e

5. Simplify the following and rewrite as powers of e. For each, explain your work, citing any theorems you use.
   a) $\exp(2x + 3)$
   b) $(\exp(x))^2$
   c) $\exp(17x)/\exp(5x)$

6. Use the second property in Theorem 3 to explain why

$$\lim_{t \to -\infty} \exp(t) = 0.$$

7. The purpose of this exercise is to prove the fourth property listed in Theorem 3: $\exp(rs) = (\exp(s))^r$, for all real numbers r and s. The idea of the proof is the same as for Theorem 1: show that two different-looking functions solve the same initial value problem, thus demonstrating that the functions must be the same. The initial value problem is

$$y' = ry \qquad y(0) = 1.$$

   a) Show that $P(t) = \exp(rt)$ solves the initial value problem. (You need to use the chain rule.)
   b) Show that $Q(t) = (\exp(t))^r$ solves the initial value problem. [Here use the chain $Q = u^r$, where $u = \exp(t)$. There is a bit of algebra involved.]
   c) From parts (a) and (b), and the fact that an initial value problem has a unique solution, it follows that $P(t) = Q(t)$, for every t. Explain how this establishes the result.

Solving $y' = ky$ Using e^{kt}

8. *Poland and Afghanistan.* Refer to problem 25 in chapter 1, section 1.2.
   a) Write out the initial value problems that summarize the information about the populations P and A given in parts (a) and (b) of problem 21.
   b) Write formulas for the solutions P and A of these initial value problems.
   *c) Use your formulas in part (b) (and a calculator) to find the population of each country in the year 2005. What were the populations in 1965?

9. *Bacterial growth.* Refer to problem 26 in section 1.2.
   a) Assuming that we begin with the colony of bacteria weighing 32 grams, write out the initial value problem that summarizes the information about the weight P of the colony.
   b) Write a formula for the solution P of this initial value problem.
   c) How much does the colony weigh after 30 minutes? After 2 hours?

10. *Radioactivity*. Refer to problem 27 in section 1.2.
    a) Assuming that when we begin the sample of radium weighs 1 gram, write out the initial value problem that summarizes the information about the weight R of the sample.
    b) How much did the sample weigh 20 years ago? How much will it weigh 200 years hence?

11. *Intensity of radiation*. As gamma rays travel through an object, their intensity I decreases with the distance x that they have traveled. This is called **absorption**. The absorption rate dI/dx is proportional to the intensity. For some materials the multiplier in this proportion is large; they are used as radiation shields.
    a) Write down a differential equation which models the intensity of gamma rays $I(x)$ as a function of distance x.
    b) Some materials, such as lead, are better shields than others, such as air. How would this difference be expressed in your differential equation?
    c) Assume the unshielded intensity of the gamma rays is I_0. Write a formula for the intensity I in terms of the distance x and verify that it gives a solution of the initial value problem.

12. In this problem you will find a solution for the initial value problem $y' = ky$ and $y(t_0) = C$. (Notice that this isn't the original initial value problem, because t_0 was 0 originally.)
    a) Explain why you may assume $y = Ae^{kt}$ for some constant A.
    b) Find A in terms of k, C, and t_0.

Solving Other Differential Equations

13. a) *Newton's law of cooling*. Verify that

$$Q(t) = 70e^{-.1t} + 20$$

is a solution to the initial value problem $Q'(t) = -.1(Q - 20)$, $Q(0) = 90$. What is the relationship between this formula and the one found in problem 11 in section 4.2?

b) Verify that

$$Q(t) = (Q_0 - A)e^{-kt} + A$$

is a solution to the the initial value problem $Q'(t) = -k(Q - A)$, $Q(0) = A$. What is the relationship between this formula and the one found in problem 11 in section 4.2?

14. In *An Essay on the Principle of Population*, written in 1798, the British economist Thomas Robert Malthus (1766–1834) argued that food supplies grow at a constant rate, while human populations naturally grow at a constant *per capita* rate. He therefore predicted that human

populations would inevitably run out of food (unless population growth was suppressed by unnatural means).

a) Write differential equations for the size P of a human population and the size F of the food supply that reflect Malthus's assumptions about growth rates.

b) Keep track of the population in millions, and measure the food supply in millions of units, where one unit of food feeds one person for one year. Malthus's data suggested to him that the food supply in Great Britain was growing at about .28 million units per year and the per capita growth rate of the population was 2.8% per year. Let $t = 0$ be the year 1798, when Malthus estimated the population of the British Isles was $P = 7$ million people. He assumed his countrymen were on average adequately nourished, so he estimated that the food supply was $F = 7$ million units of food. Using these values, write formulas for the solutions $P = P(t)$ and $F = F(t)$ of the differential equations in (a).

c) Use the formulas in (b) to calculate the amount of food and the population at 25-year intervals for 100 years. Use these values to help you sketch graphs of $P = P(t)$ and $F = F(t)$ on the same axes.

d) The per capita food supply in any year equals the ratio $F(t)/P(t)$. What happens to this ratio as t grows larger and larger? [Use your graphs in (c) to assist your explanation.] Do your results support Malthus's prediction? Explain.

15. a) *Falling bodies.* Using the base e instead of the base 2, modify the solution $v(t)$ to the initial value problem

$$\frac{dv}{dt} = -g - bv \qquad v(0) = 0$$

that appears in exercise 21 on page 197. Show that the modified expression is still a solution.

b) If an object that falls against air resistance is $x(t)$ meters above the ground after t seconds, and it started x_0 meters above the ground, then it is the solution of the initial value problem

$$\frac{dx}{dt} = v(t) \qquad x(0) = x_0,$$

where $v(t)$ is the velocity function from the previous exercise. Find a formula for $x(t)$ using the exponential function with base e. (Compare this formula with the one in exercise 22a, page 198.)

c) Suppose the coefficient of air resistance is .2 per second. If a body is held motionless 200 meters above the ground, and then released, how far will it fall in 1 second? In 2 seconds? Use your formula from part (b). Compare these values with those you obtained in exercise 22b, page 198.

Interest Rates

Bank advertisements sometimes look like this:

Civic Bank and Trust

- Annual rate of interest 6%.
- Compounded monthly.
- Effective rate of interest 6.17%.

The first item seems very straightforward. The bank pays 6% interest per year. Thus if you deposit $100.00 for one year, at the end of the year you would expect to have $106.00. Mathematically this is the simplest way to compute interest; each year add 6% to the account. The biggest problem with this is that people often make deposits for odd fractions of a year. So if interest were paid only once each year, then a depositor who withdrew her money after 11 months would receive no interest. To avoid this problem banks usually compute and pay interest more frequently. The Civic Bank and Trust advertises interest **compounded** monthly. This means that the bank computes interest each month and credits it (that is, adds it) to the account.

Since this particular account pays interest at the rate of 6% per year and there are 12 months in a year, the interest rate is 6%/12 = 0.5% per month. The following table shows the interest computations for one year for a bank account earning 6% annual interest compounded monthly.

Month	Start	Interest	End
1	$100.0000	.5000	$100.5000
2	$100.5000	.5025	$101.0025
3	$101.0025	.5050	$101.5075
4	$101.5075	.5075	$102.0151
5	$102.0151	.5101	$102.5251
6	$102.5251	.5126	$103.0378
7	$103.0378	.5152	$103.5529
8	$103.5529	.5178	$104.0707
9	$104.0707	.5204	$104.5911
10	$104.5911	.5230	$105.1140
11	$105.1140	.5256	$105.6396
12	$105.6396	.5282	$106.1678

Notice that at the end of the year the account contains $106.17. It has effectively earned 6.17% interest. This is the meaning of the advertised *effective* rate of interest. The reason that the effective rate of interest is higher than the original rate of interest is that the interest earned each month itself earns interest in each succeeding month. (We first encountered this phenomenon when we were trying to follow the values of S, I, and R into the future.) The difference between the original rate of interest and the effective rate can be very significant. Banks routinely advertise the effective rate to attract depositors. Of course, banks do the same computations for loans. They rarely advertise the effective rate of interest for loans because customers might be repelled by the true cost of borrowing.

The effective rate of interest can be computed much more quickly than we did in the previous table. Let R denote the annual interest rate as a decimal. For example, if the interest rate is 6%, then $R = 0.06$. If interest is compounded n times per year, then each time it is compounded the interest rate is R/n. Thus each time you compound the interest you compute

$$V + \left(\frac{R}{n}\right)V = \left(1 + \frac{R}{n}\right)V,$$

where V is the value of the current deposit. This computation is done n times during the course of a year. So, if the original deposit has value V, after one year it will be worth

$$\left(1 + \frac{R}{n}\right)^n V.$$

For our example above this works out to

$$\left(1 + \frac{0.06}{12}\right)^{12} V = 1.061678\, V$$

and the effective interest rate is 6.1678%.

Many banks now compound interest daily. Some even compound interest *continuously*. The value of a deposit in an account with interest compounded continuously at the rate of 6% per year, for example, grows according to the differential equation

$$V' = 0.06\, V.$$

16. Many credit cards charge interest at an annual rate of 18%. If this rate were compounded monthly, what would the effective annual rate be?

17. In fact many credit cards compound interest daily. What is the effective rate of interest for 18% interest compounded daily? Assume that there are 365 days in a year.

18. The assumption that a year has 365 days is, in fact, *not* made by banks. They figure every one of the 12 months has 30 days, so their year is 360 days long. This practice stems from the time when interest

computations were done by hand or by tables, so simplicity won out over precision. Therefore when banks compute interest they find the daily rate of interest by dividing the annual rate of interest by 360. For example, if the annual rate of interest is 18%, then the daily rate of interest is .05%. Find the effective rate of interest for 18% compounded 360 times per year.

19. In fact, once they've obtained the daily rate as 1/360-th of the annual rate, banks then compute the interest *every* day of the year. They compound the interest 365 times. Find the effective rate of interest if the annual rate of interest is 18% and the computations are done by banks. First, compute the daily rate by dividing the annual rate by 360 and then compute interest using this daily rate 365 times.

20. Consider the following advertisement.

Civic Bank and Trust

- Annual rate of interest 6%.
- Compounded daily.
- Effective rate of interest 6.2716%.

Find the effective rate of interest for an annual rate of 6% compounded daily in the straightforward way—using 1/365-th of the annual rate 365 times. Then do the computations the way they are done in a bank. Compare your two answers.

21. There are two advertisements in the newspaper for savings accounts in two different banks. The first offers 6% interest compounded quarterly (that is, four times per year). The second offers 5.5% interest compounded continuously. Which account is better? Explain.

▷ 4.4 THE LOGARITHM FUNCTION

Suppose a population is growing at the net rate of 3 births per thousand persons per year. If there are 100,000 persons now, how many will there be 37 years from now? How long will it take the population to double?

Translating into mathematics, we want to find the function $P(t)$ that solves the initial value problem

$$P'(t) = .003P(t) \quad \text{and} \quad P(0) = 100000.$$

Using the results of section 4.3 we know that the solution is the exponential function

$$P(t) = 100000\, e^{.003t}.$$

The size of the population 37 years from now will therefore be

$$P(37) = 100000 \, e^{.111}$$
$$= 100000 \times 1.117395$$
$$\approx 111740 \text{ people}$$

To find out how long it will take the population to double, we want to find a value for t so that $P(t) = 200000$. In other words, we need to solve for t in the equation

The doubling time of a population

$$100000 \, e^{.003 \, t} = 200000.$$

Dividing both sides by 100,000, we have

$$e^{.003 \, t} = 2.$$

We can't proceed because one side is expressed in exponential form while the other isn't. One remedy is to express 2 in exponential form. In fact, $2 = e^{.693147}$, as you should verify with a calculator. Then

$$e^{.003 \, t} = 2 = e^{.693147} \qquad \text{implies} \qquad .003 \, t = .693147,$$

so $t = .693147/.003 = 231.049$. Thus it will take about 231 years for the population to double.

To determine the doubling time of the population we had to know the number b for which

$$\exp(b) = e^{b} = 2.$$

This is an aspect of a very general question: given a positive number a, find a number b for which

$$e^{b} = a.$$

Solving an exponential equation

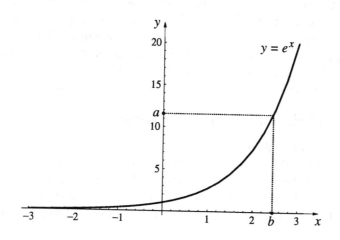

A glance at the graph of the exponential function at the bottom of page 217 shows that, by working backwards from any point $a > 0$ on the vertical axis, we can indeed find a unique point b on the horizontal axis which gives us $\exp(b) = a$.

The natural logarithm function

This process of obtaining the number b that satisfies $\exp(b) = a$ for any given positive number a is a clear and unambiguous rule. Thus, it defines a function. This function is called **the natural logarithm,** and it is denoted $\ln(a)$, or sometimes $\log(a)$. That is,

$$\ln(a) = \log(a) = \{\text{the number } b \text{ for which } \exp(b) = a\}.$$

In other words, the two statements

$$\ln(a) = b \quad \text{and} \quad \exp(b) = a$$

express exactly the same relation between the quantities a and b.

The question that led to the introduction of the logarithm function was: What number gives the exponent to which e must be raised in order to produce the value 2? This number is $\ln(2)$, and we verified that $\ln(2) = .693147$. Quite generally we can say that the number $\ln(x)$ gives the exponent to which e must be raised in order to produce the value x:

$$e^{\ln(x)} = x.$$

If we set $y = \ln(x)$, then $x = e^y$ and we can restate the last equation as a pair of companion equations:

$$e^{\ln(x)} = x \quad \text{and} \quad \ln(e^y) = y.$$

The logarithm and exponential functions are inverses

The first equation says *the exponential function "undoes" the effect of the logarithm function* and the second one says *the logarithm function "undoes" the effect of the exponential function.* For this reason the exponential and logarithm functions are said to be **inverses** of each other.

Many of the other pairs of functions—sine and arcsine, squareroot and squaring—that share a key on a calculator have this property. There are even functions (at least one can be found on any calculator) that are their own inverses—apply such a function to any number, then apply this same function to the result, and you're back at the original number. What functions do this? We will say more about inverse functions later in this section.

Properties of the Logarithm Function

The inverse relationship allows us to translate each of the properties of the exponential function into a corresponding statement about the logarithm function. We list the major pairs of properties in the following table.

For each pair, we can use the exponential property and the inverse relationship between exp and ln to establish the logarithmic property. As an example,

Exponential version	Logarithmic version
$e^0 = 1$	$\ln(1) = 0$
$e^{a+b} = e^a \cdot e^b$	$\ln(m \cdot n) = \ln(m) + \ln(n)$
$e^{a-b} = e^a / e^b$	$\ln(m/n) = \ln(m) - \ln(n)$
$(e^a)^s = e^{as}$	$\ln(m^s) = s \cdot \ln(m)$
Range of e^x is all positive reals	Domain of $\ln(x)$ is all positive reals
Domain of e^x is all real numbers	Range of $\ln(x)$ is all real numbers
$e^x \to 0$ as $x \to -\infty$	$\ln(x) \to -\infty$ as $x \to 0$
e^x grows faster than x^n, any $n > 0$	$\ln(x)$ goes to infinity slower than $x^{1/n}$, any $n > 0$

we will establish the second property. You should be able to demonstrate the others.

Proof of the second property.   Remember that to show $\ln(a) = b$, we need to show $e^b = a$. In our case a and b are more complicated. We have

$$a = m \cdot n$$
$$b = \ln(m) + \ln(n).$$

Thus, we need to show

$$e^{\ln(m)+\ln(n)} = m \cdot n.$$

But, by the exponential version of property 2,

$$e^{\ln(m)+\ln(n)} = e^{\ln(m)} \cdot e^{\ln(n)} = m \cdot n,$$

and our proof is complete.

The Derivative of the Logarithm Function

Since the natural logarithm is a function in its own right, it is reasonable to ask: What is the derivative of this function? Since the derivative describes the slope of the graph, let us begin by examining the graph of ln. Can we take advantage of the relationship between ln and exp—a function whose graph we know well—as we do this? Indeed we can, by making the following observations.

The graph of ln

- We know the point (a, b) is on the graph of $y = \ln(x)$ if and only if $b = \ln(a)$.

- We know $b = \ln(a)$ says the same thing as $a = e^b$.

- Finally, we know $a = e^b$ is true if and only if the point (b, a) is on the graph of $y = e^x$.

Putting our observations together, we have

(a, b) is on the graph of $y = \ln(x)$

if and only if

(b, a) is on the graph of $y = e^x$.

Reflection across the 45° line   The picture below demonstrates that the point (a, b) and the point (b, a) are reflections of each other about the 45° line. (Remember that points on the 45° line have the same x and y coordinates.) This is because these two points are the endpoints of the diagonal of a square whose other diagonal is the line $y = x$.

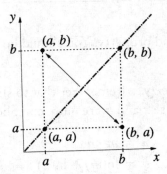

Since we have just seen that every point (a, b) on the graph of $y = \ln(x)$ corresponds to a point (b, a) on the graph of $y = e^x$, we see that *the graphs of $y = \ln(x)$ and $y = e^x$ are the reflections of each other about the line $y = x$.*

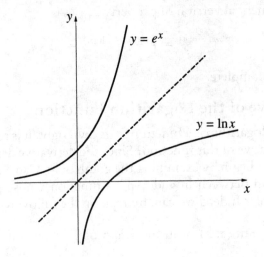

Microscopic views at mirror image points   Finally, since the two graphs are reflections of one another, a microscopic view of $\ln(x)$ at any point (b, a) will be the mirror image of the microscopic view of of e^x at the point (a, b). Any change in the y-value on one of these lines will correspond to an equal change in the x-value in its mirror image, and vice versa. The figure below shows what microscopic views of a pair of corresponding points look like, showing how a vertical change in one line equals the horizontal change in the other, and conversely.

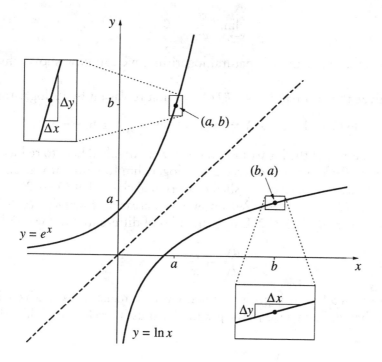

It follows that the slopes of the two lines must be reciprocals of each other. This says that the rate of change of $\ln(x)$ at $x = b$ is just the reciprocal of the rate of change of e^x at $x = a$, where $a = \ln(b)$. But the rate of change of e^x at $x = \ln(b)$ is just $e^{\ln(b)} = b$. Therefore the rate of change of $\ln(x)$ at $x = b$ is the reciprocal of this value, namely, $1/b$. We have thus proved the following result:

The slopes are reciprocals

THEOREM 1. $(\ln(x))' = 1/x$.

Note that one interpretation of this theorem is that the function $\ln(x)$ is the solution to a certain initial value problem, namely,

$$\frac{dy}{dt} = \frac{1}{x} \qquad y(1) = 0.$$

As was the case with the exponential function, we can now apply Euler's method to this differential equation as an effective way to compute values of $\ln(x)$. Applications of this idea can be found in the exercises.

Exponential Growth

The logarithm gives us a useful tool for comparing the growth rates of exponential and power functions. In the last section we claimed that e^x grows faster than any power x^p of x, as $x \to +\infty$. We interpreted that to mean

$$\lim_{x \to +\infty} \frac{x^p}{e^x} = 0,$$

for any number p. Using the natural logarithm, we can now show why it is true.

To analyze the quotient $Q = x^p/e^x$, we first replace it by its logarithm

$$\ln Q = \ln(x^p/e^x) = \ln(x^p) - \ln(e^x) = p \ln x - x.$$

Several properties of the logarithm function were invoked here to reduce $\ln Q$ to $p \ln x - x$. By another property of the logarithm function, if we can show $\ln Q \to -\infty$ we will have established our original claim that $Q \to 0$.

Let $y = \ln Q = p \ln x - x$. We know y is increasing when $dy/dx > 0$ and decreasing where $dy/dx < 0$. Using the rules of differentiation, we find

$$\frac{dy}{dx} = \frac{p}{x} - 1.$$

The expression $p/x - 1$ is positive when x is less than p and negative when x is greater than p. For x near p, the graph of y must therefore look like this:

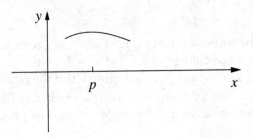

Since dy/dx remains negative as x gets large, y will continue to decrease. This does not, in itself, imply that $y \to -\infty$, however. It's conceivable that y might "level off" even as it continues to decrease—as it does in the next graph.

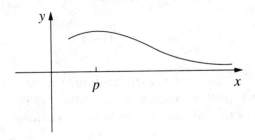

However, we can show that y does *not* "level off" in this way; it continues to plunge down to $-\infty$. We start by assuming that x has already become larger than $2p$: $x > 2p$. Then $1/x < 1/2p$ (the bigger number has the smaller reciprocal), and thus $p/x < p/2p = 1/2$. Thus, when $x > 2p$,

$$\frac{dy}{dx} = \frac{p}{x} - 1 < \frac{1}{2} - 1 = -\frac{1}{2}.$$

In other words, the slope of the graph of y is more negative than $-1/2$. The graph of y must therefore lie *below* the straight line with slope $-1/2$ that we see below:

y lies below a line that slopes down to $-\infty$

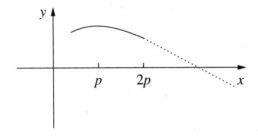

This guarantees that $y = \ln Q \to -\infty$ as $x \to \infty$. Hence $Q \to 0$, and since $Q = x^p/e^x$, we have shown that e^x grows faster than any power of x.

The Exponential Functions b^x

We have come to adopt the exponential function $\exp(x) = e^x$ as the natural one for calculus, and especially for dealing with differential equations of the form $dy/dx = ky$. Initially, though, all exponential functions b^x were on an equal footing. With the natural logarithm function, however, a *single* exponential function will meet our needs. Let's see why.

One exponential function is enough

If b is any positive real number, then $b = e^{\ln b}$. Consequently,

$$b^x = (e^{\ln b})^x = e^{\ln b \cdot x}.$$

In other words, $b^x = e^{cx}$, where $c = \ln b$. Thus, every exponential function can be expressed in terms of exp in a simple way. This is, in fact, the way computers evaluate exponents, since the computer can raise any number to any power so long as it has a way to evaluate the functions ln and exp. For instance, when you ask a computer or calculator to evaluate 2^5 (2 to the 5th power in most computer languages), it will first calculate $\ln 2$, then multiply this number by 5, then apply exp to the result. That is, it evaluates 2^5 by thinking $e^{5 \ln 2}$. While this may seem a roundabout way to come up with 32, its virtue is that the computer needs only one algorithm to calculate any base to any power, without having to consider different cases.

$b^x = e^{\ln b \cdot x}$

How to calculate 2^5

This expression gives us a new way to find the derivative of b^x. We already know that

$$(e^{cx})' = c \cdot e^{cx},$$

for any constant c. This follows from the chain rule. When $c = \ln b$, we get

$$(b^x)' = (e^{\ln(b) \cdot x})' = \ln(b) \cdot e^{\ln(b) \cdot x} = \ln(b) \cdot b^x.$$

Thus, $y = b^x$ is a solution to the differential equation

$$\frac{dy}{dx} = \ln(b) \cdot y.$$

In chapter 3, we wrote this differential equation as

$$\frac{dy}{dx} = k_b \cdot y.$$

$k_b = \ln(b)$

We see now that $k_b = \ln(b)$.

We can use the connection between k_b and the natural logarithm, and between the natural logarithm and the exponential function, to gain new insights. For example, on page 187 we argued that there must be a value of b for which $k_b = .02$. This simply means

$$\ln b = .02 \quad \text{or} \quad b = e^{.02}.$$

In other words, we now have an explicit formula that tells us the value of b for which $k_b = .02$:

$$b = e^{.02} = 1.02020134\ldots$$

Inverse Functions

Most of what we have said about the exponential and logarithm functions carries over directly to *any* pair of inverse functions. We begin by saying precisely what it means for two functions f and g to be inverses of each other.

DEFINITION.   Two functions f and g are **inverses** if

$$f(g(a)) = a \quad \text{and} \quad g(f(b)) = b$$

for every a in the domain of g and every b in the domain of f.

Observe that if f and g are inverses of each other, then each one "undoes" the effect of the other by sending any value back to the number it came from via the other function. One implication of this is that neither function can have two different input values going to the same output value. For instance, suppose b_1 and b_2 get sent to the same value by f: $f(b_1) = f(b_2)$. Applying g to both sides of this equation we would get $b_1 = g(f(b_1)) = g(f(b_2)) = b_2$, so b_1 and b_2 were actually the same number. This is an important enough property that there is a name for it:

DEFINITION. We say that a function f is **one-to-one**, usually written as **1–1**, if it is true that whenever $x_1 \neq x_2$ then it is also the case that $f(x_1) \neq f(x_2)$. Equivalently, f is 1–1 if whenever $f(x_1) = f(x_2)$, then it must be true that $x_1 = x_2$.

We have thus seen that only functions which are one-to-one can have inverses. This means that to establish inverses for some functions, we will need to restrict their domains to regions where they are one-to-one. Let's reexamine the examples we mentioned earlier to see how they fit this definition.

EXAMPLE 1.   Suppose $f(x) = \exp(x)$ and $g(x) = \ln(x)$. Then the equations

$$f(g(a)) = \exp(\ln(a)) = e^{\ln(a)} = a \quad \text{for} \quad a > 0$$

and

$$g(f(b)) = \ln(\exp(b)) = \ln(e^b) = b$$

hold for all real numbers b and for all *positive* real numbers a. The domain of the exponential function is all real numbers and the domain of the natural logarithm function is all positive real numbers.

EXAMPLE 2.   Suppose $f(x) = x^2$ and $g(x) = \sqrt{x}$. The squaring function is not invertible on its natural domain because it is not one-to-one. Since a number and its negative have the same square, we wouldn't know which one to send the square back to when we took the square root. We can't avoid the problem by saying that $g(4) = \pm 2$, since a function has to have only one output for each input. The squaring function is invertible, though, if we restrict it to nonnegative real numbers. Then

x^2 is invertible on $x \geq 0$

$$f(g(a)) = (\sqrt{a})^2 = a \quad \text{(for } a \geq 0)$$

and

$$g(f(b)) = \sqrt{b^2} = b \quad \text{(for } b \geq 0).$$

The domain of the square root function is all $b \geq 0$.

Note that we could have restricted the domain of f in another way to make it one-to-one by considering only nonpositive real numbers. Now g is no longer the inverse of this restricted f. For instance, $g(f(-3)) = g(9) = 3 \neq -3$. What would the inverse of f be in this case?

EXAMPLE 3. Suppose $f(x) = \sin(x)$ and $g(x) = \arcsin(x)$. Since f is not one-to-one on its natural domain, we again need to restrict it in order for it to have an inverse. By convention, the domain of $\sin(x)$ is taken to be $-\pi/2 \leq x \leq \pi/2$.

sin x is
invertible on
$-\pi/2 \leq x \leq \pi/2$

$$f(g(a)) = \sin(\arcsin(a)) = a \quad (\text{for } -1 \leq a \leq 1)$$

and

$$g(f(b)) = \arcsin(\sin(b)) = b \quad (\text{for } -\pi/2 \leq b \leq \pi/2).$$

Each pair of inverse functions shares corresponding properties, just as the logarithm and exponential functions do—the particular properties depending on the particular functions. But two they all share are

- The range of f is the domain of g;
- The domain of f is the range of g.

The exercises check this for examples 2 and 3.

Finally, the graphs—and therefore the derivatives—of a function and of its inverse are mirror images, exactly like those of the exponential and logarithm functions. We begin with the same list of observations.

- We know the point (a, b) is on the graph of $y = g(x)$ if and only if $b = g(a)$.
- We know $b = g(a)$ says the same thing as $a = f(b)$.
- Finally, we know $a = f(b)$ is true if and only if the point (b, a) is on the graph of $y = f(x)$.

As before, putting our observations together, we have

(a, b) is on the graph of $y = g(x)$
if and only if
(b, a) is on the graph of $y = f(x)$.

The graphs of inverse
functions are mirror
images . . .

Exactly as before, we have that the point (a, b) and the point (b, a) are reflections of each other about the line $y = x$. Since we have just seen that every point (a, b) on the graph of $y = g(x)$ corresponds to a point (b, a) on the graph of $y = f(x)$, we again see that *the graphs of $y = g(x)$ and $y = f(x)$ are the reflections of each other about the line $y = x$.*

Finally, since the two graphs are reflections of one another, the local linear approximation of $g(x)$ at any point (a, b) will be the mirror image of the local linear approximation of $f(x)$ at the point (b, a). Any change in the y-value on one of these local lines will correspond to an equal change in the x-value in its mirror image, and vice versa. Just as before, it follows that the slopes of the two lines must be reciprocals of each other. This says that the rate of change of $g(x)$ at $x = b$ is the reciprocal of the rate of change of $f(x)$ at $x = a$, where $a = g(b)$. But the rate of change of $f(x)$ at $x = g(b)$ is just $f'(g(b))$. Therefore the rate of change of $g(x)$ at $x = b$ is the reciprocal of this value, namely $1/f'(g(b))$. We have thus proved the following result:

THEOREM 2. *If the functions f and g are inverses, then g is locally linear at (b, a) if and only if f is locally linear at (a, b). When local linearity holds,*

... and their derivatives are reciprocals

$$g'(b) = \frac{1}{f'(a)}.$$

Exercises

1. Determine the numerical value of each of the following.

a) $\ln(2e)$ *e) $\ln e^3$ i) e^{-1} m) $\ln(\sqrt{e})$

b) $e^{\ln 2}$ *f) $e^{3\ln 2}$ j) $(e^{\ln 2})^3$ n) $e^{2\ln 3}$

c) $\ln 10$ g) $\ln 10^3$ k) $e^{\ln 10}$ o) $e^{\ln 1000}$

d) $\ln(1/e)$ h) $\ln(1/2)$ *l) $e^{-\ln 2}$ p) $e^{-3\ln 2}$

2. a) In the text we noted that the function $\ln x$ is the solution to the initial value problem

$$\frac{dy}{dt} = \frac{1}{x} \quad y(1) = 0,$$

so that we can use Euler's method to compute values for $\ln x$. Use this method to evaluate $\ln 2$ to 3 decimal places. What value of Δx gives the desired accuracy?

b) If you now wanted to calculate $\ln 6$ to 3 decimals, can you think of a better way to do it than simply starting at $x = 1$ and running Euler's method out to $x = 6$? Remember the basic properties of logarithms, and figure out a way to use the results of part (a).

c) Suppose you had figured out that $\ln 2 = .693147....$ How would you use Euler's method to calculate $\ln 1300$ quickly? You might find the fact that $2^{10} = 1024$ helpful.

3. The rate of growth of the population of a particular country is proportional to the population. The last two censuses determined that the population in 1980 was 40,000,000, and in 1985 it was 45,000,000. What will the population be in 1995?

4. Find the derivatives of the following functions.

    *a) $\ln(3x)$

    b) $17\ln(x)$

    c) $\ln(e^w)$

    d) $\ln(2^t)$

    e) $\pi\ln(3e^{4s})$

5. Suppose a bacterial population grows so that its mass is

$$P(t) = 200e^{.12t} \quad \text{grams}$$

after t hours. Its initial mass is $P(0) = 200$ grams. When will its mass double, to 400 grams? How much longer will it take to double again, to 800 grams? After the population reaches 800 grams, how long will it take for yet another doubling to happen? What is the *doubling time* of this population?

*6. Suppose a beam of X-rays whose intensity is A rads (the "rad" is a unit of radiation) falls perpendicularly on a heavy concrete wall. After the rays have penetrated s feet of the wall, the radiation intensity has fallen to

$$R(s) = Ae^{-.35s} \quad \text{rads.}$$

What is the radiation intensity 3 inches inside the wall? 18 inches? (Your answers will be expressed in terms of A.) How far into the wall must the rays travel before their intensity is cut in half, to $A/2$? How much further before the intensity is $A/4$?

7. Virtually all living things take up carbon as they grow. This carbon comes in two principal forms: normal, stable carbon—C^{12}—and radioactive carbon—C^{14}. C^{14} decays into C^{12} at a rate proportional to the amount of C^{14} remaining. While the organism is alive, this lost C^{14} is continually replenished. After the organism dies, though, the C^{14} is no longer replaced, so the percentage of C^{14} decreases exponentially over time. It is found that after 5730 years, half the original C^{14} remains. If an archaeologist finds a bone with only 20% of the original C^{14} present, how old is it?

8. The human population of the world appears to be growing exponentially. If there were 2.5 billion people in 1960, and 3.5 billion in 1980, how many will there be in 2010?

9. If bacteria increase at a rate proportional to the current number, how long will it take 1000 bacteria to increase to 10,000 if it takes them 17 minutes to increase to 2000?

10. Suppose sugar in water dissolves at a rate proportional to the amount left undissolved. If 40 lb of sugar reduces to 12 lb in 4 hours, how long will you have to wait until 99% of the sugar is dissolved?

11. Atmospheric pressure is a function of altitude. Assume that at any given altitude the rate of change of pressure with altitude is proportional to the pressure there. If the barometer reads 30 psi (pounds per square inch) at sea level and 24 psi at 6000 feet above sea level, how high are you when the barometer reads 20 psi?

12. a) An important concept in many economic analyses is the idea of **present value**. It is used to compare the values of different possible payments made at different times. As a simple example, suppose you had a small wood lot and had the choice of selling the timber on it now for $5,000 or waiting 10 years for the trees to get larger, at which point you estimate the timber could be sold for $8,000. To compare these two options, you need to convert the prospect of $8,000 ten years from now into an equivalent amount of money now—its present value. This is the amount of money you would need to invest now to have $8,000 in 10 years. Suppose you thought you could invest money at an annual interest rate of 4% compounded continuously. If you invested $5,000 now at this rate, then in 10 years you would have $5000\,e^{.4} = \$7,459.12$. That is, $5,000 now is worth $7,459.12 in 10 years—both amounts have the same present value. Clearly $8,000 in 10 years must have a slightly greater present value under the assumption of a 4% annual interest rate. What is it?

    b) On the other hand, if you can get a higher interest rate than 4%, the present value of the $8,000 will be much less. What is the present value of a payment of $8,000 ten years from now if the annual interest rate is 8%?

    c) At what interest rate do $5,000 now and $8,000 in ten years have the same present value?

13. Use properties of exp to prove the following properties of the logarithm. (Remember that $\ln a = b$ means $a = \exp b$.)
    a) $\ln(1) = 0$.
    b) $\ln(m/n) = \ln(m) - \ln(n)$.
    c) $\ln(m^n) = n \ln(m)$.

14. a) Use a graphing program to find a good numerical approximation to $(\ln x)'$ at $x = 2$. Make a short table, for decreasing interval sizes Δx, of the quantity $\Delta(\ln x)/\Delta x$.

    b) Use a graphing program to find a good numerical approximation to $(e^x)'$ at $x = \ln(2) = .6931\ldots$. Make a short table for decreasing interval sizes Δx, of the quantity $\Delta(e^x)/\Delta x$.

    c) What is the relationship between the values you got in parts (a) and (b)?

15. Find a solution (using $\ln x$) to the differential equation

$$f'(x) = 3/x \quad \text{satisfying} \quad f(1) = 2 \, .$$

16. a) Find a formula using the natural logarithm function giving the solution of $y' = a/x$ with $y(1) = b$.
    b) Solve $P' = 2/t$ with $P(1) = 5$.

17. Find the domain and range of each of the following pairs of inverse functions.
    a) $f(x) = x^2$ (restricted to $x \geq 0$) and $g(x) = \sqrt{x}$.
    b) $f(x) = \sin(x)$ (restricted to $-\pi/2 \leq x \leq \pi/2$) and $g(x) = \arcsin(x)$.

18. Show that $f(x) = 1/x$ equals its own inverse. What are the domain and range of f?

19. Let n be a positive integer, and let $f(x) = x^n$. What is an inverse of f? How do we need to restrict the domain of f for it to have an inverse? Caution: The answer depends on n.

20. a) What is the inverse g of the function $f(x) = 1 - 3x$?
    b) Do f and g satisfy Theorem 2?

21. What is an inverse of $f(x) = x^2 - 4$?

22. Use the relationship between the derivatives of a function and its inverse to find the indicated derivatives.
    a) $g'(100)$ for $g(x) = \sqrt{x}$.
    b) $g'(\sqrt{2}/2)$ for $g(x) = \arcsin(x)$.
    c) $g'(1/2)$ for $g(x) = 1/x$.

23. a) Use Theorem 2 and the fact that $(x^2)' = 2x$ to derive the formula for the derivative of $\sqrt{x}$.
    b) Use Theorem 2 and the fact that $(x^n)' = nx^{n-1}$ to derive the formula for the derivative of $\sqrt[n]{x}$.

24. Compare the rates of growth of e^x and b^x for both $e < b$ and $1 < b < e$.

⊳ 4.5 THE EQUATION $y' = f(t)$

Most differential equations we have encountered express the rate of growth of a quantity *in terms of the quantity itself*. The simplest models for biological growth had this form: $y' = ky$ and $y' = ky^p$. Even when several variables were present—as in the S-I-R model and the predator–prey models—it was most natural to express the rates at which those variables change in terms of the variables themselves. Even the motion of a spring (pages 198–199) was described that way: the rate of change of position equaled the velocity, and the rate of change of velocity was proportional to the position.

The motion of a falling body ...

Sometimes, though, a differential equation will express the rate of change of a variable *directly in terms of the input variable*. For example, on page 197 we saw that the velocity dx/dt of a body falling under the sole influence of gravity is a linear function of the time:

$$\frac{dx}{dt} = -gt + v_0.$$

Here x is the height of the body above the ground, g is the acceleration due to gravity, and v_0 is the velocity at time $t = 0$. This equation has the general form

$$\frac{dx}{dt} = f(t),$$

where $f(t)$ is a given function of t. We will now consider special methods that can be used to study differential equations of this special form.

Antiderivatives

To solve the equation of motion of a body falling under gravity, we must find a function $x(t)$ whose derivative is given as

$$x' = -gt + v_0.$$

We can call upon our knowledge of the rules of differentiation to find x. Consider $-gt$ first. What function has $-gt$ as its derivative? We can start with t^2, whose derivative is $2t$. Since we want the derivative to turn out to be $-gt$, we can reason this way:

$$-gt = -\frac{g}{2} \cdot 2t = -\frac{g}{2} \times \text{ the derivative of } t^2.$$

This leads us to identify $-gt^2/2$ as a function whose derivative is $-gt$. Check for yourself that this is correct by differentiating $-gt^2/2$.

Now consider v_0, the other part of dx/dt. What function has the constant v_0 as its derivative? A derivative is a rate of growth, and we know that the linear functions are precisely the ones that have constant growth rates. Furthermore, the rate is the multiplier for a linear function, so we conclude that any linear function of the form $v_0 t + b$ has derivative v_0.

If we put the two pieces together, we find that

$$x(t) = -\frac{g}{2} t^2 + v_0 t + b$$

is a solution to the differential equation, for any value of b. (Recall from section 4.2 that a differential equation can have many solutions.) We constructed this formula for $x(t)$ by "undoing" the process of differentiation, a process

... and its solution

$-gt$ is the derivative of $-gt^2/2$

v_0 is the derivative of $v_0 t$ and of $v_0 t + b$

The antiderivative of a function

sometimes called **antidifferentiation**. The function produced is called an **antiderivative**. Thus:

$$-\frac{g}{2}t^2 + v_0 t + b \text{ is an } \textit{antiderivative} \text{ of } -gt + v_0$$

because

$$-gt + v_0 \text{ is the } \textit{derivative} \text{ of } -\frac{g}{2}t^2 + v_0 t + b.$$

All the functions $F(x)$ + C are antiderivatives of $F'(x)$

Note that a function has only one derivative, but it has many antiderivatives. If $F(t)$ is an antiderivative of $f(t)$, then so is $F(t) + C$, where C is any constant.

The list of functions and their derivatives that we compiled in chapter 3 (see page 131) can be "turned around" to become a list of functions and their antiderivatives. Note that the antiderivative column should really be labeled "an antiderivative" since we could add a constant to any of the listed functions and still have an antiderivative for the function in the first column.

Function	Antiderivative
0	c
x^p	$\frac{1}{p+1}x^{p+1}$ (if $p \neq -1$)
x^{-1}	$\ln x$
$\sin x$	$-\cos x$
$\cos x$	$\sin x$
$\exp x = e^x$	$\exp x = e^x$
b^x	$\frac{1}{\ln b}b^x$

Every power of x has an antiderivative

Notice the formula for the antiderivative of x^p requires $p + 1 \neq 0$, that is, $p \neq -1$. This leaves out x^{-1}. However, the antiderivative of x^{-1} is $\ln x$, so no power of x is excluded from the table.

We also had differentiation rules that told us how to deal with different *combinations* of functions. Each of these rules has an analogue in antidifferentiation. The simplest combinations are a sum and a constant multiple.

Function	Antiderivative
$f(x)$	$F(x)$
$g(x)$	$G(x)$
$c \cdot f(x)$	$c \cdot F(x)$
$f(x) + g(x)$	$F(x) + G(x)$

We defer a discussion of the analogue of the chain rule to chapter 11.

With just these rules we can find the antiderivative of any polynomial, for instance. (Recall that a polynomial is a sum of constant multiples of powers of the input variable.) Here is a collection of sample antiderivatives that illustrate the various rules. To emphasize the fact that antiderivatives are determined only up to an additive constant, various constants have been tacked on—any other constant would work just as well. You should compare this table with the one on page 132.

Function	Antiderivative
$5x^4 - 2x^3$	$x^5 - \frac{1}{2}x^4 + 17$
$5x^4 - 2x^3 + 17x$	$x^5 - \frac{1}{2}x^4 + \frac{17}{2}x^2 - 243.77$
$6 \cdot 10^z + 17/z^7$	$6 \cdot 10^z / \ln 10 - 17/6z^6 + .002$
$3 \sin t - 2t^3$	$-3 \cos t - \frac{1}{2}t^4 + 5 \ln 7$
$\pi \cos x + \pi^2$	$\pi \sin x + \pi^2 x - 12e^{7.21}$

Euler's Method Revisited

If we know the formula for an antiderivative of $f(t)$, then we can write down a solution to the differential equation $dy/dt = f(t)$. For example, the general solution to

$$\frac{dy}{dt} = 12t^2 + \sin t$$

is $y = 4t^3 - \cos t + C$. In such a case we have a shortcut to solving the differential equation without needing to use Euler's method. Often, though, there is no formula for an antiderivative of $f(t)$—even when $f(t)$ itself has a simple formula. There is no formula for the antiderivative of $\cos(t^2)$, or $\sin t/t$, or $\sqrt{1 + t^3}$, for instance. In other cases, $f(t)$ may not even be given by a formula. It may be a data function, given as a graph made by a pen tracing on a moving sheet of graph paper.

Whether we can find a formula for an antiderivative of $f(t)$ or not, we can still solve the differential equation $dy/dt = f(t)$ by Euler's method. It turns out that Euler's method takes on a relatively simple form in such cases. Let's investigate this in the following context.

Let V be the volume of water in a reservoir serving a small town, measured in millions of gallons. Then V is a function of the time t, measured in days. Rainfall adds water to the reservoir, while evaporation and consumption by the townspeople take it away. Let f be the *net rate* at which water is flowing into the reservoir, in millions of gallons per day. Sometimes f will be positive—when rainfall exceeds evaporation and consumption—and sometimes f will be

The volume of a reservoir varies over time

negative. The net inflow rate varies from day to day; that is, f is a function of time: $f = f(t)$. Our model of the reservoir is the differential equation

$$\frac{dV}{dt} = f(t) \quad \text{millions of gallons per day.}$$

The net inflow rate

Suppose $f(t)$ is measured every two days, and those measurements are recorded in the following table.

Time t (days)	Rate $f(t)$ ($10^6 \times$ gals. per day)
0	.34
2	.11
4	−.07
6	−.23
8	−.14
10	.03
12	.08

Note that in this table we are able to write down the rate for all values of t immediately, without having to calculate the intermediate values of the dependent variable V. This is in marked contrast with most of the examples we've looked at in this course where we had to know the values of all the variables for any time t before we could calculate the new rate value at that time. It is this simplification that gives differential equations of the form $y' = f(t)$ their special structure.

If we assume the value of $f(t)$ remains constant for the two days after each measurement is made, we can approximate the total change in V over these 14 days.

The following table tells us two things: first, how much V changes over each two-day period; second, the *total* change in V that has accumulated by the end of each period. Since $\Delta t = 2$ days, we calculate ΔV by

The accumulated change in V

$$\Delta V = V' \cdot \Delta t = f(t) \cdot \Delta t = 2 \cdot f(t).$$

Starting t	Current ΔV	Accumulated ΔV	Ending t
0	.68	.68	2
2	.22	.90	4
4	−.14	.76	6
6	−.46	.30	8
8	−.28	.02	10
10	.06	.08	12
12	.16	.24	14

At the end of the 14 days, V has accumulated a total change of .24 million gallons. Notice that this does not depend on the initial size of V. If V had been 92.64 million gallons at the start, it would be $92.64 + .24 = 92.88$ million gallons at the end. If it had been only 2 million gallons at the start, it would be $2 + .24 = 2.24$ million gallons at the end. Other models do not behave this way: in two weeks, a rabbit population of 900 will change much more than a population of 90. The total change in V is independent of V because the *rate* at which V changes is independent of V.

We can therefore use Euler's method to solve any differential equation of the form $dy/dt = f(t)$ *independently* of an initial value for y. We just calculate the total accumulated change in y, and add that total to any given initial y. Here is how it works when the initial value of t is a, and the time step is Δt.

> Euler's method can calculate just the accumulated change in y

Starting t	Current Δy	Accumulated Δy	Ending t
a	$f(a) \cdot \Delta t$	Previously	$a + \Delta t$
$a + \Delta t$	$f(a + \Delta t) \cdot \Delta t$	accumulated	$a + 2\Delta t$
$a + 2\Delta t$	$f(a + 2\Delta t) \cdot \Delta t$	Δy	$a + 3\Delta t$
$a + 3\Delta t$	$f(a + 3\Delta t) \cdot \Delta t$	+	$a + 4\Delta t$
$\vdots$	$\vdots$	current	$\vdots$
$a + (n-1)\Delta t$	$f(a + (n-1)\Delta t) \cdot \Delta t$	Δy	$a + n\Delta t$

The third column is too small to hold the values of "accumulated Δy." Instead, it contains the instructions for obtaining those values. It says: To get the current value of "accumulated Δy," add the "current Δy" to the previous value of "accumulated Δy."

> The accumulated change in y

Let's use Euler's method to find the accumulated Δy when $t = 4$, given that

$$\frac{dy}{dt} = \cos(t^2)$$

and t is initially 0. If we use 8 steps, then $\Delta t = .5$ and we obtain the following:

Starting t	Current Δy	Accumulated Δy	Ending t
0	.5000	.5000	.5
.5	.4845	.9845	1.0
1.0	.2702	1.2546	1.5
1.5	−.3141	.9405	2.0
2.0	−.3268	.6137	2.5
2.5	.4997	1.1134	3.0
3.0	−.4556	.6579	3.5
3.5	.4752	1.1330	4.0

The following program generated the last three columns of this table.

Program: TABLE

```
DEF fnf (t) = COS(t ^ 2)
tinitial = 0
tfinal = 4
numberofsteps = 2 ^ 3
deltat = (tfinal - tinitial) / numberofsteps
t = tinitial
accumulation = 0
FOR k = 1 TO numberofsteps
    deltay = fnf(t) * deltat
    accumulation = accumulation + deltay
    t = t + deltat
    PRINT deltay, accumulation, t
NEXT k
```

TABLE is a modification of the program SIRVALUE (page 56). To empha-size the fact that it is the accumulated change that matters rather than the actual value of y, we have modified the program accordingly. Note that `accumulation` always starts at 0, no matter what the initial value of y is. The first line of the program takes advantage of a capacity most programming languages have to define functions which can then be referred to elsewhere in the program.

As usual, to find the exact value of the accumulated Δy, it is necessary to recalculate, using more steps and smaller step sizes Δt. If we use TABLE to do this, we find

Number of steps	Accumulated Δy
2^3	1.13304
2^6	.65639
2^9	.60212
2^{12}	.59542
2^{15}	.59458
2^{18}	.59448

Thus we can say that if $dy/dt = \cos(t^2)$, then y increases by .594 ... when t increases from 0 to 4.

In the same way we changed SIRVALUE to produce the program SIRPLOT (page 60), we can change the program TABLE into one that will *plot* the values of y. In the following program all those changes are made, and one more besides: we have increased the number of steps to 400 to get a closer approximation to the true values of y.

Program: PLOT

```
Set Up GRAPHICS
DEF fnf (t) = COS(t ^ 2)
tinitial = 0
tfinal = 4
numberofsteps = 400
deltat = (tfinal - tinitial) / numberofsteps
t = tinitial
accumulation = 0
FOR k = 1 TO numberofsteps
    deltay = fnf(t) * deltat
    Plot the line from (t, accumulation)
        to (t + deltat, accumulation + deltay)
    accumulation = accumulation + deltay
    t = t + deltat
NEXT k
```

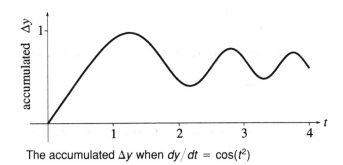

The accumulated Δy when $dy/dt = \cos(t^2)$

Let's compare our reservoir model with population growth. The rate at which a population grows depends, in an obvious way, on the size of the population. By contrast, the rate at which the reservoir fills does *not* depend on how much water there is in the reservoir. It depends on factors *outside* the reservoir: rainfall and consumption. These factors are said to be **exogenous** (from the Greek *exo-*, "outside" and *-gen*, "produced," or "born"). The opposite is called an **endogenous** factor (from the Greek *endo-*, "within"). Evaporation is an endogenous factor for the reservoir model; population size is certainly an endogenous factor for a population model.

Precisely because exogenous factors are "outside the system," we need to be *given* the information on how they vary over time. In the reservoir model, this information appears in the function $f(t)$ that describes the rate at which

Exogenous and endogenous factors

V changes. In general, if y depends on exogenous factors that vary over time, we can expect the differential equation for y to involve a function of time:

$$\frac{dy}{dt} = f(t)$$

Thus, we can view this section as dealing with models that involve exogenous factors.

> The differential equation of motion for a falling body, $dx/dt = -gt + v_0$, indicates that gravity is an exogenous factor. In Greek and medieval European science, the reason an object fell to the ground was assumed to lie within the object itself—it was the object's "heaviness." By making the cause of motion exogenous, rather than endogenous, Galileo and Newton started a scientific revolution.

Exercises

1. Find a formula $y = F(t)$ for a solution to the differential equation $dy/dt = f(t)$ when $f(t)$ is

   a) $5t - 3$
   *b) $t^6 - 8t^5 + 22\pi^3$
   c) $5e^t - 3\sin t$
   d) $12\sqrt{t}$
   e) $2^t + 7/t^9$
   f) $5e^{4t} - 1/t$

*2. Find $G(5)$ if $y = G(x)$ is the solution to the initial value problem

$$\frac{dy}{dx} = \frac{1}{x^2} \qquad y(2) = 3.$$

3. Find $F(2)$ if $y = F(x)$ is the solution to the initial value problem

$$\frac{dy}{dx} = \frac{1}{x} \qquad y(1) = 5.$$

4. Find $H(3)$ if $y = H(x)$ is the solution to the initial value problem

$$\frac{dy}{dx} = x^3 - 7x^2 + 19 \qquad y(-1) = 5.$$

5. Find $L(-2)$ if $y = L(x)$ is the solution to the initial value problem

$$\frac{dy}{dx} = e^{3x} \qquad y(1) = 6.$$

6. a) Sketch the graph of the solution to the initial value problem

$$\frac{dy}{dx} = \sin x \qquad y(0) = 1$$

over the interval $0 \le x \le 4\pi$.

   b) By finding a suitable antiderivative, evaluate $y(2)$.

7. **a)** Sketch the graph of the solution to the initial value problem

$$\frac{dy}{dx} = \sin(x^2) \qquad y(0) = 0$$

over the interval $0 \le x \le 5$. (This one can't be done by finding an antiderivative.)

b) What is the slope of the solution graph at $x = 0$? Does your graph show this?

c) How many peaks (local maxima) does the solution have on the interval $0 \le x \le 5$?

d) What is the maximum value that the solution achieves on the interval $0 \le x \le 5$? For which value of x does this happen?

e) What is $y(6)$?

8.***a)** What is the accumulated change in y if $dy/dt = 3t^2 - 2t$ and t increases from 0 to 1? What if t increases from 1 to 2? What if t increases from 0 to 2?

b) Sketch the graph of the accumulated change in y as a function of t. Let $0 \le t \le 2$.

9. **a)** Here's another problem for which there is no formula for an antiderivative. Sketch the graph of the solution to the initial value problem

$$\frac{dy}{dx} = \frac{\sin x}{x} \qquad y(0) = 0$$

on the interval $0 \le x \le 40$. [Note: $\sin x / x$ is not defined when $x = 0$, so take the initial value of x to be .00001. That is, use $y(.00001) = 0$.]

b) How many peaks (local maxima) does the solution have on the interval $0 \le x \le 40$?

c) What is the maximum value of the solution on the interval $0 \le x \le 40$? For which x is this maximum achieved?

> ## 4.6 CHAPTER SUMMARY

The Main Ideas

- A **system of differential equations** expresses the derivatives of a set of functions in terms of those functions and the input variable.

- An **initial value problem** is a system of differential equations together with values of the functions for some specified value of the input variable.

- Many processes in the physical, biological, and social sciences are **modeled** as initial value problems.
- A **solution to a system of differential equations** is a set of functions which make the equations *true* when they and their derivatives are substituted into the equations.
- A **solution to an initial value problem** is a set of functions that solve the differential equations and satisfy the initial conditions. Typically, a solution is unique.
- **Euler's method** provides a recipe to find the solution to an initial value problem.
- In special circumstances it is possible to find **formulas** for the solution to a system of differential equations. If the differential equations involve **parameters**, the solutions will, too.
- Systems of differential equations define functions as their solutions. Among the most important are the **exponential** and **logarithm functions**.
- The **natural logarithm** function is the inverse of the exponential function.
- The **graphs** and the **derivatives** of a function and its inverse are connected geometrically to each other by **reflection**.
- Exponential functions b^x **grow to infinity** faster than any power of x.
- The solution to $dy/dx = f(x)$ is an **antiderivative** of f—that is, a function whose derivative is f.

Expectations

- You should be able to use computer programs to produce tables and graphs of solutions to initial value problems.
- You should be able to check whether a system of differential equations reflects the hypotheses being made in constructing a model of a process.
- You should be able to verify whether a set of functions given by formulas is a solution to a system of differential equations.
- You should be familiar with the basic properties of the exponential and logarithm functions.
- You should be able to express solutions to initial value problems involving exponential growth or decay in terms of the exponential function.
- You should be able to solve $dy/dx = f(x)$ by antidifferentiation when $f(x)$ is a basic function or a simple combination of basic functions.
- You should be able to analyze and graph the inverse of a given function.

CHAPTER **5**

TECHNIQUES OF DIFFERENTIATION

In this chapter we focus on functions given by *formulas*. The derivatives of such functions are then also given by formulas. In chapter 4 we used information about the derivative of a function to recover the function itself; now we go *from* the function *to* its derivative. We develop the rules for *differentiating* a function: computing the formula for its derivative from the formula for the function. Then we use differentiation to investigate the properties of functions, especially their *extreme values*. Finally we examine a powerful method for solving equations that depends on being able to find a formula for a derivative.

➤ 5.1 THE DIFFERENTIATION RULES

There are three kinds of differentiation rules. First, any basic function has a specific rule giving its derivative. Second, the *chain rule* will find the derivative of a *chain* of functions. Third, there are general rules that allow us to calculate the derivatives of algebraic combinations—for example, sums, products, and quotients—of any functions provided we know the derivatives of each of the component functions. To obtain all three kinds of rules we will typically start with the analytic definition of the derivative as the limit of a quotient of differences:

> The **derivative** of the function f at x is the value of the limit
> $$\lim_{\Delta x \to 0} \frac{f(x + \Delta x) - f(x)}{\Delta x} = f'(x).$$

In this chapter we will look at the cases where this limit can be evaluated exactly. Although using this definition of derivative usually leads to many algebraic manipulations, the other interpretations of derivatives as slopes, rates, and multipliers will still be helpful in visualizing what's going on. The process of calculating the derivative of a function is called **differentiation**. For this reason, functions which are locally linear and not locally vertical (so they do have slopes, and hence derivatives at every point) are called **differentiable** functions. Our goal in this chapter is to differentiate functions given by formulas.

Derivatives of Basic Functions

Functions given by formulas have derivatives given by formulas

When a function is given by a *formula*, there is in fact a formula for its derivative. We have already seen several examples in chapters 3 and 4. These examples include all of what we may consider the **basic functions**. We collect these formulas in the following table.

Derivatives of Basic Functions

Function	Derivative
$mx + b$	m
x^r	rx^{r-1}
$\sin x$	$\cos x$
$\cos x$	$-\sin x$
e^x	e^x
$\ln x$	$1/x$

In the case of the linear function $mx + b$, we obtained the derivative by using its geometric description as the *slope* of the graph of the function. The derivatives of the exponential and logarithm functions came from the definition of the exponential function as the solution of an initial value problem. To find the derivatives of the other functions we will need to start from the definition.

An Example: $f(x) = x^3$

We begin by examining the calculation of the derivative of $f(x) = x^3$ using the definition. The change Δy in $y = f(x)$ corresponding to a change Δx in x is given by

$$\Delta y = f(x + \Delta x) - f(x)$$

$$= (x + \Delta x)^3 - x^3$$

$$= 3x^2 \cdot \Delta x + 3x(\Delta x)^2 + (\Delta x)^3.$$

From this we get

$$f'(x) = \lim_{\Delta x \to 0} \frac{\Delta y}{\Delta x}$$

$$= \lim_{\Delta x \to 0} 3x^2 + 3x \cdot \Delta x + (\Delta x)^2.$$

To see what's happening with this expression, let's consider the specific value $x = 2$ and evaluate the corresponding values of $\Delta y / \Delta x$ for successively smaller Δx.

Δx	$2^2 + 6\Delta x + (\Delta x)^2$	$\Delta y / \Delta x$
.1	$12 + .6 + .01$	12.61
.01	$12 + .06 + .0001$	12.0601
.001	$12 + .006 + .000001$	12.006001
.0001	$12 + .0006 + .00000001$	12.00060001
.00001	$12 + .00006 + .0000000001$	12.0000600001

The value of $\Delta y / \Delta x$ gets closer and closer to 12 as Δx gets smaller and smaller

It is clear from this table that we can make $\Delta y / \Delta x$ as close to 12 as we like by making Δx small enough. Therefore $f'(2) = 12$.

Note that in the table above we have used positive values of Δx. You should check to convince yourself that if we had used negative values of Δx we would have come up with a different set of approximations $\Delta y / \Delta x$, but that the limit would still be the same, namely, 12—it doesn't matter whether we use positive or negative values for Δx, or a mixture of the two, so long as $\Delta x \to 0$.

In general, for any given x, the second and third terms in the expansion for $\Delta y / \Delta x$ become vanishingly small as $\Delta x \to 0$, so that $\Delta y / \Delta x$ can be made as close to $3x^2$ as we like by making Δx small enough. For this reason, we say that the derivative $f'(x)$ is *exactly* $3x^2$:

$$f'(x) = \lim_{\Delta x \to 0} 3x^2 + 3x \cdot \Delta x + (\Delta x)^2 = 3x^2.$$

In other words, given the function f specified by the formula $f(x) = x^3$ we have found the formula for its derivative function f': $f'(x) = 3x^2$. Note that this general formula agrees with the specific value $f'(2) = 12$ we have already obtained.

Notice the difference between the statements

$$f'(x) \approx \Delta y / \Delta x \qquad \text{and} \qquad f'(x) = 3x^2.$$

For a particular value of Δx, the corresponding value of $\Delta y / \Delta x$ is an approximation of $f'(x)$. We can obtain another, better approximation by computing $\Delta y / \Delta x$ for a smaller Δx. The successively better approximations differ from one another by less and less. In particular, they differ less and less from the *limit value* $3x^2$. The value of the derivative $f'(x)$ is *exactly* $3x^2$.

More generally, for any function $y = f(x)$, a particular difference quotient $\Delta y / \Delta x$ is an approximation of $f'(x)$. Successively smaller values of Δx give successively better approximations of $f'(x)$. Again $f'(x)$ *exactly* equals the limiting value of these successive approximations. In some cases, however, we are only able to approximate that limiting value, as we often did in chapter 3, and for many purposes the approximation is entirely satisfactory. In this chapter we will concentrate on the exact statements that are possible for functions given by formulas.

The Other Basic Functions

Our formula for the derivative of the function $f(x) = x^3$ is one instance of the general rule for the derivative of $f(x) = x^r$.

The rule for the derivative of a power function

> For every real number r, the derivative of $f(x) = x^r$ is $f'(x) = rx^{r-1}$.

We can prove this rule for the case when r is a positive integer using algebraic manipulations very like the ones carried out for x^3; see the exercises for verifications of this and the other differentiation rules in this section. Using a rule for quotients of functions (coming later in this section), we can show that this rule also holds for *negative* integer exponents. Further arguments using the chain rule show that the pattern still holds for *rational* exponents. We can eliminate this case-by-case approach, though, by recalling the approach developed in chapter 4. We saw that we can give meaning to b^r for any positive base b and any real number r by defining

$$b^r = e^{r \ln(b)}.$$

Using the formulas for the derivatives of e^x and $\ln x$ together with the chain rule, we can prove the rule for $x > 0$ and for arbitrary real exponent r directly, without first proving the special cases for integer or rational exponents. See the exercises for details. Arguments justifying the formulas for the derivatives of the trigonometric functions are also in the exercises.

Combining Functions

We can form new functions by combining functions. We have already studied one of the most useful ways of doing this in chapter 3 when we looked at forming "chains" of functions and developed the **chain rule** for taking the derivative of such a chain. Suppose $u = f(x)$ and $y = g(u)$. Chaining these two functions together we have y as a function of x:

Functions combined by chains...

$$y = h(x) = g(f(x)).$$

The chain rule tells us how to find the derivative of y with respect to x. In function notation it takes the form

$$h'(x) = g'(f(x)) \cdot f'(x).$$

In Leibniz notation, using $f(x) = u$ we can write the chain rule as

$$\frac{dy}{dx} = \frac{dy}{du} \cdot \frac{du}{dx}.$$

The chain rule

We also saw in chapter 3 that the polynomial $5x^3 - 7x^2 + 3$ can be thought of as an *algebraic* combination of simple functions. We can build an even more complicated function by forming a quotient with this polynomial in the numerator and the difference of the functions $\sin x$ and e^x in the denominator. The result is

...and algebraically

$$\frac{5x^3 - 7x^2 + 3}{\sin x - e^x}.$$

The derivative of this function, as well as of other functions formed by adding, subtracting, multiplying, and dividing simpler functions, is obtained by use of the following rules for the derivatives of algebraic combinations of differentiable functions.

Algebraic Combinations of Functions

Function	Derivative
$f(x) + g(x)$	$f'(x) + g'(x)$
$f(x) - g(x)$	$f'(x) - g'(x)$
$cf(x)$	$cf'(x)$
$f(x) \cdot g(x)$	$f'(x) \cdot g(x) + f(x) \cdot g'(x)$
$\dfrac{f(x)}{g(x)}$	$\dfrac{g(x) \cdot f'(x) - f(x) \cdot g'(x)}{[g(x)]^2}$

Combining functions by adding, subtracting, multiplying, and dividing

Notice the signs in the
rules
Notice carefully that the product rule has a plus sign but the quotient rule has a
minus sign. You can remember these formulas better if you think about where
these signs come from. Increasing *either* factor increases a (positive) product,
so the derivative of *each* factor appears with a plus sign in the formula for the
derivative of a product. Similarly, increasing the numerator increases a positive
quotient, so the derivative of the numerator appears with a plus sign in the
formula for the derivative of a quotient. However, increasing the denominator
decreases a positive quotient, so the derivative of the denominator appears with
a minus sign.

Let's now use the rules to differentiate the quotient

$$\frac{5x^3 - 7x^2 + 3}{\sin x - e^x}.$$

First, the derivative of the numerator $5x^3 - 7x^2 + 3$ is

$$5(3x^2) - 7(2x) + 0 \qquad \text{which equals} \qquad 15x^2 - 14x.$$

Similarly the derivative of $\sin x - e^x$ is $\cos x - e^x$. Finally, the derivative of the
quotient function is obtained by using the rule for quotients:

$$\frac{(\sin x - e^x)(15x^2 - 14x) - (5x^3 - 7x^2 + 3)(\cos x - e^x)}{(\sin x - e^x)^2}.$$

The following examples further illustrate the use of the rules for algebraic
combinations of functions.

Function	Derivative
$-3e^t + \sqrt[3]{t}$	$-3e^t + (1/3)t^{-2/3}$
$\dfrac{5}{x^3} - 7x^4 + \ln x$	$5(-3)x^{-4} - 7(4x^3) + 1/x$
$7\sqrt{x}\cos x$	$7\left(\dfrac{1}{2\sqrt{x}}\right)\cos x + 7\sqrt{x}(-\sin x)$
$\left(\dfrac{4}{3}\right)\pi r^3$	$\left(\dfrac{4}{3}\right)\pi 3r^2$
$\dfrac{3s^6}{s^2 - s}$	$\dfrac{(s^2 - s)3(6s^5) - 3s^6(2s - 1)}{(s^2 - s)^2}$

For another kind of example, suppose the per capita daily energy consumption in a country is currently 800,000 BTU, and, due to energy conservation
efforts, it is falling at the rate of 1,000 BTU per year. Suppose too that the
population of the country is currently 200,000,000 people and is rising at the

rate of 1,000,000 people per year. Is the total daily energy consumption of this country rising or falling? By how much?

Three different quantities vary with time in this example: daily per capita energy consumption, population, and total daily energy consumption. We can model this situation with three functions $C(t)$, $P(t)$, and $E(t)$.

$C(t)$:   per capita consumption at time t
$P(t)$:   population at time t
$E(t)$:   total energy consumption at time t

Since the per capita consumption times the number of people in the population gives the total energy consumption, these three functions are related algebraically:

$$E(t) = C(t) \cdot P(t).$$

If $t = 0$ represents today, then we are given the two rates of change

$$C'(0) = -1{,}000 = -10^3 \text{ BTU per person per year, and}$$
$$P'(0) = 1{,}000{,}000 = 10^6 \text{ persons per year.}$$

Using the product rule we can compute the current rate of change of the total daily energy consumption:

$$\begin{aligned}
E'(0) &= C(0) \cdot P'(0) + C'(0) \cdot P(0) \\
&= (8 \times 10^5) \cdot (10^6) + (-10^3) \cdot (2 \times 10^8) \\
&= (8 \times 10^{11}) - (2 \times 10^{11}) \\
&= 6 \times 10^{11} \text{ BTU per year.}
\end{aligned}$$

So the total daily energy consumption is currently rising at the rate of 6×10^{11} BTU per year. Thus the growth in the population more than offsets the efforts to conserve energy.

Finally, it is a useful exercise to check that the units make sense in this computation. Recall that $C(t)$ represents *per capita* daily energy consumption, so the units for $C(0) \cdot P'(0)$ are

Checking units

$$\frac{\text{BTU}}{\text{person}} \cdot \frac{\text{persons}}{\text{year}} = \frac{\text{BTU}}{\text{year}}$$

and, similarly, the units for $C'(0) \cdot P(0)$ are

$$\frac{\text{BTU}}{\text{person}} \cdot \frac{1}{\text{year}} \cdot \text{persons} = \frac{\text{BTU}}{\text{year}}$$

Informal Arguments

All of the rules for differentiating algebraic combinations of functions can be proved by using the algebraic definition of the derivative as a limit of a difference quotient. In fact, we will examine such a formal proof below. However, informal arguments based on geometric ideas or other intuitive insights are also valuable aids to understanding. Here are three examples of such arguments.

Stretching y coordinates

- If a new function g is obtained from f by multiplying by a positive constant c, so $g(x) = cf(x)$, what is the relationship between the graphs of $y = f(x)$ and of $y = g(x)$? Stretching (or compressing, if c is less than 1) the y coordinates of the points of the graph of f by a factor of c yields the graph of g.

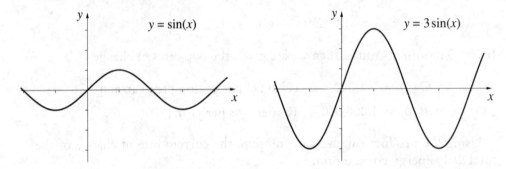

What then is the relationship between the slopes $f'(x)$ and $g'(x)$? If the y coordinates are tripled, the slope will be three times as great. If they are halved, the slope will also be half as much. More generally, the elongated (or compressed) graph of g has a slope equal to c times the slope of the original graph of f. In other words, $g(x) = cf(x)$ implies $g'(x) = cf'(x)$.

Shifting y coordinates

- Now suppose instead that g is obtained from f by adding a constant b, so $g(x) = f(x) + b$. This time the graph of $y = g(x)$ is obtained from the graph of $y = f(x)$ by shifting up or down (according to the sign of c) by $|c|$ units. What is the relationship between the slopes $f'(x)$ and $g'(x)$? The shifted graph has exactly the same slope as the original graph, so in this case, $g(x) = f(x) + b$ implies $g'(x) = f'(x)$.

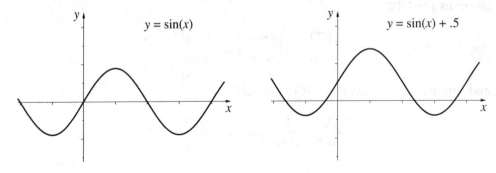

There is a similar pattern when the coordinates of the input variable are stretched or shifted—that is when $y = f(u)$ and u is *rescaled* by the linear relation $u = mx + b$. These results depend on the chain rule and appear in the exercises.

The fact that the derivative of $f(x) + b$ is the same as the derivative of $f(x)$ is a special case of the general *addition rule*, which says *the derivative of a sum is the sum of the derivatives*. In the special case, the derivative of the constant function b is zero, so adding a constant leaves the derivative unchanged. To see how natural it is to add rates in the general case, consider the following example.

- Suppose we are diluting concentrated orange juice by mixing it with water in a big tub. We may let $f(t)$ be the amount (in gallons) of concentrate in the tub, and $g(t)$ the amount of water in the tub at time t. Then $f'(t)$ is the rate at which concentrate is being added at time t (measured in gallons per minute), and $g'(t)$ is the rate at which water is flowing into the tub. $F(t) = f(t) + g(t)$ then gives the total amount of liquid in the tub at time t, and $F'(t)$ is the rate by which that total amount of liquid is changing at time t. Clearly that rate is the sum of the rates of flow of concentrate and water into the tank. If at some particular moment we are adding concentrate at the rate of 3.2 gal/min and water at the rate of 1.1 gal/min, the liquid in the tub is increasing by 4.3 gal/min at that moment.

Adding flows

A Formal Proof: The Product Rule

We include here the algebraic calculations yielding the rule for the derivative of the product of two arbitrary functions—just to give the flavor of these arguments. Algebraic arguments for the rest of these rules may be found in the exercises.

The Product Rule

$$F(x) = f(x) \cdot g(x) \qquad implies \qquad F'(x) = f'(x) \cdot g(x) + f(x) \cdot g'(x)$$

To save some writing, let

$$\Delta F = F(x + \Delta x) - F(x),$$
$$\Delta f = f(x + \Delta x) - f(x),$$

and

$$\Delta g = g(x + \Delta x) - g(x).$$

Rewrite the last two equations as

$$f(x + \Delta x) = f(x) + \Delta f$$
$$g(x + \Delta x) = g(x) + \Delta g.$$

Now we can write

$$F(x + \Delta x) = f(x + \Delta x) \cdot g(x + \Delta x)$$
$$= (f(x) + \Delta f) \cdot (g(x) + \Delta g)$$
$$= f(x) \cdot g(x) + f(x) \cdot \Delta g + \Delta f \cdot g(x) + \Delta f \cdot \Delta g$$

A simple expression for ΔF

This gives us a simple expression for

$$\Delta F = F(x + \Delta x) - F(x),$$

namely,

$$\Delta F = f(x) \cdot \Delta g + \Delta f \cdot g(x) + \Delta f \cdot \Delta g.$$

Interpret Δf, Δg, and ΔF as areas

These quantities all have nice geometric interpretations. First, think of the numbers $f(x)$ and $g(x)$ as lengths that depend on x; then $F(x)$ naturally stands for the area of the rectangle whose sides are $f(x)$ and $g(x)$. If the sides of the rectangle grow by the amounts Δf and Δg, then the area F grows by ΔF. As the following diagram shows, ΔF has three parts, corresponding to the three terms in the expression we derived algebraically for ΔF.

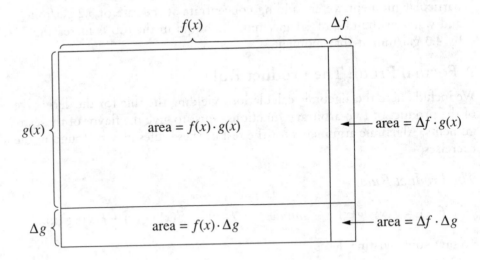

Now we divide ΔF by Δx and finish the argument:

$$\frac{\Delta F}{\Delta x} = \frac{f(x) \cdot \Delta g + \Delta f \cdot g(x) + \Delta f \cdot \Delta g}{\Delta x}$$

$$= f(x) \cdot \frac{\Delta g}{\Delta x} + \frac{\Delta f}{\Delta x} \cdot g(x) + \frac{\Delta f \cdot \Delta g}{\Delta x}$$

Consider what happens to each of the three terms as Δx gets smaller and smaller. In the first term, the second factor $\Delta g / \Delta x$ approaches $g'(x)$—by the *definition* of the derivative. The first factor, $f(x)$, doesn't change at all as Δx

shrinks. So the first term approaches $f(x) \cdot g'(x)$. Similarly, in the second term, the quotient $\Delta f / \Delta x$ approaches $f'(x)$, and the second term approaches $f'(x) \cdot g(x)$.

Finally, look at the third term. We would know what to expect if we had another factor of Δx in the denominator. We can put ourselves in familiar territory by the "trick" of multiplying the third term by $\Delta x / \Delta x$:

$$\frac{\Delta f \cdot \Delta g}{\Delta x} = \frac{\Delta f}{\Delta x} \cdot \frac{\Delta g}{\Delta x} \cdot \Delta x$$

Thus we can see that as Δx approaches zero, the third term itself approaches $f'(x) \cdot g'(x) \cdot 0 = 0$.

We may summarize our calculation by writing

$$\lim_{\Delta x \to 0} \frac{\Delta F}{\Delta x} = f(x) \cdot \left(\lim_{\Delta x \to 0} \frac{\Delta g}{\Delta x} \right) + \left(\lim_{\Delta x \to 0} \frac{\Delta f}{\Delta x} \right) \cdot g(x)$$

$$+ \left(\lim_{\Delta x \to 0} \frac{\Delta f}{\Delta x} \right) \cdot \left(\lim_{\Delta x \to 0} \frac{\Delta g}{\Delta x} \right) \cdot \left(\lim_{\Delta x \to 0} \Delta x \right)$$

from which we have

$$\lim_{\Delta x \to 0} \frac{\Delta F}{\Delta x} = f(x)g'(x) + f'(x) \cdot g(x) + f'(x) \cdot g'(x) \cdot 0$$

$$= f(x) \cdot g'(x) + f'(x) \cdot g(x).$$

This completes the proof of the product rule. Other formal arguments are left to the exercises.

Exercises

Finding Derivatives

1. Find the derivative of each of the following functions.
   a) $3x^5 - 10x^2 + 8$
   *b) $(5x^{12} + 2)(\pi - \pi^2 x^4)$
   c) $\sqrt{u} - 3/u^3 + 2u^7$
   d) $mx + b \ (m, b \text{ constant})$
   e) $.5 \sin x + \sqrt[3]{x} + \pi^2$
   *f) $\dfrac{\pi - \pi^2 x^4}{5x^{12} + 2}$
   g) $2\sqrt{x} - \dfrac{1}{\sqrt{x}}$
   h) $\tan z \ (\sin z - 5)$
   i) $\dfrac{\sin x}{x^2}$
   j) $x^2 e^x$
   k) $\cos x + e^x$
   l) $\sin x / \cos x$
   m) $e^x \ln x$
   n) $\dfrac{2^x}{10 + \sin x}$
   o) $\sin(e^x \cos x)$
   p) $6e^{\cos t} / 5\sqrt[3]{t}$
   q) $\ln(x^2 + xe^x)$
   r) $\dfrac{5x^2 + \ln x}{7\sqrt{x} + 5}$

2. Suppose f and g are functions and that we are given

$$f(2) = 3 \qquad g(2) = 4 \qquad g(3) = 2$$

$$f'(2) = 2 \qquad g'(2) = -1 \qquad g'(3) = 17$$

Evaluate the derivative of each of the following functions at $t = 2$:

a) $f(t) + g(t)$

f) $\sqrt{g(t)}$

b) $5f(t) - 2g(t)$

g) $t^2 f(t)$

c) $f(t)g(t)$

h) $(f(t))^2 + (g(t))^2$

d) $\dfrac{f(t)}{g(t)}$

i) $\dfrac{1}{f(t)}$

e) $g(f(t))$

j) $f(3t - (g(1 + t))^2)$

k) What additional piece of information would you need to calculate the derivative of $f(g(t))$ at $t = 2$?

l) Estimate the value of $f(t)/g(t)$ at $t = 1.95$.

3. a) Extend the product rule to express $(f(t)g(t)h(t))'$ in terms of f, g, and h.

b) If the length, width, and height of a rectangular box are changing at the rates of 3, 6, and -5 inches/minute at the moment when all three dimensions happen to be 10 inches, at what rate is the volume of the box changing then?

c) If the length, width, and height of a box are 10 inches, 12 inches, and 8 inches, respectively, and if the length and height of the box are changing at the rates of 3 inches/minute and -2 inches/minute, respectively, at what rate must the width be changing to keep the volume of the box constant?

4. In this problem we examine the effect of stretching or shifting the coordinates of the input variable of a function. Your answers should address both the *algebra* and the *geometry* of the problem to show how the algebraic relations between the functions are manifested in their graphs.

a) Suppose $f(x) = \sin(x)$ and $g(x) = \sin(mx)$, where m is a constant stretching factor. What is the relation between $f'(x)$ and $g'(x)$?

b) As in (a), suppose $f(x) = \sin(x)$, but this time $g(x) = \sin(x + b)$ where b is the size of a (constant) shift. What is the relation between $f'(x)$ and $g'(x)$ this time?

c) Now consider the general case: $f(x)$ is an unspecified differentiable function and $g(x) = f(mx + b)$, where the input variable is stretched by the constant factor m and shifted by the constant amount b. What is the relation between $f'(x)$ and $g'(x)$ in this general case?

5. Which of the following functions has a derivative which is always positive (except at $x = 0$, where neither the function nor its derivative is defined)?

$$1/x \qquad -1/x \qquad 1/x^2 \qquad -1/x^2$$

6. a) As a function of its radius r, the volume of a sphere is given by the formula $V(r) = \frac{4}{3}\pi r^3$. If r is measured in centimeters, what are the units for $V'(r)$?

b) Explain why square cm are *not* the appropriate units for $V'(r)$, even though dimensionally correct.

7. Do the following.

a) Show that $\dfrac{1}{1-x^2}$ and $\dfrac{x^2}{1-x^2}$ have the same derivative.

b) If $f'(x) = g'(x)$ for every x, what can be concluded about the relationship between f and g? [Hint: What is $(f(x) - g(x))'$?]

c) Show that $\dfrac{1}{1-x^2} = \dfrac{x^2}{1-x^2} + C$ by finding C.

8. Suppose that the current total daily energy consumption in a particular country is 16×10^{13} BTU and is rising at the rate of 6×10^{11} BTU per year. Suppose that the current population is 2×10^8 people and is rising at the rate of 10^6 people per year. What is the current daily per capita energy consumption? Is it rising or falling? By how much?

9. The population of a particular country is 15,000,000 people and is growing at the rate of 10,000 people per year. In the same country the per capita yearly expenditure for energy is $1,000 per person and is growing at the rate of $8 per year. What is the country's current total yearly energy expenditure? How fast is the country's total yearly energy expenditure growing?

10. The population of a particular country is 30 million and is rising at the rate of 4,000 people per year. The total yearly personal income in the country is 20 billion dollars, and it is rising at the rate of 500 million dollars per year. What is the current per capita personal income? Is it rising or falling? By how much?

11. An explorer is marooned on an iceberg. The top of the iceberg is shaped like a square with sides of length 100 feet. The length of the sides is shrinking at the rate of 2 feet per day. How fast is the area of the top of the iceberg shrinking? Assuming the sides continue to shrink at the rate of 2 feet per day, what will be the dimensions of the top of the iceberg in five days? How fast will the area of the top of the iceberg be shrinking then?

12. Suppose the iceberg of problem 11 is shaped like a cube. How fast is the volume of the cube shrinking when the sides have length 100 feet? How fast after five days?

Deriving Differentiation Rules

13. In this problem we calculate the derivative of $f(x) = x^4$.

a) Expand $f(x + \Delta x) = (x + \Delta x)^4 = (x + \Delta x)(x + \Delta x)(x + \Delta x)(x + \Delta x)$ as a sum of 12 terms. (Don't collect "like" terms yet.)

b) How many terms in part (a) involve *no* Δx's? What form do such terms have?

c) How many terms in part (a) involve exactly *one* Δx? What form do such terms have?

d) Group the terms in part (a) so that $f(x + \Delta x)$ has the form

$$Ax^4 + B\,\Delta x + R(\Delta x)^2,$$

where there are no Δx's among the terms in A or B, but R has several terms, some involving Δx. Use part (b) to check your value of A; use part (c) to check your value of B.

e) Compute the quotient $\dfrac{f(x + \Delta x) - f(x)}{\Delta x}$, taking advantage of part (d).

f) Now find

$$\lim_{n \to 0} \frac{f(x + \Delta x) - f(x)}{\Delta x}.$$

This is the derivative of x^4. Is your result here compatible with the rule for the derivative of x^n?

14. In this problem we calculate the derivative of $f(x) = x^n$, where n is any positive integer.

a) First show that you can write

$$f(x + \Delta x) = x^n + nx^{n-1}\Delta x + R(\Delta x)^2$$

by developing the following line of argument. Write $(x + \Delta x)^n$ as a product of n identical factors:

$$(x + \Delta x)^n = \underbrace{(x + \Delta x)}_{\text{1st}}\underbrace{(x + \Delta x)}_{\text{2nd}}\underbrace{(x + \Delta x)}_{\text{3rd}} \ldots \underbrace{(x + \Delta x)}_{n\text{th}}.$$

But now, before tackling this general case, look at the following examples. In the examples we use notation to help us keep track of which factors are contributing to the final result.

i) Consider the product $(a + b)(\underline{a} + \underline{b}) = a\underline{a} + a\underline{b} + b\underline{a} + b\underline{b}$. There are four individual terms. Each term contains one of the entries in the first factor (namely, a or b) and one of the entries in the second factor (namely, $\underline{a}$ or $\underline{b}$). The four terms represent thereby all possible ways of choosing one entry in the first factor and one entry in the second factor.

ii) Multiply out the product $(a + b)(\underline{a} + \underline{b})(A + B)$. (Don't combine like terms yet.) Does each term contain one entry from the first factor, one from the second, and one from the third? How many terms did you get? In fact there are two ways to choose an entry from the first factor, two ways to choose an entry from the second factor, and two ways to choose an entry from the third factor. Therefore, how many ways can you make a choice consisting of one entry from the first, one from the second, and one from the third?

Now return to the general case:

$$(x + \Delta x)^n = \underbrace{(x + \Delta x)}_{\text{1st}}\underbrace{(x + \Delta x)}_{\text{2nd}}\underbrace{(x + \Delta x)}_{\text{3rd}}\dots\underbrace{(x + \Delta x)}_{n\text{th}}.$$

How many ways can you choose an entry from each factor and *not* get any Δx's? Multiply these chosen entries together; what does the product look like (apart from having no Δx's in it)?

How many ways can you choose an entry from each factor in such a way that the resulting product has *precisely one* Δx? Describe all the various choices which give that result. What does a product that contains precisely one Δx factor look like? What do you obtain for the sum of *all* such terms with precisely one Δx factor?

What is the minimum number of Δx factors in any of the remaining terms in the full expansion of $(x + \Delta x)^n$?

Do your calculations agree with this summary:

$$(x + \Delta x)^n = x^n + nx^{n-1}\Delta x + R(\Delta x)^2?$$

b) Now find the value of

$$\frac{f(x + \Delta x) - f(x)}{\Delta x}.$$

c) Finally, find

$$\lim_{\Delta x \to 0} \frac{f(x + \Delta x) - f(x)}{\Delta x}.$$

Do you get nx^{n-1}?

15. In this problem we give another derivation of the power rule based on writing

$$x^r = e^{r \ln(x)}.$$

Use the chain rule to differentiate $e^{r \ln(x)}$. Explain why your answer equals rx^{r-1}.

16. Does the rule for the derivative of x^r hold for $r = 0$? Why or why not?

17. In this exercise we prove the addition rule: $F(x) = f(x) + g(x)$ implies $F'(x) = f'(x) + g'(x)$.
  a) Show $F(x + \Delta x) - F(x) = f(x + \Delta x) - f(x) + g(x + \Delta x) - g(x)$.
  b) Divide by Δx and finish the argument.

18. In this exercise we prove the quotient rule: $F(x) = f(x)/g(x)$ implies

$$F'(x) = \frac{g(x)f'(x) - f(x)g'(x)}{[g(x)]^2}.$$

  a) Rewrite $F(x) = f(x)/g(x)$ as $f(x) = g(x)F(x)$. Pretend for the moment that you know what $F'(x)$ is and apply the product rule to find $f'(x)$ in terms of $F(x)$, $g(x)$, $F'(x)$, $g'(x)$.
  b) Replace $F(x)$ by $f(x)/g(x)$ in your expression for $f'(x)$ in part (a).
  c) Solve the equation in part (b) for $F'(x)$ in terms of $f(x)$, $g(x)$, $f'(x)$, and $g'(x)$.

19. In this problem we calculate the derivative of $f(x) = x^n$ when n is a negative integer. First write $n = -m$, so m is a positive integer. Then $f(x) = x^{-m} = 1/x^m$.
  a) Use the quotient rule and this new expression for f to find $f'(x)$.
  b) Do the algebra to reexpress $f'(x)$ as nx^{n-1}.

20. In this problem we calculate the derivatives of $\sin x$ and $\cos x$. We will need the **addition formulas**:

$$\sin(A + B) = \sin A \cos B + \cos A \sin B$$
$$\cos(A + B) = \cos A \cos B - \sin A \sin B.$$

First tackle $f(x) = \sin x$.
  a) Use the addition formula for $\sin(A + B)$ to rewrite $f(x + \Delta x)$ in terms of $\sin(x)$, $\cos(x)$, $\sin(\Delta x)$, and $\cos(\Delta x)$.

  b) The quotient $\dfrac{f(x + \Delta x) - f(x)}{\Delta x}$ can now be written in the form

$$P(\Delta x) \cdot \sin x + Q(\Delta x) \cdot \cos x,$$

  where P and Q are specific functions of Δx. What are the formulas for those functions?
  c) Use a calculator or computer to estimate the limits

$$\lim_{\Delta x \to 0} P(\Delta x) \quad \text{and} \quad \lim_{\Delta x \to 0} Q(\Delta x).$$

  (Try $\Delta x = .1, .01, .001, .0001$. Be sure your calculator is set on radians, not degrees.) Using part (b) you should now be able to determine the limit

$$\lim_{\Delta x \to 0} \frac{f(x + \Delta x) - f(x)}{\Delta x}$$

by writing it in the form

$$\left(\lim_{\Delta x \to 0} P(\Delta x)\right) \cdot \sin x + \left(\lim_{\Delta x \to 0} Q(\Delta x)\right) \cdot \cos x.$$

d) What is $f'(x)$?

e) Proceed similarly to find the derivative of $g(x) = \cos x$.

21. In this problem we calculate the derivatives of the other circular functions. Use the quotient rule together with the derivatives of $\sin x$ and $\cos x$ to verify that the derivatives of the other four circular functions are as given in the table below:

Function	Derivative
$\tan x = \dfrac{\sin x}{\cos x}$	$\sec^2 x$
$\csc x = \dfrac{1}{\sin x}$	$-\cot x \csc x$
$\sec x = \dfrac{1}{\cos x}$	$\sec x \tan x$
$\cot x = \dfrac{1}{\tan x}$	$-\csc^2 x$

Differential Equations

22. If $y = f(x)$, then the **second derivative** of f is just the derivative of the derivative of f; it is denoted $f''(x)$ or d^2y/dx^2. Find the second derivative of each of the following functions.

*a) $f(x) = e^{3x-2}$

b) $f(x) = \sin \omega x$, where ω is a constant

c) $f(x) = x^2 e^x$

23. Show that e^{3x} and e^{-3x} both satisfy the (*second-order*) differential equation

$$f''(x) = 9f(x).$$

Furthermore, show that *any* function of the form $g(x) = \alpha e^{3x} + \beta e^{-3x}$ satisfies this differential equation. Here α and β are arbitrary constants. Finally, choose α and β so that $g(x)$ also satisfies the two conditions $g(0) = 12$ and $g'(0) = 15$.

24. Show that $y = \sin x$ satisfies the differential equation $y'' + y = 0$. Show that $y = \cos x$ also satisfies the differential equation. Show that,

in fact, $y = a \sin x + b \cos x$ satisfies the differential equation for any choice of constants a and b. Can you find a function $g(x)$ that satisfies these three conditions:

$$g''(x) + g(x) = 0$$
$$g(0) = 1$$
$$g'(0) = 4?$$

25. Show that $\sin \omega x$ satisfies the differential equation $y'' + \omega^2 y = 0$. What other solutions can you find to this differential equation? Can you find a function $L(x)$ that satisfies these three conditions:

$$L''(x) + 4L(x) = 0$$
$$L(0) = 36$$
$$L'(0) = 64?$$

The Colorado River Problem

Make your answer to this sequence of questions an essay. Identify all the variables you consider (for example, "A stands for the area of the lake"), and indicate the functional relationships between them ("A depends on time t, measured in weeks from the present"). Identify the derivatives of those functions, as necessary.

   The Colorado River—which excavated the Grand Canyon, among others—used to empty into the Gulf of California. It no longer does. Instead, it runs into a marshy area some miles from the Gulf and stops. One of the major reasons for this change is the construction of dams—notably the Hoover Dam. Every dam creates a lake behind it, and every lake increases the total surface area of the river. Since the rate at which water evaporates is proportional to the area of the water surface exposed to air, the lakes along the Colorado have increased the loss of river water through evaporation. Over the years, these losses (in conjunction with other factors, like increased usage by a rapidly growing population) have been significant enough to dry up the river at its mouth.

26. Let us analyze the evaporation rate along a river that was recently dammed. Suppose the lake is currently 50 yards wide, and getting wider at a rate of 3 yards per week. As the lake fills, it gets longer, too. Suppose it is currently 950 yards long, and it is extending upstream at a rate of 15 yards per week. Assuming the lake remains approximately rectangular as it grows, find
   a) the current area of the lake, in square yards;
   b) the rate at which the surface of the lake is currently growing, in square yards per week.

27. Suppose the lake continues to spread sideways at the rate of 3 yards per week, and it continues to extend upstream at the rate of 15 yards per week.

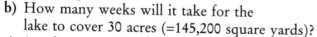
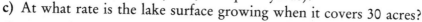

   a) Express the area of the lake as a (quadratic!) function of time, where time is measured from the present, in weeks, and where the lake's area is as given in problem 25.
   b) How many weeks will it take for the lake to cover 30 acres (=145,200 square yards)?
   c) At what rate is the lake surface growing when it covers 30 acres?

28. Compare the rates at which the surface of the lake is growing in problem 25 (which is the "current" rate) and in problem 26 (which is the rate when the lake covers 30 acres). Are these rates the same? If they are not, how do you account for the difference? In particular, the width and length grow at fixed rates, so why doesn't the area? Use what you know about derivatives to answer the question.

29. Suppose the local climate causes water to evaporate from the surface of the lake at the rate of .22 cubic yard per week, for each square yard of surface. Write a formula that expresses total evaporation per week in terms of area. Use E to denote total evaporation.

30. The lake is fed by the river, and that in turn is fed by rainwater and groundwater from its watershed. (The **watershed**, or basin, of a river is that part of the countryside containing the ponds and streams which drain into the river.) Suppose the watershed provides the lake, on average, with 25,000 cubic yards of new water each week.

   Assuming, as we did in problem 26, that the lake widens at the constant rate of 3 yards per week, and lengthens at the rate of 15 yards per week, will the time ever come that the water being added to the lake from its watershed balances the water being removed by evaporation? In other words, will the lake ever stop filling?

➤ 5.2 FINDING PARTIAL DERIVATIVES

We know from chapter 3 that no additional formulas are needed to calculate *partial* derivatives. We simply use the usual differentiation formulas, treating all the variables except one—the one with respect to which the partial derivative is formed—as if they were constants. If we do this we get new techniques for analyzing rates of change in problems that involve functions of several variables.

Some Examples

Here are two examples to illustrate the technique for calculating partial derivatives:

Finding formulas for partial derivatives

1. Suppose $f(x, y) = x^2y + 5x^3 - \sqrt{x + y}$. Then

$$f_x(x, y) = 2xy + 15x^2 - \frac{1}{2\sqrt{x + y}}, \text{ and}$$

$$f_y(x, y) = x^2 - \frac{1}{2\sqrt{x + y}}.$$

2. Suppose $g(u, v) = e^{uv} + \frac{u}{v}$. Then

$$g_u(u, v) = ve^{uv} + \frac{1}{v}, \text{ and}$$

$$g_v(u, v) = ue^{uv} - \frac{u}{v^2}.$$

Eradication of Disease

Controlling—or, better still, eradicating—a communicable disease depends first on the development of a vaccine. But even after this step has been accomplished, public health officials must still answer important questions, including:

- What proportion of the population must be vaccinated in order to eliminate the disease?
- At what age should people be vaccinated?

In their 1982 article, "Directly Transmitted Infectious Diseases: Control by Vaccination" (*Science*, Vol. 215, 1053–1060), Roy Anderson and Robert May formulate a model for the spread of disease that permits them to answer these and other questions. For a particular disease in a particular environment, the important variables in their model are

1. The average human life expectancy L, in years;
2. The average age A at which individuals catch the disease, in years;
3. The average age V at which individuals are vaccinated against the disease, in years.

Anderson and May deduce from their model that in order to eradicate the disease, the proportion of the population that is vaccinated must exceed p, where p is given by

$$p = \frac{L + V}{L + A}.$$

For a disease like measles, public health officials can directly affect the variable V, for example, by the recommendations they make to physicians about immunization schedules for children. They may also indirectly affect the variables A and L, because public health policy influences factors which can modify the age at which children catch the disease or the overall life expectancy of the population. (Many other factors affect these variables as well.) Which of these three variables has the greatest effect on the proportion of the population that must be vaccinated?

Partial derivatives can tell us which variables are most significant

In other words, which is largest: $\partial p / \partial L$, $\partial p / \partial A$, or $\partial p / \partial V$?

Using the rules, we compute:

$$\frac{\partial p}{\partial L} = \frac{1 \cdot (L + A) - 1 \cdot (L + V)}{(L + A)^2} = \frac{A - V}{(L + A)^2},$$

$$\frac{\partial p}{\partial A} = \frac{-(L + V)}{(L + A)^2}, \text{ and}$$

$$\frac{\partial p}{\partial V} = \frac{1}{L + A}.$$

For measles in the United States, reasonable values of the variables are $L = 70$ years, $A = 5$ years, and $V = 1$ year. Using these values, the crucial proportion of the population needing to be vaccinated is $p = 71/75 = .947$, and the partial derivatives are

$$\frac{\partial p}{\partial L} = \frac{4}{(75)^2} = .0007,$$

$$\frac{\partial p}{\partial A} = \frac{-71}{(75)^2} = -.0126,$$

$$\frac{\partial p}{\partial V} = \frac{1}{75} = .0133.$$

A comment is in order here on units. While the input variables L, A, and V are all measured in years—so the rates are *per year*, the output variable p is dimensionless: it is the ratio of persons vaccinated to persons not vaccinated. It would be reasonable to write p as a percentage. Then we can attach the units *percent per year* to each of the three partial derivatives. Thus we have:

Determining units

$$\frac{\partial p}{\partial L} = .07\% \text{ per year}$$

$$\frac{\partial p}{\partial A} = -1.26\% \text{ per year}$$

$$\frac{\partial p}{\partial V} = 1.33\% \text{ per year.}$$

It is not surprising that a change in average life expectancy has a negligible effect on the proportion p of the population that must be vaccinated in order to eradicate measles. Nor is it surprising that changing the age of vaccination has the greatest effect on p. But it is not obvious ahead of time that changing the age at which children catch the disease has nearly as large an effect on p:

- Decreasing the age of vaccination decreases the proportion p by 1.33% per year of decrease.
- Increasing the age at which children catch measles decreases the proportion p by 1.26% per year of increase.

Changes can also go the "wrong" way. For example, in an area where use of communal child care facilities is growing, contact among very young children increases, and the age at which children are exposed to—and can catch—communicable diseases like measles falls. The Anderson–May model tells us that immunization practices must change to compensate: either the age of vaccination must drop a like amount, or the fraction of the population that is vaccinated must grow by 1.26% per year of decrease in the average age at infection.

Exercises

Finding Partial Derivatives

1. Find the partial derivatives of the following functions.
   a) $x^2 y$

   b) $\sqrt{x + y}$

   c) e^{xy}

   d) $\dfrac{y}{x}$

   *e) $\dfrac{x + y}{y + z}$

   f) $\sin \dfrac{y}{x}$

2 **a)** Suppose $f(x, y) = e^{-(x+2y)}(2x - 5y)$. Find $f_x(x, y)$ and $f_y(x, y)$.

b) Find a point (a, b) at which $f_x(a, b) = 0$. At such a point a small change in x leaves the value of f virtually unchanged.

c) Find a point (a, b) at which a small *increase* in the x-value would produce the same change in $f(a, b)$ as would the same-sized *decrease* in the y-value.

3. Suppose $g(u, v) = \dfrac{\sin u + v^2 + 7uv}{1 + u^2 + v^4}$. Find $g_u(u, v)$ and $g_v(u, v)$.

4. The **second partial derivatives** of $z = f(x, y)$ are the partial derivatives of $\partial f / \partial x$ and $\partial f / \partial y$, namely:

$$\frac{\partial^2 f}{\partial x^2} = \frac{\partial}{\partial x}\left(\frac{\partial f}{\partial x}\right)$$

$$\frac{\partial^2 f}{\partial x \partial y} = \frac{\partial}{\partial x}\left(\frac{\partial f}{\partial y}\right)$$

$$\frac{\partial^2 f}{\partial y \partial x} = \frac{\partial}{\partial y}\left(\frac{\partial f}{\partial x}\right)$$

$$\frac{\partial^2 f}{\partial y^2} = \frac{\partial}{\partial y}\left(\frac{\partial f}{\partial y}\right)$$

Find the four second partial derivatives of the following functions.

a) $x^2 y$

b) $\sqrt{x + y}$

c) e^{xy}

*d) $\dfrac{y}{x}$

e) $\sin \dfrac{y}{x}$

Eradication of Disease

5. Suppose you were dealing with measles in a developing country where $L = 50$ years, $A = 4$ years, and $V = 2$ years. Discuss the impact on measles control if increased public health efforts increase L to 55 years, A to 5 years, and decrease V to 1.5 years.

Partial Differential Equations

6. Show that the function $z = \dfrac{1}{\sqrt{t}} \exp \dfrac{-x^2}{4t}$ satisfies the **partial differential equation**

$$\frac{\partial^2 z}{\partial x^2} = \frac{\partial z}{\partial t}.$$

7. Show that *every* linear function of the form $z = px + qy + c$ satisfies the partial differential equation

$$\frac{\partial^2 z}{\partial x^2} + \frac{\partial^2 z}{\partial y^2} = 0.$$

Here p, q, and c are arbitrary constants.

8. Show that the function $z = e^x \sin y$ also satisfies the partial differential equation

$$\frac{\partial^2 z}{\partial x^2} + \frac{\partial^2 z}{\partial y^2} = 0.$$

➤ 5.3 THE SHAPE OF THE GRAPH OF A FUNCTION

We know from chapter 3 that the derivative gives us qualitative information about the shape of the graph of a differentiable function.

Function	Derivative
increasing	positive
decreasing	negative
level	zero
steep (rising or falling)	large (positive or negative)
gradual (rising or falling)	small (positive or negative)
straight	constant

Having a formula for the derivative of a function will thus give us a great deal of information about the behavior of the function itself. In particular we will be interested in using the derivative to solve **optimization problems**—finding maximum or minimum values of a function. Such problems occur frequently in many fields.

Contexts for optimization problems

- Economists actually define human rationality in terms of optimization. Each person is assumed to have a *utility function*, a function that assigns to each of many possible outcomes its *utility*, a numerical measure of its value to the person. (Different people may have different utility functions, depending on their personal value systems.) A rational person is one who acts to maximize his or her utility. Sometimes utility is expressed in terms of money. For example, a rational manufacturer will seek to maximize her profit (in dollars). Her profit will depend on—that is, be a function of—such variables as the cost of her raw materials and the unit price she charges for her product.

- Many physical laws are expressed as minimum principles. Ordinary soap bubbles exhibit one of these principles. A soap film has a *surface energy* which is proportional to its surface area. For almost any physical system, its stable state is one which minimizes its energy. Stable soap films are thus examples of *minimal surfaces*. Interfaces involving crystals also have surface energies, leading to the study of crystalline minimal surfaces.

- Statisticians develop mathematical summaries for data—in other words, mathematical models. For example, a relationship between two numerical variables may be summarized by a linear function, say $y = mx + b$, where x and y are the variables of interest. It would be very rare to find data for which such a linear relation holds exactly. In a particular case, the statistician chooses the linear model that minimizes the *discrepancy* between the actual values of y and the theoretical values obtained from the linear function. Statisticians frequently measure this discrepancy by summing the squares of the differences between the actual and the theoretical values of y for each data point. The *best-fitting line* or *regression line* is the graph of the linear function which is optimal in this sense.

- Psychologists who study decision making have found that some people are "risk averse"; they make their decisions primarily to avoid risks. If we regard risk as a function of the various outcomes under consideration (a bit like a utility function), such a person acts to minimize this function.

The derivative is the key tool here. We will develop a general procedure for using the derivative of a function to locate the extremes of the functions.

Language

Here is a graph of what we might consider a "generic" differentiable function.

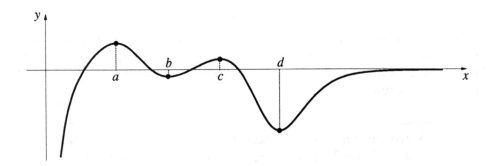

The most distinctive features are the tops and bottoms, points where the graph levels and the derivative is zero. We distinguish between **local** extremes, like those occurring at the points $x = b$, $x = c$, and $x = d$ and a **global** extreme, like the global maximum at the point $x = a$. The function has a **local minimum** at $x = b$ because $f(x) \geq f(b)$ for all x *sufficiently near* b. The function has a **global maximum** at $x = a$ because $f(x) \leq f(a)$ for *all* x. Notice that this particular function does not have a global minimum. What kinds of local extremes does the function have at $x = c$ and $x = d$? The convention is to say that *all* extremes are local extremes, and a local extreme may or may not also be a global extreme.

Local extremes and global extremes

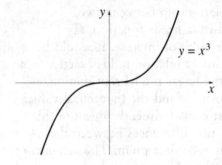

Examining the graph of as simple a function as $f(x) = x^3$ shows us that a function need not have any extremes at all. Moreover, since $f'(0) = 0$ for this function, a zero derivative doesn't necessarily identify a point where an extreme occurs.

Can a function have an extreme at a point *other* than where the derivative is zero? Consider the graph of $f(x) = x^{2/3}$ below.

Some functions have no extremes, even if their derivative equals zero

A minimum can occur at a cusp

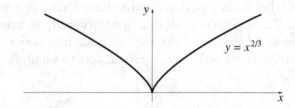

This function is differentiable everywhere except at the point $x = 0$. And it is at this very point, where

$$f'(x) = \frac{2}{3}x^{-1/3} = \frac{2}{3\sqrt[3]{x}}$$

is undefined, that the function has its global minimum. For this reason, points where the derivative fails to exist (or is infinite) are as important as points where the derivative equals zero. All of these kinds of points are called **critical points** for the function.

> A **critical point** for a function f is a point on the graph of f where f' equals zero or infinity or fails to exist.

How can we recognize, from looking at its graph, that a function fails to be differentiable at a point? We know that when the graph has a sharp corner or **cusp** it isn't locally linear at that point, and so has no derivative there. Thus critical points occur at these places. When the graph is locally linear but vertical, the slope cannot be given a finite numerical value. Critical points also occur at these vertical places where the slope is infinite. The graph at the right is an example of such a curve.

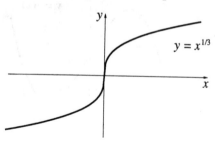

A weaker condition than differentiability, but one that is useful, especially in this context, is **continuity**. We say that a function f is **continuous at a point** $x = a$ if

- it is defined at the point, and

- we can achieve changes in the output that are arbitrarily small by restricting changes in the input to be sufficiently small.

This second condition can also be expressed in the following form:

Given any positive number ϵ (the proposed bound on the change in the output is traditionally designated by the Greek letter ϵ, pronounced "epsilon"), there is always a positive number δ (the Greek letter "delta"), such that whenever the change in the input is less than δ, then the corresponding change in the output will be less than ϵ.

A function is continuous on a set of real numbers if it is continuous at each point of the set. A natural way to think of (and recognize) a function which is continuous on an interval is by its graph on that interval, which is continuous in the usual sense: it has no gaps or jumps in it. You can draw it without lifting your pencil from the page. Of the four functions whose graphs appear on the next page, f is continuous (and differentiable); g is continuous, but not differentiable (because of the cusp at $x = a$); h is not continuous, because h is undefined at $x = b$; and k is not continuous, because of the "jump" at $x = c$.

Graphs of continuous functions have no gaps or jumps

Luckily, the functions that we are likely to encounter are continuous on their natural domains. Among functions given by formulas, the only exceptions we have to worry about are quotients, like $f(x) = 1/x$, which have gaps in their natural domains at points where the denominator vanishes [at $x = 0$ in the case of $f(x) = 1/x$], so their domains are not single intervals. For convenience, we usually confine our attention to functions continuous on an interval.

A quotient isn't continuous where its denominator is zero

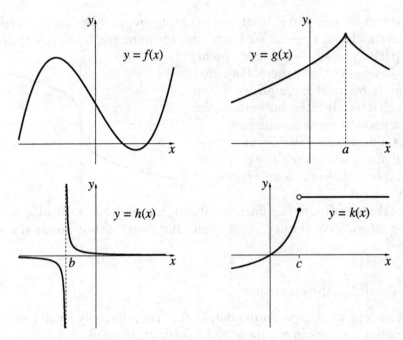

The Existence of Extremes

Not every function has extremes, as the example of $f(x) = x^3$ shows. We are, however, guaranteed extremes for certain functions.

A continuous function on a finite closed interval has extremes

Principle I. We are guaranteed that a function has a global maximum and a global minimum if we know:

- the domain of the function is a *finite closed interval*, and
- the function is *continuous* on this interval.

The domain restriction in most optimization problems is likely to come from the physical constraints of the problem, not the mathematics.

A **finite closed interval**, written $[a, b]$, is the set of all real numbers between a and b, including the endpoints a and b. Other kinds of finite intervals are $(a, b]$ (which excludes the endpoint a), $[a, b)$, and (a, b). Infinite intervals include open and closed "rays" like $(a, +\infty)$, $[a, +\infty)$, $(-\infty, b)$ and $(-\infty, b]$, and the entire real line.

For example, Principle I does not apply to $f(x) = 1/x$ on the finite closed interval $[-1, 2]$, because this function isn't continuous at every point on this interval—in fact, the function isn't even defined for $x = 0$.

However, Principle I *does* apply to the same function if we change the finite closed interval to one that doesn't include $x = 0$. In the figure below we use $[1, 3]$.

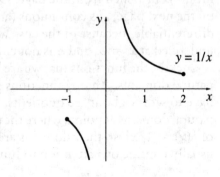

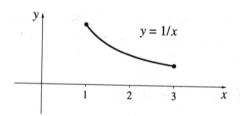

A function which fails to satisfy the first condition of Principle I can still have global extremes. For any function continuous on an interval we can apply the following principle.

Principle II. If a function f is continuous on an interval and has a local extreme at a point $x = c$ of the interval, then either $x = c$ is a critical point for the function, or $x = c$ is an endpoint of the interval. In other words, we are guaranteed that one of the following three conditions holds:

- $f'(c) = 0$,
- $f'(c)$ is undefined,
- $x = c$ is an endpoint of the interval.

Continuous functions have extremes only at endpoints or critical points

The following graphs illustrate three local maxima, each satisfying one of these three conditions.

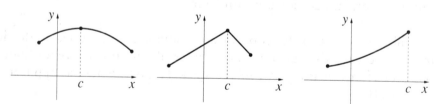

When we apply Principle II to optimization problems, an important part of the task will be ascertaining which, if any, of the critical points or endpoints we find actually gives the extreme we're looking for. We'll examine a variety of techniques, graphical and analytical, for locating critical points and determining what kind of extreme point (if any) they are.

Finding Extremes
Using a Graphical Approach

If we can use a computer to examine the graph of the function of interest, we can determine the existence and location of extremes by inspection. However, every graphing utility requires the user to specify the interval on which the function will be graphed, and careful analysis may be required in order to choose an interval that contains all the extremes of interest.

Computer graphing can be easy if the general location of the extremes is known

For functions given by data, whose graphs have only finitely many points, we can zoom in to find the exact coordinates of the extreme datapoints. For a function given by a formula, we can estimate the coordinates of an extreme to arbitrary accuracy by zooming in on the point as closely as desired. This is the method we used in some of the exercises of chapter 1, and it is quite satisfactory in many situations.

Using the Formula for the Derivative

Formulas can give exact answers and can handle parameters

In this chapter we are concentrating on functions given by formulas. For these functions we may want a method other than the approximation using a graphing utility described above.

- For some functions, the determination of extremes using a formula for the derivative is at least as easy as using the computer.
- Some functions are described in terms of a *parameter*, a constant whose value may vary from one problem to another. For example, the rate equation from the *S-I-R* model for change in the number of infected, $I' = aSI - bI$, involves two parameters a and b, the transmission and recovery coefficients. For such a function, we cannot use the computer unless we specify a numerical value for each parameter, thus limiting the generality of our results.
- The computer gives only an approximation, while an exact answer may convey important additional information.

We will assume that the function we are studying is continuous on the domain of interest and that its domain is an interval. Our procedure for finding extremes of a function given by a formula $y = f(x)$ is thus a direct application of Principle II.

The Search for Local Extremes

1. Determine the domain of the function and identify its endpoints, if any. Keep in mind that in an applied context, the domain may be determined by physical or other restrictions.
2. Find a formula for $f'(x)$.
3. Find any roots of $f'(x) = 0$ in the domain. An estimation procedure, for example, Newton's method (see section 5.5), may be used for complicated equations.
4. Find any points in the domain where $f'(x)$ is undefined.
5. Determine the shape of the graph to locate any local extremes. Find the shape either by looking directly at the graph of $y = f(x)$ or by analyzing the sign of $f'(x)$ on either side of each critical point $x = c$. [Sometimes the second derivative $f''(c)$ aids the analysis; see the exercises for details.]

The Search for Global Extremes

1. Find the local extremes, as above.
2. If the domain is a finite closed interval, it is only necessary to compare the values of f at each of the critical points and at the endpoints to determine which is the global maximum and which is the global minimum.
3. More generally, use the shape of the graph to ascertain whether the desired global extreme exists and to identify it.

In the succeeding sections and in the exercises we will carry out these search procedures in a variety of situations. Since there are often substantial algebraic difficulties in analyzing the sign of the derivative, or even determining when it is equal to zero, in realistic problems, in section 5.6 we will develop some numerical methods for handling such complications.

Exercises

Describing Functions

1. For each of the following graphs, is the function continuous? Is the function differentiable?

a)

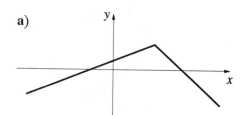

b)

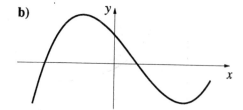

(c)

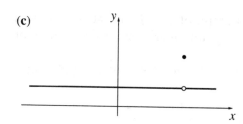

d)

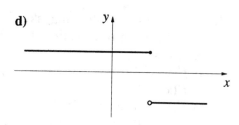

2. For each of the following graphs of a function $y = f(x)$, is f' increasing or decreasing? At the indicated point, what is the sign of f'? Is f' (*not* f) increasing or decreasing at the point? What does this then say about the sign of f'' (the derivative of f') at the point?

a)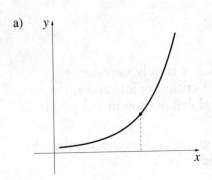

b)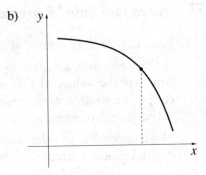

c)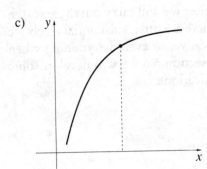

d)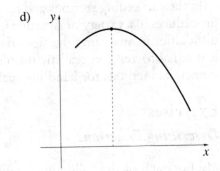

e)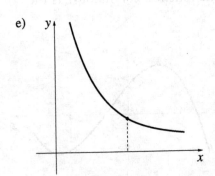

f)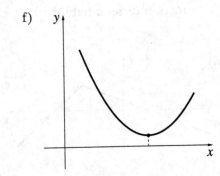

3. For each of the following, sketch a graph of $y = f(x)$ that is consistent with the given information. On each graph, mark any critical points or extremes.

   a) $f'(x) > 0$ for $x < 1$; $f'(1) = 0$; $f'(x) < 0$ for $1 < x < 2$; $f'(2) = 0$; $f'(x) > 0$ for $x > 2$.

   b) $f'(x) > 0$ for $x < 2$; $f'(2) = 0$; $f'(x) > 0$ for $x > 2$.

   c) $f'(x) > 0$ for $x < 2$; $f'(2) = 0$; $f'(x) < 0$ for $x > 2$.

   d) $f'(3) = 0$; $f''(3) > 0$.

4. *The geometric meaning of the second derivative.* If f is any function and if $f''(x)$ is positive over some interval, then f' is increasing over that interval, and we say the curve is **concave upward** over that interval. If $f''(x)$ is negative over some interval, then f' is decreasing over that interval, and we say the curve is **concave downward** over

that interval. You should study the graphs in problem 2 until you are clear what this means geometrically.

a) Suppose there were four functions f, g, h, and k, and that $f(0) = g(0) = h(0) = k(0) = 0$, and $f'(0) = g'(0) = h'(0) = k'(0) = 1$. Suppose, moreover, that $f''(0) = 1$, $g''(0) = 5$, $h''(0) = -1$, and $k''(0) = -5$. Sketch possible graphs of these functions near the origin.

b) We know that the magnitude of the first derivative tells us how steep the curve is—the greater the value of f', positive or negative, the steeper the curve. What does the magnitude of the second derivative tell us geometrically about the shape of the curve? Complete the sentence: "The greater the value of f'', the _____."

5. a) Here is the graph you saw back on page 265:

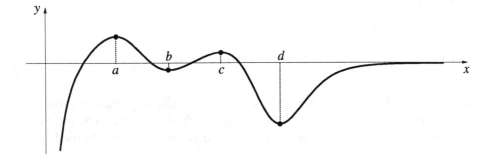

At which point is the second derivative greater, b or d, and why?

b) Reproduce a sketch of this curve and indicate where the curve is concave up and where it is concave down.

c) What must be true about the second derivative at the points where the curve changes concavity from up to down, or vice versa? Give a clear justification for your answer.

d) What must be true about the second derivative near the right-hand end of the graph, and why?

e) Put all this together to sketch the graph of the second derivative of this function. Label the values a, b, c, d on your sketch.

6. *Second derivative test for maxima and minima.* Explain why the following test works.

- If $f'(c) = 0$ and $f''(c) > 0$, then f has a *local minimum* at $x = c$.
- If $f'(c) = 0$ and $f''(c) < 0$, then f has a *local maximum* at $x = c$.

[Hint: If $f'(c) = 0$ and $f''(c) > 0$, what can you say about the value of $f'(x)$—and hence the geometry of the graph of f—on either side of c?]

Finding Critical Points

7. For each of the following functions, find the critical points, if any, without using a computer or calculator. Can you sketch the graph of the function near the critical point? Use the second derivative test if you can't figure out the behavior from a simpler inspection.

*a) $f(x) = x^{1/3}$

b) $f(x) = x^3 + \frac{3}{2}x^2 - 6x + 5$

c) $f(x) = \dfrac{2x + 1}{x - 1}$

d) $f(x) = \sin x$

e) $f(x) = \sqrt{1 - x^2}$

f) $f(x) = \dfrac{e^x}{x}$

g) $f(x) = x \ln x$

h) $f(x) = x^c + \dfrac{1}{x^c}$, where c is some constant

8. For each of the following graphs, mark any critical points or extremes. Indicate which extremes are local and which are global. (Assume that at their ends the curves continue in the direction they are headed.)

a)

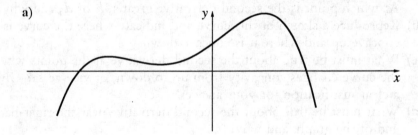

b)

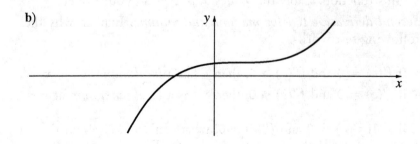

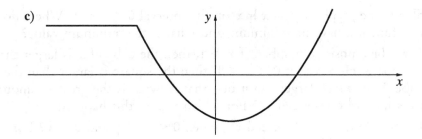

Finding Extremes

Except where indicated, you should not use a computer or calculator to solve the following problems.

*9. For what positive value of x does $f(x) = x + (7/x)$ attain its minimum value? Explain how you found this value.

10. For what value of x in the interval $[1, 2]$ does $f(x) = x + (7/x)$ attain its minimum value? Explain how you found this value.

11. Use a graphing program to make a sketch of the function $y = f(x) = x^2 2^{-x}$ on the interval $0 \le x \le 10$. From the graph, estimate the value of x which makes y largest, accurate to four decimal places. Then find where y takes on its maximum by setting the derivative $f'(x)$ equal to 0. You should find that the maximum occurs at $x = 2/\ln 2$. Finally, what is the numerical value of this estimate of $2/\ln 2$, accurate to seven decimal places?

12. Use a graphing program to sketch the graph of $y = (1/x^2) + x$ on the interval $0 < x < 4$ and estimate the value of x that makes y smallest on this interval. Then use the derivative of y to find the exact value of x that makes y smallest on the same interval. Compare the estimated and exact values.

13. What is the smallest value $y = (4/x^2) + x$ takes on when x is a positive number? Explain how you found this value.

14. The function $y = x^4 - 42x^2 - 80x$ has two local minima. Where are they (that is, what are their x coordinates?), and what are their values? Which is the lower of the two minima? In this problem you will have to solve a cubic equation. You can use an estimation procedure, but it is also possible to solve by factoring the cubic. (The roots are integers.)

15. The function $y = x^4 - 6x^2 + 7$ has two local minima. Where are they, and which of the two is lower? In this problem you will have to solve a cubic equation. Do this by factoring and then approximating the x values to four decimal places.

16. Sketch the graph of $y = x \ln x$ on the interval $0 < x < 1$. Where does this function have its minimum, and what is the minimum value?

17. Let x be a positive number; if $x > 1$, then the cube of x is larger than its square. However, if $0 < x < 1$, then the square is larger than the cube. How *much* larger can it be—that is, what is the greatest amount by which x^2 can exceed x^3? For which x does this happen?

18. By how much can x^p exceed x^q, when $0 < p < q$ and $0 < x$? For which x does this happen?

▷ 5.4 OPTIMAL SHAPES

The Problem of the Optimal Tin Can

What is the minimum surface area?

Suppose you are a tin can manufacturer. You must make a can to hold a certain volume V of canned tomatoes. Naturally, the can will be a cylinder, but the proportions, the height h and the radius r, can vary. Your task is to choose the proportions so that you use the least amount of tin to make the can. In other words, you want the surface area of the can to be as small as possible.

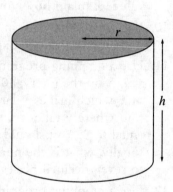

The Solution

The surface area is the sum of the areas of the two circles at the top and bottom of the can, plus the area of the rectangle that would be obtained if the top and bottom were removed and the side cut vertically.

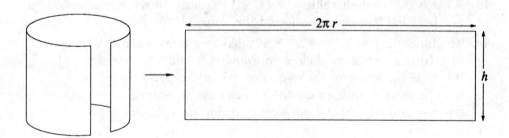

Thus A depends on r and h:

$$A = 2\pi r^2 + 2\pi rh.$$

However, r and h cannot vary independently. Because the volume V is fixed, r and h are related by

Finding A as a function of r

$$V = \pi r^2 h.$$

Solving the equation above for h in terms of r,

$$h = \frac{V}{\pi r^2},$$

we may express A as a function of r alone:

$$A = f(r) = 2\pi r^2 + 2\pi r \frac{V}{\pi r^2} = 2\pi r^2 + \frac{2V}{r}.$$

Notice that the formula for this function involves the parameter V. The mathematical description of our task is to find the value of r that makes $A = f(r)$ a minimum.

V is a parameter

Following the procedure of the previous section, we first determine the domain of the function. Clearly this problem makes physical sense only for $r > 0$. Looking at the equation

Finding the domain of $f(r)$

$$h = \frac{V}{\pi r^2},$$

we see that although V is fixed, r can be arbitrarily large provided h is sufficiently small (resulting in a can that looks like an elephant stepped on it). Thus the domain of our function is $r > 0$, which is not a closed interval, so we have no guarantee that a minimum exists.

Next we compute $f'(r)$, keeping in mind that the symbols V and π represent constants and that we are differentiating with respect to the variable r.

$$f'(r) = 4\pi r - \frac{2V}{r^2} = \frac{4\pi r^3 - 2V}{r^2}.$$

The derivative is undefined at $r = 0$, which is outside the domain under consideration. So now we set the derivative equal to zero and solve for any possible critical points.

Looking for critical points

$$f'(r) = \frac{4\pi r^3 - 2V}{r^2}$$

$$0 = \frac{4\pi r^3 - 2V}{r^2}$$

$$0 = 4\pi r^3 - 2V$$

$$r = \sqrt[3]{V/2\pi}.$$

Thus $r = \sqrt[3]{V/2\pi}$ is the only critical point.

Finding the shape of
the graph of A

We can actually sketch the shape of the graph of A versus r based on this analysis of $f'(r)$. The sign of $f'(r)$ is determined by its numerator, since the denominator r^2 is always positive.

- When $r < \sqrt[3]{V/2\pi}$, the numerator is negative, so the graph of A is falling.
- When $r > \sqrt[3]{V/2\pi}$, the numerator is positive, so the graph is rising.

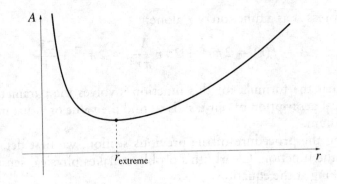

Obviously, A has a *minimum* at

$$r_{\text{extreme}} = \sqrt[3]{V/2\pi}.$$

A has a minimum but
no maximum

It is also obvious that there is no *maximum* area, since

$$\lim_{r \to 0} A = \infty \quad \text{and} \quad \lim_{r \to \infty} A = \infty.$$

Thus we see again that not *every* optimization problem has a solution.

The Mathematical Context: Optimal Shapes

It is interesting to find the value of the height $h = h_{\text{extreme}}$ when the area is a minimum,

$$h_{\text{extreme}} = \frac{V}{\pi r_{\text{extreme}}^2}.$$

When we replace r_{extreme} by $\sqrt[3]{V/2\pi}$ and simplify this expression, we obtain

$$h_{\text{extreme}} = 2r_{\text{extreme}}.$$

In other words, the height of the optimal tin can exactly equals its diameter. Campbell soup cans are far from optimal, but a can of Progresso plum tomatoes has diameter 4 inches and height 4.5 inches. Does someone at Progresso know calculus?

Problems like the one above—as well as others that you will find in the exercises—have been among the mainstays of calculus courses for generations. However, if you really had to face the problem of optimizing the shape of a tin can as a practical matter, you would probably have a numerical value for V specified in advance. In that case, your first impulse might be to use a graphing utility to approximate $r_{extreme}$ as accurately as your needs warrant. This is entirely appropriate. Using a graphing utility gives us the answer as quickly as taking derivatives, and you are less likely to make mistakes in algebra. Moreover, the functions encountered in many other real problems are often complicated and messy, and their derivatives are hard to analyze. A graphing utility may provide the only practical course open to you.

Graphing utilities are often best for solving practical problems

So why insist that you use derivatives on these problems? It is because the actual context of these examples is not really saving money for food canners. The real context is geometry. If we had used a graphing utility to solve the can problem, we might have noticed that the optimal radius was about half the height, but we wouldn't have known that the relationship is exact. Nor would we have recognized that the relationship holds for cylinders of arbitrary volume.

So why do things the hard way?

Using a graphing utility can be mechanical, which is part of its virtue if we just need a specific answer to a particular problem. Part of the purpose of a calculus course, though, is to develop the concepts, tools, and the geometric vocabulary for thinking about problems in general, to see the connecting threads between apparently disparate settings. The more clearly you can think within a general framework, not just that of the specific problem being addressed, the better your chances of seeing unsuspected relationships in the problem at hand or thinking more creatively about other problems in other settings. The ability to give precise expression to our intuitions can lead to deeper insights. You should thus try to see the exercises that follow as not just about fences and storage bins, but as opportunities to observe that, often, the geometric regularity that pleases the eye is also optimal.

Symmetric shapes are often optimal

Exercises

For each of the following problems, find a function relating the variables and use differentiation to find the optimal value specified. Some of the problems (1, 3, 4, and 5) are expressed in terms of a general parameter—P, L, A, and L again—rather than specific numerical values. If this gives you trouble, try doing the problem using the specific parameter value given at the end of the problem for computer verification. Use other specific values as needed until you can do the problem in terms of the general parameter, which should behave exactly like any of the specific values you tried.

1. Show that the rectangle of perimeter P whose area is a maximum is a square. Use a graphing utility to check your answer for the special case when $P = 100$ feet.

2. An open rectangular box is to be made from a piece of cardboard 8 inches wide and 15 inches long by cutting a square from each corner and bending up the sides. Find the dimensions of the box of largest volume. Use a graphing utility to check your answer.

3. One side of an open field is bounded by a straight river. A farmer has L feet of fencing. How should the farmer proportion a rectangular plot along the river in order to enclose as great an area as possible? Use a graphing utility to check your answer for the special case when $L = 100$ feet.

4. An open storage bin with a square base and vertical sides is to be constructed from A square feet of wood. Determine the dimensions of the bin if its volume is to be a maximum. (Neglect the thickness of the wood and any waste in construction.) Use a graphing utility to check your answer for the special case when $A = 100$ square feet.

5. A roman window is shaped like a rectangle surmounted by a semicircle. If the perimeter of the window is L feet, what are the dimensions of the window of maximum area? Use a graphing utility to check your answer for the special case when $L = 100$ feet.

6. Suppose the roman window of problem 5 has clear glass in its rectangular part and colored glass in its semicircular part. If the colored glass transmits only half as much light per square foot as the clear glass does, what are the dimensions of the window that transmits the most light? Use a graphing utility to check your answer for the special case when $L = 100$ feet.

7. A cylindrical oil can with radius r inches and height h inches is made with a steel top and bottom and cardboard sides. The steel costs 3 cents per square inch, the cardboard costs 1 cent per square inch, and rolling the crimp around the top and bottom edges costs 1/2 cent per linear inch. (Both crimps are done at the same time, so only count the contribution of one circumference.)
   a) Express the cost C of the can as a function of r and h.
   b) Find the dimensions of the cheapest can holding 100 cubic inches of oil. (You'll need to solve a cubic equation to find the critical point. Use a graphing utility or an estimation procedure to approximate the critical point to three decimal places.)

▷ 5.5 NEWTON'S METHOD

Finding Critical Points

When we solve optimization problems for functions given by formulas, we begin by calculating the derivative and using the derivative formula to find critical points. Almost always the derivative is defined for all elements of the domain, and we find the critical points by determining the roots of the equation obtained by setting the derivative equal to zero.

Finding the roots of this equation often requires an estimation procedure. For example, consider the function

$$f(x) = x^4 + x^3 + x^2 + x + 1.$$

The derivative of f is

$$f'(x) = 4x^3 + 3x^2 + 2x + 1,$$

which is certainly defined for all x.

In order to use a graphing utility to find the roots of $f'(x) = 0$, we need to choose an interval that will contain the roots we seek. Since $f'(0) = 1 > 0$ and $f'(-1) = -2 < 0$, we know f' has at least one root on $[-1, 0]$. (Why is this?) But might there be other roots outside this interval? *Solving $f'(x) = 0$*

It is easy to see that $f'(x)$ is positive for all $x > 0$, so there are no roots to the right of $[-1, 0]$. What about $x < -1$? Rewriting the derivative as

$$f'(x) = (2x + 1)(2x^2 + 1) + x^2$$

lets us see that $f'(x)$ is negative for all $x < -1$ (check this for yourself), so there are no roots to the left of $[-1, 0]$ either.

Examining the graph of $y = 4x^3 + 3x^2 + 2x + 1$ on $[-1, 0]$, we see that it crosses the x-axis exactly once, so there is a unique critical point. Progressively shrinking the interval, we find that, to eight decimal places, this critical point is

$$c = -.60582958\ldots.$$

Notice that in this method it requires roughly the same amount of work to get each additional digit in the answer.

There is, however, another approximation procedure we can use, called Newton's method. This has wide applicability and usually converges very rapidly to solutions—additional digits are much easier to get than in the method we just looked at. It also has the virtue of being algorithmic, so that we can write a single computer program which can then be used for any root-finding problem we may encounter without a great deal of thought and decision making. Newton's method is also an interesting application of both local linearity and successive approximation, two important themes of this course. *A good algorithm is a convenient tool*

Local Linearity and the Tangent Line

Let's give the derivative function the new name g to emphasize that we now want to consider it as a function in its own right, out of the context of the function f from which it was derived. Look at the following graph of the function *The root is an x-intercept of the graph*

$$y = g(x) = 4x^3 + 3x^2 + 2x + 1.$$

We are seeking the number r such that $g(r) = 0$. The root r is the x-coordinate of the point where the graph crosses the x-axis.

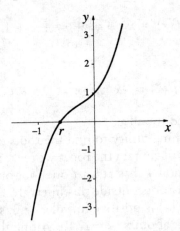

The basic plan of attack in Newton's method is to replace the graph of $y = g(x)$ by a straight line that looks reasonably like the graph near the root r. Then, the x-intercept of that line will be a reasonably good estimate for r. The graph below includes such a line.

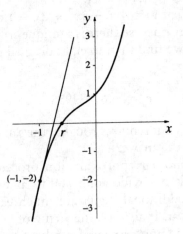

The tangent line extends the local linear approximation

The function g is locally linear at $x = -1$, and we have drawn an extension of the local linear approximation of g at this point. This line is called the **tangent line** to the graph of $y = g(x)$ at $x = -1$, by analogy to the tangent line to a circle. To find the x-intercept of the tangent line, we must know its equation. Clearly the line passes through the **point of tangency** $(-1, g(-1)) = (-1, -2)$. What is its slope? It is the same as the slope of the local linear approximation at $(-1, g(-1))$, namely

$$g'(-1) = 12(-1)^2 + 6(-1) + 2 = 8.$$

Thus the equation of the tangent line is

The equation of the tangent line

$$y + 2 = 8(x + 1).$$

To find the x-intercept of this line, we must set y equal to zero and solve for x: $0 + 2 = 8(x + 1)$ gives us $x = -.75$. Of course, this x-intercept is not equal to r, but it's a better approximation than, say, -1. To get an even better approximation, we repeat this process, starting with the line tangent to the graph of g at $x = -0.75$ instead of at $x = -1$.

Finding the x-intercept of the tangent line

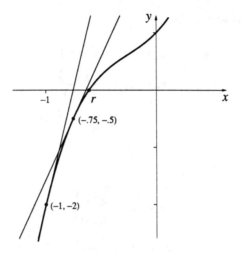

The slope of this new tangent line is $g'(-.75) = 4.25$, and it passes through $(-.75, -.5)$, so its equation is

$$y + .5 = 4.25(x + .75).$$

Setting $y = 0$ and solving for x gives a new x-intercept equal to $-.6323529$, closer still to r.

It seems reasonable to repeat the process yet again, using the tangent line at $x = -0.6323529$. Let's first introduce some notation to keep track of our computations. Call the original point of tangency x_0, so $x_0 = -1$. Let x_1 be the x-intercept of the tangent line at $x = x_0$. Draw the tangent line at $x = x_1$, and call its x-intercept x_2. Continuing in this way, we get a sequence of points $x_0, x_1, x_2, x_3, \ldots$ which appear to approach nearer and nearer to r. That is, they appear to approach r as their limit.

Successive approximations get closer to the root r

The Algorithm

This process of using one number to determine the next number in the sequence is the heart of Newton's method—it is an iterative method. Moreover, it turns out to be quite simple to calculate each new estimate in terms of the previous one. To see how this works, let's compute x_1 in terms of x_0. We know

The general equation
of the tangent line

x_1 is the x-intercept of the line tangent to the graph of g at $(x_0, g(x_0))$. The slope of this line is $g'(x_0)$, so

$$y - g(x_0) = g'(x_0)(x - x_0)$$

is the **equation of the tangent line**. Since this line crosses the x-axis at the point $(x_1, 0)$, we set $x = x_1$ and $y = 0$ in the equation to obtain

$$0 - g(x_0) = g'(x_0)(x_1 - x_0).$$

Now it is easy to solve for x_1:

$$g'(x_0)(x_1 - x_0) = -g(x_0)$$

$$x_1 - x_0 = \frac{-g(x_0)}{g'(x_0)}$$

$$x_1 = x_0 - \frac{g(x_0)}{g'(x_0)}.$$

In the same way we get

$$x_2 = x_1 - \frac{g(x_1)}{g'(x_1)},$$

$$x_3 = x_2 - \frac{g(x_2)}{g'(x_2)},$$

and so on.

To summarize, suppose that x_0 is given some value START. Then Newton's method is the computation of the sequence of numbers determined by

$$x_0 = \text{START}$$

$$x_{n+1} = x_n - \frac{g(x_n)}{g'(x_n)}, \quad n = 0, 1, 2, 3, \ldots$$

As we have seen many times, the sequence

$$x_1, x_2, x_3, \ldots, x_n, \ldots$$

The limit of
the successive
approximations is
the root

is a list of numbers to which we can always add a new term—by iterating our method yet again. For most functions, if we begin with an appropriate starting value of x_0, there is another number r that is the *limit* of this list of numbers, in the sense that the difference between x_n and r becomes as small as we wish as n increases without bound,

$$r = \lim_{n \to \infty} x_n.$$

The numbers $x_1, x_2, x_3, \ldots, x_n, \ldots$ constitute a sequence of *successive approximations* for the root r of the equation $g(x) = 0$. We can write a computer program to carry out this algorithm for as many steps as we choose. The program NEWTON does just that for $g(x) = 4x^3 + 3x^2 + 2x + 1$.

Program: NEWTON
Newton's method for solving $g(x) = 4x^3 + 3x^2 + 2x + 1 = 0$

```
start = -1
numberofsteps = 8
x = start
FOR n = 0 to numberofsteps
    print n,x                    {This prints x_n}
    g = 4*x^3+3*x^2+2*x+1
    gprime = 12*x^2+6*x+2
    x = x - g/gprime
NEXT n
```

If we program a computer using this algorithm with start = -1, then we get

$$x_0 = -1.00000000000$$
$$x_1 = -.75000000000$$
$$x_2 = -.63235294118$$
$$x_3 = -.60687911790$$
$$x_4 = -.60583128240$$
$$x_5 = -.60582958619$$
$$x_6 = -.60582958619$$
$$x_7 = -.60582958619$$
$$x_8 = -.60582958619$$

Thus we have found the root of $g(x) = 0$—the critical point we were looking for. In fact, after only six steps we could see that the value of the critical point was specified to at least ten decimal places. Also at the sixth step, we had the eight decimal places obtained with the use of the graphing utility. In fact, it turns out that the number of decimal places fixed roughly doubles with each round. In the above list, for instance, x_2 fixed one decimal, x_3 fixed two decimals, x_4 fixed four, x_5 fixed ten (the eleventh digit of the root is really an 8, which gets rounded to a 9 in $x_6 - x_8$), and x_6 would have fixed at least twenty if we had printed them all out! Moreover, by changing only three lines of the program—the first, sixth, and seventh—we can use NEWTON for any other function. With the use of the program NEWTON, we will see that in most cases we can obtain results more quickly and to a higher degree of accuracy with Newton's method than by using a graphing utility, although it is still sometimes helpful to use a graphing utility to get a reasonable starting value.

Examples

EXAMPLE 1.   Start with $\cos x = x$. The solution(s) to this equation (if any) will be the x coordinates of any points of intersection of the graphs of $y = \cos x$ and $y = x$. Sketch these two graphs and convince yourself that there is one solution, between 0 and $\pi/2$. The equation $\cos x = x$ is not in the form $g(x) = 0$, so rewrite it as $\cos x - x = 0$. Now we can apply Newton's method with $g(x) = \cos x - x$. Try starting with $x_0 = 1$. This gives the iteration scheme

$$x_0 = 1$$

$$x_{n+1} = x_n - \frac{\cos x_n - x_n}{-\sin x_n - 1}, \quad n = 0, 1, 2, \ldots.$$

The numbers we get are

$$x_0 = 1.000000000$$
$$x_1 = .750363868\ldots$$
$$x_2 = .739112891\ldots$$
$$x_3 = .739085133\ldots$$
$$x_4 = .739085133\ldots.$$

We have the solution to nine decimal places in only four steps. Not only does Newton's method work, it works fast!

EXAMPLE 2.   Suppose we continue with the equation $\cos x = x$, but this time choose $x_0 = 0$. What will we find? The numbers we get are

$$x_0 = .000000000$$
$$x_1 = 1.000000000$$
$$x_2 = .750363868.$$

There's no need to continue; we can see that we will obtain $r = .739085133\ldots$ as in Example 1. Look again at your sketch and see why you might have predicted this result.

EXAMPLE 3.   Next, let's find the roots of the polynomial $x^5 - 3x + 1$. This means solving the equation $x^5 - 3x + 1 = 0$. We know the necessary derivative, so we're ready to apply Newton's method, except for one thing: Which starting value x_0 do we pick? This is the part of Newton's method that leaves us on our own.

Finding the starting value x_0 can be hard

Assuming that some graphing software is available, the best thing to do is graph the function. But for most graphing utilities, we need to choose an interval. How do we choose one which is sure to include all the roots of the polynomial? The derivative $5x^4 - 3$ of this polynomial is simple enough that we can use it to get an idea of the shape of the graph of $y = x^5 - 3x + 1$ before we turn to the computer. Clearly the derivative is zero only for $x = \pm \sqrt[4]{3/5}$, and the derivative is positive except between these two values of x. In other words, we know the shape of the graph of $y = g(x)$—it is increasing, then decreases for a bit, then increases from there on out. This still is not enough information to tell us how many roots g has, though; its graph might lie in any one of the following configurations and so have 1, 2, or 3 roots (there are two other possibilities not shown—one has 2 roots and one has 1 root).

Using the derivative to find the shape of the graph

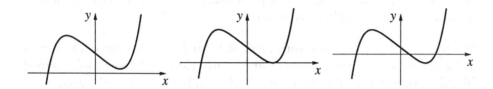

We can thus say that $g(x) = 0$ has at least one and at most three real roots. However, if we further observe that $g(-2) = -25$, $g(-1) = 3$, $g(0) = 1$, $g(1) = -1$, and $g(2) = 27$, we see that the graph of g must cross the x-axis at some value of x between -2 and -1, between 0 and 1, and again between 1 and 2. Therefore the right-hand sketch above must be the correct one.

Or we can almost as easily turn to a graphing utility. If we try the interval $[-5, 5]$, we see again that the graph crosses the x-axis in exactly three points.

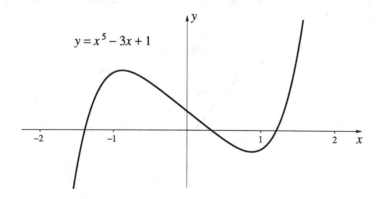

$$y = x^5 - 3x + 1$$

One of the roots is between -24 and -1, one is between 0 and 1, and the third is between 1 and 2. To find the first, we apply Newton's method with $x_0 = -2$. Then we get

Finding the root between -2 and -1

$$x_0 = -2.000000000$$
$$x_1 = -1.675324675\ldots$$
$$x_2 = -1.478238029\ldots$$
$$x_3 = -1.400445373\ldots$$
$$x_4 = -1.389019863\ldots$$
$$x_5 = -1.388792073\ldots$$
$$x_6 = -1.388791984\ldots$$
$$x_7 = -1.388791984\ldots$$

This took a few more steps than the other examples, but not a lot. Notice again that once there are any decimals fixed at all, the number of decimals fixed roughly doubles in the next approximation. In the exercises you will be asked to compute the other two roots.

EXAMPLE 4. Let's use Newton's method to find the obvious solution $r = 0$ of $x^3 - 5x = 0$. If we choose x_0 sufficiently close to 0, Newton's method should work just fine. But what does "sufficiently close" mean? Suppose we try $x_0 = 1$. Then we get

$$x_0 = 1$$
$$x_1 = -1$$
$$x_2 = 1$$
$$x_3 = -1$$
$$x_4 = 1$$
$$\vdots$$

Newton's method can fail

The x_n's oscillate endlessly, never getting close to 0. Going back to the geometric interpretation of Newton's method, this oscillation can be explained by the graph below.

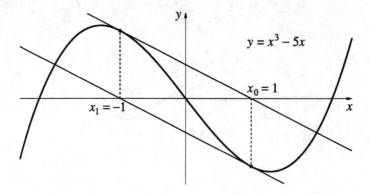

Using more advanced methods, it is possible to get precise estimates for how close x_0 needs to be to r in order for Newton's method to succeed. For now, we'll just have to rely on common sense and trial and error.

One important thing to note is the relation between algebra and Newton's method. Although we can now solve many more equations than we could earlier, this doesn't mean that we can abandon algebra. In fact, given a new equation, you should first try to solve it algebraically, for exact solutions are often better. Only when this fails should you look for approximate solutions using Newton's method. So don't forget algebra—you'll still need it!

Exercises

1. *The Babylonian algorithm.* Show that the Babylonian algorithm of chapter 2 is the same as Newton's method applied to the equation $x^2 - a = 0$.

*2. When Newton introduced his method, he did so with the example $x^3 - 2x - 5 = 0$. Show that this equation has only one root, and find it.

> *This example appeared in 1669 in an unpublished manuscript of Newton's (a published version came later, in 1711). The interesting fact is that Newton's method differs from the one presented here: his scheme was more complicated, requiring a different formula to get each approximation. In 1690, Joseph Raphson transformed Newton's scheme into the one used above. Thus, "Newton's method" is more properly called the "Newton–Raphson method," and many modern texts use this more accurate name.*

3. Use Newton's method to find a solution of $x^3 + 2x^2 + 10x = 20$ near the point $x = 1$.

> *The approximate solution 1;22,7,42,33,4,40 of this equation appears in a book written in 1228 by Leonardo of Pisa (also known as Fibonacci). This number looks odd because it's written in sexagesimal notation: it translates into*

$$1 + \frac{22}{60} + \frac{7}{60^2} + \frac{42}{60^3} + \frac{33}{60^4} + \frac{4}{60^5} + \frac{40}{60^6}.$$

> *This solution is accurate to ten decimal places, which is not bad for 750 years ago. In the Middle Ages, there was a lot of interest in solving equations. There were even contests, with a prize going to the person who could solve the most. The quadratic formula, which expresses algebraically the roots of any second-degree equation, had been known for thousands of years, but there were no general methods for finding roots of higher-degree equations in the thirteenth century. We don't know how Leonardo found his solution—why give away your secrets to your competitors!*

4. Use Newton's method to find a solution of $x^3 + 3x^2 = 5$.

> *In 1530, Nicolo Tartaglia was challenged to solve this equation algebraically. Five years later, in 1535, he found the solution*

$$x = \sqrt[3]{\frac{3 + \sqrt{5}}{2}} + \sqrt[3]{\frac{3 - \sqrt{5}}{2}} - 1.$$

> *Initially, Tartaglia could only solve certain types of cubic equations, but this was enough to let him win some famous contests with other mathematicians of the time. By 1541, he knew the general solution, but he made the mistake of telling Geronimo Cardano. Cardano published the solution in 1545 and the resulting formulas are called "Cardan's formulas."*

The above solution of $x^3 + 3x^2 = 5$ is called a **solution by radicals** because it is obtained by extracting various roots or radicals. Similarly, some time before 1545, Luigi Ferrari showed that any fourth-degree equation can be solved by radicals. This led to an intense interest in the fifth-degree equation. To see what happens in this case, read the next problem.

5. In Example 3, we saw that one root of $x^5 - 3x + 1$ was $-1.39887919\ldots$. Use Newton's method to find the other two roots.

> In 1826, Niels Henrik Abel proved that the general polynomial of degree 5 or greater cannot be solved by radicals. Using the work of Evariste Galois (done around 1830, but not understood until many years later), it can be shown that the roots of the equation $x^5 - 3x + 1 = 0$ cannot be expressed by any combination of radicals. Thus algebra can't solve this equation—some kind of successive approximation technique is unavoidable!

*6. One of the more surprising applications of Newton's method is to compute reciprocals. To make things more concrete, we will compute 1/3.4567. Note that this number is the root of the equation $1/x = 3.4567$.

a) Show that the formula of Newton's method gives us

$$x_{n+1} = 2x_n - 3.4567x_n^2.$$

b) Using $x_0 = .5$ and the formula from (a), compute 1/3.4567 to a high degree of accuracy.

c) Try starting with $x_0 = 1$. What happens? Explain graphically what goes wrong.

This method for computing reciprocals is important because it involves only *multiplication* and *subtraction*. Since $a/b = a \cdot (1/b)$, this implies that division can likewise be built from multiplication and subtraction. Thus, when designing a computer, the division routine doesn't need to be built from scratch—the designer can use the method illustrated here. There are some computers that do division this way.

7. In this problem we will determine the maximum value of the function

$$f(x) = \frac{x+1}{x^4+1}.$$

a) Graph $f(x)$ and convince yourself that the maximum value occurs somewhere around $x = .5$. Of course, the exact location is where the slope of the graph is zero, that is, where $f'(x) = 0$. So we need to solve this equation.

b) Compute $f'(x)$.

c) Since the answer to (b) is a fraction, it vanishes when its numerator does. Setting the numerator equal to 0 gives a fourth-degree equation. Use Newton's method to find a solution near $x = .5$.

d) Compute the maximum value of $f(x)$.

8. Consider the hyperbola $y = 1/x$ and the circle $x^2 - 4x + y^2 + 3 = 0$.
   a) By graphing the circle and the hyperbola, convince yourself that there are two points of intersection.
   b) By substituting $y = 1/x$ into the equation of the circle, obtain a fourth-degree equation satisfied by the x coordinate of the points of intersection.
   c) Solve the equation from (b) by Newton's method, and then determine the points of intersection.

9. Sometimes Newton's method doesn't work so nicely. For example, consider the equation $\sin x = 0$.
   a) Compute x_1 using Newton's method for each of the four starting values $x_0 = 1.55, 1.56, 1.57$, and 1.58.
   b) The answers you get are wildly different. Using the basic formula

$$x_{n+1} = x_n - \frac{g(x_n)}{g'(x_n)},$$

   explain why.

The Epidemic Runs Its Course

We return to the epidemiology example we have studied since chapter 1. Recall that our S-I-R model keeps track of three subgroups of the population: the susceptible, the infected, and the recovered. One of the interesting features of the model is that the larger the initial susceptible population, the more rapidly the epidemic runs its course. We observe this by choosing fixed values of $R_0 = R(0)$ and $I_0 = I(0)$ and looking at graphs of $S(t)$ versus t for various values of $S_0 = S(0)$.

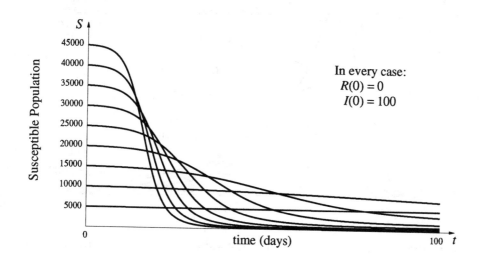

We see in each case that for sufficiently large t the graph of S levels off, approaching a value we'll call S_∞:

$$S_\infty = \lim_{t \to \infty} S(t).$$

What we mean by the epidemic "running its course" is that $S(t)$ reaches this limit value. We can see from the graphs that the value of S_0 affects the number S_∞ of individuals who escape the disease entirely. It turns out that we can actually find the value of S_∞ if we know the values of S_0, I_0, and the parameters a and b.

Recall that a is the *transmission coefficient*, and b is the *recovery coefficient* for the disease. The differential equations of the S-I-R model are

$$S' = -aSI$$
$$I' = aSI - bI$$
$$R' = bI.$$

10. Use the differentiation rules together with these differential equations to show that

$$(I + S - (b/a) \cdot \ln S)' = 0$$

11. Explain why the result of problem 10 means that $I + S - (b/a) \cdot \ln S$ has the same value—call it C—for every value of t.

12. Look at the graphs of the solutions $I(t)$ for various values of $S(0)$ below.

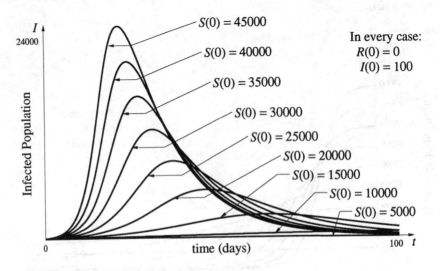

Write $\lim_{t \to \infty} I(t) = I_\infty$. What is the value of I_∞ for all values of $S(0)$?

13. Use the results of problems 11 and 12 to explain why

$$S_\infty - \frac{b}{a}\ln(S_\infty) = I_0 + S_0 - \frac{b}{a}\ln(S_0).$$

This equation determines S_∞ implicitly as a function of I_0 and S_0. For particular values of I_0 and S_0 (and of the parameters), you can use Newton's method to find S_∞.

14. Use the values

$$a = .00001 \text{ (person-days)}^{-1}$$
$$b = .08 \text{ day}^{-1}$$
$$I_0 = 100 \text{ persons}$$
$$S_0 = 35,000 \text{ persons}.$$

Writing x instead of S_∞ gives

$$x - 8000\ln(x) = -48,605.$$

Apply Newton's method to find S_∞. Judging from the graph for $S_0 = 35,000$, it looks like a reasonable first estimate for S_∞ might be 100.

15. Using the same values of a, b, and I_0 as in problem 14, determine the value of S_∞ for each of the following initial population sizes:
    a) $S_0 = 45,000$.
    b) $S_0 = 25,000$.
    c) $S_0 = 5,000$.

➤ 5.6 CHAPTER SUMMARY

The Main Ideas

- **Formulas** for **derivatives** can be calculated for functions given by formulas using the definition of the derivative as the limit of difference quotients

$$\lim_{\Delta x \to 0} \frac{f(x + \Delta x) - f(x)}{\Delta x}.$$

- Each particular difference quotient is an **approximation** to the derivative. Successive approximations, for smaller and smaller Δx, approach the derivative as a limit. The formula for the derivative gives its *exact* value.

- Formulas for **partial derivatives** are obtained using the formulas for derivatives of functions of a single variable by simply treating all variables other than the one of interest as if they were constants.
- **Optimization** problems occur in many different contexts. For instance, we seek to maximize benefits, minimize energy, and minimize error.
- The **sign** of the derivative indicates where a graph rises and where it falls.
- Functions which are **continuous** on a **finite closed interval** have **global extremes**.
- For a function continuous on an interval, its **local extremes** occur at **critical points**—points where the derivative equals zero or fails to exist—or at endpoints.
- Local linearity permits us to replace the graph of a function $y = g(x)$ by its **tangent line** at a point near a root of $g(x) = 0$, and then the x-intercept of the tangent line is a better approximation to the root. Successive approximations, obtained by iterating this procedure, yield **Newton's method** for solving the equation $g(x) = 0$.

Expectations

- You should be able to **differentiate** a function given by a formula.
- You should be able to use differentiation formulas to **calculate** partial derivatives.
- In most cases, you should be able to determine from a graph of a function on an interval whether that function is **continuous** and/or **differentiable** on that interval.
- You should be able to find **critical points** for a function of one or several variables given by a formula.
- You should be able to use the formula for the derivative of a function of a single variable to find **local and global extremes**.
- You should be able to use **Newton's method** to solve an equation of the form $g(x) = 0$.

Chapter Exercises

Prices, Demand, and Profit

Suppose the demand D (in units sold) for a particular product is determined by its price p (in dollars), $D = f(p)$. It is reasonable to assume that when the price is low, the demand will be high, but as the price rises, the demand will fall. In other words, we assume that the slope of the demand function is negative. If the manufacturing cost for each unit of the product is c dollars, then the profit per unit at price p is $p - c$. Finally the total profit T gained at the unit price p will be the number of units sold at the price p [that is, the demand $D(p)$] multiplied by the profit per unit $p - c$.

$$T = g(p) = D(p)\,\text{units} \times (p - c)\,\frac{\text{dollars}}{\text{unit}}.$$

In this series of problems we will determine the effect of the demand function and of the unit manufacturing cost on the maximum total profit.

1. Suppose the demand function is linear

$$D = f(p) = 1000 - 500p \text{ units,}$$

and the unit manufacturing cost is .20, so the total profit is

$$T = g(p) = (1000 - 500p)(p - .20) \text{ dollars.}$$

Find the "best" price—that is, find the price that yields the maximum total profit.

2. Suppose the demand function is the same as in problem 1, but the unit manufacturing cost rises to .30. What is the "best" price now? How much of the rise in the unit manufacturing cost is passed on to the consumer if the manufacturer charges this best price?

3. Suppose the demand function for a particular product is $D(p) = 2000 - 500p$, and that the unit manufacturing cost is .30. What price should the manufacturer charge to maximize profit? Suppose the unit manufacturing cost rises to .50. What price should the manufacturer charge to maximize profit now? How much of the rise in the unit manufacturing cost should be passed on to the consumer?

4. If the demand function for a product is $D(p) = 1500 - 100p$, compare the "best" price for unit manufacturing costs of .30 and .50. How much of the rise in cost should the manufacturer pass on to the consumer?

5. As you may have noticed, problems 2–4 illustrate an interesting phenomenon. In each case, exactly half of the rise in the unit manufacturing cost should be passed on to the consumer. Is this a coincidence?
   a) Consider the most general case of a linear demand function

$$D = f(p) = a - mp$$

and unit cost c. What is the "best" price? Is exactly half the unit manufacturing cost passed on to the consumer? Explain your answer.

b) Now consider a nonlinear demand function

$$D = f(p) = \frac{1000}{1 + p^2}.$$

Find the "best" price for unit costs
  i) .50 dollar per unit;
  ii) 1.00 dollar per unit;
  iii) 1.50 dollars per unit.
How much of the price increase is passed on to the consumer in cases
(ii) and (iii)?

CHAPTER 6

THE INTEGRAL

There are many contexts—work, energy, area, volume, distance traveled, and profit and loss are just a few—where the quantity in which we are interested is a product of known quantities. For example, the electrical energy needed to burn three 100-watt light bulbs for Δt hours is $300 \cdot \Delta t$ watt-hours. In this example, though, the calculation becomes more complicated if lights are turned off and on during the time interval Δt. We face the same complication in any context in which one of the factors in a product varies. To describe such a product we will introduce the **integral**.

As you will see, the integral itself can be viewed as a variable quantity. By analyzing the rate at which that quantity changes, we will find that every integral can be expressed as the solution to a particular differential equation. We will thus be able to use all our tools for solving differential equations to determine integrals.

➤ 6.1 MEASURING WORK

Human Work

Let's measure the work done by the staff of an office that processes catalog orders. Suppose a typical worker in the office can process 10 orders an hour. Then we would expect 6 people to process 60 orders an hour; in 2 hours, they could process 120 orders.

Processing catalog orders

$$10 \frac{\text{orders per hour}}{\text{person}} \times 6 \text{ persons} \times 2 \text{ hours} = 120 \text{ orders}.$$

Notice that a staff of 4 people working 3 hours could process the same number of orders:

$$10 \frac{\text{orders per hour}}{\text{person}} \times 4 \text{ persons} \times 3 \text{ hours} = 120 \text{ orders} .$$

Human work is measured as a product

It is natural to say that 6 persons working 2 hours do the same amount of work as 4 persons working 3 hours. This suggests that we use the product

$$\text{number of workers} \times \text{elapsed time}$$

to measure **human work**. In these terms, it takes 12 "person-hours" of human work to process 120 orders.

Another name that has been used in the past for this unit of work is the "man-hour." If the task is large, work can even be measured in "man-months" or "man-years." The term we will use most of the time is "staff-hour."

Measuring the work in terms of person-hours or staff-hours may seem a little strange at first—after all, a typical manager of our catalog order office would be most interested in the number of orders processed; that is, the **production** of the office. Notice, however, that we can rephrase the rate at which orders are processed as 10 orders per staff-hour. This is sometimes called the **productivity rate**. The productivity rate allows us to translate human work into production:

Productivity rate

$$\text{production} = \text{productivity rate} \times \text{human work}$$

$$120 \text{ orders} = 10 \frac{\text{orders}}{\text{staff-hour}} \times 12 \text{ staff-hours}.$$

As this equation shows, production varies linearly with work and the productivity rate serves as multiplier (see our discussion of the multiplier on pages 28–29).

If we modify the productivity rate in a suitable way, we can use this equation for other kinds of jobs. For example, we can use it to predict how much mowing a lawn mowing crew will do. Suppose the productivity rate is .7 acre per staff-hour. Then we expect that a staff of S working for H hours can mow

Mowing lawns

$$.7 \frac{\text{acre}}{\text{staff-hour}} \times SH \text{ staff-hours} = .7SH \text{ acres}$$

of lawn altogether.

Staff-hours provide a common measure of work in different jobs

Production is measured differently in different jobs—as orders processed, or acres mowed, or houses painted. However, in all these jobs human work is measured in the *same* way, as *staff-hours*, which gives us a common unit that can be translated from one job to another.

A staff of S working steadily for H hours does SH staff-hours of work. Suppose, though, the staffing level S is not constant, as in the graph below. Can we still find the total amount of work done?

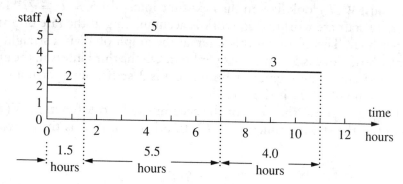

The basic formula works only when the staffing level is constant. But staffing *is* constant over certain time intervals. Thus, to find the total amount of work done, we should simply use the basic formula on each of those intervals, and then add up the individual contributions. These calculations are done in the following table. The total work is 42.5 staff-hours. So if the productivity rate is 10 orders per staff-hour, 425 orders can be processed.

$$
\begin{array}{lll}
2 \text{ staff} & \times\ 1.5 \text{ hours} & =\ \ 3.0 \ \text{ staff-hours} \\
5 & \times\ 5.5 & =\ 27.5 \\
3 & \times\ 4.0 & =\ \underline{12.0} \\
& & \ \ 42.5 \ \text{ staff-hours}
\end{array}
$$

Accumulated Work

The last calculation tells us how much work got done over an entire day. What can we tell an office manager who wants to know how work is progressing *during* the day?

At the beginning of the day, only two people are working, so after the first T hours (where $0 \le T \le 1.5$)

work done up to time T = 2 staff $\times\ T$ hours = $2T$ staff-hours.

Even before we consider what happens after 1.5 hours, this expression calls our attention to the fact that *accumulated work is a function*—let's denote it $W(T)$. According to the formula, for the first 1.5 hours $W(T)$ is a linear function whose multiplier is

$$
W' = 2 \frac{\text{staff-hours}}{\text{hour}}.
$$

Nonconstant staffing

The work done is a *sum* of products

Work *accumulates* at
a rate equal to the
number of staff

This multiplier is the **rate** at which work is being accumulated. It is also the **slope** of the graph of $W(T)$ over the interval $0 \leq T \leq 1.5$. With this insight, we can determine the rest of the graph of $W(T)$.

What must $W(T)$ look like on the next time interval $1.5 \leq T \leq 7$? Here 5 members of staff are working, so work is accumulating at the rate of 5 staff-hours per hour. Therefore, on this interval the graph of W is a straight line segment whose slope is 5 staff-hours per hour. On the third interval, the graph is another straight line segment whose slope is 3 staff-hours per hour. The complete graph of $W(T)$ is shown below.

S is the derivative of
W, so ...

As the graphs show, the *slope* of the accumulated work function $W(T)$ is the *height* of the staffing function $S(T)$. In other words, S is the *derivative* of W:

$$W'(T) = S(T).$$

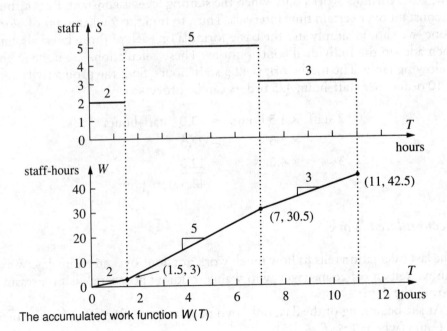

The accumulated work function $W(T)$

Notice that the units for W' and for S are compatible: the units for W' are "staff-hours per hour," which we can think of as "staff," the units for S.

We can describe the relation between S and W another way. At the moment, we have explained S in terms of W. However, since we started with S, it is really more appropriate to reverse the roles, and explain W in terms of S. Section 4.5 gives us the language to do this: W is an **antiderivative** of S. In other words, $y = W(T)$ is a solution to the differential equation

... W is an
antiderivative of S

$$\frac{dy}{dT} = S(T).$$

As we find accumulation functions in other contexts, this relation will give us crucial information.

Before leaving this example we note some special features of S and W. The staffing function S is said to be **piecewise constant**, or a **step function**. The graphs illustrate the general fact that the derivative of a piecewise-*linear* function (W, in this case) is piecewise *constant*.

The derivative of a piecewise-linear function

Summary

The example of human work illustrates the key ideas we will meet, again and again, in different contexts in this chapter. Essentially, we have two functions $W(t)$ and $S(t)$ and two different ways of expressing the relation between them: On the one hand,

$$W(t) \text{ is an accumulation function for } S(t),$$

while on the other hand,

$$S(t) \text{ is the derivative of } W(t).$$

Exploring the far-reaching implications of functions connected by such a twofold relationship will occupy the rest of this chapter.

Electrical Energy

Just as humans do work, so does electricity. A power company charges customers for the work done by the electricity it supplies, and it measures that work in a way that is strictly analogous to the way we measure human work. The work done by electricity is usually referred to as **(electrical) energy**.

For example, suppose we illuminate two light bulbs—one rated at 100 watts, the other at 60 watts. It will take the same amount of electrical energy to burn the 100-watt bulb for 3 hours as it will to burn the 60-watt bulb for 5 hours. Both will use 300 watt-hours of electricity. The power of the light bulb— measured in watts—is analogous to the number of staff working (and, in fact, workers have sometimes been called man*power*). The time the bulb burns is analogous to the time the staff works. Finally, the product

The analogy between electrical energy and human work

$$\text{energy} = \text{power} \times \text{elapsed time}$$

for electricity is analogous to the product

$$\text{work} = \text{number of staff} \times \text{elapsed time}$$

for human effort.

Electric *power* is measured in watts, in kilowatts (=1,000 watts), and in megawatts (=1,000,000 watts). Electric *energy* is measured in watt-hours, in kilowatt-hours (abbreviated kWh), and in megawatt-hours (abbreviated

MWh). Since an individual electrical appliance has a power demand of about 1 kWh, kilowatt-hours are suitable units to use for describing the energy consumption of a house, while megawatt-hours are more natural for a whole town.

Suppose the power demand of a town over a 24-hour period is described by the following graph:

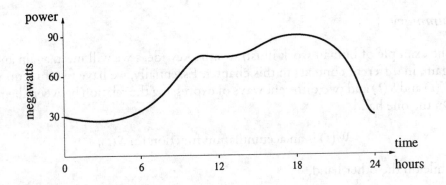

Power is analogous to staffing level

Since this graph describes *power*, its vertical height over any point t on the time axis tells us the total wattage of the light bulbs, dishwashers, computers, etc., that are turned on in the town at that instant. This demand fluctuates between 30 and 90 megawatts, roughly. The problem is to determine the total amount of *energy* used in a day—how many megawatt-hours are there in this graph? Although the equation

$$\text{energy} = \text{power} \times \text{elapsed time},$$

gives the basic relation between energy and power, we can't use it directly because the power demand isn't constant.

The staffing function $S(t)$ we considered earlier wasn't constant, either, but we were still able to compute staff-hours because $S(t)$ was *piecewise* constant. This suggests that we should replace the power graph by a piecewise-constant graph that **approximates** it. Here is one such approximation:

A piecewise-constant approximation

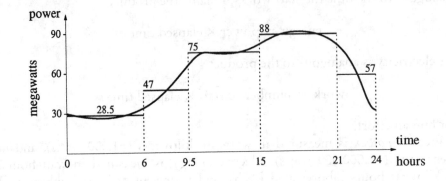

As you can see, the step function has five steps, so our approximation to the total energy consumption of the town will be a sum of five individual products:

$$\text{energy} \approx 28.5 \times 6 + 47 \times 3.5 + \cdots + 57 \times 3 = 1447 \text{ MWh}.$$

This value is only an *estimate*, though. How can we get a better estimate? The answer is clear: start with a step function that approximates the power graph *more closely*. In principle, we can get as good an approximation as we might desire this way. We are limited only by the precision of the power graph itself. As our approximation to the power graph improves, so does the accuracy of the calculation that estimates energy consumption.

Better estimates

In summary, we determine the energy consumption of the town by a sequence of successive approximations. The steps in the sequence are listed in the box below.

---

1. **Approximate** the power demand by a step function.
2. **Estimate** energy consumption from this approximation.
3. **Improve** the energy estimate by choosing a new step function that follows power demand more closely.

Accumulated Energy Consumption

Energy is being consumed steadily over the entire day; can we determine how much energy has been used through the first T hours of the day? We'll denote this quantity $E(T)$ and call it the **energy accumulation function**. For example, we already have the estimate $E(24) = 1447$ MWh; can we estimate $E(3)$ or $E(17.6)$?

Energy accumulation

Once again, the earlier example of human effort can guide us. We saw that work accumulates at a rate equal to the number of staff present:

$$W'(T) = S(T).$$

Since $S(T)$ was piecewise constant, this rate equation allowed us to determine $W(T)$ as a piecewise-linear function.

We claim that there is an analogous relation between accumulated energy consumption and power demand—namely,

$$E'(T) = p(T).$$

Unlike $S(T)$, the function $p(T)$ is not piecewise constant. Therefore, the argument we used to show that $W'(T) = S(T)$ will not work here. We need another argument.

To explain why the differential equation $E'(T) = p(T)$ should be true, we will start by analyzing the derivative $E'(T)$. We have the standard approximation

$$E'(T) \approx \frac{\Delta E}{\Delta T} = \frac{E(T + \Delta T) - E(T)}{\Delta T}.$$

Assume we have made ΔT so small that, to the level of precision we require, the approximation $\Delta E / \Delta T$ agrees with $E'(T)$. The numerator ΔE is, by definition, the total energy used up to time $T + \Delta T$, minus the total energy used up to time T. This is just the energy used during the time interval ΔT that runs from time T to time $T + \Delta T$:

$$\Delta E = \text{energy used between times } T \text{ and } T + \Delta T.$$

Since the elapsed time ΔT is small, the power demand should be nearly constant, so we can get a good estimate for energy consumption from the basic equation

$$\text{energy used} = \text{power} \times \text{elapsed time}.$$

In particular, if we represent the power by $p(T)$, which is the power demand at the beginning of the time period from T to $T + \Delta T$, then we have

$$\Delta E \approx p(T) \cdot \Delta T.$$

Using this value in our approximation for the derivative $E'(T)$, we get

$$E'(T) \approx \frac{\Delta E}{\Delta T} \approx \frac{p(T) \cdot \Delta T}{\Delta T} = p(T).$$

That is, $E'(T) \approx p(T)$, and the approximation becomes more and more exact as the time interval ΔT shrinks to 0. Thus,

$$E'(T) = \lim_{\Delta T \to 0} \frac{\Delta E}{\Delta T} = p(T).$$

Here is another way to arrive at the same conclusion. Our starting point is the basic formula

$$\Delta E \approx p(T) \cdot \Delta T,$$

which holds over a small time interval ΔT. This formula tells us how E responds to small changes in T. But that is exactly what the **microscope equation** tells us:

$$\Delta E \approx E'(T) \cdot \Delta T.$$

Since these equations give the same information, their multipliers must be the same:

$$p(T) = E'(T).$$

In words, the differential equation $E' = p$ says that *power is the rate at which energy is consumed*. In purely mathematical terms:

> The energy accumulation function $y = E(t)$ is a solution to the differential equation $dy/dt = p(t)$.

In fact, $y = E(t)$ is *the* solution to the **initial value problem**

$$\frac{dy}{dt} = p(t) \qquad y(0) = 0.$$

We can use all the methods described in section 4.5 to solve this problem.

The relation we have explored between power and energy can be found in an analogous form in many other contexts, as we will see in the next two sections. In section 6.4 we will turn back to accumulation functions and investigate them as solutions to differential equations. Then, in chapter 11, we will look at some special methods for solving the particular differential equations that arise in accumulation problems.

Exercises

Human Work

1. House-painting is a job that can be done by several people working simultaneously, so we can measure the amount of work done in "staff-hours." Consider a house-painting business run by some students. Because of class schedules, different numbers of students will be painting at different times of the day. Let $S(T)$ be the number of staff present at time T, measured in hours from 8 A.M., and suppose that during an 8-hour work day, we have

$$S(T) = \begin{cases} 3 & 0 \le T < 2 \\ 2 & 2 \le T < 4.5 \\ 4 & 4.5 \le T \le 8. \end{cases}$$

   *a) Draw the graph of the step function defined here, and compute the total number of staff hours.

   b) Draw the graph that shows how staff-hours *accumulate* on this job. This is the graph of the **accumulated work** function $W(T)$. (Compare the graphs of staff and staff-hours on page 300.)

   c) Determine the derivative $W'(T)$. Is $W'(T) = S(T)$?

2. Suppose that there is a house-painting job to be done, and by past experience the students know that four of them could finish it in 6 hours. But for the first 3.5 hours, only two students can show up, and after that, five will be available.

   *a) How long will the whole job take?

   b) Draw a graph of the staffing function for this problem. Mark on the graph the time that the job is finished.

   c) Draw the graph of the accumulated work function $W(T)$.

   d) Determine the derivative $W'(T)$. Is $W'(T) = S(T)$?

Average staffing. Suppose a job can be done in 3 hours when 6 people work the first hour and 9 work during the last 2 hours. Then the job takes 24 staff-hours of work, and the **average staffing** is

$$\text{average staffing} = \frac{24 \text{ staff-hours}}{3 \text{ hours}} = 8 \text{ staff.}$$

This means that a *constant* staffing level of 8 persons can accomplish the job in the same time that the given variable staffing level did. Note that the average staffing level (8 persons) is *not* the average of the two numbers 9 and 6!

*3. What is the average staffing of the jobs considered in exercises 1 and 2 above?

4. a) Draw the graph that shows how work would accumulate in the job described in exercise 1 if the workforce was kept at the *average* staffing level instead of the varying level described in the exercise. Compare this graph to the graph you drew in exercise 1b.

   b) What is the derivative $W'(T)$ of the work accumulation function whose graph you drew in part (a)?

5. What is the average staffing for the job described by the graph on page 299?

Electrical Energy

6. On Monday evening, a 1500-watt space heater is left on from 7 until 11 P.M. How many kilowatt-hours of electricity does it consume?

7. a) That same heater also has settings for 500 and 1000 watts. Suppose that on Tuesday we put it on the 1000-watt setting from 6 to 8 P.M., then switch to 1500 watts from 8 till 11 P.M., and then on the 500-watt setting through the night until 8 A.M., Wednesday. How much energy is consumed (in kilowatt-hours)?

   b) Sketch the graphs of power demand $p(T)$ and accumulated energy consumption $E(T)$ for the space heater from Tuesday evening to Wednesday morning. Determine whether $E'(T) = p(T)$ in this case.

c) The **average power demand** of the space heater is defined by:

$$\text{average power demand} = \frac{\text{energy consumption}}{\text{elapsed time}}.$$

If energy consumption is measured in kilowatt-hours, and time in hours, then we can measure average power demand in kilowatts—the same as power itself. (Notice the similarity with average staffing.) What is the average power demand from Tuesday evening to Wednesday morning? If the heater could be set at this average power level, how would the energy consumption compare to the actual energy consumption you determined in part (a)?

8. The graphs on page 302 describe the power demand of a town over a 24-hour period. Give an estimate of the average power demand of the town during that period. Explain what you did to produce your estimate.

Work as Force × Distance

The effort it takes to move an object is also called work. Since it takes *twice* as much effort to move the object twice as far, or to move another object that is twice as heavy, we can see that the work done in moving an object is proportional to both the force applied and to the distance moved. The simplest way to express this fact is to define

$$\text{work} = \text{force} \times \text{distance}.$$

For example, to lift a weight of 20 pounds straight up it takes 20 pounds of force. If the vertical distance is 3 feet, then

$$20 \text{ pounds} \times 3 \text{ feet} = 60 \text{ foot-pounds}$$

of work is done. Thus, once again the quantity we are interested in has the form of a product. The *foot-pound* is one of the standard units for measuring work.

9. Suppose a tractor pulls a loaded wagon over a road whose steepness varies. If the first 150 feet of road are relatively level and the tractor has to exert only 200 pounds of force, while the next 400 feet are inclined and the tractor has to exert 550 pounds of force, how much work does the tractor do altogether?

10. A motor on a large ship is lifting a 2000-pound anchor that is already out of the water at the end of a 30-foot chain. The chain weighs 40 pounds per foot. As the motor lifts the anchor, the part of the chain

that is hanging gets shorter and shorter, thereby reducing the weight the motor must lift.

a) What is the combined weight of anchor and hanging chain when the anchor has been lifted x feet above its initial position?

b) Divide the 30-foot distance that the anchor must move into three equal intervals of 10 feet each. Estimate how much work the motor does lifting the anchor and chain over each 10-foot interval by multiplying the combined weight at the *bottom* of the interval by the 10-foot height. What is your estimate for the total work done by the motor in raising the anchor and chain 30 feet?

c) Repeat all the steps of part (b), but this time use 30 equal intervals of 1 foot each. Is your new estimate of the work done larger or smaller than your estimate in part (b)? Which estimate is likely to be more accurate? On what do you base your judgment?

d) If you ignore the weight of the chain entirely, what is your estimate of the work done? How much *extra* work do you therefore estimate the motor must do to raise the heavy chain along with the anchor?

➤ 6.2 RIEMANN SUMS

In the last section we estimated energy consumption in a town by replacing the power function $p(t)$ by a step function. Let's pause to describe that process in somewhat more general terms that we can adapt to other contexts. The power graph, the approximating step function, and the energy estimate are shown below.

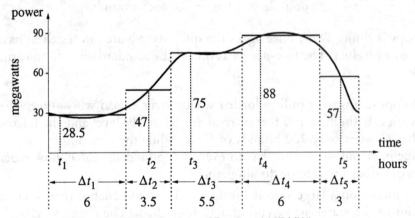

Energy $\approx 28.5 \times 6 + 47 \times 3.5 + \cdots + 57 \times 3 = 1447$ MWh.

The height of the first step is 28.5 megawatts. This is the *actual* power level at the time t_1 indicated on the graph. That is, $p(t_1) = 28.5$ megawatts. We found a power level of 28.5 megawatts by **sampling** the power function at the time t_1. The height of the first step could have been different if we had sampled the power function at a different time. In general, if we sample the power function $p(t)$ at the time t_1 in the interval Δt_1, then we would estimate the energy used during that time to be

Sampling the power function

$$\text{energy} \approx p(t_1) \cdot \Delta t_1 \text{ MWh.}$$

Notice that t_1 is not in the middle, or at either end, of the first interval. It is simply a time when the power demand is representative of what's happening over the entire interval. Furthermore, t_1 is not even unique; there is another sampling time (near $t = 5$ hours) when the power level is again 28.5 megawatts.

We can describe what happens in the other time intervals the same way. If we sample the kth interval at the point t_k, then the height of the kth power step will be $p(t_k)$ and our estimate for the energy used during that time will be

$$\text{energy} \approx p(t_k) \cdot \Delta t_k \text{ MWh.}$$

We now have a general way to construct an approximation for the power function and an estimate for the energy consumed over a 24-hour period. It involves these steps.

1. Choose any number n of subintervals, and let them have arbitrary widths $\Delta t_1, \Delta t_2, \ldots, \Delta t_n$, subject only to the condition

A procedure for approximating the power level and energy use

$$\Delta t_1 + \cdots + \Delta t_n = 24 \text{ hours.}$$

2. Sample the kth subinterval at any point t_k, and let $p(t_k)$ represent the power level over this subinterval.

3. Estimate the energy used over the 24 hours by the sum

$$\text{energy} \approx p(t_1) \cdot \Delta t_1 + p(t_2) \cdot \Delta t_2 + \cdots + p(t_n) \cdot \Delta t_n \text{ MWh.}$$

The expression on the right is called a **Riemann sum** for the power function $p(t)$ on the interval $0 \le t \le 24$ hours.

The work of Bernhard Riemann (1826–1866) has had a profound influence on contemporary mathematicians and physicists. His revolutionary ideas about the geometry of space, for example, are the basis for Einstein's theory of general relativity.

The enormous range of choices in this process means there are innumerable ways to construct a Riemann sum for $p(t)$. However, we are not really interested in *arbitrary* Riemann sums. On the contrary, we want to build Riemann

Choices that lead to good estimates

sums that will give us good estimates for energy consumption. Therefore, we will choose each subinterval Δt_k so small that the power demand over that subinterval differs only very little from the sampled value $p(t_k)$. A Riemann sum constructed with *these* choices will then differ only very little from the total energy used during the 24-hour time interval.

Essentially, we use a Riemann sum to resolve a dilemma. We know the basic formula

$$\text{energy} = \text{power} \times \text{time}$$

The dilemma

works when power is constant, but in general power *isn't* constant—that's the dilemma. We resolve the dilemma by using instead a *sum* of terms of the form *power* × *time*. With this sum we get an estimate for the energy.

In this section we will explore some other problems that present the same dilemma. In each case we will start with a basic formula that involves a product of two constant factors, and we will need to adapt the formula to the situation where one of the factors varies. The solution will be to construct a Riemann sum of such products, producing an estimate for the quantity we were after in the first place. As we work through these problems, you should pause to compare each to the problem of energy consumption.

Calculating Distance Traveled

It is easy to tell how far a car has traveled by reading its odometer. The problem is more complicated for a ship, particularly a sailing ship in the days before electronic navigation was common. The crew always had instruments that could measure—or at least estimate—the velocity of the ship at any time. Then, during any time interval in which the ship's velocity is constant, the distance traveled is given by the familiar formula

Estimating velocity and distance

$$\text{distance} = \text{velocity} \times \text{elapsed time.}$$

If the velocity is *not* constant, then this formula does not work. The remedy is to break up the long time period into several short ones. Suppose their lengths are $\Delta t_1, \Delta t_2, \ldots, \Delta t_n$. By assumption, the velocity is a function of time t; let's denote it $v(t)$. At some time t_k during each time period Δt_k measure the velocity: $v_k = v(t_k)$. Then the Riemann sum

Sampling the velocity function

$$v(t_1) \cdot \Delta t_1 + v(t_2) \cdot \Delta t_2 + \cdots + v(t_n) \cdot \Delta t_n$$

is an estimate for the total distance traveled.

For example, suppose the velocity is measured five times during a 15-hour trip—once every 3 hours—as shown in the table on the next page. Then the basic formula

$$\text{distance} = \text{velocity} \times \text{elapsed time}$$

gives us an estimate for the distance traveled during each 3-hour period, and the sum of these distances is an estimate of the total distance traveled during the 15 hours. These calculations appear in the right-hand column of the table. (Note that the first measurement is used to calculate the distance traveled between hours 0 and 3, while the last measurement, taken 12 hours after the start, is used to calculate the distance traveled between hours 12 and 15.)

Sampling Time (hours)	Elapsed Time (hours)	Velocity (miles/hour)	Distance Traveled (miles)
0	3	1.4	$3 \times 1.4 = 4.20$
3	3	5.25	$3 \times 5.25 = 15.75$
6	3	4.3	$3 \times 4.3 = 12.90$
9	3	4.6	$3 \times 4.6 = 13.80$
12	3	5.0	$3 \times 5.0 = 15.00$
			61.65

Thus we estimate the ship has traveled 61.65 miles during the 15 hours. The number 61.65, obtained by adding the numbers in the rightmost column, is a Riemann sum for the velocity function.

The estimated distance is a Riemann sum for the velocity function

Consider the specific choices that we made to construct this Riemann sum:

$$\Delta t_1 = \Delta t_2 = \Delta t_3 = \Delta t_4 = \Delta t_5 = 3$$
$$t_1 = 0, \quad t_2 = 3, \quad t_3 = 6, \quad t_4 = 9, \quad t_5 = 12.$$

These choices differ from the choices we made in the energy example in two notable ways. First, all the subintervals here are the same size. This is because it is natural to take velocity readings at regular time intervals. By contrast, in the energy example the subintervals were of different widths. Those widths were chosen in order to make a piecewise-constant function that followed the power demand graph closely. Second, all the sampling times lie at the beginning of the subintervals in which they appear. Again, this is natural and convenient for velocity measurements. In the energy example, the sampling times were chosen with an eye to the power graph. Even though we can make arbitrary choices in constructing a Riemann sum, we will do it systematically whenever possible. This means choosing subintervals of equal size and sampling points at the "same" place within each interval.

Intervals and sampling times are chosen in a systematic way

Let's turn back to our estimate for the total distance. Since the velocity of the ship could have fluctuated significantly during each of the 3-hour periods we used, our estimate is rather rough. To improve the estimate we could measure the velocity more frequently—for example, every 15 minutes. If we did,

Improving the distance estimate

the Riemann sum would have 60 terms (four distances per hour for 15 hours). The individual terms in the sum would all be much smaller, though, because they would be estimates for the distance traveled in 15 minutes instead of in 3 hours. For instance, the first of the 60 terms would be

$$1.4 \frac{\text{miles}}{\text{hour}} \times .25 \text{ hour} = .35 \text{ mile}.$$

Of course it may not make practical sense to do such a precise calculation. Other factors, such as water currents or the inaccuracy of the velocity measurements themselves, may keep us from getting a good estimate for the distance. Essentially, the Riemann sum is only a *model* for the distance covered by a ship.

Calculating Areas

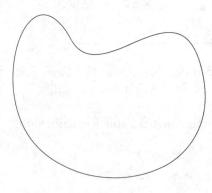

The area of a rectangle is just the product of its length and its width. How can we measure the area of a region that has an irregular boundary, like the one at the left? We would like to use the basic formula

area = length × width.

However, since the region doesn't have straight sides, there is nothing we can call a "length" or a "width" to work with.

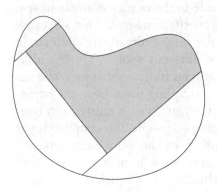

We can begin to deal with this problem by breaking up the region into smaller regions that *do* have straight sides—with, at most, only one curved side. This can be done many different ways. The lower figure shows one possibility. The sum of the areas of all the little regions will be the area we are looking for. Although we haven't yet solved the original problem, we have at least **reduced** it to another problem that looks simpler and may be easier to solve. Let's now work on the reduced problem for the shaded region.

Following is the shaded region, turned so that it sits flat on one of its straight sides. We would like to calculate its area using the formula

width × height,

but this formula applies only to rectangles. We can, however, approximate the region by a collection of rectangles, as shown at the right. The formula *does* apply to the individual rectangles and the sum of their areas will approximate the area of the whole region.

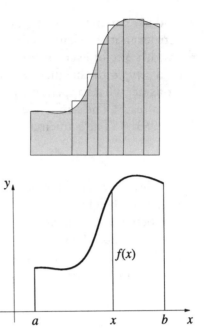

To get the area of a rectangle, we must measure its width and height. Their heights vary with the height of the curved top of the shaded region. To describe that height in a systematic way, we have placed the shaded region in a coordinate plane so that it sits on the x-axis. The other two straight sides lie on the vertical lines $x = a$ and $x = b$. The curved side defines the graph of a function $y = f(x)$. Therefore, at each point x, the vertical height from the axis to the curve is $f(x)$.

By introducing a coordinate plane we gain access to mathematical tools—such as the language of functions—to describe the various areas.

The kth rectangle has been singled out on the left, below. We let Δx_k denote the width of its base. By **sampling** the function f at a properly chosen point x_k in the base, we get the height $f(x_k)$ of the rectangle. Its area is therefore $f(x_k) \cdot \Delta x_k$. If we do the same thing for all n rectangles shown on the right, we can write their total area as

Calculating the areas of the rectangles

$$f(x_1) \cdot \Delta x_1 + f(x_2) \cdot \Delta x_2 + \cdots + f(x_n) \cdot \Delta x_n.$$

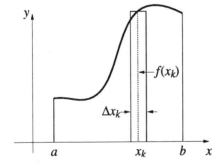

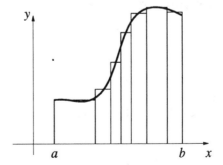

Notice that our estimate for the area has the form of a Riemann sum for the height function $f(x)$ over the interval $a \leq x \leq b$. To get a better estimate, we

The area estimate is a Riemann sum

should use narrower rectangles, and more of them. In other words, we should construct another Riemann sum in which the number of terms, n, is larger and the width Δx_k of every subinterval is smaller. Putting it yet another way, we should *sample* the height more often.

Consider what happens if we apply this procedure to a region whose area we know already. The semicircle of radius $r = 1$ has an area of $\pi r^2/2 = \pi/2 = 1.5707963\ldots$. The semicircle is the graph of the function

$$f(x) = \sqrt{1 - x^2},$$

which lies over the interval $-1 \le x \le 1$. To get the figure on the left, we sampled the height $f(x)$ at 20 evenly spaced points, starting with $x = -1$. In the better approximation on the right, we increased the number of sample points to 50. The values of the shaded areas were calculated with the program RIEMANN, which we will develop later in this section. Note that with 50 rectangles the Riemann sum is within .005 of $\pi/2$, the exact value of the area.

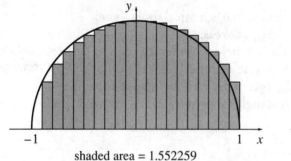

shaded area = 1.552259

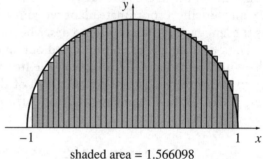

shaded area = 1.566098

Calculating Lengths

It is to be expected that products—and ultimately, Riemann sums—will be involved in calculating areas. It is more surprising to find that we can use them to calculate *lengths*, too. In fact, when we are working in a coordinate plane, using a product to describe the length of a straight line is even quite natural.

The length of a straight line

To see how this can happen, consider a line segment in the x, y plane that has a known slope m. If we also know the horizontal separation between the two ends, we can find the length of the segment. Suppose the horizontal separation is Δx and the vertical separation Δy. Then the length of the segment is

$$\sqrt{\Delta x^2 + \Delta y^2}$$

by the Pythagorean theorem (see page 78). Since $\Delta y = m \cdot \Delta x$, we can rewrite this as

$$\sqrt{\Delta x^2 + (m \cdot \Delta x)^2} = \Delta x \cdot \sqrt{1 + m^2}.$$

In other words, if a line has slope m and it is Δx units wide, then its length is the product

$$\sqrt{1 + m^2} \cdot \Delta x.$$

Suppose the line is *curved*, instead of straight. Can we describe its length the same way? We'll assume that the curve is the graph $y = g(x)$. The complication is that the slope $m = g'(x)$ now varies with x.

The length of a curved line

Suppose $g'(x)$ doesn't vary too much over an interval of length Δx. Then the curve is nearly straight. If we select a single point x_* in the interval and sample the slope $g'(x_*)$ there, we would expect the length of the curve to be approximately

$$\sqrt{1 + (g'(x_*))^2} \cdot \Delta x.$$

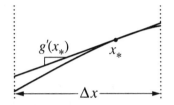

As the figure shows, this is the exact length of the straight line segment that lies over the same interval Δx and is tangent to the curve at the point $x = x_*$.

If the slope $g'(x)$ varies appreciably over the interval, we should subdivide the interval into small pieces $\Delta x_1, \Delta x_2, \ldots, \Delta x_n$, over which the curve is nearly straight. Then, if we sample the slope at the point x_k in the kth subinterval, the sum

$$\sqrt{1 + (g'(x_1))^2} \cdot \Delta x_1 + \cdots + \sqrt{1 + (g'(x_n))^2} \cdot \Delta x_n$$

will give us an estimate for the total length of the curve.

Once again, we find an expression that has the form of a Riemann sum. There is, however, a new ingredient worth noting. The estimate is a Riemann sum not for the *original* function $g(x)$ but for *another* function

The length of a curve is estimated by a Riemann sum

$$f(x) = \sqrt{1 + (g'(x))^2}$$

that we constructed using g. The important thing is that the length is estimated by a Riemann sum for *some* function.

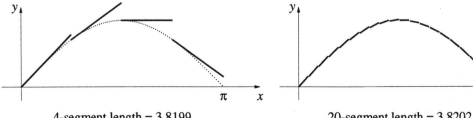

4-segment length = 3.8199   20-segment length = 3.8202

The figure at the bottom of page 315 shows two estimates for the length of the graph of $y = \sin x$ between 0 and π. In each case, we used subintervals of equal length and we sampled the slope at the left end of each subinterval. As you can see, the four segments approximate the graph of $y = \sin x$ only very roughly. When we increase the number of segments to 20, on the right, the approximation to the shape of the graph becomes quite good. Notice that the graph itself is not shown on the right; only the 20 segments.

To calculate the two lengths, we constructed Riemann sums for the function $f(x) = \sqrt{1 + \cos^2 x}$. We used the fact that the derivative of $g(x) = \sin x$ is $g'(x) = \cos x$, and we did the calculations using the program RIEMANN. By using the program with still smaller subintervals you can show that

$$\text{the exact length} = 3.820197789.\ldots$$

Thus, the 20-segment estimate is already accurate to four decimal places.

We have already constructed estimates for the length of a curve in chapter 2 (pages 77–79). Those estimates were sums, too, but they were not Riemann sums. The terms had the form $\sqrt{\Delta x^2 + \Delta y^2}$; they were not products of the form $\sqrt{1 + m^2} \cdot \Delta x$. The sums in chapter 2 may seem more straightforward. However, we are developing Riemann sums as a powerful general tool for dealing with many different questions. By expressing lengths as Riemann sums we gain access to that power.

Definition

The Riemann sums that appear in the calculation of power, distance, and length are instances of a general mathematical object that can be constructed for *any* function whatsoever. We pause now to describe that construction apart from any particular context. In what follows it will be convenient for us to write an interval of the form $a \le x \le b$ more compactly as $[a, b]$.

Notation: [a, b]

Suppose the function $f(x)$ is defined for x in the interval $[a, b]$. Then a **Riemann sum** for $f(x)$ on $[a, b]$ is an expression of the form

$$f(x_1) \cdot \Delta x_1 + f(x_2) \cdot \Delta x_2 + \cdots + f(x_n) \cdot \Delta x_n.$$

The interval $[a, b]$ has been divided into n subintervals whose lengths are $\Delta x_1, \ldots, \Delta x_n$, and for each k from 1 to n, x_k is some point in the kth subinterval.

Data for a Riemann sum

Notice that once the function and the interval have been specified, a Riemann sum is determined by the following data:

- A **decomposition** of the original interval into subintervals (which determines the lengths of the subintervals).

- A **sampling point** chosen from each subinterval (which determines a value of the function on each subinterval).

A Riemann sum for $f(x)$ is a sum of products of values of Δx and values of $y = f(x)$. If x and y have units, then so does the Riemann sum; its units are the units for x times the units for y. When a Riemann sum arises in a particular context, the notation may look different from what appears in the definition just given: the variable might not be x, and the function might not be $f(x)$. For example, the energy approximation we considered at the beginning of the section is a Riemann sum for the power demand function $p(t)$ on $[0, 24]$. The length approximation for the graph $y = \sin x$ is a Riemann sum for the function $\sqrt{1 + \cos^2 x}$ on $[0, \pi]$.

Units

It is important to note that, from a mathematical point of view, a Riemann sum is just a number. It's the *context* that provides the meaning: Riemann sums for a power demand that varies over time approximate total energy consumption; Riemann sums for a velocity that varies over time approximate total distance; and Riemann sums for a height that varies over distance approximate total area.

A Riemann sum is just a number

To illustrate the generality of a Riemann sum, and to stress that it is just a number arrived at through arbitrary choices, let's work through an example without a context. Consider the function

$$f(x) = \sqrt{1 + x^3} \quad \text{on} \quad [1, 3].$$

We will break up the full interval $[1, 3]$ into three subintervals $[1, 1.6]$, $[1.6, 2.3]$, and $[2.3, 3]$. Thus

The data

$$\Delta x_1 = .6 \qquad \Delta x_2 = \Delta x_3 = .7.$$

Next we'll pick a point in each subinterval, say $x_1 = 1.3$, $x_2 = 2$, and $x_3 = 2.8$. Here are the data laid out on the x-axis.

With these data we get the following Riemann sum for $\sqrt{1 + x^3}$ on $[1, 3]$:

$$f(x_1) \cdot \Delta x_1 + f(x_2) \cdot \Delta x_2 + f(x_3) \cdot \Delta x_3$$
$$= \sqrt{1 + 1.3^3} \times .6 + \sqrt{1 + 2^3} \times .7 + \sqrt{1 + 2.8^3} \times .7$$
$$= 6.5263866.$$

In this case, the choice of the subintervals, as well as the choice of the point x_k in each subinterval, was haphazard. Different data would produce a different value for the Riemann sum.

Keep in mind that an individual Riemann sum is not especially significant. Ultimately, we are interested in seeing what happens when we recalculate Riemann sums with smaller and smaller subintervals. For that reason, it is helpful to do the calculations systematically.

Calculating a Riemann sum algorithmically.   As we have seen with our contextual problems, the data for a Riemann sum are not usually chosen in a haphazard fashion. In fact, when dealing with functions given by formulas, such as the function $f(x) = \sqrt{1-x^2}$ whose graph is a semicircle, it pays to be systematic. We use subintervals of equal size and pick the "same" point from each subinterval (e.g., always pick the midpoint or always pick the left endpoint). The benefit of systematic choices is that we can write down the computations involved in a Riemann sum in a simple algorithmic form that can be carried out on a computer.

Systematic data

Let's illustrate how this strategy applies to the function $\sqrt{1+x^3}$ on $[1, 3]$. Since the whole interval is $3-1 = 2$ units long, if we construct n subintervals of equal length Δx, then $\Delta x = 2/n$. For each $k = 1, \ldots, n$, we choose the sampling point x_k to be the left endpoint of the kth subinterval. Here is a picture of the data:

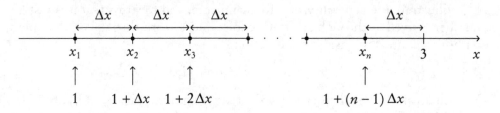

In this systematic approach, the space between one sampling point and the next is Δx, the same as the width of a subinterval. This puts the kth sampling point at $x = 1 + (k-1)\Delta x$.

In the following table, we add up the terms in a Riemann sum S for $f(x) = \sqrt{1+x^3}$ on the interval $[1, 3]$. We used $n = 4$ subintervals and always sampled f at the left endpoint. Each row shows the following:

1. The current sampling point;
2. The value of f at that point;
3. The current term $\Delta S = f \cdot \Delta x$ in the sum;
4. The accumulated value of S.

Left Endpoint	Current $\sqrt{1+x^3}$	Current ΔS	Accumulated S
1	1.4142	.7071	.7071
1.5	2.0917	1.0458	1.7529
2	3	1.5	3.2529
2.5	4.0774	2.0387	5.2916

The Riemann sum S appears as the final value 5.2916 in the fourth column.

The program RIEMANN, below, will generate the last two columns in the table on the previous page. The statement x = a on the sixth line determines the position of the first sampling point. Within the FOR–NEXT loop, the statement x = x + deltax moves the sampling point to its next position.

Program: RIEMANN
Left endpoint Riemann sums

```
DEF fnf (x) = SQR(1 + x ^ 3)
a = 1
b = 3
numberofsteps = 4
deltax = (b - a) / numberofsteps
x = a
accumulation = 0
FOR k = 1 TO numberofsteps
        deltaS = fnf(x) * deltax
        accumulation = accumulation + deltaS
        x = x + deltax
        PRINT deltaS, accumulation
NEXT k
```

By modifying RIEMANN, you can calculate Riemann sums for other sampling points and for other functions. For example, to sample at midpoints, you must start at the midpoint of the first subinterval. Since the subinterval is Δx units wide, its midpoint is $\Delta x / 2$ units from the left endpoint, which is $x = a$. Thus, if you change the statement on the sixth line to x = a + deltax / 2, the program will then generate midpoint Riemann sums.

Summation Notation

Because Riemann sums arise frequently and because they are unwieldy to write out in full, we now introduce a method—called **summation notation**—that allows us to write them more compactly. To see how it works, look first at the sum

$$1^2 + 2^2 + 3^2 + \cdots + 50^2.$$

Using summation notation, we can express this as

$$\sum_{k=1}^{50} k^2.$$

For a somewhat more abstract example, consider the sum

$$a_1 + a_2 + a_3 + \cdots + a_n,$$

which we can express as

$$\sum_{k=1}^{n} a_k.$$

Sigma notation

We use the capital letter *sigma* $\sum$ from the Greek alphabet to denote a sum. For this reason, summation notation is sometimes referred to as **sigma notation**. You should regard $\sum$ as an instruction telling you to **sum** the numbers of the indicated form as the index k runs through the integers, starting at the integer displayed below the $\sum$ and ending at the integer displayed above it. Notice that changing the index k to some other letter has no effect on the sum. For example,

$$\sum_{k=1}^{20} k^3 = \sum_{j=1}^{20} j^3,$$

since both expressions give the sum of the cubes of the first 20 positive integers. Other aspects of summation notation will be covered in the exercises.

Summation notation allows us to write the Riemann sum

$$f(x_1) \cdot \Delta x_1 + \cdots + f(x_n) \cdot \Delta x_n$$

more efficiently as

$$\sum_{k=1}^{n} f(x_k) \cdot \Delta x_k.$$

Be sure not to get tied into one particular way of using these symbols. For example, you should instantly recognize

$$\sum_{i=1}^{m} \Delta t_i \, g(t_i)$$

as a Riemann sum. In what follows we will commonly use summation notation when working with Riemann sums. The important thing to remember is that summation notation is only a "shorthand" way to express a Riemann sum in a more compact form.

Exercises

Making Approximations

*1. Estimate the average velocity of the ship whose motion is described on page 311. The voyage lasts 15 hours.

2. The aim of this question is to determine how much electrical energy was consumed in a house over a 24-hour period, when the power demand p was measured at different times to have these values:

Time (24-hour clock)	Power (watts)
1:30	275
5:00	240
8:00	730
9:30	300
11:00	150
15:00	225
18:30	1880
20:00	950
22:30	700
23:00	350

Notice that the time interval is from $t = 0$ hours to $t = 24$ hours, but the power demand was not sampled at either of those times.

a) Set up an estimate for the energy consumption in the form of a Riemann sum $p(t_1)\Delta t_1 + \cdots + p(t_n)\Delta t_n$ for the power function $p(t)$. To do this, you must identify explicitly the value of n, the sampling times t_k, and the time intervals Δt_k that you used in constructing your estimate. (Note: The sampling times come from the table, but there is wide latitude in how you choose the subintervals Δt_k.)

b) What is the estimated energy consumption, using your choice of data? There is no single "correct" answer to this question. Your estimate depends on the choices you made in setting up the Riemann sum.

c) Plot the data given in the table in part (a) on a t, p-coordinate plane. Then draw on the same coordinate plane the step function that represents your estimate of the power function $p(t)$. The width of the kth step should be the time interval Δt_k that you specified in part (a). Is it?

d) Estimate the *average* power demand in the house during the 24-hour period.

Waste production.   A colony of living yeast cells in a vat of fermenting grape juice produces waste products—mainly alcohol and carbon dioxide—as it consumes the sugar in the grape juice. It is reasonable to expect that another yeast colony, twice as large as this one, would produce twice as much waste over the same time period. Moreover, since waste accumulates over time, if we double the time period, we would expect our colony to produce twice as much waste.

These observations suggest that waste production is proportional to both the size of the colony and the amount of time that passes. If P is the size of the colony, in grams, and Δt is a short time interval, then we can express waste production W as a function of P and Δt:

$$W = k \cdot P \cdot \Delta t \text{ grams.}$$

If Δt is measured in hours, then the multiplier k has to be measured in units of grams of waste per hour per gram of yeast.

The preceding formula is useful only over a time interval Δt in which the population size P does not vary significantly. If the time interval is large, and the population size can be expressed as a function $P(t)$ of the time t, then we can estimate waste production by breaking up the whole time interval into a succession of smaller intervals $\Delta t_1, \Delta t_2, \ldots, \Delta t_n$ and forming a Riemann sum

$$kP(t_1)\,\Delta t_1 + \cdots + kP(t_n)\,\Delta t_n \approx W \text{ grams.}$$

The time t_k must lie within the time interval Δt_k, and $P(t_k)$ must be a good approximation to the population size $P(t)$ throughout that time interval.

*3. Suppose the colony starts with 300 grams of yeast (i.e., at time $t = 0$ hours) and it grows exponentially according to the formula

$$P(t) = 300\, e^{0.2\,t}.$$

If the waste production constant k is 0.1 gram per hour per gram of yeast, estimate how much waste is produced in the first four hours. Use a Riemann sum with four hour-long time intervals and measure the population size of the yeast in the middle of each interval—that is, "on the half-hour."

Using RIEMANN

4. a) Calculate left endpoint Riemann sums for the function $\sqrt{1 + x^3}$ on the interval $[1, 3]$ using 40, 400, 4000, and 40000 equally spaced subintervals. How many decimal points in this sequence have stabilized?

b) The left endpoint Riemann sums for $\sqrt{1 + x^3}$ on the interval $[1, 3]$ seem to be approaching a limit as the number of subintervals increases without bound. Give the numerical value of that limit, accurate to four decimal places.

c) Calculate left endpoint Riemann sums for the function $\sqrt{1 + x^3}$ on the interval $[3, 7]$. Construct a sequence of Riemann sums using more and more subintervals, until you can determine the limiting value of these sums, accurate to three decimal places. What is that limit?

d) Calculate left endpoint Riemann sums for the function $\sqrt{1 + x^3}$ on the interval $[1, 7]$ in order to determine the limiting value of the sums to three decimal place accuracy. What is that value? How are the limiting values in parts (b), (c), and (d) related? How are the corresponding *intervals* related?

5. Modify RIEMANN so it will calculate a Riemann sum by sampling the given function at the *midpoint* of each subinterval, instead of the left endpoint. Describe exactly how you changed the program to do this.

6. a) Calculate *midpoint* Riemann sums for the function $\sqrt{1 + x^3}$ on the interval $[1, 3]$ using 40, 400, 4000, and 40000 equally spaced subintervals. How many decimal points in this sequence have stabilized?

b) Roughly how many subintervals are needed to make the midpoint Riemann sums for $\sqrt{1 + x^3}$ on the interval $[1, 3]$ stabilize out to the first four digits? What is the stable value? Compare this to the limiting value you found earlier for left endpoint Riemann sums. Is one value larger than the other? Could they be the same?

c) Comment on the relative "efficiency" of midpoint Riemann sums versus left endpoint Riemann sums (at least for the function $\sqrt{1 + x^3}$ on the interval $[1, 3]$). To get the same level of accuracy, an *efficient* calculation will take fewer steps than an *inefficient* one.

7. a) Modify RIEMANN to calculate *right endpoint* Riemann sums, and use it to calculate right endpoint Riemann sums for the function $\sqrt{1 + x^3}$ on the interval $[1, 3]$ using 40, 400, 4000, and 40000 equally spaced subintervals. How many digits in this sequence have stabilized?

b) Comment on the efficiency of right endpoint Riemann sums as compared to left endpoint and to midpoint Riemann sums—at least as far as the function $\sqrt{1 + x^3}$ is concerned.

8. Calculate left endpoint Riemann sums for the function

$$f(x) = \sqrt{1 - x^2} \quad \text{on the interval } [-1, 1].$$

Use 20 and 50 equally spaced subintervals. Compare your values with the estimates for the area of a semicircle given on page 314.

9. **a)** Calculate left endpoint Riemann sums for the function

$$f(x) = \sqrt{1 + \cos^2 x} \quad \text{on the interval } [0, \pi].$$

Use 4 and 20 equally spaced subintervals. Compare your values with the estimates for the length of the graph of $y = \sin x$ between 0 and π, given on pages 315–316.

b) What is the limiting value of the Riemann sums, as the number of subintervals becomes infinite? Find the limit to 11 decimal places accuracy.

10. Calculate left endpoint Riemann sums for the function

$$f(x) = \cos(x^2) \quad \text{on the interval } [0, 4],$$

using 100, 1000, and 10000 equally spaced subintervals.

11. Calculate left endpoint Riemann sums for the function

$$f(x) = \frac{\cos x}{1 + x^2} \quad \text{on the interval } [2, 3],$$

using 10, 100, and 1000 equally spaced subintervals. The Riemann sums are all negative. Why? (A suggestion: Sketch the graph of f. What does that tell you about the signs of the terms in a Riemann sum for f?)

12. **a)** Calculate midpoint Riemann sums for the function

$$H(z) = z^3 \quad \text{on the interval } [-2, 2],$$

using 10, 100, and 1000 equally spaced subintervals. The Riemann sums are all zero. Why? (On some computers and calculators, you may find that there will be a nonzero digit in the fourteenth or fifteenth decimal place—this is due to "round-off error.")

b) Repeat part (a) using *left endpoint* Riemann sums. Are the results still zero? Can you explain the difference, if any, between these two results?

Volume as a Riemann Sum

If you slice a rectangular parallelepiped (e.g., a brick or a shoebox) parallel to a face, the area A of a **cross section** does not vary. The same is true for a cylinder (e.g., a can of spinach or a coin). For *any* solid that has a constant cross section (e.g., the object on the right, below), its volume is just the product of its cross-sectional area with its thickness.

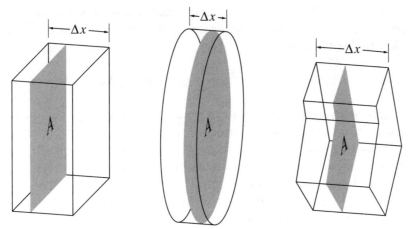

Volume = area of cross section × thickness = $A \cdot \Delta x$

Most solids don't have such a regular shape. They are more like the one shown below. If you take cross-sectional slices perpendicular to some fixed line (which will become our x-axis), the slices will not generally have a regular shape. They may be roughly oval, as shown below, but they will generally vary in area. Suppose the area of the cross section x inches along the axis is $A(x)$ square inches. Because $A(x)$ varies with x, you cannot calculate the volume of this solid using the simple formula above. However, you can *estimate* the volume as a Riemann sum for A.

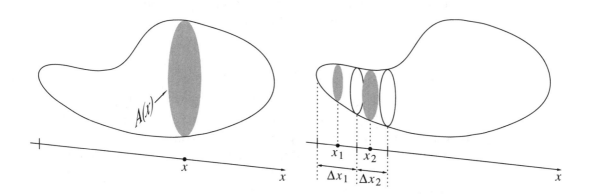

The procedure should now be familiar to you. Subdivide the x-axis into segments of length $\Delta x_1, \Delta x_2, \ldots, \Delta x_n$ inches, respectively. The solid piece that lies over the first segment has a thickness of Δx_1 inches. If you slice this piece at a point x_1 inches along the x-axis, the area of the slice is $A(x_1)$ square inches, and the volume of the piece is approximately $A(x_1) \cdot \Delta x_1$ cubic inches. The second piece is Δx_2 inches thick. If you slice it x_2 inches along the x-axis, the slice has an area of $A(x_2)$ square inches, so the second piece has an approximate volume of $A(x_2) \cdot \Delta x_2$ cubic inches. If you continue in this way and add up the n volumes, you get an estimate for the total volume that has the form of a Riemann sum for the area function $A(x)$:

$$\text{volume} \approx A(x_1)\,\Delta x_1 + A(x_2)\,\Delta x_2 + \cdots + A(x_n)\,\Delta x_n \text{ cubic inches.}$$

One place where this approach can be used is in medical diagnosis. The X-ray technique known as a CAT scan provides a sequence of precisely spaced cross-sectional views of a patient. From these views much information about the state of the patient's internal organs can be gained without invasive surgery. In particular, the volume of a specific piece of tissue can be estimated, as a Riemann sum, from the areas of individual slices and the spacing between them. The next exercise gives an example.

13. A CAT scan of a human liver shows us X-ray "slices" spaced 2 centimeters apart. If the areas of the slices are 72, 145, 139, 127, 111, 89, 63, and 22 square centimeters, estimate the volume of the liver.

14. The volume of a sphere whose radius is r is exactly $V = 4\pi r^3/3$.
   a) Using the formula, determine the volume of the sphere whose radius is 3. Give the numerical value to four decimal places accuracy. One way to get a sphere of radius 3 is to rotate the graph of the semicircle

   $$r(x) = \sqrt{9 - x^2} \qquad -3 \le x \le 3$$

   around the x-axis. Every cross section perpendicular to the x-axis is a circle. At the point x, the radius of the circle is $r(x)$, and its area is

   $$A = \pi r^2 = \pi(r(x))^2 = A(x).$$

   You can thus get estimates for the volume of the sphere by constructing Riemann sums for $A(x)$ on the interval $[-3, 3]$.
   b) Calculate a sequence of estimates for the volume of the sphere that use more and more slices, until the value of the estimate stabilizes out to four decimal places. Does this value agree with the value given by the formula in part (a)?

15. a) Rotate the graph of $r(x) = .5x$, with $0 \le x \le 6$ around the x-axis. What shape do you get? Describe it precisely, and find its volume using an appropriate geometric formula.
   b) Calculate a sequence of estimates for the volume of the same object by constructing Riemann sums for the area function $A(x) = \pi(r(x))^2$. Continue until your estimates stabilize out to four decimal places. What value do you get?

Summation Notation

16. Determine the numerical value of each of the following:

a) $\displaystyle\sum_{k=1}^{10} k$ *b) $\displaystyle\sum_{k=1}^{5} k^2$ c) $\displaystyle\sum_{j=0}^{4} 2j + 1$

17. Write "the sum of the first five positive even integers" in summation notation.

18. Determine the numerical value of

a) $\displaystyle\sum_{n=1}^{5} \left(\frac{1}{n} - \frac{1}{n+1}\right)$ b) $\displaystyle\sum_{n=1}^{500} \left(\frac{1}{n} - \frac{1}{n+1}\right)$

19. Express the following sums using summation notation.
    a) $1^2 + 2^2 + 3^2 + \cdots + n^2$.
    b) $2^1 + 2^2 + 2^3 + \cdots + 2^m$.
    c) $f(s_1)\Delta s + f(s_2)\Delta s + \cdots + f(s_{12})\Delta s$.
    d) $y_1^2\,\Delta y_1 + y_2^2\,\Delta y_2 + \cdots + y_n^2\,\Delta y_n$.

20. Express each of the following as a sum written out term by term. (There is no need to calculate the numerical value, even when that can be done.)

a) $\displaystyle\sum_{l=3}^{n-1} a_l$ b) $\displaystyle\sum_{j=0}^{4} \frac{j+1}{j^2+1}$ c) $\displaystyle\sum_{k=1}^{5} H(x_k)\,\Delta x_k$.

21. Acquire experimental evidence for the claim

$$\left(\sum_{k=1}^{n} k\right)^2 = \sum_{k=1}^{n} k^3$$

by determining the numerical values of both sides of the equation for $n = 2, 3, 4, 5,$ and 6.

22. Let $g(u) = 25 - u^2$ and suppose the interval $[0, 2]$ has been divided into four equal subintervals Δu and u_j is the left endpoint of the jth interval. Determine the numerical value of the Riemann sum

$$\sum_{j=1}^{4} g(u_j)\,\Delta u.$$

Length and Area

23. Using Riemann sums with equal subintervals, estimate the length of the parabola $y = x^2$ over the interval $0 \le x \le 1$. Obtain a sequence of estimates that stabilize to four decimal places. How many subintervals

did you need? (Compare your result here with the earlier result on page 83.)

24. Using Riemann sums, obtain a sequence of estimates for the area under each of the following curves. Continue until the first four decimal places stabilize in your estimates.

a) $y = x^2$ over $[0, 1]$ **c)** $y = x \sin x$ over $[0, \pi]$

b) $y = x^2$ over $[0, 3]$

25. What is the area under the curve $y = \exp(-x^2)$ over the interval $[0, 1]$? Give an estimate that is accurate to four decimal places. Sketch the curve and shade the area.

26. a) Estimate, to four decimal place accuracy, the length of the graph of the natural logarithm function $y = \ln x$ over the interval $[1, e]$.

     b) Estimate, to four decimal place accuracy, the length of the graph of the exponential function $y = \exp(x)$ over the interval $[0, 1]$.

27. a) What is the length of the hyperbola $y = 1/x$ over the interval $[1, 4]$? Obtain an estimate that is accurate to four decimal places.

     b) What is the area under the hyperbola over the same interval? Obtain an estimate that is accurate to four decimal places.

28. The graph of $y = \sqrt{4 - x^2}$ is a semicircle whose radius is 2. The circumference of the whole circle is 4π, so the length of the part of the circle in the first quadrant is exactly π.

     a) Using left endpoint Riemann sums, estimate the length of the graph $y = \sqrt{4 - x^2}$ over the interval $[0, 2]$ in the first quadrant. How many subintervals did you need in order to get an estimate that has the value $3.14159\ldots$?

     b) There is a technical problem that makes it impossible to use *right* endpoint Riemann sums. What is the problem?

$\succ$ **6.3 THE INTEGRAL**

Refining Riemann Sums

In the last section, we estimated the electrical energy a town consumed by constructing a Riemann sum for the power demand function $p(t)$. Because we sampled the power function only five times in a 24-hour period, our estimate was fairly rough. We would get a better estimate by sampling more frequently—that is, by constructing a Riemann sum with more terms and shorter subintervals. The process of refining Riemann sums in this way leads to the mathematical object called the **integral**.

To see what an integral is—and how it emerges from this process of refining Riemann sums—let's return to the function

$$\sqrt{1 + x^3} \quad \text{on} \quad [1, 3]$$

Refining Riemann sums with equal subintervals

we analyzed at the end of the last section. What happens when we refine Riemann sums for this function by using smaller subintervals? If we systematically choose n equal subintervals and evaluate $\sqrt{1 + x^3}$ at the left endpoint of each subinterval, then we can use the program RIEMANN (page 319) to produce the values in the following table. For future reference we record the size of the subinterval $\Delta x = 2/n$ as well.

Left Endpoint Riemann
Sums for $\sqrt{1 + x^3}$ on $[1, 3]$

n	Δx	Riemann Sum
100	.02	6.191 236 2
1 000	.002	6.226 082 6
10 000	.0002	6.229 571 7
100 000	.00002	6.229 920 6

The first four digits have stabilized, suggesting that these Riemann sums, at least, approach the limit 6.229 . . . .

It's too soon to say that *all* the Riemann sums for $\sqrt{1 + x^3}$ on the interval $[1, 3]$ approach this limit, though. There is such an enormous diversity of choices at our disposal when we construct a Riemann sum. We haven't seen what happens, for instance, if we choose midpoints instead of left endpoints, or if we choose subintervals that are not all of the same size. Let's explore the first possibility.

To modify RIEMANN to choose midpoints, we need only change the line

Midpoints versus left endpoints

```
x = a
```

that determines the position of the first sampling point to

```
x = a + deltax / 2.
```

With this modification, RIEMANN produces the following data.

Midpoint Riemann Sums
for $\sqrt{1 + x^3}$ on $[1, 3]$

n	Δx	Riemann Sum
10	.2	6.227 476 5
100	.02	6.229 934 5
1 000	.002	6.229 959 1
10 000	.0002	6.229 959 4

This time, the first *seven* digits have stabilized, even though we used only 10,000 subintervals—ten times fewer than we needed to get four digits to stabilize using left endpoints! This is further evidence that the Riemann sums converge to a limit, and we can even specify the limit more precisely as 6.229959....

These tables also suggest that midpoints are more "efficient" than left endpoints in revealing the limiting value of successive Riemann sums. This is indeed true. In Chapter 11, we will look into this further.

We still have another possibility to consider: What happens if we choose subintervals of different sizes—as we did in calculating the energy consumption of a town? By allowing variable subintervals, we make the problem messier to deal with, but it does not become conceptually more difficult. In fact, we can still get all the information we really need in order to understand what happens when we refine Riemann sums.

To see what form this information will take, look back at the two tables for midpoints and left endpoints, and compare the values of the Riemann sums that they report for the same size subinterval. When the subinterval was fairly large, the values differed by a relatively large amount. For example, when $\Delta x = .02$ we got two sums (namely 6.2299345 and 6.1912362) that differ by more than .038. As the subinterval got smaller, the difference between the Riemann sums got smaller, too. (When $\Delta x = .0002$, the sums differ by less than .0004.) This is the general pattern. That is to say, for subintervals with a given maximum size, the various Riemann sums that can be produced will still differ from one another, but those sums will all lie within a certain range that gets smaller as the size of the largest subinterval gets smaller.

Refining Riemann sums with *unequal* subintervals

The connection between the range of Riemann sums and the size of the largest subinterval is subtle and technically complex; this course will not explore it in detail. However, we can at least see what happens concretely to Riemann sums for the function $\sqrt{1 + x^3}$ over the interval [1, 3]. The following table shows the smallest and largest possible Riemann sum that can be produced when no subinterval is larger than the maximum size Δx_k given in the first column.

Maximum Size of Δx_k	Riemann Sums Range from	to	Difference between Extremes
.02	6.113690	6.346328	.232638
.002	6.218328	6.241592	.023264
.0002	6.228796	6.231122	.002326
.00002	6.229843	6.230076	.000233
.000002	6.229948	6.229971	.000023

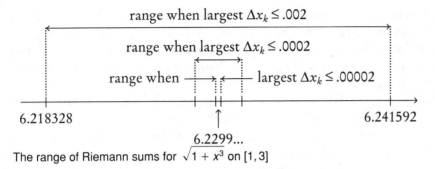

The range of Riemann sums for $\sqrt{1 + x^3}$ on $[1, 3]$

This table provides the most compelling evidence that there is a single num- The integral
ber 6.2299...that *all* Riemann sums will be arbitrarily close to, if they are
constructed with sufficiently small subintervals Δx_k. This number is called the
integral of $\sqrt{1 + x^3}$ on the interval $[1, 3]$, and we will express this by writing

$$\int_1^3 \sqrt{1 + x^3}\, dx = 6.2299.\ldots$$

Each Riemann sum approximates this integral, and in general the approximations get better as the size of the largest subinterval is made smaller. Moreover,
as the subintervals get smaller, the location of the sampling points matters less
and less.

The unusual symbol $\int$ that appears here reflects the historical origins of the
integral. We'll have more to say about it after we consider the definition.

Definition

The purpose of the following definition is to give a name to the number to
which the Riemann sums for a function converge, *when those sums do indeed
converge.*

> Suppose all the Riemann sums for a function $f(x)$ on an
> interval $[a, b]$ get arbitrarily close to a single number when
> the lengths $\Delta x_1, \ldots, \Delta x_n$ are made small enough. Then this
> number is called the **integral** of $f(x)$ on $[a, b]$ and it is denoted
>
> $$\int_a^b f(x)\, dx.$$

The function f is called the **integrand**. The definition begins with a *Suppose*...because there are functions whose Riemann sums don't converge. We'll
look at an example on page 333. However, that example is quite special. All

A typical function has
an integral

the functions that typically arise in context, and nearly all the functions we study in calculus, *do* have integrals. In particular, every continuous function has an integral, and so do many noncontinuous functions—such as the step functions with which we began this chapter. (Continuous functions are discussed on page 267.)

Notice that the definition doesn't speak about the choice of sampling points. The condition that the Riemann sums be close to a single number involves only the subintervals $\Delta x_1, \Delta x_2, \ldots, \Delta x_n$. This is important; it says *once the subintervals are small enough, it doesn't matter which sampling points x_k we choose—all of the Riemann sums will be close to the value of the integral.* (Of course, some will still be closer to the value of the integral than others.)

How an integral
expresses a product

The integral allows us to resolve the dilemma we stated at the beginning of the chapter. Here is the dilemma: How can we describe the product of two quantities when one of them varies? Consider, for example, how we expressed the energy consumption of a town over a 24-hour period. The basic relation

$$\text{energy} = \text{power} \times \text{elapsed time}$$

cannot be used directly, because power demand varies. Indirectly, though, we can use the relation to build a Riemann sum for power demand p over time. This gives us an *approximation*:

$$\text{energy} \approx \sum_{k=1}^{n} p(t_k)\,\Delta t_k \quad \text{megawatt-hours.}$$

As these sums are refined, two things happen. First, they converge to the true level of energy consumption. Second, they converge to the integral—by the definition of the integral. Thus, energy consumption is described *exactly* by the integral

$$\text{energy} = \int_0^{24} p(t)\,dt \quad \text{megawatt-hours}$$

of the power demand p. In other words, *energy is the integral of power over time*.

On page 310 we asked how far a ship would travel in 15 hours if we knew its velocity was $v(t)$ miles per hour at time t. We saw the distance could be estimated by a Riemann sum for the v. Therefore, reasoning just as we did for energy, we conclude that the *exact* distance is given by the integral

$$\text{distance} = \int_0^{15} v(t)\,dt \quad \text{miles.}$$

The units for an
integral

The energy integral has the same units as the Riemann sums that approximate it. Its units are the product of the megawatts used to measure p and the hours used to measure dt (or t). The units for the distance integral are the product of the miles per hour used to measure velocity and the hours used to measure time. In general, the units for the integral

$$\int_a^b f(x)\, dx$$

are the product of the units for f and the units for x.

Because the integral is approximated by its Riemann sums, we can use summation notation (introduced in the previous section) to write

$$\int_1^3 \sqrt{1+x^3}\, dx \approx \sum_{k=1}^n \sqrt{1+x_k^3}\, \Delta x_k.$$

Why $\int$ is used to denote an integral

This expression helps reveal where the rather unusual-looking notation for the integral comes from. In seventeenth century Europe (when calculus was being created), the letter "s" was written two ways: as "s" and as "$\int$". The $\int$ that appears in the integral and the $\sum$ that appears in the Riemann sum both serve as abbreviations for the word *sum.* While we think of the Riemann sum as a sum of products of the form $\sqrt{1+x_k^3}\cdot \Delta x_k$, in which the various Δx_k are small quantities, some of the early users of calculus thought of the integral as a sum of products of the form $\sqrt{1+x^3}\cdot dx$, in which dx is an "infinitesimally" small quantity.

Now we do not use infinitesimals or regard the integral as a sum directly. On the contrary, for us the integral is a *limit* of Riemann sums as the subinterval lengths Δx_k all shrink to 0. In fact, we can express the integral directly as a limit:

$$\int_a^b f(x)\, dx = \lim_{\Delta x_k \to 0} \sum_{k=1}^n f(x_k)\, \Delta x_k.$$

The process of calculating an integral is called **integration**. Integration means "putting together." To see why this name is appropriate, notice that we determine energy consumption over a long time interval by putting together a lot of energy computations $p\cdot \Delta t$ over a succession of short periods.

Integration "puts together" products of the form $f(x)\,\Delta x$

A Function That Does Not Have an Integral

Riemann sums converge to a single number for many functions—but not all. For example, Riemann sums for

$$J(x) = \begin{cases} 0 & \text{if } x \text{ is rational} \\ 1 & \text{if } x \text{ is irrational} \end{cases}$$

do not converge. Let's see why.

A rational number is the quotient p/q of one integer p by another q. An irrational number is one that is not such a quotient; for example, $\sqrt{2}$ is irrational. The values of J are continually jumping between 0 and 1.

Why Riemann sums for $J(x)$ do not converge

Suppose we construct a Riemann sum for J on the interval $[0, 1]$ using the subintervals $\Delta x_1, \Delta x_2, \ldots, \Delta x_n$. Every subinterval contains both rational and irrational numbers. Thus, we could choose all the sampling points to be rational numbers $r_1, r_2, \ldots, r_n$. In that case,

$$J(r_1) = J(r_2) = \cdots = J(r_n) = 0,$$

so the Riemann sum has the value

$$\sum_{k=1}^{n} J(r_k)\,\Delta x_k = \sum_{k=1}^{n} 0 \cdot \Delta x_k = 0.$$

But we could also choose all the sampling points to be irrational numbers $s_1, s_2, \ldots, s_n$. In that case,

$$J(s_1) = J(s_2) = \cdots = J(s_n) = 1,$$

and the Riemann sum would have the value

$$\sum_{k=1}^{n} J(s_k)\,\Delta x_k = \sum_{k=1}^{n} 1 \cdot \Delta x_k = \sum_{k=1}^{n} \Delta x_k = 1.$$

(For any subdivision Δx_k, $\Delta x_1 + \Delta x_2 + \cdots + \Delta x_n = 1$ because the subintervals together form the interval $[0, 1]$.) If some sampling points are rational and others irrational, the value of the Riemann sum will lie somewhere between 0 and 1. Thus, the Riemann sums range from 0 to 1, *no matter how small the subintervals Δx_k are chosen.* They cannot converge to any single number.

The function J shows us that not every function has an integral. The definition of the integral (page 331) takes this into account. It doesn't guarantee that an arbitrary function will have an integral. It simply says that *if* the Riemann sums converge to a single number, then we can give that number a name—the integral.

Can you imagine what the graph of $y = J(x)$ would look like? It would consist of two horizontal lines with gaps; in the upper line, the gaps would be at all the rational points, in the lower at the irrational points. This is impossible to draw! Roughly speaking, any graph you can draw you can integrate.

Visualizing the Integral

The eye plays an important role in our thinking. We visualize concepts whenever we can. Given a function $y = f(x)$, we visualize it as a graph in the x, y-coordinate plane. We visualize the derivative of f as the slope of its graph at any point. We can visualize the integral of f, too. We will view it as the area under that graph. Let's see why we can.

We have already made a connection between areas and Riemann sums, in the last section. Our starting point was the basic formula

$$\text{area} = \text{height} \times \text{width}.$$

Since the height of the graph $y = f(x)$ at any point x is just $f(x)$, we were tempted to say that

$$\text{area} = f(x) \cdot (b - a).$$

Of course we couldn't do this, because the height $f(x)$ is variable. The remedy was to slice up the interval $[a, b]$ into small pieces Δx_k, and assemble a collection of products $f(x_k) \Delta x_k$:

$$\sum_{k=1}^{n} f(x_k) \Delta x_k = f(x_1) \Delta x_1 + f(x_2) \Delta x_2 + \cdots + f(x_n) \Delta x_n.$$

This is a Riemann sum. It represents the total area of a row of side-by-side rectangles whose tops approximate the graph of f. As the Riemann sums are refined, the tops of the rectangles approach the shape of the graph, and their areas approach the area under the graph. But the process of refining Riemann sums leads to the integral, so the integral must be the area under the graph.

Riemann sums converge to the area . . .

. . . and to the integral

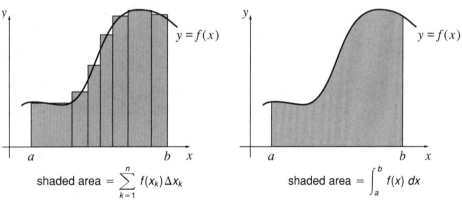

$$\text{shaded area} = \sum_{k=1}^{n} f(x_k) \Delta x_k \qquad \text{shaded area} = \int_{a}^{b} f(x)\, dx$$

Every integral we have encountered can be visualized as the area under a graph. For instance, since

$$\text{energy use} = \int_{0}^{24} p(t)\, dt,$$

we can now say that the energy used by a town is just the area under its power demand graph.

Energy used is the area under the power graph

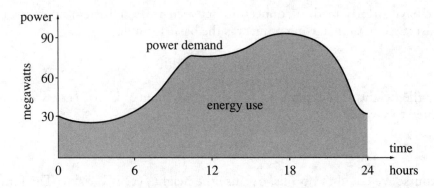

The area interpretation
is not always helpful!

Although we can always visualize an integral as an area, it may not be very enlightening in particular circumstances. For example, in the last section (page 315) we estimated the length of the graph $y = \sin x$ from $x = 0$ to $x = \pi$. Our estimates came from Riemann sums for the function $f(x) = \sqrt{1 + \cos^2 x}$ over the interval $[0, \pi]$. These Riemann sums converge to the integral

$$\int_0^\pi \sqrt{1 + \cos^2 x}\, dx,$$

which we can now view as the area under the graph of $\sqrt{1 + \cos^2 x}$.

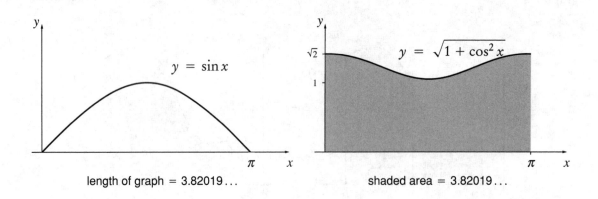

length of graph = 3.82019...

shaded area = 3.82019...

More generally, the length of $y = g(x)$ will always equal the area under $y = \sqrt{1 + (g'(x))^2}$.

The Integral of a Negative Function

Up to this point, we have been dealing with a function $f(x)$ that is never negative on the interval $[a, b]$: $f(x) \geq 0$. Its graph therefore lies entirely above the x-axis. What happens if $f(x)$ does take on negative values? We'll first consider an example.

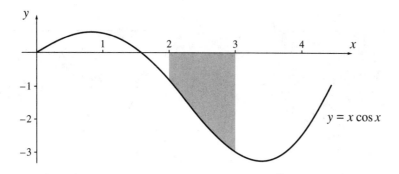

The graph of $f(x) = x \cos x$ is shown above. On the interval $[2, 3]$, it lies entirely below the x-axis. As you can check,

$$\int_2^3 x \cos x \, dx = -1.969080.$$

The integral is negative, but areas are positive. Therefore, it seems we can't interpret *this* integral as an area. But there is more to the story. The shaded region is 1 unit wide and varies in height from 1 to 3 units. If we say the average height is about 2 units, then the area is about 2 square units. Except for the negative sign, our rough estimate for the area is almost exactly the value of the integral.

The integral is the *negative* of the area

In fact, the integral of a *negative* function is always the *negative* of the area between its graph and the x-axis. To see why this is always true, we'll look first at the simplest possibility—a constant function. The integral of a constant function is just the product of that constant value by the width of the interval. For example, suppose $f(x) = -7$ on the interval $[1, 3]$. The region between the graph and the x-axis is a rectangle whose area is $7 \times 2 = 14$. However,

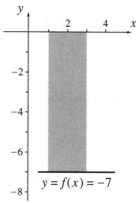

$$f(x) \cdot \Delta x = -7 \times 2 = -14.$$

This is the *negative* of the area of the region.

Let's turn now to an arbitrary function $f(x)$ whose values vary but remain negative over the interval $[a, b]$. Each term in the Riemann sum on the left is the *negative* of one of the shaded rectangles. In the process of refinement, the total area of the rectangles approaches the shaded area on the right. At the same time, the Riemann sums approach the integral. Thus, the integral must be the negative of the shaded area.

An arbitrary function that takes only negative values

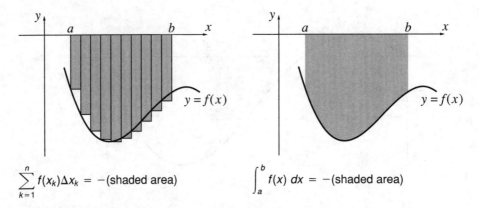

$$\sum_{k=1}^{n} f(x_k)\Delta x_k = -(\text{shaded area})$$

$$\int_a^b f(x)\, dx = -(\text{shaded area})$$

Functions with Both Positive and Negative Values

The final possibility to consider is that $f(x)$ takes both positive and negative values on the interval $[a, b]$. In that case its graph lies partly above the x-axis and partly below. By considering these two parts separately we can see that

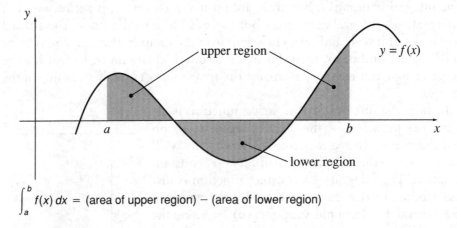

$$\int_a^b f(x)\, dx = (\text{area of upper region}) - (\text{area of lower region})$$

The graph of $y = \sin x$ on the interval $[0, \pi]$ is the mirror image of the graph on the interval $[\pi, 2\pi]$. The first half lies above the x-axis, the second half below. Since the upper and lower areas are equal, it follows that

$$\int_0^{2\pi} \sin x\, dx = 0.$$

Signed Area

There is a way to simplify the geometric interpretation of an integral as an area. It involves introducing the notion of *signed area*, by analogy with the notion of signed length.

Consider the two points 2 and −2 on the y-axis at the right. Although the line that goes up from 0 to 2 has the same length as the line that goes down from 0 to −2, we customarily attach a sign to those lengths to take into account the *direction* of the line. Specifically, we assign a positive length to a line that goes up and a negative length to a line that goes down. Thus the line from 0 to −2 has **signed length** −2.

To adapt this pattern to areas, just assign to any area that goes up from the x-axis a positive value and to any area that goes down from the x-axis a negative value. Then the **signed area** of a region that is partly above and partly below the x-axis is just the sum of the areas of the parts—taking the signs of the different parts into account.

A region below the x-axis has negative area

Consider, for example, the graph of $y = x$ over the interval $[-2, 3]$. The upper region is a triangle whose area is 4.5. The lower region is another triangle; its area is 2, and its *signed* area is −2. Thus, the total signed area is +2.5, and it follows that

$$\int_{-2}^{3} x \, dx = 2.5.$$

You should confirm that Riemann sums for $f(x) = x$ over the interval $[-2, 3]$ converge to the value 2.5 (see exercise 13).

Now that we can describe the signed area of a region in the x, y plane, we have a simple and uniform way to visualize the integral of any function:

$$\int_{a}^{b} f(x) \, dx = \text{the } \textbf{signed area}$$
between the graph of $f(x)$ and the x-axis.

Error Bounds

A Riemann sum determines the value of an integral only approximately. For example,

The error in a Riemann sum

$$\int_{1}^{3} \sqrt{1 + x^3} \, dx = 6.229959\ldots,$$

but a left endpoint Riemann sum with 100 equal subintervals Δx gives

$$\sum_{k=1}^{100} \sqrt{1 + x^3}\, \Delta x = 6.191236.$$

If we use this sum as an estimate for the value of the integral, we make an **error** of

$$6.229959 - 6.191236 = .038723.$$

By increasing the number of subintervals, we can reduce the size of the error. For example, with 100,000 subintervals, the error is only .000038. (This information comes from page 329.) The fact that the first four digits in the error are now 0 means, roughly speaking, that the first four digits in the new estimate are correct.

Finding the error without knowing the exact value

In this example, we could measure the error in a Riemann sum because we knew the value of the integral. Usually, though, we *don't* know the value of the integral—that's why we're calculating Riemann sums! We will describe here a method to decide how inaccurate a Riemann sum is *without first knowing the value of the integral*. For example, suppose we estimate the value of

$$\int_0^1 e^{-x^2}\, dx$$

using a left endpoint Riemann sum with 1000 equal subintervals. The value we get is .747140. Our method will tell us that this differs from the true value of the integral by *no more than* .000633. So the method does not tell us the exact size of the error. It says only that the error is not larger than .000633. Such a

Error bounds

number is called an **error bound**. The actual error—that is, the true difference between the value of the integral and the value of the Riemann sum—may be a lot less than .000633. (That is indeed the case. In the exercises you are asked to show that the actual error is about half this number.)

We have two ways to indicate that .747140 is an estimate for the value of the integral, with an error bound of .000633. One is to use a "plus-minus" sign ($\pm$):

$$\int_0^1 e^{-x^2}\, dx = .747140 \pm .000633.$$

Since $.747140 - .000633 = .746507$ and $.747140 + .000633 = .747773$, this is the same as

$$.746507 \leq \int_0^1 e^{-x^2}\, dx \leq .747773.$$

The number .746507 is called a **lower bound** for the integral, and .747773 is called an **upper bound**. Thus, the true value of the integral is .74..., and the third digit is either a 6 or a 7.

Upper and lower bounds

Our method will tell us even more. In this case, it will tell us that the original Riemann sum is *already* larger than the integral. In other words, we can drop the upper bound from .747773 to .747140:

$$.746507 \le \int_0^1 e^{-x^2}\, dx \le .747140.$$

The Method

We want to get a bound on the difference between the integral of a function and a Riemann sum for that function. By visualizing both the integral and the Riemann sum as areas, we can visualize their difference as an area, too. We'll assume that all subintervals in the Riemann sum have the same width Δx. This will help keep the details simple. Thus

Visualize the error as an area

$$\text{error} = \left| \int_a^b f(x)\, dx - \sum_{k=1}^{n} f(x_k)\, \Delta x \right|.$$

(The absolute value $|u - v|$ of the difference tells us how far apart u and v are.)

We'll also assume the function $f(x)$ is *positive* and *increasing* on an interval $[a, b]$. We say $f(x)$ is **increasing** if its graph rises as x goes from left to right.

Let's start with left endpoints for the Riemann sum. Because $f(x)$ is increasing, the rectangles lie entirely below the graph of the function:

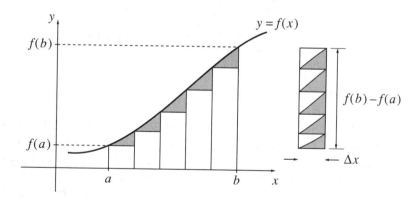

The integral is the area under the graph, and the Riemann sum is the area of the rectangles. Therefore, the error is the area of the shaded region that lies above the rectangles and below the graph. This region consists of a number of

Slide the little errors
into a stack

separate pieces that sit on top of the individual rectangles. Slide them to the right and stack them on top of one another, as shown in the preceding figure. They fit together inside a single rectangle of width Δx and height $f(b) - f(a)$. Thus the error (which is the area of the shaded region) is no greater than the area of this rectangle:

$$\text{error} \leq \Delta x \ (f(b) - f(a)).$$

The number $\Delta x \ (f(b) - f(a))$ is our error bound. It is clear from the figure that this is not the *exact* value of the error. The error is smaller, but it is difficult to say exactly how much smaller. Notice also that we need very little information to find the error bound—just Δx and the function values $f(a)$ and $f(b)$. We do *not* need to know the exact value of the integral!

The error bound is
proportional to Δx

The error bound is proportional to Δx. If we cut Δx in half, that will cut the error bound in half. If we make Δx a tenth of what it had been, that will make the error bound a tenth of what *it* had been. In the figure below, Δx is one–fifth its value in the previous figure. The rectangle on the right shows how much smaller the error bound has become as a result. It demonstrates how Riemann sums converge as the size of the subinterval Δx shrinks to zero.

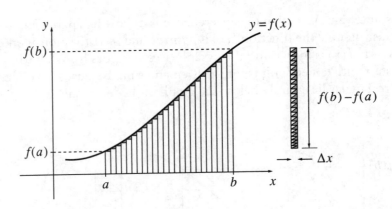

Right endpoints

Let's go back to the original Δx, but switch to *right* endpoints. Then the tops of the rectangles lie above the graph shown at the top of page 343. The error is the vertically hatched region between the graph and the tops of the rectangles. Once again, we can slide the little errors to the right and stack them on top of one another. They fit inside the same rectangle we had before. Thus, whether the Riemann sum is constructed with left endpoints or right endpoints, we find the same error bound:

$$\text{error} \leq \Delta x \ (f(b) - f(a)).$$

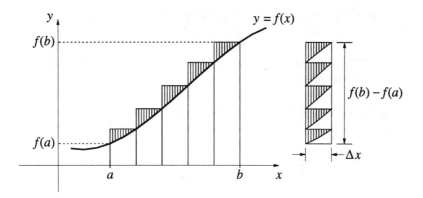

Because $f(x)$ is increasing, the left endpoint Riemann sum is smaller than the integral, while the right endpoint Riemann sum is larger. On a number line, these three values are arranged as follows:

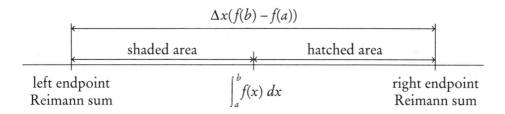

The distance from the left endpoint Riemann sum to the integral is represented by the shaded area, and the distance from the right endpoint by the hatched area. Notice that these two areas exactly fill the rectangle that gives us the error bound. Thus the distance between the two Riemann sums on the number line is exactly $\Delta x \left(f(b) - f(a) \right)$, as shown.

Finally, suppose that the Riemann sum has *arbitrary* sampling points x_k:

Arbitrary sampling points

$$\sum_{k=1}^{n} f(x_k)\, \Delta x.$$

Since f is increasing on the interval $[a, b]$, its values get larger as x goes from left to right. Therefore, on the kth subinterval,

$$f(\text{left endpoint}) \leq f(x_k) \leq f(\text{right endpoint}).$$

In other words, the rectangle built over the left endpoint is the shortest, and the one built over the right endpoint is the tallest. The one built over the sampling point x_k lies somewhere in between.

The areas of these three rectangles are arranged in the same order:

$$f(\text{left endpoint})\, \Delta x \leq f(x_k)\, \Delta x \leq f(\text{right endpoint})\, \Delta x.$$

If we add up these areas, we get Riemann sums. The Riemann sums are arranged in the same order as their individual terms:

$$\left\{ \begin{array}{c} \text{left endpoint} \\ \text{Riemann sum} \end{array} \right\} \leq \sum_{k=1}^{n} f(x_k) \, \Delta x \leq \left\{ \begin{array}{c} \text{right endpoint} \\ \text{Riemann sum} \end{array} \right\}.$$

Thus, the left endpoint and the right endpoint Riemann sums are extremes: *every* Riemann sum for f that uses a subinterval size of Δx lies between these two.

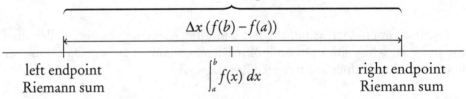

It follows that $\Delta x \, (f(b) - f(a))$ is an error bound for all Riemann sums whose subintervals are Δx units wide.

Error bounds for a decreasing function

If $f(x)$ is a positive function but is *decreasing* on the interval $[a, b]$, we get essentially the same result. Any Riemann sum for f that uses a subinterval of size Δx differs from the integral by no more than $\Delta x \, (f(a) - f(b))$. When f is decreasing, however, $f(a)$ is larger than $f(b)$, so we write the height of the rectangle as $f(a) - f(b)$. To avoid having to pay attention to this distinction, we can use absolute values to describe the error bound:

$$\text{error} \leq \Delta x \, \left| f(b) - f(a) \right|.$$

Also, when f is decreasing, the left endpoint Riemann sum is larger than the integral, while the right endpoint Riemann sum is smaller. The difference between the right and the left endpoint Riemann sums is still $\Delta x \, |f(b) - f(a)|$.

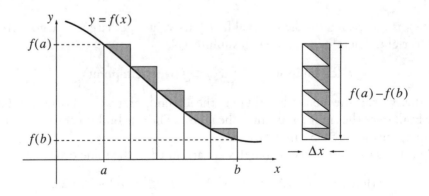

Up to this point, we have assumed $f(x)$ was either always increasing, or else always decreasing, on the interval $[a, b]$. Such a function is said to be **monotonic**. If $f(x)$ is *not* monotonic, the process of getting an error bound for Riemann sums is only slightly more complicated. Monotonic functions

Here is how to get an error bound for the Riemann sums constructed for a nonmonotonic function. First break up the interval $[a, b]$ into smaller pieces on which the function *is* monotonic. Error bounds for nonmonotonic functions

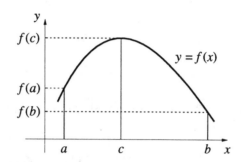

In this figure, there are two such intervals: $[a, c]$ and $[c, b]$. Suppose we construct a Riemann sum for f by using rectangles of width Δx_1 on the first interval, and Δx_2 on the second. Then the total error for this sum will be no larger than the sum of the error bounds on the two intervals: The monotonic pieces

$$\text{total error} \le \Delta x_1 \left| f(c) - f(a) \right| + \Delta x_2 \left| f(b) - f(c) \right|.$$

By making Δx_1 and Δx_2 sufficiently small, we can make the error as small as we wish.

This method can be applied to any nonmonotonic function that can be broken up into monotonic pieces. For other functions, more than two pieces may be needed.

Using the Method

Earlier we said that when we use a left endpoint Riemann sum with 1000 equal subintervals to estimate the value of the integral An example

$$\int_0^1 e^{-x^2}\, dx,$$

the error is no larger than .000633. Let's see how our method would lead to this conclusion.

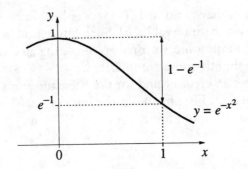

On the interval $[0, 1]$, $f(x) = e^{-x^2}$ is a decreasing function. Furthermore,

$$f(0) = 1 \qquad f(1) = e^{-1} \approx .3679.$$

If we divide $[0, 1]$ into 1000 equal subintervals Δx, then $\Delta x = 1/1000 = .001$. The error bound is therefore

$$.001 \times |.3679 - 1| = .001 \times .6321 = .0006321.$$

Any number *larger* than this one will also be an error bound. By "rounding up," we get .000633. This is slightly shorter to write, and it is the bound we claimed earlier. An even shorter bound is .00064.

Furthermore, since $f(x) = e^{-x^2}$ is decreasing, any Riemann sum constructed with *left* endpoints is larger than the actual value of the integral. Since the left endpoint Riemann sum with 1000 equal subdivisions has the value .747140, upper and lower bounds for the integral are

$$.746507 = .747140 - .000633 \le \int_0^1 e^{-x^2}\,dx \le .747140.$$

Integration Rules

Just as there are rules that tell us how to find the derivative of various combinations of functions, there are other rules that tell us how to find the integral. Here are three that are exactly analogous to differentiation rules:

$$\int_a^b f(x) + g(x)\,dx = \int_a^b f(x)\,dx + \int_a^b g(x)\,dx$$

$$\int_a^b f(x) - g(x)\,dx = \int_a^b f(x)\,dx - \int_a^b g(x)\,dx$$

$$\int_a^b c\,f(x)\,dx = c \int_a^b f(x)\,dx.$$

Let's see why the third rule is true. A Riemann sum for the integral on the left looks like

$$\sum_{k=1}^{n} cf(x_k)\,\Delta x_k = cf(x_1)\,\Delta x_1 + \cdots + cf(x_n)\,\Delta x_n.$$

Since the factor c appears in every term in the sum, we can move it outside the summation:

$$c\left(f(x_1)\,\Delta x_1 + \cdots + f(x_n)\,\Delta x_n\right) = c\left(\sum_{k=1}^{n} f(x_k)\,\Delta x_k\right).$$

The new expression is c times a Riemann sum for

$$\int_a^b f(x)\,dx.$$

Since the Riemann sum expressions are equal, the integral expressions they converge to must be equal, as well. You can use similar arguments to show why the other two rules are true.

Here is one example of the way we can use these rules:

$$\int_1^3 4\sqrt{1+x^3}\,dx = 4\int_1^3 \sqrt{1+x^3}\,dx = 4 \times 6.229959 = 24.919836.$$

(The value of the second integral is given on page 339.) Here is another example:

$$\int_2^9 5x^7 - 2x^3 + 24x\,dx = 5\int_2^9 x^7\,dx - 2\int_2^9 x^3\,dx + 24\int_2^9 x\,dx.$$

Of course, we must still determine the value of various integrals of the form

$$\int_a^b x^n\,dx.$$

However, the example shows us that, once we know the value of these special integrals, we can determine the value of the integral of any polynomial.

Here are two more rules that have no direct analogue in differentiation. The first says that if $f(x) \le g(x)$ for every x in the interval $[a, b]$, then

$$\int_a^b f(x)\,dx \le \int_a^b g(x)\,dx.$$

In the second, c is a point somewhere in the interval $[a, b]$:

$$\int_a^b f(x)\,dx = \int_a^c f(x)\,dx + \int_c^b f(x)\,dx.$$

If you visualize an integral as an area, it is clear why these rules are true.

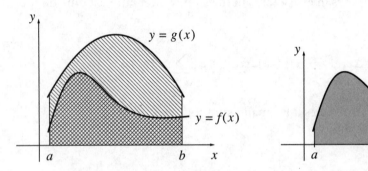

The second rule can be used to understand the results of exercise 4 on pages 322–323. It concerns the three integrals

$$\int_1^3 \sqrt{1 + x^3}\,dx = 6.229959, \qquad \int_3^7 \sqrt{1 + x^3}\,dx = 45.820012,$$

and

$$\int_1^7 \sqrt{1 + x^3}\,dx = 52.049971.$$

Since the third interval $[1, 7]$ is just the first $[1, 3]$ combined with the second $[3, 7]$, the third integral is the sum of the first and the second. The numerical values confirm this.

Exercises

1. Determine the values of the following integrals.

*a) $\displaystyle\int_2^{15} 3\,dx$ c) $\displaystyle\int_{-5}^{-3} 7\,dx$ e) $\displaystyle\int_{-4}^{9} -2\,dz$

b) $\displaystyle\int_{15}^{2} 3\,dx$ d) $\displaystyle\int_{-3}^{-5} 7\,dx$ f) $\displaystyle\int_{9}^{-4} -2\,dz$

2. a) Sketch the graph of

$$g(x) = \begin{cases} 7 & \text{if } 1 \le x < 5, \\ -3 & \text{if } 5 \le x \le 10 \end{cases}$$

b) Determine $\displaystyle\int_1^7 g(x)\,dx$, $\displaystyle\int_7^{10} g(x)\,dx$, and $\displaystyle\int_1^{10} g(x)\,dx$.

Refining Riemann Sums

3. a) By refining Riemann sums, find the value of the following integral to four decimal place accuracy. Do the computations twice: first, using left endpoints; second, using midpoints.

$$\int_0^1 \frac{1}{1+x^3}\,dx.$$

b) How many subintervals did you need to get four decimal place accuracy when you used left endpoints and when you used midpoints? Which sampling points gave more efficient computations—left endpoints or midpoints?

4. By refining appropriate Riemann sums, determine the value of each of the following integrals, accurate to four decimal places. Use whatever sampling points you wish, but justify your claim that your answer is accurate to four decimal places.

a) $\int_1^4 \sqrt{1+x^3}\,dx$ **b)** $\int_4^7 \sqrt{1+x^3}\,dx$ ***c)** $\int_0^3 \frac{\cos x}{1+x^2}\,dx$

5. Determine the value of the following integrals to four decimal places accuracy.

a) $\int_1^2 e^{-x^2}\,dx$ **c)** $\int_0^4 \sin(x^2)\,dx$

b) $\int_0^4 \cos(x^2)\,dx$ **d)** $\int_0^1 \frac{4}{1+x^2}\,dx$

6. a) What is the length of the graph of $y = \sqrt{x}$ from $x = 1$ to $x = 4$?
b) What is the length of the graph of $y = x^2$ from $x = 1$ to $x = 2$?
c) Why are the answers in parts (a) and (b) the same?

7. Both of the curves $y = 2^x$ and $y = 1 + x^{3/2}$ pass through $(0, 1)$ and $(1, 2)$. Which is the shorter one? Can you decide simply by looking at the graphs?

8. A pyramid is 30 feet tall. The area of a horizontal cross section x feet from the top of the pyramid measures $2x^2$ square feet. What is the area of the base? What is the volume of the pyramid, to the nearest cubic foot?

Error Bounds

9. A left endpoint Riemann sum with 1000 equally spaced subintervals gives the estimate .135432 for the value of the integral

$$\int_1^2 e^{-x^2}\,dx.$$

a) Is the true value of the integral larger or smaller than this estimate? Explain.

*b) Find an error bound for this estimate.

c) Using the information you have already assembled, find lower and upper bounds A and B:

$$A \le \int_1^2 e^{-x^2}\, dx \le B.$$

*d) The lower and upper bounds allow you to determine a certain number of digits in the *exact* value of the integral. How many digits do you know, and what are they?

10. A left endpoint Riemann sum with 100 equally spaced subintervals gives the estimate .342652 for the value of the integral

$$\int_0^{\pi/4} \tan x\, dx.$$

a) Is the true value of the integral larger or smaller than this estimate? Explain.

b) Find an error bound for this estimate.

c) Using the information you have already assembled, find lower and upper bounds A and B:

$$A \le \int_0^{\pi/4} \tan x\, dx \le B.$$

d) The lower and upper bounds allow you to determine a certain number of digits in the *exact* value of the integral. How many digits do you know, and what are they?

11. **a)** In the next section you will see that

$$\int_0^{\pi/2} \sin x\, dx = 1$$

exactly. Here, estimate the value by a Riemann sum using the left endpoints of 100 equal subintervals.

b) Find an error bound for this estimate, and use it to construct the best possible lower and upper bounds

$$A \le \int_0^{\pi/2} \sin x\, dx \le B.$$

12. In the text (page 329), a Riemann sum using left endpoints on 1000 equal subintervals produces an estimate of 6.226083 for the value of

$$\int_1^3 \sqrt{1 + x^3}\, dx.$$

a) Is the true value of the integral larger or smaller than this estimate? Explain your answer, and do so without referring to the fact that the true value of the integral is known to be 6.229959....

b) Find an error bound for this estimate.

c) Find the upper and lower bounds for the value of the integral that are determined by this estimate.

d) According to these bounds, how many digits of the value of the integral are now known for certain?

The Average Value of a Function

In the exercises for section 6.1 we saw that the average staffing level for a job is

$$\text{average staffing} = \frac{\text{total staff-hours}}{\text{hours worked}}.$$

If $S(t)$ represents the number of staff working at time t, then the total staff-hours accumulated between $t = a$ and $t = b$ hours is

$$\text{total staff-hours} = \int_a^b S(t)\, dt \quad \text{staff-hours.}$$

Therefore the average staffing is

$$\text{average staffing} = \frac{1}{b - a} \int_a^b S(t)\, dt \quad \text{staff.}$$

Likewise, if a town's power demand was $p(t)$ megawatts at t hours, then its average power demand between $t = a$ and $t = b$ hours is

$$\text{average power demand} = \frac{1}{b - a} \int_a^b p(t)\, dt \quad \text{megawatts.}$$

We can define the **average value** of an arbitrary function $f(x)$ over an interval $a \le x \le b$ by following this pattern: Geometric meaning of the average

$$\text{average value of } f = \frac{1}{b - a} \int_a^b f(x)\, dx.$$

The average value of f is sometimes denoted $\overline{f}$. Since

$$\overline{f} \cdot (b - a) = \int_a^b f(x)\, dx,$$

the area under the horizontal line $y = \bar{f}$ between $x = a$ and $x = b$ is the same as the area under the graph $y = f(x)$. See the graph below.

13. a) What is the average value of $f(x) = 5$ on the interval $[1, 7]$?
    b) What is the average value of $f(x) = \sin x$ on the interval $[0, 2\pi]$? On $[0, 100\pi]$?
    c) What is the average value of $f(x) = \sin^2 x$ on the interval $[0, 2\pi]$? On $[0, 100\pi]$?

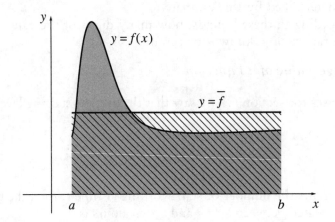

14. a) What is the average value of the function

$$H(x) = \begin{cases} 1 & \text{if } 0 \le x < 4, \\ 12 & \text{if } 4 \le x < 6, \\ 1 & \text{if } 6 \le x \le 20 \end{cases}$$

on the interval $[0, 20]$?
    b) Is the average $\bar{H}$ larger or smaller than the average of the two numbers 12 and 1 that represent the largest and smallest values of the function?
    c) Sketch the graph $y = H(x)$ along with the horizontal line $y = \bar{H}$, and show directly that the same area lies under each of these two graphs over the interval $[0, 20]$.

15. a) What are the maximum and minimum values of $f(x) = x^2 e^{-x}$ on the interval $[0, 20]$? What is the *average* of the maximum and the minimum?
    b) What is the average $\bar{f}$ of $f(x)$ on the interval $[0, 20]$?
    c) Why aren't these two averages the same?

The Integral as a Signed Area

16. By refining Riemann sums, confirm that $\displaystyle\int_{-2}^{3} x \, dx = 2.5$.

17. a) Sketch the graphs of $y = \cos x$ and $y = 5 + \cos x$ over the interval $[0, 4\pi]$.

    b) Find $\displaystyle\int_0^{4\pi} \cos x \, dx$ by visualizing the integral as a signed area.

    c) Find $\displaystyle\int_0^{4\pi} 5 + \cos x \, dx$. Why does $\displaystyle\int_0^{4\pi} 5 \, dx$ have the same value?

18. a) By refining appropriate Riemann sums, determine the value of the integral $\displaystyle\int_0^{\pi} \sin^2 x \, dx$ to four decimal places accuracy.

    b) Sketch the graph of $y = \sin^2 x$ on the interval $0 \le x \le \pi$. Note that your graph lies inside the rectangle formed by the lines $y = 0$, $y = 1$, $x = 0$, and $x = \pi$. (Sketch this rectangle.)

    c) Explain why the area under the graph of $y = \sin^2 x$ is exactly *half* of the area of the rectangle you sketched in part (b). What is the area of that rectangle?

    d) Using your observations in part (c), explain why $\displaystyle\int_0^{\pi} \sin^2 x \, dx$ is *exactly $\pi/2$.*

19. a) On what interval $a \le x \le b$ does the graph of the function $y = 4 - x^2$ lie *above* the x-axis?

    b) Sketch the graph of $y = 4 - x^2$ on the interval $a \le x \le b$ you determined in part (a).

    c) What is the area of the region that lies above the x-axis and below the graph of $y = 4 - x^2$?

20. a) What is the signed area (see page 339) between the graph of $y = x^3 - x$ and the x-axis on the interval $-1 \le x \le 2$?

    b) Sketch the graph of $y = x^3 - x$ on the interval $-1 \le x \le 2$. On the basis of your sketch, support or refute the following claim: the *signed* area between the graph of $y = x^3 - x$ and the x-axis on the interval $-1 \le x \le 2$ is exactly the same as the area between the graph of $y = x^3 - x$ and the x-axis on the interval $+1 \le x \le 2$.

➤ ## 6.4 THE FUNDAMENTAL THEOREM OF CALCULUS

Two Views of Power and Energy

In section 6.1 we considered how much energy a town consumed over 24 hours when it was using $p(t)$ megawatts of power at time t. Suppose $E(T)$ megawatt-hours of energy were consumed during the first T hours. Then the

Energy is the integral of power...

integral, introduced in section 6.3, gave us the language to describe how E depends on p:

$$E(T) = \int_0^T p(t)\, dt.$$

Because E is the integral of p, we can visualize $E(T)$ as the area under the power graph $y = p(t)$ as the time t sweeps from 0 hours to T hours:

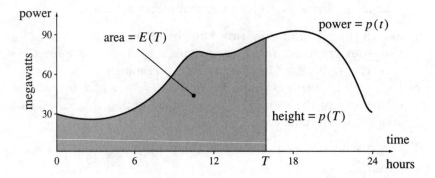

As T increases, so does $E(T)$. The exact relation between E and T is shown in the graph below. The height of the graph of E at any point T is equal to the area under the graph of p from 0 out to the point T.

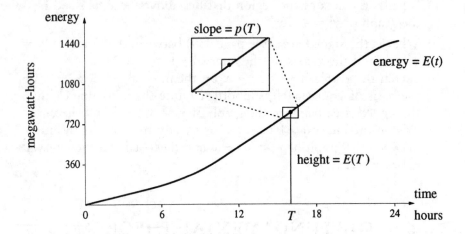

The microscope window on the graph of E reminds us that the slope of the graph at any point T is just $p(T)$:

...and power is the derivative of energy

$$p(T) = E'(T).$$

We discovered this fact in section 6.1, where we stated it in the following form: *Power is the rate at which energy is consumed.*

Pause now to study the graphs of power and energy. You should convince yourself that the *height* of the E graph at any time equals the *area* under the p graph up to that time. For example, when $T = 0$ no area has accumulated, so $E(0) = 0$. Furthermore, up to $T = 6$ hours, power demand was almost constant at about 30 megawatts. Therefore, $E(6)$ should be about

$$30 \text{ megawatts} \times 6 \text{ hours} = 180 \text{ megawatt-hours.}$$

It is. You should also convince yourself that the *slope* of the E graph at any point equals the *height* of the p graph at that point. Thus, for example, the graph of E will be steepest where the graph of p is tallest.

Notice that when we write the energy accumulation function $E(T)$ as an integral,

$$E(T) = \int_0^T p(t)\, dt,$$

we have introduced a new ingredient. The time variable T appears as one of the "limits of integration." By definition, the integral is a single number. However, that number depends on the interval of integration $[0, T]$. As soon as we treat T as a variable, the integral itself becomes a variable, too. Here the value of the integral varies with T.

A new ingredient: variable limits of integration

We now have two ways of viewing the relation between power and energy. According to the first view, the energy accumulation function E is the integral of power demand:

$$E(T) = \int_0^T p(t)\, dt.$$

According to the second view, the energy accumulation function is the solution $y = E(t)$ to an initial value problem defined by power demand:

$$y' = p(t); \qquad y(0) = 0.$$

An integral is a solution to a differential equation

If we take the first view, then we find E by refining Riemann sums—because that is the way to determine the value of an integral. If we take the second view, then we can find E by using any of the methods for solving initial value problems that we studied in chapter 4. Thus, the energy integral is a solution to a certain differential equation.

This is unexpected. Differential equations involve derivatives. At first glance, they have nothing to do with integrals. Nevertheless, the relation between power and energy shows us that there is a deep connection between derivatives and integrals. As we shall see, the connection holds for the integral of *any* function. The connection is so important—because it links together the two basic processes of calculus—that it has been called the **fundamental theorem of calculus**.

The fundamental theorem of calculus

The fundamental theorem gives us a powerful new tool to calculate integrals. Our aim in this section is to see why the theorem is true, and to begin to explore its use as a tool. In chapter 11 we will consider many specific integration techniques that are based on the fundamental theorem.

Integrals and Differential Equations

We begin with a statement of the fundamental theorem for a typical function $f(x)$.

The Fundamental Theorem of Calculus

The solution $y = A(x)$ to the initial value problem

$$y' = f(x) \qquad y(a) = 0$$

is the **accumulation function** $A(X) = \int_a^X f(x)\,dx$.

We have always been able to find the value of the integral

$$\int_a^b f(x)\,dx$$

by refining Riemann sums. The fundamental theorem gives us a new way. It says: First, find the solution $y = A(x)$ to the initial value problem $y' = f(x)$, $y(a) = 0$ using any suitable method for solving the differential equation. Then, once we have $A(x)$, we get the value of the integral by evaluating A at $x = b$.

A test case

To see how all this works, let's find the value of the integral

$$\int_0^4 \cos(x^2)\,dx$$

two ways: by refining Riemann sums and by solving the initial value problem

$$y' = \cos(x^2); \quad y(0) = 0$$

using Euler's method.

RIEMANN versus...

To estimate the value of the integral, we use the program RIEMANN from page 319. It produces the sequence of left endpoint sums shown in the table below on the left. Since the first three digits have stabilized, the value of the integral is .594.... The integral was deliberately chosen to lead to the initial value problem we first considered on page 235. We solved that problem by Euler's method, using the program TABLE, and produced the table of estimates for the value of $y(4)$ that appear in the table on the right on page 357. We see $y(4) = .594...$.

...TABLE

Program: RIEMANN
Left endpoint Riemann sums

```
DEF fnf (x) = COS(x ^ 2)
a = 0
b = 4
numberofsteps = 2 ^ 3
deltax = (b - a) / numberofsteps
x = a
accumulation = 0
FOR k = 1 TO numberofsteps
   deltaS = fnf(x) * deltax
   accumulation = accumulation + deltaS
   x = x + deltax
NEXT k
PRINT accumulation
```

Program: TABLE
Euler's method

```
DEF fnf (t) = COS(t ^ 2)
tinitial = 0
tfinal = 4
numberofsteps = 2 ^ 3
deltat = (tfinal - tinitial) / numberofsteps
t = tinitial
accumulation = 0
FOR k = 1 TO numberofsteps
   deltay = fnf(t) * deltat
   accumulation = accumulation + deltay
   t = t + deltat
NEXT k
PRINT accumulation
```

Number of Steps	Estimated Value of the Integral
2^3	1.13304
2^6	.65639
2^9	.60212
2^{12}	.59542
2^{15}	.59458
2^{18}	.59448

Number of Steps	Estimated Value of $y(4)$
2^3	1.13304
2^6	.65639
2^9	.60212
2^{12}	.59542
2^{15}	.59458
2^{18}	.59448

RIEMANN estimates the value of an integral by calculating Riemann sums. TABLE solves a differential equation by constructing estimates using Euler's method. These appear to be quite different tasks, but they lead to exactly the same results! But this is no accident. Compare the programs. (In both, the PRINT statement has been put outside the loop. This speeds up the calculations but still gives us the final outcome.) Once you make necessary modifications (e.g., change x to t, a to tinitial, etc.), you can see the two programs are the same.

The two programs are the same

The very fact that these two programs do the same thing gives us one proof of the fundamental theorem of calculus.

Graphing Accumulation Functions

Let's take a closer look at the solution $y = A(x)$ to the initial value problem

$$y' = \cos(x^2); \qquad y(0) = 0.$$

We can write it as the accumulation function

$$A(X) = \int_0^X \cos(x^2)\, dx.$$

By switching to the graphing version of TABLE (this is the program PLOT; see page 237), we can graph A. Here is the result.

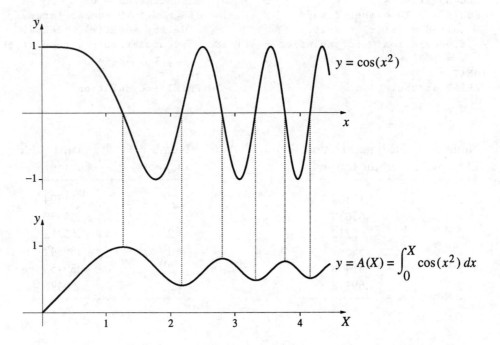

The relation between $y = \cos(x^2)$ and $y = A(x)$ is the same as the relation between power and energy.

- The *height* of the graph $y = A(x)$ at any point $x = X$ is equal to the *signed area* between the graph $y = \cos(x^2)$ and the x-axis over the interval $0 \le x \le X$.
- On the intervals where $\cos(x^2)$ is positive, $A(x)$ is increasing. On the intervals where $\cos(x^2)$ is negative, $A(x)$ is decreasing.
- When $\cos(x^2) = 0$, $A(x)$ has a maximum or a minimum.
- The *slope* of the graph of $y = A(x)$ at any point $x = X$ is equal to the *height* of the graph $y = \cos(x^2)$ at that point.

In summary, the lower curve is the integral of the upper one, and the upper curve is the derivative of the lower one.

To get a better idea of the simplicity and power of this approach to integrals, let's look at another example:

$$A(X) = \int_1^X \ln x \, dx.$$

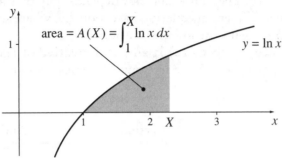

We find A, as always, by solving the initial value problem

$$y' = \ln x; \quad y(1) = 0$$

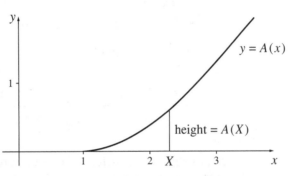

using Euler's method. The graphs $y = \ln x$ and $y = A(x)$ are shown at the right. Notice once again that the height of the graph of $A(x)$ at any point $x = X$ equals the area under the graph of $y = \ln x$ from $x = 1$ to $x = X$. Also, the graph of A becomes steeper as the height of $\ln x$ increases. In particular, the graph of A is horizontal (at $x = 1$) when $\ln x = 0$.

Antiderivatives

We have developed a novel approach to integration in this section. We start by replacing a given integral by an accumulation function:

Find an integral...

$$\int_a^b f(x) \, dx \quad \leadsto \quad A(X) = \int_a^X f(x) \, dx.$$

Then we try to find $A(X)$. If we do, then the original integral is just the value of A at $X = b$.

At first glance, this doesn't seem to be a sensible approach. We appear to be making the problem harder: instead of searching for a single number, we must now find an entire function. However, we know that $y = A(x)$ solves the initial value problem

$$y' = f(x), \quad y(a) = 0.$$

This means we can use the complete "bag of tools" we have for solving differential equations to find A. The real advantage of the new approach is that it reduces integration to the fundamental activity of calculus—solving differential equations.

...by solving a differential equation...

...using
antidifferentiation

The differential equation $y' = f(x)$ that arises in integration problems is special. The right–hand side depends only on the input variable x. We studied this differential equation in section 4.5, where we developed a special method to solve it—**antidifferentiation.**

We say that $F(x)$ is an **antiderivative** of $f(x)$ if f is the derivative of F: $F'(x) = f(x)$. Here are some examples (from page 233):

Function	Antiderivative
$5x^4 - 2x^3$	$x^5 - \frac{1}{2}x^4$
$5x^4 - 2x^3 + 17x$	$x^5 - \frac{1}{2}x^4 + \frac{17}{2}x^2$
$6 \cdot 10^z + 17/z^7$	$6 \cdot 10^z / \ln 10 - 17/6z^6$
$3 \sin t - 2t^3$	$-3 \cos t - \frac{1}{2}t^4$
$\pi \cos x + \pi^2$	$\pi \sin x + \pi^2 x$

A is an antiderivative of f

Since $y = A(x)$ solves the differential equation $y' = f(x)$, we have $A'(x) = f(x)$. Thus, $A(x)$ is an antiderivative of $f(x)$, so we can try to find A by antidifferentiating f.

Here is an example. Suppose we want to find

$$A(X) = \int_2^X 5x^4 - 2x^3 \, dx.$$

Notice that $f(x) = 5x^4 - 2x^3$ is the first function in the previous table. Since A must be an antiderivative of f, let's try the antiderivative for f that we find in the table:

$$F(x) = x^5 - \frac{1}{2}x^4.$$

Check the condition A(2) = 0

The problem is that F must also satisfy the initial condition $F(2) = 0$. However,

$$F(2) = 2^5 - \frac{1}{2} \cdot 2^4 = 32 - \frac{1}{2} \cdot 16 = 24 \neq 0,$$

so the initial condition does not hold *for this particular choice* of antiderivative. But this problem is easy to fix. Let

$$A(x) = F(x) - 24.$$

Since $A(x)$ differs from $F(x)$ only by a constant, it has the same derivative—namely, $5x^4 - 2x^3$. So $A(x)$ is still an antiderivative of $5x^4 - 2x^3$. But it also satisfies the initial condition:

$$A(2) = F(2) - 24 = 24 - 24 = 0.$$

Therefore $A(x)$ solves the problem; it has the right derivative *and* the right value at $x = 2$. Thus we have a formula for the accumulation function

$$\int_2^X 5x^4 - 2x^3\, dx = X^5 - \tfrac{1}{2}X^4 - 24.$$

The key step in finding the correct accumulation function A was to recognize that a given function f has infinitely many antiderivatives: if F is an antiderivative, then so is $F + C$, for any constant C. The general procedure for finding an accumulation function involves these two steps:

To find $A(X) = \displaystyle\int_a^X f(x)\, dx$:

1. First find *an* antiderivative $F(x)$ of $f(x)$.
2. Then set $A(x) = F(x) - F(a)$.

COMMENT. Recall that some functions f simply cannot be integrated—the Riemann sums they define may not converge. Although we have not stated it explicitly, you should keep in mind that the procedure just described applies only to functions that can be integrated.

A caution

EXAMPLE.   To illustrate the procedure, let's find the accumulation function

$$A(X) = \int_0^X \sin x\, dx.$$

The first step is to find an antiderivative for $f(x) = \sin x$. A natural choice is $F(x) = -\cos x$. To carry out the second step, note that $a = 0$. Since $F(0) = -\cos 0 = -1$, we set

$$A(x) = F(x) - F(0) = -\cos x - (-1) = 1 - \cos x.$$

The graphs $y = \sin x$ and $y = 1 - \cos x$ follow.

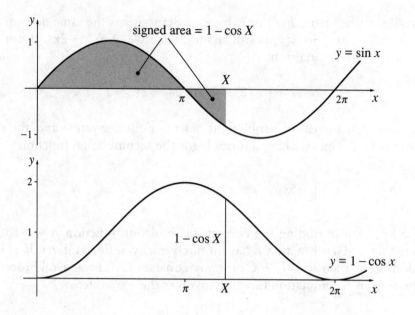

Now that we have a formula for $A(x)$ we can find the exact value of integrals involving $\sin x$. For instance,

$$\int_0^\pi \sin x \, dx = A(\pi) = 1 - \cos \pi = 1 - (-1) = 2.$$

Also,

$$\int_0^{\pi/2} \sin x \, dx = A(\pi/2) = 1 - \cos \pi/2 = 1 - (0) = 1,$$

and

$$\int_0^{2\pi} \sin x \, dx = A(2\pi) = 1 - \cos 2\pi = 1 - (1) = 0.$$

These values are *exact*, and we got them without calculating Riemann sums. However, we have already found the value of the third integral. On page 338 we argued from the shape of the graph $y = \sin x$ that the signed area between $x = 0$ and $x = 2\pi$ must be 0. That argument, however, is no help in finding the area to $x = \pi$ or to $x = \pi/2$. Before the fundamental theorem showed us we could evaluate integrals by finding antiderivatives, we could only make estimates using Riemann sums. Here, for example, are midpoint Riemann sums for

$$\int_0^{\pi/2} \sin x \, dx :$$

Subintervals	Riemann Sum
10	1.001 028 868
100	1.000 010 325
1000	1.000 000 147
10000	1.000 000 045

According to the table, the value of the integral is $1.000000\ldots$, to six decimal place accuracy. That is valuable information, and is often as accurate as we need. However, the fundamental theorem tells us that the value of the integral is 1 *exactly*! With the new approach, we can achieve absolute precision.

The fundamental theorem makes absolute precision possible

Precision is the result of having a *formula* for the antiderivative. Notice how we used that formula to express the value of the integral. Starting with an arbitrary antiderivative $F(x)$ of $f(x)$, we get

$$A(X) = \int_a^X f(x)\,dx = F(X) - F(a).$$

If we set $X = b$, we find

$$\int_a^b f(x)\,dx = F(b) - F(a).$$

In other words,

> If $F(x)$ is any antiderivative of $f(x)$,
> then $\displaystyle\int_a^b f(x)\,dx = F(b) - F(a).$

EXAMPLE.   Evaluate $\displaystyle\int_0^{\ln 2} e^{-u}\,du.$

For the antiderivative, we can choose $F(u) = -e^{-u}$. Then

$$\int_0^{\ln 2} e^{-u}\,du = F(\ln 2) - F(0)$$
$$= -e^{-\ln 2} - (-e^0)$$
$$= -(1/2) - (-1)$$
$$= 1/2.$$

EXAMPLE.   Evaluate $\int_2^3 \ln t \, dt$.

For the antiderivative, we can choose $F(t) = t \ln t - t$. Using the product rule, we can show that this is indeed an antiderivative:

$$F'(t) = t \cdot \frac{1}{t} + 1 \cdot \ln t - 1$$
$$= 1 + \ln t - 1$$
$$= \ln t.$$

Thus,

$$\int_2^3 \ln t \, dt = F(3) - F(2)$$
$$= 3 \cdot \ln 3 - 3 - (2 \cdot \ln 2 - 2)$$
$$= \ln 3^3 - \ln 2^2 - 1$$
$$= \ln 27 - \ln 4 - 1$$
$$= \ln(27/4) - 1.$$

While formulas make it possible to get exact values, they do present us with problems of their own. For instance, we need to know that $t \ln t - t$ is an antiderivative of $\ln t$. This is not obvious. In fact, there is no guarantee that the antiderivative of a function given by a formula will have a formula! The antiderivatives of $\cos(x^2)$ and $\sin(x)/x$ do not have formulas, for instance. Many techniques have been devised to find the formula for an antiderivative. In chapter 11 we will survey some of those that are most frequently used.

Parameters

When integrals depend on parameters...

In Section 4.2 we considered differential equations that involved parameters (see pages 188–192). It also happens that integrals can involve parameters. However, parameters complicate numerical work. If we calculate the value of an integral numerically, by making estimates with Riemann sums, we must first fix the value of any parameters that appear. This makes it difficult to see how the value of the integral depends on the parameters. We would have to give new values to the parameters and then recalculate the Riemann sums.

...the fundamental theorem can help make the relation clearer

The outcome is much simpler and more transparent if we are able to use the fundamental theorem to get a *formula* for the integral. The parameters just appear in the formula, so it is immediately clear how the integral depends on the parameters.

Here is an example that we shall explore further later. We want to see how the integrals

$$\int_a^b \sin(\alpha x)\, dx \qquad \text{and} \qquad \int_a^b \cos(\alpha x)\, dx$$

depend on the parameter α, and also on the parameters a and b. To begin, you should check that $F(x) = -\cos(\alpha x)/\alpha$ is an antiderivative of $\sin(\alpha x)$. Therefore,

$$\int_a^b \sin(\alpha x)\, dx = \frac{-\cos(\alpha b)}{\alpha} - \frac{-\cos(\alpha a)}{\alpha} = \frac{\cos(\alpha a) - \cos(\alpha b)}{\alpha}.$$

In a similar way, you should be able to show that

$$\int_a^b \cos(\alpha x)\, dx = \frac{\sin(\alpha b) - \sin(\alpha a)}{\alpha}.$$

Suppose the interval $[a, b]$ is exactly one-half of a full period: $[0, \pi/\alpha]$. Then

$$\int_0^{\pi/\alpha} \sin(\alpha x)\, dx = \frac{\cos(\alpha \cdot 0) - \cos(\alpha \pi/\alpha)}{\alpha}$$

$$= \frac{\cos 0 - \cos \pi}{\alpha}$$

$$= \frac{1 - (-1)}{\alpha} = \frac{2}{\alpha}.$$

Exercises

Constructing Accumulation Functions

1. a) Obtain a formula for the accumulation function

$$A(X) = \int_2^X 5\, dx$$

and sketch its graph on the interval $2 \le X \le 6$.
b) Is $A'(X) = 5$?

2. Let $f(x) = 2 + x$ on the interval $0 \le x \le 5$.
a) Sketch the graph of $y = f(x)$.
b) Obtain a formula for the accumulation function

$$A(X) = \int_0^X f(x)\, dx$$

and sketch its graph on the interval $0 \le X \le 5$.
c) Verify that $A'(X) = f(X)$ for every X in $0 \le X \le 5$.
d) By comparing the graphs of f and A, verify that, at any point X, the *slope* of the graph of A is the same as the *height* of the graph of f.

3. **a)** Consider the accumulation function

$$A(X) = \int_0^X x^3 \, dx.$$

Using the fact that $A'(X) = X^3$, obtain a formula that expresses A in terms of X.

b) Modify A so that accumulation begins at the value $x = 1$ instead of $x = 0$ as in part (a). Thus

$$A(X) = \int_1^X x^3 \, dx.$$

It is still true that $A'(X) = X^3$, but now $A(1) = 0$. Obtain a formula that expresses this modified A in terms of X. How do the formulas for A in parts (a) and (b) differ?

Using the Fundamental Theorem

4. Find $A'(X)$ when

*a) $A(X) = \int_0^X \cos(x) \, dx$

e) $A(X) = \int_0^X \sin(x^2) \, dx$

b) $A(X) = \int_0^X \sin(x) \, dx$

f) $A(X) = \int_0^X \sin^2 x \, dx$

c) $A(X) = \int_0^X \cos(x^2) \, dx$

g) $A(X) = \int_0^X \ln t \, dt$

d) $A(X) = \int_0^X \cos(t^2) \, dt$

h) $A(X) = \int_0^X x^2 - 4x^3 \, dx$

*5. Find all critical points of the function

$$A(X) = \int_0^X \cos(x^2) \, dx$$

on the interval $0 \le X \le 4$. Indicate which critical points are local maxima and which are local minima. (Critical points and local maxima and minima are discussed on pages 266–271.)

6. Find all critical points of the function

$$A(X) = \int_0^X \sin(x^2) \, dx$$

on the interval $0 \le X \le 4$. Indicate which critical points are local maxima and which are local minima.

7. Find *all* critical points of the function

$$A(X) = \int_0^X x^2 - 4x^3 \, dx.$$

Indicate which critical points are local maxima and which are local minima.

8. Express the solution to each of the following initial value problems as an accumulation function (that is, as an integral with a variable upper limit of integration).

a) $y' = \cos(x^2)$, $y(\sqrt{\pi}) = 0$
b) $y' = \sin(x^2)$, $y(0) = 0$
c) $y' = \sin(x^2)$, $y(0) = 5$
d) $y' = e^{-x^2}$, $y(0) = 0$

9. Sketch the graphs of the following accumulation functions over the indicated intervals.

a) $\int_0^X \sin(x^2) \, dx$, $0 \le X \le 4$

b) $\int_0^X \frac{\sin(x)}{x} \, dx$, $0 \le X \le 4$

Formulas for Integrals

10. Determine the exact value of each of the following integrals.

*a) $\int_3^7 2 - 3x + 5x^2 \, dx$

b) $\int_0^{5\pi} \sin x \, dx$

c) $\int_0^{5\pi} \sin(2x) \, dx$

d) $\int_0^1 e^t \, dt$

e) $\int_1^6 dx/x$

f) $\int_0^4 7u - 12u^5 \, du$

g) $\int_0^1 2^t \, dt$

h) $\int_{-1}^1 s^2 \, ds$

11. Express the values of the following integrals in terms of the parameters they contain.

a) $\int_3^7 kx \, dx$

*b) $\int_0^\pi \sin(\alpha x) \, dx$

c) $\int_1^4 px^2 - x^3 \, dx$

d) $\int_0^1 e^{ct} \, dt$

e) $\int_{\ln 2}^{\ln 3} e^{ct} \, dt$

f) $\int_1^b 5 - x \, dx$

g) $\int_0^1 a^t \, dt$

h) $\int_1^2 u^c \, du$

12. Find a formula for the solution of each of the following initial value problems.

   *a) $y' = x^2 - 4x^3$, $y(0) = 0$
   b) $y' = x^2 - 4x^3$, $y(3) = 0$
   c) $y' = x^2 - 4x^3$, $y(3) = 10$
   d) $y' = \cos(3x)$, $y(\pi) = 0$

13. Find the average value of each of the following functions over the indicated interval.

   a) $x^2 - x^3$ over $[0, 1]$
   b) $\ln x$ over $[1, e]$
   c) $\sin x$ over $[0, \pi]$

14. a) What is the average value of the function $px - x^2$ on the interval $[0, 1]$? The average depends on the parameter p.
   b) For which value of p will that average be zero?

> ## 6.5   CHAPTER SUMMARY

The Main Ideas

- A **Riemann sum** for the function $f(x)$ on the interval $[a, b]$ is a sum of the form

$$f(x_1) \cdot \Delta x_1 + f(x_2) \cdot \Delta x_2 + \cdots + f(x_n) \cdot \Delta x_n,$$

  where the interval $[a, b]$ has been subdivided into n subintervals whose lengths are $\Delta x_1, \Delta x_2, \ldots, \Delta x_n$, and each x_k is a sampling point in the kth subinterval (for each k from 1 to n).

- Riemann sums can be used to approximate a variety of quantities expressed as **products** where one factor varies with the other.

- Riemann sums give more accurate **approximations** as the lengths $\Delta x_1, \Delta x_2, \ldots, \Delta x_n$ are made small.

- If the Riemann sums for a function $f(x)$ on an interval $[a, b]$ converge, the limit is called the **integral** of $f(x)$ on $[a, b]$, and it is denoted

$$\int_a^b f(x)\, dx.$$

- The **units** of $\int_a^b f(x)\, dx$ equal the product of the units of $f(x)$ and the units of x.

- **The Fundamental Theorem of Calculus.** The solution $y = A(x)$ of the initial value problem

$$y' = f(x) \qquad y(a) = 0$$

is the **accumulation function**

$$A(X) = \int_a^X f(x)\, dx.$$

- If $F(x)$ is an **antiderivative** of $f(x)$, then

$$\int_a^b f(x)\, dx = F(b) - F(a).$$

- The integral $\int_a^b f(x)\, dx$ equals the **signed area** between the graph of $f(x)$ and the x-axis.

- If $f(x)$ is **monotonic** on $[a, b]$ and if $\int_a^b f(x)\, dx$ is approximated by a Riemann sum with subintervals of width Δx, then the **error** in the approximation is at most $\Delta x \cdot |f(b) - f(a)|$.

Expectations

- You should be able to write down (by hand) a Riemann sum to approximate a quantity expressed as a product (e.g., human effort, electrical energy, work, distance traveled, area).

- You should be able to write down an integral giving the *exact* value of a quantity approximated by a Riemann sum.

- You should be able to use **sigma notation** to abbreviate a sum, and you should be able to read sigma notation to calculate a sum.

- You should be able to use a computer program to compute the value of a Riemann sum.

- You should be able to find an error bound when approximating an integral by a Riemann sum.

- You should know and be able to use the **integration rules**.

- You should be able to use the fundamental theorem of calculus to find the value of an integral.

- You should be able to use an antiderivative to find the value of an integral.

CHAPTER 7

PERIODICITY

In seeking to describe and understand natural processes, we search for patterns. Patterns that repeat are particularly useful, because we can predict what they will do in the future. The sun rises every day and the seasons repeat every year. These are the most obvious examples of cyclic, or periodic, patterns, but there are many more of scientific interest, too. Periodic behavior is the subject of this chapter. We shall take up the questions of describing and measuring it. To begin, let's look at some intriguing examples of periodic or near-periodic behavior.

Many patterns are periodic

➤ 7.1 PERIODIC BEHAVIOR

EXAMPLE 1: POPULATIONS.   In chapter 4 we studied several models that describe how interacting populations might change over time. Two of those models—one devised by May and the other by Lotka and Volterra—predict that when one species preys on another, both predator and prey populations will fluctuate periodically over time. How can we tell if that actually happens in nature? Ecologists have examined data for a number of species. Some of the best evidence is found in the records of Hudson's Bay Company, which trapped fur-bearing animals in Canada for almost 200 years. The graph on the next page gives the data for the numbers of lynx pelts harvested in the Mackenzie River region of Canada during the years 1821 to 1934 (Finerty, *Population Cycles in Small Mammals.* New Haven, CT: Yale University Press, 1980). (The lynx is a predator; its main prey is the snowshoe hare.) Clearly the numbers go up and down every 10 years in something like a periodic pattern. There is even a more complex pattern, with one large bulge and three smaller

Predator and prey populations fluctuate periodically

ones, that repeats about every 40 years. Data sets like this appear frequently in scientific inquiries, and they raise important questions. Here is one: If a quantity we are studying really does fluctuate in a periodic way, why might that happen? Here is another: If there appear to be several periodic influences, what are they, and how strong are they? To explore these questions we will develop a language to describe and analyze periodic functions.

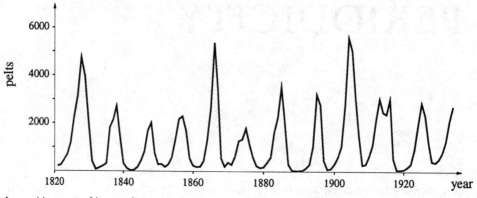

Annual harvest of lynx pelts.

EXAMPLE 2: THE EARTH'S ORBIT. The earth orbits the sun, returning to its original position after one year. This is the most obvious periodic behavior; it explains the cycle of seasons, for example. But there are other, more subtle, periodicities in the earth's orbital motion. The orbit is an ellipse which turns slowly in space, returning to its original position after about 23,000 years. This movement is called **precession**. The orbit fluctuates in other ways that have periods of 41,000 years (the **obliquity** cycle) and 95,000, 123,000, and 413,000 years (the **eccentricity** cycles).

The position and the shape of the earth's orbit both fluctuate periodically

EXAMPLE 3: THE CLIMATE. In 1941 the Serbian geophysicist Milutin Milankovitch proposed that all the different periodicities in the earth's orbit affect the climate—that is, the long-term weather patterns over the entire planet. Therefore, he concluded, there should also be periodic fluctuations in the climate, with the same periods as the earth's orbit. In fact, it is possible to test this hypothesis, because there are features of the geological record that tell us about long-term weather patterns. For example, in a year when the weather is warm and wet, rains will fill streams and rivers with mud that is eventually carried to lake bottoms. The result is a thick sediment layer. In a dry year, the sediment layer will be thinner. Over geological time, lakes dry out and their beds turn to clay or shale. By measuring the annual layers over thousands of years, we can see how the climate has varied. Other features that have been analyzed the same way are the thickness of annual ice layers in the Antarctic ice cap, the fluctuations of CO_2 concentrations in the ice caps, changes in the

Fluctuations in the climate appear in the geological record

O^{18}/O^{16} ratio in deep-sea sediments and ice caps. In chapter 12 we will look at the results of one such study.

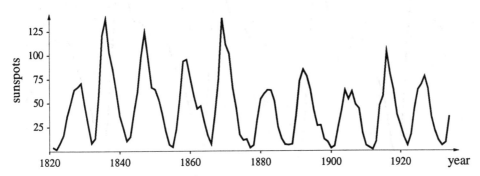

Daily average number of sunspots.

EXAMPLE 4: SUNSPOT CYCLES. The number of sunspots fluctuates, reaching a peak every 11 years or so. The graph above shows the average daily number of sunspots during each year from 1821 to 1934. Compare this with the lynx graph which covers the same years. Some earthbound events (e.g., auroras, television interference) seem to follow the same 11-year pattern. According to some scientists, other meteorological phenomena—such as rainfall, average temperature, and CO_2 concentrations in the atmosphere—are also "sunspot cycles," fluctuating with the same 11-year period.

It is difficult to get firm evidence, though, because many fluctuations with different possible causes can be found in the data. Even if there is an 11-year cycle, it may be "drowned out" by the effects of these other causes.

Data can have both periodic and random influences

The problem of detecting periodic fluctuations in "noisy" data is one that scientists often face. In chapter 12 we will introduce a mathematical tool called the **power spectrum**, and we will use it to detect and measure periodic behavior—even when it is swamped by random fluctuations.

➤ 7.2   PERIOD, FREQUENCY, AND THE CIRCULAR FUNCTIONS

We are familiar with the notions of period and frequency from everyday experience. For example, a full moon occurs every 28 days, which means that a lunar cycle has a *period* of 28 days and a *frequency* of once per 28 days. Moreover, whatever phase the moon is in today, it will be in the same phase 28 days from now. Let's see how to extend these notions to functions.

The function $y = g(x)$ whose graph is sketched at the top of the next page has a pattern that repeats. The space T between one high point and the next tells us the period of this repeating pattern. There is nothing special about the

Defining a periodic function

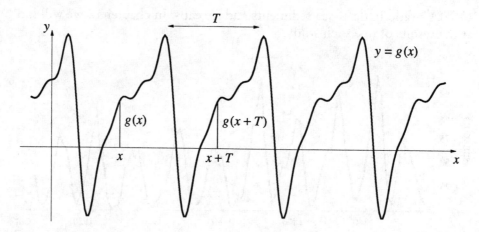

high point, though. If we take *any* two points x and $x + T$ that are spaced one period apart, we find that g has the same value at those points. (This is analogous to saying that the moon is in the same phase on any two days that are 28 days apart.) The condition $g(x + T) = g(x)$ for every x guarantees that g will be periodic. We make it the basis of our definition.

> We say that a function $g(x)$ is **periodic** if there is a positive or negative number T for which
>
> $$g(x + T) = g(x) \qquad \text{for all } x.$$
>
> We call T a **period** of $g(x)$.

A periodic function has many periods ...

Since the graph of g repeats after x increases by T, it also repeats after x increases by $2T$, or $-3T$, or any integer multiple (positive or negative) of T. This means that a periodic function always has many periods. (That's why the definition refers to "*a* period" rather than "*the* period.") The same is true of the moon; its phases also repeat after 2×28 days, or 3×28 days. Nevertheless, we think of 28 days as *the* period of the lunar cycle, because we see the entire pattern precisely once. We can say the same for any periodic function:

... but we call the smallest positive one *the* period

> The **period** of a periodic function is its smallest positive period. It is the size of a single cycle.

Frequency

Another measure of a periodic function is its frequency. Consider first the lunar cycle. Its frequency is the number of cycles—or fractions of a cycle—that occur in unit time. If we measure time in days, then the frequency is 1/28-th of a cycle per day. If we measure time in *years*, though, then the frequency is about 13 cycles per year. Here is the calculation:

$$\frac{365 \text{ days/year}}{28 \text{ days/cycle}} \approx 13 \text{ cycles/year}.$$

Using this example as a pattern, we make the following definition.

> If the function $g(x)$ is periodic, then its
> **frequency** is the number of cycles per unit x.

Notice that the period and the frequency of the lunar cycle are reciprocals: the period is 28 days—the time needed to complete one cycle—while the frequency is 1/28-th of a cycle per day. In the example below, t is measured in seconds and g has a period of .2 second. Its frequency is therefore 5 cycles per second.

The frequency of a cycle is the reciprocal of its period

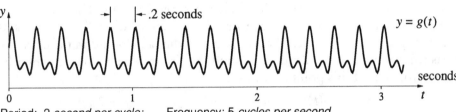

Period: *.2 second per cycle;*   Frequency: *5 cycles per second.*

In general, if f is the frequency of a periodic function $g(x)$ and T is its period, then we have

$$f = \frac{1}{T} \quad \text{and} \quad T = \frac{1}{f}.$$

The units are also related in a reciprocal fashion: if the period is measured in seconds, then the frequency is measured in cycles per second.

Because many quantities fluctuate periodically over time, the input variable of a periodic function will often be *time*. If time is measured in seconds, then frequency is measured in "cycles per second." The term **hertz** is a special unit used to measure time frequencies; it equals one cycle per second. Hertz is abbreviated Hz; thus a **kilohertz** (kHz) and a **megahertz** (MHz) are 1,000 and 1,000,000 cycles per second, respectively. This unit is commonly used to describe sound, light, radio, and television waves. For example, an orchestra tunes to an A at 440 Hz. If an FM radio station broadcasts at 88.5 MHz, this means its carrier frequency is 88,500,000 cycles per second.

The units for measuring frequency over time

Quantities may also be periodic in other dimensions. For instance, a scientist studying the phenomenon of ripple formation in a river bed might be interested in the function $h(x)$ measuring the height of a ripple as a function of its distance x along the river bed. This would lead to a function of period, say, 10 inches and corresponding frequency of .1 cycle per inch.

Functions can be periodic over other units as well

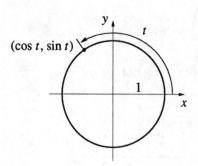

Circular functions. While there are innumerable examples of periodic functions, two in particular are considered basic: the sine and the cosine. They are called circular functions because they are defined by means of a circle. To be specific, take the circle of radius 1 centered at the origin in the x, y plane. Given any real number t, measure a distance of t units around the circumference of the circle. Start on the positive x-axis, and measure counterclockwise if t is positive, clockwise if t is negative. The coordinates of the point you reach this way are, by definition, the cosine and the sine functions of t, respectively:

$$x = \cos(t)$$
$$y = \sin(t).$$

Why cos t and sin t are periodic

The whole circumference of the circle measures 2π units. Therefore, if we add 2π units to the t units we have already measured, we will arrive back at the same point on the circle. That is, we get to the same point on the circle by measuring either t or $t + 2\pi$ units around the circumference. We can describe the coordinates of this point two ways:

$$(\cos(t), \sin(t)) \qquad \text{or} \qquad (\cos(t + 2\pi), \sin(t + 2\pi)).$$

Thus

$$\cos(t + 2\pi) = \cos(t) \qquad \sin(t + 2\pi) = \sin(t),$$

so $\cos(t)$ and $\sin(t)$ are both periodic, and they have the same period, 2π. Here are their graphs. By reading their slopes we can see $(\sin t)' = \cos t$ and $(\cos t)' = -\sin t$.

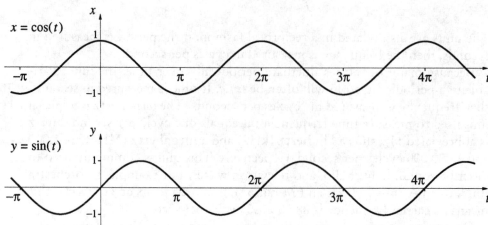

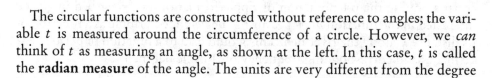

Radian measure

The circular functions are constructed without reference to angles; the variable t is measured around the circumference of a circle. However, we *can* think of t as measuring an angle, as shown at the left. In this case, t is called the **radian measure** of the angle. The units are very different from the degree measurement of an angle: an angle of 1 radian is much larger than an angle of 1 degree. The radian measure of a 90° angle is $\pi/2 \approx 1.57$, for instance. If we thought of t as an angle measured in degrees, the slope of $\sin(t)$ would equal $.017 \cos t$! (See the exercises.) Only when we measure t in radians do we get a simple result: $(\sin t)' = \cos t$. This is why we always measure angles in radians in calculus.

Compare the graph of $y = \sin(t)$ above with that of $y = \sin(4t)$, shown below. Their scales are identical, so it is clear that the frequency of $\sin(4t)$ is four times the frequency of $\sin(t)$. The general pattern is described in the table at the top of the next page.

Changing the frequency

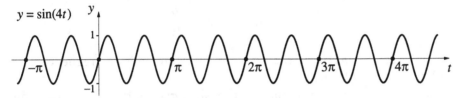

Notice that it is the *frequency*—not the period—that is increased by a factor of b when we multiply the input variable by b.

By using the information in the table, we can construct circular functions with any period or frequency whatsoever. For instance, suppose we wanted a cosine function $x = \cos(bt)$ with a frequency of 5 cycles per unit t. This means

Constructing circular function with a given frequency

$$5 = \text{frequency} = \frac{b}{2\pi},$$

which implies that we should set $b = 10\pi$ and $x = \cos(10\pi t)$. In order to see the high-frequency behavior of this function better, we magnify the graph a bit. Compare the figure below with the graph of $x = \cos(t)$ on page 376.

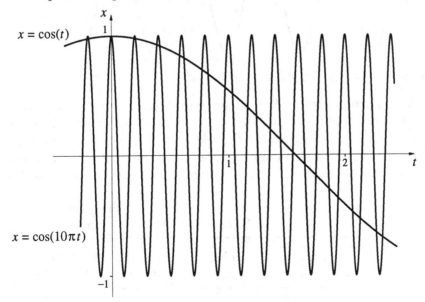

Function		Period	Frequency
$\sin(t)$	$\cos(t)$	2π	$1/2\pi$
$\sin(4t)$	$\cos(4t)$	$2\pi/4$	$4/2\pi$
$\sin(bt)$	$\cos(bt)$	$2\pi/b$	$b/2\pi$

We still have equal scales on the horizontal and vertical axes. Finally, notice that $\cos(10\pi t)$ has exactly 5 cycles on the interval $0 \le t \le 1$.

Frequency ω

We will denote the frequency by ω, the lowercase letter *omega* from the Greek alphabet. If

$$\omega = \text{frequency} = \frac{b}{2\pi},$$

then $b = 2\pi\omega$. Therefore, the basic circular functions of frequency ω are $\cos(2\pi\omega t)$ and $\sin(2\pi\omega t)$.

Amplitude

Suppose we take the basic sine function $\sin(2\pi\omega t)$ of frequency ω and multiply it by a factor A:

$$y = A\sin(2\pi\omega t).$$

The graph of this function oscillates between $y = -A$ and $y = +A$. The number A is called the **amplitude** of the function.

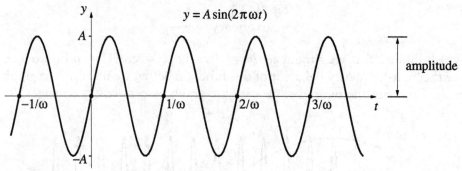

The sine function of amplitude A and frequency ω.

Physical interpretations. Sounds are transmitted to our ears as fluctuations in air pressure. Light is transmitted to our eyes as fluctuations in a more abstract medium—the electromagnetic field. Both kinds of fluctuations can be described using circular functions of time t. The amplitude and the frequency of these functions have the physical interpretations given in the following table.

	Amplitude	Frequency	Frequency range
sound	loudness	pitch	$10-15000$ Hz
light	intensity	color	$4 \times 10^{14} - 7.5 \times 10^{14}$ Hz

Exercises

Circular Functions

1. Choose ω so that the function $\cos(2\pi\omega t)$ has each of the following periods.

    a) 1   *b) 5   c) 2π d) π e) 1/3

2. Determine the period and the frequency of the following functions.

    a) $\sin(x)$, $\sin(2x)$, $\sin(x) + \sin(2x)$
    b) $\sin(2x)$, $\sin(3x)$, $\sin(2x) + \sin(3x)$
    c) $\sin(6x)$, $\sin(9x)$, $\sin(6x) + \sin(9x)$

3. Suppose a and b are positive integers. Describe how the periods of $\sin(ax)$, $\sin(bx)$, $\sin(ax) + \sin(bx)$ are related. (As the previous exercise shows, the relation between the periods depends on the relation between a and b. Make this clear in your explanation.)

4. a) What are the amplitude and frequency of $g(x) = 5\cos(3x)$?
    b) What are the amplitude and frequency of $g'(x)$?

5. a) Is the antiderivative $\int_0^x 5\cos(3t)\, dt$ periodic?
    b) If so, what are its amplitude and frequency?

6. Use the definition of the circular functions to explain why

$$\sin(-t) = -\sin(t) \qquad \sin\left(\frac{\pi}{2} - t\right) = \cos(t)$$

$$\cos(-t) = +\cos(t) \qquad \sin(\pi - t) = \sin(t).$$

    hold for all values of t. Describe how these properties are reflected in the graphs of the sine and cosine functions.

7. *a) What is the average value of the function $\sin(s)$ over the interval $0 \le s \le \pi$? (This is a half-period.)
    b) What is the average value of $\sin(s)$ over $\pi/2 \le s \le 3\pi/2$? (This is also a half-period.)
    c) What is the average value of $\sin(s)$ over $0 \le s \le 2\pi$? (This is a full period.)
    d) Let c be any number. Find the average value of $\sin(s)$ over the full period $c \le s \le c + 2\pi$.
    e) Your work should demonstrate that the average value of $\sin(s)$ over a full period does not depend on the point c where you begin the period. Does it? Is the same true for the average value over a half-period? Explain.

8. a) What is the period T of $P(t) = A\sin(bt)$?
    b) Let c be any number. Find the average value of $P(t)$ over the full period $c \le t \le c + T$. Does this value depend on the choice of c?
    c) What is the average value of $P(t)$ over the half-period $0 \le t \le T/2$?

Phase

There is still another aspect of circular functions to consider besides amplitude and frequency. It is called *phase difference*. We can illustrate this with the two functions graphed below. They have the same amplitude and frequency, but differ in phase. Specifically, the variable u in the expression $\sin(u)$ is called the **phase**. In the dotted graph the phase is $u = bt$, while in the solid graph it is $u = bt - \varphi$. They differ in phase by $bt - (bt - \varphi) = \varphi$. In the exercises you will see why a **phase difference** of φ produces a shift—which we call a **phase shift**—of φ/b in the graphs. (φ is the Greek letter *phi*.)

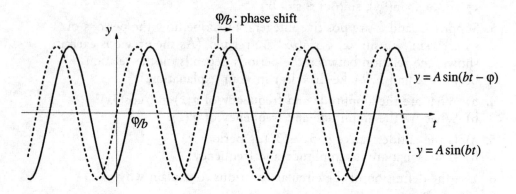

*9. The functions $\sin(x)$ and $\cos(x)$ have the same amplitude and frequency; they differ only in phase. In other words,

$$\cos(x) = \sin(x - \varphi)$$

for an appropriately chosen phase difference φ. What is φ?

10. The functions $\sin(x)$ and $-\sin(x)$ *also* differ only in phase. What is their phase difference? In other words, find φ so that

$$\sin(x - \varphi) = -\sin(x).$$

(Note: A circular function and its negative are sometimes said to be "180 degrees out of phase." The value of φ you found here should explain this phrase.)

11. What is the phase difference between $\sin(x)$ and $-\cos(x)$?

12. a) Graph $y = \sin(t)$ and $y = \sin(t - \pi/3)$ on the same plane.
   *b) What is the phase difference between these two functions?
   c) What is the phase shift between their graphs?

13. a) Graph together on the same coordinate plane $y = \cos(t)$ and $y = \cos(t + \pi/4)$.
   b) What is the phase difference between these two functions?
   c) What is the phase shift between their graphs?

14. We know $y = \cos(t)$ has a maximum at the origin. Determine the point closest to the origin where $y = \cos(t + \pi/4)$ has *its* maximum. Is the second maximum shifted from the first by the amount of the phase shift you identified in the previous question?

15. Repeat the last two exercises for the pair of functions $y = \cos(2t)$ and $\cos(2t + \pi/4)$. Is the phase difference equal to the phase shift in this case?

16. Verify that the graph of $y = A\sin(bt - \varphi)$ crosses the t-axis at the point $t = \varphi/b$. This shows that $A\sin(bt - \varphi)$ is "phase-shifted" by the amount φ/b in relation to $A\sin(bt)$. (Refer to the graph on page 380.)

17. a) At what point nearest the origin does the function $A\cos(bt - \varphi)$ reach its maximum value?
    b) Explain why this shows $A\cos(bt - \varphi)$ is "phase-shifted" by the amount φ/b in relation to $A\cos(bt)$.

18. a) Let $f(x) = \sin(x) - .7\cos(x)$. Using a graphing utility, sketch the graph of $f(x)$.
    b) The function $f(x)$ is periodic. What is its period? From your graph, estimate its amplitude.
    c) In fact, $f(x)$ can be viewed as a "phase-shifted" sine function:

$$f(x) = A\sin(bx - \varphi).$$

From your graph, estimate the phase difference φ and the amplitude A.

19. a) For each of the values $\varphi = 0, \pi/4, \pi/2, 3\pi/4, \pi$, sketch the graph $y = \sin(x) \cdot \sin(x - \varphi)$ over the interval $0 \leq x \leq 2\pi$. Put the five graphs on the same coordinate plane.
    b) For which graphs is the average value positive, for which is it negative, and for which is it 0? Estimate by eye.

∗20. The purpose of this exercise is to determine the average value

$$F(\varphi) = \frac{1}{2\pi} \int_0^{2\pi} \sin(x)\sin(x - \varphi)\,dx$$

for an *arbitrary* value of the parameter φ. To stress that the average value is actually a function of φ, we have written it as $F(\varphi)$. Here is one way to determine a formula for $F(\varphi)$ in terms of φ. First, using a "sum of two angles" formula and exercise 6, above, write

$$\sin(x - \varphi) = \cos(\varphi)\sin(x) - \sin(\varphi)\cos(x).$$

Then consider

$$\frac{1}{2\pi}\left[\cos(\varphi)\int_0^{2\pi}(\sin(x))^2\,dx - \sin(\varphi)\int_0^{2\pi}\sin(x)\cos(x)\,dx\right],$$

and determine the values of the two integrals separately.

21. **a)** Sketch the graph of the *average value function* $F(\varphi)$ you found in the previous exercise. Use the interval $0 \leq \varphi \leq \pi$.
    b) In exercise 19 you estimated the value of $F(\varphi)$ for five specific values of φ. Compare your estimates with the exact values that you can now calculate using the formula for $F(\varphi)$.

22. Sketch the graph of $y = \cos(x)\sin(x - \varphi)$ for each of the following values of φ: $0, \pi/2, 2\pi/3, \pi$. Use the interval $0 \leq x \leq 2\pi$. Estimate by eye the average value of each function over that interval.

23. **a)** Obtain a formula for the average value function

$$G(\varphi) = \frac{1}{2\pi} \int_0^{2\pi} \cos(x)\sin(x - \varphi)\,dx.$$

   and sketch the graph of $G(\varphi)$ over the interval $0 \leq \varphi \leq \pi$.
   b) Use your formula for $G(\varphi)$ to compute the average value of the function $\cos(x)\sin(x - \varphi)$ exactly for $\varphi = 0, \pi/2, 2\pi/3, \pi$. Compare these values with your estimates in the previous exercise.

*24. How large a phase difference φ is needed to make the graphs of $y = \sin(3x)$ and $y = \sin(3x - \varphi)$ coincide?

25. Sketch the graphs of the following functions.
    a) $y = 3\sin(2x - \pi/6) - 1$
    b) $y = 4\sin(2x - \pi) + 2$
    c) $y = 4\sin(2x + \pi) + 2$

$y = A\sin(bt - \varphi)$

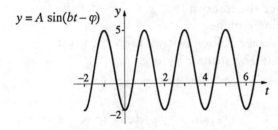

26. The function whose graph is sketched at the left has the form

$$G(x) = A\sin(bt - \varphi) + C.$$

   Determine the values of A, b, C, and φ.

27. Write equations for three different functions all having amplitude 4 and period 5, whose graphs pass through the point $(6, 7)$. Be sure the functions are really different—if $g(t)$ is one solution, then $h(t) = g(t + 5)$ would really be just the same solution.

Derivatives with Degrees

28. **a)** In this exercise measure the angle θ in degrees. Estimate the derivative of $\sin(\theta)$ at $\theta = 0°$ by calculating $\sin(\theta)/\theta$ for $\theta = 2°$, $1°, .5°$.
    b) Estimate the derivative of $\sin(\theta)$ at $\theta = 60°$ in a similar way. Is $(\sin(30°))' = \cos(30°)$?

29. a) Your calculations in the previous exercise should support the claim that $(\sin(\theta))' = k \cos(\theta)$ for a particular value of k, when θ is measured in degrees. What is k, approximately?

b) If t is the radian measure of an angle, and if θ is its degree measure, then θ will be a function of t. What is it? Now use the chain rule to get a precise expression for the constant k.

▷ 7.3   DIFFERENTIAL EQUATIONS WITH PERIODIC SOLUTIONS

The models of predator-prey interactions constructed by Lotka–Volterra and by May (see chapter 4) provide us with examples of systems of differential equations that have periodic solutions. Similar examples can be found in many areas of science. We shall analyze some of them in this section. In particular, we will try to understand how the frequency and the amplitude of the periodic solutions depend on the parameters given in the model.

Oscillating Springs

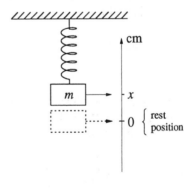

We want to study the motion of a weight that hangs from the end of a spring. First let the weight come to rest. Then pull down on it. You can feel the spring pulling it back up. If you push up on the weight, the spring (and gravity) push it back down. The force you feel is called the **spring force.** Now release the weight; it will move. We'll assume that the only influence on the motion is the spring force. (In particular, we will ignore the force of friction.) With this assumption we can construct a model to describe the motion. We'll suppose the weight has a mass of m grams, and it is x centimeters above its rest position after t seconds. (If the weight goes below the rest position, then x will be negative.)

The Linear Spring

The simplest assumption we can reasonably make is that the spring force is proportional to the amount x that the spring has been displaced:

A linear spring

$$\text{force} = -cx.$$

In this case the spring is said to be **linear.** The multiplier c is called the **spring constant.** It is a positive number that varies from one spring to another. The minus sign tells us the force pushes down if $x > 0$, and it pushes up if $x < 0$.

Because this model describes an oscillating spring governed by a linear spring force, it is called the **linear oscillator.**

Newton's laws of motion

To see how the spring force affects the motion of the weight, we use Newton's laws. In their simplest form, they say that the force acting on a body is the product of its mass and its acceleration. Suppose $v = dx/dt$ is the velocity of the weight in cm/sec, and dv/dt is its acceleration in cm/sec². Then

$$\text{force} = m\,\frac{dv}{dt} \quad \text{g-cm/sec}^2.$$

If we equate our two expressions for the force, we get

$$m\,\frac{dv}{dt} = -cx \quad \text{or} \quad \frac{dv}{dt} = -b^2 x \quad \text{cm/sec}^2,$$

where we have set $c/m = b^2$. It is more convenient to write c/m as b^2 here, because then $b = \sqrt{c/m}$ itself will be measured in units of 1/sec. (To see why, note that $-b^2 x$ is measured in units of cm/sec².)

Suppose we move the weight to the point $x = a$ cm on the scale, hold it motionless for a moment, and then release it at time $t = 0$ sec. This gives us the initial value problem

The linear oscillator

$$
\begin{aligned}
x' &= v & x(0) &= a \\
v' &= -b^2 x & v(0) &= 0.
\end{aligned}
$$

The solution with fixed parameters

If we give the parameters a and b specific values, we can solve this initial value problem using Euler's method. The figure below shows the solution $x(t)$ for two different sets of parameter values:

$$
\begin{aligned}
a &= 4 \text{ cm} & a &= -5 \text{ cm} \\
b &= 5 \text{ per sec} & b &= 9 \text{ per sec}
\end{aligned}
$$

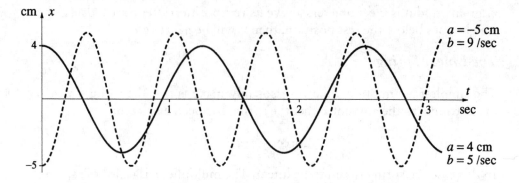

The graphs were made in the usual way, with the differential equation solver of a computer. They indicate that the weight bounces up and down in a periodic fashion. The amplitude of the oscillation is precisely a, and the frequency appears to be linked directly to the value of b. For instance, when $b = 9/\text{sec}$, the motion completes just under 3 cycles in 2 seconds. This is a frequency of slightly less than 1.5 cycles per second. When $b = 5/\text{sec}$, the motion undergoes roughly 2 cycles in 2.5 seconds, a frequency of about .8 cycle per second. If the frequency is indeed proportional to b, the multiplier must be about 1/6:

$$\text{frequency} \approx \frac{b}{6} \text{ cycles/sec.}$$

We can get a better idea how the parameters in a problem affect the solution if we solve the problem with a method that doesn't require us to fix the values of the parameters in advance. This point is discussed in section 4.2, pages 188–192. It is particularly useful if we can express the solution by a *formula*, which it turns out we can do in this case. To get a formula, let us begin by noticing that

The solution for arbitrary parameter values

$$(x')' = v' = -b^2 x.$$

In other words, $x(t)$ is a function whose second derivative is the negative of itself (times the constant b^2). This suggests that we try

$$x(t) = \sin(bt) \quad \text{or} \quad x(t) = \cos(bt).$$

You should check that $x'' = -b^2 x$ in both cases.

Turn now to the initial conditions. Since $\sin(0) = 0$, there is no way to modify $\sin(bt)$ to make it satisfy the condition $x(0) = a$. However,

$$x(t) = a\cos(bt)$$

does satisfy it. Finally, we can use the differential equation $x' = v$ to define $v(t)$:

$$v(t) = (a\cos(bt))' = -ab\sin(bt).$$

Notice that $v(0) = -ab\sin(0) = 0$, so the second initial condition is satisfied.

In summary, we have a formula for the solution that incorporates the parameters. With this formula we see that the motion is really periodic—a fact that Euler's method could only suggest. Furthermore, the parameters determine the amplitude and frequency of the solution in the following way:

The formula proves the motion is periodic

position : $a\cos(bt)$ cm from rest after t sec

amplitude : a cm

frequency : $b/2\pi$ cycles/sec.

Graph of the general
linear oscillator

We can see the relation between the motion and the parameters in the graph
below (in which we take $a > 0$).

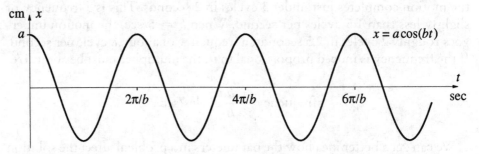

Here are some further properties of the motion that follow from our formula for the solution. Recall that the parameter b depends on the mass m of
the weight and the spring constant c: $b^2 = c/m$.

- The amplitude depends only on the initial conditions, not on the mass
  m or the spring constant c.
- The frequency depends only on the mass and the spring constant, not
  on the initial amplitude.

These properties are a consequence of the fact that the spring force is *linear*.
As we shall see, a nonlinear spring and a pendulum move differently.

The Nonlinear Spring

The harder you pull on a spring, the more it stretches. If the stretch is exactly proportional to the pull (i.e., the force), the spring is linear. In other
words, to double the stretch you must double the force. Most springs behave this way when they are
stretched only a small amount.
This is called their **linear range.**
Outside that range, the relation
is more complicated. One possibility is that, to double the
stretch, you must increase the
force by *more than* double. A
spring that works this way is
called a **hard spring.** The graph
at the left shows the relation
between the applied force and
the displacement (or stretch x)
of a hard spring.

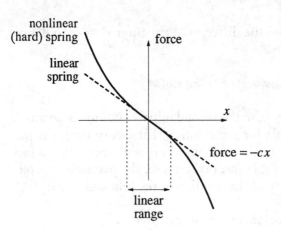

In a *nonlinear* spring, force is no longer proportional to displacement. Thus,
if we write

$$\text{force} = -cx,$$

we must allow the multiplier c to depend on x. One simple way to achieve this is to replace c by $c + \gamma x^2$. (We use x^2 rather than just x to ensure that $-x$ will have the same effect as $+x$. The multiplier γ is the Greek letter *gamma*.) Then

$$\text{force} = -cx - \gamma x^3.$$

Since force $= m\,(dv/dt)$ as well, we have

$$m\frac{dv}{dt} = -cx - \gamma x^3 \quad \text{or} \quad \frac{dv}{dt} = -b^2 x - \beta x^3 \quad \text{cm/sec}^2.$$

Here $b^2 = c/m$ and $\beta = \gamma/m$. By taking the same initial conditions as before, we get the following initial value problem:

$$\begin{aligned} x' &= v & x(0) &= a \\ v' &= -b^2 x - \beta x^3 & v(0) &= 0. \end{aligned}$$

A nonlinear oscillator

To solve this problem using Euler's method, we must fix the values of the three parameters. For the two parameters that determine the spring force, we choose:

$$b = 5 \text{ per sec} \qquad \beta = .2 \text{ per cm}^2\text{-sec}^2.$$

We have deliberately chosen b to have the same value it did for our first solution to the linear problem. In this way, we can compare the nonlinear spring to the linear spring that has the same spring constant. We do this in the figure below. The dashed graph shows the linear spring when its initial amplitude is $a = 4$ cm. The solid graph shows the hard spring when its initial amplitude is $a = 1.5$ cm. Note that the two oscillations have the same frequency.

Comparing a hard spring to a linear spring

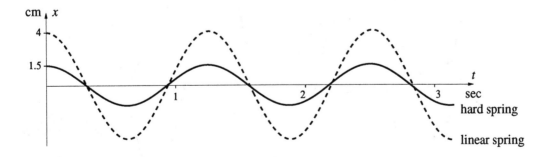

The nonlinear spring behaves like the linear one because the amplitudes are small. To understand this reason, we must compare the accelerations of the two springs. For the linear spring we have $v' = -25x$, while for the nonlinear

The effect of amplitude on acceleration

spring, $v' = -25x - .2x^3$. As the following graph shows, these expressions are approximately equal when the amplitude x lies between $+2$ cm and -2 cm. In other words, the linear range of the hard spring is $-2 \le x \le 2$ cm. Since the initial amplitude was 1.5 cm—well within the linear range—the hard spring acts like a linear one. In particular, its frequency is approximated closely by the formula $b/2\pi$ cycles per second. This is $5/2\pi \approx .8$ Hz.

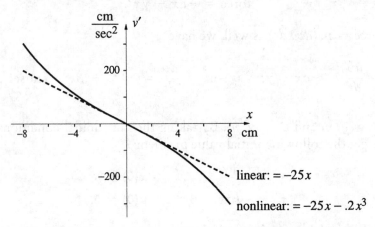

Large-amplitude oscillations

A different set of circumstances is reflected in the following graph. The hard spring has been given an initial amplitude of 8 cm. As the graph of v' shown above indicates, the hard spring experiences an acceleration about 50% greater than the linear spring at the that amplitude. As a consequence, the hard spring oscillates with a noticeably higher frequency! It completes 3 cycles in the time it takes the linear spring to complete $2\frac{1}{2}$—or 6 cycles while the linear spring completes 5. The frequency of the hard spring is therefore about 6/5-th the frequency of the linear spring, or $6/5 \times 5/2\pi = 3/\pi \approx .95$ Hz.

The frequency of a nonlinear spring depends on its amplitude

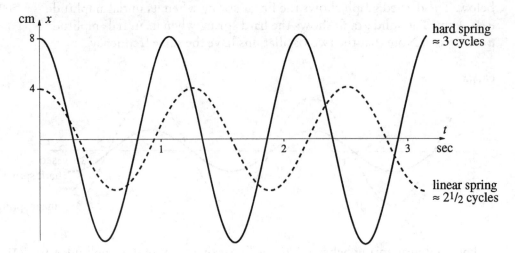

The solutions of the nonlinear spring problem still look like cosine functions, but they're not. It's easier to see the difference if we take a large-amplitude solution, and look at velocity instead of position. In the graph below you can see how the velocity of a hard spring differs from a pure sine function of the same period and amplitude. Since there are no sine or cosine functions here, we can't even yet be sure that the motion of a nonlinear spring is truly periodic! We will prove this, though, in the next section by using the notion of a **first integral.**

A mathematical comment

There are other ways we might have modified the basic equation $v' = -b^2x$ to make the spring nonlinear. The formula $v' = -b^2x - \beta x^3$ is only one possibility. Incidentally, our study of a hard spring was based on choosing $\beta > 0$ in this formula. Suppose we choose $\beta < 0$ instead. As you will see in the exercises, this is a **soft** spring: we can double the stretch in a soft spring by using *less than* double the force. The pendulum, which we will study next, behaves like a soft spring.

Other nonlinear springs

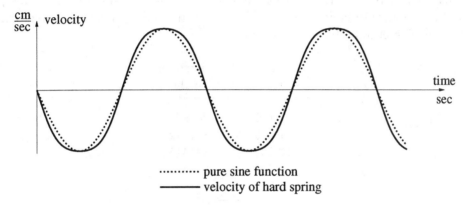

·········· pure sine function
——— velocity of hard spring

Although we can use sine and cosine functions to solve the *linear* oscillator problem, there are, in general, no formulas for the solutions to the *nonlinear* oscillator problems. We must use numerical methods to find their graphs—as we have done in the last two pages.

The basic differential equation for a linear spring is also used to model a vibrating string. Think of a tightly stretched wire, like a piano string or a guitar string. Let x be the distance the center of the string has moved from rest at any instant t. The larger x is, the more strongly the tension on the string will pull it back toward its rest position. Since x is usually very small, it makes sense to assume that this "restoring force" is a linear function of x: $-cx$. If v is the velocity of the string, then $mv' = -cx$ by Newton's laws of motion. Because of the connection between vibrating strings and music, this differential equation is called the **harmonic oscillator.**

The harmonic oscillator

The Sine and Cosine Revisited

The sine and cosine functions first appear in trigonometry, where they are defined for the acute angles of a right triangle. Negative angles and angles larger than 90° are outside their domain. This is a serious limitation. To overcome it, we redefine the sine and cosine on a circle. The main consequence of this change is that the sine and cosine become *periodic*.

However, neither circles nor triangles are particularly useful if we want to *calculate* the values of the sine or the cosine. [How would you use one of them to determine sin(1) to four—or even two—decimal places accuracy?] Our experience with the harmonic oscillator gives yet another way to define the sine and the cosine functions—a way that conveys computational power.

The idea is simple. With hindsight we know that $u = \sin(t)$ and $v = \cos(t)$ are the solutions to the initial value problem

$$u' = v \qquad u(0) = 0$$
$$v' = -u \qquad v(0) = 1.$$

Now make a fresh start with this initial value problem, and *define* $u = \sin(t)$ and $v = \cos(t)$ to be its solution! Then we can calculate sin(1), for instance, by Euler's method. Here is the result.

Number of steps	Estimate of sin(1)
100	.845671
1 000	.841892
10 000	.841513
100 000	.841475
1 000 000	.841471

So we can say $\sin(1) = .8415$ to four decimal places accuracy.

Our point of view here is that *differential equations define functions*. In chapter 10, we shall consider still another method for defining and calculating these important functions, using infinite series.

The Pendulum

We are going to study the motion of a pendulum that can swing in a full 360° circle. To keep the physical details as simple as possible, we'll assume its mass is 1 unit, and that all the mass is concentrated in the center of the pendulum bob, 1 unit from the pivot point. Assume that the pendulum is x units from its rest position at time t, where x is measured around the circular path that the bob traces out. Assume the velocity is v. Take counterclockwise positions and velocities to be positive, clockwise ones to be negative. When the pendulum is at rest we have $x = v = 0$.

When the pendulum is moving, there must be forces at work. Let's ignore friction, as we did with the spring. The force that pulls the pendulum back toward the rest position is gravity. However, gravity itself—**G** in the figure at the right—pulls straight down. Part of the pull of **G** works straight along the arm of the pendulum, and is resisted by the pivot. (If not, the pendulum would be pulled out of the pivot and fall to the floor!) It is the other part, labeled **F**, that moves the pendulum sideways.

The size of **F** depends on the position x of the pendulum. When $x = 0$, the sideways force **F** is zero. When $x = \pi/2$ (the pendulum is horizontal), the entire pull of **G** is "sideways," so **F = G**. To see how **F** depends on x in general, note first that we can think of x as the radian measure of the angle between the pendulum and the vertical (because x is measured around a circle of radius 1). In the small right triangle, the hypotenuse is **G** and the side opposite the angle x is exactly as long as **F**. By trigonometry, **F** = **G** $\sin x$.

Let's choose units which make the size of **G** equal to 1. Then the size of **F** is simply $\sin x$. Since **F** points in the clockwise (or negative) direction when x is positive, we must write **F** $= -\sin x$. According to Newton's laws of motion, the force **F** is the product of the mass and the acceleration of the pendulum. Since the mass is 1 unit and the acceleration is $v' = x''$, we finally get $x'' = -\sin x$, or

Newton's laws produce a model of the pendulum

$$x' = v \qquad v' = -\sin x.$$

Now that we have an explicit description of the restoring force, we can see that the pendulum behaves like a nonlinear spring. However, it is true that doubling the displacement x always *less than* doubles the force, as the graph below demonstrates. Thus the pendulum is like a **soft** spring.

The pendulum is a soft spring

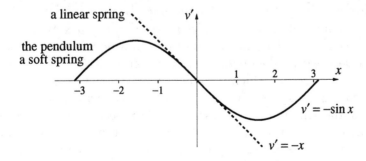

Because a swinging pendulum is used to keep time, it is important to control the period of the swing. Physics analyzes how the period depends on the pendulum's length and mass. We will confine ourselves to analyzing how the period depends on its amplitude.

Small-amplitude
oscillations

Let's draw on our experience with springs. According to the graph above, the restoring force of the pendulum is essentially linear for small amplitudes— say, for $-.5 \leq x \leq .5$ radian. Therefore, if the amplitude stays small, it is reasonable to expect that the pendulum will behave like a linear oscillator. As the graph indicates, the differential equation of the linear oscillator is $v' = -x$. This is of the form $v' = -b^2x$ with $b = 1$. The period of such a linear oscillator is $2\pi/b = 2\pi \approx 6.28$. Let's see if the pendulum has this period when it swings with a small amplitude. We use Euler's method to solve the initial value problem

$$x' = v \qquad\qquad x(0) = a$$

$$v' = -\sin x \qquad v(0) = 0$$

for several small values of a. The results appear in the graph below.

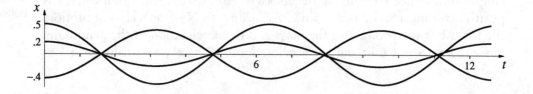

As you can see, small-amplitude oscillations have virtually the same period. Thus, we would not expect the fluctuations in the amplitude of the pendulum on a grandfather's clock to affect the timekeeping.

Large-amplitude
oscillations

What happens to the period, though, if the pendulum swings in a large arc? The largest possible initial amplitude we can give the pendulum would point it straight up. The pendulum is then 180° from the rest position, corresponding to a value of $x = \pi = 3.14159\ldots$. In the graph below we see the solution that has an initial amplitude of $x = 3$, which is very near the maximum possible. Its period is much larger than the period of the solution with $x = .5$, which has been carried over from the previous graph for comparison. Even the solution with $x = 2$ has a period which is significantly larger. We saw that the period of a hard spring got shorter (its frequency increased) when its amplitude increased. But the pendulum is a soft spring and shows motions of longer period as its initial amplitude is increased. Notice how flat the large-amplitude graph is. This means that the pendulum lingers at the top of its swing for a long time. That's why the period becomes so large. Check the graph now and confirm that the period of the large swing is about 17.

Although we can't get formulas to describe the motion of the pendulum for most initial conditions, there are two special circumstances when we can. Consider a pendulum that is initially at rest: $x = 0$ and $v = 0$ when $t = 0$.

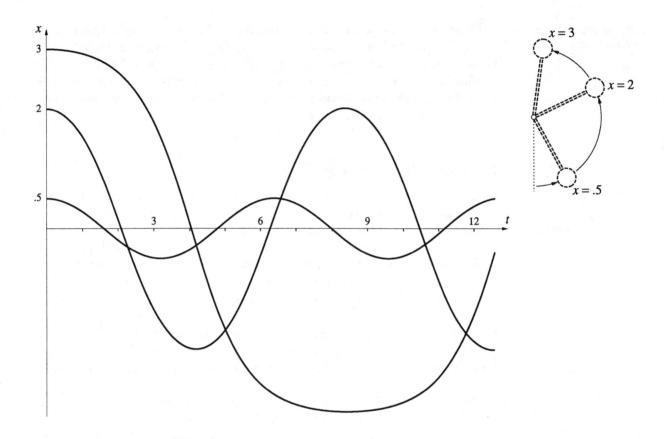

It will remain at rest forever: $x(t) = 0$, $v(t) = 0$ for all $t \geq 0$. What we really mean is that the constant functions $x(t) = 0$ and $v(t) = 0$ solve the initial value problem

The pendulum at rest

$$x' = v \qquad \qquad x(0) = 0$$

$$v' = -\sin x \qquad v(0) = 0.$$

There is another way for the pendulum to remain at rest. The key is that v must not change. But $v' = -\sin x$, so v will remain fixed if $v' = -\sin x = 0$. Now, $\sin x = 0$ if $x = 0$. This yields the rest solution we have just identified. But $\sin x$ is also zero if $x = \pi$. You should check that the constant functions $x(t) = \pi$, $v(t) = 0$ solve this initial value problem:

The pendulum balanced on end

$$x' = v \qquad \qquad x(0) = \pi$$

$$v' = -\sin x \qquad v(0) = 0.$$

Since the pendulum points straight up when $x = \pi$ radians, this motionless solution corresponds to the pendulum balancing on its end.

Stable and unstable
equilibrium

These two solutions are called **equilibrium** solutions (from the Latin *æqui-*, "equal" + *libra,* a balance scale). If the pendulum is disturbed from its rest position, it tends to return to rest. For this reason, rest is said to be a **stable** equilibrium. Contrast what happens if the pendulum is disturbed when it is balanced upright. This is said to be an **unstable** equilibrium. We will take a longer look at equilibria in chapter 8.

Predator–Prey Ecology

Many animal populations undergo nearly periodic fluctuations in size. It is even more remarkable that the period of those fluctuations varies little from species to species. This fascinates ecologists and frustrates many who hunt, fish, and trap those populations to make their livelihood. Why should there be fluctuations, and can something be done to alter or eliminate them?

Why do populations
fluctuate?

There are models of predator–prey interaction that exhibit periodic behavior. Consequently, some researchers have proposed that the fluctuations observed in a real population occur because that species is either the predator or the prey for another species. There are various models; we will study one proposed by R. May. As we did with the spring and the pendulum, we will ask how the frequency and amplitude of periodic solutions depend on the initial conditions.

May's model involves two populations that vary in size over time: the predator y and the prey x. The numbers x and y have been set to an arbitrary scale; they lie between 0 and 20. The model also has six adjustable parameters, but we will simply fix their values:

$$\text{prey:} \quad x' = .6x\left(1 - \frac{x}{10}\right) - \frac{.5xy}{x + 1}$$

$$\text{predator:} \quad y' = .1y\left(1 - \frac{y}{2x}\right).$$

These equations will be our starting point. However, if you wish to learn more about the premises behind May's model, you can refer to section 4.1 (page 168).

A predator-free
equilibrium . . .

To begin to explore the model, let's see what happens to the prey population when there are no predators ($y = 0$). Then the size of x is governed by the simpler differential equation $x' = .6x(1 - x/10)$. This is logistic growth, and x will eventually approach the carrying capacity of the environment, which in this case is 10. (See pages 161–163.) In fact, you should check that

$$x(t) = 10 \qquad y(t) = 0$$

is an equilibrium solution of May's original differential equations.

Now suppose we introduce a small number of predators: $y = .1$. Then the equilibrium is lost, and the predator and prey populations fall into cyclic patterns with the same period. Here are the graphs:

... upset by predators

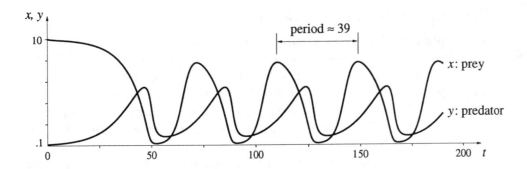

In the other models of periodic behavior we have studied, the frequency and amplitude have depended on the initial conditions. Is the same true here? The following graphs illustrate what happens if the initial populations are either

Solutions with various initial conditions ...

$$x = 8, \quad y = 2 \quad \text{or} \quad x = 1.1, \quad y = 2.2.$$

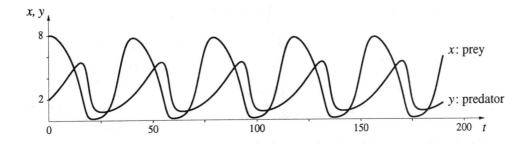

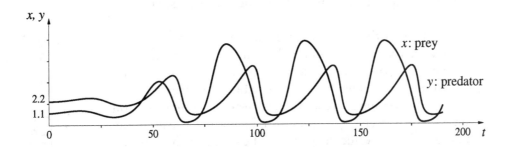

...all have the
same amplitude
and frequency

In all of these graphs, periodic behavior eventually emerges. What is most striking, though, is that it is the *same* behavior in all cases. The amplitude and the period *do not* depend on the initial conditions. Moreover, even though the populations peak at different times on the three graphs (i.e., the *phases* are different), the y peak always comes about 14 time units after the x peak.

Proving a Solution Is Periodic

Can we *prove* that
systems have periodic
oscillations?

The graphs in the last ten pages provide strong evidence that nonlinear springs, pendulums, and predator–prey systems can oscillate in a periodic way. The evidence is numerical, though. It is based on Euler's method, which gives us only *approximate* solutions to differential equations. Can we now go one step further and *prove* that the solutions to these and other systems are periodic?

The virtue
of a formula

Notice that we already have a proof in the case of a linear spring. The solutions are given by formulas that involve sines and cosines, and these are periodic by their very design as circular functions. But we have no formulas for the solutions of the other systems. In particular, we are not able to say anything about the general properties of the solutions (the way we can about sine and cosine functions). The approach we take now does not depend on having a formula for the solution.

> It may seem that what we should do is develop more methods for finding formulas for solutions. In fact, two hundred years of research was devoted to this goal, and much has been accomplished. However, it is now clear that most solutions simply have no representation "in closed form" (that is, as formulas). This isn't a confession that we can't find the solutions. It just means the formulas we have are inadequate to describe the solutions we can find.

The Pendulum—A Qualitative Approach

Let's work with the pendulum and model it by the following initial value problem:

$$
\begin{aligned}
x' &= v & x(0) &= a \\
v' &= -\sin x & v(0) &= 0.
\end{aligned}
$$

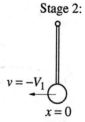

Stage 1:

We'll assume $0 < a < \pi$. Thus, at the start the pendulum is motionless and raised to the right. Call this **stage 1**. We'll analyze what happens to x and v in a qualitative sense. That is, we'll pay attention to the *signs* of these quantities, and whether they're increasing or decreasing, but not their exact numerical values.

According to the differential equations, v determines the rate at which x changes, and x determines the rate at which v changes. In particular, since we start with $0 < x < \pi$, the expression $-\sin x$ must be negative. Thus v' is negative, so v decreases, becoming more and more negative as time goes on. Consequently, x changes at an ever increasing negative rate, and eventually its value drops to 0. The moment this happens

Stage 2:

the pendulum is hanging straight down and moving left with some large negative velocity $-V_1$. This is **stage 2.**

Immediately after the pendulum passes through stage 2, x becomes negative. Consequently, $v' = -\sin x$ now has a positive value (because x is negative). So v stops decreasing and starts increasing. Since x gets more and more negative, v increases more and more rapidly. Eventually v must become 0. Suppose $x = -B_1$ at the moment this happens. The pendulum is then poised motionless and raised up B_1 units to the left. We have reached **stage 3.**

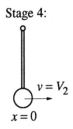

Stage 3:
$v = 0$
$x = -B_1$

The situation is now similar to stage 1, because $v = 0$ once again. The difference is that x is now negative instead of positive. This just means that v' is positive. Consequently v becomes more and more positive, implying that x changes at an ever increasing positive rate. Eventually x reaches 0. The moment this happens the pendulum is again hanging straight down (as it was at stage 2), but now it is moving to the right with some large positive velocity V_2. Let's call this **stage 4.** It is similar to stage 2.

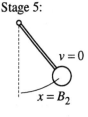

Stage 4:
$v = V_2$
$x = 0$

Immediately after the pendulum passes through stage 4, x becomes positive. This makes $v' = -\sin x$ negative, so v stops increasing and starts decreasing. Eventually v becomes 0 again (just as it did in the events that lead up to stage 3). At the moment the pendulum stops, x has reached some positive value B_2. Let's call this **stage 5.**

Stage 5:
$v = 0$
$x = B_2$

The "Trade-off" Between Speed and Height

We appear to have gone "full circle." The pendulum has returned to the right and is once again motionless—just as it was at the start. However, we don't know that the *current* position of the pendulum (which is $x = B_2$) is the same as its *initial* position ($x = a$). This is a consequence of working qualitatively instead of quantitatively. But it is also the nub of the problem. For the motion of the pendulum to repeat itself exactly we must have $B_2 = a$. Can we prove that $B_2 = a$?

Are we back where we started?

Since a and B_2 are the successive positive values of x that occur when $v = 0$, it makes sense to explore the connection between x and v. In a real pendulum there is an obvious connection. The higher the pendulum bob rises, the more slowly it moves. If you review the sequence of stages we just went through, you'll see that the same thing is true of our mathematical model. This suggests that we should focus on the height of the pendulum bob and the magnitude of the velocity. This is called the **speed;** it is just the absolute value $|v|$ of the velocity.

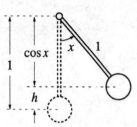

A little trigonometry shows us that when the pendulum makes an angle of x radians with the vertical, the height of the pendulum bob is

$$h = 1 - \cos x.$$

When x is a function of time t, then h is too and we have

$$h(t) = 1 - \cos(x(t)).$$

Our intuition about the pendulum tells us that every change in height is offset by a change in speed. (This is the "trade-off.") It makes sense, therefore, to compare the *rates* at which the height and the speed change over time. However, the speed $|v(t)|$ involves an absolute value, and this is difficult to deal with in calculus. (The absolute value function is not differentiable at 0.) Since we are using $|v|$ simply as a way to ignore the difference between positive and negative velocities, we can replace $|v|$ by v^2. Then we find

Changes in speed... modified

$$\frac{d}{dt}(v(t))^2 = 2 \cdot v(t) \cdot v'(t) = -2 \cdot v \cdot \sin x.$$

Notice that we needed the chain rule to differentiate $(v(t))^2$. After that we used the differential equations of the pendulum to replace v' by $-\sin x$.

Changes in height

The height of the pendulum changes at this rate:

$$\frac{d}{dt} h(t) = \sin(x(t)) \cdot x'(t) = \sin x \cdot v.$$

We needed the chain rule again, and we used the differential equations of the pendulum to replace x' by v.

The two derivatives are almost exactly the same; except for sign, they differ only by a factor of 2. If we use $\frac{1}{2}v^2$ instead of v^2, then the trade-off is exact: every increase in $\frac{1}{2}v^2$ is *exactly* matched by a decrease in h, and *vice versa*. Therefore, if we combine $\frac{1}{2}v^2$ and h to make the new quantity

$$E = \tfrac{1}{2}v^2 + h = \tfrac{1}{2}v^2 + 1 - \cos x,$$

then we can say that the value of E does not change as the pendulum moves.

Since E depends on v and h, and these are functions of the time t, E itself is a function of t. To say that E doesn't change as the pendulum moves is to say that this function is a constant—in other words, that its derivative is 0. This was, in fact, the way we constructed E in the first place. Let's remind ourselves of why this worked. Since $E = \frac{1}{2}v^2 + h$,

Showing E is a constant

$$\frac{dE}{dt} = v \cdot v' + h'$$
$$= v \cdot (-\sin x) + \sin x \cdot v$$
$$= 0.$$

To get the second line we used the fact that $v' = -\sin x$ and $x' = v$ when $x(t)$ and $v(t)$ describe pendulum motion.

E is called the **energy** of the pendulum. The fact that E doesn't change is called the **conservation of energy** of the pendulum. A number of problems in physics can be analyzed starting from the fact that the energy of many systems is constant.

Let's calculate the value of E at the five different stages of our pendulum:

Stage	v	x	h	E
1	0	a	$1 - \cos a$	$1 - \cos a$
2	$-V_1$	0	0	$\frac{1}{2}(-V_1)^2$
3	0	$-B_1$	$1 - \cos B_1$	$1 - \cos B_1$
4	V_2	0	0	$\frac{1}{2}V_2^2$
5	0	B_2	$1 - \cos B_2$	$1 - \cos B_2$

By the conservation of energy, all the quantities in the right-hand column have the same value. Looking at the value for E in stages 2 and 4, we see that $V_1 = V_2$—*whenever the pendulum is at the bottom of its swing ($x = 0$), it is moving with the same speed*, the velocity being positive when the pendulum is swinging to the right, negative when it is swinging to the left. Similarly, if we look at the value of E at stages 1, 3, and 5, we see that

$$1 - \cos a = 1 - \cos B_1 = 1 - \cos B_2.$$

We can put this another way: *whenever the pendulum is motionless, it must be back at its starting height $h = 1 - \cos a$.*

In particular, we have thus shown that B_2 (the position of the pendulum after it's gone over and back) $= a$ (the position of the pendulum at the beginning). Thus the value for x and the value for v are the same in stage 5 and in stage 1—the two stages are mathematically indistinguishable. Since the solution to an initial value problem depends only on the differential equation and the initial values, what happens after stage 5 must be identical to what happens after stage 1—the second swing of the pendulum must be identical to the first! Thus the motion is periodic, which completes our proof.

...and that the oscillations are periodic

You can also use the fact that the value of E doesn't change to determine the velocities $-V_1$ and V_2 that the pendulum achieves at the bottom of its swing. In the exercises you are asked to show that

$$V_1 = V_2 = \sqrt{2 - 2\cos a}.$$

First Integrals

Notice in what we have just done that we haven't solved the differential equation for the pendulum in the sense of finding explicit formulas giving x and v
in terms of t. Instead we found a combination of x and v that remained constant over time and used this to deduce some of the behavior of x and v. Such
a combination of the variables that remains constant is called a **first integral** of
the differential equation. A surprising amount of information about a system
can be inferred from first integrals (when they exist). They play an important
role in many branches of physics, giving rise to the basic conservation laws for
energy, momentum, and angular momentum. We will have more to say about
first integrals and conservation laws in chapter 8.

First integrals

In the exercises you are asked to explore first integrals for linear and nonlinear springs—and to prove thereby that (frictionless) nonlinear springs have
periodic motions.

Exercises

Linear Springs

In the text we always assumed that the weight on the spring was motionless at
$t = 0$ seconds. The first four exercises explore what happens if the weight is
given an initial *impulse*. For example, instead of simply releasing the weight,
you could hit it out of your hand with a hammer. This means $v(0) \neq 0$. The
general initial value problem is

$$x' = v \qquad x(0) = a$$
$$v' = -b^2 x \qquad v(0) = p.$$

The aim is to see how the period, amplitude, and phase of the solution depend
on this new condition.

1. *Pure impulse.* Take $b = 5$ per second, as in the first example in the
   text, but suppose

   $$a = 0 \text{ cm} \qquad p = 20 \text{ cm/sec}.$$

   (In other words, you strike the weight with a hammer as it sits
   motionless at the rest position $x = 0$ cm.)
   a) Use the differential equation solver on a computer to solve the
      initial value problem numerically and graph the result.
   b) From the graph, estimate the period and the amplitude of the
      solution.
   c) Find a formula for this solution, using the graph as a guide.
   d) From the formula, determine the period and amplitude of the
      solution. Does the period depend on the initial impulse p, or only
      on the spring constant b? Does the amplitude depend on p?

2. *Impulse and displacement.* Take $a = 4$ cm and $b = 5$ per second, as in the first example on page 384. But assume now that the weight is given an initial *downward* impulse of $p = -20$ cm/sec.
   a) Solve the initial value problem numerically and graph the result.
   *b) From the graph, estimate the period and the amplitude of the solution. Compare these with the period and the amplitude of the solution obtained in the text for $p = 0$ cm/sec.

3. Let a and b have the values they did in the last exercise, but change p to $+20$ cm/sec. Graph the solution, and compare the amplitude and phase of this solution with the solution of the previous exercise.

4. Let a, b, and p have arbitrary values. The last two exercises suggest that the solution to the general initial value problem for a linear spring can be given by the formula $x(t) = A \sin(bt - \varphi)$. The amplitude A and the phase difference φ depend on the initial conditions. Show that the formula for $x(t)$ is correct by expressing A and φ in terms of the initial conditions.

5. *Strength of the spring.* Take two springs, and suppose the second is twice as strong as the first. That is, assume the second spring constant is twice the first. Put equal weights on the ends of the two springs, and use the initial value $v(0) = 0$ in both cases. Which weight oscillates with the higher frequency? How are the frequencies of the two related—for example, is the frequency of the second equal to twice the frequency of the first, or should the multiplier be a different number?

6. a) *Effect of the weight.* Hang weights from two identical springs (i.e., springs with the same spring constant). Suppose the mass of the second weight is twice that of the first. Which weight oscillates with the higher frequency? How much higher—twice as high, or some other multiplier?
   b) Do this experiment in your head. Measure the frequency of the oscillations of a 200-gram weight on a spring. Suppose a second weight oscillates at twice the frequency. What is *its* mass?

A reality check. Do your results in the last two exercises agree with your intuitions about the way springs operate?

7. a) *First integral.* Show that $E = \frac{1}{2}v^2 + \frac{1}{2}b^2x^2$ is a first integral for the linear spring

$$x' = v \qquad x(0) = a$$
$$v' = -b^2x \qquad v(0) = p.$$

In other words, if the functions $x(t)$ and $v(t)$ solve this initial value problem, you must show that the combination

$$E = \tfrac{1}{2}(v(t))^2 + \tfrac{1}{2}b^2(x(t))^2$$

does not change as t varies.

*b) What value does E have in this problem?

c) If x is measured in cm and t in sec, what are the units for E?

8. a) This exercise concerns the initial value problem in the previous question. When $x = 0$, what are the possible values that v can have?

b) At a moment when the weight on the spring is motionless, how far is it from the rest position?

9. You already know that the initial value problem in exercise 7 has a solution of the form $x(t) = A\sin(bt - \varphi)$ and therefore must be periodic. Give a different proof of periodicity using the first integral from the same exercise, following the approach used in the case of the pendulum.

Nonlinear Springs

10.*a) Suppose the acceleration v' of the weight on a hard spring depends on the displacement x of the weight according to the formula $v' = -16x - x^3$ cm/sec². If you pull the weight down $a = 2$ cm, hold it motionless (so $p = 0$ cm/sec), and then release it, what will its frequency be?

b) How far must you pull the weight so that its frequency will be double the frequency in part (a)? (Assume $p = 0$ cm/sec, so there is still no initial impulse.)

11. Suppose the acceleration of the weight on a hard spring is given by $v' = -16x - .1x^3$ cm/sec². If the weight is oscillating with very small amplitude, what is the frequency of the oscillation?

12. a) Suppose a weight on a spring accelerates according to the formula

$$\frac{dv}{dt} = -\frac{25x}{1 + x^2} \quad \text{cm/sec}^2.$$

This is a soft spring. Explain why. (Graph v' as a function of x.)

b) If the initial amplitude of the weight is $a = 4$ cm, and there is no initial impulse (so $p = 0$ cm/sec), what is the frequency of the oscillation?

c) Double the initial amplitude, making $a = 8$ cm but keeping $p = 0$ cm/sec. What happens to the frequency?

d) Suppose you make the initial amplitude $a = 100$ cm. Now what happens to the frequency?

13. *First integrals.* Suppose the acceleration on a nonlinear spring is

$$v' = -b^2 x - \beta x^3, \quad \text{where} \quad v = x'.$$

Show that the function

$$E = \tfrac{1}{2}v^2 + \tfrac{1}{2}b^2x^2 + \tfrac{1}{4}\beta x^4$$

is a first integral. [See the text (page 400) and exercise 7, above.]

14. Suppose the acceleration on a nonlinear spring is $v' = -16x - x^3$ cm/sec², and initially $x = 2$ cm and $v = 0$ cm/sec.
    a) The first integral of the preceding exercise must have a fixed value for this spring. What is that value?
    b) How fast is the spring moving when it passes through the rest position?
    c) Can the spring ever be more than 2 cm away from the rest position? Explain your answer.

15. Construct a first integral for the initial value problem

$$x' = v \qquad\qquad x(0) = a$$
$$v' = -b^2x - \beta x^3 \qquad v(0) = p$$

and use it to show that the solution to the problem is periodic.

16. a) Show that the function

$$E = \tfrac{1}{2}v^2 + \tfrac{25}{2}\ln(1 + x^2)$$

is a first integral for the soft spring in exercise 12.
    b) If the initial amplitude is $a = 4$ cm and the initial velocity is 0 cm/sec, what is the speed of the weight as it moves past the rest position?
    c) Prove that the motion of this spring is periodic.

17. Suppose the acceleration on a nonlinear spring has the general form $v' = -f(x)$. Can you find a first integral for this spring? In other words, you are being asked to show that a first integral always exists whenever the rate of change of the velocity depends only on the position x (and not, for instance, on v itself, or on the time t).

The Pendulum

These questions deal with the initial value problem

$$x' = v \qquad\qquad x(0) = a$$
$$v' = -\sin x \qquad v(0) = p.$$

In particular, we want to allow an initial impulse $p \neq 0$.

18. Take $a = 0$ and give the pendulum three different initial impulses: $p = .05, p = .1, p = .2$. Use the differential equation solver on a

computer to graph the three motions that result. Determine the period of the motion in each case. Are the periods noticeably different?

19. What is the period of the motion if $p = 1$? If $p = 2$?

20. By experiment, find how large an initial impulse p is needed to knock the pendulum "over the top," so it spins around its axis instead of oscillating? Assume $x(0) = 0$. (Note: When the pendulum spins, x just keeps getting larger and larger.) Of course any enormous value for p will guarantee that the pendulum spins. Your task is to find the *threshold*; this is the smallest initial impulse that will cause spinning.

21. a) Suppose the initial position is horizontal: $a = +\pi/2$. If you give the pendulum an initial impulse p in the same direction (that is, $p > 0$), find by experiment how large p must be to cause the pendulum to spin. Once again, the challenge is to find the threshold value.

    b) Reverse the direction of the initial impulse: $p < 0$, and choose p so the pendulum spins. What is the smallest $|p|$ that will cause spinning?

22. *First integrals*. Consider the initial value problem described in the text:

$$x' = v \qquad\qquad x(0) = a$$
$$v' = -\sin x \qquad v(0) = 0.$$

Use the first integral for this problem found on page 398 to show that $v = \sqrt{2 - 2\cos a}$ when $x = 0$.

23. a) Suppose the pendulum described in the previous exercise is at rest [$x(0) = 0$], but given an initial impulse $v(0) = p$. What value does the first integral have in this case?

    b) Redo exercise 20 using the information the first integral gives you. You should be able to find the exact threshold value of the impulse that will push the pendulum "over the top."

*24. Redo exercise 21 using an appropriate first integral. Find the threshold value exactly.

Predator–Prey Ecology

25. a) *The May model*. The differential equations for this model are on page 394. Show that the constant functions

$$x(t) = 10 \qquad y(t) = 0$$

are a solution to the equations. This is an equilibrium solution, as defined in the discussion of the pendulum (page 394).

    b) Is $x(t) = 0$, $y(t) = 0$ an equilibrium solution?

c) Here is yet another equilibrium solution:

$$x(t) = \frac{-23 \pm \sqrt{889}}{6} \qquad y(t) = \frac{-23 \pm \sqrt{889}}{3}.$$

Either verify that it *is* an equilibrium, or explain how it was derived.

26. a) Use a computer differential equation solver to graph the solution to the May model that is determined by the initial conditions

$$x(0) = 1.13 \qquad y(0) = 2.27.$$

These initial conditions are very close to the equilibrium solution in part (c) of the previous exercise. Does the solution you've just graphed suggest that this equilibrium is *stable* or that it is *unstable* (as described on page 394)?

b) Change the initial conditions to

$$x(0) = 5 \qquad y(0) = 5$$

and graph the solution. Compare this solution to those determined by the initial conditions used in the text. In particular, compare the shapes of the graphs, their periods, and the time interval between the peak of x and the peak of y.

27. Consider this scenario. Imagine that the prey species x is an agricultural pest, while the predator y does not harm any crops. Farmers would like to eliminate the pest, and they propose to do so by bringing in a large number of predators. Does this strategy work according to the May model? Suppose that we start with a relatively large number of predators:

$$x(0) = 5 \qquad y(0) = 50.$$

What happens? In particular, does the pest disappear?

*28. *The Lotka–Volterra model.* We use the differential equations found in chapter 4, page 170, modified so that relevant values of x and y will be roughly the same size:

$$x' = .1x - .005xy$$

$$y' = .004xy - .04y.$$

Take $x(0) = 20$ and $y(0) = 10$. Use a computer differential equation solver to graph the solution to this initial value problem. The solutions are periodic. What is the period? Which peaks first, the prey x or the predator y? How much sooner?

29. Solve the Lotka–Volterra model with $x(0) = 10$ and $y(0) = 5$. What is the period of the solutions, and what is the difference between the times when the two populations peak? Compare these results with those of the previous exercise.

30. Show that $x(t) = 0$, $y(t) = 0$ is an equilibrium solution of the Lotka–Volterra equations. To test the stability of this solution, take these nearby initial conditions:

$$x(0) = .1 \qquad y(0) = .1$$

and find the solution. Does it remain near the equilibrium? If so, the equilibrium is stable; if not, it is unstable.

31. Show that $x(t) = 10$, $y(t) = 20$ is another equilibrium solution of these Lotka–Volterra equations. Is this equilibrium stable? (We will have more to say about stability of equilibria in chapter 8.)

32. This is a repeat of the biological pest control scenario you treated above, using the May model. Solve the Lotka–Volterra model when the initial populations are

$$x(0) = 5 \qquad y(0) = 50.$$

What happens? In particular, does the pest disappear?

33. *First integrals.* As remarkable as it may seem, the Lotka–Volterra model has a first integral. Show that the function

$$E = a \ln y + d \ln x - by - cx$$

is a first integral of the Lotka–Volterra model given in the general form

$$x' = ax - bxy$$
$$y' = cxy - dy.$$

34. Prove that the solutions of the Lotka–Volterra equations are periodic.

The van der Pol Oscillator

One of the essential functions of the electronic circuits in a television or radio transmitter is to generate a periodic "signal" that is stable in amplitude and period. One such circuit is described by the van der Pol differential equations. In this circuit $x(t)$ represents the current, and $y(t)$ the voltage, at time t. These functions satisfy the differential equations

$$x' = y \qquad y' = Ay - By^3 - x \qquad \text{with } A, B > 0.$$

35. Take $A = 4$, $B = 1$. Make a sketch of the solution whose initial values are $x(0) = .1$, $y(0) = 0$. Your sketch should show that this solution is *not* periodic at the outset, but becomes periodic after some time has passed. Determine the (eventual) period and amplitude of this solution.

36. Obtain the solution whose initial values are $x(0) = 2$, $y(0) = 0$, and then the one whose initial values are $x(0) = 4$, $y(0) = 0$. What are the periods and amplitudes of these solutions? What effect does the initial current $x(0)$ have on the period or the amplitude?

➤ 7.4 CHAPTER SUMMARY

The Main Ideas

- There are many phenomena which exhibit **periodic** and near-periodic behavior. They are modeled by differential equations with periodic solutions.

- A periodic function repeats: the smallest number T for which $g(x + T) = g(x)$ for all x is the **period** of the function g. Its **frequency** is the reciprocal of its period, $\omega = 1/T$.

- The **circular functions** are periodic; they include the sine, cosine, and tangent functions. The period of $\sin(t)$ and $\cos(t)$ is 2π and the frequency is $1/2\pi$. The frequency of $A\sin(bt)$ and $A\cos(bt)$ is $b/2\pi$, and the **amplitude** is A. In $A\sin(bt + \varphi)$, the **phase** is shifted by φ.

- A **linear spring** is one for which the spring force is proportional to the amount that the spring has been displaced. The motion of a linear spring is periodic. Its amplitude depends only on the initial conditions, and its frequency only on the mass and the spring constant.

- In a **nonlinear spring,** the force is no longer proportional to the displacement. The motion of a nonlinear spring can still be periodic, although it is no longer described simply by sines and cosines. Its frequency depends on its amplitude. A pendulum is a nonlinear spring. It has two **equilibria,** one **stable** and one **unstable.**

- Many quantities oscillate periodically, or nearly so. Frequently the behavior of these quantities can be modeled by systems of differential equations. Pendulums, electronic components, and animal populations are some examples.

- In some initial value problems, it may still be possible to find a **first integral**—a combination of the variables that remains constant—even when we can't find formulas for the variables separately. We can often derive important properties of the system (such as periodicity) from these constant combinations.

Expectations

- You should be able to find the **period, frequency,** and **amplitude** of sine and cosine functions.
- You should be able to convert between **radian** measure and degrees.
- You should be able to find a formula for the solution of the differential equation describing a **linear spring.**
- You should be able to use Euler's method to describe the motion of a nonlinear spring.
- You should be able to analyze oscillations of various kinds to determine their periodicity.

CHAPTER 8

DYNAMICAL SYSTEMS

A recurring theme in this book is the use of mathematical models consisting of a set of differential equations to explore the behavior of physical systems as they evolve over time. Some examples we have encountered are the S-I-R epidemiological model, predator–prey systems, and the motion of a pendulum. We call such a set of differential equations a **dynamical system.** Dynamical systems play important roles in all branches of science. In this chapter we will develop some general tools for thinking about them, with particular emphasis on the kinds of geometric insight provided by the concepts of **state space** and **vector field.**

➢ 8.1 STATE SPACES AND VECTOR FIELDS

If you look back at the examples we've considered, many of them take the following form: we have two (or more) variable quantities x and y that are functions of time, and we want to find the nature of these functions. What we have to work with is a model for the way the functions $x(t)$ and $y(t)$ are changing—that is, we are told how to calculate $x'(t)$ and $y'(t)$ whenever we know the values of x and y, and possibly t. From a given starting point, we typically used something like Euler's method to get values for x and y at times on either side of the starting value. We then graphed the solutions as functions of time—x against t and y against t.

The standard way to graph solutions

In many instances, the rules determining $x'(t)$ and $y'(t)$ depend only on the current values of x and y, but not on the value of t, so that knowing the current state of the system (as specified by its x and y values) is sufficient to determine the future and past states of the system. Such systems are said to be **autonomous.** These are the only systems we will be considering in this chapter.

A new way to graph solutions: as trajectories in state space

In autonomous systems there is another way of visualizing the solutions that can be very powerful. Instead of plotting values of x and y as functions of time, we view these values as coordinates of a point in the x, y plane. As the system changes, the point (x, y) will trace out a curve in this plane. The point (x, y) is called a **state,** and the portion of the plane corresponding to physically possible states is called the **state space** of the system. The solution curves that get traced out in state space are called **trajectories.** By looking at three examples, we will see how this method of analysis can help us understand the overall behavior of a system.

There are a number of effective software packages available which can perform efficiently all the operations we will be considering, and one of them would probably be the most useful tool for exploring the ideas in this chapter. On the other hand, the basic numerical operations are quite simple, and it is easy to modify the programs developed earlier in the text to perform these operations as well. For those of you who enjoy programming, we will from time to time point out some of these modifications. It can be instructive to implement them in your own programs, and we urge you to do so.

Predator–Prey Models

In section 4.1, we looked at several models for the dynamics of a simple system consisting of foxes (F) and rabbits (R). Our first model was

$$R' = .1R\left(1 - \frac{R}{10000}\right) - .005RF \qquad \text{rabbits per month}$$

$$F' = .00004RF - .04F \qquad \text{foxes per month.}$$

When we started with the initial values $R(0) = 2000$ rabbits and $F(0) = 10$ foxes, Euler's method produced the following solutions for the first 250 months:

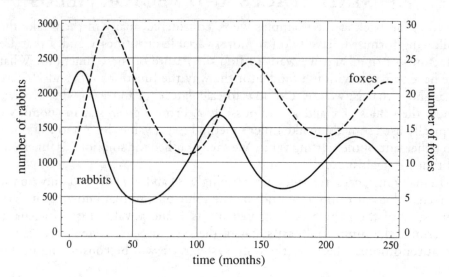

Let's see how the same model looks when we express it in the language of
state spaces. The state space consists of points in the R, F plane. For physical reasons our state space consists only of points (R, F) satisfying $R \geq 0$ and
$F \geq 0$. That is, our state space is the first quadrant of the R, F plane together
with the bounding portions of the R-axis and the F-axis. We can easily modify
the program used to obtain the curves in the previous picture to plot the corresponding trajectory in the R, F plane. We only need change the specification
of the dimensions of the viewing window and change the PLOT command to
plot points with coordinates (R, F) instead of (t, R) and (t, F); all the rest of the
calculations are unchanged. Here's what the same solution looks like when we
do this:

Two ways of
representing the same
solution graphically

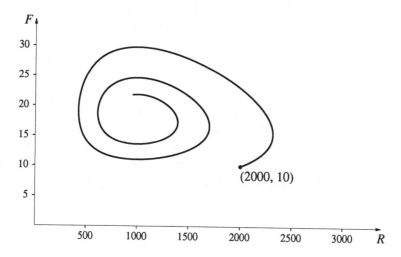

You should notice several things here:

- The trajectory looks like a spiral, moving in toward, but never reaching,
  some point at its center. We will see later (see page 416) how to
  determine the coordinates of this limit state.

- If we had started at any other initial state with $R > 0$ and $F > 0$, we
  would have gotten another spiral converging to the same limit (try it
  and see).

- From the trajectory alone, there is no way of determining the time at
  which the system passes through the different states. In part, this simply
  emphasizes that the succession of states the system moves through does
  not depend on the initial value of t, nor does it depend on the units in
  which t is measured—if t were measured in days or years, rather than in
  months, the trajectory would be unchanged.

If we wanted to include some information about time, one way would be
to label some points on the trajectory with the associated time value. If we label

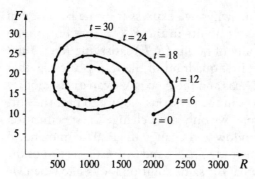

the points every 6 months, say, we would get the picture at the left. Note that the points are not uniformly spaced along the trajectory: the spacing is largest between points relatively far from the origin, where the values of R and F are largest. Moreover, the closer we come to the limit state, the tighter the spacing becomes.

The differential equations indicate how the state changes

Could we have foreseen some of this behavior by looking at the original differential equations? Since the differential equations give R' and F' as functions of R and F alone, for each point (R, F) in the state space we can calculate the associated values for R' and F'. Knowing these values, we can in turn tell in what direction and with what speed a trajectory would be moving as it passed through the point (R, F). Using our (by now) standard argument, in time Δt the change in R would be $\approx R' \Delta t$, while F would change by $\approx F' \Delta t$. We can convey this information graphically by choosing a number of points in the state space, and from each point (R, F) drawing an arrow to the point $(R + R' \Delta t, F + F' \Delta t)$. We would typically choose a value for Δt that keeps the arrows a reasonable size. Here's what we get in our current example when we choose a 16 × 16 grid of points in the region $0 \leq R \leq 3000$ and $0 \leq F \leq 30$, with $\Delta t = 1$.

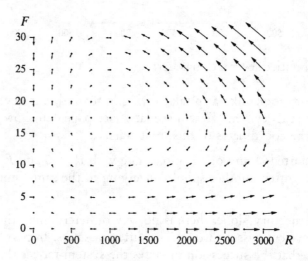

Arrows indicate the change in state...

Several things are immediately clear from this picture: the arrows suggest a general counterclockwise flow in the plane; change is most rapid in the upper right corner; near the limit point of the flow and near the origin change is so slow that arrows don't even show up there.

Moreover, since the method used to construct the arrows is exactly the way Euler's method calculates the trajectories themselves, the solution trajectory through a given initial state is a curve in the state space which at every point is tangent to the arrow at that point. For instance, if we superpose the trajectory graphed on page 411 on the figure on page 412, we get the following:

. . . and trajectories are tangent to the arrows

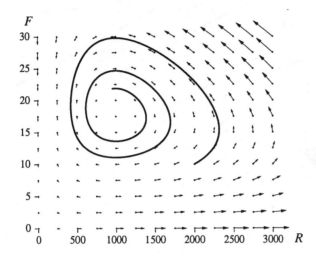

The net effect of this construction is thus to transform a problem in *analysis*—solving a system of differential equations—into a problem in *geometry*—finding a curve which is tangent everywhere to a prescribed set of arrows. This correspondence between the analytical and the geometrical ways of formulating a problem is very powerful. Let's sum up the way this correspondence was established:

How to construct a geometrical visualization of a dynamical system

- We set up a **state space** for the system being studied. Each point—called a **state**—in the space corresponds to a possible pair of values the system could have.

- There is a rule which assigns to each point in the state space a **velocity vector**—which can be visualized as an arrow in the space based at the given point—specifying the rates at which the coordinates of the point are changing. The rule itself, which is just our original set of rate equations, is called a **vector field.** Geometrically, we can visualize the vector field as the state space with all the associated arrows.

- Solutions to the dynamical system correspond to **trajectories** in the state space. At every point on a trajectory the associated velocity vector specified by the vector field will be tangent to the trajectory. The existence and uniqueness principle for the solutions of differential equations—there is a unique solution for each set of initial values—is geometrically expressed by the property that every point in the state space lies on exactly one trajectory. The set of all possible trajectories is called the **phase portrait** of the system. For instance, part of the phase portrait of the system we have been considering appears below. We have drawn only a few trajectories—if we had drawn them all, we would have seen only a black rectangle since there is a trajectory through every point.

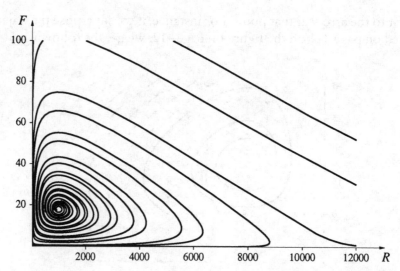

Simplifying the picture

There is almost too much detail in the picture of the vector field and the phase portrait. One way to see the underlying simplicity is to notice that the space is divided into four regions according to whether F' and R' are positive or negative. The *signs* of F' and R' in turn determine the *direction* of the associated velocity vector. For instance, if F' and R' are both positive, then F and R must both be increasing, which means the velocity vector will be pointing up and to the right, while if $F' > 0$ and $R' < 0$, the velocity vector will be pointing up ($F' > 0$) and to the left ($R' < 0$). Let's see which states correspond to which behaviors. Here are the original rate equations:

$$R' = .1R\left(1 - \frac{R}{10000}\right) - .005RF \qquad \text{rabbits per month}$$

$$F' = .00004RF - .04F \qquad\qquad\qquad \text{foxes per month.}$$

The equation for F' is slightly simpler, so we'll start there. We see that $F' = 0$ in exactly two cases:

1. When $F = 0$, or
2. When $.00004R - .04 = 0$, which is equivalent to saying $R = 1000$.

The first case simply says that if we are ever on the R-axis ($F = 0$), then we stay there—a trajectory starting on the R-axis must move horizontally. (If you start with no foxes, you will never have any at a later time.) The second case says that the value of F isn't changing whenever $R = 1000$. The set of points satisfying $R = 1000$ is just a vertical line in the state space. The condition that $F' = 0$ on this line can be expressed geometrically by saying that any trajectory crossing this line must do so horizontally (why?).

Divide the state space into regions

The remainder of the quadrant consists of two regions: one consists of all points (R, F) with $0 \le R < 1000$ and $F > 0$; the other consists of all points (R, F) with $R > 1000$ and $F > 0$. Moreover, since we've already accounted for all the points where $F' = 0$, it must be true that at every point of these two regions F' must be > 0 or < 0; F' can't equal 0 in either region. Further, within any one region F' must be always positive or always negative. If it were positive at some points and negative at others in a single region, there would have to be transition points where it took on the value 0, which we have just observed can't happen. (Be sure you see why this is so!) Thus, to determine the sign of F' in an entire region, we only need to see what the sign is at one point in that region. For instance, if we let $R = 2000$ and $F = 1$, we see that $F' = .08 - .04$, which is positive. Therefore we will have $F' > 0$ (fox population increasing) for any other state (R, F) with $R > 1000$. Similarly, we can show that $F' < 0$ (fox population decreasing) if $0 \le R < 1000$—the test point $R = 0$ and $F = 1$ is easy to evaluate. We could, of course, have arrived at the same conclusions through more formal algebraic arguments, which are fairly straightforward in this instance. In other problems, though, the "test point" approach may be the more convenient.

In exactly the same way, if we look at the first rate equation, we find that $R' = 0$ in two cases:

1. When $R = 0$, or
2. When $.1(1 - R/10000) - .005F = 0$. This is just the equation of a line, which can be rewritten as $F = 20 - .002R$.

The interpretations of these two cases are similar to the preceding analysis: any trajectory starting on the F-axis must stay on the F-axis; any trajectory crossing the line $F = 20 - .002R$ must cross it vertically, since $R' = 0$ there, so the R-value isn't changing. Further, for any other state (R, F) we have $R' > 0$ if the point is below this line (the point $R = 1$ and $F = 0$ is a convenient test point where it's easy to see without doing any arithmetic that $R' > 0$), and $R' < 0$ if the point is above the line.

We can combine all this information into the following picture. We have drawn a number of velocity vectors along the lines where $R' = 0$ and $F' = 0$, with one or two others in each region.

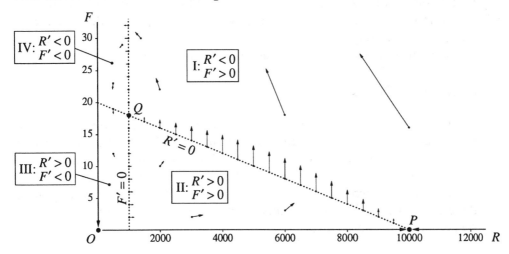

We see that the entire state space is divided into four regions:

1. Region I, above the line $F = 20 - .002R$ and to the right of the line $R = 1000$. Here $R' < 0$, and $F' > 0$, so all velocity vectors are pointing up and to the left.

2. Region II, below the line $F = 20 - .002R$ and to the right of the line $R = 1000$. Here $R' > 0$, and $F' > 0$, and all velocity vectors are pointing up and to the right.

3. Region III, below the line $F = 20 - .002R$ and to the left of the line $R = 1000$. Here $R' > 0$, and $F' < 0$, and all velocity vectors are pointing down and to the right.

4. Region IV, above the line $F = 20 - .002R$ and to the left of the line $R = 1000$. Here $R' < 0$, and $F' < 0$, and all velocity vectors are pointing down and to the left.

There are simple
trajectories: three are
just points . . .

Notice that this diagram makes it clear what the limit state of the spirals is: it is the point $Q = (1000, 18)$ where the line $R = 1000$ and the line $F = 20 - .002R$ intersect. Notice that at Q both $R' = 0$ and $F' = 0$, so that if we are ever at Q, we never leave—the point Q is a trajectory all by itself. The points $O = (0, 0)$ and $P = (10\,000, 0)$ are the two other such point trajectories. While the typical trajectory looks like a spiral coming into the point Q, note that this picture contains three other "special" trajectories in addition to the point trajectories:

- The F-axis for $F > 0$. The point $(0, 0)$ is *not* part of this trajectory.

- The portion of the R-axis with $0 < R < 1000$. Here the flow is to the right, toward the point P.

. . . and three are
straight line segments

- The portion of the R-axis with $1000 < R$. Flow is to the left, toward P, with movement being slower and slower as P is approached. Note that this is entirely separate from the preceding trajectory—you can't start at any point on one of them and get to any point on the other.

Equilibrium Points

There are different
kinds of equilibrium
points

The three points O, P, and Q in the previous figure—single points which are also trajectories—are called **equilibrium points** for the system. If the system is ever in such a state, it stays in it forever. Moreover, the system can't reach such a state from any other state (although it may be able to come very close). Nevertheless, the behavior of the system is not the same near the three points. If we zoom in on each of these points and draw some of the nearby trajectories, we get the following pictures:

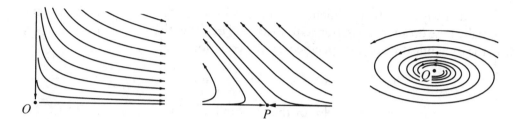

Points O and P look fairly similar—they would look even more alike if we crossed over into the negative R and negative F regions and included the trajectories there as well (impossible to do in the real world, but elementary in mathematics!). In both cases there is one direction from which trajectories come straight toward the point (in the case of O, this is the F-axis; for P, this is the R-axis), and one direction in which trajectories move directly away from the point (the R-axis in the case of point O; and the line of slope $-.0092$, as we'll see how to find this later, in the case of P). The remaining trajectories look sort of like hyperbolas asymptotic to these two lines. Equilibrium points of this sort are called **saddle points.** They are characterized by the property that there is exactly one direction along which the system can be displaced and still move back toward the equilibrium point. Displacements in any other direction get amplified, with the state eventually moving even further away.

Saddle point equilibrium

Point Q is quite different. If the state experiences a small displacement away from Q in any direction, over time it will move back toward Q. Such equilibrium points are called **attractors,** and Q is an example of a particular kind of attractor called a **spiral attractor.** In this example, Q is an attractor for almost the entire space—if we start with any point (R, F) with $R > 0$ and $F > 0$, the trajectory through (R, F) will eventually come arbitrarily close to Q and stay there. We will shortly see examples (see page 423, for instance) of attractors that draw from more limited portions of the state space.

Spiral equilibrium

For future reference, we define here the concept of **repellor** and **spiral repellor.** Their vector fields look just like those for the attractors, but with all the arrows reversed. If the state experiences a small displacement from a repellor, over time this displacement will increase. We will see examples of a repellor in section 8.3. It turns out that there is a relatively small number of kinds of equilibrium points that a system can have, and we will meet most of them in the next several examples. We will turn more systematically to the problem of identifying the kinds of equilibrium points in section 8.2.

Attractors and repellors

The Pendulum Revisited

In chapter 7 we analyzed the motion of a pendulum. Let's see how this analysis looks when translated into the language of state space. We first need to figure out what the appropriate coordinates are, which means deciding what information we need in order to specify the state of a pendulum. If you look back at the model in the last chapter, you will recall that the two variables we

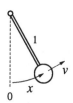

needed were the displacement x and the velocity v. Since x and v can potentially take on any values, our state space will be the entire x, v plane. As before, the dynamical system is specified by the equations

$$x' = v \qquad v' = -\sin x.$$

Here is what the vector field for this system looks like:

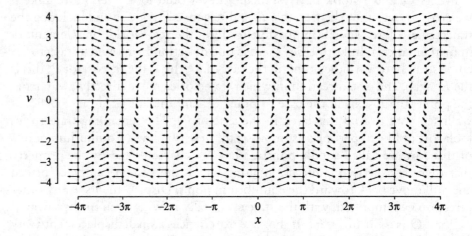

We have included in this diagram the lines where $v' = 0$ (the vertical lines at every multiple of π) and the line where $x' = 0$ (the horizontal line at $v = 0$). Note that the velocity vectors are horizontal on the lines corresponding to $v' = 0$ and are vertical on the line corresponding to $x' = 0$. The points where these two sets of lines intersect—all points of the form $(k\pi, 0)$ for k an integer—are the equilibrium points of the system. Let's sketch the phase portrait of this system to see more clearly what's going on:

The equilibrium points

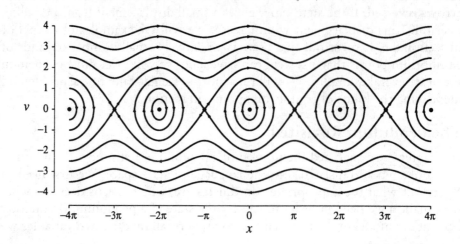

We see that there are several different kinds of trajectories:

- There are the wavy trajectories moving from left to right across the
  top of the state space. Note that for these trajectories the value of
  v is always positive, and x just keeps increasing. These trajectories
  correspond to the cases where the velocity is great enough that the
  pendulum can go over the top, continuing to loop around
  counterclockwise (since x is increasing and x is measured in a
  counterclockwise direction) forever. Notice that v takes on its minimum
  value when x is an odd multiple of π, which is what we would expect,
  since the pendulum is at the top of its arc then. Similarly, v takes on its
  maximum value at the bottom of its arc—x an even multiple of π.

  Trajectories from left to right in the state plane correspond to $v > 0$

- The wavy trajectories moving from right to left across the bottom
  are similar, except that v is always negative. This corresponds to the
  pendulum spinning around in a clockwise direction.

- There are the closed loops. Here x oscillates back and forth between
  some maximum and minimum value symmetrically placed about an even
  multiple of π. These trajectories correspond to a pendulum swinging
  back and forth. The fact that some are centered at x-values other than
  0 is due to the fact that the same position of the pendulum can be
  specified by an infinite number of values of x, all differing from each
  other by multiples of 2π.

  Oscillations of the pendulum correspond to closed loops in the state plane

- There are the equilibrium points $(k\pi, 0)$, with k an even integer. This
  corresponds to the pendulum hanging straight down. If we perturb
  the system to a state slightly away from such a point, the pendulum
  swings back and forth, and the corresponding trajectory loops around
  the equilibrium point forever. The system neither comes back to the
  equilibrium point—the condition for an attractor—nor does it go
  wandering off even further away—the condition for a repellor. Such an
  equilibrium point is called a **center.** Notice that it is neither an attractor
  nor a repellor; it is said to be a **neutral equilibrium.**

  A center: a neutral equilibrium

- There are the equilibrium points $(k\pi, 0)$, with k an odd integer,
  corresponding to the pendulum balanced vertically. These are saddle
  points—if we perturb the system slightly with *exactly* the right v-value
  for the given x-value, the system will move back toward the vertical
  position; any other combination, though, will cause the pendulum to
  spin around and around forever or to oscillate back and forth forever,
  depending on whether the v-value is greater than or less than the critical
  value.

- There are the trajectories connecting the saddle points. These correspond
  to cases where the pendulum has just enough velocity so that it keeps
  moving closer to the vertical position without either overshooting and
  spinning around, or coming to a stop and reversing direction. In fact,

A connected curve in the phase portrait may be composed of more than one trajectory

these trajectories divide the state space: on one side of such a trajectory are points corresponding to states where the pendulum will spin, and on the other side are points corresponding to states where the pendulum will swing back and forth. Note that the saddle points are *not* part of these trajectories, and that each arc between saddle points is a separate trajectory—you can't get from a point on one of them to a point on another.

First Integrals Again

In the case of the pendulum, we have another way of thinking about the trajectories. Recall that in section 7.3 we saw that, for any given initial conditions, the quantity $E = \frac{1}{2}v^2 + 1 - \cos x$ was constant over time. In the vocabulary of this chapter, if (x, v) is any state on the trajectory through (x_1, v_1), then it must be true that $\frac{1}{2}v^2 + 1 - \cos x = \frac{1}{2}v_1^2 + 1 - \cos x_1$. But this relation determines a curve in the x, v plane. We thus have an algebraic condition for each of the trajectories, which will depend on the initial values. In this example we could actually get an equation for each trajectory by first using the initial values to determine the energy $E = \frac{1}{2}v_1^2 + 1 - \cos x_1$. We could then solve for

First integrals can give equations of trajectories

v in terms of x by $v = \pm\sqrt{2(E - 1 + \cos x)}$ and plot the resulting function. (Whether we took the plus sign or the minus sign would depend on whether the pendulum was moving counterclockwise or clockwise.) Let's return to our previous sketch of the phase portrait and label some of the trajectories by their corresponding values of E:

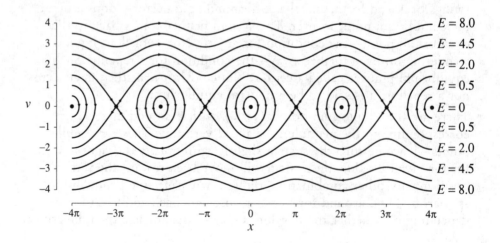

Note that for each value of $E \geq 0$ there is more than one trajectory having that value as its energy.

We can now characterize the different kinds of trajectories by their associated energy E:

- If $E > 2$, we get a trajectory extending from $x = \infty$ to $x = -\infty$ (or vice versa).
- If $0 < E < 2$, the trajectory is a closed loop.
- If $E = 0$, we get a neutral equilibrium point.
- If $E = 2$, we get either a saddle point equilibrium or a trajectory connecting two such saddle points.

A Model for the Acquisition of Immunity

One of the roles of mathematical modeling is to allow researchers to explore possible mechanisms to explain an observed phenomenon. As an example of this, consider the phenomenon of immunity: for many infections, particularly those due to viruses, once you've been exposed to the disease your body continues to produce high levels of antibodies to the disease for the rest of your life, even in the absence of any further stimulation from the virus.

A mathematical model can be used to think about the feasibility of a proposed explanation

> *A capsule summary of the immune response: Vertebrates have a wide variety of specialized cells called* lymphocytes *circulating in their blood streams and lymphatic systems at all times. Each lymphocyte has the ability to recognize and bind with a specific kind of invading organism. The invader is called an* antigen, *and the neutralizing molecules produced by the responding lymphocytes are called* antibodies. *Prior to infection, the concentration level of a particular antibody is typically so low as to be undetectable, but the appearance of the antigen causes the system to respond by producing large quantities of the appropriate antibody. If the body can continue to produce high levels of antibodies, it will be immune to reinfection.*

In their book *Infectious Diseases of Humans* (New York: Oxford University Press, 1991), Roy Anderson and Robert May propose the following model as a possible mechanism for how antibody levels are sustained. Suppose that there are two kinds of lymphocytes (called *effector cells*) whose densities at time t are denoted by $E_1(t)$ and $E_2(t)$, with the type 2 cells being the potential antibodies for the disease in question. They assume further that new cells of type i ($i = 1$ or $i = 2$) are produced by the bone marrow at constant rates Λ_i and they die at per capita rates of μ_i. They assume that each cell type is an antigen for the other—that is, contact with cell type 2 triggers cell type 1 to proliferate, and vice versa. They further assume that this proliferation response saturates to a maximum net rate which is dependent on the product of their respective densities. The following equations express this behavior:

$$dE_1/dt = \Lambda_1 - \mu_1 E_1 + a_1 E_1 E_2/(1 + b_1 E_1 E_2),$$
$$dE_2/dt = \Lambda_2 - \mu_2 E_2 + a_2 E_1 E_2/(1 + b_2 E_1 E_2).$$

Here the parameters Λ_i, μ_i, a_i, and b_i would have to be determined by experimental means. At this stage, though, when we are simply exploring to see if such a mechanism might account for the phenomenon of permanent immunity, we can try a range of values for the parameters to see how they affect the behavior of the model.

If we take values for the parameters in this equation of $\Lambda_1 = \Lambda_2 = 8000$, $\mu_1 = \mu_2 = 1000$, $a_1 = a_2 = 10$, and $b_1 = b_2 = 10^{-6}$, we get the following picture for the vector field:

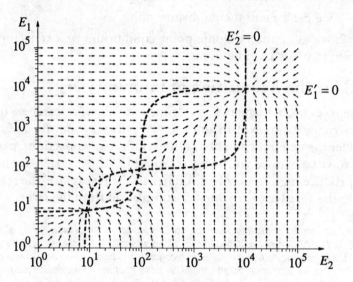

There are a couple of features to notice about this graph:

1. First, since the velocity vectors differ so much in their size, we have recorded only the direction of the velocity vectors, drawing all the arrows to be the same length. Thus we don't really show the vector field, but its close relative, the **direction field**. This is often a useful substitute.

Knowing the direction of the trajectories is often sufficient

2. Second, since the range of values we want to represent is so great, we have employed a common device from the sciences of plotting the values on a **log–log scale.** That is, we have plotted the values so that each interval spanning a power of 10—from 10^0 to 10^1, or from 10^3 to 10^4—gets the same space. This is equivalent to plotting the logarithms of the values on ordinary graph paper. This allows us to see effects that take place at different scales. If we hadn't done this, but had plotted this information on regular graph paper with the values running from 0 to 10^5, then some of our most interesting behavior—from 10^0 to 10^2—would be compressed into the lower left-hand corner of the graph, occupying only .001 of the vertical and horizontal scales.

How to express graphical information acting on several different orders of magnitude

We have included in the graph the two curves corresponding to all points satisfying $E_1' = 0$ and $E_2' = 0$ (note that these curves are *not* trajectories). These curves intersect at the three points $P_1 = (8.7689, 8.7689)$, $P_2 = (92.0869, 92.0869)$, and $P_3 = (9907.14, 9907.14)$, which are then the equilibrium points of this system. The points P_1 and P_3 appear to be attractors, while the point P_2

is a saddle point. In the next section (see page 432) we will see how to zoom in and look at the trajectories near each of these points to confirm this impression. Here is a picture of the phase portrait for this system.

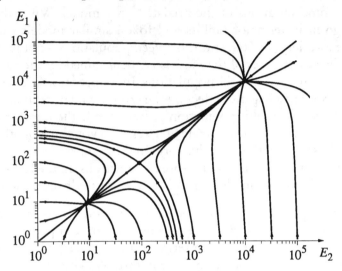

Note that neither P_1 nor P_3 is an attractor for the entire system. The **basin of attraction** for P_1 appears to be a region in the lower left of the graph, while the basin of attraction for P_3 is everything else. The boundary separating these two basins is formed by the two heavily shaded trajectories which come toward the point P_2 (since P_2 is a saddle point, there are only two such trajectories—every other trajectory eventually veers off and heads toward either P_1 or P_3).

We can now interpret this system in the following way. State P_1 represents the **virgin** or **resting state** of the system, with coordinate values on the order of magnitude of $E_i \approx \Lambda_i / \mu_i$, which would just be the steady-state values we would have if there were no interactions between the two kinds of cells (why?). Note that after small perturbations (i.e., anything roughly less than a 10-fold increase of type 1 or type 2 cells) from P_1, the system will settle back to this resting state.

Now, though, suppose a viral pathogen appears which possesses an antigen which is identical to that expressed by cell type 1. This has an effect equivalent to moving vertically in the E_1, E_2 plane to a state which is now in the basin for P_3. As a result, the system immediately starts producing large quantities of type 2 cells (which are antibodies for the virus) very rapidly, the virus is wiped out, and the system settles into a new state—the **immune state, P_3**—and remains there. There are now so many type 2 cells permanently floating around the body that no further infection by the viral pathogen is possible. The only way the system can be switched back to state P_1 is if some other agent, such as radiation therapy or infection with an HIV virus, for instance, kills off large numbers of *both* the type 1 and type 2 cells, moving the system back into the basin of attraction for P_1. Just killing off large numbers of one type of cell won't move the system back to state P_1—do you see why?

Exercises

Two Species Interactions

We look at some variations of the predator–prey model. While the original context is given in terms of rabbits and foxes, similar models can be constructed for a variety of interactions between populations—not just predator and prey. The key features of the models are determined by the nature of the **feedback structure** between the populations. In the predator–prey models, the number of foxes has a negative effect on the growth rate of rabbits—the more foxes, the slower the rabbit population grows—while the number of rabbits has a positive effect on the growth rate of foxes. Can you think of other pairs of quantities whose interaction is of this sort? In the first problem we will look at several different models for predator–prey interactions. In the following three problems we will look at models for other kinds of feedback structures.

1. Below are four predator–prey models. In each model all the letters other than R and F are constant parameters. You can perform a general analysis, giving your answers in terms of the unspecified parameters a, b, c, and so on, or, if you are more comfortable with specific values, you can perform the analysis using $a = .1$, $b = .005$, $c = .00004$, $d = g = .04$, $e = .001$, $f = .05$, $h = .004$, and $K = 10,000$. For each model you should carry out the following steps to sketch the vector field for the model in the first quadrant of the R, F plane. Compare your work with the steps that led up to the analysis of the vector field on page 415.

    • Write down in words a justification for each rate equation. Why is the model a reasonable one? What is it saying about the way rabbit and fox populations change?

    • Draw (in red) the set of points where $R' = 0$, and mark the regions where $R' > 0$ and $R' < 0$.

    • Draw (in green) the set of points where $F' = 0$, and mark the regions where $F' > 0$ and $F' < 0$.

    • Mark the equilibrium points. What color are they?

    • Sketch representative vectors of the vector field, and then sketch a couple of trajectories that follow these vectors. You might use a computer to verify your sketches.

    • On the basis of your sketches make a conjecture about the stability of the equilibrium points.

    a) The original Lotka–Volterra model, proposed independently in the mid-1920s by Lotka and Volterra. This model stimulated much of the subsequent development of mathematical population biology.

$$R' = aR - bRF$$
$$F' = cRF - dF.$$

b) The Leslie–Gower model

$$R' = aR - bRF$$

$$F' = \left(e - f\frac{F}{R}\right)F.$$

c) Leslie–Gower with carrying capacity for rabbits

$$R' = aR\left(1 - \frac{R}{K}\right) - bRF$$

$$F' = \left(e - f\frac{F}{R}\right)F.$$

d) Another combination

$$R' = aR\left(1 - \frac{R}{K}\right) - bRF$$

$$F' = cRF + gF - hF^2.$$

2. *Symbiosis and mutualism.* Many flowers cannot pollinate themselves; instead insects like bees transport pollen from one flower to another. For their part, bees collect nectar from flowers and make honey to feed new bees. This sort of feedback structure in which the presence of each element has a positive effect on the growth rate of the other is called **symbiosis** or **mutualism**. (There is a distinction made between these two interactions, but mathematically they are similar.) Here is a model: B is the number of bees per acre, measured in hundreds of bees, while C is the weight of clover per acre, in thousands of pounds. Assume time to be measured in months.

$$B' = .1(1 - .01B + .005C)B$$
$$C' = .03(1 + .04B - .1C)C.$$

a) Do these equations describe symbiosis? What terms account for symbiosis?

b) Each equation has a negative term in it. What aspect of reality is this term capturing?

c) Sketch the vector field for this system in the B, C plane. Find the equilibrium points, and mark them on your sketch.

d) Draw some trajectories on your sketch, and use them to determine the stability of the equilibrium points.

e) Suppose an acre of land has 10,000 pounds of clover on it, and a hive of 2,000 bees is introduced. [What are the values of $B(0)$ and $C(0)$ in this case?] What happens? Answer this question both by drawing a trajectory and by describing the situation in words.

f) Let a couple of years pass after the situation in part (e) has stabilized. Suppose the field is now mowed so only 2,000 pounds of clover remain on it. The bee–clover system is now at what point on the B, C plane? What happens now? Does the bee population drop? Does it stay down, or does it recover? Does the clover grow back?

g) This scenario is an alternative to part (f); it is also played out a couple of years after the situation in part (e) has stabilized. Suppose an insecticide applied to the clover field kills two-thirds of the bees. The insecticide is then washed away by rain, leaving the remaining bees unaffected. What happens?

3. *Competition.* As a third kind of feedback structure, consider two species X and Y competing for the same food or territory. In this case each has a negative impact on the growth rate of the other. If we let x and y be the number of individuals of species X and Y, respectively, then the larger y is, the less rapidly x increases—and vice versa. Here is a specific model to consider:

$$x' = .15(1 - .005x - .010y)x$$
$$y' = .03(1 - .004x - .005y)y .$$

The term $-.010y$ in the first equation shows explicitly how an increase in y reduces the growth rate x'. In the second equation $-.004x$ tells us how much X affects the growth of Y. Notice that Y affects X more strongly than X affects Y.

If x and y are both small, then the parenthetical terms are approximately 1, so the equations reduce to

$$x' = .15x$$
$$y' = .03y .$$

Thus, in these circumstances X's per capita growth rate is five times as large as Y's.

In the competition for resources, will the growth rate advantage permit X to win the competition and drive out Y, or will the more adverse effect that Y has on the growth of X permit Y to win? Perhaps the two species will both survive and share the resources for which they compete. The purpose of this exercise is to decide these questions.

a) Suppose we start with $x = y = 10$. What are the two growth rates x' and y'? Is x' about five times as large as y' in this case? What are the approximate values of x and y after .5 time units have elapsed? Is X growing significantly more rapidly than Y?

*b) How many equilibrium points does this system have, and where are they?

c) Sketch *and label* in the x, y plane the points where $x' = 0$ and where $y' = 0$. The vector field typically points in one of four directions:

up and to the right; up and to the left; down and to the right; or, down and to the left. Indicate on your sketch the zones where these different directions occur and draw representative vectors in each zone.

(Note: Only three of the zones actually occur in the first quadrant; no vectors there point down and to the right.)

d) Sketch on the x, y plane the trajectory that starts at the point $(x, y) = (10, 10)$. Now answer the question: What happens to a population of 10 individuals each from species X and from species Y? In particular, does X gain an early lead? Does X keep its lead? Does either X or Y eventually vanish?

e) Is the outcome of part (d) typical, or is it not? Try several other starting points: $(x, y) = (150, 25)$, $(300, 10)$, $(200, 200)$, $(50, 200)$. Do these starting points lead to the same *eventual* outcome, or are there different outcomes? Use a computer to confirm your analysis.

f) Describe the type of each equilibrium point you found in part (a). Is any equilibrium an attractor?

4. *Fairer competition.* The vector field in exercise 3 shows that species X didn't have a chance: all trajectories in the first quadrant flow to the equilibrium at $(0, 200)$. We can attribute this to the strength of the adverse effect Y has on X—that is, to the size of the term $-.010y$ in the first equation when compared to the corresponding term $-.004x$ in the second equation. Let's try to give X a better chance by increasing this term to $-.006x$. The equations become

$$x' = .15(1 - .005x - .010y)x$$
$$y' = .03(1 - .006x - .005y)y .$$

*a) Sketch and label in the x, y plane the points where $x' = 0$ and where $y' = 0$. Sketch representative vectors for the vector field. Mark all equilibrium points.

b) What happens to a population consisting of 10 individuals each from species X and species Y? Is the outcome significantly different from what it was in exercise 3? To get quantitatively precise results you will probably find a computer helpful.

*c) What happens to a population consisting of 150 individuals from species X and 25 individuals from species Y? Is this outcome significantly different from what it was in exercise 3?

*d) Is it possible for X and Y to coexist? What must x and y be? Is that coexistence *stable;* that is, if x and y are changed slightly, will the original values be restored?

e) Sometimes X wins the competition, sometimes Y. Mark in the x, y plane the dividing line between those starting points that lead to X winning and those that lead to Y winning.

*f) Identify the type of each equilibrium point.

g) An often-articulated concept in ecology is the *principle of competitive exclusion,* which states that you can't have a stable situation in which two species compete for the same resource—one of them will eventually crowd out the other. Is the model you've been exploring in this problem consistent with such a principle?

5. *More on the Lotka–Volterra model.* The Lotka–Volterra model

$$R' = aR - bR\,F$$
$$F' = cR\,F - dF,$$

while it had a major impact on the development of mathematical biology, was found to be flawed in several important ways. The chief problem is that the equilibrium point $(d/c, a/b)$ is a neutral equilibrium point—given any starting state, the system would follow a closed trajectory. This in itself was all right and, in fact, stimulated a great many important investigations on whether or not cycles were an intrinsic feature of many populations. The difficulty was that there were so many possible closed trajectories—which one the system followed depended on where it started. A second difficulty, related to the first, is that there is a first integral for the Lotka–Volterra model. What is seen as a virtue in a physical system like the pendulum—since it is equivalent to the conservation of energy—is unrealistic in an ecological system, where there are almost certainly too many outside forces at work for any quantity to be conserved there. In the following exercises we will explore some of these behaviors. As before, you can either perform a general analysis of the model or use the specific parameter values $a = .1$, $b = .005$, $c = .00004$, and $d = .04$.

a) Sketch the vector field, together with some typical trajectories, in the rest of the R, F plane, including negative values. What happens to any trajectory starting at a state with a negative R or negative F value?

b) For this exercise you will need to go back to a computer program that implements Euler's method of approximating the trajectory by drawing a straight line segment from a point in the direction indicated by the velocity vector (commercial packages use fancier routines which accommodate for the kind of phenomena you are about to see!). Using the specific values for a, b, c, and d suggested above, starting from the point $(2000, 10)$ in the R, F plane, and using a time step $\Delta t = 1$, draw the first 500 segments of Euler's approximation to the trajectory. What does the trajectory look like? Would you think the trajectory was a closed loop on the basis of this result? How small does Δt have to be before the trajectory looks like it closes? Can you explain this phenomenon?

c) Using the same values for a, b, c, and d as in the preceding part, start at the point $(2000, 1)$ and use $\Delta t = 2$. This time calculate the first 1000 segments of Euler's approximation. What happens? (Your computer will probably give you some sort of overflow message.) Can you explain this? [Think about your answer to part (a).]

d) *Getting a first integral for the system.* Show that the Lotka–Volterra equations imply that

$$\frac{R'}{R}(cR - d) = \frac{F'}{F}(a - bF).$$

Integrate this equation and show that the expression

$$cR + bF - d \ln R - a \ln F$$

must be a constant for all points on a given trajectory. If we know one point on the trajectory (such as the starting point), we can evaluate the constant.

e) Show that the function $f(R) = cR - d \ln R$ is decreasing for $0 < R < d/c$ and is increasing for $d/c < R < \infty$. Hence argue that for any given value of F there are at most two values of R giving the same value for the expression $cR + bF - d \ln R - a \ln F$. Hence conclude that the trajectories for the Lotka–Volterra equations can't be spirals, but must then be closed loops.

The Pendulum

6. Suppose instead of an idealized frictionless pendulum, we wanted to model a pendulum that "ran down." One approach we might try is to throw in a term for air resistance. Let's see what happens when we add a term to the expression for v' which suggests that there is a drag effect which is proportional to the value of v—the larger v is, the greater will be the drag. Here are equations that do this:

$$x' = v$$
$$v' = -\sin x - .1v.$$

Perform a vector field analysis of this model, indicating the regions where the velocity vectors are pointing in the various combinations of up, down, right, and left. Try sketching in some trajectories. Where are the equilibrium points? What kinds are they?

The Anderson–May Model

7. Consider $dE_1/dt = \Lambda_1 - \mu_1 E_1 + a_1 E_1 E_2/(1 + b_1 E_1 E_2)$. For what values of E_1 is it possible to find a value for E_2 making $dE_1/dt = 0$?

Express your answer in terms of the parameters Λ_1, μ_1, a_1, and b_1. Is your answer consistent with the graph on page 422?

8. In the same book, *Infectious Diseases of Humans*, containing the previous model, Anderson and May propose another model to explain the acquisition of (apparently) permanent immunity. In this model there is just the virus and the lymphocyte cells (effector cells) that kill the virus. We denote their populations at time t by $V(t)$ and $E(t)$. They propose the model

$$\frac{dE}{dt} = \Lambda - \mu E + \varepsilon V E$$

$$\frac{dV}{dt} = rV - \sigma V E$$

Here Λ is the (constant) rate of background production of the lymphocytes by the bone marrow, μ is the per capita death rate of such cells, and r is the intrinsic growth rate of the virus if none of the specific lymphocytes was present. Both the increased production of the lymphocytes and the death of the virus are assumed to proceed at rates proportional to the number of their interactions, determined by their product.

a) Show that in the absence of any virus, the effector cells have a stable equilibrium of Λ/μ.

b) Perform a state space analysis of the vector field. Note that there will be two very different cases, depending on whether $\Lambda/\mu > r/\sigma$ or $\Lambda/\mu < r/\sigma$. In each case say what you can about the equilibrium points and the expected long-term behavior of the system.

*c) Using parameter values $\Lambda = 1$, $\mu = r = .5$, and $\varepsilon = \sigma = .01$, and starting values $E = V = 1$, find the resulting trajectory.

*d) How long, approximately, will it take the spiral to make one revolution? If this time, call it T, is roughly the same length as the lifetime of the infected individual, what will appear to be happening? It might help to plot both E and V as functions of time over the interval $[0, T]$.

> ## 8.2 LOCAL BEHAVIOR OF DYNAMICAL SYSTEMS

A Microscopic View

Phase portraits under the microscope

One of the themes of this book has been the concept of the "microscope." When we zoom in on some part of a geometrical object, the structure typically becomes much simpler. In chapter 3 we used this approach to think about the behavior of functions. In this section we will use the same idea to analyze the behavior of a vector field and its phase portrait. There are two parts to this process:

1. We shift the origin of the coordinate system to center on the point we are interested in—we **localize**—and

2. We approximate the vector field and its phase portrait by linear approximations—we **linearize**.

To get a feel for how this works, let's go back and look at exercise 4 on page 427 of the previous section. There we had two species X and Y competing for the same food source. We modeled the dynamics of this system by the equations

$$x' = .15(1 - .005x - .010y)x$$
$$y' = .03(1 - .006x - .005y)y.$$

The phase portrait for this system looks like

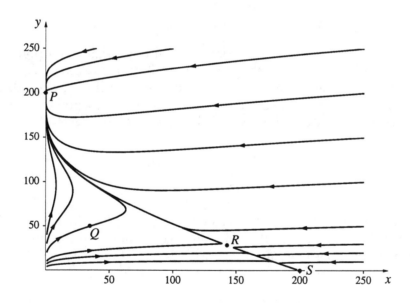

The three equilibrium points—$P = (0, 200)$, $R = (\frac{1000}{7}, \frac{200}{7})$, and $S = (0, 200)$—are indicated, together with a generic point $Q = (35, 50)$. Note that P and S are attractors and that R is a saddle point. As was the case with the Anderson–May model, there is a trajectory flowing away from R to each of the attractors. There are also two trajectories (not shown) flowing directly toward R and forming the boundary between the basins of attraction for P and S. We will see how to construct this boundary shortly (page 440).

Let's first zoom in on the point Q and see what the phase portrait looks like there. If we take the region ± 1 unit on either side of Q, we get the following phase portrait:

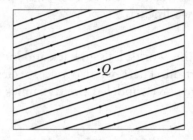

At this level, all the trajectories appear to be parallel straight lines. How could we have anticipated this picture? The first step in analyzing this phase portrait is to observe that since we are interested in its behavior near $Q = (35, 50)$, instead of working with the variables $x(t)$ and $y(t)$, we introduce new variables $r(t)$ and $s(t)$ which measure how far we are from Q:

$$r(t) = x(t) - 35$$
$$s(t) = y(t) - 50.$$

Shifting the origin

The effect of this transformation is simply to shift the origin to the point Q—the location of every point in the plane is now measured relative to Q rather than to the x, y origin. A point is close to the point Q if its r, s coordinates are small. Further, if we are given the r, s coordinates of a point, we can always recover the x, y coordinates, and vice versa—we can transform in either direction:

$$r = x - 35 \Longleftrightarrow x = r + 35$$
$$s = y - 50 \Longleftrightarrow y = s + 50.$$

Next, note that $r'(t) = x'(t)$ and $s'(t) = y'(t)$ so that the new variables change at the same rates as the old ones. We can now express our original differential equations in terms of the variables r and s by replacing x' by r', x by $r + 35$, y' by s', and y by $s + 50$. When we do this, we get

$$r' = .15(1 - .005(r + 35) - .010(s + 50))(r + 35)$$
$$= 1.70625 + .0225r - .0525s - .00075r^2 - .0015rs$$
$$s' = .03(1 - .006(r + 35) - .005(s + 50))(s + 50)$$
$$= .81 - .009r + .0087s - .00018rs - .00015s^2.$$

What we have accomplished by this is to transform a problem about trajectories near the point $(35, 50)$ in the x, y plane into a problem about trajectories near the origin in the r, s plane—we have **localized** the problem to the point we are interested in.

The second step comes in analyzing the r, s system: since we are only interested in its behavior near the origin, we will be looking at values of r and s that are small. Under these circumstances, the contributions of the constant terms will far outweigh the contributions of any of the terms involving r and s. For instance, in our current example we are looking at a window that is ± 1 unit wide and ± 1 unit high around Q. In this window, the terms involving r or s are at most 3% of the constant term in the case of r', and a little over 1% in the case of s'. If we had used a smaller window, the contributions of the nonconstant terms would be even less significant. This means that *near the r, s origin* the vector field for this system is well-approximated by the behavior of the related **constant linear system:**

Near an ordinary point, a vector field is almost constant

$$r' = 1.70625$$
$$s' = .81.$$

Note that 1.70625 and .81 are just the values of x' and y' at Q.

In this linearized system, any change Δt in the time produces a change $\Delta r = 1.70625\Delta t$ in r, and a change $\Delta s = .81\Delta t$ in s. Thus the velocity vectors in the vector field near Q would all have the same length and would be pointing in the same direction, with slope $\Delta s / \Delta r = .81 / 1.70625 = .4747$. This in turn means that near Q all trajectories have the same slope and are traversed at the same speed.

Near an ordinary point all trajectories look the same

We would see a similar picture—a family of parallel straight lines—whenever we zoom in on the phase portrait near any other ordinary (i.e., nonequilibrium) point (x_*, y_*). The vector field near such a point can always be approximated by a constant linear system of the form

$$r' = e$$
$$s' = f,$$

where e and f are the values of x' and y' at (x_*, y_*). The trajectories of this approximating linear system will be lines of slope f / e.

Near an equilibrium point, the picture is more complicated. No matter how far in we zoom, the phase portrait never looks like a family of straight lines. For instance, here's what the picture looks like when we zoom in on $R = (1000/7, 200/7) \approx (142.857, 28.571)$:

Equilibrium points are different

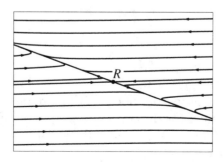

If we zoomed in to a window one-hundredth the size of this one, the picture would be indistinguishable from this one.

Here we see four trajectories that look almost like straight lines—two coming directly toward R and two going directly away. All the other trajectories appear to be asymptotic to these two sets. On page 439 in the next section you will see how to find the equations of these asymptotes.

What happens when we linearize the vector field at R? As before, we first shift the origin so that it is centered at R by changing to coordinates r and s, where

$$r(t) = x(t) - 1000/7$$
$$s(t) = y(t) - 200/7.$$

When we then write the differential equations in terms of r and s, we get as before that $x' = r'$ and $y' = s'$ and

$$r' = .15(1 - .005(r + 1000/7) - .010(s + 200/7))(r + 1000/7)$$
$$= -.107143r - .214286s - .00075r^2 - .0015rs$$
$$s' = .03(1 - .006(r + 1000/7) - .005(s + 200/7))(s + 200/7)$$
$$= -.00514286r - .00428571s - .00018rs - .00015s^2.$$

The vector field is not constant locally

This time, though, the constant term in the expression for both r' and s' is 0. This is because the point R was an equilibrium point, which meant that both x' and y', and hence r' and s', were 0 there. If we are considering only small values of r and s, though, say much smaller than 1, then the terms involving r^2 or s^2 or rs will be much smaller than the terms involving r and s alone. We

It is linear, however

can therefore simplify our equations at R by taking only the first powers of r and s, getting for the linearized system

$$r' = -.107143r - .214286s$$
$$s' = -.00514286r - .00428571s.$$

In a similar fashion we could hope to explore the behavior of any other dynamical system about any of its equilibrium points by approximating the vector field there by a linear system of the form

$$r' = ar + bs$$
$$s' = cr + ds,$$

for suitable constants a, b, c, and d.

We will see in section 8.3 how to use this linearized form of the vector field to discover many of the properties of equilibrium points.

How can we find values for the constants a, b, c, and d? If the differential equations specifying the rates of change of the variables are polynomials, then we can proceed as above:

- Shift the origin to the point we're interested in.
- Express the rate equations in terms of the new local variables.
- Throw away all the terms except the first-degree terms.

This process requires some fairly tedious algebra. Moreover, what if the differential equations are not polynomials? Suppose, for instance, we wanted to study the local behavior of the Anderson–May model (page 422) at the saddle point $P_2 = (92.0869, 92.0869)$. Note that the differential equations are of the form

$$dE_1/dt = f_1(E_1, E_2)$$
$$dE_2/dt = f_2(E_1, E_2),$$

where f_1 and f_2 are the functions given in the text. But f_1 and f_2 are just functions, and we learned in chapter 3 how to construct locally linear approximations to them. This was, in fact, how we defined derivatives in the first place. Thus if E_1 changes by a small amount $\Delta E_1 = E_1 - 92.0869$, the function f_i will change by approximately $\partial f_i / \partial E_1 \times \Delta E_1$. Similarly, a small change $\Delta E_2 = E_2 - 92.0869$ will produce a change of approximately $\partial f_i / \partial E_2 \times \Delta E_2$ in the function f_i. The total change in the function f_i can then be approximated by the sum of these changes:

> To linearize a vector field, linearize the functions that determine it

$$\Delta f_1(E_1, E_2) \approx \frac{\partial f_1}{\partial E_1} \Delta E_1 + \frac{\partial f_1}{\partial E_2} \Delta E_2$$

$$\Delta f_2(E_1, E_2) \approx \frac{\partial f_2}{\partial E_1} \Delta E_1 + \frac{\partial f_2}{\partial E_2} \Delta E_2.$$

But since P_2 is an equilibrium point, we have by definition that f_1 and f_2 are both zero there, so $\Delta f_i(E_1, E_2) = f_i(E_1, E_2) - f_i(P_2)$ is just $f_i(E_1, E_2)$. Further, if you look closely you will see that the quantity $\Delta E_1 = E_1 - 92.0869$ is identical with what we have been calling the local coordinate r, and $\Delta E_2 = E_2 - 92.0869$ is just the other local coordinate s. Thus, since $E_1' = r'$ and $E_2' = s'$, we have

> The general form for the local linearization at an equilibrium point...

$$r' = \frac{\partial f_1}{\partial E_1} r + \frac{\partial f_1}{\partial E_2} s$$

$$s' = \frac{\partial f_2}{\partial E_1} r + \frac{\partial f_2}{\partial E_2} s,$$

where the partial derivatives are evaluated at P_2. Notice that there is nothing in this expression that is specific to this particular problem. The local linearization of any vector field at any equilibrium point will be in this form.

Finally, using the values given for the different parameters back on page 422, we can evaluate all the partial derivatives to get the specific local linearization for the point P_2:

$$r' = -94.5525\, r + 905.448\, s$$
$$s' = 905.448\, r - 94.5525\, s.$$

We will see in the next section how knowing this form will allow us to find the boundary between the two basins of attraction.

...and at a generic point

For completeness, let's remind ourselves of what the local linearization would look like at a nonequilibrium point in the current formulation. The result is immediate and simple, using the analysis we used before. If Q is a generic point, then the local linearization consists of parallel lines, whose slopes are given by the constant rate equations

$$r' = f_1(Q)$$
$$s' = f_2(Q).$$

Exercises

1. Find the local linearizations at all the equilibrium points in exercises 2–4 at the end of the previous section.

2. **a)** Show that the Lotka–Volterra equations

$$R' = aR - bRF$$
$$F' = cRF - dF$$

have an equilibrium point at $(d/c, a/b)$.

   *b) What is the local linearization there?

   c) What is a striking feature of this linearization, and what is its physical significance?

   d) The trajectories for the local linearizations turn out to be ellipses. If r and f are the local variables, find constants α and β such that the expression $\alpha r^2 + \beta f^2$ is constant on any trajectory.

3. Find the local linearization at the point P_1 in the Anderson–May model for the acquisition of immunity discussed in the last section, using the parameter values given in the text on page 422.

4. Go back to the second Anderson–May model analyzed in exercise 8 of the previous section (page 430). Using the parameter values given in part (c), find the local linearizations at all equilibrium points.

> ## 8.3   A TAXONOMY OF EQUILIBRIUM POINTS

An intuitive classification of equilibrium points

In the exercises and examples we have seen so far in this chapter, there have been several kinds of trajectories near equilibrium points: spirals toward and spirals away from the equilibrium, closed loops about the equilibrium, trajectories that looked vaguely like hyperbolas, and trajectories that seemed to arc more or less directly into or away from the equilibrium. It turns out that this rough classification covers virtually all the equilibrium behaviors we might encounter in a two-dimensional state space. There are many ways to demonstrate this, but we can accomplish almost everything with a couple of simple insights. We begin with a summary of the different kinds of equilibrium points, then turn to the question of devising ways to figure out from the equations what kind we are dealing with.

Suppose, then, that we are studying a two-dimensional dynamical system and that we have linearized the system at an equilibrium point. The point is either an attractor, repellor, saddle point, or neutral point. Attractors and repellors can be further subdivided according to whether they have one or two straight line trajectories, or whether their trajectories are spirals. Note that any attractor can be converted into a repellor simply by reversing the arrows, and vice versa. (How do you accomplish this arrow reversal at the level of the defining differential equations?) If you reverse all the arrows at a saddle point, you get another saddle point. If you reverse the arrows at a neutral point, you get the same closed loops, but they are traversed in the opposite direction.

Here, then, is a listing of all the kinds of equilibrium points. There are five generic types. (*Generic* here means "general"; if you generate a random equilibrium point, it will almost certainly be one of these.) They are most easily categorized by whether or not they have **fixed line** trajectories—that is, trajectories which are straight lines going directly toward or directly away from the equilibrium point.

The existence or not of straight line trajectories and how to find them when they do exist is an instance of the so-called eigenvector *problem. Analogous problems occur elsewhere in many parts of mathematics, physics, and even population biology. Being able to find such eigenvectors efficiently is an important problem in computational mathematics.*

Nodes.   Two pairs of fixed lines, all trajectories flowing toward the equilibrium (attractors) or away from it (repellors).

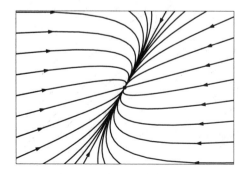
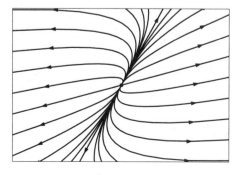

Spirals.   No fixed lines, all trajectories spiraling toward the equilibrium (attractors) or away from it (repellors).

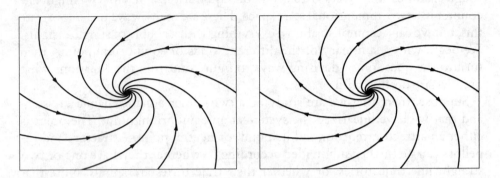

Saddle point.   Two pairs of fixed lines, with the flow along one pair being toward the equilibrium, and the flow along the other pair away from it. All other trajectories are asymptotic to these lines.

In addition to these five generic cases, there are three more types that arise under more specialized conditions:

Special nodes.   One pair of fixed lines, all trajectories flowing toward the equilibrium (attractors) or away from it (repellors).

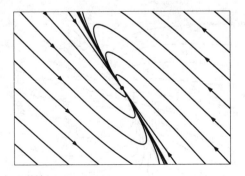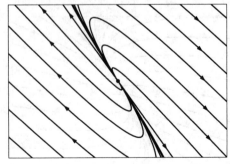

Center.   No fixed lines, all trajectories flowing around the equilibrium in closed loops.

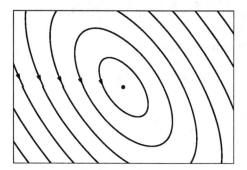

Except for a variety of highly specialized (or *degenerate,* in mathematical terminology) cases, examples of which are given in the exercises, the region near every equilibrium point will look like one of the above (although the exact shape may vary).

Clearly, it would be helpful to have an efficient way to determine whether or not fixed lines exist, and what their equations are if they do.

Straight Line Trajectories

Given a dynamical system

$$r' = ar + bs$$
$$s' = cr + ds,$$

how can we tell whether or not it has any straight line trajectories? Note that the line $s = mr$ will be a trajectory for this system provided the slope of the line—namely, m—equals the slope of the vector field at every point (r, s) on the line. But the slope of the vector field at any point (r, s) is just s'/r', which in turn is equal to $(cr + ds)/(ar + bs)$. Since every point on the line of slope m is of the form (r, mr), what we are really asking, then, is whether there are any values of m which satisfy the equation

The condition for a fixed line trajectory

$$m = \frac{cr + dmr}{ar + bmr} = \frac{c + dm}{a + bm}.$$

To see how this works, let's return to the example of two competing species which we last looked at on page 431. There we zoomed in on the saddle point $R = (1000/7, 200/7)$ and found that the local linear approximation was

$$r' = -.1071r - .2143s$$
$$s' = -.0051r - .0043s.$$

If this system has a straight line trajectory of slope m, then m must satisfy

$$m = \frac{-.0051 - .0043m}{-.1071 - .2143m},$$

which leads to the quadratic equation

$$.2143m^2 + .1028m - .0051 = 0,$$

which has roots

$$m = .0454 \quad \text{and} \quad m = -.5250.$$

Thus the lines $s = .0454r$ and $s = -.5250r$ are trajectories of the linear system. To be more exact, each of these lines is made up of three distinct trajectories: the portion of the line consisting of all points with $r > 0$, the portion with $r < 0$, and the origin (which is the saddle point R) by itself, which is always a trajectory in any linear system. To see whether flow along these trajectories is toward the origin or away from it, we could look to see where the lines lie in the state plane. It is just as simple, though, to try a test point. For instance, a typical point on the line $s = .0454r$ is $(1, .0454)$. When we substitute these values into the original rate equations, we find that

$$r' = -.1071 \times 1 - .2143 \times .0454$$

$$s' = -.0051 \times 1 - .0043 \times .0454.$$

We don't even need to do the arithmetic to be able to tell that both r' and s' are negative at this point; hence both r and s are decreasing, which means that on the line of slope .0454 movement is toward the origin. Similarly, on the line of slope $-.5250$ the flow is away from the origin. Finally, it turns out (as is the case with every linear system with straight line trajectories) that every other trajectory is asymptotic to these lines.

The crux of this approach was the use of the quadratic formula. Of course, it may happen—and we will see examples in the exercises—that when we try the same approach on another system, we find there are no real roots to the equation. This means that there are no fixed lines, so that trajectories must be spirals or closed loops.

Attractors and Basins of Attraction

One byproduct of the analysis in the previous section is that it gives us a technique for sketching the boundary separating two basins of attraction. Let's continue with the previous example to illustrate how this is done. We observed that the boundary between the two basins was formed by the two trajectories coming directly into the saddle point R between the two attractors P and S.

We have just seen that near R these two trajectories looked like the straight line of slope .0454. We can therefore take a point on this line on each side of R and run the system backward (if we go forward, we simply approach R) in time to reconstruct the trajectories, and hence get the boundary of the basins of attraction.

Exercises

1. In this exercise we look at a number of different linear systems to see what kinds of trajectories we get. In each case you should sketch the trajectories. Do this as before by first identifying the regions in the plane where $r' = 0, r' > 0$, and $r' < 0$, and similarly for s'. Then sketch trajectories consistent with this information. You might want to use a graphing program to check any answer you're unsure of.
   a) $r' = 4r + s$ $s' = 2r + 3s.$
   b) $r' = 4r + s$ $s' = -2r + 3s.$
   c) $r' = 2r + 3s$ $s' = 4r + s.$
   d) $r' = -4r + 4s$ $s' = 2r + s.$
   e) $r' = -.4r - 4s$ $s' = 2r - .5s.$
   f) $r' = -.4r - 4s$ $s' = 2r + .4s.$
   g) Make up and analyze four more linear systems.

2. If you start with a given linear system and consider the related system in which all the coefficients are four times as big, how do the trajectories change?

3. If you start with a given linear system and consider the related system in which all the coefficients have their signs reversed, how do the trajectories change?

4.*a) Use the quadratic formula to find the general solution to the equation

$$m = \frac{c + dm}{a + bm}.$$

   b) In exercises 1, 3, and 4 in the previous section you found local linearizations at the equilibrium points of a number of examples discussed earlier. Determine which of these have straight line trajectories and which do not. For those that do, find the equations of the lines and determine for each line whether the flow is toward the origin or away from it.

   c) What is the general condition for a linear dynamical system to have straight line trajectories?

5. Make up a system that has the lines of slope ± 1 as trajectories.

6. What is the condition for a system to have exactly one fixed line? Construct a couple of systems that have only one fixed line and sketch their phase portraits.

7. *Degeneracy.* The analysis developed in this section implicitly assumed that in the local linearization at least one of the coefficients in each of the expressions for r' and s' was nonzero. If this is not true, then many more possibilities open up. The following two systems have the origin as their only equilibrium point. In each case, write down the local linearization and draw in the trajectory pattern for the linearized system. Notice that the linearized systems have more than one equilibrium point. Then do the standard phase plane analysis for the original system—identify the regions in the plane where $r' = 0$ and where $s' = 0$, and specify what the direction field is doing in the rest of the plane, as usual. Sketch in some typical trajectories. Comment on the connections between the linearized and unlinearized forms.

   a) $r' = r^2$, $s' = -s$. You should see a sort of hybrid between a saddle point and an attractor here.

   b) $r' = r^2 + s^2$, $s' = r$.

8. a) Use the technique presented at the end of this section (page 440) to graph the boundary between the two basins of attraction.

   b) In the same way, construct the boundary between the two basins in the competing species model we've been discussing—exercise 4 on page 427.

Distance from the Origin

Another way to distinguish between different kinds of trajectories is to see how their distance from the origin varies over time. For saddle points the distance will first decrease and then increase. For spiral attractors and nodal attractors the distance may be always decreasing, or it may fluctuate, depending on how flat the trajectory is.

Again, let's look at a general linear system

$$r' = ar + bs$$
$$s' = cr + ds.$$

Consider the system moving along some trajectory in r, s space. At time t it will be at a point $(r(t), s(t))$, situated at a distance $d(t) = \sqrt{r(t)^2 + s(t)^2}$. We would like to know how the function $d(t)$ behaves. Is it always increasing? Always decreasing? Or does it have local maxima and minima? To answer this we need to know if $d'(t)$ is ever $= 0$, or if it is always positive or always negative. We can simplify our calculations if we look at the square of the distance: $D(t) = d(t)^2 = r(t)^2 + s(t)^2$. The function D will be increasing and decreasing at exactly the same points as the function d, and it's easier to work with.

9. a) Show that

$$D'(t) = 2r(t)r'(t) + 2s(t)s'(t)$$
$$= 2[r(ar + bs) + s(cr + ds)]$$
$$= 2[ar^2 + (b + c)rs + ds^2].$$

b) Show that if we look at points on the line of slope m, so that $s = mr$, we will have $D'(t) = 0$ there if and only if

$$a + (b + c)m + dm^2 = 0.$$

c) Use the binomial theorem to conclude that this happens precisely where

$$m = \frac{-(b + c) \pm \sqrt{(b + c)^2 - 4ad}}{2d}.$$

d) Show in particular, if $(b + c)^2 - 4ad < 0$, there are no solutions to $D'(t) = 0$, and the distance must always be strictly increasing along all trajectories, or strictly decreasing along all trajectories.

e) Return to the example

$$r' = -.1071r - .2143s$$
$$s' = -.0051r - .0043s,$$

and determine the point where each trajectory passes closest to the origin. These points all lie on two lines. Find the equations of these two lines. These lines will not be trajectories themselves. However, the "vertices" of all the trajectories will lie on them.

10. Choose four of the exercises in the first part of this section and analyze them to see where (and whether) trajectories have a closest approach to the origin.

11. a) Use the results of this section to construct a dynamical system whose trajectories are spirals that are always moving away from the origin.

b) Use the results to construct a dynamical system whose trajectories are flattened spirals, so that the distance from the origin, while increasing overall, has local maxima and minima.

*12. It turns out that trajectories which form closed loops should really be considered as a special kind of spiral. In fact, a flattened spiral will close up precisely when the two directions in which the distance is a maximum or minimum are perpendicular to each other. Express this as a condition on the coefficients $a, b, c,$ and d in the dynamical system.

13. Write down the equations of some dynamical systems that will have closed orbits.

▷ 8.4 LIMIT CYCLES

With this analysis of the behavior of vector fields near equilibrium points, we now know most of the possibilities for the long-term behavior of trajectories. The one important phenomenon we haven't discussed is **limit cycles.** To see an example of this, let's return to May's predator–prey model which we first encountered in chapter 4. If $x(t)$ and $y(t)$ are the prey and predator populations, respectively, at time t, then the general form of May's model is

$$x' = ax\left(1 - \frac{x}{b}\right) - \frac{cxy}{x + d}$$

$$y' = ey\left(1 - \frac{y}{fx}\right).$$

The parameters a, b, c, d, e, and f are all positive.

Using parameter values of $a = .6, b = 10, c = .5, d = 1, e = .1$, and $f = 2$, let's take several different starting values and sketch the resulting trajectories. Here's what we find:

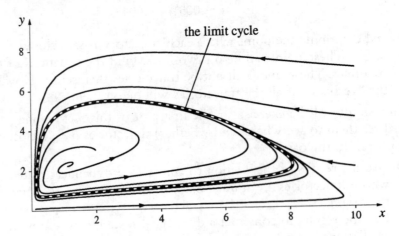

Notice that no matter where we start, the trajectory is apparently always drawn to the closed loop shown as a dashed curve above. This loop is an example of an **attracting limit cycle.** As usual, we could reverse all the arrows in our vector field, in which case this example would be converted to a **repelling limit cycle.**

A limit cycle is very different from the kind of behavior we saw in the neighborhood of a neutral equilibrium point called a center. Around a center there is a closed loop trajectory through every point: displace the state slightly, and it would move happily along the new loop. If the state is on an attracting limit cycle, though, and you displace it, it will move back toward the cycle it started from. For this reason limit cycles make very good models for cyclic behavior, whether it is in the firing of neurons or population cycles of mammals.

Limit cycles give models for cyclic behavior

The size of the limit cycle, and even its very existence, depends on the specific values of the parameters in the model. If you change the parameters, you change the limit cycle. If you change the parameters enough, the limit cycle may disappear altogether. (See the exercises.)

A result proved early in this century is the **Poincaré–Bendixson theorem,** which says that equilibrium points and limit cycles are as complicated as dynamical systems in two variables can get. Once we pass to three variables, the situation becomes much more complicated. Many of the phenomena associated with such systems have been discovered only within the past 50 years, and their exploration is a subject of continuing research. In the next section we will give a brief introduction to some of the new behaviors that can arise.

Exercises

1. Using the parameter values given in the text, find the coordinates of the equilibrium point at the center of the limit cycle and show that it is, in fact, a repellor.

2. May's model is interesting in that it exhibits a phenomenon known as **Hopf bifurcation:** the existence of a limit cycle depends on the values of the parameters. Choose one of the parameters in May's model and try a range of values both larger and smaller than in the example we've worked out. At what value does the limit cycle disappear? When this happens, the equilibrium point inside the cycle has become an attractor. Can you work out analytically when this happens?

➢ 8.5 BEYOND THE PLANE: THREE-DIMENSIONAL SYSTEMS

Up to now we have worked with dynamical systems in which there are only two interacting quantities. We have thought of the two quantities as specifying a point in the state space, which we think of as some subset of the plane. The dynamical system defined a vector field on the state space. These geometric notions carry over to dynamical systems involving more than two interacting quantities.

In particular, if we have a dynamical system consisting of three interacting quantities, then we think of the values of the three quantities as specifying a point (or state) in three-dimensional space. So, for instance, if we have an ecological system consisting of three species, then we think of the numbers x, y, z

of each of the three species as specifying a point (x, y, z) in space. The set of all possible points or "states" consists of the set $\{(x, y, z) : x \geq 0, y \geq 0, z \geq 0\}$, the "first octant" in x, y, z space. We think of the dynamical system as a vector field, that is, as a rule which assigns to each point of the state space a vector. As in the case of the plane, we can define equilibrium points, trajectories, limit cycles, attractors, and the like.

In three-dimensional space there is a much wider range of behavior possible, even in the case of equilibrium points. We do, of course, have point attractors and repellors: all trajectories near a point attractor flow toward the attractor and all trajectories near a point repellor flow away from the repellor. However, a greater range of combinations is possible: an equilibrium point can attract all points along some plane, but repel all other points. Or the equilibrium point could be a center, surrounded by closed orbits lying in some plane which attract trajectories off the plane.

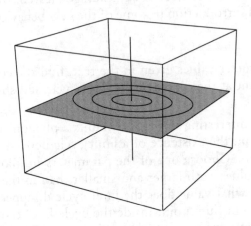

It is worth pointing out that we can represent the two-dimensional systems we've been exploring so far in this chapter in three dimensions by introducing a time axis. This has the effect of "unwinding" the trajectories by stretching them out in the t-direction: closed trajectories become endless coils, and equilibrium points become straight lines parallel to the t-axis.

The analytic tools we introduced to find and explore the nature of equilibrium points in two-dimensional systems carry over to three dimensions. In particular, in investigating the nature of an equilibrium point analytically, we first localize the system at the equilibrium point and linearize. The behavior of the linearized system can then be explored using analogues of the techniques introduced in the previous section (or using simple linear algebra).

There are also, of course, limit cycles, which can be attractors, repellors, or a mix of the two (attracting, for example, all trajectories on a plane, but repelling all trajectories off the plane), in three-dimensional systems. As in the case of dynamical systems in the plane, attracting limit cycles in a three-dimensional system signals stable periodic behavior.

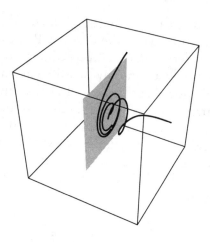

However, more complicated types of periodic behavior are possible in the three-dimensional case: we could, for example, have an attracting torus in the state space.

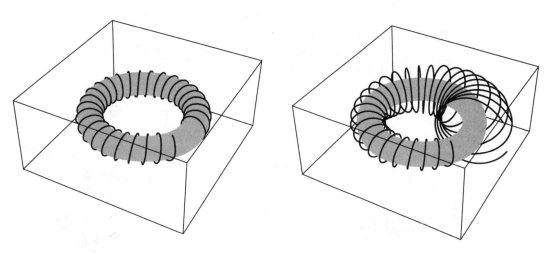

In this case, the behavior of the states does not settle down to periodic behavior, but to a behavior which is approximately periodic (often called **quasi-periodic**). In the plane, there is a well-studied phenomenon called Hopf bifurcation in which changing the parameters in a dynamical system can cause an attracting fixed point to become a repellor surrounded by a stable attracting limit cycle. Such dynamical systems arise in modeling situations in which a state begins to oscillate. In three dimensions we also see the same sort of phenomenon in which an attractor can give birth to an attracting limit cycle. However, there are also three-dimensional systems in which varying the parameters results in an attracting limit cycle becoming a repelling limit cycle enclosed by an attracting torus (this is also called Hopf bifurcation and is frequently encountered in applications).

These sorts of behavior are relatively straightforward generalizations of behavior in the plane. At the turn of the century, Poincaré realized that simple three-dimensional systems could have exceedingly complicated trajectories which exhibit behavior totally unlike any two-dimensional trajectory. Discoveries in the last three decades have made it clear that qualitatively new types of *attractors* (not just trajectories) can exist in three-dimensional systems with even very simple equations. The most famous such attractor was discovered by a meteorologist, Edward Lorenz, in the course of using dynamical systems to model weather patterns. He discovered a class of simple systems with an attractor that corresponded to behavior which was in no sense periodic. An example of such a system is

$$x' = -3x - 3y$$
$$y' = -xz + 30x - y$$
$$z' = xy - z.$$

All trajectories of the system entered a bounded region of the state space and tended toward a clearly defined geometrical object (resembling a butterfly). But along the attractor, nearby points followed trajectories which rapidly diverged from one another. Below, we have sketched two views of a trajectory beginning at $(0,1,0)$ of the system above.

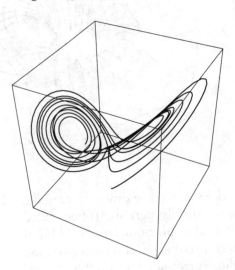 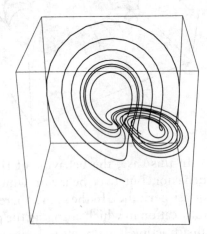

As Lorenz noted in the paper describing his discovery [Deterministic Nonperiodic Flow, *Journal of Atmospheric Science*, **20**:130 (1963)], this divergence of trajectories along an attractor has astonishing practical implications. It means that the trajectories of nearby points in state space could (and would) wind up following very different paths along the attractor. Since we never know initial conditions exactly (and even if we did, a computer truncates decimal expansions of the coordinates of any point, effectively replacing the point with a nearby point), this means that long-term predictions using a model possessing such an attractor are impossible. In other words, although the future is completely determined by a dynamical system given an initial state, it is unknowable in systems of the sort discovered by Lorenz, because initial states are never known exactly in practice. Such systems are called **chaotic**, and attractors which are not points, limit cycles, or tori are called **strange attractors**. These systems have been intensively studied in the last 30 years and are still far from completely understood. Chaotic systems have been used to attempt to model a wide variety of real situations which exhibit unpredictable behavior: business cycles, turbulence, heart attacks, among others. Although fascinating and philosophically provocative, most of this work is still very speculative.

Systems involving more than three variables can still be treated geometrically—we think of the space of states as a higher-dimensional space (one dimension for each quantity) and the dynamical system as defining a vector field on the state space. Of course, we cannot visualize such spaces directly, but the geometrical insight we gain in dimensions 2 and 3 very frequently allows us to handle such systems.

Exercises

In the next two exercises, we look at some three-dimensional systems which arise in ecology. These questions are challenging and you will probably find it helpful to work them out with a friend.

1. Consider a system consisting of three species: giant carnivorous reptiles, vegetarian mammals, and plants. Suppose that the populations of these are given by x, y, and z, respectively. The reptiles eat the mammals, the mammals eat the plants, and the plants compete among themselves. Explain why the following system is consistent with these hypotheses:

$$x' = -.2x + .0001xy$$
$$y' = -.05y - .001xy + .000001yz$$
$$z' = z - .00001z^2 - .0001yz.$$

a) Find all equilibrium points of the system. There are five, one of which is physically impossible. Describe the significance of the other four.

b) The most interesting equilibrium is the one in which all three species are present. Localize the system at this equilibrium, using local variables u, v, and w. Linearize. Show that the linearized system has the form

$$u' = .003v$$

$$v' = -2u + .002w$$

$$w' = -8v - .8w.$$

Can you determine whether the equilibrium is an attractor? This is a hard question—it *is* an attractor. One way to show this is a generalization of the technique that we used in the preceding section is to examine the distance of points on a trajectory from the origin over time. For the current problem we use a **generalized distance function**

$$D(t) = 8 \cdot 10^6 u^2 + 12000 v^2 + 3w^2.$$

Show, using arguments like those we used when we looked at ordinary distance, that as we move along a trajectory, the value of D must decrease. Hence conclude that the equilibrium point must be an attractor.

2. The system of equations

$$x' = x - .001x^2 + .002xy - .1xz$$
$$y' = y - .01y^2 + .001xy$$
$$z' = -z + .001xz$$

arises in a general family of models proposed in 1980 by Heithaus, Culver, and Beattie [Models of Some Ant-Plant Mutualisms, *American Naturalist*, 116:347–361 (1980)] for investigating the interactions of three species: violets, ants, and mice. Violets produce seeds with density x (per square meter, say). The ants take some of the seeds and use the seed covering for food. But they leave the remainder, which is still a perfectly good seed, in their refuse piles, which happily turn out to be good sites for germination. The ants have density y. Finally, the seeds are also taken by mice, who use the whole seed for food (destroying both the cover and the seed within). The mice have density z.

a) Explain why these equations are consistent with the hypotheses we made on the interactions between the violets, ants, and mice.

*b) Find all equilibrium points for the system. Don't forget the points where one or more of the variables equals 0.

$$x' = x - .001x^2 + .002xy - .1xz$$
$$y' = y - .01y^2 + .001xy$$
$$z' = -z + .001xz.$$

*c) Localize the model at each of these equilibria, using local coordinates u, v, and w as before, and linearize.

d) In the case of the equilibrium point (1000, 200, 4) the local linearization is

$$u' = -u + 2v - 100w$$
$$v' = .2u - 2v$$
$$w' = .004u.$$

As in the preceding exercise, show that this point is an attractor by examining the generalized distance function $R(t) = u(t)^2 + v(t)^2 + 25000w(t)^2$ and showing that the value of R decreases as you move along a trajectory.

➢ 8.6 CHAPTER SUMMARY

The Main Ideas

- A dynamical system can be viewed as a geometrical object. The possible values of the dependent variables are then the coordinates of a point—called a **state.** The set of all possible points is called the **state space** for the system.

- The differential equations become a rule, assigning a **velocity vector** to each state. Thought of in this way, the equations are called a **vector field.**

- Solutions to the differential equations correspond to **trajectories** in the state space. At every point a trajectory is tangent to the corresponding velocity vector, and is changing at the rate given by the length of the vector. The set of all possible trajectories is called the **phase portrait** of the system.

- **Equilibrium points** are points where the velocity vector is 0. An equilibrium point is a trajectory consisting of a single point. A dynamical system is conveniently analyzed by examining the nature of its equilibrium points—whether they are **attractors, repellors, saddle points,** or **centers.**

- To study the nature of an equilibrium point it is helpful to look at the **local linearization** of the vector field near the point.

- Determining whether **fixed line trajectories** exist is a crucial part of analyzing the nature of an equilibrium point.

- In addition to equilibrium points, dynamical systems in two dimensions may also have **limit cycles** that shape the asymptotic behavior of the system.

- In higher-dimensional state spaces, there are not only the obvious extensions of point attractors and limit cycles, but it is possible to have an **attracting torus** as well. There are even more complicated attracting objects called **strange attractors.**

Expectations

- You should be able to describe the assumptions embodied in a particular dynamical system modeling the interaction between two (or three) species and evaluate whether the assumptions seem reasonable.

- For a dynamical system with two dependent variables, you should be able to determine the regions where each variable is zero or has a constant sign, find equilibrium points, sketch representative vectors of the vector field, and draw trajectories that are consistent with this information.

- You should be able to determine whether a linear system of differential equations with two dependent variables has **fixed line** trajectories—that is, trajectories which are straight lines going directly toward or directly away from an equilibrium point.

- You should be able to **localize** and **linearize** a dynamical system in two dependent variables to explore its behavior near an equilibrium point.

- You should be able to recognize the five generic types of equilibrium points: attracting and repelling **nodes**, attracting and repelling **spirals**, and **saddle points**.

- Using a differential equation solver, you should be able to recognize when a dynamical system has a **limit cycle**.

- You should be able to analyze a dynamical system with three dependent variables.

CHAPTER 9

FUNCTIONS OF SEVERAL VARIABLES

Functions that depend on several input variables first appeared in the S-I-R model at the beginning of the course. Usually, the number of variables has not been an issue for us. For instance, when we introduced the derivative in chapter 3, we used partial derivatives to treat functions of several variables in a parallel fashion. However, when there are questions of visualization and geometric understanding, the number of variables *does* matter. Every variable adds a dimension to the problem—one way or another. For example, if a function has two input variables instead of one, we will see that its graph is a surface rather than a curve.

The problem of visualizing a function of several variables

This chapter deals with the geometry of functions of two or more variables. We start with graphs and level sets. These are the basic tools for visualization. Then we turn to microscopic views, and see what form the microscope equation takes. Finally, we consider optimization problems using both direct visual methods and dynamical systems.

➤ 9.1 GRAPHS AND LEVEL SETS

The graph at the right comes from a model that describes how the average daily temperature at one place varies over the course of a year. It shows the temperature A in °F, and the time t in months from January. As we would expect, the temperature is a periodic function [which we can write as $A(t)$], and its period is 12 months. Furthermore,

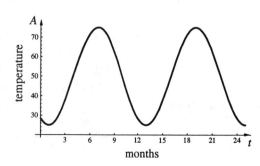

the lowest temperature occurs in February (when $t \approx 1$ or 13) and the highest in July (when $t \approx 7$ or 19). This is about what we would expect.

Underground temperatures fluctuate less

However, all these temperature fluctuations disappear a few feet underground. Below a depth of 6 or 8 feet, the temperature of the soil remains about 55°F year-round! Between ground level and that depth, the temperature still fluctuates, but the range from low to high decreases with the depth. Here is what happens at some specific depths. Notice how the time at which

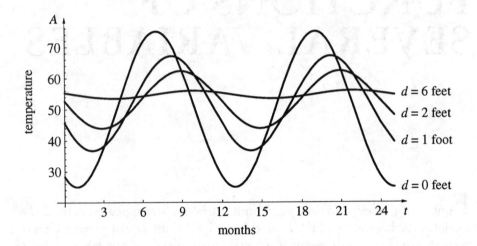

the temperature peaks gets later and later as we go farther and farther underground. For example, at $d = 2$ feet the highest temperature occurs in September ($t \approx 9$), not July. It literally takes time for the heat to sink in. In chapter 7 we called this a phase shift. The lowest temperature shifts in just the same way. At a depth of 2 feet, it is colder in March than in January.

The phase shifts with the depth

Thus A is really a function of *two* variables, the depth d as well as the time t. To reflect this addition, let's change our notation for the function to $A(t, d)$. In the figure above, d plays the role of a parameter: it has a fixed value for each graph. We can reverse these roles and make t the parameter. This is done in the figure on the top of the next page. It shows us how the temperature varies with the depth at fixed times of year. Notice that, in April and October, the extreme temperature is not found on the surface. In October, for example, the soil is warmest at a depth of about 9 inches.

Graphing temperature as a function of depth

The lower figure on the page is a single graph that combines all the information in these two sets of graphs. Each point on the bottom of the box corresponds to a particular depth and a particular time. The height of the surface above that point tells us the temperature at that depth and time. For example, suppose you want to find the temperature 4 feet below the surface at the beginning of July. Working from the bottom front corner of the box, move 4 feet to

Reading a surface graph

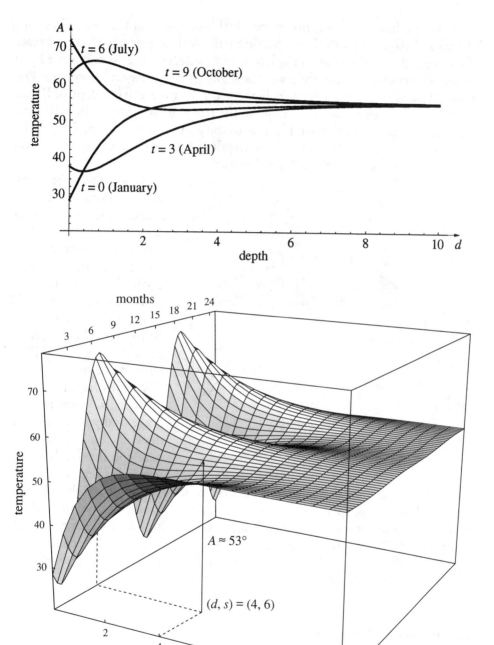

the right and then 6 months toward the back. This is the point $(d, s) = (4, 6)$. The height of the graph above this point is the temperature A that we want.

Grid lines are slices
of the surface that
show how the function
depends on each
variable separately

There is a definite connection between this surface and the two collections of curves. Imagine that the box containing the surface graph is a loaf of bread. If you slice the loaf parallel to the left or right side, this slice is taken at a fixed depth. The cut face of the slice will look like one of the graphs on page 454. These show how the temperature depends on the time at fixed depths. If you slice the loaf the other way—parallel to the front or back face—then the time is fixed. The cut face will look like one of the graphs on page 455. They show how the temperature depends on the depth at fixed times. The grid lines on the surface are precisely these "slice" marks.

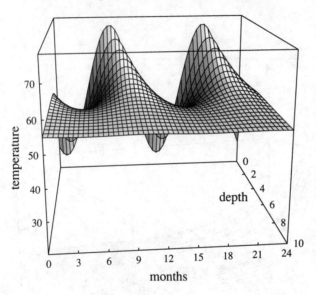

Here is the same graph seen from a different viewpoint. Now time is measured from the left, while depth is measured from the back. The temperature is still the height, though. One advantage of this view is that it shows more clearly how the peak temperature is phase-shifted with the depth.

We now have two ways to visualize how the average daily temperature depends on the time of year and the depth below ground. One is the surface graph itself, and the other is a collection of curves that are slices of the surface. The surface gives us an overall view, but it is not so easy to read the surface graph to determine the temperature at a specific time and depth. Check this yourself: What is the temperature 2 feet underground at the beginning of April? The slices are much more helpful here. You should be able to read from either collection of slices that $A \approx 44°$F.

Comparing the surface
to its slices

Examples of Graphs

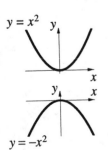

The purpose of this section is to get some experience constructing and interpreting surface graphs. To work in a context, look first at the functions $y = x^2$ and $y = -x^2$. They provide us with standard examples of a minimum and a maximum when there is just one input variable. Let's consider now the corresponding examples for two input variables. Besides an ordinary maximum and an ordinary minimum, we will find a *third* type—called a minimax—that is completely new. It arises because a function can have a minimum with respect to one of its input variables and a maximum with respect to the other.

A Minimum: $z = x^2 + y^2$

At the origin $(x, y) = (0, 0)$, $z = 0$. At any other point, either x or y is nonzero. Its square is positive, so $z > 0$. Consequently, z has a minimum at the origin. The graph of this function is a parabolic **bowl** whose lowest point sits on the origin. As always, the grid lines are slices, made by fixing the value of x or y. For example, if $y = c$, then the slice is $z = x^2 + c^2$. This is an ordinary parabolic curve.

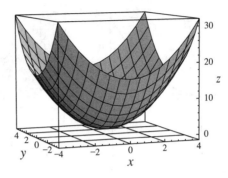

A Maximum: $z = -x^2 - y^2$

For any x and y, the value of z in this example is the opposite of its value in the previous one. Thus, z is everywhere negative, except at the origin, where its value is 0. Thus z has a maximum at the origin. Its graph is an upside-down bowl, or **peak**, whose highest point reaches up and touches the origin. Grid lines are the curves $z = -x^2 - c^2$ and $z = -c^2 - y^2$. These are parabolic curves that open downward.

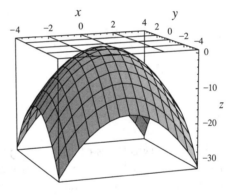

A Minimax: $z = x^2 - y^2$

Suppose we fix y at $y = 0$. This slice has the equation $z = x^2$, so it is an ordinary parabola (that opens upward). Thus, as far as the input x is concerned, z has a *minimum* at the origin. Suppose, instead, that we fix x at $x = 0$. Then we get a slice whose equation is $z = -y^2$. It is also an ordinary parabola, but this one opens downward. As far as y is concerned, z has a *maximum* at the origin. It is clear from the graph how upward-opening

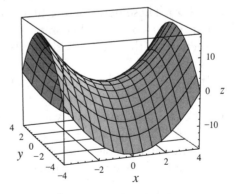

slices in the x-direction fit together with downward-opening slices in the y-direction. Because of the shape of the surface, a minimax is commonly called a **saddle**, or a **saddle point**.

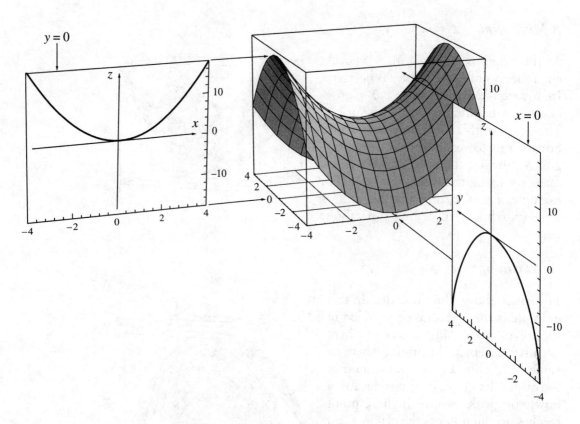

The points where $y = c$ form a vertical plane parallel to the x, z plane

Here are two slices of $z = x^2 - y^2$ shown in more detail. Points in the box have three coordinates: (x, y, z). If we set $y = 0$ we are selecting the points of the form $(x, 0, z)$. These make up the x, z plane. On this plane the equation $z = x^2 - y^2$ becomes simply $z = x^2$. The graph of *this* equation is a curve in the x, z plane—specifically, the parabola shown. The situation is similar if y is given some other fixed value. For example, $y = -4$ specifies the points $(x, -4, z)$. These describe the plane that forms the front face of the box. The equation $z = x^2 - y^2$ becomes $z = x^2 - 16$. The curve tracing out the intersection of the saddle with the front of the box is precisely the graph of $z = x^2 - 16$.

If $x = 0$ we get the points $(0, y, z)$ that make up the y, z plane. On this plane the equation simplifies to $z = -y^2$, and its graph is the parabolic curve shown. Giving x a different fixed value leads to similar results. A good example is $x = -4$. The points $(-4, y, z)$ lie on the plane that forms the left side of the box. The equation becomes $z = 16 - y^2$ there, and this is the parabolic curve marking the intersection of the saddle with the left side of the box.

The points where $x = c$ form a vertical plane parallel to the y, z plane

As you can see, it is valuable for you to be able to generate surface graphs yourself. There are now a number of computer utilities which will do the job. Some can even rotate the surface while you watch, or give you a stereo view. However, even without one of these powerful utilities, you should try to generate the slicing curves that make up the grid lines of the surface.

A Cubic: $z = x^3 - 4x - y^2$

Slices of this graph are downward-opening parabolas (when $x = c$) and are cubic curves that have the same shape (when $y = c$). Notice that each cubic curve has a maximum and a minimum, and each parabola has a maximum. The

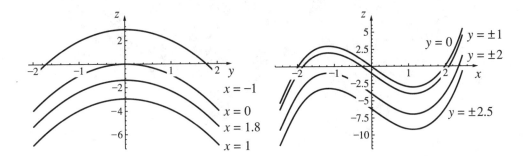

surface graph itself has a *peak* where the cubics have their maximum, but it has a *saddle* where the cubics have a minimum. Do you see why? The saddle point is a minimax for $z = x^3 - 4x - y^2$: z has a minimum there *as a function of x alone* but a maximum *as a function of y alone*.

The surface has a saddle

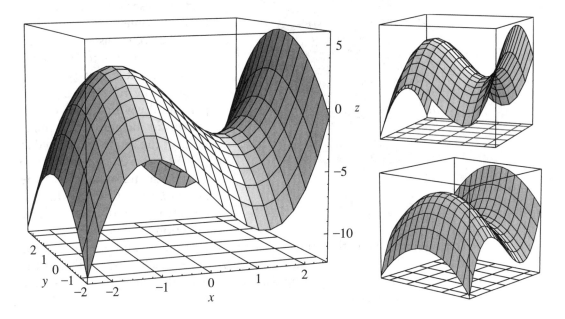

The small figures on the right show the same surface as the large figure; they just show it from different viewpoints. As a practical matter, you should look at these surfaces the way you would look at sculpture: "walk around them" by generating diverse views.

See the graph from different viewpoints

Energy of the Pendulum: $E = 1 - \cos \theta + \frac{1}{2}v^2$

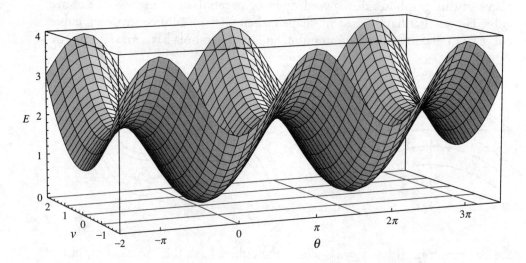

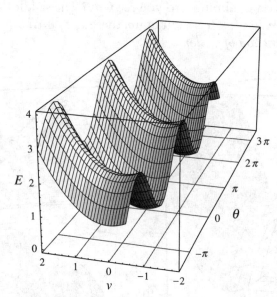

This function first came up in chapter 7, where it was used to demonstrate that a dynamical system describing the motion of a frictionless pendulum had periodic solutions. It was used again in chapter 8 to clarify the phase portrait of that dynamical system. The function E varies periodically with θ, and you can see this in the graph. The minimum at the origin is repeated at $(\theta, v) = (2\pi, 0)$, and so on. The graph also has a saddle at the point $(\theta, v) = (-\pi, 0)$. This too repeats with period 2π in the θ direction.

The figure above is the same surface with part cut away by a slice of the form $v = c$. These slices are sine curves: $E = 1 - \cos \theta + \frac{1}{2}c^2$. Slices of the form $\theta = c$ are upward-opening parabolas. From this viewpoint, the saddle points show up clearly.

One way to describe what happens to a real pendulum—that is, one governed by frictional forces as well as gravity—is to say that its energy "runs down" over time. Now, at any moment the pendulum's energy is a point on this graph. As the energy runs down, that point must work its way down the graph. Ultimately, it must reach the bottom of the graph— the minimum energy point at the origin $(\theta, v) = (0, 0)$. This is the stable equilibrium point. The *pendulum hangs straight down ($\theta = 0$) and is motionless ($v = 0$). The graph gives us an abstract—but still vivid and concrete—way of thinking of the dissipation of energy.*

From Graphs to Levels

There is still another way to picture a function of two variables. To see how it works we can start with an ordinary graph. On the right is the graph of

$$z = f(x, y) = x^3 - 4x - y^2,$$

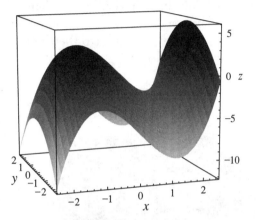

the cubic function we considered on page 459. This graph looks different, though. The difference is that points are shaded according to their height. Points at the bottom are lightest, points at the top are darkest.

Notice that the flat x, y plane is shaded exactly like the graph above it. For instance, the dark spot centered at the point $(x, y) = (-1, 0)$ is directly under the peak on the graph. The other dark patch, near the right edge of the plane, is under

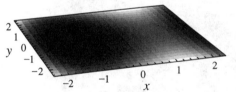

the highest visible part of the surface. Consequently, the shading on the x, y plane gives us the same information as the graph. In other words, *the intensity of shading at (x, y) is proportional to the value of the function $f(x, y)$.*

The figure in the x, y plane is called a **density plot**. Think of the intensity of shading as the *density* of ink on the page. Here are density plots of the standard minimum, maximum, and minimax. Compare these with the graphs

Density plots

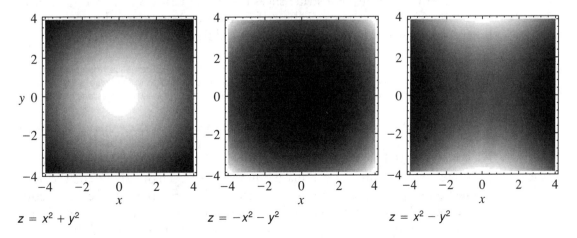

$z = x^2 + y^2$ $z = -x^2 - y^2$ $z = x^2 - y^2$

on page 457. The third density plot is the most interesting. From the center of the x, y plane, the shading increases to the right and left. Therefore, z has a minimum in the horizontal direction. However, the shading *decreases* above and below the center. Therefore, z has a *maximum* in the vertical direction. Thus, you really can see there is a minimax at the origin.

A sample plot

Try your hand at reading the density plot on the left below. You should see two maxima (directly above and below the origin), a minimum (at the origin itself), and two saddles (to the right and the left of the origin). The function defining the plot is

$$f(x, y) = (x^2 + (y - 1)^2 - 3)(3 - x^2 - (y + 1)^2).$$

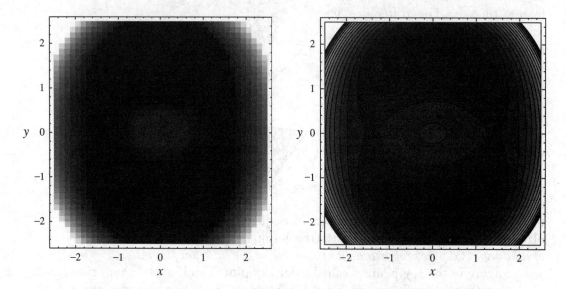

Can you visualize what the graph looks like? This density plot should help you, and you can also construct slices by setting $x = c$ and $y = c$. The slices $x = 0$ and $y = 0$ are especially useful. With them you could determine the exact coordinates of the maxima and the saddles.

> *These density plots show a "checkerboard" pattern because Mathematica (the computer program that produces them) shades each little square according to the value of the function at the center of the square. This pattern is an artifact; it is not inherent to a density plot.*

In a density plot, the shading varies smoothly with the value of the function. This is accurate, but it may be a bit difficult to read. On the right you see a modified density plot. There is still shading, but there are now just a few distinct shades. This makes a sharp boundary between one shade and the next. The boundary is called a **contour**, or a **level**. The figure itself is called a **contour plot**. The two maxima on the vertical line $x = 0$ stand out more clearly on the contour plot. Also, the contour lines around the two saddles help us

From densities to contours

see that the function has a minimum in the vertical direction and a maximum in the horizontal direction.

Once we have contour lines to separate one density level from the next, we can even dispense with the shading. The figure on the right is just the contour plot from the opposite page, minus the shading. The contour lines, or level curves, now stand out clearly. On each contour, the value of the function is constant. This is also called a **contour plot**.

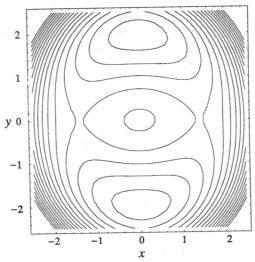

There is some loss of information here, however. For example, we can't tell where the value of the function is large and where it is small. Nevertheless, the nested ovals on the vertical line $x = 0$ *do* tell us that there is either a maximum or a minimum at the center of each nest.

For reference purposes, here are the contour plots for the standard minimum, maximum, and saddle. In the first two cases, the contours are concentric ovals. These look the same, so only one is illustrated. The other two pictures show a saddle. In general, the contours around a saddle are a family of hyperbolas. However, it is possible for one of the contour lines to pass exactly through the minimax point. That contour is a pair of crossed lines, as shown in the version on the far right. You should compare these contour plots with the density plots of the same functions on page 461, and with their graphs on page 457.

Contours of the standard functions

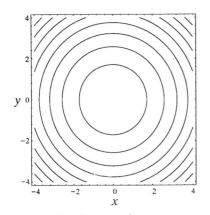

two functions but one plot
$z = x^2 + y^2$
$z = -x^2 - y^2$

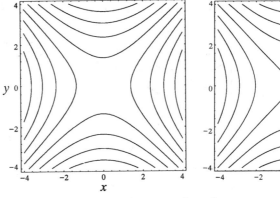

two plots of a single function $z = x^2 = y^2$

Contours are horizontal slices of a graph

There is a direct connection between the contour plot of a function and its graph. Contours are horizontal slices of the graph, just as grid lines are vertical slices. Below, we use the standard functions $z = x^2 - y^2$ and $z = x^2 + y^2$ to illustrate the connection. Notice that every contour down in the x, y plane lies exactly below, and has the same shape as, a horizontal slice of the graph. This picture explains why contours are called *level* curves.

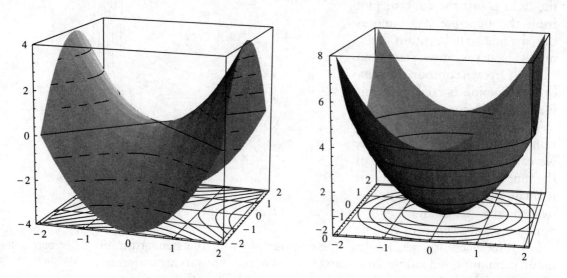

Energy of the pendulum, again

To get some more experience with contour plots, we return to the energy function of the pendulum:

$$E(\theta, v) = 1 - \cos \theta + \tfrac{1}{2}v^2.$$

From the contour plot alone you should be able to see that E has either a minimum or a maximum at $(\theta, v) = (0, 0)$, and another at $(2\pi, 0)$. The contours also provide evidence that there is a saddle (minimax) near $(\theta, v) = (-\pi, 0)$ and $(\pi, 0)$. It is also apparent that E is a *periodic* function of θ.

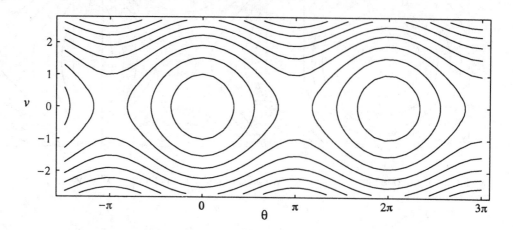

What you should find most striking about this plot, however, is the way it resembles the phase portrait of the pendulum (chapter 8, pages 417–421). Every level curve here looks like a trajectory of the dynamical system. This is no accident. We know from chapter 8 that the energy is a first integral for the dynamics. In other words, energy is constant along each trajectory—this is the law of conservation of energy. But each level curve shows where the energy function has some fixed value. Therefore, each trajectory must lie on a single energy level.

Energy contours are trajectories of the dynamics

We can carry the connection between contours and trajectories even further. Closed trajectories correspond to *oscillations* of the pendulum. But the closed trajectories are the closed contours, and these are the ones that surround the minimum. In particular, they are *low*-energy levels. By contrast, at higher energies ($E > 2$, in fact), the pendulum will just continue to spin

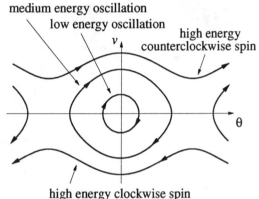

in whatever direction it was moving initially. Thus, each high-energy level is occupied by *two* trajectories—one for clockwise spinning and one for counterclockwise.

Technical Summary

The examples we have seen so far were meant to introduce some of the common ways of visualizing a function $z = f(x, y)$. To use them most effectively, though, you need to know more precisely how each is defined. We review here the definition of a graph, a density plot, a contour plot, and a terraced density plot.

Graph. The graph of $z = f(x, y)$ lies in the three-dimensional space with coordinates (x, y, z). To construct it, take any input (x, y). Identify this with the point $(x, y, 0)$ in the x, y plane (which is defined by the condition $z = 0$). The corresponding point on the graph lies at the height $z = f(x, y)$

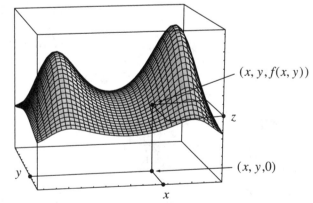

above the x, y plane. This point has coordinates $(x, y, f(x, y))$. **The graph is the set of all points of the form $(x, y, f(x, y))$.** This is a two-dimensional surface.

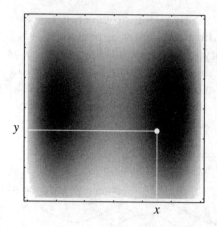

Density plot. The density plot of $z = f(x, y)$ lies in the two-dimensional x, y plane. Choose any rectangle where the function is defined, and let m and M be the minimum and maximum values, respectively, of $f(x, y)$ on the rectangle. Define

$$\rho(x, y) = \frac{f(x, y) - m}{M - m}.$$

Then ρ (greek rho) satisfies $0 \le \rho(x, y) \le 1$ on the rectangle; it is called a **density function. In the density plot, the density of ink—or darkness—at (x, y) is $\rho(x, y)$.**

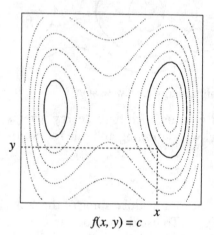

$f(x, y) = c$

Contour plot. A **contour** of $z = f(x, y)$ is the set of points in the x, y plane where f has some fixed value:

$$f(x, y) = c.$$

That fixed value c is called the **level** of the contour. (The two solid ovals in the figure at the left make up a single contour.) A contour is also called a level curve. **A contour plot of f is a collection of curves $f(x, y) = c_j$ in the x, y plane.** In the plot it is customary to use constants $c_1, c_2, \ldots$ that are equally spaced; that is, the interval between one c_j and the next always has the same value Δc.

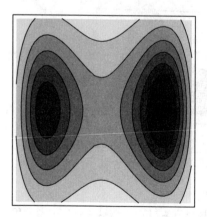

Terraced density plot. This is a contour plot in which the region between two adjacent contours is shaded with ink of a single density. If the contours are at levels c_1 and c_2, then the density that is typically chosen is the one for the level half-way between these two—that is, for their average $(c_1 + c_2)/2$. Each region is called a **terrace.** Often, a terraced density plot is drawn in color, using different colors for each terrace. Television weather programs use terraced density plots to describe the temperature forecast for a large region.

We find density plots everywhere. A photograph is a density plot of the light that fell on the film when it was exposed. A newspaper "halftone" illustration is also a density plot of an image.

The pros and cons.   Each of these modes of visualization has advantages and disadvantages. All are reasonably good at indicating the extremes (the maxima and minima) of a function. A contour plot needs some additional information—for example, a label on each contour to indicate its level—to distinguish between maxima and minima. However, if you want to know the numerical value of $f(x, y)$ at a particular point (x, y), a contour plot with labels offers more precision than a density plot. It's usually better than a graph, too.

Overall, a graph has the biggest visual impact, but there is a cost. It takes three dimensions to represent the graph of a function of two variables, but only two to represent a plot. The cost is that extra dimension. It means that we cannot draw the graph of a function of three variables. That would take four mutually perpendicular axes—an impossibility in our three-dimensional space. However, we can produce a contour plot.

Plots are visually economical in comparison to graphs

Contours of a Function of Three Variables

We pause here for a brief glimpse of a large subject. By analogy with the definition for a function of two variables, we say that a **contour** of the function $f(x, y, z)$ is the set of points (x, y, z) that satisfy the equation

Contours and levels

$$f(x, y, z) = c,$$

for some fixed number c. We call c the **level** of the contour.

Let's find the contours of $w = x^2 + y^2 + z^2$. This is completely analogous to the function $x^2 + y^2$ with two inputs. (What do the contours of $x^2 + y^2$ look like?) In particular, w has a minimum when $(x, y, z) = (0, 0, 0)$. As the following diagram shows, $w = x^2 + y^2 + z^2$ is the square of the distance from the origin to (x, y, z). (We use the Pythagorean theorem twice: once for p^2 and once for q^2.) Consequently,

The standard minimum

$$p^2 = x^2 + y^2$$
$$q^2 = p^2 + z^2$$
$$= x^2 + y^2 + z^2$$
$$= w,$$

all points (x, y, z) where w has a fixed value lie a fixed distance from the origin. Specifically, $w = x^2 + y^2 + z^2 = c$ is the following set:

- $c > 0$: the sphere of radius $\sqrt{c}$ entered at the origin;
- $c = 0$: the origin itself;
- $c < 0$: the empty set.

The contours are spheres

The contour plot of $w = x^2 + y^2 + z^2$ is thus a nest of concentric spheres, as shown in the illustration below. The value of w is constant on each sphere. (The tops of the spheres have been cut away so you can see how the spheres nest; the whole thing resembles an onion.)

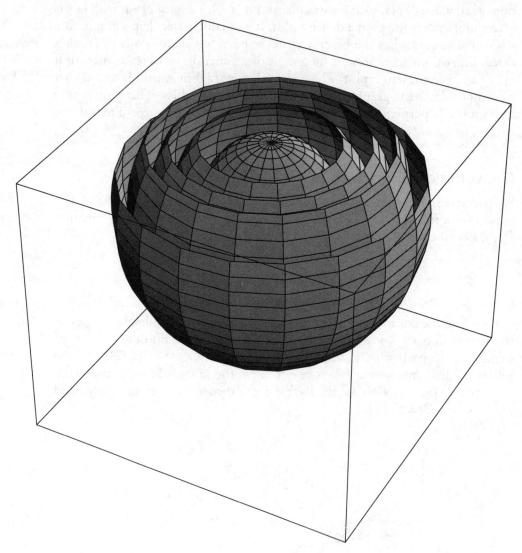

A standard minimax

On the next page is the contour plot of another standard function with three input variables:

$$w = f(x, y, z) = x^2 + y^2 - z^2.$$

A quarter of each surface has been cut away so you can see how the surfaces nest together. Note that $w = 0$ is a cone, and every surface with $w < 0$ consists of two disconnected (but congruent) pieces—an upper half and a lower half.

You should compare this function to the standard minimax $x^2 - y^2$ in two variables. The three-variable function w has a minimum with respect to *both* of the variables x and y, while it has a maximum with respect to z. (Do you see why? The arguments are exactly the same as they were for two input variables on page 457.) Furthermore, the contours of $x^2 - y^2$ are a family of hyperbolas, and the contours of $x^2 + y^2 - z^2$ are surfaces obtained by rotating these hyperbolas about a common axis.

The contours are hyperbolic shapes

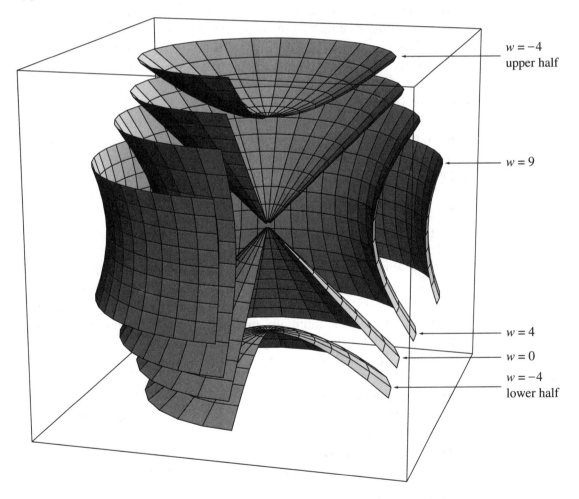

$w = -4$
upper half

$w = 9$

$w = 4$

$w = 0$

$w = -4$
lower half

It is a general fact—and our two examples provide good evidence for it—that a single contour of a function of three variables is a *surface*. Thus a contour is a curve or a surface, depending on the number of input variables. We often use the term **level set** (rather than a level *curve* or a level *surface*) as a generic name for a contour.

When there are three input variables, the contours are surfaces

Exercises

In many of these exercises it will be essential to have a computer program to make graphs, terraced density plots, and contour plots of functions of two variables.

1.*a) Use a computer to obtain a graph of the function $z = \sin x \sin y$ on the domain $0 \le x \le 2\pi$, $0 \le y \le 2\pi$. How many maximum points do you see? How many minimum points? How many saddles?

 b) Determine, as well as you can from the graph, the location of the maximum, minimum, and saddle points.

2. (Continuation.) Make the domain $-2\pi \le x \le 2\pi$, $-2\pi \le y \le 2\pi$ and answer the same questions you did in the previous exercise. (Does the graph look like an egg carton?)

3. Obtain a terraced density plot (or a contour plot) of $z = \sin x \sin y$ on the domain $-2\pi \le x \le 2\pi$, $-2\pi \le y \le 2\pi$. Locate the maximum, minimum, and saddle points of the function. Do these results agree with those from the previous exercise?

4. Obtain the graph of $z = \sin x \cos y$ on the domain $0 \le x \le 2\pi$, $0 \le y \le 2\pi$. How does this graph differ from the one in exercise 1? In what ways is it similar?

5. Obtain the graph of $z = 2x + 4x^2 - x^4 - y^2$ when $-2 \le x \le 2$, $-4 \le y \le 4$. Locate all the minimum, maximum, and saddle points in this domain. (Note: The minimum is on the boundary!)

6. (Continuation.) Obtain a terraced density plot (or contour plot) for the function in the previous exercise, using the same domain. Use the plot to locate all the minimum, maximum, and saddle points. Compare your results with those of the previous exercise.

7. a) Obtain the graph of $z = 2x - y$ on the domain $-2 \le x \le 2$, $-2 \le y \le 2$. What is the shape of the graph?

 b) Graph the same function of the domain $2 \le x \le 6$, $0 \le y \le 4$. What is the shape of the graph? How does this graph compare to the one in part (a)?

8. a) (Continuation.) Sketch three different slices of the graph of $z = 2x - y$ in the y-direction. What do the slices have in common? How are they different?

 b) Answer the same questions for slices in the x-direction.

*9. a) Obtain the graph of $z = .3x + .8y + 2.3$; choose the domain yourself. Where does the graph intercept the z-axis?

 b) Describe the vertical slices of this graph in the y-direction and in the x-direction.

10. Describe the vertical slices of the graph of $z = px + qy + r$ in the y-direction and in the x-direction.

11. a) Compare the contours of the function $z = x^2 + 2y^2$ to those of $z = x^2 + y^2$.
    b) What is the shape of the graph of $z = x^2 + 2y^2$? Decide this first using only the information you have about the contours. Then use a computer to obtain the graph.

12. a) Compare the contours of the function $z = x^2 - 2y^2$ to those of $z = x^2 - y^2$.
    b) What is the shape of the graph of $z = x^2 - 2y^2$? Decide this first using only the information you have about the contours. Then use a computer to obtain the graph.

13. a) Obtain a contour plot of the function $z = x^2 + xy + y^2$.
    b) What is the shape of the graph of $z = x^2 + xy + y^2$? Decide this first using only the information you have about the contours. Then use a computer to obtain the graph.

14. a) Obtain a contour plot of the function $z = x^2 + 3xy + y^2$.
    b) What is the shape of the graph of $z = x^2 + 3xy + y^2$? Decide this first using only the information you have about the contours. Then use a computer to obtain the graph.

15. a) Obtain a contour plot of the function $z = x^2 + 2xy + y^2$.
    b) What is the shape of the graph of $z = x^2 + 2xy + y^2$? Decide this first using only the information you have about the contours. Then use a computer to obtain the graph.

16. Complete this statement: The function $f(x, y; p) = x^2 + pxy + 4y^2$, which depends on the parameter p, has a minimum at the origin when _____ and a minimax when _____.

17. a) Obtain the graph and a terraced density plot of the function $z = 3x^2 + 17xy + 12y^2$. What is the shape of the graph?
    b) What is the shape of the contours? Indicate how the contours fit on the graph.

18. a) Obtain the graph and a terraced density plot of the function $z = 3x^2 + 7xy + 12y^2$. What is the shape of the graph?
    b) What is the shape of the contours? Indicate how the contours fit on the graph.

19. a) Obtain the graph and a terraced density plot of the function $z = 3x^2 + 12xy + 12y^2$. What is the shape of the graph?
    b) What is the shape of the contours? Indicate how the contours fit on the graph.

20. Obtain the graph of $z = f(x, y) = xy$ on the domain $-3 \le x \le 3$, $-3 \le y \le 3$. Does this function have a maximum or a minimum or a saddle point? Where?

21. **a)** (Continuation.) Sketch slices of the graph of $z = xy$ in the y-direction, for each of the values $x = -2, -1, 0, 1,$ and 2. What is the general shape of each of these slices?

b) Repeat part (a), but make the five slices in the x-direction—that is, fix y instead of x.

22. (Continuation.) Show how the slices you obtained in the previous exercise fit (or appear) on the graph you obtained in the exercise just before that one.

23. **a)** (Continuation.) Let $x = u + v$ and $y = u - v$. Express z in terms of u and v, using the fact that $z = xy$. Then obtain the graph of z as a function of the new variables u and v.

b) What is the shape of the graph you just obtained? Compare it to the graph of $z = xy$ you obtained earlier.

24. Can you draw a network of straight lines on the saddle surface $z = x^2 - y^2$?

25. Obtain a terraced density plot of $z = xy$. How do the contours of this plot fit on the graph of $z = xy$ you obtained in a previous exercise?

26. The graphs of $z = x^2 + 5xy + 10y^2$ and $z = 3$ intersect in a curve. What is the shape of that curve?

27. The graphs of $z = x^2 + y^3$ and $z = 0$ intersect in a curve. What is the shape of that curve?

28. The graphs of $z = 2x - y$ and $z = .3x + .8y + 2.3$ intersect in a curve. What is the shape of that curve?

First Integrals

29. A hard spring described by the dynamical system

$$\frac{dx}{dt} = v \qquad \frac{dv}{dt} = -cx - \beta x^3$$

has a first integral of the form

$$E(x, v) = \tfrac{1}{2}cx^2 + \tfrac{1}{4}\beta x^4 + \tfrac{1}{2}v^2.$$

This is the **energy** of the spring. (See section 7.3, especially exercise 13, page 402.)

*a) Let $c = 16$ and $\beta = 1$. Obtain the graph of $E(x, v)$ on a domain that has the origin at its center. Locate all the minimum, maximum, and saddle points in this domain.

b) What is the state of the spring (that is, its position x and its velocity v) when it has minimum energy?

30. A soft spring described by the dynamical system

$$\frac{dx}{dt} = v \qquad \frac{dv}{dt} = -\frac{25x}{1 + x^2}$$

has an energy integral of the form

$$E(x, v) = \frac{25}{2}\ln(1 + x^2) + \frac{1}{2}v^2.$$

(See exercise 16, page 403.)

a) Obtain the graph of $E(x, v)$. Experiment with different possibilities for the domain until you get a good representation.

b) Obtain a terraced density plot of $E(x, v)$ over the same domain you chose in part (a). Compare the two representations of E.

c) Does the spring have a state of minimum energy? If so, where is it?

d) Does the spring have a state of *maximum* energy? Explain your answer.

31. a) *The Lotka–Volterra equations.* According to exercise 33 of section 7.3 (page 406), the function

$$E(x, y) = .1 \ln y + .04 \ln x - .005y - .004x$$

is a first integral of the dynamical system

$$x' = .1x - .005xy$$
$$y' = .004xy - .04y.$$

Obtain the graph of E on the domain $1 \le x \le 50$, $1 \le y \le 50$. (Why not enlarge the domain to $0 \le x \le 50$, $0 \le y \le 50$?)

a) Find all maximum, minimum, and saddle points on this graph. What is the connection between the maximum of E and the equilibrium point of the dynamical system?

b) Obtain a contour plot of E on the same domain as in part (a). Compare the contours of E and the trajectories of the dynamical system. (This reveals a *conservation of "energy"* for the solutions of the Lotka–Volterra equations.)

32. a) (Continuation.) Here is another first integral of the same dynamical system as in the previous exercise:

$$H(x, y) = x^{.04}y^{.1} / \exp(.005y + .004x).$$

Obtain the graph of H and compare it to the graph of E in the previous exercise.

b) Obtain a contour plot of H, and compare the contours to the trajectories of the dynamical system.

➤   9.2 LOCAL LINEARITY

Local linearity is the central idea of chapter 3: it says that a graph looks straight when viewed under a microscope. Using this observation we were able to give a precise meaning to the *rate of change* of a function and, as a consequence, to see why Euler's method produces solutions to differential equations. At the time we concentrated on functions with a single input variable. In this section we explore local linearity for functions with two or more input variables.

Microscopic Views

Magnifying a graph

Consider the cubic $f(x, y) = x^3 - 4x - y^2$ that we used as an example in the previous section. We'll examine both the graph and the plot of f under a microscope. In the figure below we see successive magnifications of the graph near the point where $(x, y) = (1.5, -1)$. The initial graph, in the left rear, is drawn over the square

$$-2.5 \le x \le 2.5 \qquad -2.5 \le y \le 2.5.$$

With each magnification, the portion of the surface we see bends less and less. **The graph approaches the shape of a flat plane.**

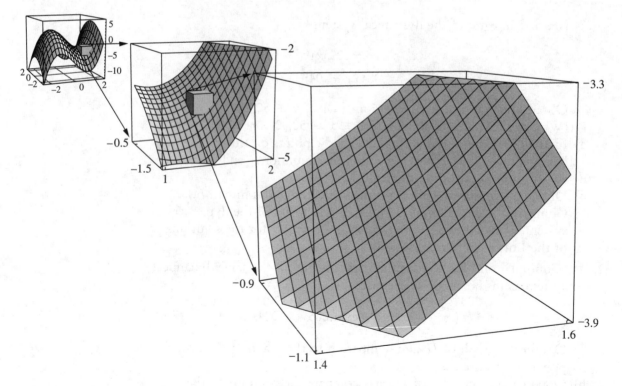

Contour plots for $f(x, y) = x^3 - 4x - y^2$ appear on the next page. Again, we magnify near the point $(x, y) = (1.5, -1)$. Each window below is a small part of the window to its left. In the large-scale plot, which is the first one on the left, the contours are quite variable in their direction and spacing. With each magnification, that variability decreases. **The contours become straight, parallel, and equally spaced.**

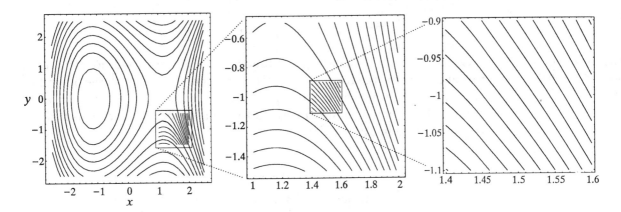

The process of magnification thus leads us to functions whose graphs are flat and whose contours are straight, parallel, and equally spaced. As we shall now see, these are the linear functions.

Linear Functions

A linear function is defined by the way its output responds to *changes* in the input. Specifically, in chapter 1 we said

$$y = f(x) \quad \text{is linear if} \quad \Delta y = m \cdot \Delta x.$$

Responses to changes in input

This is the simplest possibility: changes in output are strictly proportional to changes in input. The multiplier m is both the *rate* at which y changes with respect to x and the *slope* of the graph of f.

Exactly the same idea defines a linear function of two or more variables: the change in output is strictly proportional to the change in any one of the inputs.

The definition

> The function $z = f(x_1, x_2, \ldots, x_n)$ is **linear** if
> there are multipliers $p_1, p_2, \ldots, p_n$ for which
>
> $$\Delta z = p_1 \cdot \Delta x_1, \quad \Delta z = p_2 \cdot \Delta x_2, \quad \ldots, \quad \Delta z = p_n \cdot \Delta x_n.$$

There is one multiplier for each input variable. The multipliers are constants and they are, in general, all different.

Partial and total changes

The definition describes how z responds to each input separately. We call each $p_j \cdot \Delta x_j$ a **partial change**. The multiplier $p_j = \Delta z / \Delta x_j$ is the corresponding **partial rate of change**. Of course, several input variables may change simultaneously. In that case, the **total change** in z will just be the sum of the individual changes produced by the several variables:

$$\Delta z = p_1 \cdot \Delta x_1 + p_2 \cdot \Delta x_2 + \cdots + p_n \cdot \Delta x_n.$$

Another way to describe a linear function

Of course, if the total change of a function satisfies this condition, then each partial change has the form $p_j \cdot \Delta x_j$. (If only x_j changes, then all the other Δx_k must be 0. So Δz becomes simply $p_j \cdot \Delta x_j$.) Consequently, the function must be linear. In other words, we can use the formula for the total change as another way to define a linear function.

> The function $z = f(x_1, x_2, \ldots, x_n)$ is **linear** if there are multipliers $p_1, p_2, \ldots, p_n$ for which
>
> $$\Delta z = p_1 \cdot \Delta x_1 + p_2 \cdot \Delta x_2 + \cdots + p_n \cdot \Delta x_n.$$

Formulas for Linear Functions

From the definition to a formula

When $z = f(x_1, x_2, \ldots, x_n)$ is a linear function, we know how Δz depends on the changes Δx_j, but that doesn't tell us explicitly how z itself is related to the input variables x_j. There are several ways to express this relation as a formula, depending on the nature of the information we have about the function. For the sake of clarity, we'll develop these formulas first for a function of two variables: $z = f(x, y)$.

Given the partial rates of change and an initial point

- **The initial value form.** Suppose we know the value of a linear function at some given point—called the **initial point**—and we also know its partial rates of change. Can we construct a formula for the function? Suppose $z = z_0$ when $(x, y) = (x_0, y_0)$, and suppose the partial rates of change are

$$p = \frac{\Delta z}{\Delta x} \quad \text{and} \quad q = \frac{\Delta z}{\Delta y}.$$

If we let

$$\Delta x = x - x_0 \qquad \Delta y = y - y_0 \qquad \Delta z = z - z_0,$$

then we can write

$$z - z_0 = \Delta z$$
$$= p \cdot \Delta x + q \cdot \Delta y$$
$$= p \cdot (x - x_0) + q \cdot (y - y_0).$$

This is the initial value form of a linear function. For example, if the initial point is $(x, y) = (4, 3)$, $z = 5$, and the partial rates of change are $\Delta z / \Delta x = -\frac{1}{2}$, $\Delta z / \Delta y = +1$, the equation of the linear function can be written

$$z - 5 = -\tfrac{1}{2}(x - 4) + (y - 3).$$

- **The intercept form.** This is a special case of the initial value form, in which the initial point is the origin: $(x, y) = (0, 0)$, $z = r$. The formula becomes

$$z - r = p x + q y \qquad \text{or} \qquad z = p x + q y + r.$$

Given the partial rates of change and the z-intercept

As we shall see, the graph of this function in x, y, z space passes through the point $(x, y, z) = (0, 0, r)$ on the z-axis. This point is called the **z-intercept** of the graph. Sometimes we simply call the number r itself the z-intercept.

Notice that, with a little algebra, we can convert the previous example to the form $z = -\frac{1}{2}x + y + 4$. This is the intercept form, and the z-intercept is $z = 4$.

If there are n input variables, $x_1, x_2, \ldots, x_n$, instead of two, then a linear equation has the following forms:

The form of a linear function of n variables

$$\textbf{initial value:} \quad z - z_0 = p_1(x_1 - (x_1)_0) + p_2(x_2 - (x_2)_0) + \cdots$$
$$+ p_n(x_n - (x_n)_0)$$
$$\textbf{z-intercept:} \qquad z = p_1 x_1 + p_2 x_2 + \cdots + p_n x_n + r.$$

The Graph of a Linear Function

On the left at the top of the next page is the graph of the linear function

$$z = -\tfrac{1}{2}x + y + 4.$$

The graph is a flat plane. In particular, grid lines parallel to the x-axis (which represent vertical slices with $y = c$) are all straight lines with the same slope $\Delta z / \Delta x = -\frac{1}{2}$. The other grid lines (with $x = c$) are all straight lines with the same slope $\Delta z / \Delta y = +1$.

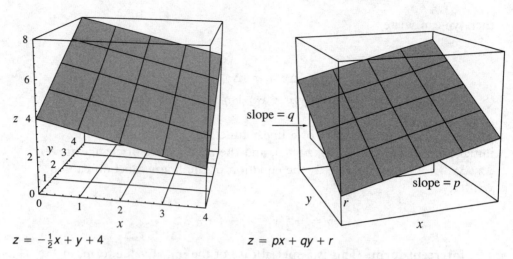

$$z = -\tfrac{1}{2}x + y + 4 \qquad\qquad z = px + qy + r$$

On the right, above, is the graph of the general linear function written in intercept form: $z = px + qy + r$. The graph is the plane that can be identified by three distinguishing features:

- It has slope p in the x-direction;
- It has slope q in the y-direction;
- It intercepts the z-axis at $z = r$.

The definition of a linear function implies that its graph is a flat plane

Let's see how we can *deduce* that the graph must be this plane. First of all, the partial rate $\Delta z / \Delta x$ tells us how z changes when y is held fixed. But if we fix $y = c$, we get a vertical slice of the graph in the x-direction. The slope of that vertical slice is $\Delta z / \Delta x = p$. Since p is constant, the slice is a straight line. The value of $y = c$ determines which slice we are looking at. Since $\Delta z / \Delta x$ doesn't depend on y, all the slices in the x-direction have the *same* slope. Similarly, all the slices in the y-direction are straight lines with the same slope q. The only surface that can be covered by a grid of straight lines in this way is a flat plane. Finally, since $z = r$ when $(x, y) = (0, 0)$, the graph intercepts the z-axis at $z = r$.

Contours of a Linear Function

A contour is a straight line

A contour of *any* function $f(x, y)$ is the set of points in the x, y plane where $f(x, y) = c$, for some given constant c. If $f = px + qy + r$, then a contour has the equation

$$px + qy + r = c \qquad \text{or} \qquad y = -\frac{p}{q}x + \frac{c - r}{q}.$$

This is an ordinary straight line in the x, y plane. Its slope is $-p/q$ and its y-intercept is $(c - r)/q$. [If $q = 0$, we can't do these divisions. However, this causes no problem; you should check that the contour is just the vertical line $x = (c - r)/p$.]

To construct a contour plot, we must give the constant c a sequence of equally spaced values c_j, with $c_{j+1} = c_j + \Delta c$. This generates a sequence of straight lines

$$px + qy + r = c_j \quad \text{or} \quad y = -\frac{p}{q}x + \frac{c_j - r}{q}.$$

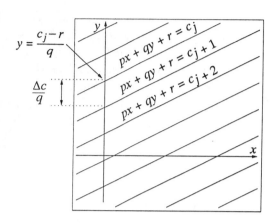

These lines all have the same slope $-p/q$, so they are parallel. (Notice the value of c doesn't affect the slope.) The y-intercept of the jth contour is $(c_j - r)/q$. Therefore, the distance along the y-axis between one intercept and the next is $\Delta c/q$. The contours are thus straight, parallel, and equally spaced. (You should check that this is still true if $q = 0$.) Note that the figure at the right, above, is drawn with $\Delta c > 0$ but $q < 0$.

Geometric Interpretation of the Partial Rates

What happens to the graph or the contour plot if you double one of the partial rates of change of a linear function? The graph on the right, below, shows the effect of doubling the partial rate with respect to x of the function $z = -\frac{1}{2}x + y + 4$. As you can see, the slope in the x-direction is dou-

Partial rates and partial slopes

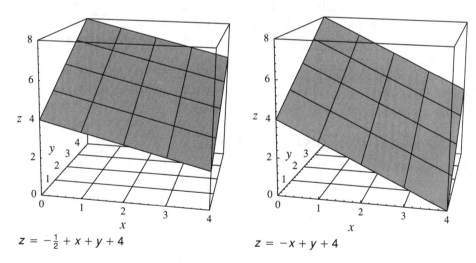

$$z = -\tfrac{1}{2} + x + y + 4 \qquad\qquad z = -x + y + 4$$

bled (from $-\frac{1}{2}$ to -1). Had we increased the partial rate by a factor of 10, the slope would have increased by a factor of 10 as well. Notice that the slope in the y-direction is not affected. Nevertheless, the overall "tilt" of the graph *has* been altered. We shall have more to say about this feature in a moment, when we introduce the **gradient** of a linear function to describe the overall tilt.

The overall tilt of a graph is altered

A change in the partial rates has a more complex effect on the contour plot. Perhaps it is more surprising, too. To make valid comparisons, we have constructed all three plots on the next page with the same spacing between

Partial rate and the spacing of contours

levels (namely, $\Delta z = 1$). Notice how the levels meet the x- and y-axes in the plot on the left ($z = -\frac{1}{2}x + y + 4$). For each unit step we take along the y-axis, the z-value increases by 1. This is the meaning of $\Delta z / \Delta y = +1$. By contrast, we have to take *two* unit steps along the x-axis to produce the same size change in z. Moreover, z *decreases* by 1 when x increases by 2. This is the meaning of $\Delta z / \Delta x = -\frac{1}{2}$. In particular, the relatively wide spacing between z-levels along the x-axis reflects the relative smallness of $\Delta z / \Delta x$.

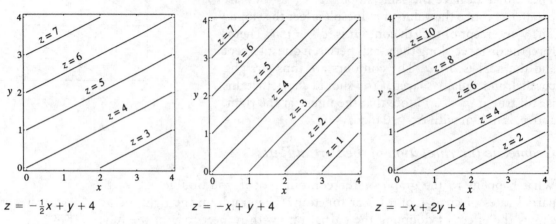

$z = -\frac{1}{2}x + y + 4$ $z = -x + y + 4$ $z = -x + 2y + 4$

The larger the partial rate, the closer the contours

Therefore, when we double the size of $\Delta z / \Delta x$—as we do in the middle plot—we should cut in half the spacing between z-levels along the x-axis. As you can see, this is exactly what happens. Notice that the spacing along the y-axis is not altered. Consequently, the contours change direction and they get packed more closely together.

Suppose we double *both* partial rates—as we do in the plot on the right. Then the spacing between contours is cut in half along both axes. Because the change is uniform, the contours keep their original direction.

The Gradient of a Linear Function

The vector of partial rates

By making use of the concept of a vector, introduced in the last chapter, we can construct still another geometric interpretation of the partial rates of a linear function. This vector is called the gradient, and it is defined in the following way.

> The **gradient** of a linear function $z = f(x, y)$ is the vector whose components are its partial rates of change:
>
> $$\text{grad } z = \nabla z = \left(\frac{\Delta z}{\Delta x}, \frac{\Delta z}{\Delta y} \right).$$

The direction of most rapid growth

The gradient is perhaps the most concise and useful tool for describing the growth of a function of several variables. To get an idea of the role that it plays, consider this question: *In what direction should we move from a given point in the x, y plane so that the value of a linear function increases most rapidly?*

Of course, the answer will depend on the linear function. Let's use $z = -\frac{1}{2}x + y + 4$ and start from the point $(x, y) = (2.4, 1.6)$. We can make z undergo a very large change simply by moving very far from this point. Therefore, to make valid comparisons, we will restrict ourselves to motions that carry us exactly one unit of distance in various directions. The vectors in the figure at the right show some of the possibilities. Their tips lie on a circle of radius 1.

Thus, to choose the direction in which z increases most rapidly, we must simply find the point on this circle where the value of z is largest. The contour line at this level must be tangent to the circle. The vector perpendicular to this contour line (see the second figure) therefore points in the direction of most rapid growth. Since perpendiculars have negative reciprocal slopes, and since all the contour lines have slope $+1/2$, it follows that the vector must have slope $-2/1$.

At the right is a magnified view of this vector. We know

$$\Delta x < 0 \qquad \frac{\Delta y}{\Delta x} = -2 \qquad \text{and} \qquad (\Delta x)^2 + (\Delta y)^2 = 1.$$

Thus $\Delta y = -2 \cdot \Delta x$, so $(\Delta x)^2 + 4(\Delta x)^2 = 1$. This implies

$$5(\Delta x)^2 = 1 \qquad \text{so} \qquad \Delta x = \frac{-1}{\sqrt{5}}, \quad \Delta y = \frac{2}{\sqrt{5}}.$$

Thus, among all the motions $(\Delta x, \Delta y)$ we have considered, we obtain the greatest change in z by choosing

$$(\Delta x, \Delta y) = \left(\frac{-1}{\sqrt{5}}, \frac{2}{\sqrt{5}}\right).$$

To determine how large this change is, we can use the alternative definition of a linear function (see page 476):

$$\Delta z = \frac{\Delta z}{\Delta x} \cdot \Delta x + \frac{\Delta z}{\Delta y} \cdot \Delta y = -\frac{1}{2} \cdot \frac{-1}{\sqrt{5}} + 1 \cdot \frac{2}{\sqrt{5}} = \frac{5}{2\sqrt{5}} = \frac{\sqrt{5}}{2}.$$

The gradient vector quickly gives us all this information. First of all, the gradient vector has the value

$$\text{grad } z = \left(\frac{\Delta z}{\Delta x}, \frac{\Delta z}{\Delta y}\right) = \left(-\frac{1}{2}, 1\right).$$

Since its slope is $1/-\frac{1}{2} = -2$, we see that it does indeed point in the direction of most rapid growth. Consequently, it is also perpendicular to the contour line. Furthermore, its *length* gives the maximum growth rate. We can see this by calculating the length using the Pythagorean theorem:

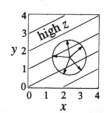

$z = -\frac{1}{2}x + y + 4$

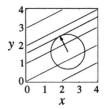

The magnitude of most rapid growth

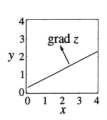

Information from the gradient

$$\text{length} = \sqrt{\left(\frac{\Delta z}{\Delta x}\right)^2 + \left(\frac{\Delta z}{\Delta y}\right)^2} = \sqrt{\frac{1}{4} + 1} = \sqrt{\frac{5}{4}} = \frac{\sqrt{5}}{2}.$$

Our findings with this example point to the following conclusion.

THEOREM. The gradient of the linear function $z = px + qy + r$ is perpendicular to its contour lines. It points in the direction in which z increases most rapidly, and its length is equal to the maximum rate of increase.

A proof

Let's see why this is true. According to the observation on the previous page, the direction of most rapid increase will be perpendicular to the contour lines. The gradient of $z = px + qy + r$ is the vector $\nabla z = (p, q)$. Its slope is q/p. On page 478 we saw that the slope of the contour lines is $-p/q$. Since these slopes are negative reciprocals, the gradient is indeed perpendicular to the contour lines.

To determine the maximum rate of increase, we must see how much z increases when we move exactly 1 unit of distance in the gradient direction. The gradient vector is (p, q), and its length is $\sqrt{p^2 + q^2}$. Therefore, the vector

$$(\Delta x, \Delta y) = \left(\frac{p}{\sqrt{p^2 + q^2}}, \frac{q}{\sqrt{p^2 + q^2}}\right)$$

is 1 unit long and in the same direction as the gradient. The increase in z along this vector is

$$\Delta z = p \cdot \Delta x + q \cdot \Delta y = p \cdot \frac{p}{\sqrt{p^2 + q^2}} + q \cdot \frac{q}{\sqrt{p^2 + q^2}} = \frac{p^2 + q^2}{\sqrt{p^2 + q^2}}$$
$$= \sqrt{p^2 + q^2}.$$

End of the proof

This *is* the length of the gradient vector, so we have confirmed that the length of the gradient is equal to the maximum rate of increase.

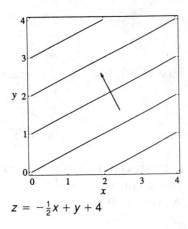

$z = -\frac{1}{2}x + y + 4$

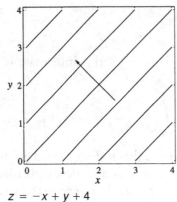

$z = -x + y + 4$

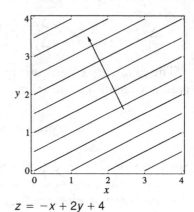

$z = -x + 2y + 4$

Shown on the previous page are the three linear functions we've already examined. In each case the gradient vector is perpendicular to the contours, and it gets longer as the space between the contours *decreases*. This is to be expected because the space between contours is also an indicator of the maximum rate of growth of the function. Widely spaced contours tell us that z changes relatively little as x and y change; closely spaced contours tell us that z changes a lot as x and y change.

Contour spacing and the length of the gradient

The connection between the gradient and the graph is particularly simple. Since the gradient (which is a vector in the x, y plane) points in the direction of greatest increase, it points in the direction in which the graph is tilted up.

The gradient points directly uphill

If we project the gradient vector onto the graph, as in the figure at right, it points directly "uphill." Putting it another way, we can say that the gradient shows us the "overall tilt" of the graph. There are two parts to this information. First, the direction of the gradient tells us which way the graph is tilted. Second, the length tells us how steep the graph is.

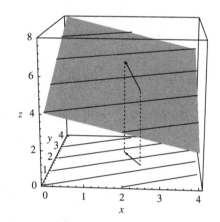

The figure at the right combines all the visual elements we have introduced to analyze a linear function: contours, graph, and gradient. Study it to see how they are related.

The Microscope Equation
Local Linearity

Let's return to arbitrary functions of two variables—that is, ones that are not necessarily linear. First we looked at magnifications of their graphs and contour plots under a microscope. We found that the graph becomes a plane, and the plot becomes a series of parallel, equally spaced lines. Next, we saw that it is precisely the linear functions which have planar graphs and uniformly parallel contour plots. Hence this function is **locally linear.**

Local linearity

Of course, not *every* function is locally linear, and even a function that *is* locally linear at most points may fail to be so at particular points. We have already seen this with functions of a single variable in chapter 3. For example, $g(x) = x^{2/3}$ is locally linear everywhere *except* the origin. It has a sharp spike there. The two-variable function $f(x, y) = (x^2 + y^2)^{1/3}$ has the same sort of spike at the origin. The two graphs at the top of the next page help make it clear that g is just a slice of f (constructed by taking $y = 0$). (Compare section 3.2, pages 100–101.) The spike is just one example; there are many other ways that a function can fail to be locally linear.

Exceptions

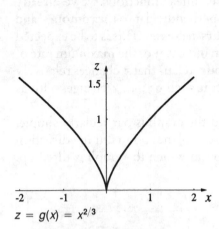

$z = g(x) = x^{2/3}$

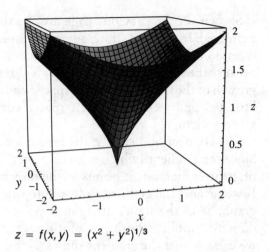

$z = f(x, y) = (x^2 + y^2)^{1/3}$

The relation between calculus and fractals

Functions that are *nowhere* locally linear are now being used in science to construct what are called **fractal** models. However, calculus does not deal with fractals. On the contrary, we remind you of the stipulation first made in chapter 3:

> **Calculus studies functions that are
> locally linear almost everywhere.**

The Microscope Equation with Two Input Variables

The equation of a microscopic view

If the function $z = f(x, y)$ is locally linear, then its graph looks like a plane when we view it under a microscope. The linear equation that describes that plane is the **microscope equation**. Since the plane is part of the graph of f, f itself must determine the form of the microscope equation. Let's see how that happens.

The idea is to reduce f to a function of one variable and then use the microscope equation for one-variable functions (described in section 3.3; see also section 3.7). Suppose the microscope is focused at the point $(x, y) = (a, b)$. If we fix y (at $y = b$), then z depends on x alone: $z = f(x, b)$. The microscope equation for this function at $x = a$ is just

The microscope equation in the x-direction...

$$\Delta z \approx \frac{\partial f}{\partial x}(a, b) \cdot \Delta x.$$

The multiplier $\partial f / \partial x$ is the rate of change of f with respect to x. We need to write it as a partial derivative because f is a function of two variables. Geometrically, the multiplier tells us the slope of a vertical slice of the graph in the x-direction.

Now reverse the roles of x and y, fixing $x = a$. The microscope equation for the function $z = f(a, y)$ at $y = b$ is

$$\Delta z \approx \frac{\partial f}{\partial y}(a, b) \cdot \Delta y.$$

...and in the
y-direction

The multiplier $\partial f / \partial y$ in this equation is the slope of a vertical slice of the graph in the y-direction. The slopes of the two vertical slices are indicated in the microscope window that appears in the foreground of the figure on the top of the opposite page.

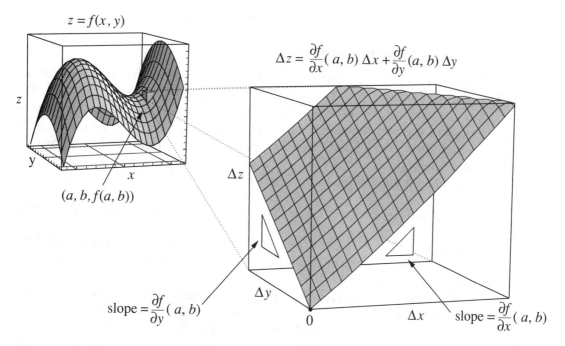

$$z = f(x, y)$$

$$\Delta z = \frac{\partial f}{\partial x}(a, b)\, \Delta x + \frac{\partial f}{\partial y}(a, b)\, \Delta y$$

$(a, b, f(a, b))$

$\text{slope} = \dfrac{\partial f}{\partial y}(a, b)$

$\text{slope} = \dfrac{\partial f}{\partial x}(a, b)$

The separate microscope equations for the x- and y-directions give us the partial changes in z. However, as we saw when we were defining linear functions (page 476), when we know all the *partial* changes, we can immediately write down the *total* change:

From partial changes
to total change

The microscope equation:

$$\Delta z \approx \frac{\partial f}{\partial x}(a, b) \cdot \Delta x + \frac{\partial f}{\partial y}(a, b) \cdot \Delta y$$

As always, the origin of the microscope window corresponds to the point $(a, b, f(a, b))$ on which the microscope is focused. The microscope coordinates

Δx, Δy, and Δz measure distances from this origin. For the sake of clarity in the figure above, we put the origin at one corner of the (three-dimensional) window.

An example

Incidentally, the function shown above is $f(x, y) = x^3 - 4x - y^2$, and the microscope is focused at $(a, b) = (1.5, -1)$. Since

$$\frac{\partial f}{\partial x} = 3x^2 - 4 = 2.75, \qquad \frac{\partial f}{\partial y} = -2y = 2$$

when $(x, y) = (1.5, -1)$, the microscope equation is

$$\Delta z \approx 2.75 \, \Delta x + 2 \, \Delta y.$$

Linear Approximation

The microscope equation gives a linear approximation

The microscope equation describes a linear function that approximates the original function near the point on which the microscope is focused. It is easy to see exactly how good the approximation is by comparing contour plots of the two functions. This is done below. In the window on the right, which shows the highest magnification, the solid contours belong to the original function $f = x^3 - 4x - y^2$. They are curved, but only slightly so. The dotted contours belong to the linear function

$$(z + 3.625) = 2.75(x - 1.5) + 2(y + 1) \qquad \text{or} \qquad z = 2.75\,x + 2\,y - 5.75.$$

[This is the microscope equation expressed in terms of the original variables x, y, and z instead of the microscope coordinates $\Delta x = x - 1.5$, $\Delta y = y - (-1)$, and $\Delta z = z - (-3.625)$.]

Comparing the contours …

The difference between the two sets of contours shows us just how good the approximation is. As you can see, the two functions are almost indistinguishable near the center of the window, which is the point $(x, y) = (1.5, -1)$. As we look farther from the center, we find the contours of f depart more and more from strict linearity.

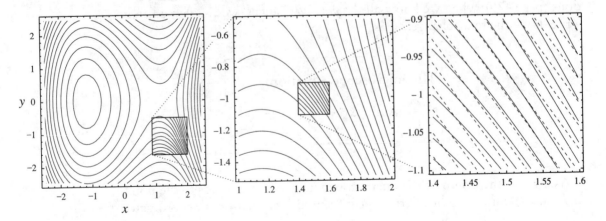

We can also compare the *graphs* of a function and its linear approximation. The graph of the linear approximation is a plane, of course. It is, in fact, the plane that is tangent to the graph of the function. In the left figure, below, we see the tangent plane to the graph of $z = f(x, y) = x^3 - 4x - y^2$ at the point where $(x, y) = (1.5, -1)$. On the right is a magnified view at the point of tangency. The graph of f is almost flat. The grid helps us distinguish it from the tangent plane, which is solid gray. Such close agreement between

...and the graphs of the function and its linear approximation

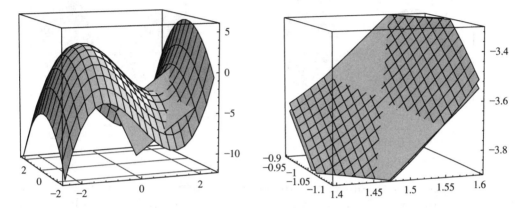

the graph and the plane demonstrates how good the linear approximation is near the point. Notice, however, that the plane diverges from the graph as we move away from the point of tangency.

At first glance you may not think that the plane in the figures above is tangent to the surface. The word "tangent" comes from the Latin *tangere*, "to touch." We sometimes take this to mean "touch at one point," like the plane in the figure at the left, below. More properly, though, two objects are **tangent** if they have the same direction at a point where they meet. The plane in the figures above *does* meet this condition—as the microscopic view helps make clear.

Tangent planes

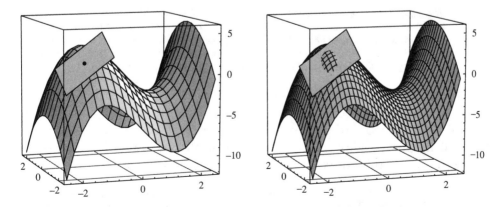

There are many different ways that a tangent plane can intersect a surface. What happens depends on the shape of the surface at the point of tangency. The

Elliptic points . . .

surface could bend the same way in all directions at that point, or it could bend up in some directions and down in others. In the first case, it will bend away from its tangent plane, so the two will meet at only one point. This is called an **elliptic point**, because the intersection turns into an ellipse if we push the plane in a bit. The right figure on the bottom of page 487 shows what happens.

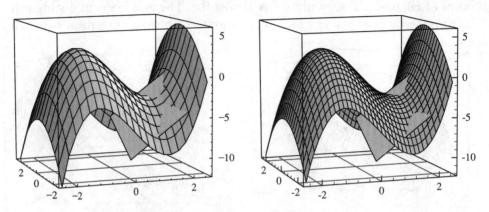

. . . and hyperbolic points

Suppose, on the other hand, that the surface bends up in some directions but down in others. Then, in some intermediate directions, it will not be bending at all. In those directions it will meet its tangent plane—which doesn't bend, either. Typically, there are two pairs of such directions. The surface and the tangent plane then intersect in an X. This always happens at a minimax (or saddle point) on a graph, and it also happens at the first point we considered on the graph of $z = x^3 - 4x - y^2$. This is shown again in the figure on the left, above. It is called a **hyperbolic point**, because the intersection turns into a hyperbola if we push the plane a bit—as on the right. The lines where the tangent plane itself intersects the surface are called **asymptotic lines**, because they are the asymptotes of the hyperbolas. (To make it easier to see the elliptical and hyperbolic intersections, we made the surface grid finer.)

> The curve of intersection between a surface and a shifted tangent plane is called the Dupin indicatrix. The Dupin indicatrix can take many forms besides the ones we have described here. However, at almost all points on almost all surfaces it turns out to be an ellipse or a hyperbola. More precisely, the indicatrix is approximately *an ellipse or a hyperbola*—in the same way that the surface itself is only approximately flat.

Parabolic points

Most points on a surface fall into one of two regions; one region consists of elliptic points, the other of hyperbolic. Points on the boundary between these two regions are said to be **parabolic**. On the graph of $z = x^3 - 4x - y^2$, if $x < 0$, the point is elliptic; if $x > 0$, it is hyperbolic; and if $x = 0$, it is parabolic. Try to confirm this yourself, just by looking at the surface.

> The classification of the points into elliptical, parabolic, and hyperbolic types is one of the first steps in studying the curvature of a surface. This is part of differential geometry; calculus provides an essential language and tool. Differential geometry is used to model the physical world at both the cosmic scale (general relativity) and the subatomic (string theory).

The Gradient

Local linearity is a powerful principle. It says that an arbitrary function looks linear when we view it on a sufficiently small scale. In particular, all these statements are approximately true in a microscope window:

Consequences of local linearity

- The contours are straight, parallel, and equally spaced;
- The graph is a flat plane (the tangent plane);
- The function has a linear formula (the microscope equation);
- The partial derivatives are the slopes of the graph in the directions of the axes.

There is one more aspect of a linear function for us to interpret—the gradient. Since partial rates become partial derivatives, we can make the following definition for an arbitrary locally linear function.

Extending the gradient to nonlinear functions

> The **gradient** of a function $z = f(x, y)$ is
> the vector whose components are the partial derivatives:
> $$\text{grad } f = \nabla f = (f_x, f_y) = \left(\frac{\partial f}{\partial x}, \frac{\partial f}{\partial y}\right).$$

The partial derivatives are functions, so the gradient varies from point to point. Thus, gradients form a **vector field** in the same sense that a dynamical system does (see chapter 8). We draw the gradient vector $(f_x(x, y), f_y(x, y))$ as an arrow whose tail is at the point (x, y).

The gradient vector field

EXAMPLE. $f(x, y) = x^3 - 4x - y^2$, $\text{grad } f = (3x^2 - 4, -2y)$.

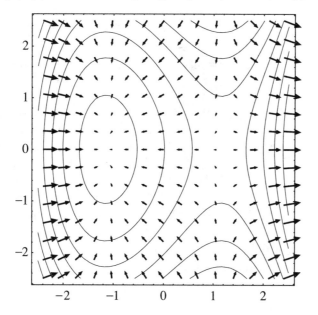

Contours and gradient vectors together

The vectors are
rescaled for clarity

There is one thing you should notice about the previous figure: we drew the gradient vectors much shorter than they actually are. For instance, at the origin grad $f = (-4, 0)$, but the arrow *as drawn* is closer to $(-.25, 0)$. The purpose of rescaling is to keep the vectors out of each other's way, so the overall pattern is easier to see.

The relation between
gradients and contours

The example shows contours as well as gradients so we can see how the two are related. The result is very striking. Even though the vectors vary in length and direction, and the contours vary in direction and spacing, the two are related the same way they were for a *linear* function (page 480 ff). First of all, each vector is perpendicular to the level curve that passes through its tail. Second, the vectors get longer where the spacing between level curves gets smaller. The similarity is no accident, of course; it is a consequence of local linearity. We can summarize and extend our observations in the following theorem. It is just a modification of the earlier theorem on the gradient of a linear function (page 482).

THEOREM.   The gradient vector field of the function $z = f(x, y)$ is perpendicular to its contour lines. At each point, the *direction* of the gradient is the direction in which z increases most rapidly; the *length* is equal to the maximum rate of increase.

A proof

To see why this theorem is true, just look in a microscope. The gradient and the contours become the gradient and the contours of the linear approximation at the point where the microscope is focused. But we already know the theorem is true for linear functions, so there is nothing more to prove.

The gradient points
uphill

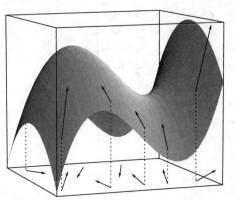

There is a direct connection between the gradient field of a function $z = f(x, y)$ and its graph. Since the gradient (which is a vector in the x, y plane) points in the direction in which z increases most rapidly, it points in the direction in which the graph is tilted up. Thus, if we project the gradient onto the graph, as we do in the figure at the left, it points directly "uphill." Since f is not a linear function, both the steepness and the uphill direction vary from point to point. The gradient vector field also varies; in this way, it keeps track of the steepness of the graph and the direction of its tilt.

The Gradient of a Function of Three Variables

Let's take a brief glance at the gradient of a function $f(x, y, z)$. It has three components:

$$\text{grad } f = \nabla f = \left(\frac{\partial f}{\partial x}, \frac{\partial f}{\partial y}, \frac{\partial f}{\partial z} \right) = (f_x, f_y, f_z),$$

and it defines a vector field in x, y, z-space. At each point, the gradient of f is perpendicular to the level set through that point, and it points in the direction in which f increases most rapidly.

EXAMPLE. In the two boxes below you can compare the gradient field of $f(x, y, z) = x^2 + y^2 - z^2$ with its level sets. At first glance, you may not find a clear pattern to the gradient vectors. After all, the picture is three-dimensional, and it is difficult to tell whether an arrow is near the front of the box or the back. However, there is a pattern: at the top and the bottom of the box, arrows point inward; closer to the middle of the box, they flare outward. The lowest

Visualizing a three-dimensional vector field

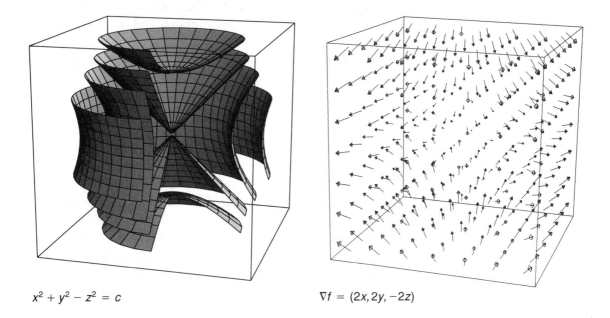

$$x^2 + y^2 - z^2 = c$$
$$\nabla f = (2x, 2y, -2z)$$

values of the function occur along the z-axis, inside the shallow bowls that sit at the top and the bottom of the box. The highest values occur outside the "equatorial belt" formed by the outermost level set. This is where the x, y plane meets the middle of the box. Notice also that the level sets are symmetric around the z-axis. The gradient field has the same symmetry, though this is harder to see.

Exercises

In many of these exercises it will be essential to have a computer program to make graphs and contour plots of functions of two variables.

1. a) Obtain the graph of $z = x^3 - 4x - y^2$ on a domain centered at the point $(x, y) = (1.5, -1)$, and magnify the graph until it looks like a plane.
   b) Estimate, by eye, the slopes in the x-direction and the y-direction of the plane you found in part (a). (You should find numbers between +2 and +3 in both cases.)

2. Obtain a contour plot of $z = x^3 - 4x - y^2$ on a domain centered at the point $(x, y) = (1.5, -1)$, and magnify the plot until the contours look straight, parallel, and equally spaced. Compare your results with the plots on page 475.

3. (Continuation.) The purpose of this exercise is to estimate the rate of change $\Delta z / \Delta x$ at $(x, y) = (1.5, -1)$, using the most highly magnified contour plot you constructed in the last exercise.

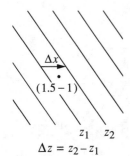

   a) What is the horizontal spacing Δx between the two contours closest to the point $(1.5, -1)$? See the illustration at the left.
   b) Find the z-levels z_1 and z_2 of those contours, and then compute $\Delta z = z_2 - z_1$.
   c) Compare the value of $\Delta z / \Delta x$ you now obtain with the slope in the x-direction that you estimated in exercise 1.

4. (Continuation.) Repeat all the work of the last exercise, this time for $\Delta z / \Delta y$.

Linear Functions

5. a) Find the z-intercept of the graph of the linear function given by the formula $z - 3 = 2(x - 4) - 3(y + 1)$.
   b. Write the formula for this linear function in intercept form.

6. a) Write, in initial value form, the formula for the linear function $z = L(x, y)$ for which $\Delta z / \Delta x = 3$, $\Delta z / \Delta y = -2$, and $L(1, 4) = 0$.
   b) Write the intercept form of the same function. What is the z-intercept of the graph of L?

7.*a) Suppose z is a linear function of x and y, and $\Delta z / \Delta x = -7$, $\Delta z / \Delta y = 12$. If (x, y) changes from $(35, 24)$ to $(33, 33)$, what is the total change in z?
   b) Suppose $z = 29$ when $(x, y) = (35, 24)$. What value does z have when $(x, y) = (33, 33)$? What value does z have when $(x, y) = (0, 0)$?
   c) Write the intercept form of the formula for z in terms of x and y.

8. Suppose z is a linear function of x and y for which we have the following information:

x	5	7	0		4	0		4
y	9	1	4	6	7	0		-1
z	2		12	-7	2		20	20

a) Fill in the blanks in this table.
b) Write the formula for z in intercept form.

9. a) Sketch the graph of $z = x - 2y + 7$ on the domain $0 \le x \le 3$, $0 \le y \le 3$.
b) Determine the slope of this graph in the x-direction, and indicate on your sketch where this slope can be found.
c) Determine the slope of this graph in the y-direction, and indicate on your sketch where this slope can be found.

10. (Continuation.) Draw the gradient vector of $z = x - 2y + 7$ in the x, y plane, and then lift it up so it sits on the graph you drew in the previous exercise. Does the gradient point directly "uphill"?

11. Sketch the graph of the linear function $z = L(x, y)$ for which

$$\frac{\Delta z}{\Delta x} = -1, \qquad \frac{\Delta z}{\Delta y} = .6, \qquad L(1, 1) = 8.$$

Be sure your graph shows clearly the slopes in the x-direction and the y-direction, and the z-intercept.

12. (Continuation.) Draw the gradient of the function L from the previous exercise, and lift it up so it sits on the graph of L you drew there. Does the gradient point directly uphill?

*13. What is the equation of the linear function $z = L(x, y)$ whose graph (a) has a slope of -4 in the x-direction, (b) a slope of $+5$ in the y-direction, and (c) passes through the point $(x, y, z) = (2, -9, 0)$?

14. Suppose $z = L(x, y)$ is a linear function whose graph contains the three points

$$(1, 1, 2) \qquad (0, 5, 4) \qquad (-3, 0, 12).$$

a) Determine the partial rates of change $\Delta z / \Delta x$ and $\Delta z / \Delta y$.
b) Where is the z-intercept of the graph?
c) If $(4, 1, c)$ is a point on the graph, what is the value of c?
d) Is the point $(2, 2, 4)$ on the graph? Explain your position.

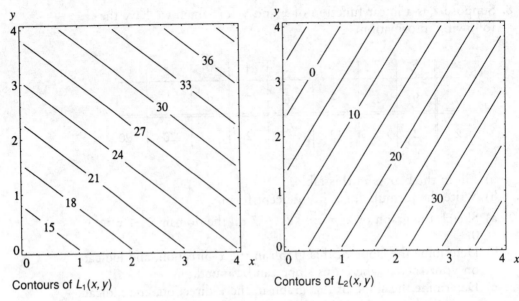

Contours of $L_1(x, y)$ Contours of $L_2(x, y)$

*15. The figure on the left, above, is the contour plot of a function
$z = L_1(x, y)$.
a) What are the values of $L_1(1, 0)$, $L_1(2, 0)$, $L_1(3, 0)$, $L_1(0, 3)$, $L_1(2, 3)$,
and $L_1(0, 0)$?
b) What are the partial rates $\Delta L_1/\Delta x$ and $\Delta L_1/\Delta y$?

16. (Continuation.) What are the values of $L_1(3, 2)$, $L_1(7, 0)$, $L_1(7, 7)$,
$L_1(1.4, 2.9)$, $L_1(-2, 9)$, and $L(-10, -100)$?
a) Find x so that $L_1(x, 0) = 0$. Find y so that $L_1(0, y) = 0$.

17. (Continuation.) Write the intercept form of the formula for $L_1(x, y)$.

18. a) Find the partial rates $\Delta L_2/\Delta x$ and $\Delta L_2/\Delta y$ of the linear function
L_2 whose contour plot is shown at the right, above.
b) Obtain the intercept form of the formula for $L_2(x, y)$.

19. The figures at the top of the next page are the graphs of L_1 and L_2.
Which is which? Explain your choice. (Note that both graphs are
shown from the same viewpoint. The x-axis is on the right, and the
y-axis is in the foreground. The z-axis has no scale on it.)

20. Determine the gradient vectors of L_1 and L_2.
a) Sketch the gradient vector of each function on the x, y plane and
on its own graph. Does the gradient point directly uphill in each
case?

21. Suppose the gradient vector of the linear function $p = L(q, r)$ is grad
$p = \nabla p = (5, -12)$. If $L(9, 15) = 17$, what is the value of $L(11, 11)$?

22. a) What is the gradient vector of the function $w = 2u + 5v$?
b) At what point on the circle $u^2 + v^2 = 1$ does w have its largest
value? What is that value?

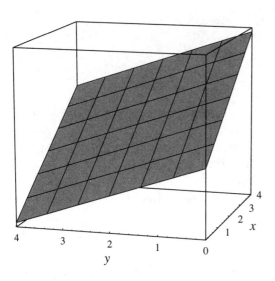
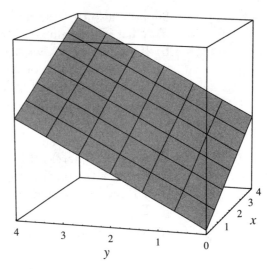

23. a) Write the formula of a linear function $z = L(x, y)$ whose gradient vector is grad $z = \nabla z = (-3, 4)$.

b) Using your formula for L, calculate the total change in L when $\Delta x = 2$, $\Delta y = 1$.

24. a) (Continuation.) In particular, continue to use your formula for $z = L(x, y)$. What is the value of $L(0, 0)$?

b) What is the maximum value of z on the circle $x^2 + y^2 = 1$? At what point (x, y) does z achieve that value?

∗c) Determine the difference between the maximum and minimum values of L on the circle $x^2 + y^2 = 1$.

25. a) What value does $z = 7x + 3y + 31$ have when $x = 5$ and $y = 2$?

b) If x increases by 2, how must y change so that the value of z doesn't change? (The change in y needed to keep z fixed when x changes is called the **trade-off**. See also chapter 3, page 154.)

The concept of a *trade-off*

c) What is the trade-off in y when x increases by α?

d) What is the trade-off in x when y increases by β?

26.∗a) Suppose $z = L(x, y)$ is a linear function for which $\Delta z / \Delta x = 5$ and $\Delta z / \Delta y = -2$. What is the trade-off in y when x increases by 50?

b) What is the trade-off in x when y increases by 1?

27. Suppose $z = L(x, y)$ is a linear function and suppose the trade-off in y when x increases by 1 is -4.

a) What is the trade-off in y when x is *decreased* by 3?

b) What is the trade-off in x when y is increased by 10? (Note that x and y are reversed here, in comparison to the earlier parts of this question.)

28. Suppose $z = L(x, y)$ is a linear function for which we know

$$L(3, 7) = -2 \qquad \frac{\Delta z}{\Delta x} = 2.$$

Suppose also that the trade-off in y when x increases by 10 is -4.

a) What is the value of $L(7, 7)$?

b) If $L(7, \beta) = -2$, what is the value of β?

c) What is the value of $\Delta z / \Delta y$?

d) Write the formula for $L(x, y)$ in intercept form.

29. **a)** Suppose the graph of the linear function $z = L(x, y)$ has a slope of -1.5 in the x-direction and -2.4 in the y-direction. What is the trade-off between x and y? That is, how much should y change when x is increased by the amount α?

b) Suppose the partial slopes become $+1.5$ and $+2.4$ in the x- and y-directions, respectively. How does that affect the trade-off? Explain.

30. **a)** Sketch in the x, y plane the set of points where $z = 2x + 3y + 7$ has the value 34.

b) If $x = 10$, then what value must y have so that the point (x, y) is on the set in part (a)?

c) If x increases from 10 to 14, how must y change so that the point (x, y) stays on the set in part (a)? In other words, what is the trade-off?

The set in the last question is called a trade-off line. *Do you see that it is just a contour line by another name?*

31. **a)** Write, in intercept form, the formula for the linear function

$$w - 4 = 3(x - 2) - 7(y + 1) - 2(z - 5).$$

b) What is the gradient vector of the linear function in part (a)?

32. Suppose the gradient of the linear function $w = L(x, y, z)$ is $\nabla w = (1, -1, 4)$. If $L(3, 0, 5) = 10$, what is the value of $L(1, 2, 3)$?

33. Describe the level sets of the function $w = f(x, y, z) = x + y + z$.

The Microscope Equation

∗34. Find the microscope equation for the function $f(x, y) = 3x^2 + 4y^2$ at the point $(x, y) = (2, -1)$.

35. **a)** (Continuation.) Use the microscope equation to estimate the values of $f(1.93, -1.05)$ and $f(2.07, -.99)$.

b) Calculate the *exact* values of the quantities in part (a), and compare those values with the estimates. In particular, indicate how many digits of accuracy the estimates have.

36. Find the microscope equation for the function $f(x, y, z) = x^2 y \sin z$ at the point $(x, y, z) = (1, 1, \pi)$.

37. Suppose $f(87, 453) = 1254$ and

$$\frac{\partial f}{\partial x}(87, 453) = -3.4 \qquad \frac{\partial f}{\partial y}(87, 453) = 4.2.$$

Estimate the following values: $f(90, 453)$, $f(87, 450)$, $f(90, 450)$, and $f(100, 500)$. Explain how you got your estimates.

38. a) (Continuation.) Find an estimate for y to solve the equation $f(87, y) = 1250$.
    b) Find an estimate for x to solve $f(x, 450) = 1275$.

39. (Continuation.) A trade-off. Go back to the starting values $x = 87$, $y = 453$, and $f(87, 453) = 1254$. If x increases from 87 to 88, how should y change to keep the value of the function fixed at 1254?

*40. a) Suppose $Q(27.3, 31.9) = 15.7$ and $Q(27.9, 31.9) = 15.2$. Estimate the value of $\partial Q / \partial x(27.6, 31.9)$.
    b) Estimate the value of $Q(27, 31.9)$.

41. Suppose $S(105, 93) = 10$, $S(110, 93) = 10.7$, $S(105, 95) = 9.3$. Estimate the value of $S(100, 100)$. Explain how you made your estimate.

42. Let P be the point $(x, y, z) = (173, -29, 553)$. Suppose $f(P) = 48$ and

$$\frac{\partial f}{\partial x}(P) = 7 \qquad \frac{\partial f}{\partial y}(P) = -2 \qquad \frac{\partial f}{\partial z}(P) = 5.$$

Estimate the value of $H(175, -30, 550)$, and explain what you did.

43. a) What is the equation of the tangent plane to the graph of $z = xy$ at the point $(x, y) = (2, -3)$?
    b) Which has a higher z-intercept: the graph or the tangent plane?

44. a) Suppose the function $H(x, y)$ has the microscope equation

$$\Delta z \approx 2.53\, \Delta x - 1.19\, \Delta y$$

at the point $(x, y) = (35, 26)$. Sketch the gradient vector ∇H at that point.
    *b) Pick the point exactly one unit away from $(35, 26)$ at which you estimate H has the largest possible value.

45. Write the microscope equation for the function $V(x, y) = x^2 y$ at the point $(x, y) = (25, 10)$.

46.*a) (Continuation.) Suppose a cardboard carton has a square base that is 25 inches on a side and a height of 10 inches. If there is an error of Δx inches in measuring the base and an error of Δy inches in measuring the height, how much error will there be in the calculated volume?

Refer to the discussion of error propagation in section 3.4

    b) Why is this a continuation of the previous question?

*47. (Continuation.) Which causes a larger percentage error in the calculated volume: a 1% error in the measurement of the length of the base, or a 1% error in the measurement of the height?

48. A large basin in the shape of
a cone is to be used as a water
reservoir. If the radius r is 186
meters and the depth h is 31
meters, how much water can the
basin hold, in cubic meters?

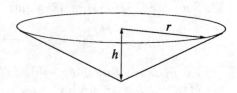

49. a) (Continuation.) If there were a 3% error in the measurement of the
radius, how much error would that lead to in the calculation of the
capacity of the basin?
b) If there were a 5% error in the measurement of the depth, how
much error would there be in the calculated capacity of the basin?
c) If *both* errors are present in the measurements, what is the total
error in the calculated capacity of the basin?

50. (Continuation.) Suppose the measured radius of the basin ($r = 186$
meters) is assumed to be accurate to within 2%. The depth has been
measured at 31 meters. Is it possible to make that measurement so
accurate that the calculated capacity is known to within 5%? How
accurate does the depth measurement have to be?

51. (Continuation.) Suppose the accuracy of the radius measurement can
only be guaranteed to be 3%. Is it still possible to measure the depth
accurately enough to guarantee that the calculated capacity is accurate
to within 5%? Explain.

52. Let's give the basin the more
realistic shape of a parabolic bowl.
If the radius r is still 186 meters
and the depth h is still 31 meters,
what is the capacity of the basin
now? Is it larger or smaller than
the conical basin of the same dimensions?

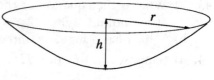

53. (Continuation.) Determine the error in the calculated capacity of the
bowl if there were a 3% error in the measurement of the radius and,
at the same time, a 5% error in the measurement of the depth.

54. (Continuation.) Is it possible for the calculated capacity to be accurate
to within 5% when the measured radius is accurate to within 2%?
How accurate does the measurement of the depth have to be to
achieve this?

55. The energy of a certain pendulum whose position is x and velocity is
v can be given by the formula

$$E(x, v) = 1 - \cos x + \tfrac{1}{2}v^2.$$

Suppose the position of the pendulum is known to be $x = \pi/2$ with
a possible error of 5% and its velocity is $v = 2$ with possible error of

10%. What is the calculated value of the energy, and how accurately is that value known?

56. **a)** A frictionless pendulum conserves energy: as the pendulum moves, the value of E does not change. Suppose $x = \pi/2$ and $v = 2$, as in the previous exercise. When x decreases by $\pi/180$ (this is 1 degree), does v increase or decrease to conserve energy?

    The conservation of energy leads to a trade-off

    b) Approximately how much does v change when x decreases by $\pi/180$?

Linear Approximations

57.*a) Write the linear approximation to $f(x, y) = \sin x \cos y$ at the point $(x, y) = (0, \pi/2)$.

   b) Write the equation of the tangent plane to the graph of $f(x, y)$ at the point $(x, y) = (0, \pi/2)$.

   c) Where does the tangent plane in part (b) meet the x, y plane?

58. Suppose $w(3, 4) = 2$ while

$$\frac{\partial w}{\partial x}(3, 4) = -1 \qquad \frac{\partial w}{\partial y}(3, 4) = 3.$$

Write the equation of the tangent plane of w at the point $(3, 4)$.

59. **a)** Write the equation of the tangent plane to the graph of the function $\varphi(x, y) = 3x^2 + 7xy - 2y^2 - 5x + 3y$ at an arbitrary point $(x, y) = (a, b)$.

   b) At what point (a, b) is the tangent plane horizontal?

   c) Magnify the graph of φ at the point you found in part (b) until it looks flat. Is it also horizontal?

The next few exercises concern the Lotka–Volterra differential equations and their linear approximations at an equilibrium point. Specifically, consider the *bounded growth* system from section 4.1 (pages 164–166):

$$R' = .1R - .00001R^2 - .005RF$$
$$F' = .00004RF - .04F.$$

60. Confirm that the system has an equilibrium point at $(R, F) = (1000, 18)$. Then obtain the phase portrait of the system near that point. What kind of equilibrium is there at $(1000, 18)$?

61. Obtain the linear approximations of the functions

$$g_1(R, F) = .1R - .00001R^2 - .005RF$$
$$g_2(R, F) = .00004RF - .04F$$

at the point $(r, F) = (1000, 18)$. Call them $\ell_1(R, F)$ and $\ell_2(R, F)$, respectively.

62. Obtain the phase portrait of the **linear** dynamical system

$$R' = \ell_1(R, F)$$
$$F' = \ell_2(R, F).$$

Does this system have an equilibrium at $(1000, 18)$? Compare this phase portrait to the phase portrait of the original *non*linear system.

The Gradient Field

63. **a)** Make a sketch of the gradient vector field of $f(x, y) = x^2 - y^2$ on the domain $-2 \le x \le 2$, $-2 \le y \le 2$.
    b) Mark on your sketch where the gradient field indicates the maximum and the minimum values of f are to be found in the domain.

64. Repeat the previous exercise, using $f(x, y) = 2x + 4x^2 - x^4 - y^2$ and the domain $-2 \le x \le 2$, $-4 \le y \le 4$. (See exercises 5 and 6 in the previous section, page 470.)

65. (Continuation.) Add to your sketch in the previous exercise a contour plot of the function $f(x, y) = 2x + 4x^2 - x^4 - y^2$ and confirm that each gradient vector is perpendicular to the contour that passes through its base. (Note: Most vectors have no contour passing through their bases, so you have to infer the shape and position of such a contour from the contours that *are* drawn.)

66. Draw a plausible set of contour lines for the function whose gradient vector field is plotted on the left below.

67. Draw a plausible gradient vector field for the function whose contour plot is shown on the right below. H marks a local maximum and L a local minimum.

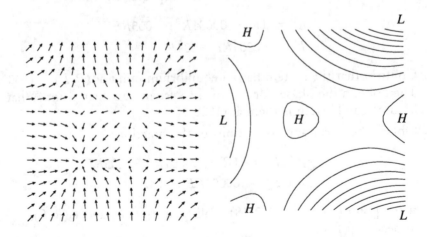

➤ 9.3 OPTIMIZATION

Optimization is the process of making the best choice from a range of possibilities. ("Optimum" is the Latin word for *best*.) We are all familiar with optimization in the economic arena: managers of an enterprise typically seek to maximize profit or minimize cost by making conscious choices. It is perhaps more surprising to learn that we sometimes use the same language to describe physical processes. For instance, the atoms in a molecule are arranged so that their total energy is minimized. A light ray travels from one point to another along the path that takes the least time. Of course atoms and photons don't make conscious choices. Nevertheless, the imagery of optimization is so vivid and useful that we try to invoke it whenever we can.

The contexts for optimization

Usually, there is a restriction—called a **constraint**—on the choices that can be made to achieve the best possible outcome. For instance, consider a factory that makes tennis rackets. We can expect that the factory managers are instructed to minimize cost *while producing a given number of rackets*. This is their constraint. Production cost is a function of many quantities that the managers can control—the number of workers, the wage scale, and the cost of the raw materials are just a few. When the managers choose values for the quantities that minimize the cost function, they must be sure the values will also satisfy the constraint.

Constrained optimization

In mathematical terms, optimization is the process of finding the minimum or maximum value of a function. The presence of constraints complicates this task, as you shall see.

Mathematical optimization

Visual Inspection

The maximum value of a function is the highest point on its graph; the minimum value is the lowest. Shown at the right is the graph of $z = x^3 - 4x - y^2$ on the domain

$$-2 \le x \le 3 \qquad -2 \le y \le 3.$$

The maximum occurs where $(x, y) = (3, 0)$, and it has the value $z = 15$. The minimum is $z \approx -12.1$, when $(x, y) \approx (1.2, 3)$. Confirm this yourself by inspecting the graph and then calculating z.

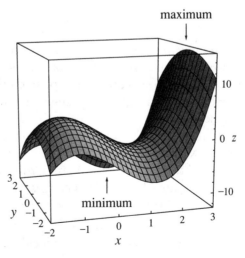

We'll use the term **extreme** to refer to either a maximum or a minimum. In the example we see several *local* extremes. These are points where the value of the function is larger or smaller than it is at any *nearby* point. There are local minima at both left-hand corners of the graph, at $(-2, -2)$ and $(-2, 3)$. There

Extremes, and local extremes

is another along the front edge, near $(1.2, -2)$. There is a local maximum in the interior, near $(-1.2, 0)$. To decide which local minimum is the *true* minimum, we must simply look. It is the same for local maxima.

The domain is a constraint

In our example, the domain of definition of the function acts as a *constraint*. If we change the domain, the positions of the extreme points can change. For example, suppose we change the position of the right-hand border in stages: first, $x \le 3$; second, $x \le 2.5$; third, $x \le 2$. (Since the graph itself doesn't change, we use a grid to show the part of the graph that satisfies the constraint in each case below.) At the start, the maximum is on the boundary. When we

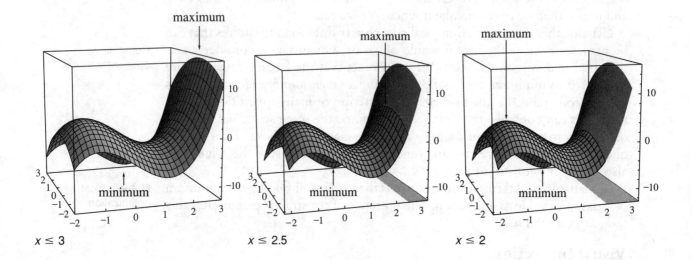

$x \le 3$ $x \le 2.5$ $x \le 2$

The maximum moves— it even jumps

first move the boundary to the left (from $x = 3$ to $x = 2.5$), the maximum just moves with it. However, when we move the boundary farther (from $x = 2.5$ to $x = 2$), the maximum jumps from the boundary to an interior point (near $(-1.2, 0)$). During these changes the minimum is not affected. It stays at the same place.

Catastrophes

The sudden jump in the maximum is called a **catastrophe**. The figures explain what happens. We impose a constraint $x \le a$, and then we reduce the value of the parameter a. At first, the maximum is at the boundary point $(a, 0)$. The position of this point changes smoothly with a, causing a gradual drop in the value of the maximum. Eventually, the maximum reaches the same value as the interior local maximum. (This happens when $a = 4/\sqrt{3}$; see the exercises.) If a continues to decrease, the local maximum at $(a, 0)$ then has a lower value than the local maximum at the interior point, so the true maximum jumps to the interior.

There are many variations on this pattern. Whenever a function depends on a parameter, the positions of its extremes do, too. There are many ways for the position of an extreme to jump suddenly *while the parameter is changing gradually*. Any such jump is called a catastrophe.

> Catastrophes make the task of optimization more interesting. If the maximum of a certain function gives the optimal solution to a problem, and that maximum jumps to a new position, then the optimal solution changes radically. For example, suppose the problem is to determine the minimum-energy configuration of atoms in a molecule. When the minimum jumps catastrophically, there is a new configuration of the same atoms, producing an isomeric form of the molecule.
>
> The quest for an optimum does not have to involve mathematical tools. A good example is John Stuart Mill's philosophy of "the greatest good for the greatest number." In politics a catastrophe is called a revolution. Even scientific research pursues an optimum in raising the question: "What is the best way to explain certain phenomena?" The consensus in the scientific community can change catastrophically, in what is called a paradigm shift or an intellectual revolution. The geological theory of plate tectonics is a familiar example. Though proposed in the 1920s, it was dismissed until the 1960s, when it was suddenly and overwhelmingly accepted.

Density Plots

We can also solve optimization problems by inspecting density plots. Suppose z is the yield from a process that is controlled by two inputs x and y, and

$$z = 3xy - 2y^2.$$

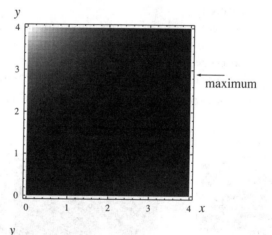

Initially we take $0 \le x \le 4$, $0 \le y \le 4$. The maximum yield is at the darkest spot in the upper density plot. It occurs on the right boundary, near $y = 3$.

The position of the maximum is subject to change if we have to impose further constraints. For instance, suppose the resource y is more limited than we first assumed, requiring us to set $y \le 2.3$. With this added constraint we see that the maximum shifts to the corner $(x, y) = (4, 2.3)$. (The points shown in a lighter gray are the ones that have been removed from consideration by the new constraint.)

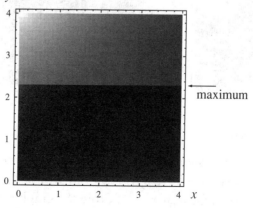

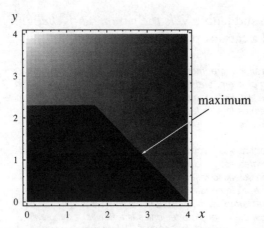

Besides limits on individual resources, we are often faced with a limit on *total* resources. In our case, let's suppose the limit has the form

$$x + y \leq 4.$$

That means all points above the line $x + y = 4$ must be removed from consideration. (They are shown in lighter gray.) This new constraint causes the maximum to shift yet again. It now appears near the point $(x, y) = (3, 1)$.

As we add constraints that force the maximum to move, the density at each new maximum is less than it was at the previous one. Thus, the value of the maximum itself decreases. In other words, each added constraint makes the optimal solution slightly "less optimal" than it had been. This is only to be expected.

Constraints reduce optimality

Extremes on the interior of a density plot

Of course the extremes may appear in the interior of the domain as well as on the boundary—and density plots can show this. Here are the density plots that correspond to the graphs of $z = x^3 - 4x - y^2$ that we saw on page 502. The maximum jumps to the interior when the value of a in the constraint $x \leq a$ drops below $a = 4/\sqrt{3}$. In all three plots the minimum appears as the

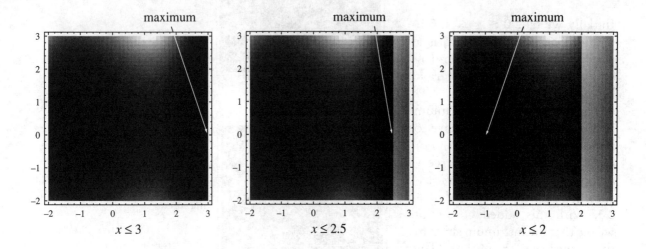

bright spot at the top of the rectangle near where $x = 1$. [The exact location is $(x, y) = (2/\sqrt{3}, 3)$.] We can also see a local minimum at the bottom of the rectangle near $x = 1$. From the graph we know there are two more at the left corners of the rectangle, but they are harder to notice in the density plots.

Contour Plots

A density plot is useful for showing the general location of the highs and lows, but we can get a more precise picture by switching to a contour plot. Let's look at the contour plots of the same problems we just analyzed using density plots.

We'll start with the function $z = 3xy - 2y^2$ we first considered on page 503 and search for its extremes subject to the constraints

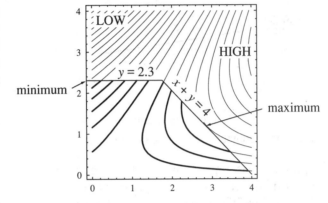

$$0 \le x$$

$$0 \le y \le 2.3$$

$$x + y \le 4$$

that we introduced earlier. From the density plot on page 503 it was obvious that the values of z increase steadily from the upper left corner of the large square to the upper part of the right side. In the contour plot, though, we need some sort of labels to show us where the low and high values of z are to be found.

To locate the extremes within the constrained region, we need to find contours that carry the lowest and the highest values of z. We can do this quite precisely. The contour that just passes through the upper left corner—and meets the region at that point alone—carries the lowest value of z. If you study the plot you can see that the contour carrying the *highest* value of z also touches the boundary at just a single point. It is the contour that is tangent to the line $x + y = 4$, and elsewhere lies outside the constrained region.

An extreme can occur where a contour is tangent to the boundary

Suppose the extreme is in the *interior* of the constrained region, rather than on the boundary. For example, the function $z = x^3 - 4x - y^2$ has an interior maximum when

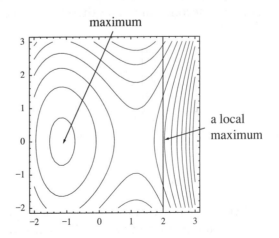

$$-2 \le x \le 2$$

$$-2 \le y \le 3.$$

Around the maximum there is a nest of concentric ovals. This pattern is characteristic for an interior extreme. Notice the local maximum where a contour is tangent to the boundary line $x = 2$.

Contour patterns at an extreme

These examples demonstrate that there are characteristic patterns that contour lines make near an extreme point. One pattern appears along a constraint curve; another pattern appears at interior points of a domain. We first met

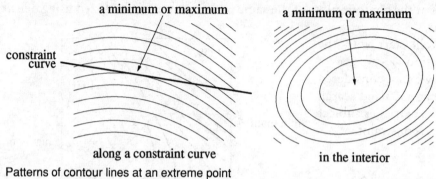

Patterns of contour lines at an extreme point

the pattern that appears at an interior extreme point when we discussed the "standard" minimum $(x^2 + y^2)$ and maximum $(-x^2 - y^2)$ in section 9.1 (page 457). **Caution:** The patterns you see here are "typical," but they do not *guarantee* the presence of an extreme point. The exercises give you a chance to explore some of the subtleties.

Dimension-Reducing Constraints

How a constraint can reduce the dimension of a problem

Constraints appear frequently in optimization problems. Thus far, however, the constraints we considered have been described by *inequalities*, like $x + y \leq 4$. Initially, x and y give us the coordinates of a point in a two-dimensional plane. The effect of the constraint $x + y \leq 4$ is to restrict the points (x, y) to just a part of that plane—but it is still a *two-dimensional* part. Sometimes, though, the constraint is given by an *equality*, like $x + y = 4$. In that case, the points (x, y) are restricted to lie on a line—which is a one-dimensional set in the plane. The second constraint therefore reduces the dimension of the problem.

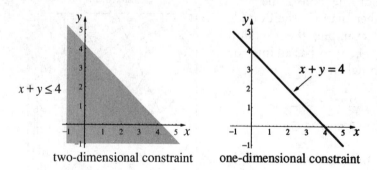

two-dimensional constraint · one-dimensional constraint

A standard form

There is a standard form for any constraint that reduces a two-dimensional optimization problem to a one-dimensional problem. The form is this:

$$\text{constraint:} \quad g(x, y) = 0.$$

For instance, the constraint $x + y = 4$ can be written this way by setting $g(x, y) = x + y - 4$. We'll look at another example in a moment.

First, though, notice that $g(x, y) = 0$ is one of the contour lines of the function $g(x, y)$, namely, the contour at level zero. Since the contours of g are curves, we often call $g(x, y) = 0$ a **constraint curve**. This curve is *one-dimensional*, even though it might twist and turn in a two-dimensional plane. We say the curve is one-dimensional because it looks like a straight line under a microscope. Likewise, a curved surface is two-dimensional because it looks like a flat plane under a microscope.

A comment on dimension

EXAMPLE.   Find the extreme values of $f(x, y) = x^2 + 8xy + 3y^2 - 5x$ subject to the constraint $g(x, y) = x^2 + y^2 - 4 = 0$. The constraint curve is a circle of radius 2, and the level curves of $z = f(x, y)$ form a set of hyperbolas. We need to find the highest and the lowest z-levels on the constraint curve.

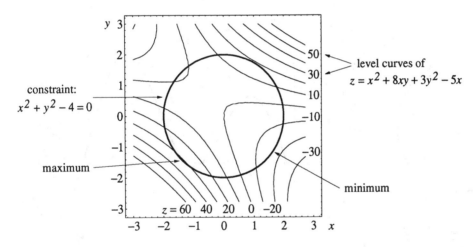

The z-levels are labeled around the right and the bottom edges of the figure. It is in the third quadrant, near the point $(-1.4, -1.5)$, that the constraint curve meets the highest z-level. At that point z is slightly more than 30. The constraint curve is evidently tangent to the contour there. The lowest z-level that the constraint curve meets is about $z = -16$, near the point $(1.6, -1.2)$. Picture in your mind how the contours between $z = -10$ and $z = -20$ fit together. Can you see that the constraint curve is tangent to the contour at the minimum?

Finding the highest and lowest levels on the constraint curve

There is still more to say about the way the constraint $x^2 + y^2 - 4 = 0$ reduces the dimension of our problem. First of all, we can describe the coordinates of any point on the constraint circle by using the circular functions:

Locate points on the constraint circle with a single variable t

$$x = 2 \cos t \qquad y = 2 \sin t.$$

We need the factor 2 because the circle has radius 2. These equations mean that x and y are now functions of a *single* variable, t. Next, consider the function

$$z = f(x, y) = x^2 + 8xy + 3y^2 - 5x$$

z becomes a function of *t*

that we seek to optimize *subject to the constraint*. But the constraint makes x and y functions of t. Thus, *when we take the constraint into account, z itself becomes a function of t:*

$$z = f(2\cos t, 2\sin t)$$
$$= (2\cos t)^2 + 8(2\cos t)(2\sin t) + 3(2\sin t)^2 - 5(2\cos t).$$

The graph of *this* function is thus just an ordinary curve in the t, z plane. It is shown on the right, below. A value of t determines a point on the circle, as shown in the contour plot on the left. (We have taken $t \approx \pi/3$.) The value of z at that point then determines the height of the graph on the right. As you can see, our chosen value of t puts z near a local maximum.

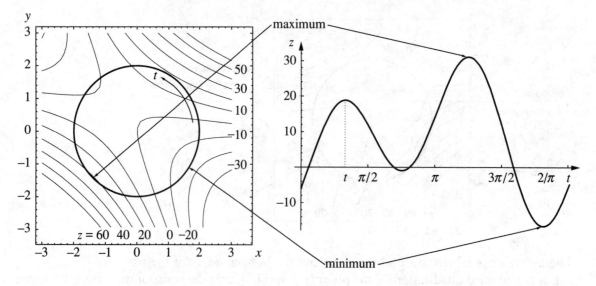

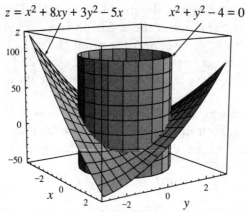

$z = x^2 + 8xy + 3y^2 - 5x$ $x^2 + y^2 - 4 = 0$

Let's look at the graph of $z = x^2 + 8xy + 3y^2 - 5x$. The constraint tells us that we should look *only* at the points on the graph that lie above the circle $x^2 + y^2 - 4 = 0$. These are the points where the graph intersects the cylinder that you see at the left. The intersection is a curve that goes up and down around the cylinder. At some point on the curve z has a maximum, and at some other point it has a minimum. (In fact, both the maximum and the minimum are visible in this view.)

Since the intersection curve lies on the cylinder, we can get a better view of the curve if we slit open the cylinder and unwrap it. Follow the sequence clockwise from the upper left. We can use coordinates to describe the curve on the flattened cylinder. The t variable takes us around the cylinder, so it becomes the horizontal coordinate. The z variable measures vertical height. The z-range in the figure above is larger than we need: it goes from -50 to $+100$. In the bottom row on the left we have rescaled the z-axis so it runs from -20 to $+35$. Compare this graph with the one on the opposite page.

Unwrap the cylinder

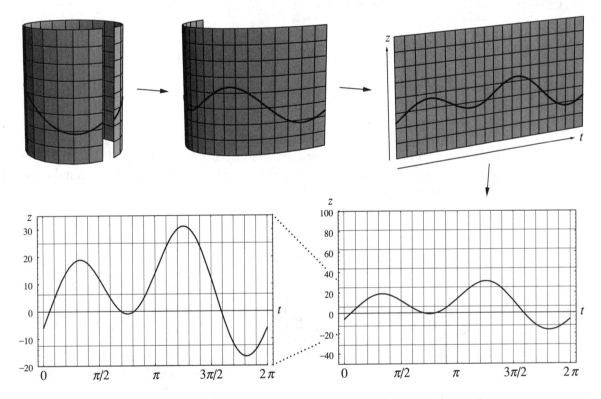

The example shows us how a constraint works to reduce the dimension of a problem in general. Suppose we want to maximize the value of the function $z = f(x, y)$, subject to the constraint $g(x, y) = 0$. Then:

A general view of the dimension-reducing effect of a constraint

- *Without* the constraint, we would look for the highest point on a *two*-dimensional surface in three-dimensional space.

- *With* the constraint, we would restrict our search to the vertical "curtain" that lies above the constraint curve (see the top of next page). The graph intersects this curtain in a curve, so we end up looking for the highest point on a *one*-dimensional curve in a two-dimensional plane. We can think of this plane as the curtain after it has been unwrapped and straightened out.

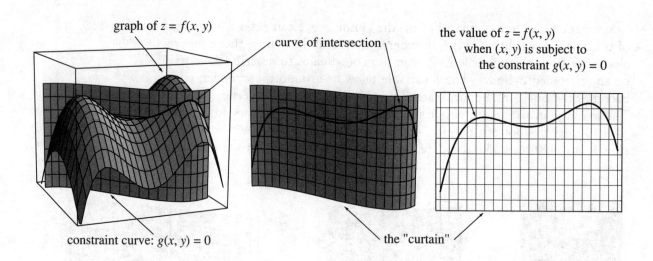

graph of $z = f(x, y)$

curve of intersection

the value of $z = f(x, y)$ when (x, y) is subject to the constraint $g(x, y) = 0$

constraint curve: $g(x, y) = 0$

the "curtain"

This is just a picture of the relation between the function and the constraint. We may still have to determine *analytically* the form that the function $f(x, y)$ takes when we impose the constraint $g(x, y) = 0$. You can find a number of possibilities in the exercises.

Extremes and Critical Points

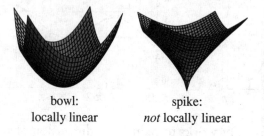

bowl: locally linear

spike: *not* locally linear

Suppose that a function $f(x, y)$ has a maximum or a minimum at an *interior* point (a, b). Suppose also that the function is *locally linear* at (a, b), so we have a "bowl" rather than a "spike" or some other irregularity in the graph. Then we must have

$$\text{grad } f(a, b) = \left(\frac{\partial f}{\partial x}(a, b), \frac{\partial f}{\partial y}(a, b) \right) = (0, 0).$$

A proof

Here is why. The gradient vector $\text{grad} f(a, b)$ tells us how the graph of $z = f(x, y)$ is tilted at the point (a, b). (We discuss the geometric meaning of the gradient on page 490.) At a maximum or a minimum, though, the graph is *not* tilted; it must be flat. Therefore, the gradient must be the zero vector.

Moving to an arbitrary number of input variables

The ideas here carry over to functions having any number of input variables. First, we give a name to a point where the gradient is zero (see page 266).

> A **critical point** of a locally linear function
> is one where the gradient vector is zero. Equivalently, all
> the first partial derivatives of the function are zero.

The observation we just made can now be restated as a theorem that connects extreme points and critical points.

THEOREM.   If a locally linear function has a maximum or a minimum at an interior point of its domain, then that point must be a critical point.

The direction of the implication in this theorem is important. Here is the theorem, written in a very abbreviated form:

A statement and its converse

$$statement: \quad extreme \implies critical.$$

When we reverse the direction of the implication, we get a new statement, abbreviated the same way:

$$converse: \quad critical \implies extreme.$$

The converse says that a critical point must be an extreme point. But that is just not true. For example, an ordinary saddle point (a minimax) is a critical point, but it is not a minimum or a maximum.

The converse of this theorem is not true

The theorem and the observation about its converse are both important in the *optimization process*—that is, the search for extremes. (Compare page 270.) Together they offer us the following guidance:

• Search for the extremes of a function among its critical points.
• A critical point may be neither a maximum nor a minimum.

Searching critical points for extremes

To see how we can find extremes by searching among the critical points of a function, we'll do a few examples. All the examples use the same basic idea. However, as the details get more complicated we bring in more powerful techniques.

EXAMPLE 1.   We'll start with the function

$$z = f(x, y) = x^3 - 4x - y^2$$

which we have frequently used as a test case in this chapter.

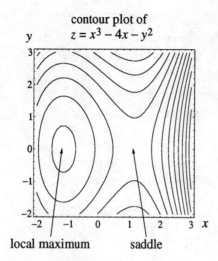

contour plot of
$z = x^3 - 4x - y^2$

local maximum   saddle

The critical points of f are the points that *simultaneously* satisfy the two equations

$$\frac{\partial f}{\partial x} = 3x^2 - 4 = 0$$

$$\frac{\partial f}{\partial y} = -2y = 0.$$

It is clear that $y = 0$ and $x = \pm \sqrt{4/3} \approx \pm 1.1547$. The two critical points are therefore

$$(+\sqrt{4/3}, 0) \quad \text{and} \quad (-\sqrt{4/3}, 0).$$

If you check the contour plot you can see that $(-\sqrt{4/3}, 0)$ is a local maximum and $(+\sqrt{4/3}, 0)$ is a saddle point.

EXAMPLE 2.   Here is a somewhat more complicated function:

$$z = g(x, y) = 2x^2y - y^2 - 4x^2 + 3y.$$

The critical points are the solutions of the equations

$$\frac{\partial g}{\partial x} = 4xy - 8x = 0 \qquad \frac{\partial g}{\partial y} = 2x^2 - 2y + 3 = 0.$$

Algebraic methods will still work, even though both variables appear in both equations. For example, we can rewrite $\partial g / \partial y = 0$ as

$$2y = 2x^2 + 3 \quad \text{or} \quad y = x^2 + \tfrac{3}{2}.$$

We can then substitute this expression for y into $\partial g / \partial x = 0$ and get

$$4x(x^2 + \tfrac{3}{2}) - 8x = 4x[x^2 + \tfrac{3}{2} - 2] = 4x[x^2 - \tfrac{1}{2}] = 0.$$

This implies $x = 0$ or $x = \pm \sqrt{1/2}$. For each x we can then find the corresponding y from the equation $y = x^2 + \tfrac{3}{2}$.

Let's work through this a second time using a geometric approach. The equations $\partial g / \partial x = 0$ and $\partial g / \partial y = 0$ both define curves in the x, y plane. The curve $\partial g / \partial y = 0$ is a parabola: $y = x^2 + \tfrac{3}{2}$. The equation $\partial g / \partial x = 0$ factors as

$$4x(y - 2) = 0 \quad \text{or} \quad x(y - 2) = 0.$$

Now a product equals 0 precisely when one of its factors equals 0, so $\partial g / \partial x = 0$ implies that *either* $x = 0$ *or* $y - 2 = 0$. In other words, the "curve" $\partial g / \partial x = 0$ consists of two lines:

$$x = 0 \quad \text{a vertical line}$$
$$y = 2 \quad \text{a horizontal line.}$$

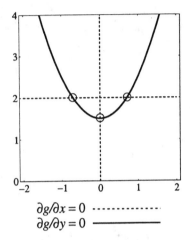

$$\partial g / \partial x = 0 \quad \text{-----------}$$
$$\partial g / \partial y = 0 \quad \text{———————}$$

The two curves are shown in the figure at the right. They intersect in three points. (The place where the horizontal and vertical lines cross is *not* one of the intersection points.)

One of the immediate benefits of the geometric approach is to make it clear that the y-coordinate of a critical point is either 2 or $\frac{3}{2}$. The critical points are therefore

$$\left(-\sqrt{1/2}, 2\right), \quad \left(0, \tfrac{3}{2}\right), \quad \left(+\sqrt{1/2}, 2\right).$$

A glance at the contour plot of g makes it clear that the first and third of these are saddle points. The middle point is a local maximum. There are several ways to determine this. One is to look at the graph of g. Another is to look at a vertical slice of the graph through the line $x = 0$. Then $z = g(0, y) = -y^2 + 3y$. Here z has a maximum when $y = \frac{3}{2}$.

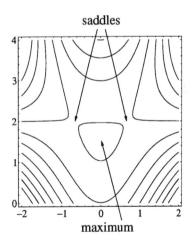

saddles

maximum

We first identified the local maximum of the function $z = x^3 - 4x - y^2$ on page 501. At the time, though, we could only estimate its position by eye. Now, however, we can specify its location exactly, because we have analytical tools for finding critical points. As the next example shows, these tools are useful even when we can't carry out the algebraic manipulations.

EXAMPLE 3. Here is a function whose critical points we *can't* find using just algebraic computations:

$$z = h(x, y) = x^2 y^2 - x^4 - y^4 - 2x^2 + 5xy + y.$$

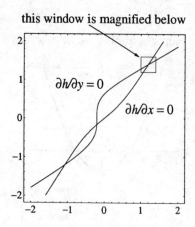

this window is magnified below

$\partial h/\partial y = 0$

$\partial h/\partial x = 0$

The equations for the critical points are

$$\frac{\partial h}{\partial x} = 2xy^2 - 4x^3 - 4x + 5y = 0$$

$$\frac{\partial h}{\partial y} = 2x^2y - 4y^3 + 5x + 1 = 0.$$

Even though we can't solve the equations algebraically, we can plot the curves they define by using the contour-plotting program of a computer. This is done at the left. The curves intersect in three points. (If you draw the plots on a large scale, you will find that these are still the only intersections.)

A contour plot of $z = h(x, y)$ itself reveals that the middle point is a saddle and the outer two are local maxima. Let's focus on the maximum in the upper right and determine its position more precisely.

We can always use a microscope. But if we magnify the contour plot, we just get a set of nested ovals. The maximum would lie somewhere inside the smallest—but we wouldn't know quite where. By contrast, the curves $\partial h/\partial x = 0$ and $\partial h/\partial y = 0$ give us a pair of "crosshairs" to focus on. Even with relatively little magnification we can see that the maximum is at $(1.202\ldots, 1.404\ldots)$.

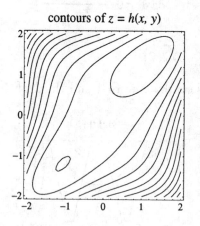

contours of $z = h(x, y)$

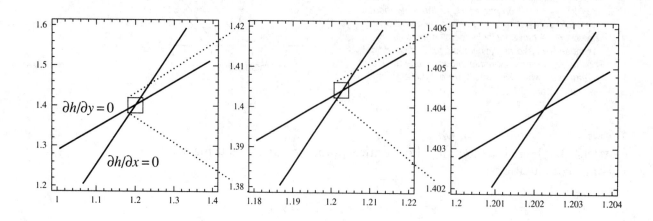

$\partial h/\partial y = 0$

$\partial h/\partial x = 0$

The Method of Steepest Ascent

Here is yet another way to find the maximum of a function $z = f(x, y)$. Imagine that the graph of f is a landscape that you're standing on. You want to get to the highest point. To do that you just walk uphill. But the uphill direction on the landscape is given by the gradient vector field of f (see page 490). So you move to higher ground by following the gradient field.

Walk uphill to get to a maximum

In fact, the gradient field defines a **dynamical system** of exactly the sort we studied in chapter 8. The differential equations are

Trajectories of the gradient dynamical system . . .

$$\frac{dx}{dt} = \frac{\partial f}{\partial x}(x, y)$$

$$\frac{dy}{dt} = \frac{\partial f}{\partial y}(x, y).$$

Because the gradient points uphill, the trajectories of this dynamical system also go uphill. Trajectories flow to the attractors of the system; these are the local maxima of the function. Furthermore, since the gradient points in the direction f increases *most rapidly*, the trajectories follow paths of **steepest ascent** to the maxima. This explains the name of the method.

. . . lead to the local maxima

EXAMPLE 1.   Let's see how the method of steepest ascent will find the local maxima of the function

$$z = h(x, y) = x^2 y^2 - x^4 - y^4 - 2x^2 + 5xy + y$$

which we considered in the previous example. The gradient field is

$$\frac{dx}{dt} = 2xy^2 - 4x^3 - 4x + 5y$$

$$\frac{dy}{dt} = 2x^2 y - 4y^3 + 5x + 1.$$

As you can see at the right, some of the trajectories flow to a local maximum near $(-1, -1)$, while others flow to the maximum whose position we determine on the opposite page. Each attractor has its own basin of attraction (as described in chapter 8). Therefore, the maximum found by the method of steepest ascent depends on the initial point of the trajectory.

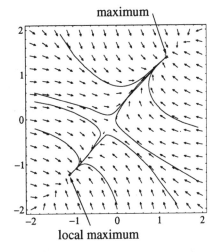

The method of steepest descent

If we replace the gradient vectors by their negatives, the new field will point directly *downhill*—to the local minima. Using the trajectories of the negative gradient field to find the minima is thus called the method of steepest descent. In the following example we use this method to investigate an economic question.

EXAMPLE 2.   Manufacturing companies ship their products to regional warehouses from large distribution centers. Suppose a company has regional warehouses at A, B, and C, as shown on the map at the left. Where should it put its distribution center X so as to minimize the total cost of supplying the three regional warehouses?

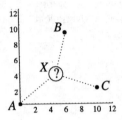

This is a complicated problem that depends on many factors. For example, X probably should be put near major roads. The managers may also want to choose a location where labor costs are lower. Certainly, the total distance between the center and the three warehouses is important. Let's simply get a *first approximation* to a solution by concentrating on the last factor. We will find the position for X that minimizes the total straight line distance (the distance "as the crow flies") from X to the three points A, B, and C. The map shows these distances as three dotted lines.

Minimize the total distance

To describe the various positions we have introduced a coordinate system in which

$$A : (0, 0) \qquad B : (6, 9) \qquad C : (10, 2).$$

The coordinates here are arbitrary. That is, they don't represent miles, or kilometers, or any of the usual units of distance—but they are *proportional* to the usual units, so we can measure with them. If we let the unknown position of X be (x, y), then we seek to minimize the function

$$S(x, y) = \sqrt{x^2 + y^2} + \sqrt{(x - 6)^2 + (y - 9)^2} + \sqrt{(x - 10)^2 + (y - 2)^2}.$$

According to the method of steepest descent, we want to find the attractor of this dynamical system:

$$\frac{dx}{dt} = -\frac{x}{\sqrt{x^2 + y^2}} - \frac{x - 6}{\sqrt{(x - 6)^2 + (y - 9)^2}} - \frac{x - 10}{\sqrt{(x - 10)^2 + (y - 2)^2}}$$

$$\frac{dy}{dt} = -\frac{y}{\sqrt{x^2 + y^2}} - \frac{y - 9}{\sqrt{(x - 6)^2 + (y - 9)^2}} - \frac{y - 2}{\sqrt{(x - 10)^2 + (y - 2)^2}}.$$

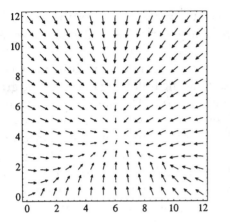

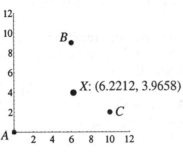

the optimum location for
the distribution center

As you can see from the vector field of the dynamical system (above left), there is a single attractor, near the point $(6, 4)$. This implies that the total distance function $S(x, y)$ has a single global minimum—which is what our intuition about the problem would lead us to expect.

> The attractor is a global minimum

To find the position of X more exactly, you can do the following. First, obtain a solution $(x(t), y(t))$ to the system of differential equations with an arbitrary initial condition—for example,

$$x(0) = 1 \qquad y(0) = 1.$$

Then, obtain the coordinates of the attractor by evaluating $(x(t), y(t))$ for larger and larger values of t, stopping when the values of $x(t)$ and $y(t)$ stabilize. You will find that

> The attractor is the limit point of a solution

$$X = \lim_{t \to \infty} (x(t), y(t)) = (6.22120\ldots, 3.96577\ldots).$$

The important point to note here is that it is not necessary to plot the vector field—or any other graphic aid, such as a contour plot or graph. You simply need to solve a system of differential equations. For example, the values above were found by modifying the computer program SIRVALUE that we introduced in chapter 2. In summary: *The method of steepest descent (or ascent) requires no graphical tools, but only a basic differential equation solver.*

> The method needs only a differential equation solver

Lagrange Multipliers

In searching for the interior extremes of a function $f(x, y)$, we have seen that it is helpful to solve the critical point equations:

$$\frac{\partial f}{\partial x} = 0 \qquad \frac{\partial f}{\partial y} = 0.$$

Equations for a constrained extreme

There is a similar set of equations we can use in the search for the *constrained* extremes of f. These equations involve a new variable called a Lagrange multiplier.

So, suppose the function $f(x, y)$ has an extreme on the constraint curve $g(x, y) = 0$ at the point (a, b). According to the following diagram, the gradient vectors ∇f and ∇g must be parallel at (a, b). (This means they are in the same direction or in opposite directions.) Here is why. We already know (see page 506) that the level curve of f that passes through the constrained maximum or minimum must be tangent to the constraint curve. Now, the gradient vector ∇f at any point is perpendicular to the level curve of f through that point—and the same is true for g. At a point where the level curves are tangent, the gradients ∇f and ∇g are perpendicular to the *same* curve, and must therefore be parallel.

∇f and ∇g must be parallel at a point where f has an extreme along the constraint curve $g = 0$

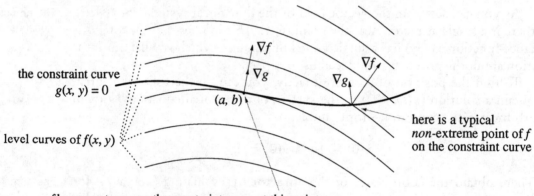

the constraint curve
$g(x, y) = 0$

∇f

∇g

∇f

∇g

(a, b)

level curves of $f(x, y)$

here is a typical *non*-extreme point of f on the constraint curve

f has an extreme on the constraint curve at this point

The multiplier equation

Parallel vectors are multiples of each other. Specifically, at a point (a, b) where ∇f and ∇g are parallel, there must be a number λ for which

$$\nabla g(a, b) = \lambda \cdot \nabla f(a, b).$$

The multiplier λ is called a **Lagrange multiplier**. In the figure above, $\lambda \approx 1/2$. If ∇g and ∇f were in *opposite* directions, then λ would be negative.

Joseph Louis Lagrange (1736–1813) was a French mathematician and a younger contemporary of Leonhard Euler. Like Euler he played an important role in making calculus the primary analytical tool for studying the physical world. He is particularly noted for his contributions to celestial mechanics, the field where Isaac Newton first applied the calculus.

If we write out the multiplier equation using the components of ∇f and ∇g, we get

$$\left(\frac{\partial g}{\partial x}(a, b), \frac{\partial g}{\partial y}(a, b) \right) = \lambda \left(\frac{\partial f}{\partial x}(a, b), \frac{\partial f}{\partial y}(a, b) \right) = \left(\lambda \frac{\partial f}{\partial x}(a, b), \lambda \frac{\partial f}{\partial y}(a, b) \right).$$

In a vector equation, the vectors are equal component by component. Thus,

$$\frac{\partial g}{\partial x}(a, b) = \lambda \frac{\partial f}{\partial x}(a, b)$$

$$\frac{\partial g}{\partial y}(a, b) = \lambda \frac{\partial f}{\partial y}(a, b).$$

Let's return to the main question, which can be stated this way: How do we determine where the function $f(x, y)$ has a maximum or a minimum, subject to the constraint $g(x, y) = 0$? If we let (a, b) denote the point we seek, then we see that a and b satisfy three equations:

> How can we find a constrained maximum or minimum?

the constraint equation : $\quad g(a, b) = 0$

the multiplier equations :
$$\begin{cases} \dfrac{\partial g}{\partial x}(a, b) = \lambda \dfrac{\partial f}{\partial x}(a, b) \\[2mm] \dfrac{\partial g}{\partial y}(a, b) = \lambda \dfrac{\partial f}{\partial y}(a, b). \end{cases}$$

In fact, there are *three* unknowns in these equations: a, b, and λ. When we solve the three equations for the three unknowns, we will determine the location of the constrained extreme. (We'll also have a piece of information we can throw away: the value of λ.)

EXAMPLE.   Find the maximum of $f(x, y) = x^p y^{1-p}$ subject to the constraint $x + y = c$. There are two parameters in this problem: p and c. We assume that $0 < p < 1$ and $0 < c$. We introduce parameters to remind you that analytic methods (such as Lagrange multipliers) are especially valuable in solving problems that depend on parameters.
  We let $g(x, y) = x + y - c$. Then

$$\nabla g = (1, 1) \qquad \nabla f = \left(p x^{p-1} y^{1-p}, (1-p) x^p y^{-p}\right),$$

so the three equations we must solve are

$$x + y - c = 0$$
$$1 = \lambda p x^{p-1} y^{1-p}$$
$$1 = \lambda(1-p) x^p y^{-p}.$$

Since the second and third equations both equal 1, we can set them equal to each other:

$$\lambda p x^{p-1} y^{1-p} = \lambda(1-p) x^p y^{-p}.$$

We can cancel the two λs and combine the powers of x and y to get

$$px^{-1}y = 1 - p \qquad \text{or} \qquad \frac{y}{x} = \frac{1-p}{p}.$$

According to the first of the three equations, $y = c - x$. If we substitute this expression for y into the last equation, we get

$$\frac{c - x}{x} = \frac{1-p}{p} \qquad \text{or} \qquad p(c - x) = (1 - p)x.$$

This equation reduces to $pc = x$, which gives us the x-coordinate of the maximum. To get the y-coordinate, we use $y = c - x = c - pc = c(1 - p)$. To sum up, the maximum is at

$$(x, y) = (cp, c(1 - p)) = c(p, 1 - p).$$

Exercises

When searching for an extreme, be sure to zoom in on the graph or plot you are using as you narrow down the location of the point you seek.

1. Inspect the graph of $z = xy$ to find the maximum value of z subject to the constraints

$$x \geq 0 \qquad y \geq 0 \qquad 3x + 8y \leq 120.$$

2. Inspect the graph of $z = 5x + 2y$ to find the minimum value of z subject to the constraints

$$x \geq 0 \qquad y \geq 0 \qquad xy \leq 10.$$

3. Inspect a contour plot of $z = 3xy - y^2$ to find the maximum value of z subject to the constraints

$$x \geq 0 \qquad y \geq 0 \qquad x + y \leq 5.$$

*4. (Continuation.) Add the constraint $x \leq a$ to the preceding three, where a is a parameter that takes values between 0 and 5. Find the maximum value of z subject to all four constraints. Describe how the position of the constraint depends on the value of the parameter a.

5. Find the maximum and minimum values of $z = 12x - 5y$ when (x, y) is exactly 1 unit from the origin.

How is the gradient involved here?

6. a) Find the maximum and minimum value of $z = px + qy$ when (x, y) is exactly 1 unit from the origin.
   b) At what point is the maximum achieved; at what point is the minimum achieved?

7. Use a graph to locate the maximum value of the function

$$z = 2xy - 5x^2 - 7y^2 + 2x + 3y.$$

There are no constraints.

8. Use a graph to find the maximum value of $z = 6x + 12y - x^3 - y^3$, subject to the constraints

$$x \geq 0 \qquad y \geq 0 \qquad x^2 + y^2 \leq 100.$$

9. a) Locate the position of the minimum of $x^4 - 2x^2 - \alpha x + y^2$ as a function of the parameter α.
   b) The position of the minimum jumps catastrophically when α passes through a certain value. At what value of α does this happen, and what jump occurs in the minimum?

10. a) Locate the maximum of $x^3 + y^3 - 3x - 3y$ subject to

$$x \leq 3 \qquad y \leq 0 \qquad x + y \leq \beta.$$

   The position of the maximum depends on the value of the parameter β, which you can assume lies between 0 and 5.
   b) The position of the maximum jumps catastrophically when β crosses a certain threshold value β_0. What is β_0?

*11. Find the maximum value of $x^2 y$ in the first quadrant, subject to the constraint $x + 5y = 10$.

12. Find the maximum and minimum values of $z = 3x + 4y$ subject to the single constraint $x^2 + 4xy + 5y^2 = 10$.

13. Find all the critical points of the following functions.
    a) $3x^2 + 7xy + 2y^2 + 5x - 6y + 3$.
    b) $\sin x \sin y$ on the domain $-4 \leq x \leq 4$, $-4 \leq y \leq 4$.
    c) $\sin xy$ on the domain $-4 \leq x \leq 4$, $-4 \leq y \leq 4$.
    d) $\exp(x^2 + y^2)$.
    e) $x^3 + y^3 - 3x - 3y$.
    f) $x^3 - 3xy^2 - x^2 - y^2$. (There are four critical points; three are saddles.)

14. a) Find the nine critical points of the function

$$C(x, y) = (x^2 + xy + y^2 - 1)(x^2 - xy + y^2 - 1).$$

   Four are minima, four are saddles, and one is a maximum.
   b) Mark the locations of the critical points on a suitable contour plot of $C(x, y)$.

15. a) Locate and classify the critical points of the energy integral of a pendulum:

$$E(x, v) = 1 - \cos x + \tfrac{1}{2} v^2.$$

   b) Compare the *critical points* of E with the *equilibrium points* of the dynamical system associated with this energy integral.

16. Use the method of steepest descent to find the minimum of the function

$$z = p(x, y) = e^{2+y-x^2-y^2} \sin x.$$

The Distribution Problem

The next two exercises are modifications of the distribution problem on page 516. In the example we assumed that deliveries from the center X to each of the regional warehouses A, B, and C happened equally often. These exercises assume that deliveries to some warehouses are more frequent than others.

17. Suppose that one truck makes deliveries to A five times each week, while a second truck is used to make three deliveries to B and two to C each week. It makes sense to locate X so that the *total weekly travel* is minimized, rather than just the total distance to the three warehouses. To get the total weekly travel, we should:

   • Multiply the distance from X to A by 5;
   • Multiply the distance from X to B by 3;
   • Multiply the distance from X to C by 2.

   (Actually, the total *round-trip* distances are twice these values, but the proportions would remain the same, so we can use these numbers.) Thus, the function to minimize is

$$T(x, y) = 5\sqrt{x^2 + y^2} + 3\sqrt{(x-6)^2 + (y-9)^2} + 2\sqrt{(x-10)^2 + (y-2)^2}.$$

   a) Use the method of steepest descent to find the minimum of T.
   b) Compare the location of the distribution center X as determined by T to its location determined by the function S of Example 2 in the text on page 517. Would you *expect* the location to change? In what direction? Does the calculated change in position agree with your intuition?

*18. Suppose that deliveries to A are twice as frequent as deliveries to either B or C. (For example, two trucks make the round-trip to A each day, but only one truck to B and one to C.) Where should the distribution center X be located in these circumstances? Explain how you got your answer.

19. A company which has four offices around the country holds an annual meeting for its top executives. The location of each office, and the number of executives at that office, are given in the following table. (The coordinates x and y of the position are given in arbitrary units.)

Office	Executives	x	y
A	32	200	300
B	17	1920	1100
C	20	2240	450
D	41	2875	1150

Where should the meeting be held if the location depends *solely* on the total travel cost for all the participants? Assume that the travel cost, per mile, is the same for every participant.

The Best-Fitting Line

Suppose we've taken measurements of two quantities x and y, and obtained the results shown in the table and graph below. We assume that y depends on x according to some rule that we don't happen to know. In particular, we'd like to know what y is when $x = 5$. We have no data. Can we *predict* what y should be?

x	y
0	1
1	3
2	4
3	3
4	5
5	?

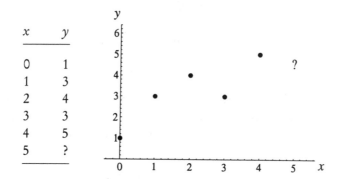

Here is a common approach to the question. We assume there is a simple underlying relation between y and x. However the measurements that give us the data contain errors or "noise" of some sort that obscure the relationship. The simplest relation is a linear function, so we assume that there is a formula $Y = mx + b$ that describes the connection between x and y.

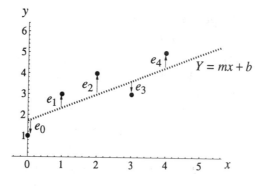

Which line should we choose? In other words, how should we choose m and b? Since the data points don't lie on a line, there is no perfect solution.

For any choices, we must expect a difference e_j between the jth data value y_j and the value $Y_j = mj + b$ predicted by the formula. These differences are the *errors* we assume are present.

A reasonable way to proceed is to **minimize** the total error. Even this involves choices. In the figure at the bottom of page 523, e_0 and e_3 are negative, so the ordinary total could be zero, or nearly so, even if the individual errors were large. We need a total that *ignores* the signs of the errors. Here is one:

$$\text{absolute error:} \quad |e_0| + |e_1| + |e_2| + |e_3| + |e_4| = \text{AE}.$$

Here is another:

$$\text{squared error:} \quad e_0^2 + e_1^2 + e_2^2 + e_3^2 + e_4^2 = \text{SE}.$$

In the following table we compare the data y_j with the calculated values Y_j and the resulting errors e_j.

x	y	Y	$e = y - Y$
0	1	b	$1 - b$
1	3	$m + b$	$3 - m - b$
2	4	$2m + b$	$4 - 2m - b$
3	3	$3m + b$	$3 - 3m - b$
4	5	$4m + b$	$5 - 4m - b$

The total errors are functions of m and b

To get the values of AE and SE, we take either the absolute values or the squares of the elements of the rightmost column, and then add. In particular, the table makes it clear that both total errors are functions of m and b. The absolute error is

$$\text{AE}(m, b) = |1 - b| + |3 - m - b| + |4 - 2m - b| + |3 - 3m - b| + |5 - 4m - b|.$$

*20. Inspect a graph and a contour plot to determine the values of m and b which minimize $\text{AE}(m, b)$.

21. a) Using a best-fitting line from the previous question, find the predicted value of y when $x = 5$.
   b) Since there is a range of best-fitting lines, there should be a range of predicted values for y when $x = 5$. What is that range?

22. Write down the function $\text{SE}(m, b)$ that describes the squared error in the fit of a straight line to the data given above.

23. a) Use a graph and a contour plot to locate the minimum of the function $\text{SE}(m, b)$ from the previous exercise. Indicate how many digits of accuracy your answer has.
   b) Use the method of steepest descent to locate the minimum. How many digits of accuracy does *this* method yield?

▷ 9.4 CHAPTER SUMMARY

The Main Ideas

- The **graph** of a function of two variables is a two-dimensional surface in a three-dimensional space.
- A function of two variables can also be viewed using a **density plot**, a **terraced density plot**, or a **contour plot**. The latter is a set of **level curves** drawn on a flat plane.
- A **contour plot** of a function of three variables is a collection of **level surfaces** in three-dimensional space.
- The graph of a **linear function** is a flat plane, and its contour plot consists of straight, parallel, and equally spaced lines.
- The **gradient** of a linear function is a vector whose components are the partial rates of change of the function.
- Under a **microscope**, the graph of a function of two variables becomes a flat plane. A contour plot turns into a set of straight, parallel, and equally spaced lines.
- The multipliers in the **microscope equation** for a function are its **partial derivatives**:

$$\Delta z = \frac{\partial f}{\partial x_1}(a, b)\, \Delta x_1 + \cdots + \frac{\partial f}{\partial x_n}(a, b)\, \Delta x_n.$$

- The **gradient** of a function is a vector whose components are the partial derivatives of the function. Its magnitude and direction give the greatest **rate of increase** of the function at each point.
- **Optimization** is a process that involves finding the **maximum** or **minimum** value of a function. There may be **constraints** present that limit the scope of the search for an **extreme**.
- Extremes can be found at **critical points**, where all partial derivatives of a locally linear function are zero.
- The **method of steepest ascent** introduces the power of dynamical systems into the optimization process.

Expectations

- Using appropriate computer software, you should be able to make a **graph**, a **terraced density plot**, and a **contour plot** of a function of two variables.
- Using appropriate graphical representations of a function of two variables, you should be able to recognize its **maxima**, **minima**, and **saddle points**.
- You should be able to estimate the **partial rates of change** of a function of two variables at a point by zooming in on a contour plot.

- You should be able to recognize the various forms of a **linear function** of several variables and transform the representation of the function from one form to another.
- You should be able to describe the geometric meaning of the partial rates of change of a linear function of two variables.
- You should be able to find the **gradient** of a function of several variables at a point.
- You should know how the gradient of a function of two variables is related to its level curves.
- You should be able to write the **microscope equation** for a function of two variables at a point.
- You should be able to use the microscope equation for a function of two variables at a point to estimate values of the function at nearby points, to find the **trade-off** in one variable when the other changes by a fixed amount, and to estimate errors.
- You should be able to find the **linear approximation** to a function of two variables at a point.
- You should be able to find the equation of the **tangent plane** to the graph of a function of two variables at a point.
- You should be able to sketch the **gradient vector field** of a function of two variables in a specified domain.
- You should be able to sketch a plausible set of contour lines for a function whose gradient vector field is given; you should be able to sketch a plausible gradient vector field for a function whose contour plot is given.
- You should be able to find the critical points of a function of two variables, and you should be able to determine whether a critical point is an extreme by inspecting a graph or a contour plot.
- You should be able to find a local maximum of a function of two variables by the method of **steepest ascent**.
- You should be able to find an extreme of a function of two variables subject to a constraint either by inspecting a graph or contour plot or by the method of **Lagrange multipliers**.

CHAPTER 10

SERIES AND APPROXIMATIONS

An important theme in this book is to give **constructive** definitions of mathematical objects. Thus, for instance, if you needed to evaluate

$$\int_0^1 e^{-x^2} dx,$$

you could set up a Riemann sum to evaluate this expression to any desired degree of accuracy. Similarly, if you wanted to evaluate a quantity like $e^{.3}$ from first principles, you could apply Euler's method to approximate the solution to the differential equation

$$y'(t) = y(t) \quad \text{with initial condition} \quad y(0) = 1,$$

using small enough intervals to get a value for $y(.3)$ to the number of decimal places you needed. You might pause for a moment to think how you would get $\sin(5)$ to seven decimal places—you wouldn't do it by drawing a unit circle and measuring the y-coordinate of the point where this circle is intersected by the line making an angle of 5 radians with the x-axis! Defining the sine function to be the solution to the second-order differential equation $y'' = -y$ with initial conditions $y = 0$ and $y' = 1$ when $t = 0$ is much better if we want to construct values of the function with more than two decimal places accuracy.

What these examples illustrate is the fact that the only functions our brains or digital computers can evaluate directly are those involving the arithmetic

Ordinary arithmetic
lies at the heart of all
calculations

operations of addition, subtraction, multiplication, and division. Anything else we or computers evaluate must ultimately be reducible to these four operations. But the only functions directly expressible in such terms are polynomials and rational functions (i.e., quotients of one polynomial by another). When you use your calculator to evaluate ln 2, and the calculator shows .69314718056, it is really doing some additions, subtractions, multiplications, and divisions to compute this 11-digit approximation to ln 2. There are no obvious connections to logarithms at all in what it does. One of the triumphs of calculus is the development of techniques for calculating highly accurate approximations of this sort quickly. In this chapter we will explore these techniques and their applications.

⊳ 10.1 APPROXIMATION NEAR A POINT AND OVER AN INTERVAL

Suppose we were interested in approximating the sine function—we might need to make a quick estimate and not have a calculator handy, or we might even be designing a calculator. In the next section we will examine a number of other contexts in which such approximations are helpful. Here is a third-degree polynomial that is a good approximation in a sense which will be made clear shortly:

$$P(x) = x - \frac{x^3}{6}.$$

(You will see in section 10.2 where $P(x)$ comes from.)

If we compare the values of $\sin(x)$ and $P(x)$ over the interval $[0, 1]$, we get the following:

x	$\sin x$	$P(x)$	$\sin x - P(x)$
0.0	0.0	0.0	0.0
.2	.198669	.198667	.000002
.4	.389418	.389333	.000085
.6	.564642	.564000	.000642
.8	.717356	.714667	.002689
1.0	.841471	.833333	.008138

The fit is good, with the largest difference occurring at $x = 1.0$, where the difference is only slightly greater than .008.

If we plot $\sin(x)$ and $P(x)$ together over the interval $[0, \pi]$, we see the ways in which $P(x)$ is both very good and not so good. Over the initial portion of the graph—out to around $x = 1$—the graphs of the two functions seem to coincide. As we move further from the origin, though, the graphs separate more and more. Thus if we were primarily interested in approximating $\sin(x)$ near the origin, $P(x)$ would be a reasonable choice. If we need to approximate $\sin(x)$ over the entire interval, $P(x)$ is less useful.

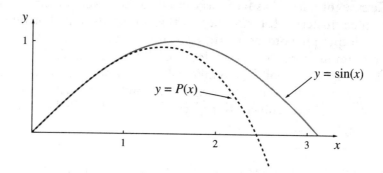

On the other hand, consider the second-degree polynomial

$$Q(x) = -.4176977x^2 + 1.312236205x - .050465497.$$

(You will see how to compute these coefficients in section 10.6.)
When we graph $Q(x)$ and $\sin(x)$ together we get the following:

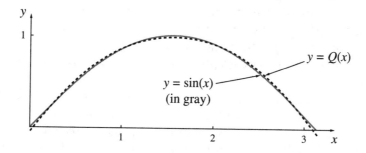

While $Q(x)$ does not fit the graph of $\sin(x)$ as well as $P(x)$ does near the origin, it is a good fit overall. In fact, $Q(x)$ exactly equals $\sin(x)$ at four values of x, and the greatest separation between the graphs of $Q(x)$ and $\sin(x)$ over the interval $[0, \pi]$ occurs at the endpoints, where the distance between the graphs is .0505 unit.

What we have here, then, are two kinds of approximation of the sine function by polynomials: we have a polynomial $P(x)$ that behaves very much like the sine function near the origin, and we have another polynomial $Q(x)$ that keeps close to the sine function over the entire interval $[0, \pi]$. Which one is the "better" approximation depends on our needs. Each solves an important problem. Since finding approximations near a point has a neater solution—Taylor polynomials—we will start with this problem. We will turn to the problem of finding approximations over an interval in section 10.6.

There's more than one way to make the "best fit" to a given curve

➤ 10.2 TAYLOR POLYNOMIALS

The general setting.   In chapter 3 we discovered that functions were locally linear at most points—when we zoomed in on them they looked more and more like straight lines. This fact was central to the development of much of the subsequent material. It turns out that this is only the initial manifestation of an even deeper phenomenon: Not only are functions locally linear, but, if we don't zoom in quite so far, they look locally like parabolas. From a little further back still they look locally like cubic polynomials, and so on. Later in this section we will see how to use the computer to visualize these "local parabolizations," "local cubicizations," etc. Let's summarize the idea and then explore its significance:

> The functions of interest to calculus look locally like polynomials at most points of their domain. The higher the degree of the polynomial, the better typically will be the fit.

Comments.   The "at most points" qualification is because of exceptions like those we ran into when we explored locally linearity. The function $|x|$, for instance, was not locally linear at $x = 0$—it's not locally like any polynomial of higher degree at that point either. The issue of what "goodness of fit" means and how it is measured is a subtle one which we will develop over the course of this section. For the time being, your intuition is a reasonable guide—one fit to a curve is better than another near some point if it "shares more phosphor" with the curve when they are graphed on a computer screen centered at the given point.

The fact that functions look locally like polynomials has profound implications conceptually and computationally. It means we can often determine the behavior of a function locally by examining the corresponding behavior of what we might call a "local polynomialization" instead. In particular, to find the values of a function near some point, or graph a function near some point, we can deal with the values or graph of a local polynomialization instead. Since we can actually evaluate polynomials directly, this can be a major simplification.

There is an extra feature to all this which makes the concept particularly attractive: not only are functions locally polynomial, it is easy to find the coefficients of the polynomials. Let's see how this works. Suppose we had some function $f(x)$ and we wanted to find the fifth-degree polynomial that best fits this function at $x = 0$. Let's call this polynomial

$$P(x) = a_0 + a_1 x + a_2 x^2 + a_3 x^3 + a_4 x^4 + a_5 x^5.$$

To determine P, we need to find values for the six coefficients a_0, a_1, a_2, a_3, a_4, a_5.

The behavior of a function can often be inferred from the behavior of a local polynomialization

We want the best fit at $x = 0$

Before we can do this, we need to define what we mean by the "best" fit to f at $x = 0$. Since we have six unknowns, we need six conditions. One obvious condition is that the graph of P should pass through the point $(0, f(0))$. But this is equivalent to requiring that $P(0) = f(0)$. Since $P(0) = a_0$, we thus must have $a_0 = f(0)$, and we have found one of the coefficients of $P(x)$. Let's summarize the argument so far:

The best fit should pass through the point $(0, f(0))$

The graph of a polynomial passes through the point $(0, f(0))$ if and only if the polynomial is of the form

$$f(0) + a_1 x + a_2 x^2 + \cdots.$$

But we're not interested in just any polynomial passing through the right point; it should be headed in the right direction as well. That is, we want the slope of P at $x = 0$ to be the same as the slope of f at this point—we want $P'(0) = f'(0)$. But

The best fit should have the right slope at $(0, f(0))$

$$P'(x) = a_1 + 2a_2 x + 3a_3 x^2 + 4a_4 x^3 + 5a_5 x^4,$$

so $P'(0) = a_1$. Our second condition therefore must be that $a_1 = f'(0)$. Again, we can summarize this as

---

The graph of a polynomial passes through the point $(0, f(0))$ and has slope $f'(0)$ there if and only if it is of the form

$$f(0) + f'(0)x + a_2 x^2 + \cdots.$$

Note that at this point we have recovered the general form for the local linear approximation to f at $x = 0$: $L(x) = f(0) + f'(0)x$.

But there is no reason to stop with the first derivative. Similarly, we would want the way in which the slope of $P(x)$ is changing—we are now talking about $P''(0)$—to behave the way the slope of f is changing at $x = 0$, etc. Each higher derivative controls a more subtle feature of the shape of the graph. We now see how we could formulate reasonable additional conditions which would determine the remaining coefficients of $P(x)$:

Say that $P(x)$ is the **best fit** to $f(x)$ at the point $x = 0$ if

$$P(0) = f(0), P'(0) = f'(0), P''(0) = f''(0), \ldots, P^{(5)}(0) = f^{(5)}(0).$$

The final criterion for
best fit at x = 0

Since $P(x)$ is a fifth-degree polynomial, all the derivatives of P beyond the fifth will be identically 0, so we can't control their values by altering the values of the a_k. What we are saying, then, is that we are using as our criterion for the best fit that all the derivatives of P as high as we can control them have the same values at $x = 0$ as the corresponding derivatives of f.

While this is a reasonable definition for something we might call the "best fit" at the point $x = 0$, it gives us no direct way to tell how good the fit really is. This is a serious shortcoming—if we want to approximate function values by polynomial values, for instance, we would like to know how many decimal places in the polynomial values are going to be correct. We will take up this question of goodness of fit later in this section; we'll be able to make measurements that allow us to to see how well the polynomial fits the function. First, though, we need to see how to determine the coefficients of the approximating polynomials and get some practice manipulating them.

Notation for higher
derivatives

Note on notation. We have used the notation $f^{(5)}(x)$ to denote the fifth derivative of $f(x)$ as a convenient shorthand for $f''''''(x)$, which is harder to read. We will use this throughout.

Finding the coefficients. We first observe that the derivatives of P at $x = 0$ are easy to express in terms of $a_1, a_2, \ldots$. We have

$$P'(x) = a_1 + 2\,a_2 x + 3\,a_3 x^2 + 4\,a_4 x^3 + 5\,a_5 x^4$$
$$P''(x) = 2\,a_2 + 3 \cdot 2\,a_3 x + 4 \cdot 3\,a_4 x^2 + 5 \cdot 4\,a_5 x^3$$
$$P^{(3)}(x) = 3 \cdot 2\,a_3 + 4 \cdot 3 \cdot 2\,a_4 x + 5 \cdot 4 \cdot 3\,a_5 x^2$$
$$P^{(4)}(x) = 4 \cdot 3 \cdot 2\,a_4 + 5 \cdot 4 \cdot 3 \cdot 2\,a_5 x$$
$$P^{(5)}(x) = 5 \cdot 4 \cdot 3 \cdot 2\,a_5.$$

Thus $P''(0) = 2\,a_2$, $P^{(3)}(0) = 3 \cdot 2\,a_3$, $P^{(4)}(0) = 4 \cdot 3 \cdot 2\,a_4$, and $P^{(5)}(0) = 5 \cdot 4 \cdot 3 \cdot 2\,a_5$.

Factorial notation

We can simplify this a bit by introducing the **factorial** notation: $n! = n \cdot (n-1) \cdot (n-2) \cdots 3 \cdot 2 \cdot 1$. This is called "$n$ factorial." Thus, for example, $7! = 7 \cdot 6 \cdot 5 \cdot 4 \cdot 3 \cdot 2 \cdot 1 = 5040$. It turns out to be convenient to extend the factorial notation to 0 by defining $0! = 1$. (Notice, for instance, that this makes the formulas below work out right.) In the exercises you will see why this extension of the notation is not only convenient, but reasonable as well!

The desired rule for
finding the coefficients

With this notation we can express compactly the equations above as $P^{(k)}(0) = k!\,a_k$ for $k = 0, 1, 2, \ldots, 5$. Finally, since we want $P^{(k)}(0) = f^{(k)}(0)$, we can solve for the coefficients of $P(x)$:

$$a_k = \frac{f^{(k)}(0)}{k!} \text{ for } k = 0, 1, 2, 3, 4, 5.$$

We can now write down an explicit formula for the fifth-degree polynomial which best fits $f(x)$ at $x = 0$ in the sense we've put forth:

$$P(x) = f(0) + f'(0)x + \frac{f^{(2)}(0)}{2!}x^2 + \frac{f^{(3)}(0)}{3!}x^3 + \frac{f^{(4)}(0)}{4!}x^4 + \frac{f^{(5)}(0)}{5!}x^5.$$

We can express this more compactly using the $\sum$ notation we introduced in the discussion of Riemann sums in chapter 6:

$$P(x) = \sum_{k=0}^{5} \frac{f^{(k)}(0)}{k!}x^k.$$

We call this the **fifth-degree Taylor polynomial for** $f(x)$. It is sometimes also called the **fifth-order Taylor polynomial**.

It should be obvious to you that we can generalize what we've done above to get a best-fitting polynomial of any degree. Thus

> The **Taylor polynomial of degree n** approximating the function $f(x)$ at $x = 0$ is given by the formula
> $$P_n(x) = \sum_{k=0}^{n} \frac{f^{(k)}(0)}{k!}x^k.$$

General rule for the Taylor polynomial at $x = 0$

We also speak of the Taylor polynomial *centered at* $x = 0$.

EXAMPLE. Consider $f(x) = \sin(x)$. Then for $n = 7$ we have

$$\begin{aligned}
f(x) &= \sin(x) & f(0) &= 0 \\
f'(x) &= \cos(x) & f'(0) &= +1 \\
f^{(2)}(x) &= -\sin(x) & f^{(2)}(0) &= 0 \\
f^{(3)}(x) &= -\cos(x) & f^{(3)}(0) &= -1 \\
f^{(4)}(x) &= \sin(x) & f^{(4)}(0) &= 0 \\
f^{(5)}(x) &= \cos(x) & f^{(5)}(0) &= +1 \\
f^{(6)}(x) &= -\sin(x) & f^{(6)}(0) &= 0 \\
f^{(7)}(x) &= -\cos(x) & f^{(7)}(0) &= -1
\end{aligned}$$

From this we can see that the pattern $0, +1, 0, -1, \ldots$ will repeat forever. Substituting these values into the formula we get that for any odd integer n, the nth-degree Taylor polynomial for $\sin(x)$ is

$$P_n(x) = x - \frac{x^3}{3!} + \frac{x^5}{5!} - \frac{x^7}{7!} + \cdots \pm \frac{x^n}{n!}.$$

Note that $P_3(x) = x - x^3/6$, which is the polynomial we met in section 10.1. We saw there that this polynomial seemed to fit the graph of the sine function only out to around $x = 1$. Now, though, we have a way to generate polynomial approximations of higher degrees, and we would expect to get better fits as the degree of the approximating polynomial is increased. To see how closely these polynomial approximations follow $\sin(x)$, here's the graph of $\sin(x)$ together with the Taylor polynomials of degrees $n = 1, 3, 5, \ldots, 17$ plotted over the interval $[0, 7.5]$:

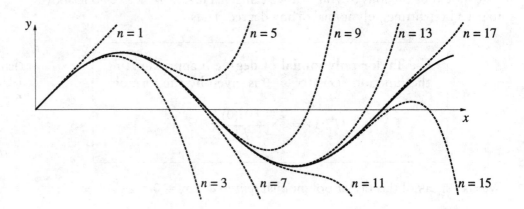

The higher the degree of the polynomial, the better the fit

Note that while each polynomial eventually wanders off to infinity, successive polynomials stay close to the sine function for longer and longer intervals—the Taylor polynomial of degree 17 is just beginning to diverge visibly by the time x reaches 2π. We might expect that if we kept going, we could find Taylor polynomials that were good fits out to $x = 100$, or $x = 1000$. This is indeed the case, although they would be long and cumbersome polynomials to work with. Fortunately, as you will see in the exercises, with a little cleverness we can use a Taylor polynomial of degree 9 to calculate $\sin(100)$ to five decimal places accuracy.

Approximating polynomials for other basic functions

Other Taylor polynomials.   In a similar fashion, we can get Taylor polynomials for other functions. You should use the general formula to verify the Taylor polynomials for the following basic functions. [The Taylor polynomial for $\sin(x)$ is included for convenient reference.]

$f(x)$	$P_n(x)$
$\sin(x)$	$x - \dfrac{x^3}{3!} + \dfrac{x^5}{5!} - \dfrac{x^7}{7!} + \cdots \pm \dfrac{x^n}{n!} \quad (n \text{ odd})$
$\cos(x)$	$1 - \dfrac{x^2}{2!} + \dfrac{x^4}{4!} - \dfrac{x^6}{6!} + \cdots \pm \dfrac{x^n}{n!} \quad (n \text{ even})$
e^x	$1 + x + \dfrac{x^2}{2!} + \dfrac{x^3}{3!} + \dfrac{x^4}{4!} + \cdots + \dfrac{x^n}{n!}$
$\ln(1-x)$	$-\left(x + \dfrac{x^2}{2} + \dfrac{x^3}{3} + \dfrac{x^4}{4} + \cdots + \dfrac{x^n}{n}\right)$
$\dfrac{1}{1-x}$	$1 + x + x^2 + x^3 + \cdots + x^n$

Taylor polynomials at points other than $x = 0$. Using exactly the same arguments we used to develop the best-fitting polynomial at $x = 0$, we can derive the more general formula for the best-fitting polynomial at any value of x. Thus, if we know the behavior of f and its derivatives at some point $x = a$, we would like to find a polynomial $P_n(x)$ which is a good approximation to $f(x)$ for values of x close to a. *(General rule for the Taylor polynomial at $x = a$)*

Since the expression $x - a$ tells us how close x is to a, we use it (instead of the variable x itself) to construct the polynomials approximating f at $x = a$:

$$P_n(x) = b_0 + b_1(x - a) + b_2(x - a)^2 + b_3(x - a)^3 + \cdots + b_n(x - a)^n.$$

You should be able to apply the reasoning we used above to derive the following:

> The **Taylor polynomial of degree n centered at $x = a$** approximating the function $f(x)$ is given by the formula
>
> $$P_n(x) = f(a) + f'(a)(x - a) + \frac{f''(a)}{2!}(x - a)^2 + \cdots + \frac{f^n(a)}{n!}(x - a)^n$$
>
> $$= \sum_{k=0}^{n} \frac{f^{(k)}(a)}{k!}(x - a)^k.$$

A computer program for graphing Taylor polynomials. Here is a program that evaluates the 17th-degree Taylor polynomial for $\sin(x)$ and graphs it over the interval $[0, 3.14]$. The first seven lines of the program constitute a subroutine for evaluating factorials. The syntax of such subroutines varies from one computer language to another, so be sure to use the format that's appropriate for you. You may even be using a language that already knows how to compute factorials, in which case you can omit the subroutine. The second set of nine lines defines the function `poly` which evaluates the 17th-degree Taylor polynomial. Note the role of the variable `Sign`—it simply changes the sign back and forth from positive to negative as each new term is added to the sum. As usual, you will have to put in commands to set up the graphics and draw lines in the format your computer language uses. You can modify this program to graph other Taylor polynomials.

Program: TAYLOR

```
Set up GRAPHICS
DEF fnfact(m)
    P = 1
    FOR r = 2 TO m
    P = P * r
    NEXT r
    fnfact = P
END DEF
DEF fnpoly(x)
    Sum = x
    Sign = -1
    FOR k = 3 TO 17 STEP 2
        Sum = Sum + Sign * x^k/fnfact(k)
        Sign = (-1) * Sign
    NEXT k
    fnpoly = Sum
END DEF
FOR x = 0 TO 3.14 STEP .01
    Plot the line from (x, fnpoly(x))
    to (x + .01, fnpoly(x + .01))
NEXT x
```

New Taylor Polynomials from Old

Given a function we want to approximate by Taylor polynomials, we could always go straight to the general formula for deriving such polynomials. On the other hand, it is often possible to avoid a lot of tedious calculation of derivatives by using a polynomial we've already calculated. It turns out that any manipulation on Taylor polynomials you might be tempted to try will probably work. Here are some examples to illustrate the kinds of manipulations that can be performed on Taylor polynomials.

Substitution in Taylor polynomials. Suppose we wanted the Taylor polynomial for e^{x^2}. We know from what we've already done that for any value of

u close to 0,

$$e^u \approx 1 + u + \frac{u^2}{2!} + \frac{u^3}{3!} + \frac{u^4}{4!} + \cdots + \frac{u^n}{n!}.$$

In this expression u can be anything, including another variable expression. For instance, if we set $u = x^2$, we get the Taylor polynomial

$$
\begin{aligned}
e^{x^2} &= e^u \\
&\approx 1 + u + \frac{u^2}{2!} + \frac{u^3}{3!} + \frac{u^4}{4!} + \cdots + \frac{u^n}{n!} \\
&= 1 + (x^2) + \frac{(x^2)^2}{2!} + \frac{(x^2)^3}{3!} + \frac{(x^2)^4}{4!} + \cdots + \frac{(x^2)^n}{n!} \\
&= 1 + x^2 + \frac{x^4}{2!} + \frac{x^6}{3!} + \frac{x^8}{4!} + \cdots + \frac{x^{2n}}{n!}.
\end{aligned}
$$

You should check to see that this is what you get if you apply the general formula for computing Taylor polynomials to the function e^{x^2}.

Similarly, suppose we wanted a Taylor polynomial for $1/(1 + x^2)$. We could start with the approximation given earlier:

$$\frac{1}{1 - u} \approx 1 + u + u^2 + u^3 + \cdots + u^n.$$

If we now replace u everywhere by $-x^2$, we get the desired expansion:

$$
\begin{aligned}
\frac{1}{1 + x^2} &= \frac{1}{1 - (-x^2)} \\
&= \frac{1}{1 - u} \\
&\approx 1 + u + u^2 + u^3 + \cdots + u^n \\
&= 1 + (-x^2) + (-x^2)^2 + (-x^2)^3 + \cdots + (-x^2)^n \\
&= 1 - x^2 + x^4 - x^6 + \cdots \pm x^{2n}.
\end{aligned}
$$

Again, you should verify that if you start with the function $f(x) = 1/(1 + x^2)$ and apply to f the general formula for deriving Taylor polynomials, you will get the preceding result. Which method is quicker?

Multiplying Taylor polynomials. Suppose we wanted the fifth-degree Taylor polynomial for $e^{3x} \cdot \sin(2x)$. We can use substitution to write down polynomial approximations for e^{3x} and $\sin(2x)$, so we can get an approximation for their product by multiplying the two polynomials:

$$
\begin{aligned}
e^{3x} \cdot \sin(2x) &\approx \left(1 + (3x) + \frac{(3x)^2}{2!} + \frac{(3x)^3}{3!} + \frac{(3x)^4}{4!} + \frac{(3x)^5}{5!}\right)\left((2x) - \frac{(2x)^3}{3!} + \frac{(2x)^5}{5!}\right) \\
&\approx 2x + 6x^2 + \frac{23}{3}x^3 + 5x^4 - \frac{61}{60}x^5.
\end{aligned}
$$

Again, you should try calculating this polynomial directly from the general rule, both to see that you get the same result, and to appreciate how much more tedious the general formula is to use in this case.

In the same way, we can also divide Taylor polynomials, raise them to powers, and chain them by composition. The exercises provide examples of some of these operations.

Differentiating Taylor polynomials.   Suppose we know a Taylor polynomial for some function f. If g is the derivative of f, we can immediately get a Taylor polynomial for g (of degree 1 less) by differentiating the polynomial we know for f. You should review the definition of Taylor polynomial to see why this is so. For instance, suppose $f(x) = 1/(1 - x)$ and $g(x) = 1/(1 - x)^2$. Verify that $f'(x) = g(x)$. It then follows that

$$\frac{1}{(1 - x)^2} = \frac{d}{dx} f(x)$$

$$\approx \frac{d}{dx}\left(1 + x + x^2 + \cdots + x^n\right)$$

$$= 1 + 2x + 3x^2 + \cdots + n x^{n-1}.$$

Integrating Taylor polynomials.   Again suppose we have functions $f(x)$ and $g(x)$ with $f'(x) = g(x)$, and suppose this time that we know a Taylor polynomial for g. We can then get a Taylor polynomial for f by antidifferentiating term by term. For instance, we find in chapter 11 that the derivative of $\arctan(x)$ is $1/(1 + x^2)$, and we have seen above how to get a Taylor polynomial for $1/(1 + x^2)$. Therefore we have

$$\arctan x = \int_0^x \frac{1}{1 + t^2}\, dt$$

$$\approx \int_0^x \left(1 - t^2 + t^4 - t^6 + \cdots \pm t^{2n}\right) dt$$

$$= \left(t - \frac{1}{3} t^3 + \frac{1}{5} t^5 - \cdots \pm \frac{1}{2n + 1} t^{2n+1}\right)\Bigg|_0^x$$

$$= x - \frac{1}{3} x^3 + \frac{1}{5} x^5 - \cdots \pm \frac{1}{2n + 1} x^{2n+1}.$$

Goodness of Fit

Graph the difference between a function and its Taylor polynomial

Let's turn to the question of **measuring** the fit between a function and one of its Taylor polynomials. The ideas here have a strong geometric flavor, so you should use a computer graphing utility to follow this discussion. Once again, consider the function $\sin(x)$ and its Taylor polynomial $P(x) = x - x^3/6$. According to the table in section 10.1, the difference $\sin(x) - P(x)$ got smaller as x got smaller. Stop now and graph the function $y = \sin(x) - P(x)$ near $x = 0$. This will show you exactly how $\sin(x) - P(x)$ depends on x. If you choose the interval $-1 \le x \le 1$ (and your graphing utility allows its vertical and horizontal scales to be set independently of each other), your graph should resemble this one.

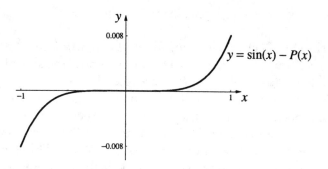

This graph looks very much like a cubic polynomial. If it really is a cubic, we can figure out its formula, because we know the value of $\sin(x) - P(x)$ is about .008 when $x = 1$. Therefore the cubic should be $y = .008x^3$ (because then $y = .008$ when $x = 1$). However, if you graph $y = .008x^3$ together with $y = \sin(x) - P(x)$, you should find a poor match (see the left-hand figure, below). Another possibility is that $\sin(x) - P(x)$ is more like a *fifth*-degree polynomial. Plot $y = .008x^5$; it's so close that it "shares phosphor" with $\sin(x) - P(x)$ near $x = 0$.

The difference looks like a power of x

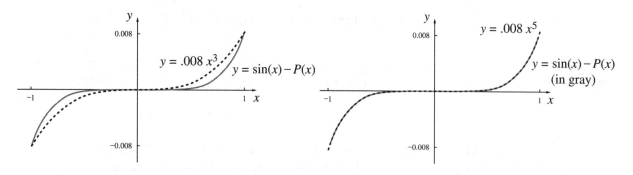

If $\sin(x) - P(X)$ were *exactly* a multiple of x^5, then $(\sin x - P(x))/x^5$ would be constant and would equal the value of the multiplier. What we actually find is this:

Finding the multiplier

x	$\dfrac{\sin x - P(x)}{x^5}$
1.0	.0081377
0.5	.0082839
0.1	.0083313
0.05	.0083328
0.01	.0083333

suggesting $\displaystyle\lim_{x \to 0} \frac{\sin x - P(x)}{x^5} = .008333....$

How $P(x)$ fits $\sin(x)$

Thus, although the ratio is not constant, it appears to converge to a definite value—which we can take to be the value of the multiplier:

$$\sin x - P(x) \approx .008333x^5 \quad \text{when} \quad x \approx 0.$$

We say that $\sin(x) - P(x)$ *has the same order of magnitude as x^5* as $x \to 0$. So $\sin(x) - P(x)$ is about as small as x^5. Thus, if we know the size of x^5, we will be able to tell how close $\sin(x)$ and $P(x)$ are to each other.

Comparing two numbers

A rough way to measure how close two numbers are is to count the number of decimal places to which they agree. But there are pitfalls here; for instance, none of the decimals of 1.00001 and 0.99999 agree, even though the difference between the two numbers is only 0.00002. This suggests that a good way to compare two numbers is to look at their difference. Therefore, we say

$A = B$ to k decimal places   *means* $A - B = 0$ to k decimal places.

Now, a number equals 0 to k decimal places precisely when it *rounds off to* 0 (when we round it to k decimal places). Since X rounds to 0 to k decimal places if and only if $|X| < .5 \times 10^{-k}$, we finally have a precise way to compare the size of two numbers:

$$\boxed{A = B \text{ to } k \text{ decimal places} \quad \textit{means} \quad |A - B| < .5 \times 10^{-k}.}$$

What the fit means computationally

Now we can say how close $P(x)$ is to $\sin(x)$. Since x is small, we can take this to mean $x = 0$ to k *decimal places*, or $|x| < .5 \times 10^{-k}$. But then,

$$|x^5 - 0| = |x - 0|^5 < (.5 \times 10^{-k})^5 < .5 \times 10^{-5k-1}$$

(since $.5^5 = .03125 < .5 \times 10^{-1}$). In other words, if $x = 0$ to k decimal places, then $x^5 = 0$ to $5k + 1$ places. Since $\sin(x) - P(x)$ has the same order of magnitude as x^5 as $x \to 0$, $\sin(x) = P(x)$ to $5k + 1$ places as well. In fact, because the multiplier in the relation

$$\sin x - P(x) \approx .008333x^5 \quad (x \approx 0)$$

is .0083..., we gain two more decimal places of accuracy. (Do you see why?) Thus, finally, we see how reliable the polynomial $P(x) = x - x^3/6$ is for calculating values of $\sin(x)$:

$$\boxed{\begin{array}{c} \text{When } x = 0 \text{ to } k \text{ decimal places of accuracy, we can use } P(x) \\ \text{to calculate the first } 5k + 3 \text{ decimal places of the value of } \sin(x). \end{array}}$$

Here are a few examples comparing $P(x)$ to the *exact* value of $\sin(x)$:

x	$P(x)$	$\sin(x)$
.0372	.0371914201920	.037194207856...
.0086	.0085998939907	.008599893991...
.0048	.0047999815680000	.0047999815680212...

The underlined digits are guaranteed to be correct, based on the number of decimal places for which x agrees with 0. (Note that, according to our rule, .0086 = 0 to *one* decimal place, not two.)

Taylor's Theorem

Taylor's theorem is the generalization of what we have just seen; it describes the goodness of fit between an arbitrary function and one of its Taylor polynomials. We'll state three versions of the theorem, gradually uncovering more information. To get started, we need a way to compare the order of magnitude of *any* two functions.

> We say that $\varphi(x)$ has the **same order of magnitude** as $q(x)$ as $x \to a$, and we write $\varphi(x) = O(q(x))$ as $x \to a$, if there is a constant C for which
>
> $$\lim_{x \to a} \frac{\varphi(x)}{q(x)} = C.$$

Order of magnitude

Now, when $\lim_{x \to a} \varphi(x)/q(x)$ is C, we have

$$\varphi(x) \approx Cq(x) \quad \text{when } x \approx a.$$

We'll frequently use this relation to express the idea that $\varphi(x)$ has the same order of magnitude as $q(x)$ as $x \to a$.

The symbol O is an uppercase "oh." When $\varphi(x) = O(q(x))$ as $x \to a$, we say $\varphi(x)$ *is "big oh"* of $q(x)$ as x approaches a. Notice that the equal sign in $\varphi(x) = O(q(x))$ does *not* mean that $\varphi(x)$ and $O(q(x))$ are equal; $O(q(x))$ isn't even a function. Instead, the equal sign and the O together tell us that $\varphi(x)$ stands in a certain relation to $q(x)$.

"Big oh" notation

> **Taylor's theorem, version 1.** If $f(x)$ has derivatives up to order n at $x = a$, then
>
> $$f(x) = f(a) + \frac{f'(a)}{1!}(x - a) + \cdots + \frac{f^{(n)}(a)}{n!}(x - a)^n + R(x),$$
>
> where $R(x) = O((x - a)^{n+1})$ as $x \to a$. The term $R(x)$ is called the **remainder**.

Informal language

This version of Taylor's theorem focuses on the general shape of the remainder function. Sometimes we just say the remainder has "order $n + 1$," using this short phrase as an abbreviation for "the order of magnitude of the function $(x - a)^{n+1}$." In the same way, we say that *a function and its nth-degree Taylor polynomial at $x = a$ agree to order $n + 1$ as $x \rightarrow a$.*

Notice that, if $\varphi(x) = O(x^3)$ as $x \rightarrow 0$, then it is also true that $\varphi(x) = O(x^2)$ (as $x \rightarrow 0$). This implies that we should take $\varphi(x) = O(x^n)$ to mean "φ has at least order n" (instead of simply "φ has order n"). In the same way, it would be more accurate (but somewhat more cumbersome) to say that $\varphi = O(q)$ means "φ has at least the order of magnitude of q."

Decimal places of accuracy

As we saw in our example, we can translate the order of agreement between the function and the polynomial into information about the number of decimal places of accuracy in the polynomial approximation. In particular, if $x - a = 0$ to k decimal places, then $(x - a)^n = 0$ to nk places, at least. Thus, as the order of magnitude n of the remainder increases, the fit increases, too. (You have already seen this illustrated with the sine function and its various Taylor polynomials, in the figure on page 534.)

A formula for the remainder

While the first version of Taylor's theorem tells us that $R(x)$ looks like $(x - a)^{n+1}$ in some general way, the next gives us a concrete formula. At least, it *looks* concrete. Notice, however, that $R(x)$ is expressed in terms of a number c_x (which depends upon x), but the formula doesn't tell us *how* c_x depends upon x. Therefore, if you want to use the formula to *compute* the value of $R(x)$, you can't. The theorem says only that c_x exists; it doesn't say how to find its value. Nevertheless, this version provides useful information, as you will see.

Taylor's theorem, version 2. Suppose f has continuous derivatives up to order $n + 1$ for all x in some interval containing a. Then, for each x in that interval, there is a number c_x between a and x for which

$$R(x) = \frac{f^{(n+1)}(c_x)}{(n + 1)!} (x - a)^{n+1}.$$

This is called **Lagrange's form of the remainder.**

Another formula

We can use the Lagrange form as an aid to computation. To see how, return to the formula

$$R(x) \approx C(x - a)^{n+1} \qquad (x \approx a)$$

that expresses $R(x) = O((x - a)^{n+1})$ as $x \rightarrow a$ (see page 541). The constant here is the limit

$$C = \lim_{x \rightarrow a} \frac{R(x)}{(x - a)^{n+1}}.$$

If we have a good estimate for the value of C, then $R(x) \approx C(x - a)^{n+1}$ gives us a good way to estimate $R(x)$. Of course, we could just evaluate the limit to determine C. In fact, that's what we did in the example; knowing $C \approx .008$ there gave us two more decimal places of accuracy in our polynomial approximation to the sine function.

But the Lagrange form of the remainder gives us another way to determine C:

Determining C from f at $x = a$

$$C = \lim_{x \to a} \frac{R(x)}{(x - a)^{n+1}} = \lim_{x \to a} \frac{f^{(n+1)}(c_x)}{(n + 1)!}$$

$$= \frac{f^{(n+1)}(\lim_{x \to a} c_x)}{(n + 1)!}$$

$$= \frac{f^{(n+1)}(a)}{(n + 1)!}.$$

In this argument, we are permitted to take the limit "inside" $f^{(n+1)}$ because $f^{(n+1)}$ is a continuous function. (That is one of the hypotheses of version 2.) Finally, since c_x lies between x and a, it follows that $c_x \to a$ as $x \to a$; in other words, $\lim_{x \to a} c_x = a$. Consequently, we get C *directly* from the function f itself, and we can therefore write

$$R(x) \approx \frac{f^{(n+1)}(a)}{(n + 1)!} (x - a)^{n+1} \qquad (x \approx a).$$

The third version of Taylor's theorem uses the Lagrange form of the remainder in a similar way to get an *error bound* for the polynomial approximation based on the size of $f^{(n+1)}(x)$.

An error bound

Taylor's theorem, version 3. Suppose that $|f^{(n+1)}(x)| \le M$ for all x in some interval containing a. Then, for each x in that interval,

$$|R(x)| \le \frac{M}{(n + 1)!} |x - a|^{n+1}.$$

With this error bound, which is derived from knowledge of $f(x)$ near $x = a$, we can determine quite precisely how many decimal places of accuracy a Taylor polynomial approximation achieves. The following example illustrates the different versions of Taylor's theorem.

EXAMPLE. Consider $\sqrt{x}$ near $x = 100$. The second-degree Taylor polynomial for $\sqrt{x}$, centered at $x = 100$, is

$$Q(x) = 10 + \frac{x - 100}{20} - \frac{(x - 100)^2}{8000}.$$

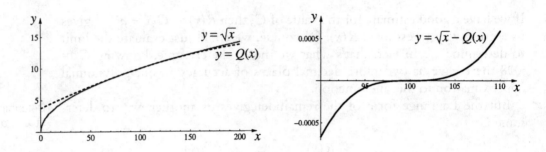

Plot $y = Q(x)$ and $y = \sqrt{x}$ together; the result should look like the figure on the left, above. Then plot the remainder $y = \sqrt{x} - Q(x)$ near $x = 100$. This graph should suggest that $\sqrt{x} - Q(x) = O((x - 100)^3)$ as $x \to 100$. In fact, this is what version 1 of Taylor's theorem asserts. Furthermore,

$$\lim_{x \to 100} \frac{\sqrt{x} - Q(x)}{(x - 100)^3} \approx 6.25 \times 10^{-7};$$

check this yourself by constructing a table of values. Thus

$$\sqrt{x} - Q(x) \approx C(x - 100)^3 \qquad \text{where } C \approx 6.25 \times 10^{-7}.$$

We can use the Lagrange form of the remainder (in version 2 of Taylor's theorem) to get the value of C another way—directly from the third derivative of $\sqrt{x}$ at $x = 100$:

$$C = \left.\frac{(x^{1/2})'''}{3!}\right|_{x = 100} = \frac{1/2 \cdot -1/2 \cdot -3/2 \cdot (100)^{-5/2}}{6} = \frac{1}{2^4 \cdot 10^5} = 6.25 \times 10^{-7}.$$

This is the *exact* value, confirming the estimate obtained above.

Let's see what the equation $\sqrt{x} - Q(x) \approx 6.25 \times 10^{-7}(x - 100)^3$ tells us about the accuracy of the polynomial approximation. If we assume $|x - 100| < .5 \times 10^{-k}$, then

$$|\sqrt{x} - Q(x)| < 6.25 \times 10^{-7} \times (.5 \times 10^{-k})^3$$

$$= .78125 \times 10^{-(3k+7)} < .5 \times 10^{-(3k+6)}.$$

Thus

$$x = 100 \text{ to } k \text{ decimal places} \implies \sqrt{x} = Q(x) \text{ to } 3k + 6 \text{ places}.$$

For example, if $x = 100.47$, then $k = 0$, so $Q(100.47) = \sqrt{100.47}$ to six decimal places. We find

$$Q(100.47) = \underline{10.0234723875},$$

and the underlined digits should be correct. In fact,

$$\sqrt{100.47} = \underline{10.0234724}521\ldots.$$

Here is a second example. If $x = 102.98$, then we can take $k = -1$, so $Q(102.98) = \sqrt{102.98}$ to $3(-1) + 6 = 3$ decimal places. We find

$$Q(102.98) = \underline{10.147}88995, \qquad \sqrt{102.98} = \underline{10.147}906187\ldots.$$

Let's see what additional light version 3 sheds on our investigation. Suppose we assume $x = 100$ to $k = 0$ decimal places. This means that x lies in the open interval $(99.5, 100.5)$. Version 3 requires that we have a bound on the size of the third derivative of $f(x) = \sqrt{x}$ over this interval. Now $f'''(x) = \frac{3}{8}x^{-5/2}$, and this is a decreasing function. (Check its graph; alternatively, note that its derivative is negative.) Its maximum value therefore occurs at the left endpoint of the (closed) interval $[99.5, 100.5]$: Version 3: an explicit error bound

$$|f'''(x)| \le f'''(99.5) = \tfrac{3}{8}(99.5)^{-5/2} < 3.8 \times 10^{-6}.$$

Therefore, from version 3 of Taylor's theorem,

$$\left| \sqrt{x} - Q(x) \right| < \frac{3.8 \times 10^{-6}}{3!} |x - 100|^3.$$

Now $|x - 100| < .5$, $|x - 100|^3 < .125$, so

$$\left| \sqrt{x} - Q(x) \right| < \frac{3.8 \times 10^{-6} \times .125}{6} = .791667 \times 10^{-7} < .5 \times 10^{-6}.$$

This proves $\sqrt{x} = Q(x)$ to six decimal places—confirming what we found earlier.

Applications

Evaluating functions. An obvious use of Taylor polynomials is to evaluate functions. In fact, whenever you ask a calculator or computer to evaluate a function—trigonometric, exponential, logarithmic—it is typically giving you the value of an appropriate polynomial (though not necessarily a Taylor polynomial). Now you can do anything your calculator can!

Evaluating integrals. The fundamental theorem of calculus gives us a quick way of evaluating a definite integral provided we can find an antiderivative for the function under the integral (cf. section 6.4). Unfortunately, many common functions, like e^{-x^2} or $(\sin x)/x$, don't have antiderivatives that can be expressed as finite algebraic combinations of the basic functions. Up until now, whenever we encountered such a function, we had to rely on a Riemann sum to estimate the integral. But now we have Taylor polynomials, and it's easy to find an antiderivative for a polynomial! Thus, if we have an awkward definite integral to evaluate, it is reasonable to expect that we can estimate it by first getting a good polynomial approximation to the integrand, and then integrating this polynomial. As an example, consider the **error function**, erf(t), defined by

The error function

$$\operatorname{erf}(t) = \frac{2}{\sqrt{\pi}} \int_0^t e^{-x^2} dx.$$

This is perhaps the most important integral in statistics. It is the basis of the so-called "normal distribution" and is widely used to decide how good certain statistical estimates are. It is important to have a way of obtaining fast, accurate approximations for erf(t). We have already seen that

$$e^{-x^2} \approx 1 - x^2 + \frac{x^4}{2!} - \frac{x^6}{3!} + \frac{x^8}{4!} - \cdots \pm \frac{x^{2n}}{(n)!}.$$

Now, if we antidifferentiate term by term:

$$\int e^{-x^2} dx \approx \int \left(1 - x^2 + \frac{x^4}{2!} - \frac{x^6}{3!} + \frac{x^8}{4!} - \cdots \pm \frac{x^{2n}}{(n)!} \right) dx$$

$$= \int 1 dx - \int x^2 dx + \int \frac{x^4}{2!} dx - \int \frac{x^6}{3!} dx + \cdots \pm \int \frac{x^{2n}}{(n)!} dx$$

$$= x - \frac{x^3}{3} + \frac{x^5}{5 \cdot 2!} - \frac{x^7}{7 \cdot 3!} + \cdots \pm \frac{x^{2n+1}}{(2n+1) \cdot n!}.$$

Thus,

$$\int_0^t e^{-x^2} dx \approx \left. x - \frac{x^3}{3} + \frac{x^5}{5 \cdot 2!} - \frac{x^7}{7 \cdot 3!} + \cdots \pm \frac{x^{2n+1}}{(2n+1) \cdot n!} \right|_0^t,$$

giving us, finally, a formula for erf(t):

A formula for
approximating the error
function

$$\operatorname{erf}(t) \approx \frac{2}{\sqrt{\pi}} \left(t - \frac{t^3}{3} + \frac{t^5}{5 \cdot 2!} - \frac{t^7}{7 \cdot 3!} + \cdots \pm \frac{t^{2n+1}}{(2n+1) \cdot n!} \right).$$

Thus if we needed to know, say, erf(1), we could quickly approximate it. For instance, letting $n = 6$, we have

$$\operatorname{erf}(1) \approx \frac{2}{\sqrt{\pi}} \left(1 - \frac{1}{3} + \frac{1}{5 \cdot 2!} - \frac{1}{7 \cdot 3!} + \frac{1}{9 \cdot 4!} - \frac{1}{11 \cdot 5!} + \frac{1}{13 \cdot 6!} \right)$$

$$\approx \frac{2}{\sqrt{\pi}} \left(1 - \frac{1}{3} + \frac{1}{10} - \frac{1}{42} + \frac{1}{216} - \frac{1}{1320} + \frac{1}{9360} \right)$$

$$\approx .746836 \frac{2}{\sqrt{\pi}} \approx .842714,$$

a value accurate to four decimals. If we had needed greater accuracy, we could simply have taken a larger value for n. For instance, if we take $n = 12$, we get the estimate $.8427007929\ldots$, where all 10 decimals are accurate (i.e., they don't change as we take larger values of n).

Evaluating limits.   Our final application of Taylor polynomials makes explicit use of the order of magnitude of the remainder. Consider the problem of evaluating a limit like

$$\lim_{x \to 0} \frac{1 - \cos(x)}{x^2}.$$

Since both numerator and denominator approach 0 as $x \to 0$, it isn't clear what the quotient is doing. If we replace $\cos(x)$ by its third-degree Taylor polynomial with remainder, though, we get

$$\cos(x) = 1 - \frac{1}{2!}x^2 + R(x),$$

and $R(x) = O(x^4)$ as $x \to 0$. Consequently, if $x \neq 0$ but $x \to 0$, then

$$\frac{1 - \cos(x)}{x^2} = \frac{1 - \left(1 - \frac{1}{2}x^2 + R(x)\right)}{x^2}$$

$$= \frac{\frac{1}{2}x^2 - R(x)}{x^2} = \frac{1}{2} - \frac{R(x)}{x^2}.$$

Since $R(x) = O(x^4)$, we know there is some constant C for which $R(x)/x^4 \to C$ as $x \to 0$. Therefore,

$$\lim_{x \to 0} \frac{1 - \cos(x)}{x^2} = \frac{1}{2} + \lim_{x \to 0} \frac{R(x)}{x^2} = \frac{1}{2} + \lim_{x \to 0} \frac{x^2 \cdot R(x)}{x^4}$$

$$= \frac{1}{2} + \lim_{x \to 0} x^2 \cdot \lim_{x \to 0} \frac{R(x)}{x^4} = \frac{1}{2} + 0 \cdot C = \frac{1}{2}.$$

There is a way to shorten these calculations—and make them more transparent—by extending the way we read the "big oh" notation. Specifically, we will read $O(q(x))$ as "some (unspecified) function that is the same order of magnitude as $q(x)$."

Extending the "big oh" notation

Then, instead of writing $\cos(x) = 1 - \frac{1}{2}x^2 + R(x)$, and then noting $R(x) = O(x^4)$ as $x \to 0$, we'll just write

$$\cos(x) = 1 - \frac{1}{2}x^2 + O(x^4) \qquad (x \to 0).$$

In this spirit,

$$\frac{1 - \cos(x)}{x^2} = \frac{1 - \left(1 - \frac{1}{2}x^2 + O(x^4)\right)}{x^2}$$

$$= \frac{\frac{1}{2}x^2 + O(x^4)}{x^2} = \frac{1}{2} + O(x^2) \qquad (x \to 0).$$

We have used the fact that $O(x^4)/x^2 = O(x^2)$. Finally, since $O(x^2) \to 0$ as $x \to 0$ (do you see why?), the limit of the last expression is just $1/2$ as $x \to 0$. Thus, once again we arrive at the result

$$\lim_{x \to 0} \frac{1 - \cos(x)}{x^2} = \frac{1}{2}.$$

Exercises

1. Find a seventh-degree Taylor polynomial centered at $x = 0$ for the indicated antiderivatives.

   *a) $\displaystyle\int \frac{\sin(x)}{x} \, dx$

   b) $\displaystyle\int e^{x^2} \, dx$

   c) $\displaystyle\int \sin(x^2) \, dx$

2. Plot the seventh-degree polynomial you found in part (a) above over the interval $[0, 5]$. Now plot the ninth-degree approximation on the same graph. When do the two polynomials begin to differ visibly?

3. Using the seventh-degree Taylor approximation

   $$E(t) \approx \int_0^t e^{-x^2} \, dx = t - \frac{t^3}{3} + \frac{t^5}{5 \cdot 2!} - \frac{t^7}{7 \cdot 3!},$$

   calculate the values of $E(.3)$ and $E(-1)$. Give only the significant digits—that is, report only those decimals of your estimates that you think are fixed. (This means you will also need to calculate the ninth-degree Taylor polynomial as well—do you see why?)

4. Calculate the values of $\sin(.4)$ and $\sin(\pi/12)$ using the seventh-degree Taylor polynomial centered at $x = 0$

   $$\sin(x) \approx x - \frac{x^3}{3!} + \frac{x^5}{5!} - \frac{x^7}{7!}.$$

   Compare your answers with what a calculator gives you.

5. Find the third-degree Taylor polynomial for $g(x) = x^3 - 3x$ at $x = 1$. Show that the Taylor polynomial is actually equal to $g(x)$—that is, the remainder is 0. What does this imply about the *fourth*-degree Taylor polynomial for g at $x = 1$?

6. Find the seventh-degree Taylor polynomial centered at $x = \pi$ for (a) $\sin(x)$; (b) $\cos(x)$; *(c) $\sin(3x)$.

7. In this problem you will compare computations using Taylor polynomials centered at $x = \pi$ with computations using Taylor polynomials centered at $x = 0$.
   a) Calculate the value of $\sin(3)$ using a seventh-degree Taylor polynomial centered at $x = 0$. How many decimal places of your estimate appear to be fixed?
   b) Now calculate the value of $\sin(3)$ using a seventh-degree Taylor polynomial centered at $x = \pi$. Now how many decimal places of your estimate appear to be fixed?

8. Write a program which evaluates a Taylor polynomial to print out $\sin(5°)$, $\sin(10°)$, $\sin(15°)$, ..., $\sin(40°)$, $\sin(45°)$ accurate to seven decimals. (Remember to convert to radians before evaluating the polynomial!)

9. *Why* $0! = 1$. When you were first introduced to exponential notation in expressions like 2^n, n was restricted to being a positive integer, and 2^n was defined to be the product of 2 multiplied by itself n times. Before long, though, you were working with expressions like 2^{-3} and $2^{1/4}$. These new expressions weren't defined in terms of the original definition. For instance, to calculate 2^{-3} you wouldn't try to multiply 2 by itself -3 times—that would be nonsense! Instead, 2^{-m} is defined by looking at the key *properties* of exponentiation for positive exponents, and extending the definition to other exponents in a way that preserves these properties. In this case, there are two such properties, one for adding exponents and one for multiplying them:

   Property A: $2^m \cdot 2^n = 2^{m+n}$ for all positive m and n.
   Property M: $(2^m)^n = 2^{mn}$ for all positive m and n.

   a) Show that to preserve property A we have to define $2^0 = 1$.
   b) Show that we then have to define $2^{-3} = 1/(2^3)$ if we are to continue to preserve property A.
   c) Show why $2^{1/4}$ must be $\sqrt[4]{2}$.
   d) In the same way, you should convince yourself that a basic property of the factorial notation is that $(n+1)! = (n+1) \cdot n!$ for any positive integer n. Then show that to preserve this property, we then have to define $0! = 1$.
   e) Show that there is no way to define $(-1)!$ which preserves this property.

10. Use the general rule to derive the fifth-degree Taylor polynomial centered at $x = 0$ for the function

$$f(x) = (1 + x)^{1/2}.$$

Use this approximation to estimate $\sqrt{1.1}$. How accurate is this?

11. Use the general rule to derive the formula for the nth-degree Taylor polynomial centered at $x = 0$ for the function

$$f(x) = (1 + x)^c \qquad \text{where } c \text{ is a constant.}$$

*12. Use the result of the preceding problem to get the sixth-degree Taylor polynomial centered at $x = 0$ for $1/\sqrt[3]{1 + x^2}$.

13. Use the result of the preceding problem to approximate

$$\int_0^1 \frac{1}{\sqrt[3]{1 + x^2}}\, dx\,.$$

*14. Calculate the first seven decimals of erf(.3). Be sure to show why you think all seven decimals are correct. What degree Taylor polynomial did you need to produce these seven decimals?

15.*a) Apply the general formula for calculating Taylor polynomials centered at $x = 0$ to the tangent function to get the fifth-degree approximation.

  b) Recall that $\tan(x) = \sin(x)/\cos(x)$. Multiply the fifth-degree Taylor polynomial for $\tan(x)$ from part (a) by the fourth-degree Taylor polynomial for $\cos(x)$ and show that you get the fifth-degree polynomial for $\sin(x)$ (discarding higher degree terms).

16. Show that the nth-degree Taylor polynomial centered at $x = 0$ for $1/(1 - x)$ is $1 + x + x^2 + \cdots + x^n$.

17. Note that

$$\int \frac{1}{1 - x}\, dx = -\ln(1 - x).$$

Use this observation, together with the result of the previous problem, to get the nth-degree Taylor polynomial centered at $x = 0$ for $\ln(1 - x)$.

18. a) Find a formula for the nth-degree Taylor polynomial centered at $x = 1$ for $\ln(x)$.

  b) Compare your answer to part (a) with the Taylor polynomial centered at $x = 0$ for $\ln(1 - x)$ you found in the previous problem. Are your results consistent?

19. a) The first-degree Taylor polynomial for e^x at $x = 0$ is $1 + x$. Plot the remainder $R_1(x) = e^x - (1 + x)$ over the interval $-.1 \le x \le .1$. How does this graph demonstrate that $R_1(x) = O(x^2)$ as $x \to 0$?

  b) There is a constant C_2 for which $R_1(x) \approx C_2 x^2$ when $x \approx 0$. Why? Estimate the value of C_2.

20. This concerns the second-degree Taylor polynomial for e^x at $x = 0$. Plot the remainder $R_2(x) = e^x - (1 + x + x^2/2)$ over the interval $-.1 \leq x \leq .1$. How does this graph demonstrate that $R_2(x) = O(x^3)$ as $x \to 0$?

    a) There is a constant C_3 for which $R_3(x) \approx C_3 x^3$ when $x \approx 0$. Why? Estimate the value of C_3.

21. Let $R_3(x) = e^x - P_3(x)$, where $P_3(x)$ is the third-degree Taylor polynomial for e^x at $x = 0$. Show $R_3(x) = O(x^4)$ as $x \to 0$.

22. At first glance, Taylor's theorem says that

$$\sin(x) = x - \tfrac{1}{6}x^3 + O(x^4) \quad \text{as } x \to 0.$$

However, graphs and calculations done in the text (pages 539–540) make it clear that

$$\sin(x) = x - \tfrac{1}{6}x^3 + O(x^5) \quad \text{as } x \to 0.$$

Explain this. Is Taylor's theorem wrong here?

23. Using a suitable formula (that is, a Taylor polynomial with remainder) for each of the functions involved, find the indicated limit.

    a) $\displaystyle \lim_{x \to 0} \frac{\sin(x)}{x}$

    b) $\displaystyle \lim_{x \to 0} \frac{e^x - (1 + x)}{x^2}$

    c) $\displaystyle \lim_{x \to 1} \frac{\ln x}{x - 1}$

    d) $\displaystyle \lim_{x \to 0} \frac{x - \sin(x)}{x^3}$

    e) $\displaystyle \lim_{x \to 0} \frac{\sin(x^2)}{1 - \cos(x)}$

24. Suppose $f(x) = 1 + x^2 + O(x^4)$ as $x \to 0$. Show that

$$(f(x))^2 = 1 + 2x^2 + O(x^4) \quad \text{as } x \to 0.$$

25. a) Using $\sin x = x - \tfrac{1}{6}x^3 + O(x^5)$ as $x \to 0$, show

$$(\sin x)^2 = x^2 - \tfrac{1}{3}x^4 + O(x^6) \quad \text{as } x \to 0.$$

    b) Using $\cos x = 1 - \tfrac{1}{2}x^2 + \tfrac{1}{24}x^4 + O(x^5)$ as $x \to 0$, show

$$(\cos x)^2 = 1 - x^2 + \tfrac{1}{3}x^4 + O(x^5) \quad \text{as } x \to 0.$$

c) Using the previous parts, show $(\sin x)^2 + (\cos x)^2 = 1 + O(x^5)$ as $x \to 0$. [Of course, you already know $(\sin x)^2 + (\cos x)^2 = 1$ *exactly.*]

26. a) Apply the general formula for calculating Taylor polynomials to the tangent function to get the fifth-degree approximation.

b) Recall that $\tan(x) = \sin(x)/\cos(x)$, so $\tan(x) \cdot \cos(x) = \sin(x)$. Multiply the fifth-degree Taylor polynomial for $\tan(x)$ from part (a) by the fifth-degree Taylor polynomial for $\cos(x)$ and show that you get the fifth-degree Taylor polynomial for $\sin(x)$ plus $O(x^6)$—that is, plus terms of order 6 and higher.

27. a) Using the formulas

$$e^u = 1 + u + \tfrac{1}{2}u^2 + \tfrac{1}{6}u^3 + O(u^4) \qquad (u \to 0)$$

$$\sin x = x - \tfrac{1}{6}x^3 + O(x^5) \qquad (x \to 0),$$

show that $e^{\sin x} = 1 + x + \tfrac{1}{2}x^2 + O(x^4)$ as $x \to 0$.

b) Apply the general formula to obtain the third-degree Taylor polynomial for $e^{\sin x}$ at $x = 0$, and compare your result with the formula in part (a).

28. Using $e^x = 1 + \tfrac{1}{2}x + \tfrac{1}{6}x^2 + \tfrac{1}{24}x^3 + O(x^4)$ as $x \to 0$, show that

$$\frac{x}{e^x - 1} = 1 - \tfrac{1}{2}x + \tfrac{1}{12}x^2 + O(x^4) \qquad (x \to 0).$$

29. Show that the following are true as $x \to \infty$.

a) $x + 1/x = O(x)$

b) $5x^7 - 12x^4 + 9 = O(x^7)$

*c) $\sqrt{1 + x^2} = O(x)$

d) $\sqrt{1 + x^p} = O(x^{p/2})$

30. a) Let $f(x) = \ln(x)$. Find the smallest bound M for which

$$|f^{(4)}(x)| \le M \qquad \text{when } |x - 1| \le .5.$$

b) Let $P_3(x)$ be the third-degree Taylor polynomial for $\ln(x)$ at $x = 1$, and let $R_3(x)$ be the remainder $R_3(x) = \ln(x) - P_3(x)$. Find a number K for which

$$|R(x)| \le K|x - 1|^4$$

for all x satisfying $|x - 1| \le .5$.

c) If you use $P_3(x)$ to approximate the value of $\ln(x)$ in the interval $.5 \le x \le 1.5$, how many digits of the approximation are correct?

  d) Suppose we restrict the interval to $|x - 1| \le .1$. Repeat parts (a) and (b), getting *smaller* values for M and K. Now how many digits of the polynomial approximation $P_3(x)$ to $\ln(x)$ are correct, if $.9 \le x \le 1.1$?

"Little oh" notation.   Similar to the "big oh" notation is another, called "little oh": if

$$\lim_{x \to a} \frac{\phi(x)}{q(x)} = 0,$$

then we write $\phi(x) = o(q(x))$ and say ϕ is *"little oh"* of q *as* $x \to a$.

31. Suppose $\phi(x) = O(x^6)$ as $x \to 0$. Show the following.
  a) $\phi(x) = O(x^5)$ as $x \to 0$.
  b) $\phi(x) = o(x^5)$ as $x \to 0$.
  c) It is false that $\phi(x) = O(x^7)$ as $x \to 0$. [One way you can do this is to give an explicit example of a function $\phi(x)$ for which $\phi(x) = O(x^6)$ but for which you can show $\phi(x) = O(x^7)$ is false.]
  d) It is false that $\phi(x) = o(x^6)$ as $x \to 0$.

32. Sketch the graph $y = x \ln(x)$ over the interval $0 < x \le 1$. Explain why your graph shows $\ln(x) = o(1/x)$ as $x \to 0$.

▷   10.3   TAYLOR SERIES

In the previous section we have been talking about approximations to functions by their Taylor polynomials. Thus, for instance, we were able to write statements like

$$\sin(x) \approx x - \frac{x^3}{3!} + \frac{x^5}{5!} - \frac{x^7}{7!},$$

where the approximation was a good one for values of x not too far from 0. On the other hand, when we looked at Taylor polynomials of higher and higher degree, the approximations were good for larger and larger values of x. We are thus tempted to write

$$\sin(x) = x - \frac{x^3}{3!} + \frac{x^5}{5!} - \frac{x^7}{7!} + \frac{x^9}{9!} - \cdots,$$

indicating that the sine function is equal to this "infinite degree" polynomial. This infinite sum is called the **Taylor series** centered at $x = 0$ for $\sin(x)$. But what would we even mean by such an infinite sum? We will explore this

You have seen infinite sums before

question in detail in section 10.5, but you should already have some intuition about what it means, for it can be interpreted in exactly the same way we interpret a more familiar statement like

$$\frac{1}{3} = .33333\ldots$$

$$= \frac{3}{10} + \frac{3}{100} + \frac{3}{1000} + \frac{3}{10000} + \frac{3}{100000} + \cdots.$$

Every decimal number is a sum of fractions whose denominators are powers of 10; 1/3 is a number whose decimal expansion happens to need an infinite number of terms to be completely precise. Of course, when a practical matter arises (for example, typing a number like 1/3 or π into a computer) just the beginning of the sum is used—the "tail" is dropped. We might write 1/3 as 0.33, or as 0.33333, or however many terms we need to get the accuracy we want. Put another way, we are saying that 1/3 is the *limit* of the finite sums of the right-hand side of the equation.

Our new formulas for Taylor series are meant to be used exactly the same way: when a computation is involved, take only the beginning of the sum, and drop the tail. Just where you cut off the tail depends on the input value x and on the level of accuracy needed. Look at what happens when we approximate the value of $\cos(\pi/3)$ by evaluating Taylor polynomials of increasingly higher degrees:

Infinite-degree polynomials are to be viewed like infinite decimals

$$1 \quad = 1.0000000$$

$$1 - \frac{1}{2!}\left(\frac{\pi}{3}\right)^2 \approx 0.4516887$$

$$1 - \frac{1}{2!}\left(\frac{\pi}{3}\right)^2 + \frac{1}{4!}\left(\frac{\pi}{3}\right)^4 \approx 0.5017962$$

$$1 - \frac{1}{2!}\left(\frac{\pi}{3}\right)^2 + \frac{1}{4!}\left(\frac{\pi}{3}\right)^4 - \frac{1}{6!}\left(\frac{\pi}{3}\right)^6 \approx 0.4999646$$

$$1 - \frac{1}{2!}\left(\frac{\pi}{3}\right)^2 + \frac{1}{4!}\left(\frac{\pi}{3}\right)^4 - \frac{1}{6!}\left(\frac{\pi}{3}\right)^6 + \frac{1}{8!}\left(\frac{\pi}{3}\right)^8 \approx 0.5000004$$

$$1 - \frac{1}{2!}\left(\frac{\pi}{3}\right)^2 + \frac{1}{4!}\left(\frac{\pi}{3}\right)^4 - \frac{1}{6!}\left(\frac{\pi}{3}\right)^6 + \frac{1}{8!}\left(\frac{\pi}{3}\right)^8 - \frac{1}{10!}\left(\frac{\pi}{3}\right)^{10} \approx 0.5000000$$

These sums were evaluated using $\pi \approx 3.141593$. As you can see, at the level of precision we are using, a sum that is six terms long gives the correct value. However, five, four, or even three terms may have been adequate for the needs at hand. The crucial fact is that these are all honest calculations using *only* the four operations of elementary arithmetic.

Note that if we had wanted to get the same six places accuracy for $\cos(x)$ for a larger value of x, we might need to go further out in the series. For instance

$\cos(7\pi/3)$ is also equal to .5, but the tenth-degree Taylor polynomial centered at $x = 0$ gives

$$1 - \frac{1}{2!}\left(\frac{7\pi}{3}\right)^2 + \frac{1}{4!}\left(\frac{7\pi}{3}\right)^4 - \frac{1}{6!}\left(\frac{7\pi}{3}\right)^6 + \frac{1}{8!}\left(\frac{7\pi}{3}\right)^8 - \frac{1}{10!}\left(\frac{7\pi}{3}\right)^{10} = -37.7302,$$

which is not even close to .5 . In fact, to get $\cos(7\pi/3)$ to six decimals, we need to use the Taylor polynomial centered at $x = 0$ of degree 30, while to get $\cos(19\pi/3)$ (also equal to .5) to six decimals we need the Taylor polynomial centered at $x = 0$ of degree 66.

The key fact, though, is that, for any value of x, if we go out in the series far enough (where what constitutes "far enough" will depend on x), we can approximate $\cos(x)$ to any number of decimal places desired. For any x, the value of $\cos(x)$ is the limit of the finite sums of the Taylor series, just as 1/3 is the limit of the finite sums of its infinite series representation.

In general, given a function $f(x)$, its Taylor series centered at $x = 0$ will be

$$f(0) + f'(0)x + \frac{f^{(2)}(0)}{2!}x^2 + \frac{f^{(3)}(0)}{3!}x^3 + \frac{f^{(4)}(0)}{4!}x^4 + \cdots = \sum_{k=0}^{\infty} \frac{f^{(k)}(0)}{k!}x^k.$$

We have the following Taylor series centered at $x = 0$ for some common functions:

$f(x)$	Taylor Series for $f(x)$
$\sin(x)$	$x - \dfrac{x^3}{3!} + \dfrac{x^5}{5!} - \dfrac{x^7}{7!} + \cdots$
$\cos(x)$	$1 - \dfrac{x^2}{2!} + \dfrac{x^4}{4!} - \dfrac{x^6}{6!} + \cdots$
e^x	$1 + x + \dfrac{x^2}{2!} + \dfrac{x^3}{3!} + \dfrac{x^4}{4!} + \cdots$
$\ln(1-x)$	$-\left(x + \dfrac{x^2}{2} + \dfrac{x^3}{3} + \dfrac{x^4}{4} + \cdots\right)$
$\dfrac{1}{1-x}$	$1 + x + x^2 + x^3 + \cdots$
$\dfrac{1}{1+x^2}$	$1 - x^2 + x^4 - x^6 + \cdots$
$(1+x)^c$	$1 + cx + \dfrac{c(c-1)}{2!}x^2 + \dfrac{c(c-1)(c-2)}{3!}x^3 + \cdots$

While it is true that $\cos(x)$ and e^x equal their Taylor series, just as $\sin(x)$ did, we have to be more careful with the last four functions. To see why this is, let's graph $1/(1 + x^2)$ and its Taylor polynomials $P_n(x) = 1 - x^2 + x^4 - x^6 + \cdots \pm x^n$ for $n = 2, 4, 6, 8, 10, 12, 14, 16, 200$, and 202. Since all the graphs are symmetric about the y-axis (why is this?), we only draw the graphs for positive x:

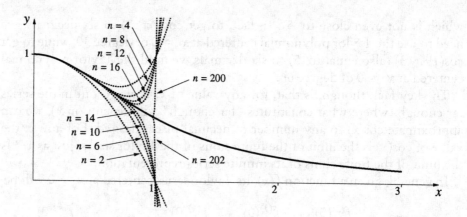

A Taylor series may not converge for all values of x

It appears that the graphs of the Taylor polynomials $P_n(x)$ approach the graph of $1/(1 + x^2)$ very nicely *so long* as $x < 1$. If $x \geq 1$, though, it looks like there is no convergence, no matter how far out in the Taylor series we go. We can thus write

$$\frac{1}{1 + x^2} = 1 - x^2 + x^4 - x^6 + \cdots \qquad \text{for } |x| < 1,$$

where the restriction on x is essential if we want to use the $=$ sign. We say that the interval $-1 < x < 1$ is the **interval of convergence** for the Taylor series centered at $x = 0$ for $1/(1 + x^2)$. Some Taylor series, like those for $\sin(x)$ and e^x, converge for all values of x—their interval of convergence is $(-\infty, \infty)$. Other Taylor series, like those for $1/(1 + x^2)$ and $\ln(1 - x)$, have finite intervals of convergence.

> *Brook Taylor (1685–1731) was an English mathematician who developed the series that bears his name in his book* Methodus incrementorum *(1715). He did not worry about questions of convergence, but used the series freely to attack many kinds of problems, including differential equations.*

REMARK.   On the one hand, it is perhaps not too surprising that a function should equal its Taylor series—after all, with more and more coefficients to fiddle with, we can control more and more of the behavior of the associated polynomials. On the other hand, we are saying that a function like $\sin(x)$ or e^x has its behavior for all values of x completely determined by the value of the function and all its derivatives at a single point, so perhaps it is surprising after all!

Exercises

1. **a)** Suppose you wanted to use the Taylor series centered at $x = 0$ to calculate $\sin(100)$. How large does n have to be before the term $(100)^n / n!$ is less than 1?

   b) If we wanted to calculate $\sin(100)$ directly using this Taylor series, we would have to go very far out before we began to approach a limit at all closely. Can you use your knowledge of the way the circular functions behave to calculate $\sin(100)$ much more rapidly (but still using the Taylor series centered at $x = 0$)? Do it.

   c) Show that we can calculate the sine of any number by using a Taylor series centered at $x = 0$ either for $\sin(x)$ or for $\cos(x)$ to a suitable value of x between 0 and $\pi/4$.

2. *a)** Suppose we wanted to calculate $\ln 5$ to seven decimal places. An obvious place to start is with the Taylor series centered at $x = 0$ for $\ln(1 - x)$:

$$-\left(x + \frac{x^2}{2} + \frac{x^3}{3} + \frac{x^4}{4} + \cdots\right),$$

with $x = -4$. What happens when you do this, and why? Try a few more values for x and see if you can make a conjecture about the interval of convergence for this Taylor series.

   b) Explain how you could use the fact that $\ln(1/A) = -\ln A$ for any real number $A > 0$ to evaluate $\ln x$ for $x > 2$. Use this to compute $\ln 5$ to seven decimals. How far out in the Taylor series did you have to go?

   c) If you wanted to calculate $\ln 1.5$, you could use the Taylor series for $\ln(1 - x)$ with either $x = -1/2$, which would lead directly to $\ln 1.5$, or you could use the series with $x = 1/3$, which would produce $\ln(2/3) = -\ln 1.5$. Which method is faster, and why?

3. We can improve the speed of our calculations of the logarithm function slightly by the following series of observations:

   *a)** Find the Taylor series centered at $u = 0$ for $\ln(1 + u)$.

   b) Find the Taylor series centered at $u = 0$ for

$$\ln\left(\frac{1 - u}{1 + u}\right).$$

   [Remember that $\ln(A/B) = \ln A - \ln B$.]

   c) Show that any $x > 0$ can be written in the form $(1 - u)/(1 + u)$ for some suitable $-1 < u < 1$.

   d) Use the preceding to evaluate $\ln 5$ to seven decimal places. How far out in the Taylor series did you have to go?

4. a) Evaluate arctan(.5) to seven decimal places.

b) Try to use the Taylor series centered at $x = 0$ to evaluate arctan(2) directly—what happens? Remembering what the arctangent function means geometrically, can you figure out a way around this difficulty?

5.*a) *Calculating π.* The Taylor series expansion for the arctangent function:

$$\arctan x = x - \tfrac{1}{3}x^3 + \tfrac{1}{5}x^5 - \cdots \pm \frac{1}{2n + 1}x^{2n+1} + \cdots$$

lies behind many of the methods for getting lots of decimals of π rapidly. For instance, since $\tan(\pi/4) = 1$, we have $\pi/4 = \arctan 1$. Use this to get a series expansion for π. How far out in the series do you have to go to evaluate π to three decimal places?

b) The reason the preceding approximations converged so slowly was that we were substituting $x = 1$ into the series, so we didn't get any help from the x^n terms in making the successive corrections get small rapidly. We would like to be able to do something with values of x between 0 and 1. We can do this by using the addition formula for the tangent function:

$$\tan(\alpha + \beta) = \frac{\tan \alpha + \tan \beta}{1 - \tan \alpha \tan \beta}.$$

Use this to show that

$$\frac{\pi}{4} = \arctan\left(\frac{1}{2}\right) + \arctan\left(\frac{1}{5}\right) + \arctan\left(\frac{1}{8}\right).$$

Now use the Taylor series for each of these three expressions to calculate π to 12 decimal places. How far out in the series do you have to go? Which series did you have to go the farthest out in before the 12th decimal stabilized? Why?

6. *Raising e to imaginary powers.* One of the major mathematical developments of the last century was the extension of the ideas of calculus to **complex numbers**—i.e., numbers of the form $r + si$, where r and s are real numbers, and i is a new symbol, defined by the property that $i \cdot i = -1$. Thus $i^3 = i^2 i = -i$, $i^4 = i^2 i^2 = (-1)(-1) = 1$, and so on. If we want to extend our standard functions to these new numbers, we proceed as we did in the previous section and look for the crucial *properties* of these functions to see what they suggest. One of the key properties of e^x as we've now seen is that it possesses a Taylor series:

$$e^x = 1 + x + \frac{x^2}{2!} + \frac{x^3}{3!} + \frac{x^4}{4!} + \cdots.$$

But this property only involves operations of ordinary arithmetic, and so makes perfectly good sense even if x is a complex number.

a) Show that if s is any real number, we must define e^{is} to be $\cos(s) + i\,\sin(s)$ if we want to preserve this property.

b) Show that $e^{\pi i} = -1$.

c) Show that if $r + si$ is any complex number, we must have

$$e^{r+si} = e^r(\cos(s) + i\,\sin(s))$$

if we want complex exponentials to preserve all the right properties.

d) Find a complex number $r + si$ such that $e^{r+si} = -5$.

7. *Hyperbolic trigonometric functions.* The hyperbolic trigonometric functions are defined by the formulas

$$\cosh(x) = \frac{e^x + e^{-x}}{2}$$

$$\sinh(x) = \frac{e^x - e^{-x}}{2}.$$

(The names of these functions are usually pronounced "cosh" and "cinch.") In this problem you will explore some of the reasons for the adjectives *hyperbolic* and *trigonometric.*

a) Modify the Taylor series centered at $x = 0$ for e^x to find a Taylor series for $\cosh(x)$. Compare your results to the Taylor series centered at $x = 0$ for $\cos(x)$.

b) Now find the Taylor series centered at $x = 0$ for $\sinh(x)$. Compare your results to the Taylor series centered at $x = 0$ for $\sin(x)$.

c) Parts (a) and (b) of this problem should begin to explain the *trigonometric* part of the story. What about the *hyperbolic* part? Recall that the familiar trigonometric functions are called *circular functions* because for any t the point $(\cos t, \sin t)$ is on the unit circle with equation $x^2 + y^2 = 1$ (cf. section 7.2). Show that the point $(\cosh t, \sinh t)$ lies on the hyperbola with equation $x^2 - y^2 = 1$.

8. Consider the Taylor series centered at $x = 0$ for $(1 + x)^c$.

a) What does the series give if you let $c = 2$? Is this reasonable?

b) What do you get if you set $c = 3$?

c) Show that if you set $c = n$, where n is a positive integer, the Taylor series will terminate. This yields a general formula—the **binomial theorem**—discovered by the twelfth-century Persian poet and mathematician, Omar Khayyam, and generalized by Newton to the form you have just obtained. Write out the first three and the last three terms of this formula.

d) Use an appropriate substitution for x and a suitable value for c to derive the Taylor series for $1/(1 - u)$. Does this agree with what we previously obtained?

e) Suppose we want to calculate $\sqrt{17}$. We might try letting $x = 16$ and $c = 1/2$ and using the Taylor series for $(1 + x)^c$. What happens when you try this?

f) We can still use the series to help us, though, if we are a little clever and write

$$\sqrt{17} = \sqrt{16 + 1} = \sqrt{16(1 + \frac{1}{16})} = \sqrt{16} \cdot \sqrt{1 + \frac{1}{16}} = 4 \cdot \sqrt{1 + \frac{1}{16}}.$$

Now apply the series using $x = 1/16$ to evaluate $\sqrt{17}$ to seven decimal places accuracy. How many terms does it take?

g) Use the same kind of trick to evaluate $\sqrt[3]{30}$.

Evaluating Taylor series rapidly.   Suppose we wanted to plot the Taylor polynomial of degree 11 associated with $\sin(x)$. For each value of x, then, we would have to evaluate

$$P_{11}(x) = x - \frac{x^3}{3!} + \frac{x^5}{5!} - \frac{x^7}{7!} + \frac{x^9}{9!} - \frac{x^{11}}{11!}.$$

Since the length of time it takes the computer to evaluate an expression like this is roughly proportional to the number of multiplications and divisions involved (additions and subtractions, by comparison, take a negligible amount of time), let's see how many of these operations are needed to evaluate $P_{11}(x)$. To calculate x^{11} requires 10 multiplications, while 11! requires 9 (if we are clever and don't bother to multiply by 1 at the end!), so the evaluation of the term $x^{11}/11!$ will require a total of 20 operations (counting the final division). Similarly, evaluating $x^9/9!$ requires 16 operations, $x^7/7!$ requires 12, on down to $x^3/3!$, which requires 4. Thus the total number of multiplications and divisions needed is

$$4 + 8 + 12 + 16 + 20 = 60.$$

This is not too bad, although if we were doing this for many different values of x, which would be the case if we wanted to graph $P_{11}(x)$, this would begin to add up. Suppose, though, that we wanted to graph something like $P_{51}(x)$ or $P_{101}(x)$. By the same analysis, evaluating $P_{51}(x)$ for a single value of x would require

$$4 + 8 + 12 + 16 + 20 + 24 + \cdots + 96 + 100 = 1300$$

multiplications and divisions, while evaluation of $P_{101}(x)$ would require 5100 operations. Thus it would take roughly 20 times as long to evaluate $P_{51}(x)$ as it takes to evaluate $P_{11}(x)$, while $P_{101}(x)$ would take about 85 times as long.

9. Show that, in general, the number of multiplications and divisions needed to evaluate $P_n(x)$ is roughly $n^2/2$.

We can be clever, though. Note that $P_{11}(x)$ can be written as

$$x\left(1 - \frac{x^2}{2\cdot 3}\left(1 - \frac{x^2}{4\cdot 5}\left(1 - \frac{x^2}{6\cdot 7}\left(1 - \frac{x^2}{8\cdot 9}\left(1 - \frac{x^2}{10\cdot 11}\right)\right)\right)\right)\right).$$

*10. How many multiplications and divisions are required to evaluate this expression?

*11. Thus this way of evaluating $P_{11}(x)$ is roughly three times as fast, a modest saving. How much faster is it if we use this method to evaluate $P_{51}(x)$?

12. Find a general formula for the number of multiplications and divisions needed to evaluate $P_n(x)$ using this way of grouping.

   Finally, we can extend these ideas to reduce the number of operations even further, so that evaluating a polynomial of degree n requires only n multiplications, as follows. Suppose we start with a polynomial

$$p(x) = a_0 + a_1 x + a_2 x^2 + a_3 x^3 + \cdots + a_{n-1} x^{n-1} + a_n x^n.$$

We can rewrite this as

$$a_0 + x(a_1 + x(a_2 + \cdots + x(a_{n-2} + x(a_{n-1} + a_n x))\cdots)).$$

You should check that with this representation it requires only n multiplications to evaluate $p(x)$ for a given x.

13. a) Write two computer programs to evaluate the 300th-degree Taylor polynomial centered at $x = 0$ for e^x, with one of the programs being the obvious, standard way, and the second program being the method given above. Evaluate $e^1 = e$ using each program, and compare the length of time required.

   b) Use these two programs to graph the 300th-degree Taylor polynomial for e^x over the interval $[0, 2]$, and compare times.

➤ 10.4 POWER SERIES AND DIFFERENTIAL EQUATIONS

So far, we have begun with functions we already know, in the sense of being able to calculate the value of the function and all its derivatives at at least one point. This in turn allowed us to write down the corresponding Taylor series. Often, though, we don't even have this much information about a function. In such cases it is frequently useful to assume that there is some infinite polynomial—called a **power series**—which represents the function, and then see if we can determine the coefficients of the polynomial.

This technique is especially useful in dealing with differential equations. To see why this is the case, think of the alternatives. If we can approximate the solution $y = y(x)$ to a certain differential equation to an acceptable degree of accuracy by, say, a 20th-degree polynomial, then the only storage space required is the insignificant space taken to keep track of the 21 coefficients. Whenever we want the value of the solution for a given value of x, we can then get a quick approximation by evaluating the polynomial at x. Other alternatives are much more costly in time or in space. We could use Euler's method to grind out the solution at x, but, as you've already discovered, this can be a slow and tedious process. Another option is to calculate lots of values and store them in a table in the computer's memory. This not only takes up a lot of memory space, but it also only gives values for a finite set of values of x, and is not much faster than evaluating a polynomial. Until 30 years ago, the table approach was the standard one—all scientists and mathematicians had a handbook of mathematical functions containing hundreds of pages of numbers giving the values of every function they might need.

To see how this can happen, let's first look at a familiar differential equation whose solutions we already know:

$$y' = y.$$

Of course, we know by now that the solutions are $y = ae^x$ for an arbitrary constant a [where $a = y(0)$]. Suppose, though, that we didn't already know how to solve this differential equation. We might see if we could find a power series of the form

$$y = a_0 + a_1 x + a_2 x^2 + a_3 x^3 + \cdots + a_n x^n + \cdots$$

which is a solution to the differential equation. Can we determine values for the coefficients $a_0, a_1, a_2, \ldots, a_n, \ldots$ that will make $y' = y$?

Using the rules for differentiation, we have

$$y' = a_1 + 2a_2 x + 3a_3 x^2 + 4a_4 x^3 + \cdots + na_n x^{n-1} + \cdots.$$

Two polynomials are equal if and only if the coefficients of corresponding powers of x are equal. Therefore, if $y' = y$, it would have to be true that

$$a_1 = a_0$$
$$2a_2 = a_1$$
$$3a_3 = a_2$$
$$\vdots$$
$$na_n = a_{n-1}$$
$$\vdots$$

Therefore the values of $a_1, a_2, a_3, \ldots$ are not arbitrary; indeed, each is determined by the preceding one. Equations like these—which deal with a sequence of quantities and relate each term to those earlier in the sequence—are called **recursion relations**. These recursion relations permit us to express every a_n in terms of a_0:

$$a_1 = a_0$$
$$a_2 = \tfrac{1}{2}a_1 \qquad\qquad\qquad\quad = \tfrac{1}{2}a_0$$
$$a_3 = \tfrac{1}{3}a_2 \quad = \tfrac{1}{3}\cdot\tfrac{1}{2}a_0 \qquad = \tfrac{1}{3!}a_0$$
$$a_4 = \tfrac{1}{4}a_3 \quad = \tfrac{1}{4}\cdot\tfrac{1}{3!}a_0 \qquad = \tfrac{1}{4!}a_0$$
$$\vdots \qquad\quad \vdots \qquad\qquad\quad \vdots$$
$$a_n = \tfrac{1}{n}a_{n-1} = \tfrac{1}{n}\cdot\frac{1}{(n-1)!}a_0 = \tfrac{1}{n!}a_0.$$
$$\vdots \qquad\quad \vdots \qquad\qquad\quad \vdots$$

Notice that a_0 remains "free": there is no equation that determines its value. Thus, without additional information, a_0 is *arbitrary*. The series for y now becomes

$$y = a_0 + a_0 x + \frac{1}{2!}a_0 x^2 + \frac{1}{3!}a_0 x^3 + \cdots + \frac{1}{n!}a_0 x^n + \cdots$$

or

$$y = a_0\left[1 + x + \frac{1}{2!}x^2 + \frac{1}{3!}x^3 + \cdots + \frac{1}{n!}x^2 + \cdots\right].$$

But the series in square brackets is just the Taylor series for e^x—we have derived the Taylor series from the differential equation alone, without using any of the other properties of the exponential function. Thus, we again find that the solutions of the differential equation $y' = y$ are

$$y = a_0 e^x,$$

where a_0 is an arbitrary constant. Notice that $y(0) = a_0$, so the value of a_0 will be determined if the initial value of y is specified.

NOTE.   In general, once we have derived a power series expression for a function, that power series will also be the Taylor series for that function. Although the two series are the same, the term Taylor series is typically reserved for those settings where we were able to evaluate the derivatives through some other means, as in the preceding section.

Bessel's Equation

For a new example, let's look at a differential equation that arises in an enormous variety of physical problems (wave motion, optics, the conduction of electricity and of heat and fluids, and the stability of columns, to name a few):

$$x^2 \cdot y'' + x \cdot y' + (x^2 - p^2) \cdot y = 0.$$

This is called the **Bessel equation of order p**. Here p is a parameter specified in advance, so we will really have a different set of solutions for each value of p. To determine a solution completely, we will also need to specify the initial values of $y(0)$ and $y'(0)$. The solutions of the Bessel equation of order p are called **Bessel functions of order p**, and the solution for a given value of p (together with particular initial conditions which needn't concern us here) is written $J_p(x)$. In general, there is no formula for a Bessel function in terms of simpler functions [although it turns out that a few special cases like $J_{1/2}(x)$, $J_{3/2}(x)$, ... can be expressed relatively simply]. To evaluate such a function we could use Euler's method, or we could try to find a power series solution.

Friedrich Wilhelm Bessel (1784–1846) was a German astronomer who studied the functions that now bear his name in his efforts to analyze the perturbations of planetary motions, particularly those of Saturn.

Consider the Bessel equation with $p = 0$. We are thus trying to solve the differential equation

$$x^2 \cdot y'' + x \cdot y' + x^2 \cdot y = 0.$$

By dividing by x, we can simplify this a bit to

$$x \cdot y'' + y' + x \cdot y = 0.$$

Let's look for a power series expansion

$$y = b_0 + b_1 x + b_2 x^2 + b_3 x^3 + \cdots.$$

We can compute

$$y' = b_1 + 2b_2 x + 3b_3 x^2 + 4b_4 x^3 + \cdots + (n + 2)(n + 1)b_{n+2} x^n + \cdots$$

and

$$y'' = 2b_2 + 6b_3 x + 12b_4 x^2 + 20b_5 x^3 + \cdots + (n + 1)b_{n+1} x^n + \cdots.$$

We can now use these expressions to calculate the series for the combination that occurs in the differential equation:

$$
\begin{aligned}
xy'' &= & 2b_2 x & + 6b_3 x^2 & + \cdots \\
y' &= b_1 + & 2b_2 x & + 3b_3 x^2 & + \cdots \\
xy &= & b_0 x & + b_1 x^2 & + \cdots \\
\hline
\end{aligned}
$$

$$
xy'' + y' + xy = b_1 + (4b_2 + b_0)x + (9b_3 + b_1)x^2 + \cdots.
$$

In general, the coefficient of x^n in the combination will be

Finding the coefficient of x^n

$$
(n + 1)nb_{n+1} + (n + 1)b_{n+1} + b_{n-1} = (n + 1)^2 b_{n+1} + b_{n-1}.
$$

If y is to be a solution to the original differential equation, this infinite series must equal 0. This in turn means that every coefficient must be 0. We thus get

$$
b_1 = 0
$$

$$
4b_2 + b_0 = 0
$$

$$
9b_3 + b_1 = 0
$$

$$
\vdots
$$

$$
n^2 b_n + b_{n-2} = 0
$$

$$
\vdots
$$

If we now solve these recursively as before, we see first off that since $b_1 = 0$, it must also be true that

$$
b_k = 0 \quad \text{for every odd } k.
$$

For the even coefficients we have

$$
b_2 = -\frac{1}{2^2} b_0
$$

$$
b_4 = -\frac{1}{4^2} b_2 = \frac{1}{2^2 4^2} b_0
$$

$$
b_6 = -\frac{1}{6^2} b_4 = -\frac{1}{2^2 4^2 6^2} b_0
$$

$$
\vdots
$$

so that in general, we have

$$
b_{2n} = \pm \frac{1}{2^2 4^2 6^2 \cdots (2n)^2} b_0 = \pm \frac{1}{2^{2n}(n!)^2} b_0.
$$

Thus any function y satisfying the Bessel equation of order 0 must be of the form

$$y = b_0 \left(1 - \frac{x^2}{2^2} + \frac{x^4}{2^4(2!)^2} - \frac{x^6}{2^6(3!)^2} + \cdots \right).$$

In particular, if we impose the initial condition $y(0) = 1$ (which requires that $b_0 = 1$), we get the zeroth-order Bessel function $J_0(x)$:

$$J_0(x) = 1 - \frac{x^2}{4} + \frac{x^4}{64} - \frac{x^6}{2304} + \frac{x^8}{147456} + \cdots.$$

The graph of the Bessel function J_0

Here is the graph of $J_0(x)$ together with the polynomial approximations of degree 2, 4, 6, ..., 30 over the interval [0, 14]:

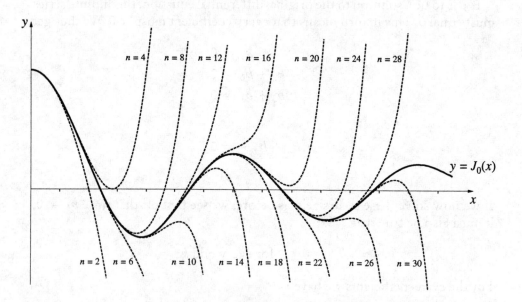

The graph of J_0 is suggestive: it appears to be oscillatory, with decreasing amplitude. Both observations are correct: it can in fact be shown that J_0 has infinitely many zeroes, spaced roughly π units apart, and that $\lim_{x \to \infty} J_0(x) = 0$.

The *S-I-R* Model One More Time

In exactly the same way, we can find power series solutions when there are several interacting variables involved. Let's look at the example we've considered at a number of points in this text to see how this works. In the *S-I-R* model we basically wanted to solve the system of equations

The *S-I-R* model

$$S' = -aSI$$
$$I' = aSI - bI$$
$$R' = bI,$$

where a and b were parameters depending on the specific situation. Let's look for solutions of the form

$$S = s_0 + s_1 t + s_2 t^2 + s_3 t^3 + \cdots$$
$$I = i_0 + i_1 t + i_2 t^2 + i_3 t^3 + \cdots$$
$$R = r_0 + r_1 t + r_2 t^2 + r_3 t^3 + \cdots.$$

If we put these series in the equation $S' = -aSI$, we get

$$s_1 + 2s_2 t + 3s_3 t^2 + \cdots = -a(s_0 + s_1 t + s_2 t^2 + \cdots)(i_0 + i_1 t + i_2 t^2 + \cdots)$$
$$= -a[s_0 i_0 + (s_0 i_1 + s_1 i_0)t + (s_0 i_2 + s_1 i_1 + s_2 i_0)t^2 + \cdots].$$

As before, if the two sides of the equation are to be equal, the coefficients of corresponding powers of t must be equal: Finding the coefficients of the power series for $S(t)$

$$s_1 = -as_0 i_0$$
$$2s_2 = -a(s_0 i_1 + s_1 i_0)$$
$$3s_3 = -a(s_0 i_2 + s_1 i_1 + s_2 i_0)$$
$$\vdots$$
$$ns_n = -a(s_0 i_{n-1} + s_1 i_{n-2} + \cdots + s_{n-2} i_1 + s_{n-1} i_0)$$

While this looks messy, it has the crucial recursive feature—each s_k is expressed in terms of previous terms. That is, if we knew all the s and the i coefficients out through the coefficients of, say, t^6 in the series for S and I, we could immediately calculate s_7. We again have a *recursion relation*. Recursion again

We could expand the equation $I' = aSI - bI$ in the same way, and get recursion relations for the coefficients i_k. In this model, though, there is a shortcut if we observe that since $S' = -aSI$, and since $I' = aSI - bI$, we have $I' = -S' - bI$. If we substitute the power series in this expression and equate coefficients, we get

$$ni_n = -ns_n - bi_{n-1},$$

which leads to

$$i_n = -s_n - \frac{b}{n} i_{n-1}.$$ Finding the power series for $I(t)$

So if we know s_n and i_{n-1}, we can calculate i_n.

We are now in a position to calculate the coefficients as far out as we like. For we will be given values for a and b when we are given the model. Moreover, since $s_0 = S(0) =$ the initial S population, and $i_0 = I(0) =$ the initial I population, we will also typically be given these values as well. But knowing

s_0 and i_0, we can determine s_1 and then i_1. But then, knowing these values, we can determine s_2 and then i_2, and so on. Since the arithmetic is tedious, this is obviously a place for a computer. Here is a program that calculates the first 50 coefficients in the power series for $S(t)$ and $I(t)$:

Program: SIRSERIES

Dimension the arrays S and I to contain 51 elements

```
a = .00001
b = 1/14
S(0) = 45400
I(0) = 2100
FOR k = 1 TO 50
    Sum = 0
    FOR j = 0 TO k - 1
        Sum = Sum + S(j) * I(k - j - 1)
    NEXT j
    S(k) = -a * SUM/k
    I(k) = -S(k) - b * I(k - 1)/k
NEXT k
```

COMMENT.   The first line in this program introduces a new feature. It notifies the computer that the variables S and I are going to be arrays—strings of numbers—and that each array will consist of 51 elements. The element S(k) corresponds to what we have been calling s_k. The integer k is called the **index** of the term in the array. The indices in this program run from 0 to 50. Different versions of BASIC express the dimensioning statement differently.

The effect of running this program is thus to create two 51-element arrays, S and I, containing the coefficients of the power series for S and I out to degree 50. Notice that the program *prints* nothing. If we just wanted to see these coefficients, we could have the computer list them. Here are the first 35 coefficients for S (read across the rows):

45400	-953.4	-172.3611	-17.982061	$-.86969127$
$5.4479852e\text{-}2$	$1.5212707e\text{-}2$	$1.4463108e\text{-}3$	$4.3532884e\text{-}5$	$-7.9100481e\text{-}6$
$-1.4207959e\text{-}6$	$-1.0846994e\text{-}7$	$-6.512610e\text{-}10$	$9.304633e\text{-}10$	$1.256507e\text{-}10$
$7.443310e\text{-}12$	$-2.191966e\text{-}13$	$-9.787285e\text{-}14$	$-1.053428e\text{-}14$	$-4.382620e\text{-}16$
$4.230290e\text{-}17$	$9.548369e\text{-}18$	$8.321674e\text{-}19$	$1.760392e\text{-}20$	$-5.533369e\text{-}21$
$-8.770972e\text{-}22$	$-6.101928e\text{-}23$	$3.678170e\text{-}25$	$6.193375e\text{-}25$	$7.6272253e\text{-}26$
$4.011923e\text{-}27$	$-1.986216e\text{-}28$	$-6.318305e\text{-}29$	$-6.271724e\text{-}30$	$-2.150100e\text{-}31$

Thus the power series for S begins

$$45400 - 953.4t - 172.3611t^2 - 17.982061t^3 - .86969127t^4$$
$$+ \cdots - 2.15010 \times 10^{-31}t^{34} + \cdots.$$

In the same fashion, we find that the power series for I begins

$$2100 + 803.4t + 143.66824t^2 + 14.561389t^3 + .60966648t^4$$
$$+ \cdots + 2.021195 \times 10^{-31}t^{34} + \cdots.$$

If we now wanted to graph these polynomials over, say, $0 \le t \le 10$, we can add some lines to the above program. We first define a couple of short subroutines SUS and INF to calculate the polynomial approximations for $S(t)$ and $I(t)$ using the coefficients we've derived in the first part of the program. (Note that these subroutines calculate polynomials in the straightforward, inefficient way. If you did the exercises in section 10.3 which developed techniques for evaluating polynomials rapidly, you might want to modify these subroutines to take advantage of the increased speed available.) Remember, too, that you will need to set up the graphics at the beginning of the program to be able to plot.

```
DEF SUS(x)
    Sum = S(0)
    FOR j = 1 TO 50
        Sum = Sum + S(j) * x^j
    NEXT j
    SUS = Sum
END DEF
DEF INF(x)
    Sum = I(0)
    FOR j = 1 TO 50
        Sum = Sum + I(j) * x^j
    NEXT j
    INF = Sum
END DEF
FOR x = 0 TO 10 STEP .01
    Plot the line from (x, SUS(x))
        to (x + .01, SUS(x + .01))
    Plot the line from (x, INF(x))
        to (x + .01, INF(x + .01))
NEXT x
```

Here is the graph of $I(t)$ over a 25-day period, together with the polynomial approximations of degree 5, 20, 30, and 70.

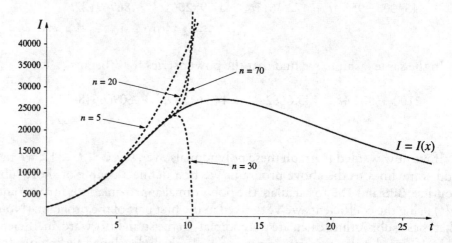

Note that these polynomials appear to converge to $I(t)$ only out to values of t around 10. If we needed polynomial approximations beyond that point, we could shift to a different point on the curve, find the values of I and S there by Euler's method, then repeat the above process. For instance, when $t = 12$, we get by Euler's method that $S(12) = 7670$ and $I(12) = 27,136$. If we now shift our clock to measure time in terms of $\tau = t - 12$, we get the following polynomial of degree 30:

$$27136 + 143.0455\tau - 282.0180\tau^2 + 23.5594\tau^3$$
$$+ .4548\tau^4 + \cdots + 1.2795 \times 10^{-25}\tau^{30}.$$

Here is what the graph of this polynomial looks like when plotted with the graph of I. On the horizontal axis we list the t-coordinates with the corresponding τ-coordinates underneath.

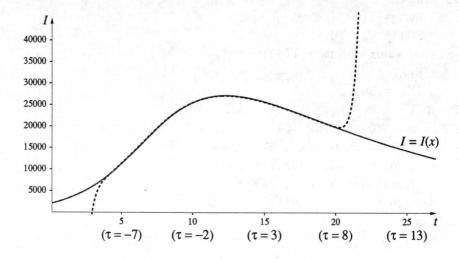

The interval of convergence seems to be approximately $4 < t < 20$. Thus if we combine this polynomial with the 30th-degree polynomial from the previous graph, we would have very accurate approximations for I over the entire interval $[0, 20]$.

Exercises

1. Find power series solutions of the form

$$y = a_0 + a_1 x + a_2 x^2 + a_3 x^3 + \cdots + a_n x^n + \cdots$$

for each of the following differential equations.

*a) $y' = 2xy$
b) $y' = 3x^2 y$
c) $y'' + xy = 0$
d) $y'' + xy' + y = 0$

2. a) Find power series solutions to the differential equation $y'' = -y$. Start with

$$y = a_0 + a_1 x + a_2 x^2 + a_3 x^3 + \cdots + a_n x^n + \cdots.$$

Notice that, in the recursion relations you obtain, the coefficients of the even terms are completely independent of the coefficients of the odd terms. This means you can get two separate power series, one with only even powers, with a_0 as arbitrary constant, and one with only odd powers, with a_1 as arbitrary constant.

b) The two power series you obtained in part (a) are the Taylor series centered at $x = 0$ of two familiar functions. Which ones? Verify that these functions do indeed satisfy the differential equation $y'' = -y$.

3. a) Find power series solutions to the differential equation $y'' = y$. As in the previous problem, the coefficients of the even terms depend only on a_0, and the coefficients of the odd terms depend only on a_1. Write down the two series, one with only even powers and a_0 as an arbitrary constant, and one with only odd powers, with a_1 as an arbitrary constant.

b) The two power series you obtained in part (a) are the Taylor series centered at $x = 0$ of two *hyperbolic trigonometric functions* (see the exercises in section 10.3). Verify that these functions do indeed satisfy the differential equation $y'' = y$.

4. a) Find power series solutions to the differential equation $y' = xy$, starting with

$$y = a_0 + a_1 x + a_2 x^2 + a_3 x^3 + \cdots + a_n x^n + \cdots.$$

What recursion relations do you get? Is $a_1 = a_3 = a_5 = \cdots = 0$?

b) Verify that

$$y = e^{x^2/2}$$

satisfies the differential equation $y' = xy$. Find the Taylor series for this function and compare it with the series you obtained in (a) using the recursion relations.

5. *The Bessel equation*
   *a) Take $p = 1$. The solution satisfying the initial condition $y' = 1/2$ when $x = 0$ is defined to be the first-order Bessel function $J_1(x)$. (It will turn out that y has to be 0 when $x = 0$, so we don't have to specify the initial value of y—we have no choice in the matter.) Find the first five terms of the power series expansion for $J_1(x)$. What is the coefficient of x^{2n+1}?

   b) Show by direct calculation from the series for J_0 and J_1 that

   $$J_0' = -J_1.$$

   c) To see, from another point of view, that $J_0' = -J_1$, take the equation

   $$x \cdot J_0'' + J_0' + x \cdot J_0 = 0$$

   and differentiate it. By doing some judicious canceling and rearranging of terms, show that

   $$x^2 \cdot (J_0')'' + x \cdot (J_0')' + (x^2 - 1)(J_0') = 0.$$

   This demonstrates that J_0' is a solution of the Bessel equation with $p = 1$.

6. a) When we found the power series expansion for solutions to the zeroth-order Bessel equation, we found that all the odd coefficients had to be 0. In particular, since b_1 is the value of y' when $x = 0$, we are saying that all solutions have to be flat at $x = 0$. This should bother you a bit. Why can't you have a solution, say, that satisfies $y = 1$ and $y' = 1$ when $x = 0$?

   b) You might get more insight on what's happening by using Euler's method, starting just a little to the right of the origin and moving left. Use Euler's method to sketch solutions with the initial values

i.	$y = 2$	$y' = 1$	when	$x = 1$
ii.	$y = 1.1$	$y' = 1$	when	$x = .1$
iii.	$y = 1.01$	$y' = 1$	when	$x = .01$.

   What seems to happen as you approach the y-axis?

7. *Legendre's differential equation*

   $$(1 - x^2)y'' - 2xy' + \ell(\ell + 1)y = 0$$

   arises in many physical problems—for example, in quantum mechanics, where its solutions are used to describe certain orbits of the electron in a hydrogen atom. In that context, the parameter ℓ is called the *angular momentum* of the electron; it must be either an integer or a "half-integer" (i.e., a number like 3/2). Quantum theory gets its name from the fact that numbers like the angular momentum of the electron in the hydrogen atom are "quantized," that is, they cannot have just any value, but must be a multiple of some "quantum"—in this case, the number 1/2.

a) Find power series solutions of Legendre's equation.

b) Quantization of angular momentum has an important consequence. Specifically, when ℓ is an integer it is possible for a series solution to stop—that is, to be a polynomial. For example, when $\ell = 1$ and $a_0 = 0$, the series solution is just $y = a_1 x$—all higher order coefficients turn out to be zero. Find polynomial solutions to Legendre's equation for $\ell = 0, 2, 3, 4,$ and 5 (consider $a_0 = 0$ or $a_1 = 0$). These solutions are called, naturally enough, **Legendre polynomials**.

8. It turns out that the power series solutions to the S-I-R model have a finite interval of convergence. By plotting the power series solutions of different degrees against the solutions obtained by Euler's method, estimate the interval of convergence.

9. **a)** *Logistic growth.* Find the first five terms of the power series solution to the differential equation

$$y' = y(1 - y).$$

Note that this is just the logistic equation, where we have chosen our units of time and of quantities of the species being studied so that the carrying capacity is 1 and the intrinsic growth rate is 1.

b) Using the initial condition $y = .1$ when $x = 0$, plot this power series solution on the same graph as the solution obtained by Euler's method. How do they compare?

c) Do the same thing with initial conditions $y = 2$ when $x = 0$.

▷ 10.5 CONVERGENCE

We have written expressions like

$$\sin(x) = x - \frac{x^3}{3!} + \frac{x^5}{5!} - \frac{x^7}{7!} + \frac{x^9}{9!} - \cdots,$$

meaning that for any value of x the series on the right will converge to $\sin(x)$. There are a couple of issues here. First, what do we mean when we say the series "converges," and how do we prove it converges to $\sin(x)$? If x is small, we can convince ourselves that the statement is true just by trying it. If x is large, though, say $x = 100^{100}$, it would be convenient to have a more general method for proving the stated convergence. Further, we have the example of the function $1/(1 + x^2)$ as a caution—it seemed to converge for small values of x (namely, for $|x| < 1$), but not for large values.

Convergence
essentially means
"decimals stabilizing"

Let's first clarify what we mean by convergence. It is, essentially, the intuitive notion of "decimals stabilizing" that we have been using all along. To make explicit what we've been doing, let's write a "generic" series

$$b_0 + b_1 + b_2 + \cdots = \sum_{m=0}^{\infty} b_m.$$

When we evaluate such a series, we look at the **partial sums**

$$S_1 = b_0 + b_1 \qquad\qquad = \sum_{m=0}^{1} b_m$$

$$S_2 = b_0 + b_1 + b_2 \qquad\qquad = \sum_{m=0}^{2} b_m$$

$$S_3 = b_0 + b_1 + b_2 + b_3 \qquad = \sum_{m=0}^{3} b_m$$

$$\vdots \qquad\qquad\qquad\qquad\qquad \vdots$$

$$S_n = b_0 + b_1 + b_2 + \ldots + b_n = \sum_{m=0}^{n} b_m$$

$$\vdots \qquad\qquad\qquad\qquad\qquad \vdots$$

Typically, when we calculate a number of these partial sums, we notice that beyond a certain point they all seem to agree on, say, the first 8 decimal places. If we keep on going, the partial sums would agree on the first 9 decimals, and, further on, on the first 10 decimals, and so forth. This is precisely what we mean by convergence:

The infinite series

$$b_0 + b_1 + b_2 + \cdots = \sum_{m=0}^{\infty} b_m$$

converges if, no matter how many decimal places are specified, it is always the case that all the partial sums eventually agree to at least this many decimal places.
Put more formally, we say the series converges if, given any number D of decimal places, it is always possible to find an integer N_D such that if k and n are both greater than N_D, then S_k and S_n agree to at least D decimal places.
The number defined by these stabilizing decimals is called the **sum** of the series.
If a series does not converge, we say it **diverges**.

In other words, for me to prove to you that the Taylor series for $\sin(x)$ converges at $x = 100^{100}$, you would specify a certain number of decimal places, say 5000, and I would have to be able to prove to you that if you took partial sums with enough terms, they would all agree to at least 5000 decimals. Moreover, I would have to be able to show the same thing happens if you specify *any* number of decimal places you want agreement on.

What it means for an infinite sum to converge

How can this be done? It seems like an enormously daunting task to be able to do for any series. We'll tackle this challenge in stages. First we'll see what goes wrong with some series that don't converge—divergent series. Then we'll look at a particular convergent series—the **geometric series**—that's relatively easy to analyze. Finally, we will look at some more general rules that will guarantee convergence of series like those for the sine, cosine, and exponential functions.

Divergent Series

Suppose we have an infinite series

$$b_0 + b_1 + b_2 + \cdots = \sum_{m=0}^{\infty} b_m,$$

and consider two successive partial sums, say

$$S_{122} = b_0 + b_1 + b_2 + \ldots + b_{122} = \sum_{m=0}^{122} b_m$$

and

$$S_{123} = b_0 + b_1 + b_2 + \ldots + b_{122} + b_{123} = \sum_{m=0}^{123} b_m.$$

Note that these two sums are exactly the same, except that the sum for S_{123} has one more term, b_{123}, added on. Now suppose that S_{122} and S_{123} agree to 19 decimal places. Recall from section 10.2 that we defined this to mean $|S_{123} - S_{122}| < .5 \times 10^{-19}$. But since $S_{123} - S_{122} = b_{123}$, this means that $|b_{123}| < .5 \times 10^{-19}$. To phrase this more generally,

> Two successive partial sums, S_n and S_{n+1}, agree out to k decimal places if and only if $|b_{n+1}| < .5 \times 10^{-k}$.

But since our definition of convergence required that we be able to fix any specified number of decimals provided we took partial sums lengthy enough, it must be true that *if the series converges*, the individual terms b_k must become arbitrarily small if we go out far enough. Intuitively, you can think of the

A necessary condition for convergence

partial sums S_k as being a series of approximations to some quantity. The term b_{k+1} can be thought of as the "correction" which is added to S_k to produce the next approximation S_{k+1}. Clearly, if the approximations are eventually becoming good ones, the corrections made should become smaller and smaller. We thus have the following necessary condition for convergence:

$$
\text{If the infinite series } b_0 + b_1 + b_2 + \cdots = \sum_{m=0}^{\infty} b_m \text{ converges, then}
$$

$$
\lim_{k \to \infty} b_k = 0.
$$

Necessary and *sufficient* mean different things

REMARK.   It is important to recognize what this criterion does and does not say. It is a **necessary** condition for convergence (i.e., every convergent sequence has to satisfy the condition $\lim_{k \to \infty} b_k = 0$), but it is not a **sufficient** condition for convergence (i.e., there are some divergent sequences that also have the property that $\lim_{k \to \infty} b_k = 0$). The criterion is usually used to detect some divergent series, and is more useful in the following form (which you should convince yourself is equivalent to the preceding):

$$
\text{If } \lim_{k \to \infty} b_k \neq 0 \text{ (either because the limit doesn't exist at all,}
$$
$$
\text{or it equals something besides 0), then the infinite series}
$$

$$
b_0 + b_1 + b_2 + \cdots = \sum_{m=0}^{\infty} b_m
$$

$$
\text{diverges.}
$$

Detecting divergent series

This criterion allows us to detect a number of divergent series right away. For instance, we saw earlier that the statement

$$
\frac{1}{1 + x^2} = 1 - x^2 + x^4 - x^6 + \cdots
$$

appeared to be true only for $|x| < 1$. Using the remarks above, we can see why this series has to diverge for $|x| \geq 1$. If we write $1 - x^2 + x^4 - x^6 + \cdots$ as $b_0 + b_1 + b_2 + \cdots$, we see that $b_k = (-1)^k x^{2k}$. Clearly b_k does not go to 0 for $|x| \geq 1$—the successive "corrections" we make to each partial sum just become larger and larger, and the partial sums will alternate more and more wildly from a huge positive number to a huge negative number. Hence the series converges *at most* for $-1 < x < 1$. We will see in the next subsection how to prove that it really does converge for all x in this interval.

Using exactly the same kind of argument, we can show that the following series also diverge for $|x| > 1$:

$f(x)$	Taylor Series for $f(x)$
$\ln(1-x)$	$-\left(x + \dfrac{x^2}{2} + \dfrac{x^3}{3} + \dfrac{x^4}{4} + \cdots\right)$
$\dfrac{1}{1-x}$	$1 + x + x^2 + x^3 + \cdots$
$(1+x)^c$	$1 + cx + \dfrac{c(c-1)}{2!}x^2 + \dfrac{c(c-1)(c-2)}{3!}x^3 + \cdots$
$\arctan x$	$x - \dfrac{1}{3}x^3 + \dfrac{1}{5}x^5 - \cdots \pm \dfrac{1}{2n+1}x^{2n+1}$

The details are left to the exercises. While these common series all happen to diverge for $|x| > 1$, it is easy to find other series that diverge for $|x| > 2$ or $|x| > 17$ or whatever—see the exercises for some examples.

The Harmonic Series

We stated earlier in this section that simply knowing that the individual terms b_k go to 0 for large values of k does not guarantee that the series

$$b_0 + b_1 + b_2 + \cdots$$

will converge. Essentially what can happen is that the b_k go to 0 slowly enough that they can still accumulate large values. The classic example of such a series is the **harmonic series**:

An important counterexample

$$1 + \frac{1}{2} + \frac{1}{3} + \frac{1}{4} + \cdots = \sum_{i=1}^{\infty} \frac{1}{i}.$$

It turns out that this series just keeps getting larger as you add more terms. It is eventually larger than 1000, or 1 million, or 100^{100}, or .... This fact is established in the exercises. A suggestive argument, though, can be quickly given by observing that the harmonic series is just what you would get if you substituted $x = 1$ into the power series

$$x + \frac{x^2}{2} + \frac{x^3}{3} + \frac{x^4}{4} + \cdots.$$

But this is just the Taylor series for $-\ln(1-x)$, and if we substitute $x = 1$ into this we get $-\ln 0$, which isn't defined. Also, $\lim_{x \to 0} -\ln x = +\infty$.

The Geometric Series

A series occurring frequently in a wide range of contexts is the **geometric series**

$$G(x) = 1 + x + x^2 + x^3 + x^4 + \cdots.$$

This is also a sequence we can analyze completely and rigorously in terms of its convergence. It will turn out that we can then reduce the analysis of the convergence of several other sequences to the behavior of this one.

To avoid divergence, $|x|$ must be less than 1

By the analysis we performed above, if $|x| \geq 1$ the individual terms of the series clearly don't go to 0, and the series therefore diverges. What about the case where $|x| < 1$?

The starting point is the partial sums. A typical partial sum looks like:

$$S_n = 1 + x + x^2 + x^3 + \cdots + x^n.$$

This is a finite number; we must find out what happens to it as n grows without bound. Since S_n is finite, we can calculate with it. In particular,

$$xS_n = x + x^2 + x^3 + \cdots + x^n + x^{n+1}.$$

Subtracting the second expression from the first, we get

$$S_n - xS_n = 1 - x^{n+1}$$

and thus (if $x \neq 1$)

A simple expression for the partial sum S_n

$$S_n = \frac{1 - x^{n+1}}{1 - x}.$$

(What is the value of S_n if $x = 1$?)

This is a handy, compact form for the partial sum. Let us see what value it has for various values of x. For example, if $x = 1/2$, then

n :	1	2	3	4	5	6	$\cdots$	$\rightarrow$	∞
S_n :	1	$\dfrac{3}{2}$	$\dfrac{7}{4}$	$\dfrac{15}{8}$	$\dfrac{31}{16}$	$\dfrac{63}{32}$	$\cdots$	$\rightarrow$	2.

Finding the limit of S_n as $n \rightarrow \infty$

It appears that as $n \rightarrow \infty$, $S_n \rightarrow 2$. Can we see this algebraically?

$$
\begin{aligned}
S_n &= \frac{1 - (1/2)^{n+1}}{1 - \frac{1}{2}} \\
&= \frac{1 - (1/2)^{n+1}}{1/2} \\
&= 2 \cdot (1 - (1/2)^{n+1}) \\
&= 2 - (1/2)^n.
\end{aligned}
$$

As $n \rightarrow \infty$, $(1/2)^n \rightarrow 0$, so the values of S_n become closer and closer to 2. Clearly the series converges, and its sum is 2.

Summing another geometric series

Similarly, when $x = -1/2$, the partial sums are

$$S_n = \frac{1 - (-1/2)^{n+1}}{3/2} = \frac{2}{3}\left(1 \pm \frac{1}{2^{n+1}}\right).$$

The presence of the $\pm$ sign does not alter the outcome: since $(1/2^{n+1}) \to 0$, the partial sums converge to $2/3$. Therefore, we can say the series converges and its sum is $2/3$.

In exactly the same way, though, for any x satisfying $|x| < 1$ we have

$$S_n = \frac{1 - x^{n+1}}{1 - x}$$

$$= \frac{1}{1 - x} (1 - x^{n+1})$$

and as $n \to \infty$, $x^{n+1} \to 0$. Therefore, $S_n \to 1/(1 - x)$. Thus the series converges, and its sum is $1/(1 - x)$.

To summarize, we have thus proved that

The geometric series

$$G(x) = 1 + x + x^2 + x^3 + x^4 + \cdots$$

converges for all x such that $|x| < 1$. In such cases the sum is

$$\frac{1}{1 - x}.$$

The series diverges for all other values of x.

Convergence and divergence of the geometric series

As a final comment, note that the formula

$$S_n = \frac{1 - x^{n+1}}{1 - x}$$

is valid for all x (except $x = 1$). Even though the partial sums aren't converging to any limit if $x > 1$, the formula can still be useful as a quick way for summing powers. Thus, for instance

$$1 + 3 + 9 + 27 + 81 + 243 = \frac{1 - 3^6}{1 - 3} = \frac{1 - 729}{-2} = \frac{-728}{-2} = 364,$$

and

$$1 - 5 + 25 - 125 + 625 - 3125 = \frac{1 - (-5)^6}{1 - (-5)} = \frac{1 - 15625}{6} = -264.$$

Alternating Series

A large class of common power series consists of the **alternating series**—series in which the terms are alternately positive and negative. The behavior of such series is particularly easy to analyze, as we shall see in this section. Here are

Many common series are alternating, at least for some values of x

some examples of alternating series we've already encountered:

$$\sin x = x - \frac{x^3}{3!} + \frac{x^5}{5!} - \frac{x^7}{7!} + \cdots$$

$$\cos x = 1 - \frac{x^2}{2!} + \frac{x^4}{4!} - \frac{x^6}{6!} + \cdots$$

$$\frac{1}{1 + x^2} = 1 - x^2 + x^4 - x^6 + \cdots \qquad \text{(for } |x| < 1\text{).}$$

Other series may be alternating for some, but not all, values of x. For instance, here are two series that are alternating for negative values of x, but not for positive values:

$$e^x = 1 + x + \frac{x^2}{2!} + \frac{x^3}{3!} + \frac{x^4}{4!} + \cdots$$

$$\ln(1 - x) = -(x + x^2 + x^3 + x^4 + \cdots) \qquad \text{(for } |x| < 1\text{).}$$

Convergence criterion for alternating series. Let us write a generic alternating series as

$$b_0 - b_1 + b_2 - b_3 + \cdots + (-1)^m b_m + \cdots,$$

where the b_m are positive. It turns out that an alternating series converges if the terms b_m both consistently shrink in size and approach zero:

$$b_0 - b_1 + b_2 - b_3 + \cdots + (-1)^m b_m + \cdots \quad \text{converges if}$$
$$0 < b_{m+1} \le b_m \text{ for all } m \text{ and } \lim_{m \to \infty} b_m = 0.$$

It is this property that makes alternating series particularly easy to deal with. Recall that this is *not* a property of series in general, as we saw by the example of the harmonic series. The reason it is true for alternating series becomes clear if we view the behavior of the partial sums geometrically:

Alternating series are easy to test for convergence

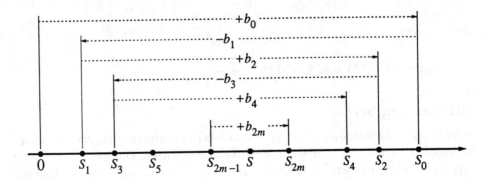

We mark the partial sums S_n on a number line. The first sum $S_0 = b_0$ lies to the right of the origin. To find S_1 we go to the left a distance b_1. Because $b_1 \le b_0$, S_1 will lie between the origin and S_0. Next we go to the right a distance b_2, which brings us to S_2. Since $b_2 \le b_1$, we will have $S_2 \le S_0$. The next move is to the left a distance b_3, and we find $S_3 \ge S_1$. We continue going back and forth in this fashion, each step being less than or equal to the preceding one, since $b_{m+1} \le b_m$. We thus get

$$0 \le S_1 \le S_3 \le S_5 \le \cdots \le S_{2m-1} \le \cdots \le S_{2m} \le \cdots \le S_4 \le S_2 \le S_0.$$

The partial sums oscillate back and forth, with all the odd sums on the left increasing and all the even sums on the right decreasing. Moreover, since $|S_n - S_{n-1}| = b_n$, and since $\lim_{n\to\infty} b_n = 0$, the difference between consecutive partial sums eventually becomes arbitrarily small—the oscillations take place within a smaller and smaller interval. Thus given any number of decimal places, we can always go far enough out in the series so that S_k and S_{k+1} agree to that many decimal places. But if n is any integer greater than k, then, since S_n lies between S_k and S_{k+1}, S_n will also agree to that many decimal places— those decimals will be fixed from k on out. The series therefore converges, as claimed—the sum is the unique number S that is greater than all the odd partial sums and less than all the even partial sums.

The partial sums oscillate, with the exact sum trapped between consecutive partial sums

For a convergent alternating series, we also have a particularly simple bound for the error when we approximate the sum S of the series by partial sums.

A simple estimate for the accuracy of the partial sums

If

$$S_n = b_0 - b_1 + b_2 - \cdots \pm b_n,$$

and if

$$0 < b_{m+1} \le b_m \quad \text{for all} \quad m \quad \text{and} \quad \lim_{m \to \infty} b_m = 0$$

(so the series converges),
then

$$|S - S_n| < b_{n+1}.$$

In words, the error in approximating S by S_n is less than the next term in the series.

PROOF.   Suppose n is odd. Then we have, as above, that $S_n < S < S_{n+1}$. Therefore $0 < S - S_n < S_{n+1} - S_n = b_{n+1}$. If n is even, a similar argument shows $0 < S_n - S < S_n - S_{n+1} = b_{n+1}$. In either case, we have $|S - S_n| < b_{n+1}$, as claimed.

Note further that we also know whether S_n is too large or too small, depending on whether n is even or odd.

Estimating the error in approximating cos(.7)

EXAMPLE.   Let's apply the error estimate for an alternating series to analyze the error if we approximate cos(.7) with a Taylor series with three terms:

$$\cos(.7) \approx 1 - \frac{1}{2!}(.7)^2 + \frac{1}{4!}(.7)^4 = 0.765004166\ldots.$$

Since the last term in this partial sum was an addition, this approximation is too big. To get an estimate of how far off it might be, we look at the next term in the series:

$$\frac{1}{6!}(.7)^6 = .0001634\ldots.$$

We thus know that the correct value for cos(.7) is somewhere in the interval

$$.76484 = .76500 - .00016 \le \cos(.7) \le .76501,$$

so we know that cos(.7) begins .76 . . . and the third decimal is either a 4 or a 5.
    If we use the partial sum with four terms, we get

$$\cos(.7) \approx 1 - \frac{1}{2!}(.7)^2 + \frac{1}{4!}(.7)^4 - \frac{1}{6!}(.7)^6 = .764840765\ldots$$

and the error would be less than

$$\frac{1}{8!}(.7)^8 = .0000014\ldots < .5 \times 10^{-5},$$

so we could now say that cos(.7) = .76484 . . . .

How many terms are needed to obtain accuracy to 12 decimal places?

If we wanted to know in advance how far out in the series we would have to go to determine cos(.7) to, say, 12 decimals, we could do it by finding a value for n such that

$$b_n = \frac{1}{n!}(.7)^n \le .5 \times 10^{-12}.$$

With a little trial and error, we see that $b_{12} \approx .3 \times 10^{-10}$, while $b_{14} < 10^{-13}$. Thus if we take the value of the 12th-degree approximation for cos(.7), we can be assured that our value will be accurate to 12 places.
    We have met this capability of getting an error estimate in a single step before, in version 3 of Taylor's theorem. It is in contrast to the approximations made in dealing with general series, where we typically had to look at the pattern of stabilizing digits in the succession of improving estimates to get a sense of how good our approximation was, and even then we had no guarantee.

Computing e.   Because of the fact that we can find sharp bounds for the accuracy of an approximation with alternating series, it is often desirable to convert a given problem to this form where we can. For instance, suppose we wanted a good value for *e*. The obvious thing to do would be to take the Taylor series for e^x and substitute $x = 1$. If we take the first 11 terms of this series, we get the approximation

$$e = e^1 \approx 1 + 1 + \frac{1}{2!} + \frac{1}{3!} + \cdots + \frac{1}{10!} = 2.718281801146\ldots,$$

but we have no way of knowing how many of these digits are correct.

Suppose instead, that we evaluate e^{-1}:

$$e^{-1} \approx 1 - 1 + \frac{1}{2!} - \frac{1}{3!} + \cdots + \frac{1}{10!} = .367879464286\ldots.$$

Since $1/(11!) = .000000025\ldots$, we know this approximation is accurate to at least seven decimals. If we take its reciprocal we get

$$1/.3678794624286\ldots = 2.718281657666\ldots,$$

which will then be accurate to six decimals (in the exercises you will show why the accuracy drops by one decimal place), so we can say $e = 2.718281\ldots.$

The Radius of Convergence

We have seen examples of power series that converge for all x (like the Taylor series for $\sin x$) and others that converge only for certain x (like the series for $\arctan x$). How can we determine the convergence of an arbitrary power series of the form

$$a_0 + a_1 x + a_2 x^2 + a_3 x^3 + \cdots + a_n x^n + \cdots ?$$

We must suspect that this series *may not* converge for all values of x. For example, does the Taylor series

$$1 + x + \frac{1}{2!} x^2 + \frac{1}{3!} x^3 + \cdots$$

converge for all values of x, or only some? When it converges, does it converge to e^x or to something else? After all, this Taylor series is designed to look like e^x only near $x = 0$; it remains to be seen how well the function and its series match up far from $x = 0$.

The question of convergence has a definitive answer. It goes like this: If the power series

$$a_0 + a_1 x + a_2 x^2 + a_3 x^3 + \cdots + a_n x^n + \cdots$$

It may be possible to convert a given problem to one involving alternating series

The answer to the convergence question

converges for a particular value of x—say $x = s$—then it automatically converges for any *smaller* value of x (meaning any x that is closer to the origin than s is; i.e., any x for which $|x| < |s|$). Likewise, if the series *diverges* for a particular value of x, then it also diverges for any value farther from the origin. In other words, the values of x where the series converges are not interspersed with the values where it diverges. On the contrary, within a certain distance R from the origin there is only convergence, while beyond that distance there is only divergence. The number R is called the **radius of convergence** of the series, and the range where it converges is called its **interval of convergence**.

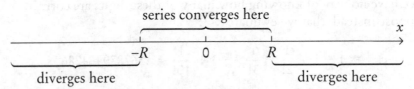

The radius of convergence of a power series

An obvious example of the radius of convergence is given by the geometric series

$$\frac{1}{1-x} = 1 + x + x^2 + x^3 + \cdots.$$

The radius of convergence of the geometric series is 1

We know that this converges for $|x| < 1$ and diverges for $|x| > 1$. Thus the radius of convergence is $R = 1$ in this case.

It is possible for a power series to converge for all x; if that happens, we take R to be ∞. At the other extreme, the series may converge only for $x = 0$. (When $x = 0$ the series collapses to its constant term a_0, so it certainly converges *at least* when $x = 0$.) If the series converges only for $x = 0$, then we take R to be 0.

At $x = R$ the series may diverge or converge; different things happen for different series. The same is true when $x = -R$. The radius of convergence tells us where the switch from convergence to divergence happens. It does not tell us what happens at the place where the switch occurs. If we know that the series converges for $x = \pm R$, then we say that $[-R, R]$ is the interval of convergence. If the series converges when $x = R$ but *not* when $x = -R$, then the interval of convergence is $(-R, R]$, and so on.

The Ratio Test

There are several ways to determine the radius of convergence of a power series. One of the simplest and most useful is by means of the **ratio test**. Because the power series to which we apply this test need not include *consecutive* powers of x (think of the Taylor series for $\cos x$ or $\sin x$), we'll write a "generic" series as

$$b_0 + b_1 + b_2 + \cdots = \sum_{m=0}^{\infty} b_m.$$

Here are three examples of the use of this notation.

1. The Taylor series for e^x is $\displaystyle\sum_{m=0}^{\infty} b_m$, where

$$b_0 = 1, \quad b_1 = x, \quad b_2 = \frac{x^2}{2!}, \quad \ldots, \quad b_m = \frac{x^m}{m!}.$$

2. The Taylor series for $\cos x$ is $\displaystyle\sum_{m=0}^{\infty} b_m$, where

$$b_0 = 1, \quad b_1 = \frac{-x^2}{2!}, \quad b_2 = \frac{x^4}{4!}, \quad \ldots, \quad b_m = (-1)^m \frac{x^{2m}}{(2m)!}.$$

3. We can even describe the series

$$17 + x + x^2 + x^4 + x^6 + x^8 + \cdots = 17 + x + \sum_{m=2}^{\infty} x^{2m-2}$$

in our generic notation, in spite of the presence of the first two terms "$17 + x$" which don't fit the pattern of later ones. We have $b_0 = 17$, $b_1 = x$, and then $b_m = x^{2m-2}$ for $m = 2, 3, 4, \ldots$. The question of convergence for a power series is unaffected by the "beginning" of the series; only the pattern in the "tail" matters. (Of course the *value* of the power series is affected by all of its terms.) So we can modify our generic notation to fit the circumstances at hand. No harm is done if we don't begin with b_0.

Convergence is determined by the "tail" of the series

Using this notation we can state the ratio test (but we give no proof).

Ratio Test: The series $b_0 + b_1 + b_2 + b_3 + \cdots + b_n + \cdots$

converges if $\displaystyle\lim_{m \to \infty} \frac{|b_{m+1}|}{|b_m|} < 1.$

Let's see what the ratio test says about the geometric series:

$$1 + x + x^2 + x^3 + \cdots .$$

Applying the ratio test to the geometric series ...

We have $b_m = x^m$, so the ratio we must consider is

$$\frac{|b_{m+1}|}{|b_m|} = \frac{|x^{m+1}|}{|x^m|} = \frac{|x|^{m+1}}{|x|^m} = |x|.$$

(Be sure you see why $|x^m| = |x|^m$.) Obviously, this ratio has the same value for all m, so the limit

$$\lim_{m \to \infty} |x| = |x|$$

exists and is less than 1 precisely when $|x| < 1$. Thus the geometric series converges for $|x| < 1$—which we know is true. This means that the radius of convergence of the geometric series is $R = 1$.

... and to e^x

Look next at the Taylor series for e^x:

$$1 + x + \frac{x^2}{2!} + \frac{x^3}{3!} + \frac{x^4}{4!} + \cdots = \sum_{m=0}^{\infty} \frac{x^m}{m!}.$$

For negative x this is an alternating series, so by the criterion for convergence of alternating series we know it converges for all $x < 0$. The radius of convergence should then be ∞. We will use the ratio test to show that in fact this series converges for *all* x.

In this case

$$b_m = \frac{x^m}{m!},$$

so the relevant ratio is

$$\frac{|b_{m+1}|}{|b_m|} = \left| \frac{x^{m+1}}{(m+1)!} \right| \cdot \left| \frac{m!}{x^m} \right|$$

$$= \frac{|x^{m+1}|}{|x^m|} \cdot \frac{m!}{(m+1)!}$$

$$= |x| \cdot \frac{1}{m+1} = \frac{|x|}{m+1}.$$

Unlike the example with the geometric series, the value of this ratio depends on m. For any particular x, as m gets larger and larger the numerator stays the same and the denominator grows, so this ratio gets smaller and smaller. In other words,

$$\lim_{m \to \infty} \frac{|b_{m+1}|}{|b_m|} = \lim_{m \to \infty} \frac{|x|}{m+1} = 0.$$

Since this limit is less than 1 for any value of x, the series converges for all x, and thus the radius of convergence of the Taylor series for e^x is $R = \infty$, as we expected.

One of the uses of the theory developed so far is that it gives us a new way of specifying functions. For example, consider the power series

$$\sum_{m=0}^{\infty} (-1)^m \frac{2^m}{m^2+1} x^m = 1 - x + \frac{4}{5}x^2 - \frac{8}{10}x^3 + \frac{16}{17}x^4 + \cdots.$$

In this case $b_m = (-1)^m \dfrac{2^m}{m^2 + 1} x^m$, so to find the radius of convergence, we compute the ratio

$$\frac{|b_{m+1}|}{|b_m|} = \frac{2^{m+1}|x|^{m+1}}{(m + 1)^2 + 1} \cdot \frac{m^2 + 1}{2^m|x|^m} = 2|x| \frac{m^2 + 1}{m^2 + 2m + 2}.$$

To figure out what happens to this ratio as m grows large, it is helpful to rewrite the factor involving the m's as

Finding the limit of $|b_{m+1}|/|b_m|$ may require some algebra

$$\frac{m^2 \cdot (1 + 1/m^2)}{m^2 \cdot (1 + 2/m + 2/m^2)} = \frac{1 + 1/m^2}{1 + 2/m + 2/m^2}.$$

Now we can see that

$$\lim_{m \to \infty} \frac{|b_{m+1}|}{|b_m|} = 2|x| \frac{1}{1} = 2|x|.$$

The limit value is less than 1 precisely when $2|x| < 1$, or, equivalently, $|x| < 1/2$, so the radius of convergence of this series is $R = 1/2$. It follows that for $|x| < 1/2$, we have a new function $f(x)$ defined by the power series:

$$f(x) = \sum_{m=0}^{\infty} (-1)^m \frac{2^m}{m^2 + 1} x^m.$$

We can also discuss the radius of convergence of a power series

$$a_0 + a_1(x - a) + a_2(x - a)^2 + \cdots + a_m(x - a)^m + \cdots$$

centered at a number a other than the origin. The radius of convergence of a power series of this form can be found by the **ratio test** in exactly the same way it was when $a = 0$.

EXAMPLE.   Let's apply the ratio test to the Taylor series centered at $a = 1$ for $\ln(x)$:

$$\ln(x) = \sum_{m=1}^{\infty} \frac{(-1)^{m-1}}{m} (x - 1)^m.$$

We can start our series with b_1, so we can take

$$b_m = \frac{(-1)^{m-1}}{m} (x - 1)^m.$$

Then the ratio we must consider is

$$\frac{|b_{m+1}|}{|b_m|} = \frac{|x - 1|^{m+1}}{m + 1} \cdot \frac{m}{|x - 1|^m} = |x - 1| \cdot \frac{m}{m + 1} = |x - 1| \cdot \frac{1}{1 + 1/m}.$$

Then

$$\lim_{m \to \infty} \frac{|b_{m+1}|}{|b_m|} = |x - 1| \cdot 1 = |x - 1|.$$

From this we conclude that this series converges for $|x - 1| < 1$. This inequality is equivalent to

$$-1 < x - 1 < 1,$$

which is an interval of "radius" 1 about $a = 1$, so the radius of convergence is $R = 1$ in this case. We may also write the interval of convergence for this power series as

$$0 < x < 2.$$

More generally, using the ratio test we find that a power series centered at a converges in an interval of "radius" R (and width $2R$) around the point $x = a$ on the x-axis. Ignoring what happens at the endpoints, we say the **interval of convergence** is

$$a - R < x < a + R.$$

Here is a picture of what this looks like:

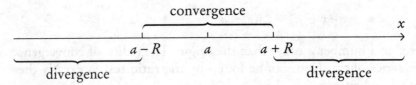

The convergence of a power series centered at a

Exercises

1. Find a formula for the sum of each of the following power series by performing suitable operations on the geometric series and the formula for its sum.

   a) $1 - x^3 + x^6 - x^9 + \cdots$.

   b) $x^2 + x^6 + x^{10} + x^{14} + \cdots$.

   c) $1 - 2x + 3x^2 - 4x^3 + \cdots$.

   d) $x + 2x^2 + 3x^3 + 4x^4 + \cdots$.

   e) $x + \dfrac{x^2}{2} + \dfrac{x^3}{3} + \dfrac{x^4}{4} + \cdots$.

2. Determine the value of each of the following infinite sums. (Each of these sums is a geometric or related series evaluated at a particular value of x.)

a) $\dfrac{1}{4} + \dfrac{1}{16} + \dfrac{1}{64} + \dfrac{1}{256} + \cdots.$

b) $.02020202.\ldots$

c) $-\dfrac{5}{2} + \dfrac{5}{4} - \dfrac{5}{8} + \cdots.$

d) $\dfrac{1}{1} - \dfrac{2}{2} + \dfrac{3}{4} - \dfrac{4}{8} + \dfrac{5}{16} - \dfrac{6}{32} + \cdots.$

e) $\dfrac{1}{1 \cdot 10} + \dfrac{1}{2 \cdot 10^2} + \dfrac{1}{3 \cdot 10^3} + \dfrac{1}{4 \cdot 10^4} + \cdots.$

3. *The multiplier effect.* Economists know that the effect on a local economy of tourist spending is greater than the amount actually spent by the tourists. The *multiplier effect* quantifies this enlarged effect. In this problem you will see that calculating the multiplier effect involves summing a geometric series.

   Suppose that, over the course of a year, tourists spend a total of A dollars in a small resort town. By the end of the year, the townspeople are therefore A dollars richer. Some of this money leaves the town— for example, to pay state and federal taxes or to pay off debts owed to "big city" banks. Some of it stays in town but gets put away as savings. Finally, a certain fraction of the original amount is spent in town, by the townspeople themselves. Suppose 3/5-ths is spent this way. The tourists and the townspeople *together* are therefore responsible for spending

$$S = A + \frac{3}{5}A \quad \text{dollars}$$

in the town that year. The second amount—$\frac{3}{5}A$ dollars—is *recirculated* money.

   Since one dollar looks much like another, the recirculated money should be handled the same way as the original tourist dollars: some will leave the town, some will be saved, and the remaining 3/5-ths will get recirculated a *second* time. The twice-recirculated amount is

$$\frac{3}{5} \times \frac{3}{5}A \quad \text{dollars},$$

and we must revise the calculation of the total amount spent in the town to

$$S = A + \frac{3}{5}A + \left(\frac{3}{5}\right)^2 A \quad \text{dollars}.$$

But the twice-recirculated dollars look like all the others, so 3/5-ths of them will get recirculated a *third* time. Revising the total dollars spent yet again, we get

$$S = A + \frac{3}{5}A + \left(\frac{3}{5}\right)^2 A + \left(\frac{3}{5}\right)^3 A \quad \text{dollars}.$$

This process never ends: no matter how many times a sum of money has been recirculated, 3/5-ths of it is recirculated once more. The total amount spend in the town is thus given by a *series*.

a) Write the series giving the total amount of money spent in the town and calculate its sum.

b) Your answer in (a) is a certain multiple of A—what is the multiplier?

c) Suppose the recirculation rate is r instead of 3/5. Write the series giving the total amount spent and calculate its sum. What is the multiplier now?

d) Suppose the recirculation rate is 1/5; what is the multiplier in this case?

e) Does a lower recirculation rate produce a smaller multiplier effect?

4. Which of the following alternating series converge, which diverge? Why?

*a) $\displaystyle\sum_{n=1}^{\infty}(-1)^n \frac{n}{n^2 + 1}$

b) $\displaystyle\sum_{n=1}^{\infty}(-1)^n \frac{1}{\sqrt{3n + 2}}$

*c) $\displaystyle\sum_{n=1}^{\infty}(-1)^n \frac{n}{\ln n}$

d) $\displaystyle\sum_{n=1}^{\infty}(-1)^n \frac{n}{5n - 4}$

e) $\displaystyle\sum_{n=1}^{\infty}(-1)^n \frac{\arctan n}{n}$

f) $\displaystyle\sum_{n=1}^{\infty}(-1)^n \frac{(1.0001)^n}{n^{10} + 1}$

g) $\displaystyle\sum_{n=1}^{\infty}(-1)^n \frac{1}{n^{1/n}}$

h) $\displaystyle\sum_{n=1}^{\infty}(-1)^n \frac{1}{\ln n}$

i) $\displaystyle\sum_{n=1}^{\infty}(-1)^n \frac{n!}{n^n}$

j) $\displaystyle\sum_{n=1}^{\infty}(-1)^n \frac{n!}{1 \cdot 3 \cdot 5 \cdots (2n - 1)}$

*5. For each of the sums in the preceding problem that converges, use the alternating series criterion to determine how far out you have to go before the sum is determined to six decimal places. Give the sum for each of these series to this many places.

6. Find a value for n so that the nth-degree Taylor series for e^x gives at least 10 places accuracy for all x in the interval $[-3, 0]$.

7. We defined the harmonic series as the infinite sum

$$1 + \frac{1}{2} + \frac{1}{3} + \frac{1}{4} + \cdots = \sum_{i=1}^{\infty} \frac{1}{i}.$$

a) Use a calculator to find the partial sums

$$S_n = 1 + \frac{1}{2} + \frac{1}{3} + \frac{1}{4} + \cdots + \frac{1}{n}$$

for $n = 1, 2, 3, \ldots, 12$.

b) Use the following program to find the value of S_n for $n = 100$. Modify the program to find the values of S_n for $n = 500, 1000,$ and 5000.

Program: HARMONIC

```
n = 100
sum = 0
FOR i = 1 TO n
      sum = sum + 1/i
NEXT i
PRINT n, sum
```

c) Group the terms in the harmonic series as indicated by the parentheses:

$$1 + \left(\frac{1}{2}\right) + \left(\frac{1}{3} + \frac{1}{4}\right) + \left(\frac{1}{5} + \frac{1}{6} + \frac{1}{7} + \frac{1}{8}\right) +$$

$$+ \left(\frac{1}{9} + \cdots + \frac{1}{16}\right) + \left(\frac{1}{17} + \cdots + \frac{1}{32}\right) + \cdots.$$

Explain why each parenthetical grouping totals at least $1/2$.

d) Following the pattern in part (c), if you add up the terms of the harmonic series forming S_n for $n = 2^k$, you can arrange the terms as $1 + k$ such groupings. Use this fact and the result of (c) to explain why S_n exceeds $1 + k \cdot \frac{1}{2}$.

e) Use part (d) to explain why the harmonic series *diverges*.

f) You might try this problem if you've had some physics—enough to know how to locate the center of mass of a system. Suppose you had n cards and wanted to stack them on the edge of a table with the top of the pile leaning out over the edge. How far out could you get the pile to reach if you were careful? Let's choose our units so the length of each card is 1. Clearly if $n = 1$, the farthest reach you could get would be $\frac{1}{2}$. If $n = 2$, you could clearly place the top card to extend half a unit beyond the bottom card. For the system to be stable, the center of mass of the two cards must be to the left of the edge of the table. Show that for this to happen, the bottom card can't extend more than 1/4 unit beyond the edge. Thus with $n = 2$, the maximum extension of the pile is $\frac{1}{2} + \frac{1}{4} = \frac{3}{4}$. The picture at the right shows 10 cards stacked carefully.

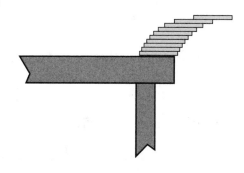

Prove that if you have n cards, the stack can be built to extend a distance of

$$\frac{1}{2} + \frac{1}{4} + \frac{1}{6} + \frac{1}{8} + \cdots + \frac{1}{2n} = \frac{1}{2}\left(1 + \frac{1}{2} + \frac{1}{3} + \frac{1}{4} + \cdots + \frac{1}{n}\right).$$

In the light of what we have just proved about the harmonic series, this shows that if you had enough cards, you could have the top card extending 100 units beyond the edge of the table!

8. *An estimate for the partial sums of the harmonic series.* You may notice in part (d) of the preceding problem that the sum $S_n = 1 + 1/2 + 1/3 + \cdots + 1/n$ grows in proportion to the exponent k of $n = 2^k$; i.e., the sum grows like the logarithm of n. We can make this more precise by comparing the value of S_n to the value of the integral

$$\int_1^n \frac{1}{x}\,dx = \ln(n).$$

a) Let's look at the case $n = 6$ to see what's going on. Consider the following picture:

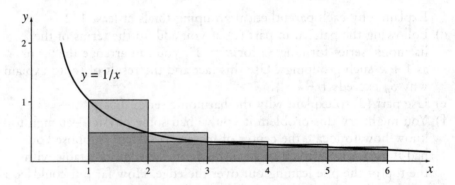

Show that the lightly shaded region plus the dark region has area equal to S_5, which can be rewritten as $S_6 - \frac{1}{6}$. Show that the dark region alone has area $S_6 - 1$. Hence prove that

$$S_6 - 1 < \int_1^6 \frac{1}{x}\,dx < S_6 - \frac{1}{6},$$

and conclude that

$$\frac{1}{6} < S_6 - \ln(6) < 1.$$

b) Show more generally that

$$\frac{1}{n} < S_n - \ln(n) < 1.$$

*c) Use part (b) to get upper and lower bounds for the value of S_{10000}.

d) Use the result of part (b) to get an estimate for how many cards you would need in part (f) of the preceding problem to make the top of the pile extend 100 units beyond the edge of the table.

Remarkably, partial sums of the harmonic series exceed $\ln(n)$ in a very regular way. It turns out that

$$\lim_{n \to \infty} \{ S_n - \ln(n) \} = \gamma,$$

where $\gamma = .5772\ldots$ is called **Euler's constant.** *(You have seen another constant named for Euler, the base $e = 2.7183\ldots$.) Although one can find the decimal expansion of γ to any desired degree of accuracy, no one knows whether γ is a rational number or not.*

9. Show that the power series for $\arctan x$ and $(1 + x)^c$ diverge for $|x| > 1$. Do the series converge or diverge when $|x| = 1$?

10. Find the radius of convergence of each of the following power series.

   a) $1 + 2x + 3x^2 + 4x^3 + \cdots$.

   b) $x + 2x^2 + 3x^3 + 4x^4 + \cdots$.

   *c) $1 + \dfrac{1}{1^2}x + \dfrac{1}{2^2}x^2 + \dfrac{1}{3^2}x^3 + \dfrac{1}{4^2}x^4 + \cdots$.

   d) $x^3 + x^6 + x^9 + x^{12} + \cdots$.

   e) $1 + (x + 1) + (x + 1)^2 + (x + 1)^3 + \cdots$.

   f) $17 + \dfrac{1}{3}x + \dfrac{1}{3^2}x^2 + \dfrac{1}{3^3}x^3 + \dfrac{1}{3^4}x^4 + \cdots$.

*11. Write out the first five terms of each of the following power series, and determine the radius of convergence of each.

   a) $\displaystyle\sum_{n=0}^{\infty} nx^n$

   b) $\displaystyle\sum_{n=0}^{\infty} \dfrac{n^2}{2^n} x^n$

   c) $\displaystyle\sum_{n=0}^{\infty} (n + 5)^2 x^n$

   d) $\displaystyle\sum_{n=0}^{\infty} \dfrac{99}{n^n} x^n$

   e) $\displaystyle\sum_{n=0}^{\infty} n! x^n$

12. Find the radius of convergence of the Taylor series for $\sin x$ and for $\cos x$. For which values of x can these series therefore represent the sine and cosine functions?

13. Find the radius of convergence of the Taylor series for $f(x) = 1/(1 + x^2)$ at $x = 0$. (See the table of Taylor series in section 10.3.) What is the radius of convergence of this series? For which values of x can this series therefore represent the function f? Do these x values constitute the *entire* domain of definition of f?

14. In the text we used the alternating series for e^x, $x < 0$, to approximate e^{-1} accurate to seven decimal places. The claim was made that in taking the reciprocal to obtain an estimate for e, the accuracy drops by one decimal place. In this problem you will see why this is true.
    a) Consider first the more general situation where two functions are reciprocals, $g(x) = 1/f(x)$. Express $g'(x)$ in terms of $f(x)$ and $f'(x)$.
    b) Use your answer in part (a) to find an expression for the relative error in g, $\Delta g/g(x) \approx g'(x)\Delta x/g(x)$, in terms of $f(x)$ and $f'(x)$. How does this compare to the relative error in f?
    c) Apply your results in part (b) to the functions e^x and e^{-x} at $x = 1$. Since e is about 3 times as large as $1/e$, explain why the error in the estimate for e should be about 3 times as large as the error in the estimate for $1/e$.

➤ 10.6 APPROXIMATION OVER INTERVALS

A powerful result in mathematical analysis is the **Weierstrass Approximation Theorem**, which states that given any continuous function $f(x)$ and given any interval $[a, b]$, there exist polynomials which fit f over this interval to any level of accuracy we care to specify. In many cases, we can find such a polynomial simply by taking a Taylor polynomial of high enough degree. There are several ways in which this is not a completely satisfactory response, however. First, some functions (like the absolute value function) have corners or other places where they aren't differentiable, so we can't even build a Taylor series at such points. Second, we have seen several functions [like $1/(1 + x^2)$] that have a finite interval of convergence, so Taylor polynomials may not be good fits no matter how high a degree we try. Third, even for well-behaved functions like $\sin(x)$ or e^x, we may have to take a very high-degree Taylor polynomial to get the same overall fit that a much lower-degree polynomial could achieve, by other means.

In this section we will develop the general machinery for finding polynomial approximations to functions over given intervals. In section 10.4 we will see how this same approach can be adapted to approximate periodic functions by **trigonometric polynomials.**

Approximation by Polynomials

EXAMPLE. Let's return to the problem introduced at the beginning of this chapter: find the second-degree polynomial which best fits the function $\sin(x)$

over the interval $[0, \pi]$. Just as we did with the Taylor polynomials, though, before we can start we need to agree on our criterion for the best fit. Here are two obvious candidates for such a criterion:

Two possible criteria
for best fit

1. The second-degree polynomial $Q(x)$ is the best fit to $\sin(x)$ over the interval $[0, \pi]$ if the *maximum* separation between $Q(x)$ and $\sin(x)$ is smaller than the maximum separation between $\sin(x)$ and any other second-degree polynomial:

$$\max_{0 \le x \le \pi} |\sin(x) - Q(x)| \text{ is the smallest possible.}$$

2. The second-degree polynomial $Q(x)$ is the best fit to $\sin(x)$ over the interval $[0, \pi]$ if the *average* separation between $Q(x)$ and $\sin(x)$ is smaller than the average separation between $\sin(x)$ and any other second-degree polynomial:

$$\frac{1}{\pi} \int_0^\pi |\sin(x) - Q(x)| \, dx \qquad \text{is the smallest possible.}$$

Unfortunately, even though their clear meanings make these two criteria very attractive, they turn out to be largely unusable—if we try to apply either criterion to a specific problem, including our current example, we are led into a maze of tedious and unwieldy calculations.

Why we don't use
either criterion

Instead, therefore, we use a criterion which, while slightly less obvious than either of the two we've already articulated, still clearly measures the same sort of property and has the added virtue of behaving well mathematically. We accomplish this by modifying criterion 2 slightly. It turns out that the major difficulty with this criterion is the presence of absolute values. If, instead of considering the average separation between $Q(x)$ and $\sin(x)$, we consider the average of the square of the separation between $Q(x)$ and $\sin(x)$, we get a criterion we can work with. (Compare this with the discussion of the best-fitting line in the exercises for section 9.3.) Since this is a definition we will be using for the rest of this section, we frame it in terms of arbitrary functions g and h, and an arbitrary interval $[a, b]$:

Given two functions g and h defined over an interval $[a, b]$, we define the **mean square separation** between g and h over this interval to be

$$\frac{1}{b - a} \int_a^b (g(x) - h(x))^2 \, dx.$$

NOTE: In this setting the word **mean** is synonymous with what we have called "average." It turns out that there is often more than one way to define the term "average"—the concepts of median and mode are two other natural ways of capturing "averageness," for instance—so we use the more technical term to avoid ambiguity.

The criterion we will use

We can now rephrase our original problem as: find the second-degree polynomial $Q(x)$ whose mean squared separation from $\sin(x)$ over the interval $[0, \pi]$ is as small as possible. In mathematical terms, we want to find coefficients a_0, a_1, and a_2 such that the integral

$$\int_0^\pi (\sin(x) - (a_0 + a_1 x + a_2 x^2))^2 dx$$

is minimized. The solution $Q(x)$ is called the quadratic **least squares approximation** to $\sin(x)$ over $[0, \pi]$.

The key to solving this problem is to observe that a_0, a_1, and a_2 can take on any values we like and that this integral can thus be considered a function of these three variables. For instance, if we couldn't think of anything cleverer to do, we might simply try various combinations of a_0, a_1, and a_2 to see how small we could make the given integral. Therefore another way to phrase our problem is

A mathematical formulation of the problem

Find values for a_0, a_1, and a_2 which minimize the function

$$F(a_0, a_1, a_2) = \int_0^\pi (\sin(x) - (a_0 + a_1 x + a_2 x^2))^2 dx.$$

We know how to find points where functions take on their extreme values—we look for the places where the partial derivatives are 0. But how do we differentiate an expression involving an integral like this? It turns out that for all continuous functions, or even functions with only a finite number of breaks in them, we can simply interchange integration and differentiation. Thus, in our example,

$$\frac{\partial}{\partial a_0} F(a_0, a_1, a_2) = \frac{\partial}{\partial a_0} \int_0^\pi (\sin(x) - (a_0 + a_1 x + a_2 x^2))^2 dx$$

$$= \int_0^\pi \frac{\partial}{\partial a_0} (\sin(x) - (a_0 + a_1 x + a_2 x^2))^2 dx$$

$$= \int_0^\pi 2(\sin(x) - (a_0 + a_1 x + a_2 x^2))(-1) \, dx \, .$$

Similarly, we have

$$\frac{\partial}{\partial a_1} F(a_0, a_1, a_2) = \int_0^\pi 2(\sin(x) - (a_0 + a_1 x + a_2 x^2))(-x) \, dx$$

$$\frac{\partial}{\partial a_2} F(a_0, a_1, a_2) = \int_0^\pi 2(\sin(x) - (a_0 + a_1 x + a_2 x^2))(-x^2) \, dx \, .$$

We now want to find values for a_0, a_1, and a_2 that make these partial derivatives simultaneously equal to 0. That is, we want

Setting the partials equal to zero gives equations for a_0, a_1, a_2

$$\int_0^\pi 2(\sin(x) - (a_0 + a_1 x + a_2 x^2))(-1) \, dx = 0$$

$$\int_0^\pi 2(\sin(x) - (a_0 + a_1 x + a_2 x^2))(-x) \, dx = 0$$

$$\int_0^\pi 2(\sin(x) - (a_0 + a_1 x + a_2 x^2))(-x^2) \, dx = 0,$$

which can be rewritten as

$$\int_0^\pi \sin(x) \, dx = \int_0^\pi (a_0 + a_1 x + a_2 x^2) dx$$

$$\int_0^\pi x \, \sin(x) \, dx = \int_0^\pi (a_0 x + a_1 x^2 + a_2 x^3) dx$$

$$\int_0^\pi x^2 \, \sin(x) \, dx = \int_0^\pi (a_0 x^2 + a_1 x^3 + a_2 x^4) dx.$$

All of these integrals can be evaluated relatively easily (see the exercises for a hint on evaluating the integrals on the left-hand side). When we do so, we are left with

Evaluating the integrals gives three linear equations

$$2 = \pi a_0 + \frac{\pi^2}{2} a_1 + \frac{\pi^3}{3} a_2$$

$$\pi = \frac{\pi^2}{2} a_0 + \frac{\pi^3}{3} a_1 + \frac{\pi^4}{4} a_2$$

$$\pi^2 - 4 = \frac{\pi^3}{3} a_0 + \frac{\pi^4}{4} a_1 + \frac{\pi^5}{5} a_2$$

But this is simply a set of three linear equations in the unknowns a_0, a_1, and a_2, and they can be solved in the usual ways. We could either replace each expression in π by a corresponding decimal approximation, or we could keep everything in terms of π. Let's do the latter; after a bit of tedious arithmetic we find

$$a_0 = \frac{12}{\pi} - \frac{120}{\pi^3} = -.050465 \ldots$$

$$a_1 = \frac{-60}{\pi^2} + \frac{720}{\pi^4} = 1.312236\ldots$$

$$a_2 = \frac{60}{\pi^3} - \frac{720}{\pi^5} = -.417697\ldots,$$

and we have

$$Q(x) = -.050465 + 1.312236\,x - .417698\,x^2,$$

which is the equation given in section 10.1 at the beginning of the chapter.

The analysis we gave for this particular case can clearly be generalized to apply to any function over any interval. When we do this we get:

How to find least-squares polynomial approximations in general

> Given a function g over an interval $[a, b]$, then the nth-degree polynomial
>
> $$P(x) = c_0 + c_1 x + c_2 x^2 + \cdots + c_n x^n$$
>
> whose mean square distance from g is a minimum has coefficients that are determined by the following $n + 1$ equations in the $n + 1$ unknowns $c_0, c_1, c_2, \ldots, c_n$:
>
> $$\int_a^b g(x)\,dx = c_0 \int_a^b dx + c_1 \int_a^b x\,dx + \cdots + c_n \int_a^b x^n\,dx$$
>
> $$\int_a^b x\,g(x)\,dx = c_0 \int_a^b x\,dx + c_1 \int_a^b x^2\,dx + \cdots + c_n \int_a^b x^{n+1}\,dx$$
>
> $$\vdots$$
>
> $$\int_a^b x^n g(x)\,dx = c_0 \int_a^b x^n\,dx + c_1 \int_a^b x^{n+1}\,dx + \cdots + c_n \int_a^b x^{2n}\,dx.$$

All the integrals on the right-hand side can be evaluated immediately. The integrals on the left-hand side will typically need to be evaluated numerically, although simple cases can be evaluated in closed form. Integration by parts is often useful in these cases. The exercises contain several problems using this technique to find approximating polynomials.

Solving the equations is a job for the computer

The real catch, though, is not in obtaining the equations—it is that solving systems of equations by hand is excruciatingly boring and subject to frequent arithmetic mistakes if there are more than two or three unknowns involved.

Fortunately, there are now a number of computer packages available which do all of this for us. Here are a couple of examples, where the details are left to the exercises.

EXAMPLE.   Let's find polynomial approximation for $1/(1 + x^2)$ over the interval $[0, 2]$. We saw earlier that the Taylor series for this function converges only for $|x| < 1$, so it will be no help. Yet with the above technique we can derive the following approximations of various degrees (see the exercises for details):

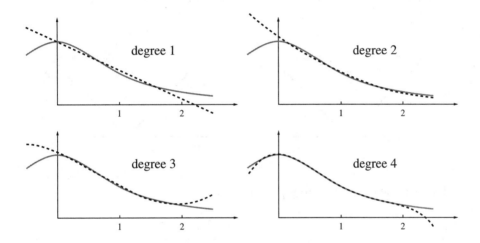

Here are the corresponding equations of the approximating polynomials:

Degree	Polynomial
1	$1.00722 - .453645x$
2	$1.08789 - .695660x + .121008x^2$
3	$1.04245 - .423017x - .219797x^2 + .113602x^3$
4	$1.00704 - .068906x - 1.01653x^2 + .733272x^3 - .154916x^4$

EXAMPLE.   We can even use this new technique to find polynomial approximations for functions that aren't differentiable at some points. For instance, let's approximate the function $h(x) = |x|$ over the interval $[-1, 1]$. Since this function is symmetric about the y-axis, and we are approximating it over an interval that is symmetric about the y-axis, only even powers of x will appear.

The technique even works when differentiability fails

(See the exercises for details.) We get the following approximations of degrees 2, 4, 6, and 8:

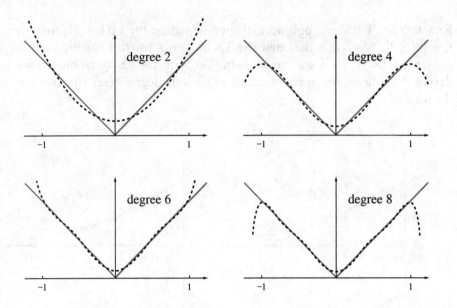

Here are the corresponding polynomials:

Degree	Polynomial
2	$.1875 + .9375x^2$
4	$.117188 + 1.64062x^2 - .820312x^4$
6	$.085449 + 2.30713x^2 - 2.81982x^4 + 1.46631x^6$
8	$.067291 + 2.960821x^2 - 6.415132x^4 - 7.698173x^6 - 3.338498x^8$

The technique is useful for data functions

A numerical example.   If we have some function which exists only as a set of data points—a numerical solution to a differential equation, perhaps, or the output of some laboratory instrument—it can often be quite useful to replace the function by an approximating polynomial. The polynomial takes up much less storage space and is easier to manipulate. To see how this works, let's return to the S-I-R model we've studied before:

$$S' = -.00001SI$$

$$I' = .00001SI - I/14$$

$$R' = I/14,$$

with initial values $S(0) = 45400$ and $I(0) = 2100$.

Let's find an eighth-degree polynomial $Q(t) = i_0 + i_1 t + i_2 t^2 + \cdots + i_8 t^8$ approximating I over the time interval $0 \leq t \leq 40$. We can do this by a minor modification of the Euler's method programs we've been using all along. Now, in addition to keeping track of the current values for S and I as we go along, we will also need to be calculating Riemann sums for the integrals

$$\int_0^{40} t^k I(t)\, dt \qquad \text{for } k = 0, 1, 2, \ldots, 8$$

as we go through each iteration of Euler's method.

Since the numbers involved become enormous very quickly, we open ourselves to various sorts of computer roundoff error. We can avoid some of these difficulties by **rescaling** our equations—using units that keep the numbers involved more manageable. Thus, for instance, suppose we measure S, I, and R in units of 10,000 people, and suppose we measure time in "decadays," where 1 decaday = 10 days. When we do this, our original differential equations become

The importance of using the right-sized units

$$S' = -SI$$
$$I' = SI - I/1.4$$
$$R' = I/1.4,$$

with initial values $S(0) = 4.54$ and $I(0) = 0.21$. The integrals we want are now of the form

$$\int_0^4 t^k I(t)\, dt \qquad \text{for } k = 0, 1, 2, \ldots, 8$$

The use of Simpson's rule (see section 11.3) will also reduce errors. It may be easiest to calculate the values of I first, using perhaps 2000 values, and store them in an array. Once you have this array of I values, it is relatively quick and easy to use Simpson's rule to calculate the nine integrals needed. If you later decide you want to get a higher-degree polynomial approximation, you don't have to rerun the program.

Using Simpson's rule helps reduce errors

Once we've evaluated these integrals, we set up and solve the corresponding system of nine equations in the nine unknown coefficients i_k. We get the following eighth-degree approximation

$$Q(t) = .3090 - .9989\, t + 7.8518\, t^2 - 3.6233\, t^3 - 3.9248\, t^4 + 4.2162\, t^5$$
$$- 1.5750\, t^6 + .2692\, t^7 - .01772\, t^8.$$

When we graph Q and I together over the interval $[0, 4]$ (decadays), we get

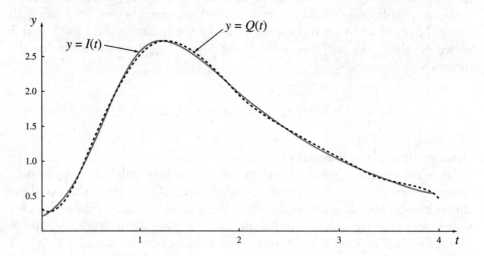

—a reasonably good fit.

A CAUTION.   Numerical least squares fitting of the sort performed in this last example fairly quickly pushes into regions where the cumulative effects of the inaccuracies of the original data, the inaccuracies of the estimates for the integrals, and the immense range in the magnitude of the numbers involved all combine to produce answers that are obviously wrong. Rescaling the equations and using good approximations for the integrals can help put off the point at which this begins to happen.

Exercises

1. To find polynomial approximations for $\sin(x)$ over the interval $[0, \pi]$, we needed to be able to evaluate integrals of the form

$$\int_0^\pi x^n \sin(x)\, dx.$$

   The value of this integral clearly depends on the value of n, so denote it by I_n.

   *a) Evaluate I_0 and I_1. (Suggestion: Use integration by parts to evaluate I_1.)

   b) Use integration by parts twice to prove the general **reduction formula**:

$$I_{n+2} = \pi^{n+2} - (n + 2)(n + 1) I_n \qquad \text{for all } n \geq 0.$$

   *c) Evaluate I_2, I_3, and I_4.

*d) If you have access to a computer package that will solve a system of equations, find the fourth-degree polynomial that best fits the sine function over the interval $[0, \pi]$. What is the maximum difference between this polynomial and the sine function over this interval?

2. To find polynomial approximations for $|x|$ over the interval $[-1, 1]$, we needed to be able to evaluate integrals of the form

$$\int_{-1}^{1} x^n |x| \, dx.$$

As before, let's denote this integral by I_n.

a) Show that

$$I_n = \begin{cases} \dfrac{2}{n+2} & \text{if } n \text{ is even} \\[2mm] 0 & \text{if } n \text{ is odd.} \end{cases}$$

b) Derive the quadratic least squares approximation to $|x|$ over $[-1, 1]$.

c) If you have access to a computer package that will solve a system of equations, find the tenth-degree polynomial that best fits $|x|$ over the interval $[-1, 1]$. What is the maximum difference between this polynomial and $|x|$ over this interval?

3. To find polynomial approximations for $1/(1 + x^2)$ over the interval $[0, 2]$, we needed to be able to evaluate integrals of the form

$$\int_{0}^{2} \frac{x^n}{1 + x^2} \, dx.$$

Call this integral I_n.

*a) Evaluate I_0 and I_1. (Suggestion: Use integration by parts to evaluate I_1.)

b) Prove the general reduction formula:

$$I_{n+2} = \frac{2^{n+1}}{n+1} - I_n \qquad \text{for all } n \geq 0.$$

*c) Evaluate I_2, I_3, and I_4.

d) If you have access to a computer package that will solve a system of equations, find the fourth-degree polynomial that best fits the sine function over the interval $[0, \pi]$. What is the maximum difference between this polynomial and the sine function over this interval?

4. Set up the equations (including evaluating all the integrals) for finding the best-fitting sixth-degree polynomial approximation to $\sin(x)$ over the interval $[-\pi, \pi]$.

5. In the S-I-R model, find the best-fitting eighth-degree polynomial approximation to $S(t)$ over the interval $0 \leq t \leq 40$.

➤ **10.7 CHAPTER SUMMARY**

The Main Ideas

- **Taylor polynomials** approximate functions at a point. The Taylor polynomial $P(x)$ of degree n is the **best fit** to $f(x)$ at $x = a$; that is, P satisfies

$$P(a) = f(a),\ P'(a) = f'(a), P''(a) = f''(a), \ldots, P^{(n)}(a) = f^{(n)}(a).$$

- **Taylor's theorem** says that a function and its Taylor polynomial of degree n agree to order $n + 1$ near the point where the polynomial is centered. Different versions expand on this idea.

- If $P(x)$ is the Taylor polynomial approximating $f(x)$ at $x = a$, then $P(u)$ approximates $f(u)$ for values of u near a; $P'(x)$ approximates $f'(x)$; and $\int P(x)\,dx$ approximates $\int f(x)\,dx$.

- A **Taylor series** is an infinite sum whose partial sums are Taylor polynomials. Some functions *equal* their Taylor series; among these are the sine, cosine, and exponential functions.

- A **power series** is an "infinite" polynomial $a_0 + a_1 x + a_x x^2 + \cdots + a_n x^n + \cdots$. If the solution of a differential equation can be represented by a power series, the coefficients a_n can be determined by **recursion relations** obtained by substituting the power series into the differential equation.

- An infinite series **converges** if, no matter how many decimal places are specified, all the partial sums eventually agree to at least this many decimal places. The number defined by these stabilizing decimals is called the **sum** of the series. If a series does not converge, we say it **diverges**.

- If the series $\sum_{m=0}^{\infty} b_m$ converges, then $\lim_{m \to \infty} b_m = 0$. The important counterexample of the **harmonic series** $\sum_{m=1}^{\infty} 1/m$ shows that $\lim_{m \to \infty} b_m = 0$ is a necessary but not sufficient condition to guarantee convergence.

- The **geometric series** $\sum_{m=0}^{\infty} x^m$ converges for all x with $|x| < 1$ and diverges for all other x.

- An **alternating series** $\sum_{m=0}^{\infty} (-1)^m b_m$ converges if $0 < b_{m+1} \leq b_m$ for all m and $\lim_{m \to \infty} b_m = 0$. For a convergent alternating series, the error in approximating the sum by a partial sum is less than the next term in the series.

- A convergent power series converges on an **interval of convergence** of width $2R$; R is called the **radius of convergence**. The **ratio test** can be used to find the radius of convergence of a power series: $\sum_{m=0}^{\infty} b_m$ converges if $\lim_{m \to \infty} |b_{m+1}|/|b_m| < 1$.

- A polynomial $P(x) = a_0 + a_1x + a_2x^2 + \cdots + a_nx^n$ is the **best fitting** approximation to a function $f(x)$ on an interval $[a, b]$ if $a_0, a_1, \ldots, a_n$ are chosen so that the **mean squared separation** between P and f

$$\frac{1}{b-a}\int_a^b (P(x) - f(x))^2 dx$$

is as small as possible. P is also called the **least squares approximation** to f on $[a, b]$.

Expectations

- Given a differentiable function $f(x)$ at a point $x = a$, you should be able to write down any of the **Taylor polynomials** or the **Taylor series** for f at a.
- You should be able to use the program TAYLOR to graph Taylor polynomials.
- You should be able to obtain new Taylor polynomials by substitution, differentiation, antidifferentiation, and multiplication.
- You should be able to use Taylor polynomials to find the value of a function to a specified degree of accuracy, to approximate integrals, and to find limits.
- You should be able to determine the order of magnitude of the agreement between a function and one of its Taylor polynomials.
- You should be able to find the power series solution to a differential equation.
- You should be able to test a series for divergence; you should be able to check a series for convergence using either the **alternating series test** or the **ratio test**.
- You should be able to find the sum of a **geometric series** and its interval of convergence.
- You should be able to estimate the error in an approximation using partial sums of an alternating series.
- You should be able to find the **radius of convergence** of a series using the ratio test.
- You should be able to set up the equations to find the **least squares** polynomial approximation of a particular degree for a given function on a specified interval. Working by hand or, if necessary, using a computer package to solve a system of equations, you should be able to find the coefficients of the least squares approximation.

CHAPTER **11**

TECHNIQUES OF INTEGRATION

Chapter 6 introduced the integral. There it was defined numerically, as the limit of approximating Riemann sums. Evaluating integrals by applying this basic definition tends to take a long time if a high level of accuracy is desired. If one is going to evaluate integrals at all frequently, it is thus important to find **techniques of integration** for doing this efficiently. For instance, if we evaluate a function at the midpoints of the subintervals, we get much faster convergence than if we use either the right or left endpoints of the subintervals.

A powerful class of techniques is based on the observation made at the end of chapter 6, where we saw that the fundamental theorem of calculus gives us a second way to find an integral, using antiderivatives. While a Riemann sum will usually give us only an approximation to the value of an integral, an antiderivative will give us the exact value. The drawback is that antiderivatives often can't be expressed in **closed form**—that is, as a **formula** in terms of named functions. Even when antiderivatives can be so expressed, the formulas are often difficult to find. Nevertheless, such a formula can be so powerful, both computationally and analytically, that it is often worth the effort needed to find it. In this chapter we will explore several techniques for finding the antiderivative of a function given by a formula.

We will conclude the chapter by developing a numerical method—Simpson's rule—that gives a good estimate for the value of an integral with relatively little computation.

▷ 11.1 ANTIDERIVATIVES

Definition

Recall that we say F is an **antiderivative** of f if $F' = f$. Here are some examples.

$$\text{FUNCTION:} \quad x^2 \quad 1/y \quad \sin u \quad 2\sin t \cos t \quad 2^z$$
$$\updownarrow \quad\; \updownarrow \quad\; \updownarrow \qquad\;\; \updownarrow \qquad\quad \updownarrow$$
$$\text{ANTIDERIVATIVE:} \quad \frac{x^3}{3} \quad \ln y \quad -\cos u \quad \sin^2 t \quad \frac{2^z}{\ln 2}$$

Undo a differentiation

Notice that you go up (↑) from the bottom row to the top by carrying out a differentiation. To go down (↓) you must "undo" that differentiation. The process of reversing, or undoing, a differentiation has come to be called **antidifferentiation**. You should differentiate each function on the bottom row to check that it is an antiderivative of the function above it.

A function has many antiderivatives

While a function can have only one derivative, it has many antiderivatives. For example, $1 - \cos u$ and $99 - \cos u$ are also antiderivatives of the function $\sin u$ because

$$(1 - \cos u)' = \sin u = (99 - \cos u)'.$$

In fact, every function $C - \cos u$ is an antiderivative of $\sin u$, for any constant C whatsoever. This observation is true in general. That is, if F is an antiderivative of a function f, then so is $F + C$, for any constant C. This follows from the addition rule for derivatives:

$$(F + C)' = F' + C' = F' + 0 = f.$$

A caution

It is tempting to claim the converse—that *every* antiderivative of f is equal to $F + C$, for some appropriately chosen value of C. In fact, you will often see this statement written. The statement is true, though, only for continuous functions. If the function f has breaks in its domain, then there will be more antiderivatives than those of the form $F + C$ for a *single* constant C—over each piece of the domain of f, F can be modified by a *different* constant and still yield an antiderivative for f. Exercises 18, 19, and 20 at the end of this section explore this for a couple of cases. If f is continuous, though, $F + C$ will cover all the possibilities, and we sometimes say that $F + C$ is *the* antiderivative of f. For the sake of keeping a compact notation, we will even write this when the domain of f consists of more than one interval. You should understand, though, that in such cases, over each piece F can be modified by a different constant.

What the "$+C$" term really means

Antiderivatives of basic functions

For future reference we collect a list of basic functions whose antiderivatives we already know. Remember that each antiderivative in the table can have an arbitrary constant added to it.

Function	Antiderivative
x^p	$\dfrac{1}{p+1}x^{p+1}$ (if $p \neq -1$)
$1/x$	$\ln x$
$\sin x$	$-\cos x$
$\cos x$	$\sin x$
e^x	e^x
b^x	$\dfrac{1}{\ln b}b^x$

All of these antiderivatives are easily verified and could have been derived with at most a little trial and error fiddling to get the right constant. You should notice one incongruity: the function $1/x$ is defined for all $x \neq 0$, but its listed antiderivative, $\ln x$, is only defined for $x > 0$. In exercise 18 (page 622) you will see how to find antiderivatives for $1/x$ over its entire domain.

Two of our basic functions—$\ln x$ and $\tan x$—do not appear in the left column of the table. This happens because there is no simple multiple of a basic function whose derivative is equal to either $\ln x$ or $\tan x$. It turns out that these functions *do* have antiderivatives, though, that can be expressed as more complicated combinations of basic functions. By differentiating $x \ln x - x$, in fact, you should be able to verify that it is an antiderivative of $\ln x$. Likewise, $-\ln(\cos x)$ is an antiderivative of $\tan x$. It would take a long time to stumble on these antiderivatives by inspection or by trial and error. It is the purpose of later sections to develop techniques which will enable us to discover antiderivatives like these quickly and efficiently. In particular, the antiderivative of $\ln x$ is derived in section 11.3 on page 636, while the antiderivative of $\tan x$ is derived in section 11.5 on page 666.

There are a couple of other functions that don't appear in the above table whose antiderivatives are needed frequently enough that they should become part of your repertoire of elementary functions that you recognize immediately:

Function	Antiderivative
$\dfrac{1}{\sqrt{1-x^2}}$	$\arcsin x$
$\dfrac{1}{1+x^2}$	$\arctan x$

The antiderivatives are inverse trigonometric functions, which we've had no need for until now. They are examples of functions that occur more often for their antiderivative properties than for themselves. Note that the derivatives

of the inverse trigonometric functions have no obvious reference to trigonometric relations. In fact, they often occur in settings where there are no triangles or periodic functions in sight. Let's see how the derivatives of these inverse
functions are derived.

Inverse Functions

We discussed inverse functions in section 4.4. Here's a quick summary of the
main points made there. Two functions f and g are **inverses** if

$$f(g(a)) = a$$

and

$$g(f(b)) = b$$

for every a in the domain of g and every b in the domain of f. It follows that
the *range* of f is the same as the *domain* of g, and vice versa.

The graphs of f and g are mirror reflections about the line $y = x$. This is a
direct translation of the definition into graphical language, since

(a, b) is on the graph of $y = g(x)$
if and only if
$g(a) = b$ (by definition of the graph of a function)
if and only if
$f(b) = a$ (by definition of inverse functions)
if and only if
(b, a) is on the graph of $y = f(x)$.

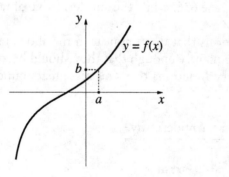
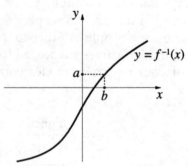

From this it follows immediately that if the graph of f is locally linear at the
point (b, a) with slope m, then the graph of g will be locally linear at the point
(a, b) with slope $1/m$. Algebraically, this is expressed as

$$g'(a) = 1/f'(b),$$

where $a = f(b)$ and $b = g(a)$.

This same result can be obtained by differentiating the expression $f(g(x)) = x$, using the chain rule:

$$1 = x' = f(g(x))' = f'(g(x))g'(x),$$

and therefore

$$g'(x) = \frac{1}{f'(g(x))}$$

for any value of x for which g is defined.

Inverse Trigonometric Functions

The arcsine function. In the discussion in chapter 4 we saw that a function has an inverse only when it is **one-to-one**, so if we want an inverse, we often have to restrict the domain of a function to a region where it is one-to-one. This is certainly the case with the sine function, which takes the same value infinitely many times. The standard choice of domain on which the sine function is one-to-one is $[-\pi/2, \pi/2]$. Over this interval the sine function increases from -1 to 1. We can then define an inverse function, which we call the **arcsine function**, written $\arcsin x$, whose domain is the interval $[-1, 1]$, and whose range is $[-\pi/2, \pi/2]$. Since the sine function is strictly increasing on its domain, the arcsine function will be strictly increasing on its domain as well—do you see why this has to be?

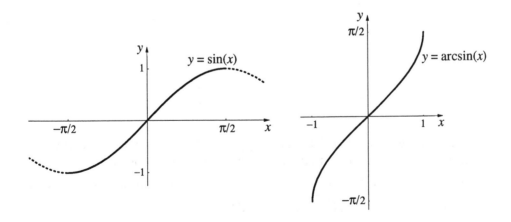

To find the derivative of the arcsine function, let $f(x) = \sin x$, and let $g(x) = \arcsin x$. Then by the remarks above, we have

$$g'(x) = \frac{1}{f'(g(x))} = \frac{1}{\cos(g(x))} = \frac{1}{\cos(\arcsin x)}.$$

To proceed, we need to figure out what $\cos(\arcsin x)$ is. We will do this in two ways, one algebraic, the other geometric. Both perspectives are useful.

The algebraic approach.   Recall that for any input t, $\sin^2 t + \cos^2 t = 1$. This can be solved for $\cos t$ as $\cos t = \pm\sqrt{1 - \sin^2 t}$. That is, the cosine of anything is the square root of 1 minus the square of the sine of that input, with a possible minus sign needed out front, depending on the context. Since the output of the arcsine function lies in the range $[-\pi/2, \pi/2]$, and the cosine function is positive (or 0) for numbers in this interval, it follows that $\cos(\arcsin x) \geq 0$ for any value of x in the domain of the arcsine function. Therefore,

$$\cos(\arcsin x) = \sqrt{1 - \sin^2(\arcsin x)} = \sqrt{1 - (\sin(\arcsin x))^2} = \sqrt{1 - x^2},$$

since $\sin(\arcsin x) = x$ by definition of inverse functions. It follows that

$$(\arcsin x)' = \frac{1}{\cos(\arcsin x)} = \frac{1}{\sqrt{1 - x^2}}$$

as claimed.

The geometric approach.   Introduce a new variable $\theta = \arcsin x$, so that $x = \sin\theta$. We can represent these relationships in the following picture:

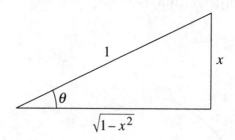

Notice that we have labeled the side opposite the angle θ as x and the hypotenuse as 1. This ensures that $\sin\theta = x$. By the Pythagorean theorem, the remaining side must then be $\sqrt{1 - x^2}$. From this picture it is then obvious that $\cos(\arcsin x) = \cos\theta = \sqrt{1 - x^2}/1 = \sqrt{1 - x^2}$, as before.

The arctangent function.   To get an inverse for the tangent function, we again need to limit the domain, and again the standard choice is to restrict it to the domain $(-\pi/2, \pi/2)$ [*not* including the endpoints this time, since $\tan(-\pi/2)$ and $\tan(\pi/2)$ aren't defined]. Over this domain the tangent function increases from $-\infty$ to $+\infty$. We can then define its inverse, called the **arctangent function**, written $\arctan x$, whose domain is the interval $(-\infty, \infty)$, and whose range is $(-\pi/2, \pi/2)$. Again, both functions are increasing over their domains.

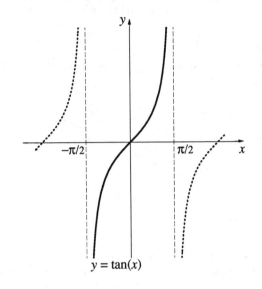

$$y = \tan(x)$$

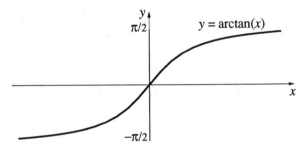

$$y = \arctan(x)$$

We proceed as before, letting $f(x) = \tan x$, and $g(x) = \arctan x$. This time we get

$$g'(x) = \frac{1}{f'(g(x))} = \frac{1}{\sec^2(g(x))} = \frac{1}{\sec^2(\arctan x)}.$$

Again, we need a trigonometric identity to proceed, and we will again derive the desired result algebraically and geometrically.

Algebraic. We start as before with the identity $\sin^2 t + \cos^2 t = 1$. Dividing through by $\cos^2 t$, we get the equivalent identity $\tan^2 t + 1 = \sec^2 t$. That is, the square of the secant of any input is just 1 plus the square of the tangent applied to the same input. In particular,

$$\sec^2(\arctan x) = \tan^2(\arctan x) + 1 = (\tan(\arctan x))^2 + 1 = x^2 + 1,$$

since $\tan(\arctan x) = x$. It follows that

$$(\arctan x)' = \frac{1}{\sec^2(\arctan x)} = \frac{1}{1 + x^2},$$

as claimed.

Geometric. Let $\theta = \arctan x$, so $x = \tan \theta$. Again we can draw a triangle reflecting these relationships:

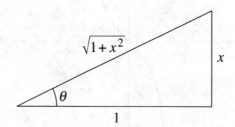

Notice that this time we have labeled the side opposite the angle θ as x and the adjacent side as 1 to ensure that $\tan \theta = x$. Again by the Pythagorean theorem, the hypotenuse must be $1 + x^2$. From this picture it is then obvious that $\sec(\arctan x) = \sec \theta = (\sqrt{1 + x^2})/1 = \sqrt{1 + x^2}$, so $\sec^2(\arctan x) = 1 + x^2$ again.

Notation

According to the fundamental theorem of calculus (see section 6.4), every accumulation function

$$A(x) = \int_a^x f(t)\, dt$$

is an antiderivative of the function f, no matter at what point $t = a$ the accumulation begins—so long as the function is defined over the entire interval from $t = a$ to $t = x$. (This is an important caution when dealing with functions like $1/x$, for instance, for which the integral from, say, -1 to 1 makes no sense.) In other words, the expression

$$\int_a^x f(t)\, dt$$

Write an antiderivative as an integral

represents an antiderivative of f. The influence of the fundamental theorem is so pervasive that this expression—with the "limits of integration" a and x omitted—is used to denote an antiderivative:

> **Notation:** An antiderivative of f is $\int f(x)\, dx$.

With this new notation the antiderivatives we have listed so far can be written in the following form.

$$\int x^p \, dx = \frac{1}{p+1} x^{p+1} + C \qquad (p \neq -1)$$

$$\int \frac{1}{x} \, dx = \ln x + C$$

$$\int \sin x \, dx = -\cos x + C$$

$$\int \cos x \, dx = \sin x + C$$

$$\int e^x \, dx = e^x + C$$

$$\int b^x \, dx = \frac{1}{\ln b} b^x + C$$

$$\int \frac{1}{\sqrt{1-x^2}} \, dx = \arcsin x + C$$

$$\int \frac{1}{1+x^2} \, dx = \arctan x + C$$

The basic antiderivatives again

The integration sign $\int$ now has two distinct meanings. Originally, it was used to describe the *number*

$$\int_a^b f(x) \, dx,$$

which was always calculated as the limit of a sequence of Riemann sums. Because this integral has a definite numerical value, it is called the **definite integral**. In its new meaning, the integration sign is used to describe the antiderivative

The definite integral is a number...

$$\int f(x) \, dx,$$

which is a *function*, not a number. To contrast the new use of $\int$ with the old, and to remind us that the new expression is a variable quantity, it is called the **indefinite integral**. The function that appears in either a definite or an indefinite integral is called the **integrand**. The terms "antiderivative" and "indefinite integral" are completely synonymous. We will tend to use the former term in general discussions, using the latter term when focusing on the process of finding the antiderivative.

...while the indefinite integral is a function

Because an indefinite integral represents an antiderivative, the process of finding an antiderivative is sometimes called **integration**. Thus the term *integration*, as well as the symbol for it, has two distinct meanings.

Using Antiderivatives

According to the fundamental theorem, we can use an *indefinite* integral to find the value of a *definite* integral—and this largely explains the importance of antiderivatives. In the language of indefinite integrals, the statement of the fundamental theorem on page 363 takes the following form.

$$\int_a^b f(x)\,dx = F(b) - F(a), \text{ where } F(x) = \int f(x)\,dx$$

EXAMPLE 1.　Find $\int_1^4 x^2\,dx$. We have that

$$\int x^2\,dx = \tfrac{1}{3}x^3 + C;$$

it follows that

$$\int_1^4 x^2\,dx = \tfrac{1}{3}4^3 + C - \left(\tfrac{1}{3}1^3 + C\right) = \frac{64}{3} + C - \frac{1}{3} - C = 21.$$

EXAMPLE 2.　Find

$$\int_0^{\pi/2} \cos t\,dt.$$

This time the indefinite integral we need is

$$\int \cos t\,dt = \sin t + C.$$

The value of the definite integral is therefore

$$\int_0^{\pi/2} \cos t\,dt = \sin\frac{\pi}{2} + C - (\sin 0 + C) = 1 + C - 0 - C = 1.$$

The calculation doesn't depend on *C*, so take *C* = 0

In each example the two appearances of C cancel each other. Thus C does not appear in the final result. This implies that it does not matter which value of C we choose to do the calculation. Usually, we just take $C = 0$.

Notation.　Because expressions like $F(b) - F(a)$ occur often when we are using indefinite integrals, we use the abbreviation

$$F(b) - F(a) = .F(x)\Big|_a^b.$$

Thus,

$$\int_1^4 x^2 \, dx = \left. \frac{x^3}{3} \right|_1^4 = \frac{64}{3} - \frac{1}{3} = 21.$$

There are clear advantages to using antiderivatives to evaluate definite integrals: we get exact values and we avoid many lengthy calculations. The difficulty is that the method works only if we can find a formula for the antiderivative.

There are several reasons why we might not find the formula we need. For instance, the antiderivative we want may be a function we have never seen before. The function $\arctan x$ is an example.

We may not recognize the antiderivative …

Even if we have a broad acquaintance with functions, we may still not be able to find the formula for a given antiderivative. The reason is simple: for most functions we can write down, the antiderivative is just not among the basic functions of calculus. For example, none of the basic functions, in any form or combination, equals

… and there may even be no formula for it

$$\int e^{x^2} \, dx \qquad \text{or} \qquad \int \frac{\sin x}{x} \, dx.$$

This does not mean that e^{x^2}, for example, has no antiderivative. On the contrary, the accumulation function

$$\int_0^x e^{t^2} \, dt$$

is an antiderivative of e^{x^2}. It can be evaluated, graphed, and analyzed like any other function. What we lack is a *formula* for this antiderivative in terms of the basic functions of calculus.

Finding Antiderivatives

In the rest of this chapter we will be deriving a number of statements involving antiderivatives. It is important to remember what such statements mean.

> The following statements are completely equivalent:
>
> $$\int f(x) \, dx = F(x) + C \qquad \text{and} \qquad F' = f.$$

In other words, a statement about antiderivatives can be verified by looking at a statement about derivatives. If it is claimed that the antiderivative of f is F, you check the statement by seeing if it is true that $F' = f$. This was how we

We differentiate to verify a statement about antiderivatives

verified the elementary antiderivatives we've considered so far. Another way of expressing these relationships is

$$\left(\int f(x)\,dx \right)' = f(x) \qquad \text{and} \qquad \int F'(x)\,dx = F(x) + C$$

for any functions f and F.

This duality between statements about derivatives and statements about antiderivatives also holds when applied to more general statements. Many of the most useful techniques for finding antiderivatives are based on converting the general rules for taking derivatives of sums, products, and chains into equivalent antiderivative form. We'll start with the simplest combinations, which involve a **function multiplied by a constant** and a **sum of two functions**.

The constant multiple and addition rules

Derivative Form	Antiderivative Form
$(k \cdot F)' = k \cdot F'$	$\int k \cdot f\,dx = k \int f\,dx$
$(F + G)' = F' + G'$	$\int (f + g)\,dx = \int f\,dx + \int g\,dx$

Let's verify these rules. Let $\int f = F$ and $\int g = G$. Then the first rule is claiming that $\int (k \cdot f) = k \cdot F$—the antiderivative of a constant times a function equals the constant times the antiderivative of the function. To verify this, we have to show that when we take the derivative of the right-hand side, we get the function under the integral on the left-hand side. But $(k \cdot F)' = k \cdot F'$ (by the derivative rule), which is just $k \cdot f$, which is what we had to show.

Similarly, to show that $\int (f + g) = F + G$—the antiderivative of the sum of two functions is the sum of their separate antiderivatives—we differentiate the right-hand side and find $(F + G)' = F' + G'$ (by the derivative rule for sums) which is just $f + g$, so the rule is true.

EXAMPLE 3.   This example illustrates the use of both the addition and the constant multiple rules.

$$\int (7e^x + \cos x)\,dx = \int 7e^x\,dx + \int \cos x\,dx$$

$$= 7 \int e^x\,dx + \int \cos x\,dx$$

$$= 7e^x + \sin x + C.$$

To verify this answer, you should take the derivative of the right-hand side to see that it equals the integrand on the left-hand side.

In the following sections we will develop the antidifferentiation rules that correspond to the product rule and to the chain rule. They are called *integration by parts* and *integration by substitution*, respectively.

While antiderivatives can be hard to find, they are easy to check. This makes "trial and error" a good strategy. In other words, if you don't immediately see what the antiderivative should be, but the function doesn't look too bad, try guessing. When you differentiate your guess, what you see may lead you to a better guess.

Trial and error

EXAMPLE 4. Find

$$F(x) = \int \cos(3x)\, dx.$$

Since the derivative of $\sin u$ is $\cos u$, it is reasonable to try $\sin(3x)$ as an antiderivative for $\cos(3x)$. Therefore:

FIRST GUESS: $F(x) = \sin(3x)$

CHECK: $F'(x) = \cos(3x) \cdot 3 \neq \cos(3x).$

We wanted $F' = \cos(3x)$ but we got $F' = 3\cos(3x)$. The chain rule gave us an extra—and unwanted—factor of 3. We can compensate for that factor by multiplying our first guess by 1/3. Then

SECOND GUESS: $F(x) = \frac{1}{3}\sin(3x)$

CHECK: $F'(x) = \frac{1}{3}\cos(3x) \cdot 3 = \cos(3x).$

Thus $\int \cos(3x)dx = \frac{1}{3}\sin(3x).$

Because indefinite integrals are difficult to calculate, reference manuals in mathematics and science often include tables of integrals. There are sometimes many hundreds of individual formulas, organized by the type of function being integrated. A modest selection of such formulas can be found at the back of this book. You should take some time to learn how these tables are arranged and get some practice using them. You should also check some of the more unlikely looking formulas by differentiating them to see that they really are the antiderivatives they are claimed to be.

Tables of Integrals

Computers are having a major impact on integration techniques. We saw in the last chapter that any continuous function—even one given by the output from some laboratory recording device or as the result of a numerical technique like Euler's method—can be approximated by a polynomial (usually using lots of computation!), and the antiderivative of a polynomial is easy to find.

Moreover, computer software packages which can find any existing formula for a definite integral are becoming widespread and will probably have a

Integration can now be done quickly and efficiently by computer software

profound impact on the importance of integration techniques over the next several years. Just as hand-held calculators have rendered obsolete many traditional arts—like using logarithms for performing multiplications or knowing how to interpolate in trig tables—there is likely to be a decreased importance placed on humans being adept at some of the more esoteric integration techniques. While some will continue to derive pleasure from becoming proficient in these skills, for most users it will generally be much faster, and more accurate, to use an appropriate software package. Nevertheless, for those going on in mathematics and the physical sciences, it will still be useful to be able to perform some of the simpler integrations by hand reasonably rapidly. The subsequent sections of this chapter develop the most commonly needed techniques for doing this.

Exercises

1. What is the inverse of the function $y = 1/x$? Sketch the graph of the function and its inverse.

*2. What is the inverse of the function $y = 1/x^3$? Sketch the graph of the the function and its inverse. Do the same for $y = 3x - 2$.

3. a) Let $\theta = \arctan x$. Then $\tan x = \theta$. Refer to the picture on page 613 showing the relationship between x and θ. Use this drawing to show that $\arctan(x) + \arctan(1/x)$ is constant—that is, its value doesn't depend on x. What is the value of the constant?

   b) Use part (a), and the derivative of $\arctan x$, to find the derivative of $\arctan(1/x)$.

   c) Use the chain rule to verify your answer to part (b).

4. The logarithm for the base b is defined as the inverse to the exponential function with base b:

$$y = \log_b x \quad \text{if} \quad x = b^y.$$

   Using only the fact that $dx/dy = \ln b \cdot b^y$, deduce the formula

$$\frac{dy}{dx} = \frac{1}{\ln b} \cdot \frac{1}{x}.$$

   (Note: This is purely an algebra problem; you don't need to invoke any differentiation rules.)

5. a) Define $\arccos x$, the inverse of the cosine function. Be sure to limit the domain of the cosine function to an interval on which it is one-to-one.

   b) Sketch the graph $y = \arccos x$. How did you limit the range of y?

   c) Determine dy/dx.

6. a) If $\theta = \arcsin x$, refer to the picture on page 611 reflecting the relation between θ and x. Using this picture, proceed as in exercise

3 to show that the sum $\arcsin x + \arccos x$ is constant. What is the value of the constant?

b) Use part (a), and the derivative of $\arcsin x$, to determine the derivative of $\arccos x$. Does this result agree with what you got in the last exercise?

*7. Find a formula for $\displaystyle\int \frac{dx}{\sqrt{1-x^2}}$.

8. Verify that the antiderivatives given in the table on page 608 are correct.

9. Find an antiderivative of each of the following functions. Don't hesitate to use the "trial-and-error" method of example 4 on page 619.

3	$5t$	$-5t$	$3 - 5t$
$7x^4$	$\dfrac{1}{y^3}$	e^{2z}	$u + \dfrac{1}{u}$
$(1 + w^3)^2$	$\cos(5v)$	$x^9 + 5x^7 - 2x^5$	$\sin t \cos t$

10. Find a formula for each of the following indefinite integrals.

$$\int 3x \, dx \qquad\qquad \int 3u \, du$$

$$\int e^z \, dz \qquad\qquad \int 5t^4 \, dt$$

$$*\int 7y + \frac{1}{y} \, dy \qquad\qquad \int 7y - \frac{4}{y^2} \, dy$$

$$*\int 5\sin w - 2\cos w \, dw \qquad\qquad \int dx$$

$$\int e^{z+2} \, dz \qquad\qquad \int \cos(4x) \, dx$$

$$\int \frac{5}{1+r^2} \, dr \qquad\qquad \int \frac{1}{\sqrt{1-4s^2}} \, ds$$

11.*a) Find an antiderivative $F(x)$ of $f(x) = 7$ for which $F(0) = 12$.
b) Find an antiderivative $G(x)$ of $f(x) = 7$ for which $G(3) = 1$.
c) Do $F(x)$ and $G(x)$ differ by a constant? If so, what is the value of that constant?

12. a) Find an antiderivative $F(t)$ of $f(t) = t + \cos t$ for which $F(0) = 3$.
b) Find an antiderivative $G(t)$ of $f(t) = t + \cos t$ for which $G(\pi/2) = -5$.
c) Do $F(t)$ and $G(t)$ differ by a constant? If so, what is the value of that constant?

13. Find an antiderivative of the function $a + by$ when a and b are fixed constants.

14. a) Verify that $(1 + x^3)^{10}$ is an antiderivative of $30x^2(1 + x^3)^9$.
    b) Find an antiderivative of $x^2(1 + x^3)^9$.
    c) Find an antiderivative of $x^2 + x^2(1 + x^3)^9$.

15. a) Verify that $x \ln x$ is an antiderivative of $1 + \ln x$.
    b) Find an antiderivative of $\ln x$. [Do you see how you can use part (a) to find this antiderivative?]

16. Recall that $F(y) = \ln(y)$ is an antiderivative of $1/y$ for $y > 0$. According to the text, *every* antiderivative of $1/y$ over this domain must be of the form $\ln(y) + C$ for an appropriate value of C.
    a) Verify that $G(y) = \ln(2y)$ is also an antiderivative of $1/y$.
    *b) Find C so that $\ln(2y) = \ln(y) + C$.

17. Verify that $-\cos^2 t$ is an antiderivative of $2 \sin t \cos t$. Since you already know $\sin^2 t$ is an antiderivative, you should be able to show

$$-\cos^2 t = \sin^2 t + C$$

for an appropriate value of C. What is C?

18. Since the function $\ln x$ is defined only when $x > 0$, the equation

$$\int \frac{1}{x}\, dx = \ln x + C$$

applies only when $x > 0$. However, the integrand $1/x$ is defined when $x < 0$ as well. Therefore, it makes sense to ask what the integral (i.e., antiderivative) of $1/x$ is when $x < 0$.
    a) When $x < 0$, then $-x > 0$, so $\ln(-x)$ is defined. In these circumstances, show that $\ln(-x)$ is an antiderivative of $1/x$.
    b) Now put these two "pieces" of antiderivative together by defining the function

$$F(x) = \begin{cases} \ln(-x) & \text{if } x < 0 \\ \ln(x) & \text{if } x > 0 \end{cases}$$

Sketch together the graphs of the functions $F(x)$ and $1/x$ in such a way that it is clear that $F(x)$ is an antiderivative of $1/x$.
    c) Explain why $F(x) = \ln|x|$. For this reason a table of integrals often contains the entry

$$\int \frac{1}{x}\, dx = \ln|x| + C.$$

d) Every function $\ln |x| + C$ is an antiderivative of $1/x$, but there are even more. As you will see, this can happen because the domain of $1/x$ is broken into two parts. Let

$$G(x) = \begin{cases} \ln(-x) & \text{if } x < 0 \\ \ln(x) + 1 & \text{if } x > 0 \end{cases}$$

Sketch together the graphs of the functions $G(x)$ and $1/x$ in such a way that it is clear that $G(x)$ is an antiderivative of $1/x$.

e) Explain why there is no value of C for which

$$\ln |x| + C = G(x).$$

This shows that the functions $\ln |x| + C$ do not exhaust the set of antiderivatives of $1/x$.

f) Construct two more antiderivatives of $1/x$ and sketch their graphs. What is the general form of the new antiderivatives you have constructed? (A suggestion: You should be able to use two separate constants C_1 and C_2 to describe the general form.)

19. On page 609 of the text there is an antiderivative for the tangent function:

$$\int \tan x \, dx = -\ln(\cos x).$$

However, this is not defined when x makes $\cos x$ either zero or negative.

a) How many separate intervals does the domain of $\tan x$ break down into?

b) For what values of x is $\cos x$ equal to zero, and for what values is it negative?

c) Modify the antiderivative $-\ln(\cos x)$ so that it *is* defined when $\cos x$ is negative. (How is this problem with the logarithm function treated in the previous question?)

d) In a typical table of integrals you will find the statement

$$\int \tan x \, dx = -\ln |\cos x| + C.$$

Explain why this does not cover all the possibilities.

e) Give a more precise expression for $\int \tan x \, dx$, modeled on the way you answered part (f) of exercise 18. How many different constants will you need?

f) Find a function G that is an antiderivative for $\tan x$ and that also satisfies the following conditions:

$$G(0) = 5$$
$$G(\pi) = -23$$
$$G(17\pi) = 197.$$

20. In the table on page 609 the antiderivative of x^p is given as $\frac{1}{p+1}x^{p+1} + C$. For some values of p this is correct, with only a single constant C needed. For other values of p, though, the domain of x^p will consist of more than one piece, and $\frac{1}{p+1}x^{p+1}$ can be modified by a different constant over each piece. For what values of p does this happen?

21. Find $F'(x)$ for the following functions. In parts (a), (b), and (d) do the problems two ways: by finding an antiderivative, and by using the fundamental theorem to get the answer without evaluating an antiderivative. Check that the answers agree.

*a) $F(x) = \displaystyle\int_0^x (t^2 + t^3)\,dt$

b) $F(x) = \displaystyle\int_1^x \frac{1}{u}\,du$

c) $F(x) = \displaystyle\int_1^x \frac{v}{1+v^3}\,dv$

*d) $F(x) = \displaystyle\int_0^{x^2} \cos t\,dt$

e) $F(x) = \displaystyle\int_1^{x^2} \frac{v}{1+v^3}\,dv$ (Hint: Let $u = x^2$ and use the chain rule.)

Comment: It may seem that parts (c) and (e) are more difficult than the others. However, there is a way to apply the fundamental theorem of calculus here to get answers to parts (c) and (e) quickly and with little effort.

22. Consider the two functions

$$F(x) = \sqrt{1+x^2} - 1 \quad \text{and} \quad G(x) = \int_0^x \frac{t}{\sqrt{1+t^2}}\,dt.$$

a) Show that F and G both satisfy the initial value problem

$$y' = \frac{x}{\sqrt{1+x^2}}, \qquad y(0) = 0.$$

*b) Since an initial value problem typically has a *unique* solution, F and G should be equal. Assuming this, determine the exact value of the following definite integrals.

$$\int_0^1 \frac{t}{\sqrt{1+t^2}}\, dt \qquad \int_0^2 \frac{t}{\sqrt{1+t^2}}\, dt \qquad \int_0^5 \frac{t}{\sqrt{1+t^2}}\, dt$$

23. The connection between integration and differentiation that is provided by the fundamental theorem of calculus makes it possible to determine an integral by solving a differential equation. For example, the accumulation function

$$A(x) = \int_0^x e^{-t^2}\, dt$$

is the solution to the initial value problem

$$y' = e^{-x^2}, \qquad y(0) = 0.$$

Therefore, $A(x)$ can be found by solving the differential equation. As you have seen, Euler's method is a useful way to solve differential equations.

a) Use either a program (e.g., PLOT) or a differential equation solver on a computer to get a graphical solution $A(x)$ to the initial value problem above.

b) Sketch the graph of $y = A(x)$ over the domain $0 \le x \le 4$.

c) Your graph should increase from left to right. How can you tell this even before you see the computer output?

d) Your graph should level off as x increases. Determine $A(5)$, $A(10)$, $A(30)$. (Approximations provided by the computer are adequate here.)

*e) Estimate $\lim_{x\to\infty} A(x)$.

f) Determine

$$\int_0^1 e^{-t^2}\, dt \quad \text{and} \quad \int_0^2 e^{-t^2}\, dt.$$

g) Determine

$$\int_1^2 e^{-t^2}\, dt.$$

*24. Find the area under the curve $y = x^3 + x$ for x between 1 and 4. (See section 6.3.)

25. Find the area under the curve $y = e^{3x}$ for x between 0 and $\ln 3$.

26. The **average value** of the function $f(x)$ on the interval $a \le x \le b$ is the integral

$$\frac{1}{b-a} \int_a^b f(x)\, dx.$$

(See the discussion of average value in chapter 6, pages 351–352.)
a) Find the average value of each of the functions $y = x$, x^2, x^3, and x^4 on the interval $0 \le x \le 1$.
b) Explain, using the graphs of $y = x$ and $y = x^2$, why the average value of x^2 is less than the average value of x on the interval $[0, 1]$.

27. a) What are the maximum, minimum, and average values of the function

$$f(x) = x + 2e^{-x}$$

on the interval $[0, 3]$?
b) Sketch the graph of $y = f(x)$ on the interval $[0, 3]$. Draw the line $y = \mu$, where μ is the average value you found in part (a).
*c) For which x does the graph of $y = f(x)$ lie above the line $y = \mu$, and for which x does it lie below the line? The region between the graph and the line has two parts—one is above the line (and below the graph) and the other is below the line (and above the graph). Shade these two regions and compare their areas: Which is larger?

▷ 11.2 INTEGRATION BY SUBSTITUTION

In the preceding section we converted a couple of general rules for differentiation—the rule for the derivative of a constant times a function and the rule for the derivative of the sum of two functions—into equivalent rules in integral form. In this section we will develop the integral form of the chain rule and see some of the ways this can be used to find antiderivatives.

Suppose we have functions F and G, with corresponding derivatives f and g. Then the chain rule says

$$(F(G(x)))' = F'(G(x))\, G'(x) = f(G(x))\, g(x).$$

If we now take the indefinite integral of these equations, we get

$$F(G(x)) + C = \int (F(G(x)))'\, dx = \int f(G(x))\, g(x)\, dx,$$

where C can be any constant.

Reversing these equalities to get a statement about integrals, we obtain:

The integral form of the chain rule

$$\int f(G(x))\, g(x)\, dx = F(G(x)) + C.$$

This somewhat unpromising expression turns out to be surprisingly useful. Here's how: Suppose we want to find an indefinite integral, and see in the integrand a *pair* of functions G and g, where $G' = g$ and where $g(x)$ is a factor of the integrand. We then find a function f so that the integrand can be written in the form $f(G(x))g(x)$. Now we only have to find an antiderivative for f. Once we have such an antiderivative, call it F, then the solution to our original problem will be $F(G(x))$. Thus, while our original antiderivative problem is not yet solved, it has been reduced to a different, simpler antiderivative problem that, when all goes well, will be easier to evaluate. Such **reduction methods** are typical of many integration techniques. We will see other examples in the remainder of this chapter.

Reduction methods transform problems into equivalent, simpler problems

EXAMPLE 1.   Suppose we try to find a formula for the integral

$$\int 3x^2(1 + x^3)^7 dx.$$

One way would be to multiply out the expression $(1 + x^3)^7$, making the integrand a polynomial with many separate terms of different degrees. (The highest degree would be 23; do you see why?) We could then carry out the integration "term by term," using the rules for sums and constant multiples of powers of x that were given in the previous section. But this is tedious—even excruciating.

Instead, notice that the expression $3x^2$ is the derivative of $1 + x^3$. If we let $G(x) = 1 + x^3$ and $g(x) = 3x^2$, we can then write the integrand as

$$3x^2(1 + x^3)^7 = (G(x))^7 g(x).$$

Now $(G(x))^7$ is clearly just $f(G(x))$ where f is the function which raises its input to the 7th power—$f(x) = x^7$ for any input x. But we recognize f as an elementary function whose antiderivative we can write down immediately as $F(x) = \frac{1}{8}x^8$. Thus the solution to our original problem will be

$$\int 3x^2(1 + x^3)^7 dx = F(G(x)) + C = \frac{1}{8}(1 + x^3)^8 + C$$

As with any integration problem, we can check our answer by taking the derivative of the right-hand side to see if it agrees with the integrand on the left. You should do this whenever you aren't quite sure of your technique (or your arithmetic!).

Note that the term $3x^2$ that appeared in the integrand above was essential for the procedure to work. The integral

$$\int (1 + x^3)^7 dx$$

cannot be found by substitution, even though it appears to have a simpler form. (Of course, the integral *can* be found by multiplying out the integrand.)

Using differential notation.   So far the symbol dx—the **differential** of x—under the integral sign has simply been an appendage, tagging along to suggest the Δx portion of the Riemann sums approximating definite integrals. It turns out we can take advantage of this notation to use the integral form of the chain

A compact notation for expressing the integral form of the chain rule rule more compactly. Instead of naming the functions G and g as above, we introduce a new variable $u = G(x)$. Then

$$\frac{du}{dx} = G'(x) = g(x),$$

and it is suggestive to multiply out this "quotient" to get $du = g(x)\,dx$. While this is reminiscent of the microscope equation we met in chapter 3, and eighteenth-century mathematicians took this equation seriously as a relation between two "infinitesimally small" quantities dx and du, we will view it only as a convenient mnemonic device. To see how this simplifies computations, reconsider the previous example. If we let $u = 1+x^3$, then $du = 3x^2 dx$, so we can write

$$3x^2(1 + x^3)^7\, dx = \underbrace{(1 + x^3)^7}_{u}\ \underbrace{3x^2\, dx}_{du} = u^7\, du.$$

It follows that

$$\int 3x^2(1 + x^3)^7 dx = \int u^7\, du$$
$$= \tfrac{1}{8}u^8 + C$$
$$= \tfrac{1}{8}(1 + x^3)^8 + C,$$

as before. We have arrived at the same answer without having to introduce the cumbersome language of all the auxiliary functions—we simply **substituted** the variable u for a certain expression in x [which we called $G(x)$ before], and replaced $G'(x)dx$ by du. For this reason this technique is called **integration by substitution.** You should always be clear, though, that integration by substitution is just the integral form of the chain rule, a relationship that becomes clear whenever you check the answer substitution gives you.

EXAMPLE 2.   Can we use the method of substitution to find

$$\int \frac{e^{5x}}{6 + e^{5x}} \, dx?$$

The numerator is almost the derivative of the denominator. This suggests we let $G(x) = 6 + e^{5x}$, giving $g(x) = G'(x) = 5e^{5x}$. Since we need to be able to factor $g(x)$ out of the integrand, we multiply numerator and denominator by 5 to get

$$\int \frac{e^{5x}}{6 + e^{5x}} \, dx = \int \frac{1}{5} \frac{1}{6 + e^{5x}} \, 5e^{5x} \, dx$$

$$= \int \frac{1}{5} \frac{1}{G(x)} \, g(x) \, dx$$

$$= \frac{1}{5} \int f(G(x))g(x) \, dx,$$

where f is just the reciprocal function—$f(x) = 1/x$. But an antiderivative for f is just $\ln x$, so the desired antiderivative is just

$$\frac{1}{5} F(G(x)) + C = \frac{1}{5} \ln(6 + e^{5x}) + C.$$

As usual, you should check this answer by differentiating to see that you really do get the original function.

Now let's see how this works using differential notation. If we set $u = 6 + e^{5x}$, then

$$\frac{du}{dx} = 5e^{5x} \qquad \text{and} \qquad du = 5e^{5x} \, dx.$$

Again we insert a factor of 5 in the numerator and an identical one in the denominator to balance it. Substitutions for u and du then yield the following:

$$\int \frac{1}{5} \cdot \frac{5e^{5x}}{6 + e^{5x}} dx = \frac{1}{5} \int \frac{5e^{5x} \, dx}{6 + e^{5x}}$$

$$= \frac{1}{5} \int \frac{du}{u}$$

$$= \frac{1}{5} \ln(u) + C$$

$$= \frac{1}{5} \ln(6 + e^{5x}) + C$$

as before.

The basic structure

The two examples above have the same structure. In both, a certain function of x is selected and called u; part of the integrand, namely $u'\,dx$, becomes du; and the rest becomes one of the basic functions of u. Specifically:

Integrand	u	du	Function of u
$3x^2(1 + x^3)^7\,dx$	$1 + x^3$	$3x^2\,dx$	u^7
$\dfrac{e^{5x}}{6 + e^{5x}}\,dx$	$6 + e^{5x}$	$5e^{5x}\,dx$	$\dfrac{1}{5} \cdot \dfrac{1}{u}$

Note that you may have to do a bit of algebraic reshaping of the integrand to cast it in the proper form. For example, we had to insert a factor of 5 to the numerator of the second example to make the numerator be the derivative of the denominator. There is no set routine to be followed to find an antiderivative most efficiently, or even any way to know whether a particular method will work until you try it. Success comes with experience and a certain amount of intelligent fiddling until something works out.

EXAMPLE 3.   The method of substitution is useful in simple problems, too. Consider

$$\int \cos(3t)\,dt.$$

If we set $u = 3t$, then $du = 3\,dt$ and

$$\int \cos(3t)\,dt = \int \cos(u) \cdot \tfrac{1}{3}\,du$$

$$= \tfrac{1}{3} \int \cos(u)\,du$$

$$= \tfrac{1}{3} \sin(u) + C$$

$$= \tfrac{1}{3} \sin(3t) + C.$$

Substitution in Definite Integrals

Until now we have been using the technique of substitution to find antiderivatives—that is, to evaluate *indefinite integrals*. Refer back to the integral form of the chain rule given in the box on page 627, and see what happens when we use this equation to evaluate a *definite integral*. Suppose we want to evaluate

$$\int_a^b f(G(x))\,g(x)\,dx.$$

We know that $F(G(x))$ is an antiderivative, so by the fundamental theorem we have

$$\int_a^b f(G(x))g(x)\,dx = F(G(x))\Big|_a^b = F(G(b)) - F(G(a)).$$

Now suppose we make the substitution $u = G(x)$ and $du = g(x)dx$. Then as x goes from a to b, u will go from $G(a)$ to $G(b)$. Moreover, we have

$$\int_{G(a)}^{G(b)} f(u)\,du = F(u)\Big|_{G(a)}^{G(b)} = F(G(b)) - F(G(a)),$$

so the two definite integrals have the same value. In other words,

---

If we make the substitution $u = G(x)$, then

$$\int_a^b f(G(x))g(x)\,dx = F(G(b)) - F(G(a)) = \int_{G(a)}^{G(b)} f(u)\,du.$$

This means that to evaluate a definite integral by substitution, we can do everything in terms of u. We don't ever need to find an antiderivative for the original integrand in terms of x or use the original limits of integration.

EXAMPLE 4. Consider the definite integral

$$\int_0^{\pi/2} \frac{\cos x\,dx}{1 + \sin x}.$$

We can evaluate this integral by making the substitution $u = 1 + \sin x$, and $du = \cos x\,dx$. Moreover, as x goes from 0 to $\pi/2$, u goes from 1 to 2. Therefore,

$$\int_0^{\pi/2} \frac{\cos x\,dx}{1 + \sin x} = \int_1^2 \frac{du}{u} = \ln u\Big|_1^2 = \ln 2.$$

Check that this is the same answer you would have gotten if you had expressed the antiderivative $\ln u$ in terms of x and evaluated the result at the limits on the original integral.

EXAMPLE 5. Evaluate

$$\int_0^1 6x^2(1 + x^3)^4\,dx.$$

With the substitution $u = 1 + 2x^3$ and $du = 6x^2 \, dx$, as x goes from 0 to 1, u goes from 1 to 3. Our integral thus becomes

$$\int_0^1 3x^2(1 + x^3)^4 \, dx = \int_0^3 u^4 \, du = \tfrac{1}{5}u^5 \Big|_1^3 = \tfrac{242}{5}.$$

Exercises

1. Evaluate the following using substitution. Do parts (a) through (e) in two ways: First, write the integrand in the form $f(G(x))g(x)$ (replace x by the appropriate variable as necessary) for appropriate functions f, G, and g, with $G' = g$, and then find $F = \int f$. Second, use differential notation. Do the remaining parts in the way you feel most confident.

a) $\displaystyle\int 2y(y^2 + 1)^{50} \, dy$

b) $\displaystyle\int \sin(5z) \, dz$

*c) $\displaystyle\int \frac{e^{\sqrt{x}}}{\sqrt{x}} \, dx$

d) $\displaystyle\int (5t + 7)^{50} \, dt$

e) $\displaystyle\int 3u^2 \sqrt[3]{u^3 + 8} \, du$

f) $\displaystyle\int \frac{1}{2v + 1} \, dv$

*g) $\displaystyle\int \tan x \, dx$

h) $\displaystyle\int \tan^2(x) \sec^2(x) \, dx$

i) $\displaystyle\int \sec(x/2) \tan(x/2) \, dx$

j) $\displaystyle\int \sin(w) \sqrt{\cos(w)} \, dw$

k) $\displaystyle\int \frac{\sin(\sqrt{s})}{\sqrt{s}} \, ds$

l) $\displaystyle\int \sqrt{3 - x} \, dx$

m) $\displaystyle\int \frac{dr}{r \ln r}$

n) $\int e^x \sin(1 + e^x)\, dx$

o) $\int \dfrac{y}{1 + y^2}\, dy$

p) $\int \dfrac{w}{\sqrt{1 - w^2}}\, dw$

q) $\int \dfrac{1}{1 + 4y^2}\, dy$

r) $\int \dfrac{1}{\sqrt{1 - 9w^2}}\, dw$

2. Use integration by substitution to find the numerical value of the following. In four of these you should get your answer in two ways: First, find an antiderivative for the given integrand. Second, use the observation in the box on page 631. Then compare the results. You should also check your results for three of the problems by finding numerical estimates for the integrals using RIEMANN.

a) $\displaystyle\int_0^1 \dfrac{e^s}{e^s + 1}\, ds$

*b) $\displaystyle\int_0^{\ln e} \dfrac{e^s}{e^s + 1}\, ds$

c) $\displaystyle\int_1^3 \dfrac{1}{2x + 1}\, dx$

*d) $\displaystyle\int_{-3}^{-1} \dfrac{1}{2x + 1}\, dx$

e) $\displaystyle\int_0^1 \dfrac{t}{\sqrt{1 + t^2}}\, dt$

f) $\displaystyle\int_0^1 \dfrac{\sin(\pi \sqrt{t})}{\sqrt{t}}\, dt$

g) $\displaystyle\int_0^2 \dfrac{1}{1 + \dfrac{x^2}{4}}\, dx$

h) $\displaystyle\int_0^{1/3} \dfrac{1}{\sqrt{1 - 9y^2}}\, dy$

3. This question concerns the integral $I = \int \sin x \, \cos x \, dx$.
   a) Find I by using the substitution $u = \sin x$.
   b) Find I by using the substitution $u = \cos x$.
   c) Compare your answers to (a) and (b). Are they the same? If not, how do they differ? Since both answers are antiderivatives of

$\sin x \cos x$, they should differ only by a constant. Is that true here? If so, what is the constant?

d) Now calculate the value of the *definite* integral

$$\int_0^{\pi/2} \sin x \, \cos x \, dx$$

twice, using the two *indefinite* integrals you found in (a) and (b). Do the two values agree, or disagree? Is your result consistent with what you expect?

4. a) Find all functions $y = F(x)$ that satisfy the differential equation

$$\frac{dy}{dx} = x^2(1 + x^3)^{13}.$$

*b) From among the functions $F(x)$ you found in part (a), select the one that satisfies $F(0) = 4$.

c) From among the functions $F(x)$ you found in part (a), select the one that satisfies $F(-1) = 4$.

5. Find a function $y = G(t)$ that solves the initial value problem

$$\frac{dy}{dt} = te^{-t^2} \qquad y(0) = 3.$$

6. a) What is the average value of the function $f(x) = x/\sqrt{1 + x^2}$ on the interval $[0, 2]$?

b) Show that the average value of the $f(x)$ on the interval $[-2, 2]$ is 0. Sketch a graph of $y = f(x)$ on this interval, and explain how the graph also shows that the average is 0.

7. a) Sketch the graph of the function $y = xe^{-x^2}$ on the interval $[0, 5]$.

b) Find the area between the graph of $y = xe^{-x^2}$ and the x-axis for $0 \le x \le 5$.

c) Find the area between the graph of $y = xe^{-x^2}$ and the x-axis for $0 \le x \le b$. Express your answer in terms of the quantity b, and denote it $A(b)$. Is $A(5)$ the same number you found in part (b)? What are the values of $A(10)$, $A(100)$, $A(1000)$?

d) It is possible to argue that the area between the graph of $y = xe^{-x^2}$ and the *entire* positive x-axis is $1/2$. Can you develop such an argument?

8. a) Use a computer graphing utility to confirm that

$$\sin^2 x = \frac{1 - \cos(2x)}{2}.$$

Sketch these graphs.

b) Find a formula for $\int \sin^2 x \, dx$. [Suggestion: Replace $\sin^2 x$ by the expression involving $\cos(2x)$, above, and integrate by substitution.]

c) What is the average value of $\sin^2 x$ on the interval $[0, \pi]$? What is its average value on any interval of the form $[0, k\pi]$, where k is a whole number?

d) Explain your results in part (c) in terms of the graph of $\sin^2 x$ you drew in part (a).

e) Here's a differential equations proof of the identity in part (a). Let $f(x) = \sin^2 x$, and let $g(x) = (1 - \cos(2x))/2$. Show that both of these functions satisfy the initial value problem

$$y'' = 2 - 4y \qquad \text{with } y(0) = 0 \text{ and } y'(0) = 0.$$

Hence conclude the two functions must be the same.

11.3   INTEGRATION BY PARTS

As in the previous section, suppose we have functions F and G, with corresponding derivatives f and g. If we use the product rule to differentiate $F(x) \cdot G(x)$, we get:

$$(F \cdot G)' = F \cdot G' + F' \cdot G = F \cdot g + f \cdot G.$$

We can turn this into a statement about indefinite integrals:

$$\int (F \cdot g + f \cdot G)\, dx = \int (F \cdot G)'\, dx = F(x) \cdot G(x) + C.$$

Unfortunately, in this form the statement is not especially useful; it applies only when the integrand has two terms of the special form $f \cdot g' + f' \cdot g$. However, if we rewrite the statement in the form

An integral form of the product rule

$$\boxed{\int F \cdot g\, dx = F \cdot G - \int f \cdot G\, dx.}$$

it becomes very useful.

EXAMPLE 1.   We will use the formula in the box to find

$$\int x \cdot \cos x\, dx.$$

If we label the parts of this integrand as follows:

$$F(x) = x \qquad g(x) = \cos x,$$

then we have

$$f(x) = 1 \quad \text{and} \quad G(x) = \sin x.$$

According to the formula,

$$\int x \cdot \cos x \, dx = x \cdot \sin x - \int \sin x \, dx$$

$$= x \cdot \sin x + \cos x + C.$$

Integrate only part of the integrand

The integrand is first broken into two parts—in this case, x and $\cos x$. One part is differentiated while the other part is integrated. [The part we integrated is $g(x) = \cos x$, and we got $G(x) = \sin x$.] For this reason, the rule described in the box is called **integration by parts**.

As with integration by substitution, integration by parts exchanges one integration task for another: Instead of finding an antiderivative for $F \cdot g$, we must find one for $f \cdot G$. The idea is to "trade in" one integration problem for a more readily solvable one.

EXAMPLE 2.   Use integration by parts to find

$$\int \ln(x) \, dx.$$

At first glance we can't integrate by parts, because there aren't two parts! But note that we can write

$$\ln(x) = \ln(x) \cdot 1,$$

and then set

$$F(x) = \ln(x) \quad g(x) = 1.$$

This implies

$$f(x) = \frac{1}{x} \quad \text{and} \quad G(x) = x,$$

and the integration-by-parts formula now gives us

$$\int \ln(x) \, dx = x \cdot \ln(x) - \int x \cdot \frac{1}{x} \, dx$$

$$= x \cdot \ln(x) - \int 1 \, dx$$

$$= x \cdot \ln(x) - x + C.$$

Example 2 shows that integration by parts—like integration by substitution—is an art rather than a set routine. If integration by parts is to work, several things must happen. First, you need to see that the method might actually apply. (In Example 2 this wasn't obvious.) Next, you need to identify the parts of the integrand that will be differentiated and integrated, respectively. The wrong choices can lead you away from a solution, rather than toward one. (See Example 3 below for a cautionary tale.) Finally, you need to be able to carry out the integration of the new integral $\int f \cdot G \, dx$. As you work you may have to reshape the integrand algebraically. Technique comes with practice, and luck is useful, too.

The ingredients of a successful integration by parts

EXAMPLE 3.   Use integration by parts to find

$$\int t \cdot e^t \, dt.$$

Set

$$F(t) = e^t \qquad g(t) = t.$$

Then

$$f(t) = e^t \qquad G(t) = \frac{t^2}{2}.$$

The integration by parts formula then gives

$$\int t \cdot e^t \, dt = \frac{t^2}{2} e^t - \int \frac{t^2}{2} e^t \, dt.$$

While this is a true statement, we are not better off—the new integral is *not* simpler than the original. A solution is eluding us here. You will have a chance to do this problem properly in the exercises.

What went wrong?

Exercises

1. Use integration by parts to find a formula for each of the following integrals.

   *a) $\int x \sin x \, dx$

   b) $\int t e^t \, dt$

   c) $\int w e^{-w} \, dw$

   d) $\int x \ln x \, dx$

e) $\int \arcsin x \, dx$

f) $\int \arctan x \, dx$

g) $\int x^2 e^{-x} dx$ (Suggestion: Apply integration by parts twice. After
the first application you should have an integral that can itself be
evaluated using integration by parts.)

h) $\int u^2 \cos u \, du$

i) $\int x \sec^2 x \, dx$

j) $\int e^{2x} (x + e^x) \, dx$

*2. Use integration by parts to obtain a formula for

$$\int (\ln x)^2 \, dx.$$

Choose $F(x) = \ln x$ and also $G'(x) = \ln x$. To continue you need to
find g, the antiderivative of $\ln x$, but this has already been obtained in
the text.

3. a) Find $\int x^2 e^x \, dx$.

b) Find $\int x^3 e^x \, dx$. [Reduce this to part (a)].

c) Find $\int x^4 e^x \, dx$.

d) What is the general pattern here? Find a formula for $\int x^n e^x \, dx$,
where n is any positive integer.

e) Find $\int e^x (5x^2 - 3x + 7) \, dx$.

4. a) Draw the graph of $y = \arctan x$ over the interval $0 \le x \le 1$. You
could have gotten the same graph by thinking of x as a function
of y—write down this relationship and the corresponding y
interval.

b) Evaluate

$$\int_0^1 \arctan x \, dx$$

and show on your graph the area this corresponds to.

c) Evaluate

$$\int_0^{\pi/4} \tan y \, dy$$

and show on your graph the area this corresponds to.
d) If we add the results of part (b) and part (c), what do you get?
From the geometry of the picture, what should you have gotten?

5. Repeat the analysis of the preceding problem by calculating the value of

$$\int_0^2 x^3 \, dx + \int_0^8 y^{1/3} \, dy$$

and seeing if it agrees with what you would predict by looking at the graphs.

6. Generalize the preceding two problems to the case where f and g are any two functions that are inverses of each other whose graphs pass through the origin.

7. **a)** What is the average value of the function $\ln x$ on the interval $[1, e]$?
   b) What is the average value of $\ln x$ on $[1, b]$? Express this in terms of b. Discuss the following claim: The average value of $\ln x$ on $[1, b]$ is approximately $\ln(b) - 1$ when b is large.

8. **a)** Sketch the graph of $f(x) = xe^{-x}$ on the interval $[0, 4]$.
   *b) What is the area between the graph of $y = f(x)$ and the x-axis for $0 \le x \le 4$?
   c) What is the area between the graph of $y = f(x)$ and the x-axis for $0 \le x \le b$? Express your answer in terms of b, and denote it $A(b)$. What is $A(100)$?

9. Find three solutions $y = f(t)$ to the differential equation

$$\frac{dy}{dt} = 5 - 2\ln t.$$

10. **a)** Find the solution $y = \varphi(t)$ to the initial value problem

$$\frac{dy}{dt} = te^{-t^2/2}, \qquad y(0) = 2.$$

   b) The function $\varphi(t)$ increases as t increases. Show this first by sketching the graph of $y = \varphi(t)$. Show it also by referring to the differential equation that $\varphi(t)$ satisfies. (What is true about the derivative of an increasing function?)
   c) Does the value of $\varphi(t)$ increase without bound as $t \to \infty$? If not, what value does $\varphi(t)$ approach?

11. a) *The differential form of integration by parts.*   If u and v are expressions in x, then the product rule can be written as

$$\frac{d}{dx}(u \cdot v) = \frac{du}{dx} \cdot v + u \cdot \frac{dv}{dx}.$$

Explain carefully how this leads to the following statement of integration by parts, and why it is equivalent to the form in the text:

$$\int u \, dv = uv - \int v \, du.$$

b) Solve a couple of the preceding problems using this notation.

Sine and Cosine Integrals

The purpose of the remaining exercises is to establish integral formulas that we will use to analyze Fourier polynomials and the power spectrum in chapter 12. In the first three, α is a constant:

$$\int \sin^2 \alpha x \, dx = \frac{x}{2} - \frac{1}{4\alpha} \sin 2\alpha x + C$$

$$\int \cos^2 \alpha x \, dx = \frac{x}{2} + \frac{1}{4\alpha} \sin 2\alpha x + C$$

$$\int \sin \alpha x \cos \alpha x \, dx = -\frac{1}{4\alpha} \cos 2\alpha x + C$$

In the remaining four, α and β are *different* constants:

$$\int \sin \alpha x \sin \beta x \, dx = \frac{1}{\beta^2 - \alpha^2} (\alpha \cos \alpha x \sin \beta x - \beta \sin \alpha x \cos \beta x) + C$$

$$\int \cos \alpha x \cos \beta x \, dx = \frac{1}{\beta^2 - \alpha^2} (\beta \cos \alpha x \sin \beta x - \alpha \sin \alpha x \cos \beta x) + C$$

$$\int \sin \alpha x \cos \beta x \, dx = \frac{1}{\beta^2 - \alpha^2} (\beta \sin \alpha x \sin \beta x + \alpha \cos \alpha x \cos \beta x) + C$$

$$\int \cos \alpha x \sin \beta x \, dx = \frac{1}{\beta^2 - \alpha^2} (-\alpha \sin \alpha x \sin \beta x - \beta \cos \alpha x \cos \beta x) + C.$$

12. a) In the later exercises we shall make frequent use of the following "trigonometric identities":

$$2 \sin \alpha x \cos \alpha x = \sin 2\alpha x$$

$$\cos^2 \alpha x - \sin^2 \alpha x = \cos 2\alpha x$$

$$\sin^2 \alpha x + \cos^2 \alpha x = 1.$$

Using a graphing package on a computer, graph together the functions

$$2 \sin \alpha x \cos \alpha x \qquad \text{and} \qquad \sin 2\alpha x$$

to show that they seem to be identical. (That is, show that they "share phosphor.") Then do the same for the pairs of functions in the other two identities.

b) We can give a different argument for the identities above using the ideas we have developed in studying initial value problems. To prove the first identity, for instance, let $f(x) = 2 \sin \alpha x \cos \alpha x$, and let $g(x) = \sin 2\alpha x$. Show that both functions satisfy

$$y'' = -4\alpha^2 y \quad \text{with } y(0) = 0 \text{ and } y'(0) = 2\alpha.$$

Hence conclude the two functions must be the same.

c) Find an initial value problem that is satisfied by both $f(x) = \cos^2 \alpha x - \sin^2 \alpha x$ and by $g(x) = \cos 2\alpha x$.

13. *Evaluating* $\displaystyle\int \sin^2 x \, dx$

a) Using integration by parts, show that

$$\int \sin^2 x \, dx = -\sin x \cos x + \int \cos^2 x \, dx.$$

b) Using the identity $\sin^2 \alpha x + \cos^2 \alpha x = 1$, show that the new integral can be written as

$$x - \int \sin^2 x \, dx.$$

c) Combining (a) and (b) algebraically, show that

$$2 \int \sin^2 x \, dx = -\sin x \cos x + x + C.$$

d) Using algebra and a trigonometric identity, conclude that

$$\int \sin^2 x \, dx = \frac{x}{2} - \frac{1}{4} \sin 2x + C.$$

14. Modify the argument of exercise 9 to show

$$\int \sin^2 \alpha x \, dx = \frac{x}{2} - \frac{1}{4\alpha} \sin 2\alpha x + C.$$

and

$$\int \cos^2 \alpha x \, dx = \frac{x}{2} + \frac{1}{4\alpha} \sin 2\alpha x + C.$$

15. *Evaluating* $\int \sin^2 \alpha x \, dx$

Determine this integral anew, without using integration by parts, by carrying out the following steps.

a) From the trigonometric identities on page 640, deduce that

$$2 \sin^2 \alpha x = 1 - \cos 2\alpha x.$$

b) Using the formula in (a), conclude that

$$\int \sin^2 \alpha x \, dx = \frac{x}{2} - \frac{1}{4\alpha} \sin 2\alpha x + C.$$

16. *Evaluating* $\int \cos^2 \alpha x \, dx$

Using only algebra and the identity $\sin^2 \alpha x + \cos^2 \alpha x = 1$, show that the previous exercise implies

$$\int \cos^2 \alpha x \, dx = \frac{x}{2} + \frac{1}{4\alpha} \sin 2\alpha x + C.$$

17. *Evaluating* $\int \sin \alpha x \cos \alpha x \, dx$

a) Using the identity $\sin 2\alpha x = 2 \sin \alpha x \cos \alpha x$, deduce the following formula

$$\int \sin \alpha x \cos \alpha x \, dx = -\frac{1}{4\alpha} \cos 2\alpha x + C.$$

b) Using integration by substitution, obtain the alternative formula

$$\int \sin \alpha x \cos \alpha x \, dx = \frac{1}{2\alpha} \sin^2 \alpha x + C.$$

c) Show that your results in (a) and (b) are compatible. [For example, use exercise 11 (a).]

18. *Evaluating* $\int \sin \alpha x \sin \beta x \, dx$

a) Use integration by parts to show that

$$\int \sin \alpha x \sin \beta x \, dx = -\frac{1}{\alpha} \cos \alpha x \sin \beta x + \frac{\beta}{\alpha} \int \cos \alpha x \cos \beta x \, dx.$$

b) Using integration by parts again, show that the new integral in part (a) can be written as

$$\int \cos \alpha x \cos \beta x \, dx = \frac{1}{\alpha} \sin \alpha x \cos \beta x + \frac{\beta}{\alpha} \int \sin \alpha x \sin \beta x \, dx.$$

c) Let $J = \int \sin \alpha x \sin \beta x \, dx$; show that combining (a) and (b) gives

$$J = -\frac{1}{\alpha} \cos \alpha x \sin \beta x + \frac{\beta}{\alpha^2} \sin \alpha x \cos \beta x + \frac{\beta^2}{\alpha^2} J.$$

d) Solve (c) for J to find

$$\int \sin \alpha x \sin \beta x \, dx = \frac{1}{\beta^2 - \alpha^2} (\alpha \cos \alpha x \sin \beta x - \beta \sin \alpha x \cos \beta x) + C.$$

19. Imitate the methods of exercise 14 to deduce

$$\int \cos \alpha x \cos \beta x \, dx = \frac{1}{\beta^2 - \alpha^2} (\beta \cos \alpha x \sin \beta x - \alpha \sin \alpha x \cos \beta x) + C.$$

and

$$\int \sin \alpha x \cos \beta x \, dx = \frac{1}{\beta^2 - \alpha^2} (\beta \sin \alpha x \sin \beta x + \alpha \cos \alpha x \cos \beta x) + C.$$

20. *Evaluating* $\int \cos \alpha x \sin \beta x \, dx$

This integral is the same as the one in exercise 17, if you exchange the factors α and β. Do that, and obtain the formula

$$\int \cos \alpha x \sin \beta x \, dx = \frac{1}{\alpha^2 - \beta^2} (\alpha \sin \alpha x \sin \beta x + \beta \cos \alpha x \cos \beta x) + C.$$

21. Determine the following.

a) $\displaystyle\int_0^{2\pi} \sin^2 x \, dx.$

*b) $\displaystyle\int_0^{n\pi} \cos^2 x \, dx$, where n is a positive integer.

c) $\displaystyle\int_0^{2\pi} \sin x \sin \beta x \, dx$, where $\beta \neq 1$.

d) $\displaystyle\int_0^{\pi} \sin x \sin \beta x \, dx$, $\beta \neq 1$.

> ## 11.4 SEPARATION OF VARIABLES AND PARTIAL FRACTIONS

One of the principal uses of integration techniques is to find closed form solutions to differential equations. If you look back at the methods we have developed so far in this chapter, they are all applicable to differential equations of the form $y' = f(t)$ for some function f—that is, the rate at which y changes is a function of the independent variable only. In such cases we only need to find an antiderivative F for f, choose the constant C to satisfy the initial value, and we have our solution. As we saw in the early chapters, though, the behavior of y' often depends on the values of y rather than on t—think of the S-I-R model or the various predator–prey problems. In this section we will see how our earlier techniques can be adapted to apply to problems of this sort as well.

The Differential Equation $y' = y$

As you know, the exponential functions $y = Ce^t$ are the solutions to the differential equation $dy/dt = y$. Let's put aside this knowledge for a moment and rediscover these solutions using a new method. The method involves the connection between inverse functions and their derivatives. With it we will be able to explore a variety of problems that had been beyond our reach.

The idea behind the new method is quite simple: Instead of thinking of y as a function of t, convert to thinking of t as a function of y, thereby looking for the **inverse function.** We know that the derivative of the inverse function is the reciprocal of the derivative of the original function, so we can rewrite the given differential equation by using its reciprocal:

$$\frac{dy}{dt} = y \qquad \text{becomes} \qquad \frac{dt}{dy} = \frac{1}{y}.$$

Then solve the new differential equation $dt/dy = 1/y$. While this may not look very different, it has the property that the rate of change of the *dependent variable*—now t—is expressed as a function of the *independent* variable—now y. But this is just the form we have been considering in the earlier sections of this chapter. A solution to the new equation is a function $t = g(y)$ whose derivative is $1/y$. This is one of the basic antiderivatives listed in the table on page 609:

$$t = g(y) = \ln y + k,$$

where k is an arbitrary constant.

The solution to the original differential equation $dy/dt = y$ is the inverse of $t = \ln y + k$. We find it by solving this equation for y:

$$t - k = \ln y$$

A new method for solving $y' = y$

Find the inverse function instead

The new differential equation . . .

. . . and its solution

The inverse function . . .

$$e^{t-k} = y$$
$$e^t \cdot e^{-k} = y.$$

...solves the original
problem

Thus $y(t) = Ce^t$, where we have replaced the constant e^{-k} by C.

Indefinite Integrals

The language of indefinite integrals and differentials again provides a convenient mnemonic for this new method. First, we use the original differential equation to relate the differentials dy and dt:

The differential equation expressed using differentials

$$dy = \frac{dy}{dt} \, dt = y \, dt.$$

The use of differentials is introduced on page 628. At this point the equation makes sense for either t or y being the independent variable. Now if we try to integrate this equation with respect to t, we get

$$y = \int dy = \int y(t) \, dt.$$

We can't find the last integral, because we don't know what y is as a function of t. Remember, an indefinite integral is an antiderivative, so an expression of the form

$$\int f \, dt$$

represents a function $F(t)$ whose derivative is $f(t)$. If f is *not* given by a formula in t, there is no way to get a formula for $F(t)$.

Suppose, though, that we divide both sides of the differential equation $dy = y \, dt$ by y, and integrate with respect to y. The equation takes the form

$$\frac{dy}{y} = dt.$$

Now, if we introduce indefinite integrals, we have

$$\int \frac{dy}{y} = \int dt.$$

This time the variables y and t have been separated from each other, and we *can* find the integrals. In fact,

The variables are now separated

$$\ln y = \int \frac{dy}{y} = \int dt = t + b,$$

where b is an arbitrary constant. (We could just as easily have added the constant to the left side instead—do you see why we don't have to add a constant to both sides?) To complete the work we solve for y:

The solution once
again

$$y = e^{t+b} = e^t \cdot e^b = Ce^t,$$

where $C = e^b$.

Summary

The first time we went through the method, we replaced

$$\frac{dy}{dt} = y \qquad \text{by} \qquad \frac{dt}{dy} = \frac{1}{y}.$$

These differential equations express the same relation between y and t. Each is just the reciprocal of the other. In the first, y depends on t; in the second, though, t depends on y. The second time we went through the method, using indefinite integrals, we replaced

$$dy = y\,dt \qquad \text{by} \qquad dt = \frac{dy}{y}.$$

Separate the variables
to integrate

This change was also algebraic, and it had the same effect: the dependent variable changed from y to t. More important, in the new differential equation (using the differentials themselves!) the variables are separated. That allows us to do the integration. We get a solution in the form $t = g(y)$. The solution to the original problem is the *inverse* $y = f(t)$ of $t = g(y)$.

Separation of Variables

With the method of **separation of variables,** introduced in the previous pages, we can obtain formulas for solutions to a number of differential equations that were previously accessible only by Euler's method. Recall that one of the clear advantages of a *formula* is that it allows us to see how the parameters in the problem affect the solution. We'll look at two problems. First we'll show how the method can explain the rather baffling formula for supergrowth that we gave in chapter 4. Then, using the method of **partial fractions,** to be discussed next, we'll give a formula for logistic growth.

Supergrowth

In section 4.2 we modeled the growth of a population Q by the initial value problem

$$\frac{dQ}{dt} = kQ^{1.2} \qquad Q(0) = A.$$

To get a formula for the solution, transform the differential equation in the following way:

Separate the variables

$$\frac{dQ}{dt} = kQ^{1.2} \quad \rightsquigarrow \quad dQ = kQ^{1.2}\, dt \quad \rightsquigarrow \quad \frac{dQ}{Q^{1.2}} = k\, dt.$$

Now integrate:

$$\int \frac{dQ}{Q^{1.2}} = \int k\, dt.$$

Because the variables have been separated, the integrals can be found:

$$\int \frac{dQ}{Q^{1.2}} = \int Q^{-1.2}\, dQ = \frac{1}{-.2} Q^{-.2}$$

$$\int k\, dt = kt + C$$

Therefore, $(-1/.2)\, Q^{-.2} = kt + C$. Now we must solve this equation for Q. Here is one possible approach. First, we can write

Solve for Q

$$Q^{-.2} = -.2(kt + C) = C_1 - .2kt.$$

To simplify the expression, we have replaced $-.2C$ by a *new* constant C_1. Since

$$(Q^{-.2})^{-5} = Q^{-.2 \times -5} = Q^1 = Q,$$

we'll raise both sides of the previous equation to the power -5:

$$Q(t) = (Q^{-.2})^{-5} = (C_1 - .2kt)^{-5}.$$

The last step is to incorporate the initial condition $Q(0) = A$. According to the new formula for $Q(t)$,

Bring in the initial condition

$$Q(0) = (C_1 - .2k \cdot 0)^{-5} = C_1^{-5} = A.$$

Solving $A = C_1^{-5}$ for C_1, we get

$$C_1 = A^{-1/5} = \frac{1}{\sqrt[5]{A}}.$$

We are now done:

$$Q(t) = \left(\frac{1}{\sqrt[5]{A}} - .2kt \right)^{-5}.$$

Interpreting the formula

This is the formula that appears on page 183. It shows how the parameters k and A affect the solution. In particular, we called this *supergrowth* because the model predicts that the population Q becomes infinite when

$$\frac{1}{\sqrt[5]{A}} - .2kt = 0 \qquad \text{that is, when } t = \frac{1}{.2k\sqrt[5]{A}}.$$

Partial Fractions

Using separation of variables with a partial fraction's decomposition (to be described below), we will obtain a formula for the solution to the logistic equation (section 4.1). The method of **partial fractions** is a useful tool for solving many integration problems.

Logistic Growth

Consider this initial value problem associated with the logistic differential equation:

$$\frac{dP}{dt} = kP\left(1 - \frac{P}{C}\right) \rightsquigarrow P(0) = A.$$

We will find a formula for the solution that incorporates the growth parameter k and the carrying capacity C.

Step 1: Separate the variables

The first step is to transform the equation into one where the variables are separated:

$$\frac{dP}{dt} = kP\left(1 - \frac{P}{C}\right) \rightsquigarrow \frac{dP}{P(1 - P/C)} = k\,dt.$$

Integrating this equation we get

$$\int \frac{dP}{P(1 - P/C)} = \int k\,dt.$$

The denominator has an unfamiliar form

We are stuck now, because the integral on the left doesn't appear in our table of integrals (page 609). If the denominator had *only* P or *only* $1 - P/C$, we could use the natural logarithm. The difficulty is that the denominator is the product of both terms.

There is a way out of the difficulty. We will use algebra to transform the integrand into a form we can work with. The first step is to simplify the denominator a bit:

$$\frac{1}{P(1 - P/C)} = \frac{1}{P(C/C - P/C)} = \frac{1}{P(C - P)/C} = \frac{C}{P(C - P)}.$$

(This wasn't essential; it just makes later steps easier to write.) The next step will be the crucial one. To understand why we take it, consider the rule for adding two fractions:

$$\frac{\alpha}{x + a} + \frac{\beta}{x + b} = \frac{\alpha(x + b) + \beta(x + a)}{(x + a)(x + b)}.$$

The denominator is a product—very much like the product $P(C - P)$ in our integrand! Perhaps we can write *that* as a sum of two simpler fractions:

$$\frac{C}{P(C - P)} = \frac{\alpha}{P} + \frac{\beta}{C - P}.$$

Step 2: Write the integrand as a sum of simple fractions

What values should α and β have? According to the rule for adding fractions,

$$\frac{\alpha}{P} + \frac{\beta}{C - P} = \frac{\alpha(C - P) + \beta P}{P(C - P)},$$

and this should equal the original integrand:

$$\frac{\alpha(C - P) + \beta P}{P(C - P)} = \frac{C}{P(C - P)}.$$

Since the denominators are equal, the numerators must also be equal:

Determining α and β

$$\alpha(C - P) + \beta P = C.$$

In fact, they must be equal *as polynomials in the variable P*. If we rewrite the last equation, collecting terms that involve the same power of P, we get

$$(\beta - \alpha)P + \alpha C = 0 \cdot P + 1 \cdot C.$$

Since two polynomials are equal precisely when their coefficients are equal, it follows that

$$\beta - \alpha = 0 \qquad \alpha = 1.$$

Thus $\alpha = \beta = 1$, and we have

$$\frac{C}{P(C - P)} = \frac{1}{P} + \frac{1}{C - P}.$$

The partial fractions decomposition

The simpler expressions on the right are called **partial fractions.** Their denominators are the different *parts* of the denominator of the integrand. The equation that expresses the integrand as a sum of partial fractions is called a **partial fractions decomposition.**

We can now return to the integral equation we are trying to solve:

$$\int \frac{C}{P(C-P)} \, dP = \int k \, dt.$$

Step 3: Evaluate the integrals

The right-hand side equals $kt + b$, where b is the usual constant of integration. Thanks to the partial fractions decomposition, the left-hand side can be written

$$\int \frac{C}{P(C-P)} \, dP = \int \frac{1}{P} \, dP + \int \frac{1}{C-P} \, dP.$$

The first integral on the right is straightforward:

$$\int \frac{1}{P} \, dP = \ln P.$$

The second can be solved by using the substitution $C - P = u$, with $dP = -du$:

$$\int \frac{1}{C-P} \, dP = \int \frac{-du}{u} = -\ln u = -\ln(C-P).$$

Putting everything together we find

$$\ln P - \ln(C-P) = \ln\left(\frac{P}{C-P}\right) = kt + b.$$

Step 4: Solve for P

As we have seen, separation of variables usually leaves us with an inverse function to find. This problem is no different. We must solve the last equation for P. The first step is to exponentiate both sides:

$$\frac{P}{C-P} = e^{kt+b} = e^b \cdot e^{kt} = Be^{kt}.$$

To simplify the expression a bit, we have replaced e^b by $B = e^b$. Multiplying both sides by $C - P$ gives

$$P = Be^{kt}(C-P) = CBe^{kt} - PBe^{kt}.$$

Now bring the last term over to the left, and then factor out P:

$$P + PBe^{kt} = (1 + Be^{kt})P = CBe^{kt}.$$

The final step is to divide by the coefficient $1 + Be^{kt}$:

$$P(t) = \frac{CBe^{kt}}{1 + Be^{kt}}.$$

The formula for $P(t)$

Lastly, we must see how the initial condition $P(0) = A$ affects the solution. We could substitute $t = 0, P = A$ into the last formula, but that produces an algebraic mess. We want to know how A affects the constant B, and we can see that directly by making our substitutions into the equation above:

Step 5: Incorporate the initial condition

$$\frac{P}{C - P} = Be^{kt} \quad \rightsquigarrow \quad \frac{A}{C - A} = Be^0 = B.$$

Now replace B by $A/(C - A)$ in our formula for $P(t)$. This yields

$$P(t) = \frac{CAe^{kt}/(C - A)}{1 + Ae^{kt}/(C - A)} = \frac{CAe^{kt}}{C - A + Ae^{kt}}.$$

If we write $-A + Ae^{kt} = A(e^{kt} - 1)$, we get one of the standard forms of the solution to the logistic equation:

$$P(t) = \frac{CAe^{kt}}{C + A(e^{kt} - 1)}.$$

The complete solution

REMARK. The method of partial fractions can be used to evaluate integrals of the form

$$\int \frac{dx}{(x + a_1)(x + a_2) \cdots (x + a_n)} \quad \text{or} \quad \int \frac{P(x)}{Q(x)} \, dx,$$

where $P(x)$ and $Q(x)$ are arbitrary polynomials. Tables of integrals and calculus references describe how the method works in these cases. As one example, though, let's compute the antiderivative of the cosecant function, since we will need it in the next section.

EXAMPLE—THE ANTIDERIVATIVE OF THE COSECANT. We first use a trigonometric identity to transform the integral slightly:

$$\int \csc x \, dx = \int \frac{1}{\sin x} \, dx$$

$$= \int \frac{\sin x}{\sin^2 x} \, dx$$

$$= \int \frac{\sin x \, dx}{1 - \cos^2 x}.$$

If we now make the substitution $u = \cos x$, with $du = -\sin x \, dx$, this becomes

$$\int \csc x \, dx = \int \frac{-du}{1 - u^2}$$

$$= \int \frac{-1}{2}\left(\frac{1}{1 + u} + \frac{1}{1 - u}\right) du$$

$$= \frac{-1}{2}\int \frac{du}{1 + u} - \frac{1}{2}\int \frac{du}{1 - u}$$

$$= \frac{-1}{2}\ln(1 + u) + \frac{1}{2}\ln(1 - u) + C$$

$$= \frac{1}{2}\ln\frac{1 - u}{1 + u} + C$$

$$= \frac{1}{2}\ln\frac{1 - \cos x}{1 + \cos x} + C.$$

We can simplify this slightly by multiplying both numerator and denominator by $1 + \cos x$ to get

The final form of the antiderivative of the cosecant

$$\int \csc x \, dx = \frac{1}{2}\ln\frac{1 - \cos^2 x}{(1 + \cos x)^2} + C = \ln\left|\frac{\sin x}{1 + \cos x}\right| + C$$

$$= -\ln|\csc x + \cot x| + C.$$

Note that in the next to last line we used the general fact about logarithms that $n \ln A = \ln(A^n)$ for any value of n and any $A > 0$. Note also that since the domain of the secant function consists of infinitely many separate intervals, the "$+C$" at the end of the antiderivative needs to be interpreted as potentially a different value of C over each interval.

In the same fashion we can obtain the antiderivative for the secant function:

The antiderivative of the secant

$$\int \sec x \, dx = \ln|\sec x + \tan x| + C.$$

Exercises

Separation of Variables

*1. Use the method of separation of variables to find a formula for the solution of the differential equation $dy/dt = y + 5$. Your formula should contain an arbitrary constant to reflect the fact that many functions solve the differential equation.

2. Use the method of separation of variables to find formulas for the solutions to the following differential equations. In each case your formula should be expressed in terms of the input variable that is indicated [e.g., in part (a) it is t].

a) $dy/dt = 1/y$

b) $dz/dx = 3/(z - 2)$

*c) $dy/dx = x/y$

d) $dy/dx = y/x$

e) $du/dv = u/(u - 1)$

f) $dv/dt = -\sqrt{v}$

3. *A cooling liquid.* According to Newton's law of cooling (see section 4.1), in a room where the ambient temperature is C, the temperature Q of a hot object will change according to the differential equation

$$\frac{dQ}{dt} = -k(Q - C).$$

The constant k gives the rate at which the object cools.

a) Find a formula for the solution to this equation using the method of separation of variables. Your formula should contain an arbitrary constant.

*b) Suppose C is 20°C and k is .1° per minute per °C. If time t is measured in minutes, and $Q(0) = 90°C$, what will Q be after 20 minutes?

c) How long does it take for the temperature to drop to 30°C?

4. a) Suppose a cold drink at 36°F is sitting in the open air on a summer day when the temperature is 90°F. If the drink warms up at a rate of .2°F per minute per °F of temperature difference, write a differential equation to model what will happen to the temperature of the drink over time.

b) Obtain a formula for the temperature of the drink as a function of the number of minutes t that have passed since its temperature was 36°F.

c) What will the temperature of the drink be after 5 minutes? After 10 minutes?

d) How long will it take for the drink to reach 55°F?

5. *A leaking tank.* In chapter 4 we used the differential equation

$$\frac{dV}{dt} = -k\sqrt{V}$$

to model the volume $V(t)$ of water in a leaking tank after t hours (see section 4.2, pages 195–196).

a) Use the method of separation of variables to show that

$$V(t) = \frac{k^2}{4}(C - t)^2$$

is a solution to the differential equation, for any value of the constant C.

b) Explain why the function

$$V(t) = \begin{cases} \dfrac{k^2}{4}(C - t)^2 & \text{if } 0 \le t \le C \\ 0 & \text{if } C < t \end{cases}$$

is *also* a solution to the differential equation. Why is *this* solution more relevant to the leaking tank problem than the solution in part (a)?

6. *A falling body with air resistance.* We have used the differential equation

$$\frac{dv}{dt} = -g - bv$$

to model the motion of a body falling under the influence of gravity (g) and air resistance (bv). Here v is the velocity of the body at time t. (See pages 197–198.)

a) Solve the differential equation by separating variables, and obtain

$$v(t) = \frac{1}{b}(Ce^{-bt} - g),$$

where C is an arbitrary constant.

b) Now impose the initial condition $v(0) = 0$ (so the body starts its fall from rest) to determine the value of C. What is the formula for $v(t)$ now?

c) Exercise 21 on page 197 gives the solution to the initial value problem as

$$v(t) = \frac{g}{b}(2^{-bt/.69} - 1).$$

Reconcile this expression with the one you obtained in part (b) of this exercise.

d) The distance $x(t)$ that the body has fallen by time t is given by the integral

$$x(t) = \int_0^t v(t)\, dt, \qquad \text{because } \frac{dx}{dt} = v \text{ and } x(0) = 0.$$

Use your formula for $v(t)$ from part (b) to find $x(t)$.

7. a) *Supergrowth.* We have analyzed the differential equation

$$\frac{dQ}{dt} = kQ^p$$

when $p = 1.2$ (and, of course, when $p = 1$). Find a formula for the solution $Q(t)$ when $p = 2$. Your formula should contain an arbitrary constant C.

b) Add the initial condition $Q(0) = A$. This will fix the value of the constant C. What is the formula for $Q(t)$ when the initial condition is incorporated?

c) Your formula in part (b) should demonstrate that Q becomes infinite at some finite time $t = \tau$. When is τ? Your answer should be expressed in terms of the growth constant k and the initial population size A.

d) Suppose the values of k and A are known only imprecisely, and they could be in error by as much as 5%. That makes the value of τ uncertain. Which error causes the greater uncertainty, the error in k or the error in A? (See the discussion of error analysis for the supergrowth model on pages 190–191.)

*8. *General supergrowth.* Find the solution to the initial value problem

$$\frac{dQ}{dt} = kQ^p \qquad Q(0) = A$$

for *any* value of the power p. For which values of p does Q blow up to ∞ at a finite time $t = \tau$? What is τ?

Partial Fractions

9. Use the method of partial fractions to determine the values of α, β, and γ in the following equations.

a) $\dfrac{1}{(x-1)(x+2)} = \dfrac{\alpha}{x-1} + \dfrac{\beta}{x+2}$

b) $\dfrac{x}{(x-1)(x+2)} = \dfrac{\alpha}{x-1} + \dfrac{\beta}{x+2}$

c) $\dfrac{1}{x(x^2-1)} = \dfrac{\alpha}{x} + \dfrac{\beta}{x-1} + \dfrac{\gamma}{x+1}$

d) $\dfrac{x}{2x^2+3x+1} = \dfrac{\alpha}{2x+1} + \dfrac{\beta}{x+1}$

e) $\dfrac{1}{x(x^2+1)} = \dfrac{\alpha}{x} + \dfrac{\beta x+\gamma}{x^2+1}$ \qquad (Note that x^2+1 can't be factored.)

10. Find a formula for each of these indefinite integrals.

*a) $\displaystyle\int \frac{3\,dx}{(x-1)(x+2)}$

d) $\displaystyle\int \frac{x\,dx}{1-x^2}$

b) $\displaystyle\int \frac{5x+3}{(x-1)(x+2)}\,dx$

e) $\displaystyle\int \frac{1-u}{u^2-4}\,du$

c) $\displaystyle\int \frac{dt}{t(t^2-1)}$

f) $\displaystyle\int \frac{x^2+2x+1}{x(x^2+1)}\,dx$

11. Determine

a) $\displaystyle\int_2^3 \frac{3\,dx}{(x-1)(x+2)}$

c) $\displaystyle\int_0^{\pi/4} \frac{x\,dx}{1-x^2}$

b) $\displaystyle\int_2^4 \frac{dt}{t(t^2-1)}$

d) $\displaystyle\int_1^{\sqrt{3}} \frac{x^2+2x+1}{x(x^2+1)}\,dx$

12. Mirror the derivation of $\displaystyle\int \csc x\,dx$ to find $\displaystyle\int \sec x\,dx$.

13. Consider the particular logistic growth model defined by

$$\frac{dP}{dt} = .2P\left(1-\frac{P}{10}\right) \text{ lb/hr} \qquad P(0) = .5 \text{ lb}$$

(Compare this with the fermentation problems, pages 172–174.)
a) Obtain the formula for the solution to this initial value problem.
b) How large will P be after 3 hours? After 10 hours?
c) When will P reach one-half the carrying capacity—that is, for which t is $P = 5$ lb?

14. Derive the formula for $\int \sec x\,dx$ given on page 652, using methods similar to those used to find an antiderivative for the cosecant function.

⊳ 11.5 TRIGONOMETRIC INTEGRALS

The preceding sections have covered the main integration techniques and concepts likely to be needed by most users of calculus. These techniques, together with the numerical methods discussed in section 11.6, should be part of the basic tool kit of every practitioner of calculus. For those going on in physics or mathematics, there are additional methods, largely involving trigonometric functions in various ways, that are sometimes useful. The purpose of this section is to develop the most commonly used of these techniques.

Recall that there are only a few simple antiderivatives we can write down immediately by inspection. All nonnumerical integration techniques consist of finding transformations that will reduce some new class of integration problems to a class we already know how to solve. Once we have a new class of solvable problems, then we look for other classes of problems that can be reduced to this new class, and so on. The techniques we will be developing in this section involve ways of making such transformations through the use of basic trigonometric identities, typically in conjunction with integration by parts or by substitution. Before we proceed with the integration techniques, it will be helpful to list the trigonometric identities used.

Review of trigonometric identities.   The most frequently used identity is

$$\sin^2 x + \cos^2 x = 1,$$

and the equivalent form obtained by dividing through by $\cos^2 x$:

$$\tan^2 x + 1 = \sec^2 x,$$

and by $\sin^2 x$:

$$1 + \cot^2 x = \csc^2 x.$$

The only other identities you will need have already been encountered:

$$\sin 2x = 2 \sin x \cos x \quad \text{and} \quad \cos 2x = \cos^2 x - \sin^2 x$$

plus the two other forms of the second of these identities

$$\cos^2 x = \tfrac{1}{2}(1 + \cos 2x) \quad \text{and} \quad \sin^2 x = \tfrac{1}{2}(1 - \cos 2x).$$

Inverse Substitution

The method of substitution outlined in section 11.2 worked by taking a complicated integrand and breaking it down into simpler components, reducing the problem of finding an antiderivative for something in the form $f(G(x))g(x)$ to the problem of finding an antiderivative for f. In some cases, though, we go in the opposite direction: we have an integral $\int f(x)\,dx$ we want to find but can't evaluate directly. Instead, we can find a function $G(u)$ with derivative $g(u)$ such that we can find an antiderivative for $f(G(u))g(u)$. Since we know this integral is $F(G(u))$, we can now figure out what the desired function F must be. As with the earlier substitution techniques, this **inverse substitution** is conveniently expressed using differential notation.

> Success sometimes comes by making things more complicated

EXAMPLE 1.   Suppose we want to evaluate

$$\int \sqrt{4 - x^2}\,dx.$$

If we substitute $x = 2 \sin u$, so that $dx = 2 \cos u\, du$, look what happens:

$$\int \sqrt{4 - x^2}\, dx = \int \sqrt{4 - (2 \sin u)^2}\, 2 \cos u\, du$$

$$= \int \sqrt{4 - 4 \sin^2 u}\, 2 \cos u\, du$$

$$= \int 2 \sqrt{1 - \sin^2 u}\, 2 \cos u\, du$$

$$= \int 2 \cos u\, 2 \cos u\, du$$

$$= 4 \int \cos^2 u\, du.$$

But this is just an antiderivative we have already found in the exercises in section 11.3, namely,

$$\int \cos^2 u\, du = \frac{u}{2} + \frac{1}{4} \sin 2u + C$$

$$= \frac{u}{2} + \frac{1}{4} \cdot 2 \sin u \cos u + C$$

$$= \frac{1}{2}(u + \sin u \cos u) + C.$$

To find the desired antiderivative for the original function of x, we now replace u by its expression in terms of x by inverting the relationship: If $x = 2 \sin u$, then $\sin u = x/2$, and $u = \arcsin(x/2)$. As we found in section 11.1, drawing a picture expressing the relationship between x and u makes it easy to visualize the other trigonometric functions:

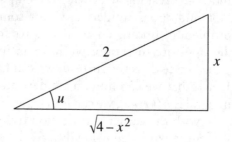

From the picture we see that

$$\cos u = \frac{\sqrt{4 - x^2}}{2} \qquad \text{and} \qquad \tan u = \frac{x}{\sqrt{4 - x^2}}.$$

We can now find an expression for the desired antiderivative in terms of x:

$$\int \sqrt{4 - x^2}\, dx = 2(u + \sin u \cos u) + C$$

$$= 2\left(\arcsin \frac{x}{2} + \frac{x}{2}\frac{\sqrt{4 - x^2}}{2}\right) + C$$

$$= 2\arcsin \frac{x}{2} + \frac{x\sqrt{4 - x^2}}{2} + C.$$

As usual, you should check this result by differentiating the right-hand side to see that you do obtain the integrand on the left.

Similar substitutions allow us to evaluate other integrals involving square roots of quadratic expressions. Here is a summary of useful substitutions. In each case, a is a positive real number.

$$
\begin{array}{lll}
\text{To transform} & a^2 - x^2 & \text{let} \quad x = a\sin u; \\
\text{To transform} & a^2 + x^2 & \text{let} \quad x = a\tan u; \\
\text{To transform} & x^2 - a^2 & \text{let} \quad x = a\sec u.
\end{array}
$$

EXAMPLE 2.   Integrate

$$\int \frac{dx}{\sqrt{x^2 + 9}}.$$

If we set $x = 3\tan u$, then $dx = 3\sec^2 u\, du$, and the integral becomes

$$\int \frac{dx}{\sqrt{x^2 + 9}} = \int \frac{3\sec^2 u\, du}{\sqrt{9\sec^2 u}}$$

$$= \int \sec u\, du$$

$$= \ln|\sec u + \tan u| + C\,(\text{as we saw in section 11.4}).$$

To express this in terms of x, we again draw a picture showing the relation between u and x:

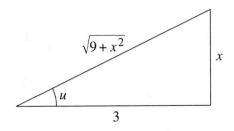

From this picture we see that $\sec u = \sqrt{9 + x^2}/3$. Therefore

$$\int \frac{dx}{\sqrt{x^2 + 9}} = \ln \left| \frac{\sqrt{9 + x^2}}{3} + \frac{x}{3} \right| + C$$

$$= \ln \left| \frac{\sqrt{9 + x^2} + x}{3} \right| + C = \ln \left| \sqrt{9 + x^2} + x \right| + C',$$

where $C' = C - \ln 3$ is a new constant. As usual, you should differentiate to check that this really is the claimed antiderivative.

EXAMPLE 3.   Evaluate

$$\int \frac{dx}{\sqrt{9x^2 - 16}}.$$

We first write $\sqrt{9x^2 - 16}$ as $\sqrt{9}\sqrt{x^2 - (16/9)} = 3\sqrt{x^2 - (16/9)}$. Use the substitution $x = (4/3)\sec u$, with $dx = (4/3)\sec u \tan u \, du$. This gives

$$\int \frac{dx}{\sqrt{9x^2 - 16}} = \int \frac{(4/3)\sec u \, \tan u \, du}{3\sqrt{(16/9)\sec^2 u - (16/9)}}$$

$$= \int \frac{(4/3)\sec u \, \tan u \, du}{3 \cdot (4/3)\tan u} = \frac{1}{3}\int \sec u \, du$$

$$= \frac{1}{3}\ln|\sec u + \tan u| + C.$$

Again we need a picture to relate x and u:

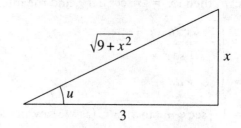

Thus $\tan u = \sqrt{9x^2 - 16}/4$, which gives

$$\int \frac{dx}{\sqrt{9x^2 - 16}} = \frac{1}{3}\ln|\sec u + \tan u| + C$$

$$= \frac{1}{3}\ln\left|\frac{3x}{4} + \frac{\sqrt{9x^2 - 16}}{4}\right| + C$$

$$= \frac{1}{3}\ln|3x + \sqrt{9x^2 - 16}| + C',$$

where $C' = C - (\ln 4)/3$.

As usual, you should differentiate this final expression to confirm that it really is the desired antiderivative.

Inverse Substitution and Definite Integrals

We saw on page 631 in section 11.2 how to use substitution to evaluate a definite integral. When we transformed an integral originally expressed in terms of a variable x into one expressed in terms of a variable u, the two integrals had the same numerical value. The same can be done with the inverse substitution technique we have just been considering. Let's see how this works. Suppose we start with a function $f(x)$ to be integrated over an interval $[a, b]$. If we only knew an antiderivative F for f, we could easily write

$$\int_a^b f(x)\, dx = F(b) - F(a),$$

as usual. In the examples we've just been considering, we found the antiderivative for f by making a substitution $x = G(u)$ for some function G and then finding an antiderivative for $f(G(u))g(u)$, where $G' = g$. This antiderivative we know is $F(G(u))$, where F is the function we are trying to find. We were able to obtain F by replacing u by its expression in x. To do this we needed to find the inverse function G^{-1} for G, so that $x = G(u)$ was equivalent to $u = G^{-1}(x)$, and $F(x) = F(G(G^{-1}(x)))$. It is this last step we can eliminate in calculating definite integrals.

If we want x to go from a to b, what must u do? What interval of u values will get transformed to this interval of x values by G? The value of u such that $G(u) = a$ is $A = G^{-1}(a)$. Similarly, the u value that gets transformed to b is $B = G^{-1}(b)$. Thus under the substitution $x = G(u)$, as u goes from A to B, x will go from a to b. Now look at the corresponding definite integral:

$$\int_A^B f(G(u))\, g(u)\, du = F(G(u)) \Big|_A^B$$
$$= F(G(B)) - F(G(A))$$
$$= F(G(G^{-1}(b))) - F(G(G^{-1}(a)))$$
$$= F(b) - F(a),$$

which is just the desired value of the original definite integral. To summarize:

> If we make the substitution $x = G(u)$, then
> $$\int_a^b f(x)\, dx = F(b) - F(a) = \int_A^B f(G(u))\, g(u)\, du,$$
> where $A = G^{-1}(a)$ and $B = G^{-1}(b)$.

Let's look back at a couple of the preceding examples to see how this works.

EXAMPLE 4.   Evaluate

$$\int_{-2}^{2} \sqrt{4 - x^2}\,dx.$$

In Example 1 we found an antiderivative for this function by making the substitution $x = 2\sin u = G(u)$. The inverse function is then $G^{-1}(x) = \arcsin(x/2)$. To get x to go from -2 to 2, u must go from $G^{-1}(-2) = \arcsin(-1) = -\pi/2 = A$ to $G^{-1}(2) = \arcsin 1 = \pi/2 = B$. Then to evaluate the integral from $x = -2$ to $x = 2$, we only need to evaluate the antiderivative we found for the u integral between $u = -\pi/2$ and $u = \pi/2$.

$$\int_{-2}^{2} \sqrt{4 - x^2}\,dx = 2(u + \sin u \cos u)\Big|_{-\pi/2}^{\pi/2} = \pi - (-\pi) = 2\pi.$$

(Note that this is just half the area of a circle of radius 2. How could we have anticipated this result from the form of the problem?)

EXAMPLE 5.   Suppose we wanted the integral

$$\int_{0}^{3} \frac{dx}{\sqrt{x^2 + 9}}.$$

In Example 2 we found an antiderivative by letting $x = 3\tan u$. Here $G^{-1}(x) = \arctan(x/3)$. For x to go from 0 to 3, u must go from 0 to $\arctan 1 = \pi/4$. Using the u-antiderivative we found in Example 2, we have

$$\int_{0}^{3} \frac{dx}{\sqrt{x^2 + 9}} = \ln|\sec u + \tan u|\Big|_{0}^{\pi/4}$$

$$= \ln|\sqrt{2} + 1| - \ln|1 + 0| = \ln(\sqrt{2} + 1).$$

Completing the Square

Integrands involving terms of the form $Ax^2 + Bx + C$ can always be put in the form $A(u^2 \pm b^2)$ for a suitable variable u and constant b. The technique for doing this is the standard method of **completing the square**:

$$Ax^2 + Bx + C = A\left(x^2 + \frac{B}{A}x\right) + C$$

$$= A\left(x^2 + \frac{B}{A}x + \frac{B^2}{4A^2}\right) + C - \frac{B^2}{4A^2}$$

$$= A\left(x + \frac{B}{2A}\right)^2 + \frac{4AC - B^2}{4A^2}.$$

The substitutions

$$u = x + \frac{B}{2A} \quad \text{and} \quad b = \frac{\sqrt{|4AC - B^2|}}{2A}$$

then transform the problem to a form where we can use the techniques already developed. The following examples should make this clear.

EXAMPLE 6.   Consider the integral

$$\int \frac{dx}{x^2 + 4x + 5}.$$

This may not immediately remind us of anything we've seen before. But if we rewrite it in the form

$$\int \frac{dx}{(x^2 + 4x + 4) + 1} = \int \frac{dx}{(x + 2)^2 + 1},$$

it now begins to resemble something involving an arctangent. In fact, if we make the substitution $u = x + 2$, so $du = dx$, we can write

$$\int \frac{dx}{x^2 + 4x + 5} = \int \frac{du}{u^2 + 1}$$
$$= \arctan u + C$$
$$= \arctan(x + 2) + C.$$

EXAMPLE 7.   The technique of completing the square even works for expressions we could have factored directly, if we had noticed:

$$\int \frac{dx}{x^2 + 4x + 3} = \int \frac{dx}{(x + 2)^2 - 1}$$
$$= \int \frac{dx}{(x + 2 - 1)(x + 2 + 1)}$$
$$= \int \frac{dx}{(x + 1)(x + 3)}$$
$$= \frac{1}{2} \int \frac{dx}{x + 1} - \frac{1}{2} \int \frac{dx}{x + 3}$$
$$= \frac{1}{2} \ln \left| \frac{x + 1}{x + 3} \right| + C.$$

EXAMPLE 8.   Evaluate

$$\int \frac{dx}{\sqrt{6x - x^2}}.$$

Note that $6x - x^2 = -(x^2 - 6x) = -(x - 3)^2 + 9 = 9 - (x - 3)^2$. If we now substitute $x - 3 = 3u$, with $dx = 3\,du$, we get

$$
\int \frac{dx}{\sqrt{6x - x^2}} = \int \frac{dx}{\sqrt{9 - (x - 3)^2}}
$$

$$
= \int \frac{3\,du}{\sqrt{9 - 9u^2}} = \int \frac{du}{\sqrt{1 - u^2}}
$$

$$
= \arcsin u + C
$$

$$
= \arcsin \frac{x - 3}{3} + C.
$$

Trigonometric Polynomials

A **trigonometric polynomial** is any sum of constant multiples of products of trigonometric functions. The preceding techniques have shown some cases where such trigonometric polynomials can arise, even though the original problem had no apparent reference to trigonometric functions. There are many different ways of breaking trigonometric polynomials down into special cases which can then be integrated. We will develop one way which has the virtue of using few special cases, so that it can be used fairly automatically. It also introduces a powerful tool—that of **reduction formula**—which can be used to generate mathematical results interesting in their own right. One example is the striking representation of π derived in section 12.1. Other examples are developed in the exercises at the end of this section.

Since every trigonometric function is expressible in terms of sines and cosines, any trigonometric polynomial can be written as a sum of terms of the form $c \sin^m x \cos^n x$ where c is a constant and m and n are integers—positive, negative, or 0. For instance, $5 \sec^2 x \tan^5 x$ can be rewritten as $5 \sin^5 x \cos^{-7} x$. To find antiderivatives for trigonometric polynomials, it therefore suffices to be able to evaluate integrals of the form

$$
\int \sin^m x \cos^n x \, dx.
$$

We will see how to find antiderivatives for functions of this sort by breaking the problem into a series of special cases:

Category I   Either $m \geq 0$ or $n \geq 0$ (or both)

    Case 1 $m = 1$ or $n = 1$

    Case 2 $m = 0$ or $n = 0$

Category II m and n both negative

Category I: Either m ≥ 0 or n ≥ 0 (or both)

Assume for the sake of explicitness that $m \geq 0$. We can then use the identity $\sin^2 x = 1 - \cos^2 x$ to replace $\sin^m x$ entirely by cosine terms if m is even, or to replace all but one of the sine terms by cosines if m is odd. A similar replacement can be made if $n \geq 0$.

EXAMPLE 9

$$\begin{aligned}
\sin^4 x \, \cos^6 x &= (1 - \cos^2 x)^2 \cdot \cos^6 x \\
&= (1 - 2 \cos^2 x + \cos^4 x) \cdot \cos^6 x \\
&= \cos^6 x - 2 \cos^8 x + \cos^{10} x.
\end{aligned}$$

(Note that in this example we could just as well have expressed $\cos^6 x$ entirely in terms of $\sin x$.)

EXAMPLE 10

$$\begin{aligned}
\sin^3 x \, \cos^{-8} x &= \sin x \cdot (1 - \cos^2 x) \cdot \cos^{-8} x \\
&= \sin x \, \cos^{-8} x - \sin x \, \cos^{-6} x.
\end{aligned}$$

EXAMPLE 11

$$\begin{aligned}
\sin^{-7} x \, \cos^7 x &= \sin^{-7} x \cdot (1 - \sin^2 x)^3 \cdot \cos x \\
&= \sin^{-7} x \, \cos x - 3 \sin^{-5} x \, \cos x + 3 \sin^{-3} x \, \cos x \\
&\quad - \sin^{-1} x \, \cos x.
\end{aligned}$$

We can thus reduce any problem in Category I to one of two special cases:

Case 1 $m = 1$ or $n = 1$
Case 2 $m = 0$ or $n = 0$

We will now see how to find antiderivatives for these cases.

CASE 1: $m = 1$ OR $n = 1$.   Since the two possibilities are analogous, we will consider the case with $n = 1$. Then m can be any real number at all, not necessarily an integer. We make the substitution $u = \sin x$, so that $du = \cos x \, dx$, and

$$\int \sin^m x \, \cos x \, dx = \int u^m \, du = \begin{cases} \dfrac{1}{m+1} u^{m+1} + C & \text{if } m \neq -1 \\[2mm] \ln|u| + C & \text{if } m = -1. \end{cases}$$

Replacing u by its expression in x we have the antiderivative:

$$\int \sin^m x \cos x \, dx = \begin{cases} \dfrac{1}{m+1}\sin^{m+1} x + C & \text{if } m \neq -1 \\[2ex] \ln|\sin x| + C & \text{if } m = -1. \end{cases}$$

The antiderivative of the cotangent

REMARK. The instance $m = -1$ in this case is worth singling out, as it gives us an antiderivative for $\cot x$:

$$\int \cot x \, dx = \int \frac{\cos x}{\sin x} dx = \ln|\sin x| + C.$$

Integrals where $m = 1$ are handled in a completely analogous fashion. You should check that

$$\int \cos^n x \sin x \, dx = \begin{cases} \dfrac{-1}{n+1}\cos^{n+1} x + C & \text{if } n \neq -1 \\[2ex] -\ln|\cos x| + C & \text{if } n = -1. \end{cases}$$

The antiderivative of the tangent

REMARK. Notice that $n = -1$ gives us an antiderivative for $\tan x$:

$$\int \tan x \, dx = \int \frac{\sin x}{\cos x} dx = -\ln|\cos x| + C.$$

CASE II: $m = 0$ OR $n = 0$. Again the two possibilities are analogous, so we will look at instances where $n = 0$. There are a number of clever ways for dealing with antiderivatives of functions of this form, many of them depending on special subcases according to whether m is even or odd, positive or negative, etc. We will develop a single method which deals with all cases in the same way.

Think of $\sin^n x$ as $\sin^{n-1} x \cdot \sin x$ and use integration by parts, with

$$F(x) = \sin^{n-1} x \qquad \text{and} \qquad g(x) = \sin x.$$

Then

$$f(x) = (n-1)\sin^{n-2} x \cos x \qquad \text{and} \qquad G(x) = -\cos x.$$

Therefore

$$\int \sin^n x \, dx = -\sin^{n-1} x \cos x + (n-1)\int \sin^{n-2} x \cos^2 x \, dx.$$

Now since $\cos^2 x = 1 - \sin^2 x$, we can rewrite the integral on the right-hand side as

$$\int \sin^{n-2} x \, \cos^2 x \, dx = \int \sin^{n-2} x \, dx - \int \sin^n x \, dx$$

—an expression involving the original integral we are trying to evaluate! If we now substitute this expression in our original equation and bring all the terms involving $\sin^n x$ over to the left-hand side, we have

$$n \int \sin^n x \, dx = -\sin^{n-1} x \, \cos x + (n-1) \int \sin^{n-2} x \, dx,$$

so that

$$\int \sin^n x \, dx = \frac{-1}{n} \sin^{n-1} x \, \cos x + \frac{n-1}{n} \int \sin^{n-2} x \, dx.$$

We thus have a **reduction formula** which reduces the problem of finding an antiderivative for $\sin^n x$ to the problem of finding an antiderivative for $\sin^{n-2} x$. This in turn can be reduced to finding an antiderivative for $\sin^{n-4} x$, and so on, until we get down to having to find an antiderivative for $\sin x$ (if n is odd), or for 1 (if n is even).

EXAMPLE 12

$$\int \sin^5 x \, dx = \frac{-1}{5} \sin^4 x \, \cos x + \frac{4}{5} \int \sin^3 x \, dx$$

$$= \frac{-1}{5} \sin^4 x \, \cos x + \frac{4}{5} \left(\frac{-1}{3} \sin^2 x \, \cos x + \frac{2}{3} \int \sin x \, dx \right)$$

$$= \frac{-1}{5} \sin^4 x \, \cos x - \frac{4}{15} \sin^2 x \, \cos x - \frac{8}{15} \cos x + C.$$

Check this answer by taking the derivative of the right-hand side. To show that this derivative really is equal to the integrand on the left, you will need to express all the cosines in terms of sines.

EXAMPLE 13.   If we let $n = 2$, we quickly get the antiderivative for $\sin^2 x$ that we've needed at several points already:

$$\int \sin^2 x \, dx = \frac{-1}{2} \sin x \, \cos x + \frac{1}{2} \int 1 \, dx$$

$$= \frac{x}{2} - \frac{1}{2} \sin x \, \cos x.$$

In its current form, the reduction formula works best for $n > 0$

The reduction formula as stated is most convenient for $n > 0$, although it is true for any number $n \neq 0$. For if $n < 0$, though, we want to *increase* the exponent, replacing a problem of finding an antiderivative for $\sin^n x$ by a problem where the exponent is less negative. We can do this by rearranging the formula as

$$\int \sin^{n-2} x \, dx = \frac{n}{n-1} \int \sin^n x \, dx + \frac{1}{n-1} \sin^{n-1} x \cos x.$$

Since we are interested in negative exponents, call $n - 2$ by a new name, $-k$. But if $n - 2 = -k$, then $n = -k + 2$, and we can rewrite our formula as

The reduction formula for negative exponents

$$\int \sin^{-k} x \, dx = -\frac{1}{k-1} \sin^{-(k-1)} x \cos x + \frac{k-2}{k-1} \int \sin^{-(k-2)} x \, dx.$$

With this formula we can reduce the problem of finding an antiderivative for $\sin^{-k} x$ to the problem of finding an antiderivative for $\sin^{-k+2} x$. This in turn can be reduced to finding an antiderivative for $\sin^{-k+4} x$, and so on, until we get up to having to find an antiderivative for $\sin^{-1} x$ (if k is odd), or for 1 (if k is even). All we need, then, is an antiderivative for $\sin^{-1} x$. But $\sin^{-1} x = \csc x$, and in section 11.4 (pages 651–652) we found that

$$\int \csc x \, dx = \int \sin x \, dx = -\ln|\csc x + \cot x| + C.$$

We can now handle antiderivatives for any negative integer exponent of the sine function.

EXAMPLE 14.   We can check this formula by trying $k = 2$, which will give us the antiderivative of $\csc^2 x$:

$$\int \csc^2 x \, dx = \int \sin^{-2} x \, dx$$

$$= -\frac{1}{1} \sin^{-1} x \cos x + \frac{0}{1} \int \sin^0 x \, dx$$

$$= -\sin^{-1} x \cos x + C = -\cot x + C,$$

as it should.

EXAMPLE 15

$$\int \sin^{-3} x \, dx = -\frac{1}{2} \sin^{-2} x \cos x + \frac{1}{2} \int \sin^{-1} x \, dx$$

$$= \frac{1}{2}(-\sin^{-2} x \cos x - \ln|\csc x + \cot x|) + C.$$

In the exercises you are asked to derive the following reduction formulas for the cosine function:

$$\int \cos^m x \, dx = \frac{1}{m} \cos^{m-1} x \, \sin x + \frac{m-1}{m} \int \cos^{m-2} x \, dx.$$

and

$$\int \cos^{-m} x \, dx = \frac{1}{m-1} \cos^{-m+1} x \, \sin x + \frac{m-2}{m-1} \int \cos^{-m+2} x \, dx.$$

Category II: Both $m < 0$ and $n < 0$

If we divide the identity $\sin^2 x + \cos^2 x = 1$ by $\sin^2 x \cos^2 x$, we get the identity

$$\sin^{-2} x + \cos^{-2} x = \sin^{-2} x \cos^{-2} x.$$

Next we will see how to use this identity to express anything of the form $\cos^{-r} x \sin^{-s} x$ (where $r > 0$ and $s > 0$) as a sum of terms of the form $\sin^{-h} x$, or $\cos^{-i} x$, or $\sin x \cos^{-j} x$, or $\sin^{-k} x \cos x$. Since we learned how to find antiderivatives for expressions like these in the previous cases, we will then be done.

The trick in transforming $\cos^{-r} x \sin^{-s} x$ to the desired form is to multiply by $(\cos x \cos^{-1} x)$ or $(\sin x \sin^{-1} x)$ as needed so that both the sine and the cosine terms appear to be *even* negative exponents. Then simply keep using the identity above until there's nothing left to use it on. The following three examples should make clear how the reduction then works.

EXAMPLE 16   (*r* and *s* both even already)

$$
\begin{aligned}
\sin^{-4} x \cos^{-6} x &= (\sin^{-2} x \cos^{-2} x)^2 \cos^{-2} x \\
&= (\sin^{-2} x + \cos^{-2} x)^2 \cos^{-2} x \\
&= (\sin^{-4} x + 2 \sin^{-2} x \cos^{-2} x + \cos^{-4} x) \cos^{-2} x \\
&= (\sin^{-4} x + 2(\sin^{-2} x + \cos^{-2} x) + \cos^{-4} x) \cos^{-2} x \\
&= \sin^{-4} x \cos^{-2} x + 2 \sin^{-2} x \cos^{-2} x + 2 \cos^{-4} x \\
&\quad + \cos^{-6} x \\
&= \sin^{-2} x (\sin^{-2} x + \cos^{-2} x) + 2(\sin^{-2} x + \cos^{-2} x) \\
&\quad + 2 \cos^{-4} x + \cos^{-6} x \\
&= \sin^{-4} x + \sin^{-2} x \cos^{-2} x + 2 \sin^{-2} x + 2 \cos^{-2} x \\
&\quad + 2 \cos^{-4} x + \cos^{-6} x \\
&= \sin^{-4} x + (\sin^{-2} x + \cos^{-2} x) + 2 \sin^{-2} x \\
&\quad + 2 \cos^{-2} x + 2 \cos^{-4} x + \cos^{-6} x \\
&= \sin^{-4} x + 3 \sin^{-2} x + 3 \cos^{-2} x + 2 \cos^{-4} x + \cos^{-6} x.
\end{aligned}
$$

While this process is tedious, it requires little thought—you simply replace $\sin^{-2}x \cos^{-2}x$ with $\sin^{-2}x + \cos^{-2}x$ at every opportunity until there is no negative-exponent sine term multiplying any negative-exponent cosine term. We will use this result to demonstrate how to deal with cases where either r or s (or both) is odd.

EXAMPLE 17 (r even and s odd)

$$\sin^{-4}x \cos^{-5}x = \sin^{-4}x \cos^{-6}x \cos x$$
$$= \sin^{-4}x \cos x + 3 \sin^{-2}x \cos x + 3 \cos^{-1}x$$
$$+ 2 \cos^{-3}x + \cos^{-5}x$$

EXAMPLE 18 (both r and s odd)

$$\sin^{-3}x \cos^{-5}x = \sin x \sin^{-4}x \cos^{-6}x \cos x$$
$$= \sin^{-3}x \cos x + 3 \sin^{-1}x \cos x + 3 \sin x \cos^{-1}x$$
$$+ 2 \sin x \cos^{-3}x + \sin x \cos^{-5}x$$

Exercises

1. Find the following antiderivatives (a is a constant> 0):

*a) $\displaystyle\int \frac{dx}{\sqrt{1 - 4x^2}}$

b) $\displaystyle\int \frac{dx}{\sqrt{1 + 4x^2}}$

c) $\displaystyle\int \frac{dx}{\sqrt{4 + x^2}}$

d) $\displaystyle\int \frac{x\,dx}{\sqrt{4 + x^2}}$

e) $\displaystyle\int \frac{dx}{(a^2 - x^2)^{3/2}}$

f) $\displaystyle\int \frac{dx}{4 + x^2}$

g) $\displaystyle\int \frac{x\,dx}{4 + x^2}$

*h) $\displaystyle\int \frac{dx}{x\sqrt{4 + x^2}}$

i) $\displaystyle\int \frac{x\,dx}{\sqrt{x^2 - a^2}}$

j) $\displaystyle\int \frac{dx}{(a^2 + x^2)^2}$

2. Evaluate the following integrals:

a) $\displaystyle\int_{1}^{-1} \frac{dx}{4 - x^2}$

b) $\displaystyle\int_{1}^{2} \sqrt{x^2 - 1}\,dx$

c) $\displaystyle\int_{0}^{\pi/3} x \sec^2 x\,dx$

d) $\displaystyle\int_{0}^{1} \frac{dx}{(2 - x^2)^{3/2}}$

e) $\displaystyle\int_{0}^{\infty} \frac{dx}{9 + x^2}$

f) $\displaystyle\int_{a}^{2a} x^3 \sqrt{x^2 - a^2}\,dx$

3. Sketch the ellipse

$$\frac{x^2}{a^2} + \frac{y^2}{b^2} = 1,$$

labeling the coordinates of the points where it crosses the x-axis and the y-axis. Prove that the area of this ellipse is πab.

4. Find the following antiderivatives:

*a) $\displaystyle\int \frac{dx}{\sqrt{x^2 - 2x - 8}}$

e) $\displaystyle\int \frac{x\,dx}{\sqrt{5 + 4x - x^2}}$

b) $\displaystyle\int \frac{dx}{x^2 + 6x + 10}$

f) $\displaystyle\int \frac{(2x + 7)\,dx}{4x^2 + 4x + 5}$

c) $\displaystyle\int \frac{dx}{\sqrt{x^2 + 6x + 8}}$

g) $\displaystyle\int \frac{(4x - 3)\,dx}{\sqrt{-x^2 - 2x}}$

d) $\displaystyle\int \frac{x\,dx}{x^2 + 4x + 5}$

h) $\displaystyle\int \frac{dx}{(a^2 - x^2 - 2x)^2}$

5. a) If $x = a \sec u$, where a is a constant, draw a right triangle containing an angle u with lengths of sides specified to reflect this relation between x, a, and u.

b) What is $\sin u$?

c) What is $\cos u$?

*d)** What is $\tan u$?

6. Evaluate the following:

a) $\displaystyle\int \frac{dx}{\sin x \, \cos x}$

f) $\displaystyle\int \tan^5 x \, dx$

b) $\displaystyle\int \cos^3 x \, \sin^{-4} x \, dx$

g) $\displaystyle\int \frac{\sin^3 5x \, dx}{\sqrt[3]{\cos 5x}}$

c) $\displaystyle\int \csc^4 x \, \cot^2 x \, dx$

h) $\displaystyle\int \frac{\cos^3(\ln x)\,dx}{x}$

d) $\displaystyle\int \sin 3x \, \cot 3x \, dx$

i) $\displaystyle\int \sec^4 x \, \ln(\tan x) \, dx$

e) $\displaystyle\int_0^{\pi/2} \sin^n x \, \cos^3 x \, dx$

j) $\displaystyle\int_0^{a/2} \frac{dx}{(a^2 - x^2)^{3/2}}$

7. Use the analysis of Example 17 (page 670) to find an antiderivative for $\sin^{-4} x \, \cos^{-5} x$.

Reduction Formulas

8. Derive the reduction formulas for the cosine function given on page 669.

9. a) By writing $\tan^n x = \tan^{n-2} x (\sec^2 x - 1)$, get a reduction formula which expresses $\int \tan^n x \, dx$ in terms of $\int \tan^{n-2} x \, dx$.

b) Use this evaluation formula to find $\int \tan^6 x \, dx$.

c) Show that

$$\int_0^{\pi/4} \tan^n x \, dx = \begin{cases} \dfrac{1}{n-1} - \dfrac{1}{n-3} + \cdots \pm \dfrac{1}{3} \mp 1 \pm \pi/4 & \text{if } n \text{ is even} \\[4mm] \dfrac{1}{n-1} - \dfrac{1}{n-3} + \cdots \pm \dfrac{1}{4} \mp \dfrac{1}{2} \pm \dfrac{1}{2} \ln 2 & \text{if } n \text{ is odd.} \end{cases}$$

d) Give a clear argument why

$$\lim_{n \to \infty} \int_0^{\pi/4} \tan^n x \, dx = 0.$$

e) Prove that

$$\lim_{k \to \infty} \left(1 - \frac{1}{3} + \frac{1}{5} - \cdots \pm \frac{1}{2k+1} \right) = \frac{\pi}{4},$$

and

$$\lim_{k \to \infty} \left(1 - \frac{1}{2} + \frac{1}{3} - \cdots \pm \frac{1}{k} \right) = \ln 2$$

10. a) By writing $\sec^n x$ as $\sec^{n-2} x \, \sec^2 x$ and using integration by parts, get a reduction formula which expresses $\int \sec^n x \, dx$ in terms of $\int \sec^{n-2} x \, dx$.

b) Since $\sec x = \cos^{-1} x$, the formula you got in part (a) could also have been obtained from the reduction formula for cosines on page 669. Try it and see if the formulas are in fact the same.

11. a) Find a reduction formula expressing $\int x^n e^x \, dx$ in terms of $\int x^{n-1} e^x \, dx$.

b) Using the results of part (a), show that

$$\frac{1}{n!} \int_0^t x^n e^x \, dx = e^t \left(\frac{t^n}{n!} - \frac{t^{n-1}}{(n-1)!} + \frac{t^{n-2}}{(n-2)!} - \cdots \pm \frac{t^2}{2!} \mp t \pm 1 \right) \mp 1.$$

c) Explain why, for a fixed value of t,

$$\lim_{n \to \infty} \frac{1}{n!} \int_0^t x^n e^x \, dx = 0.$$

d) Prove that

$$\lim_{n \to \infty} \left(1 - t + \frac{t^2}{2!} - \frac{t^3}{3!} + \frac{t^4}{4!} - \cdots \pm \frac{t^n}{n!} \right) = e^{-t}.$$

12. a) Find a reduction formula expressing

$$\int \frac{dx}{(1 + x^2)^n}$$

in terms of

$$\int \frac{dx}{(1 + x^2)^{n-2}}.$$

You can do this using integration by parts, or you can use a trigonometric substitution.

b) What is the exact value of

$$\int_0^1 \frac{dx}{(1 + x^2)^{10}} ?$$

13. Our approach to integrating trigonometric polynomials was to express everything in the form $\sin^m x \cos^n x$. We can just as readily express everything in the form $\sec^j x \tan^k x$ (j and k integers—positive, negative, or 0), and develop our technique by dealing with various cases of this. See if you can work out the details, trying to parallel the approach developed in the text using sines and cosines as our basic functions.

➢ 11.6 SIMPSON'S RULE

This chapter has concentrated on formulas for antiderivatives, because a formula conveys compactly a lot of information. However, you must not lose sight of the fact that most antiderivatives cannot be found by such analytic methods. The integrand may be a data function, for instance, and thus have no formula. And even when the integrand is given by a formula, there may be no formula for the antiderivative itself. One possibility in such cases is to **approximate** such a function by a function—such as a polynomial—for which we can readily find an antiderivative. In chapter 10 we saw some methods for doing this. In chapter 12 Fourier series are introduced, providing another family of approximating functions for which antiderivatives can be readily obtained. Another approach is to find a desired definite integral using approximating rectangles, as we did in chapter 6.

In any case, numerical methods are inescapable, but accurate results require many calculations. This takes time—even on a modern high-speed computer. A numerical method is said to be **efficient** if it gets accurate results quickly, that is, with relatively few calculations. In chapter 6 we saw that *midpoint* Riemann

Return to numerical methods

Efficient numerical
integration

sums are much more efficient than left or right *endpoint* Riemann sums. We will look at these and other methods in detail in this section. The most efficient method we will develop is called Simpson's rule.

The Trapezoid Rule

We interpret the integral

$$\int_a^b f(x)\,dx$$

as the area under the graph $y = f(x)$ between $x = a$ and $x = b$. We interpret a Riemann sum as the total area of a collection of rectangles that approximate the area under the graph. The tops of the rectangles are level, and they represent the graph of a step function. Clearly, we get a better approximation to the graph by using slanted lines. They form the tops of a sequence of **trapezoids** that approximate the area under the graph.

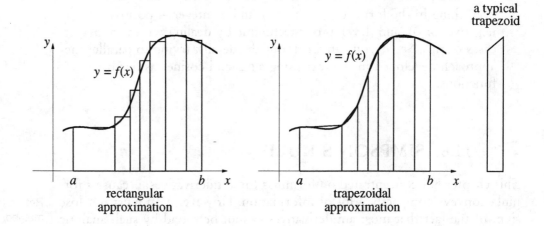

rectangular
approximation

trapezoidal
approximation

a typical
trapezoid

Replace rectangles by
trapezoids

Each rectangle is
sandwiched between
two rectangles...

Let's figure out the areas of these trapezoids. They are related in a simple way to the rectangles that we would construct at the right and left endpoints to calculate Riemann sums. To see the relation, let's take a closer look at a single trapezoid. It is sandwiched between two rectangles, one taller and one shorter. In our picture the height of the taller rectangle is f(right endpoint). We will call it the *right rectangle*. The height of the shorter rectangle is f(left endpoint). We will call it the *left rectangle*. For other trapezoids the left rectangle may be the taller one. In any case, the trapezoid is exactly half-way between the two rectangles in size, and thus its area is the *average* of the areas of the rectangles:

...whose average area
equals the area of the
trapezoid

$$\text{area trapezoid} = \tfrac{1}{2}(\text{area left rectangle} + \text{area right rectangle}).$$

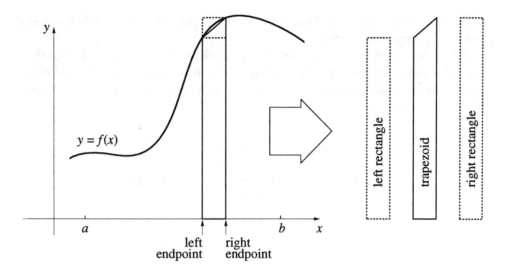

If we sum over all of the trapezoids, the areas of the right rectangles sum to the right Riemann sum, and similarly for the left rectangles. Thus

trapezoidal approximation $= \frac{1}{2}$ (left Riemann sum + right Riemann sum).

The trapezoidal approximation is the average of left and right Riemann sums

The pictures make it clear that the trapezoidal approximation should be significantly better than either a left or a right Riemann sum. To test this numerically, let's get numerical estimates for the integral

$$\int_1^3 \frac{1}{x}\, dx = \ln 3 = 1.098612288668.\ldots$$

(The relation between the trapezoid approximation and the left and right Riemann sums holds for any choice Δx_k of subintervals. However, we will use equal subintervals to make the calculations simpler.) Here is how our four main estimates compare when we use 100 subintervals.

Comparing approximations

$n = 100$	Approximation	Error
Right	1.09197525	6.63×10^{-3}
Left	1.10530858	-6.69×10^{-3}
Midpoint	1.09859747	1.48×10^{-5}
Trapezoidal	1.09864191	-2.90×10^{-5}

The figures in this table are calculated to eight decimal places, and the column marked *error* is the difference

$$1.09861228 - \text{approximation},$$

so that the error is negative if the approximation is too large. The left Riemann sum is too large, for instance.

Before we comment on the differences between the estimates, let's gather more data. Here are the calculations for 1000 subintervals. Note that the midpoint and trapezoidal approximations are more than 1000 times better than the left or right Riemann sums!

$n = 1000$	Approximation	Error
Right	1.09794591	6.663×10^{-4}
Left	1.09927925	-6.669×10^{-4}
Midpoint	1.09861214	1.4×10^{-7}
Trapezoidal	1.09861258	-2.9×10^{-7}

A surprise: the midpoint approximation is even better than the trapezoidal

We expected the trapezoidal approximation to be better than either the right or left Riemann sum. The surprising observation is that the midpoint Riemann sum is even better! In fact, it appears that the midpoint Riemann sum has only *half* the error of the trapezoidal approximation.

The figures below explain geometrically why the midpoint approximation is better than the trapezoidal. The first step is shown on the left. Take a midpoint rectangle (whose height is $f(\text{midpoint})$), and rotate the top edge around the midpoint until it is tangent to the graph of $y = f(x)$. Call this a **midpoint trapezoid.** Notice that the trapezoid has the same area as the rectangle.

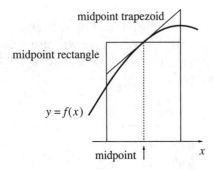
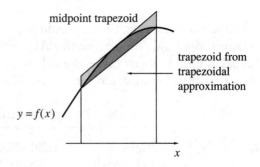

Shading depicts the errors

The second step is to compare the midpoint trapezoid to the one used in the trapezoidal approximation. This is done on the right. The error coming from the midpoint trapezoid is shaded light gray, while the error from the trapezoidal approximation is dark gray. The midpoint trapezoid is the better approximation. Since the midpoint rectangle has the same area as the midpoint trapezoid, we now see why the midpoint Riemann sum is more accurate than the trapezoidal approximation. This picture also explains why the errors of the two approximations have different signs, which we noticed first in the tables.

Simpson's Rule

Our goal is a calculation scheme whose error is as small as possible. The trapezoidal and the midpoint approximations are both good—but we can combine them to get something even better. Here is why. The tables and the figure on page 676 indicate that their *errors* have opposite signs, and the midpoint error is only about half the size of the trapezoidal error (in absolute value). Thus, if we form the sum

Combine good approximations . . .

$$2 \times \text{midpoint approximation} + \text{trapezoidal approximation},$$

then most of the error will cancel. Now this sum is approximately three times the value of the integral, because each term in it approximates the integral itself. Therefore, if we divide by three, then

. . . so that most of the error cancels

$$\tfrac{2}{3} \times \text{midpoint approximation} + \tfrac{1}{3} \times \text{trapezoidal approximation}$$

should be a superb approximation to the integral.

Let's try this approximation on our test integral

$$\int_1^3 \frac{1}{x}\,dx$$

with $n = 100$ subintervals. Using the numbers from the table on page 676, we obtain

$$\tfrac{2}{3} \times 1.09859747 + \tfrac{1}{3} \times 1.09864191 = 1.098612283,$$

which gives the integral accurate to eight decimal places. This method of approximating integrals is called **Simpson's rule.**

We can use the program RIEMANN to do the calculation. First calculate left and right Riemann sums, and take their average. That is the trapezoidal approximation. Then calculate the midpoint Riemann sum. Since

Using Riemann sums to carry out Simpson's rule

$$\text{trapezoid} = \tfrac{1}{2} \times \text{left} + \tfrac{1}{2} \times \text{right},$$

Simpson's rule reduces to this combination of left, right, and midpoint sums:

$$
\begin{aligned}
\tfrac{2}{3} \times \text{midpoint} &+ \tfrac{1}{3} \times \text{trapezoidal} \\
&= \tfrac{2}{3}\text{midpoint} + \tfrac{1}{3}\left(\tfrac{1}{2} \times \text{left} + \tfrac{1}{2} \times \text{right}\right) \\
&= \tfrac{2}{3} \times \text{midpoint} + \tfrac{1}{6} \times \text{left} + \tfrac{1}{6} \times \text{right} \\
&= \tfrac{1}{6}(4 \times \text{midpoint} + \text{left} + \text{right})
\end{aligned}
$$

> **Simpson's rule:**
>
> $$\int f(x)\,dx \approx \tfrac{1}{6}(\text{left sum} + \text{right sum} + 4 \times \text{midpoint sum})$$

The accuracy of Simpson's rule

You can get even more accuracy if you keep track of more digits in the left, right, and midpoint Riemann sums. For example, if you estimate

$$\int_1^3 \frac{1}{x}\,dx = \ln 3 = 1.098612288668\ldots$$

to 14 decimal places, you will get the following:

$$\begin{aligned}
\text{left:} \quad & 1.105\,308\,583\,647\,79 \\
\text{right:} \quad & 1.091\,975\,250\,314\,45 \\
\text{midpoint:} \quad & 1.098\,597\,475\,005\,31
\end{aligned}$$

When combined these give the estimate $1.098\,612\,288\,997\,3$, which differs from the true value by less than 3.3×10^{-10}. In other words, the calculation is actually correct to nine decimal places.

An error bound for Simpson's rule

It is possible to get a bound on the error produced by using Simpson's rule to estimate the value of

$$\int_a^b f(x)\,dx.$$

(See the discussion of error bounds in section 6.3.) Specifically,

$$\left| \int_a^b f(x)\,dx - \text{Simpson's rule} \right| \leq \frac{M(b-a)^5}{2880n^4},$$

where n is the number of subintervals used in the Riemann sums and M is a bound on the size of the fourth derivative of f:

$$|f^{(4)}(x)| \leq M \qquad \text{for all } a \leq x \leq b.$$

The crucial factor n^{-4}

The most important factor in the error bound is the n^4 that appears in the denominator. In our example $n = 100 = 10^2$, and this leads to the factor $1/n^4 = 10^{-8}$ in the error bound. As we saw, the actual error was less than 10^{-9}. Essentially, n is the number of computations we do, and error bound tells how many decimal places of accuracy we can count on. According to the error bound, a tenfold increase in the number of computations produces four more decimal places of accuracy. *That* is why Simpson's method is efficient.

Exercises

Because Simpson's rule is so efficient, we can use it to get accurate values of some of the fundamental constants of mathematics. For example, since

$$4 \cdot \int_0^1 \frac{dx}{1 + x^2} = 4\arctan(x)\Big|_0^1 = 4\arctan(1) = 4 \cdot \frac{\pi}{4} = \pi,$$

we can estimate the value of π by using Simpson's rule to approximate this integral.

1. Evaluate the expression above (including the factor of 4) using Simpson's rule with $n = 2, 4, 8,$ and 16. How accurate is each of these estimates of π; that is, how many decimal places of each estimate agree with the true value of π?

2. **a)** Over the interval $0 \le x \le 1$ it is true that

$$|f^{(4)}(x)| \le 96 \qquad \text{when} \qquad f(x) = \frac{4}{1 + x^2}.$$

(You don't need to show this, but how might you do it?) Use this bound to show that $n = 256 = 2^8$ will guarantee that you can find the first 10 decimals of π by using the method of the previous question.

   b) Show that if $n = 128 = 2^7$, then the error bound for Simpson's rule does *not* guarantee that you can find the first 10 decimals of π by the same method.

   c) Run Simpson's rule with $n = 2^7$ to estimate π. How many decimal places *are* correct? Does this surprise you? In fact the error bound is too timid: it says that the error is no larger than the bound it gives, but the actual error may be much smaller. From your work in part (a), which power of 2 is sufficient to get 10 decimal places accuracy?

3. ***a)** In section 6.3 (page 345) a left Riemann sum for

$$\int_0^1 e^{-x^2} dx$$

with 1000 equal subdivisions gave three decimal places accuracy. How many subdivisions n are needed to get that much accuracy using Simpson's rule? Let n be a power of 2. Start with $n = 1$ and increase n until three digits stabilize.

   b) If you use Simpson's rule with $n = 1000$ to estimate this integral, how many digits stabilize?

4. On page 329 a midpoint Riemann sum with $n = 10,000$ shows that

$$\int_1^3 \sqrt{1 + x^3}\, dx = 6.229959.\dots$$

How many subdivisions n are needed to get this much accuracy using Simpson's rule? Start with $n = 1$ and keep doubling it until seven digits stabilize.

11.7 IMPROPER INTEGRALS

The Lifetime of Light Bulbs

The lifetime of a light bulb is unpredictable

Ordinary light bulbs are supposed to burn about 700 hours, but of course some last longer while others burn out more quickly. It is impossible to know, in advance, the lifetime of a particular bulb you might buy, but it is possible to describe what happens to a large batch of bulbs.

Suppose we take a batch of 1000 light bulbs, start them burning at the same time, and note how long it takes each one to burn out. Let

$L(t)$ = fraction of bulbs that burn out before t hours.

Then $L(t)$ might have a graph that looks like this:

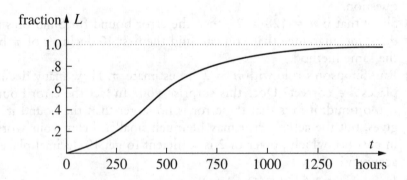

In this example, $L(400) \approx .5$, so about half the bulbs burned out before 400 hours. Furthermore, all but a few have burned out by 1250 hours.

Manufacturers are very concerned about the way the lifetime of light bulbs varies. They study the output of their factories on a regular basis. It is more common, though, for them to talk about the *rate* r at which bulbs burn out. The rate varies over time, too. In fact, in terms of L, r is just the derivative

The burnout rate

$$r(t) = L'(t) \quad \text{bulbs per hour.}$$

However, if we *start* with the rate r, then we get L as the integral

$$L(t) = \int_0^t r(s)\, ds.$$

This is yet another consequence of the fundamental theorem of calculus. The integral expression is quite handy. For example, the fraction of bulbs that burn out between $t = a$ hours and $t = b$ hours is

Lifetime is the integral of burnout rate . . .

$$L(b) - L(a) = \int_a^b r(s) \, ds.$$

We can even use the integral to say that all the bulbs burn out eventually:

$$L(t) = \int_0^t r(s) \, ds = 1 \quad \text{when } t \text{ is sufficiently large.}$$

In practice r is the average burnout rate for many batches of light bulbs, so we can't identify the precise moment when L becomes 1. All we can really say is

. . . but there is no upper limit to the lifetime

$$L(\infty) = \int_0^\infty r(s) \, ds = 1.$$

This is called an **improper integral,** because it cannot be calculated directly: its "domain of integration" is infinite. By definition, its value is obtained as a limit of ordinary integrals:

An integral is *improper* if the domain of integration is infinite

$$\int_0^\infty r(s) \, ds = \lim_{b \to \infty} \int_0^b r(s) \, ds.$$

The **normal density function** of probability theory provides us with another example of an improper integral. In a simple form, the function itself is

$$f(x) = \frac{1}{\sqrt{2\pi}} \exp\left(-\frac{x^2}{2}\right).$$

[Recall, $\exp(x) = e^x$.] If x is any *normally distributed quantity* whose mean is 0 and variance is 1, then the probability that a randomly chosen value of x lies between the numbers a and b is

$$\int_a^b f(x) \, dx.$$

Since the probability that x lies *somewhere* on the x-axis is 1, we have

The normal probability distribution involves an improper integral

$$\int_{-\infty}^\infty f(x) \, dx = 1.$$

This is an improper integral, and its value is defined by the limit

$$\int_{-\infty}^\infty f(x) \, dx = \lim_{b \to \infty} \int_{-b}^b f(x) \, dx.$$

In the exercises you will have a chance to evaluate this integral.

Evaluating Improper Integrals

An integral with an infinite domain of integration is only one kind of improper integral. A second kind has a finite domain of integration, but the integrand becomes infinite on that domain. For example,

$$\int_0^1 \frac{dx}{x} \quad \text{and} \quad \int_0^1 \ln x \, dx$$

are both improper in this sense. In both cases, the integrand becomes infinite as $x \to 0$. Because the difficulty lies at the endpoint 0, we define

$$\int_0^1 \frac{dx}{x} = \lim_{a \to 0} \int_a^1 \frac{dx}{x}.$$

More generally,

$$\int_a^b f(x) \, dx$$

is an improper integral if $f(x)$ becomes infinite at some point c in the interval $[a, b]$. In that case we define

$$\int_a^b f(x) \, dx = \lim_{q \to 0} \left(\int_a^{c-q} f(x) \, dx + \int_{c+q}^b f(x) \, dx \right).$$

In effect, we avoid the bad spot but "creep up" on it in the limit.

Indefinite integrals—that is, antiderivatives—can be a great help in evaluating improper integrals. Here are some examples.

EXAMPLE 1.   We can evaluate

$$\int_0^\infty e^{-x} \, dx$$

by noting first that $\int e^{-x} \, dx = -e^{-x}$. Therefore

$$\int_0^b e^{-x} \, dx = -e^{-x} \Big|_0^b = -e^{-b} - \left(-e^{-0} \right) = 1 - e^{-b}$$

and

$$\int_0^\infty e^{-x} \, dx = \lim_{b \to \infty} \int_0^b e^{-x} \, dx = \lim_{b \to \infty} \left(1 - e^{-b} \right) = 1.$$

EXAMPLE 2.   To evaluate $\int_0^1 \ln x \, dx$, we use the indefinite integral

$$\int \ln x \, dx = x \ln x - x.$$

Thus

$$\int_a^1 \ln x \, dx = x \ln x - x \Big|_a^1 = -1 - (a \ln a - a) = a - 1 - a \ln a.$$

By direct calculation (using a graphing package, for instance) we can find

$$\lim_{a \to 0} a \ln a = 0;$$

therefore

$$\int_0^1 \ln x \, dx = \lim_{a \to 0} \int_a^1 \ln x \, dx$$

$$= \lim_{a \to 0} a - 1 - a \ln a$$

$$= -1.$$

You should not assume that an improper integral always has a finite value, though. Consider the next example.

EXAMPLE 3.

$$\int_0^1 \frac{dx}{x} = \lim_{a \to 0} \int_a^1 \frac{dx}{x}$$

$$= \lim_{a \to 0} \ln(x) \Big|_a^1$$

$$= \lim_{a \to 0} (\ln(1) - \ln(a))$$

$$= \infty.$$

This is forced because $\lim_{a \to 0} \ln(a) = -\infty$, which you can see from the graph of the logarithm function.

Exercises

1. Find the value of each of the following improper integrals. (The value may be ∞.)

$$* \int_{-\infty}^0 e^x \, dx \qquad \int_0^\infty x e^{-x} \, dx$$

$$\int_1^\infty \frac{du}{u} \qquad \int_1^\infty \frac{du}{u^2}$$

$$* \int_0^1 \frac{dy}{y^2} \qquad \int_0^\infty \frac{x}{1 + x^2} \, dx$$

$$\int_0^{\pi/2} \tan x \, dx \qquad \int_1^3 \frac{x}{x^2 - 1} \, dx$$

2. Use the reduction formula for $\displaystyle\int \frac{dx}{(1+x^2)^n}$ you found on page 673 to find the exact value of $\displaystyle\int_0^\infty \frac{dx}{(1+x^2)^{10}}$.

The Normal Density Function

The next two questions concern the improper integral

$$\frac{1}{\sqrt{2\pi}} \int_{-\infty}^{\infty} e^{-x^2/2}\, dx$$

of the normal density function defined on page 681. The goal is to determine the value of this integral.

3. First, use RIEMANN to estimate the value of

$$\frac{1}{\sqrt{2\pi}} \int_{-b}^{b} e^{-x^2/2}\, dx$$

when b has the different values 1, 10, 100, and 1000. On the basis of these results, estimate

$$\lim_{b\to\infty} \frac{1}{\sqrt{2\pi}} \int_{-b}^{b} e^{-x^2/2}\, dx.$$

This gives one estimate of the value of the improper integral.

4. a) To construct a second estimate, begin by sketching the graph of the normal density function

$$f(x) = \frac{1}{\sqrt{2\pi}} e^{-x^2/2}$$

on an interval centered at the origin. Use the graph to argue that

$$\frac{1}{\sqrt{2\pi}} \int_{-b}^{b} e^{-x^2/2}\, dx = 2\left(\frac{1}{\sqrt{2\pi}} \int_{0}^{b} e^{-x^2/2}\, dx\right) = \sqrt{\frac{2}{\pi}} \int_{0}^{b} e^{-x^2/2}\, dx$$

and therefore

$$\frac{1}{\sqrt{2\pi}} \int_{-\infty}^{\infty} e^{-x^2/2}\, dx = \sqrt{\frac{2}{\pi}} \int_{0}^{\infty} e^{-x^2/2}\, dx.$$

b) Now consider the accumulation function

$$F(t) = \sqrt{\frac{2}{\pi}} \int_{0}^{t} e^{-x^2/2}\, dx.$$

We want to find $F(\infty) = \lim\limits_{t \to \infty} F(t)$. According to the fundamental theorem of calculus, $y = F(t)$ satisfies the initial value problem

$$\frac{dy}{dt} = \sqrt{\frac{2}{\pi}} \cdot e^{-t^2/2}, \qquad y(0) = 0.$$

Use a differential equation solver (e.g., PLOT) to graph the solution $y = F(t)$ to this problem. From the graph determine

$$F(\infty) = \lim_{t \to \infty} F(t).$$

c) Do your results in part (b) and exercise 3 agree? Do they agree with the value the text claims for the improper integral? (Remember, the value is the probability that a randomly chosen number will lie *somewhere* on the number line between $-\infty$ and $+\infty$.)

The Gamma Function

The **factorial function** is defined for a positive integer n by the formula

$$n! = n \cdot (n - 1) \cdot (n - 2) \cdots 3 \cdot 2 \cdot 1.$$

For example, $1! = 1$, $2! = 2$, $3! = 6$, $4! = 24$, and $10! = 3628800$. The factorial function is used often in diverse mathematical contexts, but its use is sometimes limited by the fact that it is defined only for positive integers. How might the function be defined on an expanded domain, so that we could deal with expressions like $\frac{1}{2}!$, for example? The *gamma function* answers this question.

The **gamma function** $\Gamma(x)$ is defined by the improper integral

$$\Gamma(x) = \int_0^\infty e^{-t} t^{x-1} \, dt.$$

(Notice that t is the active variable in this integral; while the integration is being performed, x is treated as a constant.)

5. Show that $\Gamma(1) = 1$.

6. Using integration by parts, show that $\Gamma(x + 1) = x \cdot \Gamma(x)$. You may use the fact that

$$\frac{t^p}{e^t} \to 0 \text{ as } t \to +\infty.$$

The property $\Gamma(x + 1) = x \cdot \Gamma(x)$ makes the gamma function like the factorial function, because

$$(n + 1)! = (n + 1) \cdot \underbrace{n \cdot (n - 1) \cdot (n - 2) \cdots 3 \cdot 2 \cdot 1}_{n!} = (n + 1) \cdot n!.$$

Notice there is a slight difference, though. We explore this now.

7. Using the property $\Gamma(x + 1) = x \cdot \Gamma(x)$, calculate $\Gamma(2)$, $\Gamma(3)$, $\Gamma(4)$, $\Gamma(5)$, and $\Gamma(6)$. On the basis of this evidence, fill in the blank:

$$\text{For a positive integer } n, \qquad \Gamma(n) = \underline{\hspace{1cm}}!$$

Using this relation, give a computable meaning to the expression $\frac{1}{2}!$.

8. Estimate the value of $\Gamma(1/2)$. (Exercises 3 and 4 above offer two ways to estimate the value of an improper integral.)

9. In fact, $\Gamma(1/2) = \sqrt{\pi}$ exactly. You can show this by employing several of the techniques developed in this chapter. Start with

$$\Gamma(1/2) = \int_0^\infty e^{-t} t^{-1/2} \, dt.$$

a) Make the substitution $u = (2t)^{1/2}$ and show that the integral becomes

$$\Gamma(1/2) = \sqrt{2} \int_0^\infty e^{-u^2/2} \, du.$$

b) From exercise 3 you know

$$\sqrt{2/\pi} \int_0^\infty e^{-u^2/2} \, du = 1.$$

(Check this.) Now, using some algebra, show $\Gamma(1/2) = \sqrt{\pi}$.

c) Compare your estimate for $\Gamma(1/2)$ from exercise 7 with the exact value $\sqrt{\pi}$.

10. a) Determine the exact values of $\Gamma(3/2)$ and $\Gamma(5/2)$.

b) In exercise 6 you gave a meaning to the expression $\frac{1}{2}!$; can you now give it an exact value?

▷ 11.8 CHAPTER SUMMARY

Main Ideas

- A function F is an **antiderivative** of f if $F' = f$. *Every* antiderivative of f is equal to $F + C$ for some appropriately chosen constant C. We write $\int f(x) \, dx = F(x) + C$.

- Differentiation rules for combinations of functions yield corresponding antidifferentiation rules. Among these are the **constant multiple** and **addition** rules. The chain rule for differentiation corresponds to **integration by substitution**. The product rule for differentiation corresponds to **integration by parts**.

- The derivative of a function and of its inverse are reciprocals. When $y = f(t)$ and $t = g(y)$ are inverses:

$$\frac{dt}{dy} = \frac{1}{dy/dt}.$$

- In some cases, the method of **separation of variables** can be used to find a *formula* for the solution of a differential equation.

- A numerical method for estimating an integral is **efficient** if it gets accurate results with relatively few calculations. The **trapezoidal approximation** is the average of a left and a right *endpoint* Riemann sum and is more efficient than either. *Midpoint* Riemann sums are even more efficient than trapezoidal approximations.

- The most efficient method developed in this chapter is **Simpson's rule.** Simpson's rule approximates an integral by

$$\int_a^b f(x)\,dx \approx \tfrac{1}{6}\,(\text{left sum} + \text{right sum} + 4 \times \text{midpoint sum}).$$

- An **improper integral** is one that cannot be calculated directly. The problem may be that its "domain of integration" is infinite or that the integrand becomes infinite on that domain. Its value is obtained as a limit of ordinary integrals.

Expectations

- You should be able to find antiderivatives of basic functions.
- You should be able to find antiderivatives of combinations of functions using the **constant multiple** and **addition** rules, as well as the **method of substitution** and **integration by parts.**
- You should be able to rewrite an integrand given as a quotient using the method of **partial fractions.**
- You should be able to express the derivative of an invertible function in terms of the derivative of its inverse. In particular, you should be able to differentiate the **arctangent, arcsine,** and **arccosine** functions.
- You should be able to solve a differential equation using the method of **separation of variables.**
- You should be able to adapt the program RIEMANN to approximate integrals using the **trapezoid rule** and **Simpson's rule.**
- You should be able to find the value of an **improper integral** as the limit of ordinary integrals.

CHAPTER 12

CASE STUDIES

To enable you to explore further the ways the concepts of calculus are used as analytical tools in scientific and mathematical investigations, this chapter presents four extended case studies. The four can be studied separately, although the first two and the last two are loosely linked.

Stirling's Formula: As an example of the way many of the ideas—Taylor series, numerical integration, reduction formulas, limits—developed in the earlier chapters of this book can be used in a tightly reasoned argument to produce some powerful mathematical insights, in the first section we derive a famous formula approximating $n!$. This formula is then applied to the binomial probability distribution.

The Poisson Distribution: Section 12.2 continues the probability theme by developing the Poisson distribution and using it to study the frequency of radioactive decay events.

The Power Spectrum: Section 12.3 builds on the study of periodicity begun in chapter 7. We develop the Fourier transform, a basic tool in the sciences for detecting the relative strength of periodic components in a noisy data set.

Fourier Series: Section 12.4 expands on some of the ideas in chapter 11. Here we develop tools for approximating functions over intervals using sums of sine and cosine terms. This is an extensively used method in a wide range of disciplines, from thermodynamics to music synthesis.

➤ 12.1 STIRLING'S FORMULA

Factorials in probability

Given a positive integer n, we define $n!$—pronounced n *factorial*—by the rule $n! = 1 \cdot 2 \cdot 3 \cdots (n-1) \cdot n$. This is a convenient concept which occurs in a number of settings, particularly combinatorial and probabilistic ones. For instance, the probability of getting exactly n heads out of $2n$ tosses of a coin turns out to be

$$\frac{(2n)!}{2^{2n}(n!)^2}.$$

n! is difficult to calculate

Unfortunately, evaluating $n!$ for values of n that are at all large is cumbersome at best. Although many calculators will compute factorials, few of them can handle numbers as large as 1000!. Even when we can evaluate $n!$, we are often as interested in the asymptotic behavior of a certain expression as much as in its exact value for specific n. For instance, using methods we develop below, it turns out that the above expression for the probability of n heads in $2n$ tosses is very close to $1/\sqrt{\pi n}$, with the approximation being more accurate the larger n is. In fact, for $n \geq 8$, the approximation is good to two places; for $n \geq 25$, the approximation gives three-place accuracy.

In his book *Methodus Differentialis* (1730), the British mathematician James Stirling published the following approximation, now know as **Stirling's formula,** for the factorial operator:

$$n! \sim \sqrt{2\pi}\, n^{n+1/2} e^{-n}.$$

While the right-hand side may look much more complicated than the left, think which one you would rather evaluate for, say, $n = 100$. To see how good this approximation is, here are some comparisons:

n	$n!$	Stirling's Approximation
2	2	1.9190
10	3,628,800	3,598,695.6
50	3.0414×10^{64}	3.0363×10^{64}
100	9.3326×10^{157}	9.3248×10^{157}
1000	4.02387×10^{2567}	4.02354×10^{2567}
10000	$2.84626 \times 10^{35659}$	$2.84624 \times 10^{35659}$

As an example of the way elementary ideas in calculus can be used to derive powerful and subtle results, we will outline a derivation of Stirling's approximation for $n!$. You should write up your own summary of this proof, filling in the gaps in the text below.

We will work in two stages. In the first stage, we will show that

$$n! \sim cn^{n+1/2}e^{-n}.$$

for some constant c. In the second stage we will show that this constant is actually $\sqrt{2\pi}$.

Stage One: Deriving the General Form

We first observe that

$$\ln(n!) = \ln 1 + \ln 2 + \cdots + \ln n.$$

It turns out to be easier to prove things about this logarithmic form. In fact, we will deal most easily with

$$A_n = \ln 1 + \ln 2 + \cdots + \ln(n-1) + \tfrac{1}{2}\ln n.$$

Thus $\ln(n!) = A_n + \tfrac{1}{2}\ln n$. Even though $\ln 1 = 0$, it will be useful to retain the term in the expression for A_n.

We will find upper and lower bounds for A_n [and hence for $\ln(n!)$] by approximating the area under the curve $y = \ln x$ by certain inscribed and circumscribed trapezoids. We will then use these bounds to predict the asymptotic behavior of A_n for large values of n.

The upper bound. Note that if we inscribe a trapezoid under the graph of $y = \ln x$ between $x = k - 1$ and $x = k$, its area will be $\tfrac{1}{2}(\ln(k-1) + \ln k)$. [How do we know that the straight line connecting the points $(k - 1, \ln(k - 1))$ and $(k, \ln k)$ will lie under the graph of $y = \ln x$?] The sum of the areas of all such trapezoids from $x = 1$ to $x = n$ is clearly less than the area under the curve $y = \ln x$ over the interval $[1, n]$.

graph of $y = \ln x$

$(k, \ln k)$

$(k - 1, \ln(k-1))$

$x = k - 1$

$x = k$

We therefore have the inequality

$$\tfrac{1}{2}(\ln 1 + \ln 2) + \tfrac{1}{2}(\ln 2 + \ln 3) + \cdots + \tfrac{1}{2}(\ln(n-1) + \ln n) < \int_1^n \ln x \, dx,$$

which is equivalent to

$$A_n < \int_1^n \ln x \, dx.$$

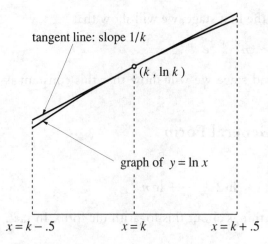

tangent line: slope $1/k$

$(k, \ln k)$

graph of $y = \ln x$

$x = k - .5$ $x = k$ $x = k + .5$

The lower bound. On the other hand, if we draw the tangent line to $y = \ln x$ at $x = k$ and form the trapezoid between $x = k - .5$ and $x = k + .5$, its area will just be $\ln k$ and will be greater than the area under the curve over the same interval. (We've used the fact—which you should check—that the area of a trapezoid equals the distance between the parallel sides times the distance between the midpoints of the other two sides.)

Adding up all such trapezoids, we get the inequality

$$\int_{3/2}^{n} \ln x \, dx < A_n.$$

Since we know that $\int \ln x \, dx = x \ln x - x$, we can evaluate these upper and lower bounds to conclude

$$n \ln n - n - \frac{3}{2} \ln \frac{3}{2} + \frac{3}{2} < A_n < n \ln n - n + 1,$$

which in turn yields

$$\left(n + \frac{1}{2}\right)\ln n - n + \frac{3}{2}\left(1 - \ln \frac{3}{2}\right) < \ln n! < \left(n + \frac{1}{2}\right)\ln n - n + 1.$$

Pause for a moment to observe that the differences

$$D_n = \left(n + \frac{1}{2}\right)\ln n - n + 1 - \ln n!$$

between the expressions in the rightmost inequality are just the accumulated error from approximating the area under $y = \ln x$ by the inscribed trapezoids. Since the error over each interval is always positive, D_n must therefore get larger as n increases. We will need this fact shortly.

Returning to our inequalities, they can finally be rewritten as

$$\frac{3}{2}\left(1 - \ln \frac{3}{2}\right) < \ln n! - \left(n + \frac{1}{2}\right)\ln n + n < 1.$$

Evaluating the constants, we thus have that for any value of n,

$$.8918 < \ln n! - \left(n + \frac{1}{2}\right)\ln n + n < 1.$$

If we exponentiate, this becomes

$$2.395 < \frac{n!}{n^{n+1/2}e^{-n}} < 2.719,$$

or

$$2.395 n^{n+1/2}e^{-n} < n! < 2.719 n^{n+1/2}e^{-n}.$$

Notice that these bounds are already quite strong and would be adequate for many estimates. Moreover, they are true for any value of n. If we are only interested in large values of n, we can do a little better. Let

These estimates are often adequate

$$\delta_n = \ln n! - \left(n + \frac{1}{2}\right)\ln n + n.$$

Then $1 - \delta_n = (n + \frac{1}{2})\ln n - n + 1 - \ln n!$ is just the expression we called D_n a moment ago and said had to be increasing as n increases. But if $D_n = 1 - \delta_n$ is increasing, it must be true that δ_n itself is decreasing as n gets larger. We thus must have $1 > \delta_1 > \delta_2 > \cdots > \delta_n \cdots > .8918$. There must therefore be some constant $d \geq .8918$ such that $\lim_{n\to\infty} \delta_n = d$. Define the constant c by $c = e^d$. Then

$$\lim_{n\to\infty} \frac{n!}{n^{n+1/2}e^{-n}} = c,$$

which is what we mean when we write

$$n! \sim c n^{n+1/2}e^{-n}.$$

This completes stage 1. In stage 2 we will see that $c = \sqrt{2\pi}$.

Stage Two: Evaluating c

We will do stage two using an interesting result of a seventeenth-century English mathematician, John Wallis (1616–1703), who showed that

$$\lim_{n\to\infty} \frac{2}{1} \times \frac{2}{3} \times \frac{4}{3} \times \frac{4}{5} \times \frac{6}{5} \times \frac{6}{7} \times \cdots \times \frac{2n}{2n - 1} \times \frac{2n}{2n + 1} = \frac{\pi}{2}.$$

Wallis's formula

Suppose for the moment that we had proved Wallis's formula. We can express it in terms of factorials by noting that we can rewrite the product of the first n even numbers—$2 \times 4 \times 6 \times \cdots \times (2n)$—by factoring a 2 out of each term, leaving us

$$2 \times 4 \times 6 \times \cdots \times (2n) = 2^n(n!).$$

Similarly, we can take the product of the first n odd integers—$1 \times 3 \times 5 \times 7 \times \cdots \times (2n - 1)$—and insert the missing even terms to get

$$1 \times 3 \times 5 \times 7 \times \cdots \times (2n - 1) = \frac{1 \times 2 \times 3 \times 4 \times \cdots \times (2n - 1) \times (2n)}{2 \times 4 \times 6 \times \cdots \times (2n)}$$

$$= \frac{(2n)!}{2^n (n!)}.$$

We can thus rewrite Wallis's formula as

$$\lim_{n \to \infty} \frac{(2^n n!)^4}{((2n)!)^2 (2n + 1)} = \frac{\pi}{2}.$$

If we now replace all the factorials by their corresponding expressions using Stirling's approximation, we get

$$\lim_{n \to \infty} \frac{2^{4n} c^4 n^{4n+2} e^{-4n}}{c^2 (2n)^{4n+1} e^{-4n} (2n + 1)} = \frac{\pi}{2},$$

which, after a great deal of cancellation, reduces to

$$\lim_{n \to \infty} \frac{c^2 n}{2(2n + 1)} = \frac{\pi}{2}.$$

Now since

$$\lim_{n \to \infty} \frac{n}{2n + 1} = \frac{1}{2},$$

this reduces to

$$\frac{c^2}{4} = \frac{\pi}{2},$$

so

$$c^2 = 2\pi,$$

and

$$c = \sqrt{2\pi},$$

as desired.

Deriving Wallis's Formula

One way to derive Wallis's formula involves the integrals

$$I_k = \int_0^{\pi/2} \sin^k x \, dx \qquad k = 0, 1, 2, 3, \ldots.$$

Note that $I_0 > I_1 > I_2 > I_3 > \cdots$. Moreover, you should verify that

$$I_0 = \frac{\pi}{2} \quad \text{and} \quad I_1 = 1.$$

Using the reduction formula derived in section 11.5 for antiderivatives of $\sin^n x$, we have a similar reduction formula for the I_k:

$$
\begin{aligned}
I_k &= \int_0^{\pi/2} \sin^k x \, dx \\
&= \frac{-1}{k} \sin^{k-1} x \cos x \Big|_0^{\pi/2} + \frac{k-1}{k} \int_0^{\pi/2} \sin^{k-2} x \, dx \\
&= \frac{k-1}{k} I_{k-2}.
\end{aligned}
$$

This in turn leads to

$$
I_k = \begin{cases}
\dfrac{2n-1}{2n} \cdot \dfrac{2n-3}{2n-2} \cdots \dfrac{1}{2} \cdot \dfrac{\pi}{2} & \text{if } k = 2n \text{ is even} \\[3mm]
\dfrac{2n}{2n+1} \cdot \dfrac{2n-2}{2n-1} \cdots \dfrac{2}{3} \cdot 1 & \text{if } k = 2n+1 \text{ is odd.}
\end{cases}
$$

Further, note that

$$\frac{I_{2n+2}}{I_{2n}} = \frac{2n+1}{2n+2},$$

which has the limit 1 for large n. Since $I_{2n} > I_{2n+1} > I_{2n+2}$, it follows that I_{2n+1}/I_{2n} also approaches 1 for large n. But this gives us

$$
\begin{aligned}
1 &= \lim_{n \to \infty} \frac{I_{2n+1}}{I_{2n}} \\
&= \lim_{n \to \infty} \left(\frac{2n}{2n+1} \cdot \frac{2n-2}{2n-1} \cdots \frac{2}{3} \cdot 1 \right) \div \left(\frac{2n-1}{2n} \cdot \frac{2n-3}{2n-2} \cdots \frac{1}{2} \cdot \frac{\pi}{2} \right) \\
&= \lim_{n \to \infty} \frac{2n}{2n+1} \cdot \frac{2n}{2n-1} \cdot \frac{2n-2}{2n-1} \cdots \frac{2n-2}{2n-3} \cdots \frac{2}{3} \cdot \frac{2}{1} \cdot \frac{2}{\pi}.
\end{aligned}
$$

If we multiply both sides of this equation by $\pi/2$, we get Wallis's formula.

Further Refinements

Using even more careful methods of analysis, it is possible to improve on Stirling's approximation and derive approximations like

$$n! \sim \sqrt{2\pi} n^{n+1/2} e^{-n+1/12n - 1/360n^3 + 1/1260n^5 - \cdots}.$$

Some refinements

If we use this expression to approximate 1000!, for instance, our result is accurate for the first 24 digits.

While this approximation and Stirling's original one are good in the sense that they give more and more accurate digits the larger n gets—so that the ratio of n! to either approximation goes to 1 as n gets large—they are bad in the sense that the difference between n! and either approximation becomes infinite as n gets large.

The Binomial Distribution

One of the most frequently encountered concepts in probability theory is the **binomial probability distribution.** Suppose we repeat a certain experiment—flipping a penny, rolling a single die, mating a pair of fruit flies, feeding cholesterol to a lab rat—over and over. Suppose further that there is some outcome we are looking for—getting heads, rolling a 2, getting a red-eyed ofspring, developing liver cancer in the rat—in each experiment. If p is the probability of obtaining the looked-for outcome in any one experiment, denote by $P(n, k, p)$ the probability of the outcome happening exactly k times in n experiments. It turns out that

$$P(n, k, p) = \frac{n!}{k!(n - k)!}p^k(1 - p)^{n-k}.$$

EXAMPLE 1.   How likely is it to get four 2's if we roll twelve dice? The probability of getting a 2 by throwing one die is $\frac{1}{6}$. Therefore the answer to the question is

$$P\left(12, 4, \frac{1}{6}\right) = \frac{12!}{4!8!}\left(\frac{1}{6}\right)^4\left(\frac{5}{6}\right)^8 = .0888281;$$

we should get exactly four 2's slightly less frequently than once out of every 11 times we roll twelve dice.

EXAMPLE 2.   What is the probability of getting exactly 47 heads if we flip 100 pennies? Since the probability of getting heads on a single toss of a penny is $\frac{1}{2}$,

$$P(100, 47, .5) = \frac{100!}{47!53!}\left(\frac{1}{2}\right)^{100} = .0665905.$$

On the average, if we flip 100 pennies, we should get 47 heads about once out of every 15 times.

The second example demonstrates the fact that calculating binomial probabilities can get very messy very quickly. Several of the exercises are designed to show how Stirling's formula can give us quick estimates that are easy to calculate and work with.

For a more complete discussion of probability, look at section 12.2.

Exercises

1. Go through the derivation in this section and find several passages that seem to you to go a bit fast or skip over details. Rewrite these sections to make them clearer and more complete.

2. Confirm the values given in the table on page 690 for the approximations of 100! and 1000! that Stirling's formula produces.

3. *Rate of growth of n!*   Factorials get very large very rapidly. The purpose of this exercise is to develop a sense of just how rapidly $n!$ grows by comparing it to exponential functions.

   Let N be some integer > 1, and consider the sequence $a_1, a_2, a_3, \ldots$ defined by

$$a_n = \frac{N^n}{n!}.$$

   a) Show that

$$a_k = \frac{N}{k} a_{k-1},$$

   and conclude that

   If $k < N$, then $a_{k-1} < a_k$
   If $k > N$, then $a_{k-1} > a_k$
   If $k = N$, then $a_{k-1} = a_k$

   We thus have a sequence that increases for a while:

$$a_1 < a_2 < \cdots < a_{N-1} = a_N,$$

   and then decreases forever after:

$$a_N > a_{N+1} > a_{N+2} > \cdots.$$

   b) If $k > 2N$, show that $a_k < .5a_{k-1}$. Hence conclude that $\lim_{n \to \infty} a_n = 0$.

   c) Use Stirling's approximation to show that

$$a_N \approx \frac{e^N}{\sqrt{2\pi N}}.$$

   Calculate the values of this expression for $N = 10$ and $N = 100$ to get an idea of how large the sequence $\{a_n\}$ can get. This shows that, *for a while*, the exponential sequence $\{N^n\}$ can get large much more rapidly than the sequence $\{n!\}$.

   d) Show that $a_n < 1$ if $n > eN$. This gives an upper bound on how long it takes the factorials to catch up with the exponentials.

4. If $n \geq 5$, then $n!$ terminates in a certain number of zeros. For instance, $5! = 120$ ends in one zero, $23! = 25852016738884976640000$ ends in four zeroes, and so on. How many zeroes are there at the end of $1000!$?

The Binomial Distribution

*5. The formula for the binomial distribution gives us that the probability of getting exactly n heads in $2n$ flips of a coin is

$$\frac{(2n)!}{(n!)^2} \left(\frac{1}{2}\right)^{2n}.$$

Using Stirling's formula show that this can be approximated by

$$\frac{1}{\sqrt{\pi n}}.$$

Use the approximation to find the probability of getting 50 heads out of 100 tosses of a coin. If you have a computer or calculator which can compute factorials, use the original binomial distribution formula to calculate the exact probability of getting 50 heads and compare the answers.

6. More generally, if we try a certain experiment n times with a probability p of success each time, the most likely number of successes is $k = np$. (Assume that p is a fraction and n is such that $n \cdot p$ is an integer.) Use Stirling's approximation to show that the probability of getting exactly np successes is

$$P(n, np, p) \approx \frac{1}{\sqrt{2\pi np(1 - p)}}.$$

Is this consistent with the answer to exercise 5?

7. *One-dimensional random walk.*   An important class of problems, including **diffusion** and **Brownian motion,** involves the long-term behavior of particles moving randomly. We will look at the simplest case of such problems. A particle starts at the origin on a line and at each stage moves one unit to the right or one unit to the left, being equally likely to do either. What can we say about where the particle will be after n steps? In this problem we will use Stirling's formula to develop some useful insights into this question.
   a) Explain why the particle will be r units to the right of the origin after n steps if and only if it has moved to the right $k = (n + r)/2$ times and to the left $n - k = (n - r)/2$ times. Explain why it could never be 3 units or 7 units to the right after 100 steps.
   b) Using the same symbols as in part (a), show that the probability of the particle's being exactly r units to the right after n steps is

$$\frac{n!}{k!(n-k)!}\left(\frac{1}{2}\right)^n.$$

c) Use Stirling's formula to show that this probability of being r units to the right after n steps is approximately

$$\frac{\sqrt{2}}{\sqrt{\pi n}(1+(r/n))^{(n+r+1)/2}(1-(r/n))^{(n-r+1)/2}}.$$

d) To simplify the denominator of this fraction, recall the Taylor series approximation for $\ln(1 + x)$:

$$\ln(1+x) = x - \frac{x^2}{2} + \cdots.$$

Hence, if r is much smaller than n, $\ln(1 + r/n)$ can be approximated by $r/n - r^2/(2n^2)$, and $\ln(1 - r/n)$ can be approximated by $-r/n - r^2/(2n^2)$. By ignoring all powers of r greater than the second, conclude that

$$(1+(r/n))^{(n+r+1)/2}(1-(r/n))^{(n-r+1)/2} \approx e^{r^2/(2n)},$$

so that the probability of being r units to the right after n steps is

$$\sqrt{\frac{2}{\pi n}}\, e^{-r^2/2n}.$$

e) Explain how we can get the answer to exercise 5 as a special case of the result just obtained in part (d).

f) Using the approximation from part (d), calculate the probability that after 100 steps the particle will be no more than 5 units away from the starting point to either the right or the left. Remember that after 100 steps it is impossible to be an odd number of units away from the starting point. The exact probability is

$$\sum_{k=48}^{52} \frac{100!}{k!(100-k)!}\left(\frac{1}{2}\right)^{100} = .382701.$$

▷ 12.2 THE POISSON DISTRIBUTION

A Linear Model for α-Ray Emission

When a radioactive element decays, we know from the study of differential equations in chapter 4 that the amount $A(t)$ of radioactive material present at time t satisfies the differential equation

$$A' = -kA,$$

where $k > 0$ is the decay constant. If A_0 is the amount present at time $t = 0$, then the solution is

$$A(t) = A_0 e^{-kt}.$$

The time T it takes for a given amount of radioactive material to decay to half the starting quantity is known as the **half-life** of the element. Since, by definition, $A(T) = .5 A(0) = .5 A_0$, we must have

$$e^{-kT} = \frac{1}{2},$$

which leads to

$$kT = \ln 2$$

The relation between the half-life and the decay constant

and therefore

$$T = \frac{\ln 2}{k}.$$

Suppose, for example, that we have a sample of polonium, which is a radioactive isotope of radium. The decay constant of polonium is $k = .500865\%$ per day, and thus its half-life is

$$T = \frac{\ln 2}{k} = \frac{\ln 2}{.00500865} = 138.39 \text{ days.}$$

By local linearity, $A(t)$ is closely approximated by a linear function for short intervals of time. Because polonium has a half-life of 138.39 days, a "short time" means several hours in this case. Thus, if we spend an afternoon in a laboratory studying the decay of polonium, we can assume that $A(t)$ is linear.

When polonium decays, it produces various sorts of radiation, including α-rays ("alpha rays"). Using a scintillation counter, one can determine the number of rays emitted in given directions:

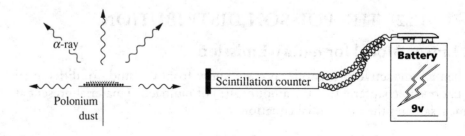

A setup like this will count a fixed percentage of the total number of α-rays emitted. Since our model of decay is linear, it follows that the number of α-rays detected should be a linear function of time. If we start counting at time $t = 0$, the number of particles observed will have a straight-line graph:

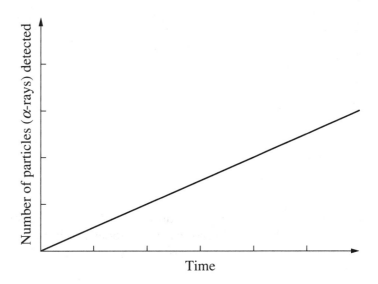

In the early twentieth century, researchers like Marie Curie and Ernest Rutherford did numerous studies of the α-rays emitted by polonium. For example, in 1911, Rutherford, Geiger, and Bateman counted the number of α-rays detected in a 7.5-second time period. They repeated their experiment 2608 times and detected a total of 10,097 α-rays. This is an average of

$$\frac{10097}{2608} = 3.8715 \ \alpha\text{-rays per 7.5-second period,}$$

so the number of α-rays per second is

$$\frac{3.8715}{7.5} = .5162 \ \alpha\text{-ray per second.}$$

Thus the straight line in the above graph has slope .5162.

This model of α-ray production has several problems. First, it predicts the existence of fractional α-rays, which makes no sense—the number detected is always a nonnegative integer. To remedy this, we can modify our model as follows:

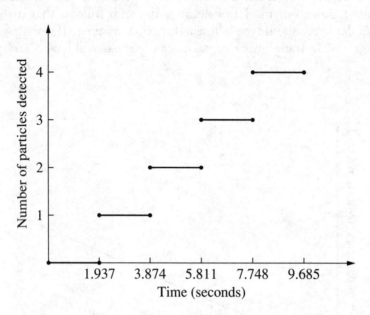

Notice that the graph is now a step function. It shows that we should see a new α-ray every $1/.5162 = 1.937$ seconds. This model also has the following consequence: if we observe the number of α-rays produced in a 7.5-second interval, then we will always see 3 or 4 particles:

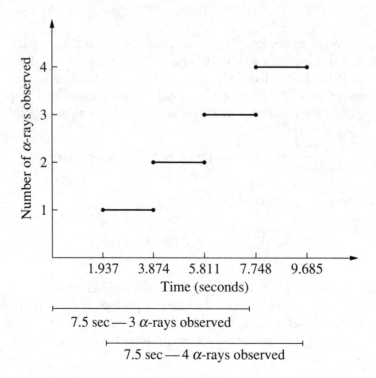

As the figure indicates, whether we get 3 or 4 depends on where the interval starts. Now comes the serious problem: this prediction is inconsistent with the experimental data collected by Rutherford and the others in 1911. For example, in 57 of the 2608 times they ran the experiment, no α-rays were observed, while in 139 cases, 7 α-rays were observed. Here are the complete data of the experiment:

Number of α-Rays Observed n	Number of Occurrences N_n
0	57
1	203
2	383
3	525
4	532
5	408
6	273
7	139
8	45
9	27
10	10
11	4
12	0
13	1
14	1
Total	2608

It follows that the linear model of α-ray emission doesn't apply to time intervals of length 7.5 seconds. This is a common occurrence—a model may work nicely over a certain range, but outside of that range, its answers may be meaningless. The problem in our case comes from the random nature of radioactive decay. In fact, there are two sources of randomness to deal with: the *time* when a polonium atom decays is random, and the *direction* in which it then emits an α-ray is also random (this affects us since the scintillation counter only detects emissions in certain directions). We need to modify our model to take the randomness into account, and this is where probability enters in.

Probability Models

The basic idea of probability theory is that the outcome of a certain event can be unpredictable in the individual instance but predictable on the average.

Randomness has structure

Throwing dice and tossing a coin are familiar examples. In this section, we will show how the Poisson probability distribution gives an excellent model of the α-ray experiment described above.

The Definition of Probability

We will let p_n denote the probability of observing exactly n α-rays in a 7.5-second time interval. By this statement, we mean the following. Suppose we run the experiment N times, where N is large. Let N_n be the number of times we observed n α-rays. Then the ratio N_n/N is the frequency with which this outcome occurs. Now imagine N getting larger and larger. Being "predictable on the average" means that the ratios N_n/N approach a fixed number, that is, the limit $\lim_{N\to\infty} N_n/N$ exists. We then define this number to be the probability p_n. Thus

$$p_n = \lim_{N\to\infty} \frac{N_n}{N}.$$

For example, the data presented on page 703 were obtained from $N = 2608$ repetitions of our experiment. From the table given there, we see that 0 α-rays were observed 57 times. This means $N_0 = 57$, and thus the probability of detecting 0 α-rays is

$$p_0 \approx \frac{N_0}{N} = \frac{57}{2608} = .0218.$$

Similarly, we can approximate p_1, p_2, and so on, using the data in the table. Our goal is to describe these probabilities $p_0, p_1, \ldots$. Ideally, we would like to have a way of determining the numbers $p_0, p_1, p_2, \ldots$ "before the fact."

Some Properties of Probabilities

In any introductory course on probability, one learns certain basic principles for working with probabilities. We will give examples to illustrate some of these principles, and more examples may be found in the exercises.

The general context

For our purposes, we will be working in the following setting. There is a certain **experiment** being performed. This might consist of flipping a coin and noting which side comes up, or running a survey asking people at random their opinions about a certain TV show, or, in our case, counting the number of α-rays detected in a 7.5-second interval. Moreover, there is a **discrete** set of possible outcomes of the experiment. That is, the possible outcomes can be listed in a sequence $O_1, O_2, O_3, \ldots$. In some cases, like throwing a pair of dice, this list might be finite. In other cases, like our α-ray experiment, the list might be infinite. What is ruled out are experiments like choosing a person at random and measuring the person's height—there is a continuum of possible outcomes here which cannot be listed in the way we've specified. Moreover, there should be a **probability** assigned to each outcome, with the outcome

O_n having probability p_n. Finally, the possible outcomes should be **disjoint**—two different outcomes can't both result from a single experiment. Thus if we are examining the attributes of a group of people, "being male" and "having green eyes" would not be acceptable outcomes in our sense unless we somehow knew in advance that there were no green-eyed males in the group.

Knowing the probabilities $p_0, p_1, \ldots$ of the possible outcomes allows us to compute other, possibly more complicated probabilities. This brings in the concept of an **event**, which is basic to probability. In the case of our α-ray experiment, here are some examples of events:

- Detecting 3 α-rays.
- Detecting 2 or 4 α-rays.
- Detecting an odd number of α-rays.

In general, an **event** is a subcollection of the possible outcomes.

> **Rule 1**   The probability of an event is simply the sum of the probabilities of its component outcomes.

The addition rule for probabilities

Thus, for the events just described, we have:

- The probability of detecting 3 α-rays is p_3;
- The probability of detecting 2 or 4 α-rays is $p_2 + p_4$;
- The probability of detecting an odd number of α-rays is the infinite sum

$$p_1 + p_3 + p_5 + p_7 + \cdots$$

(since an odd number of α-rays means that 1 or 3 or 5 or 7, etc., have been detected).

Another important property of probabilities follows directly from Rule 1:

> **Rule 2**   The sum of the probabilities of all possible outcomes is 1:
> $$\sum_{k=0}^{\infty} p_k = 1.$$

The reason for this is that the list of outcomes was stipulated to be the list of *all possible* outcomes. Hence the event consisting of all these outcomes is bound to occur every time—its probability is 1.

A third rule we will need relates the probabilities of **independent** events. Two events are independent if the occurrence or nonoccurrence of one of the events has no impact on the probability of the second event occurring. For instance, suppose we are examining a group of people. Consider the following events which may or may not occur each time we look at a person:

1. The person is female;
2. The person has green eyes;
3. The person is over 5'7" tall.

We would expect the first and second events to be independent, and also the second and third, but not the first and third.

The product rule for probabilities

> Rule 3   The probability that two or more independent events all occur is the product of their separate probabilities.

Thus, for example, suppose that in our hypothetical group of people $\frac{1}{2}$ are female, $\frac{1}{8}$ are green-eyed, and $\frac{1}{3}$ are taller than 5'7". We might then expect roughly $\frac{1}{24}$ of them to be green-eyed and over 5'7", but we would have no particular reason to expect that $\frac{1}{6}$ of them are females taller than 5'7".

A final rule that is often useful is

The probability that something doesn't happen

> Rule 4   If a certain event has a probability p of happening, then the probability that the event doesn't take place is $1 - p$.

For example, in our group of people, we would expect $\frac{2}{3}$ of them to be less than 5'7" tall, $\frac{7}{8}$ of them to have eyes colored something other than green, and so on.

The Notion of a Probability Model

A **model** is a mathematical picture of a real-life phenomenon. We have seen that dynamical systems can be used to create models of physical situations. Another type of mathematical model is a *probability model*. In general, a **probability model** for an experiment with a finite number of outcomes is a *listing of all possible outcomes and an assignment of probabilities to each outcome so that their sum is* 1. In order that the probability model be a good picture of reality, we ask that the *probability assigned to an outcome should be the* *relative frequency with which that outcome would appear if the experiment were duplicated independently a large number of times.*

As an example, a probability model for one toss of a fair die consists of a list of all possible outcomes, namely, 1, 2, 3, 4, 5, 6, and an assignment of a probability to each, namely, $\frac{1}{6}, \frac{1}{6}, \frac{1}{6}, \frac{1}{6}, \frac{1}{6}, \frac{1}{6}$, respectively. We assign the number $\frac{1}{6}$ to each outcome because we expect that if the experiment were repeated (that is, if the die were tossed) a large number of times, then any particular outcome (3, say) would occur about one-sixth of the time. Another probability model for the experiment consisting of a toss of a die might be a list of all outcomes, again 1, 2, 3, 4, 5, 6, together with an assignment of the numbers $\frac{1}{2}, 0, \frac{1}{6}, 0, 0, \frac{1}{3}$ to 1, 2, 3, 4, 5, 6, respectively. This is a probability model, because the numbers we have assigned add to 1, but it certainly does not model very well the throw of a fair die.

We would like to set up a probability model for our experiment with α-rays. The outcomes are 0, 1, 2, 3, 4, ... where, for example, the number 5 labels the outcome in which we observe 5 α-rays in our 7.5-second interval. The total number of outcomes is equal to the number of α-rays that we could conceivably see in a 7.5-second interval. Since it is conceivable (but extremely unlikely) that every atom in the sample could decay and emit an α-ray in the direction of the scintillation counter in one 7.5-second interval, we could conceivably see as many α-rays as there are atoms in the sample. This number is so large that we can think of it as infinite. To have a probability model, we need to assign numbers $p_0, p_1, p_2, \ldots$ to the outcomes 0, 1, 2, ..., respectively, so that $p_0 + p_1 + p_2 + \cdots = 1$. For the model to be reasonable, we would like each p_n to be approximately equal to the corresponding number N_n/N observed by Rutherford, Geiger, and Bateman.

The Poisson Probability Distribution

The Poisson Model of α-Ray Emission

To describe the probabilities $p_0, p_1, \ldots, p_n, \ldots$ that we will observe 0, 1, ..., $n, \ldots \alpha$-rays in a 7.5-second interval for our α-ray experiment, we use the **Poisson probability distribution**

$$p_n = \frac{\lambda^n e^{-\lambda}}{n!},$$

where λ is a number yet to be determined. The number $n!$ is called n-factorial and is defined by

$$n! = \begin{cases} n \cdot (n-1) \cdot (n-2) \cdots 3 \cdot 2 \cdot 1 & n > 0 \\ 1 & n = 0. \end{cases}$$

Thus the first few Poisson probabilities are:

$$p_0 = e^{-\lambda} \qquad p_1 = \lambda e^{-\lambda}$$

$$p_2 = \frac{\lambda^2 e^{-\lambda}}{2} \qquad p_3 = \frac{\lambda^3 e^{-\lambda}}{6}.$$

Note that this assignment does indeed give us a probability model, because

$$p_0 + p_1 + p_2 + p_3 + \cdots = \frac{\lambda^0}{0!}e^{-\lambda} + \frac{\lambda}{1!}e^{-\lambda} + \frac{\lambda^2}{2!}e^{-\lambda} + \frac{\lambda^3}{3!}e^{-\lambda} + \cdots$$

$$= e^{-\lambda}\left(1 + \frac{\lambda}{1!} + \frac{\lambda^2}{2!} + \frac{\lambda^3}{3!} + \cdots\right)$$

$$= e^{-\lambda} \cdot e^{\lambda}$$

$$= 1.$$

(The transition from the second line to the third uses the fact that the expression in parentheses is just the Taylor series for e^{λ}.)

We will shortly derive the Poisson distribution from basic principles. For the moment, though, we will assume that the probabilities $p_0, p_1, \ldots$ for α-ray emission are given by the above formulas, where we still need to choose an appropriate value for the parameter λ. The key to determining λ is the notion of **expectation**, which for us will mean the average number of α-rays observed in a 7.5-second interval.

Suppose we repeat our experiment N times. As usual, we let N_n denote the number of times exactly n α-rays were observed. Then the total number of α-rays observed in the N experiments is

$$0 \cdot N_0 + 1 \cdot N_1 + 2 \cdot N_2 + 3 \cdot N_3 + \cdots.$$

Then the "average number of α-rays observed in a 7.5-second interval" means the limit

$$E = \lim_{N \to \infty} \frac{0 \cdot N_0 + 1 \cdot N_1 + 2 \cdot N_2 + 3 \cdot N_3 + \cdots}{N}.$$

This limit is called the **expected value** or **expectation** (which explains why it is denoted E).

We claim that for the Poisson distribution, the expected value E is exactly the number λ. To see this, notice that the above limit can be written in the form

$$E = \lim_{N \to \infty}\left(0 \cdot \frac{N_0}{N} + 1 \cdot \frac{N_1}{N} + 2 \cdot \frac{N_2}{N} + 3 \cdot \frac{N_3}{N} + \cdots\right).$$

Since we defined

$$p_n = \lim_{N \to \infty} \frac{N_n}{N},$$

it follows that we get the following formula for the expectation:

$$E = 0 \cdot p_0 + 1 \cdot p_1 + 2 \cdot p_2 + 3 \cdot p_3 + \cdots = \sum_{n=0}^{\infty} n p_n.$$

The general formula for the expected value in a probability model

(Note that this equality is true for *any* probability model, not just the one we are considering.)

Substituting in the values of p_n given by the Poisson distribution, we have

$$E = 0 \cdot e^{-\lambda} + 1 \cdot \lambda e^{-\lambda} + 2 \cdot \frac{\lambda^2}{2!} e^{-\lambda} + 3 \cdot \frac{\lambda^3}{3!} e^{-\lambda} + \cdots$$

$$= \sum_{n=0}^{\infty} n \frac{\lambda^n}{n!} e^{-\lambda}$$

$$= \lambda e^{-\lambda} \sum_{n=1}^{\infty} \frac{\lambda^{n-1}}{(n-1)!},$$

where we pulled the common factor $\lambda e^{-\lambda}$ outside the summation, noted that the term in the summation corresponding to $n = 0$ is 0, and observed that

$$\frac{n}{n!} = \frac{n}{n(n-1)\cdots 2 \cdot 1} = \frac{1}{(n-1)!}.$$

Letting $k = n - 1$, we have

$$E = \lambda e^{-\lambda} \sum_{k=0}^{\infty} \frac{\lambda^k}{k!}$$

$$= \lambda e^{-\lambda} e^{\lambda}$$

$$= \lambda.$$

(We have again used the Taylor series for e^{λ}.)

This proves that the expected value is λ as claimed.

Now that we know how to interpret λ, it is easy to determine what it should be for the α-ray experiment. The data given on page 703 covered $N = 2608$ repetitions of the experiment, with

$$0 \cdot N_0 + 1 \cdot N_1 + \cdots = 10097$$

in this case. Thus

$$\frac{0 \cdot N_0 + 1 \cdot N_1 + \cdots}{N} = \frac{10097}{2608} = 3.8715$$

is an approximation of the expected value λ. However, since this is the only information about λ we have, we will let $\lambda = 3.8715$. Using this value of λ, we can then compare the frequencies predicted by the Poisson distribution to the actual data on page 703:

Number of α-Rays Observed n	Number of Occurrences N_n	Probability Approximation N_n/N	Poisson Probability p_n	Poisson Prediction $2608 p_n$
0	57	.021855	.020827	54.3
1	203	.077837	.080632	210.3
2	383	.146855	.156083	407.1
3	525	.201303	.201426	525.3
4	532	.203398	.194955	508.4
5	408	.156441	.150953	393.7
6	273	.104677	.097402	254.0
7	139	.053297	.053870	140.5
8	45	.017254	.026070	68.0
9	27	.010352	.011214	29.2
10	10	.003834	.004341	11.3
11	4	.001533	.001528	4.0
12	0	.000000	.000492	1.3
13	1	.000383	.000146	.4
14	1	.000383	.000040	.1
Totals	2608	1	1	2608

The Poisson model agrees nicely with the data since for each n, N_n/N and p_n are reasonably close. Notice that we shouldn't expect perfect agreement since N_n/N is only an approximation to p_n. We would expect these approximations to get better as we take larger values of N.

Look at the last column, labeled "Poisson prediction." The numbers here are the Poisson probabilities multiplied by $N = 2608$, and they represent the "ideal" number of occurrences. This makes it easier to compare the model to the data. For example, the graph below plots the number of occurrences, both actual and predicted.

Although the model seems to fit the data nicely, we should point out that there are statistical tests which can be used to measure the fit more precisely. These tests are part of the material covered in courses in probability and statistics.

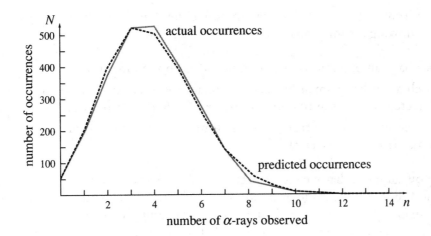

A final and very important point to make concerns the number of α-rays observed over a long period of time. Our particular Poisson model with $\lambda = 3.8715$ works only for a 7.5-second interval. What happens if we count α-rays over a longer time period? For simplicity, assume that we have a time interval of length T which is a multiple of 7.5 seconds, so that $T = 7.5N$ for some large integer N. We can regard this as running our 7.5-second experiment N consecutive times. Thus the ratio

$$\frac{\text{total number of } \alpha\text{-rays observed}}{N}$$

is an approximation to the expected value $\lambda = 3.8715$. It follows that

$$\text{total number of } \alpha\text{-rays observed} \approx 3.8715N$$
$$= \frac{3.8715}{7.5}7.5N$$
$$= .5162T.$$

This shows that for large time intervals, we recover the linear model of α-ray emissions discussed on page 701. Thus our probabilistic model is consistent with what we did earlier and yet allows us to describe what happens when the linear model breaks down.

Derivation of the Poisson Model

In the previous discussion we simply assumed that the α-ray probabilities were given by the Poisson distribution, and found that the Poisson probabilities agreed with the experimental data. Let's see where the Poisson formulas come from. It turns out that we can derive the Poisson probabilities p_n from the following assumptions:

- We have an extremely large number M of polonium atoms;
- Each atom has a small but equal probability of emitting an α-ray that is detected by our scintillation counter in a 7.5-second period;
- Observing an α-ray from a given atom is independent of observing an α-ray from any other atom.

Now suppose that we see an average of $\lambda = 3.8715$ α-rays in a 7.5-second period. Because the number of atoms M is large (in the Rutherford–Geiger–Bateman experiment $M > 10^{18}$), then the probability that a single fixed atom emits an α-ray detected by our scintillation counter in a given time period is very close to λ/M. The probability that the single atom does not emit a detected α-ray in the period is then $1 - \lambda/M$ (by Rule 4, page 706). Thus, the probability p_0 that none of the M atoms emits an α-ray in the 7.5-second period is $(1 - \lambda/M)^M$ (by Rule 3, page 706).

The fact that M is so large allows us to make a simplifying approximation. Recall that for any value of x, positive or negative,

$$e^x = \lim_{n \to \infty} \left(1 + \frac{x}{n}\right)^n.$$

Therefore

$$p_0 = \left(1 - \frac{\lambda}{M}\right)^M \approx e^{-\lambda}.$$

You might calculate some sample values for various choices of M to see how good this approximation is.

To derive the values of p_k for $k > 1$, we need a slight improvement on the estimate we just made. Let k be a relatively small number (compared to the size of M). Then

$$\left(1 - \frac{\lambda}{M}\right)^{(M-k)} = \left(\left(1 - \frac{\lambda}{M}\right)^M\right)^{(M-k)/M}.$$

Now if M is very large compared to k—we will be thinking of values of M on the order of magnitude of 10^{18} and $k < 100$—then $(M - k)/M$ will essentially equal 1. Hence

$(M - k)/M$ is essentially equal to 1

$$\left(1 - \frac{\lambda}{M}\right)^{M-k} \approx (e^{-\lambda})^1 = e^{-\lambda}.$$

We can now work out p_1. Fix your attention first on a particular atom. The probability that that atom does emit an α-ray detected by the scintillation counter while the other $M - 1$ atoms do not is (again by Rule 3)

$$\left(\frac{\lambda}{M}\right)^1 \left(1 - \frac{\lambda}{M}\right)^{M-1} \approx \frac{\lambda}{M} e^{-\lambda},$$

by the preceding approximation.

Since there are altogether M atoms which might have been responsible for the single α-ray emission, the probability that some *unspecified* atom emits an α-ray while the others do not is (Rule 1, page 705) the sum of the probability we just calculated for each of the M atoms, which is equal to M times that probability. The total probability is p_1:

$$p_1 \approx M \frac{\lambda}{M} e^{-\lambda} = \lambda e^{-\lambda}.$$

To work out p_2, note that the probability that each atom of some fixed pair of atoms emits an α-ray detected by the counter, and no other atom does, is

$$\left(\frac{\lambda}{M}\right)\left(\frac{\lambda}{M}\right)\left(1 - \frac{\lambda}{M}\right)^{M-2} \approx \frac{\lambda^2}{M^2} e^{-\lambda},$$

using our usual approximation. Since there are $\frac{1}{2}M(M - 1)$ different pairs of atoms [we can choose the first M different ways and the second $(M - 1)$ ways, but each pair gets counted twice in this scheme, so we have to divide by 2], we obtain

$$\begin{aligned}
p_2 &\approx \frac{M(M - 1)}{2} \frac{\lambda^2}{M^2} e^{-\lambda} \\
&= \left(1 - \frac{1}{M}\right) \frac{\lambda^2}{2} e^{-\lambda} \\
&\approx \frac{\lambda^2}{2} e^{-\lambda}.
\end{aligned}$$

As one can easily imagine, the computations for $p_3, p_4, \ldots$ are similar. The observant reader will note that the exact values we got for p_0, p_1, and p_2 are not the values given by the Poisson distribution. We only got the Poisson probabilities by making various approximations which were justified by the large value of M. The assumptions we have made actually lead to what is called the **binomial distribution** (see section 12.1), a distribution which tends to the

Poisson distribution in the limit $M \to \infty$. In this case, where M is large and λ relatively small, the binomial distribution is extremely close to the Poisson distribution.

Other Applications of the Poisson Distribution

The Poisson distribution can be used to model many other situations that have a random element. Examples include:

- The number of chromosome interchanges caused by exposure to X-rays for a fixed interval of time.
- The number of bacteria in a given unit of area on a Petri dish.
- The number of misprints on a page in a book.
- The number of flying-bomb hits per unit area in London during World War II.

In the exercises we will explore some examples.

Exercises

Probability Models

*1. A fair coin is tossed. If it comes up H (heads), a fair die is rolled. If the coin comes up T, the coin is tossed again. Construct a probability model for this experiment, listing the possible outcomes and their probabilities. (Hint: The list of outcomes is H1, H2, ..., H6, TT, TH.)

2. Two identical fair coins are put in a cup, shaken, and spilled out onto a table. Construct a probability model for this experiment.

3. In the disintegration of large numbers of particles Ra, it is noted that 29% of the disintegrations result in

$$Ra \longrightarrow P + A$$

and the remainder in

$$Ra \longrightarrow He^+ + B.$$

What is a model for the disintegration of a single particle of Ra?

4. In exercise 3 above, construct a probability model for the disintegration of two particles of Ra.

The Poisson Distribution

5. The purpose of this exercise is to present another way to show that the expected value E of the Poisson distribution is equal to λ. As in the

text we have

$$E = \lim_{N \to \infty} \frac{0 \cdot N_0 + 1 \cdot N_1 + 2 \cdot N_2 + 3 \cdot N_3 + \cdots}{N}.$$

The numbers np_n can be simplified as follows:

$$0 \cdot p_0 = 0$$
$$1 \cdot p_1 = 1 \cdot \lambda e^{-\lambda} = \lambda \cdot e^{-\lambda} = \lambda p_0$$
$$2 \cdot p_2 = 2 \cdot \frac{\lambda^2 e^{-\lambda}}{2} = \lambda \cdot \lambda e^{-\lambda} = \lambda p_1$$
$$3 \cdot p_3 = 3 \cdot \frac{\lambda^3 e^{-\lambda}}{6} = \lambda \cdot \frac{\lambda^2 e^{-\lambda}}{2} = \lambda p_2.$$

a) This pattern generalizes: show that

$$np_n = \lambda p_{n-1} \qquad \text{for all } n > 0.$$

b) Use part (a) to compute the expectation E (you will need to use the fact that the sum of the probabilities is $p_0 + p_1 + p_2 + \cdots = 1$).

6. A model is to be constructed for the number of rain drops that fall per square foot over a short time interval. Under what conditions would a Poisson distribution be appropriate? Under what conditions would a linear model be better?

*7. In analyzing flying-bomb hits in the south of London during World War II, investigators partitioned the area into 576 small sectors, each being $\frac{1}{4}$ of a square kilometer. There were 229 sectors with no hits, 211 sectors with exactly 1 hit, 93 sectors with exactly 2 hits, 35 sectors with 3 hits, 7 sectors with 4 hits, and 1 sector with 5 or more hits. What might lead you to expect that a Poisson distribution might be a good model for the number of hits on each sector? Fit a Poisson distribution to the data by taking λ to be the average number of hits per sector. Use this λ to compute the theoretical frequencies of 0, 1, 2, 3, 4, and 5 hits in 576 sectors.

8. A meteorite shower sprinkles a large area of the earth's surface with small meteorite hits. The average density is 5×10^{-6} hits per square meter. Set up a model assigning a probability to the number of hits per square kilometer.

9. The central processing unit (CPU) of a laptop computer will freeze if more than 10 instructions are received in a millisecond. If the average number of instructions per second received in the course of executing a large program is one per millisecond, what is the probability that the instructions received by the CPU will cause it to freeze (and, hence, the program to crash)?

▷ 12.3 THE POWER SPECTRUM

This section is an application of ideas about periodic functions and integrals to the problem of separating a signal from noise. We face this problem in our daily life. Radio and television signals have noise added to them from other radio sources we can't control. The noise sounds like hissing static on a radio and looks like "snow" on a television screen. A good receiver is designed to filter out the noise while allowing the transmitted signal to come through undistorted.

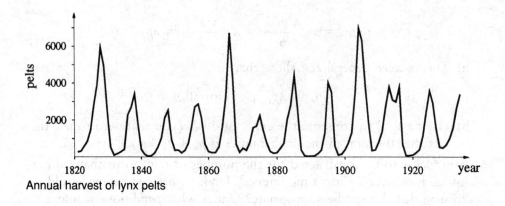

Annual harvest of lynx pelts

Scientific data and a radio broadcast have something in common: both are combinations of signal and noise. For instance, consider the annual harvest of lynx pelts by the Hudson's Bay Company. It is conceivable that the lynx population itself (the *signal*) was periodic, but various random fluctuations (the *noise*) caused the harvest (which is *signal* + *noise*) to take the form it did. If this is the case, then we should try to "filter out" the noise and find the underlying periodic signal. There is a mathematical tool to do this; it is called the **power spectrum**. We will discuss the ideas behind the power spectrum and show how it can be used to detect the underlying signal in noisy data.

Signal + Noise

To prepare for working with the power spectrum, let's first see what happens to a periodic signal that has some noise added to it. The signal we will use is a pure sine wave. The **information** that the signal carries is the frequency of that wave. The noise will also be a function, but one whose values vary in a random fashion. It can be thought of as a combination of periodic signals of all frequencies. For this reason it is sometimes called "white noise," because white light is a combination of light rays of all colors (i.e., frequencies). Here is the question we will explore: If we increase the strength of the noise, when do we lose the information contained in the original signal?

The signal and noise are shown below. As you can see, the amplitude of the signal is about 4 times as large as the amplitude of the noise. We say that the **signal-to-noise ratio** is 4:1. The combined signal + noise is no longer a pure sine wave, of course. However, it is still recognizable as a "noisy" wave with the same frequency as the original signal. The information from the signal has not yet been lost.

A signal with faint noise

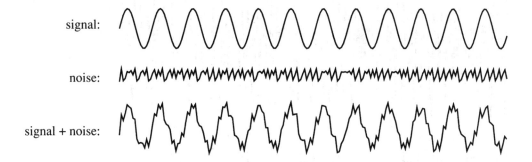

Look what happens when we increase the amplitude of the noise. In the figure below, the noise has been increased by a factor of 4, so the signal-to-noise ratio is now 1:1. The combined signal + noise is now very noisy. Would you be willing to argue that it is a wave of the same frequency as the original signal? Or would you prefer to say that it has no periodic pattern whatsoever? It appears we are close to losing the information from the original signal.

The noise level becomes stronger

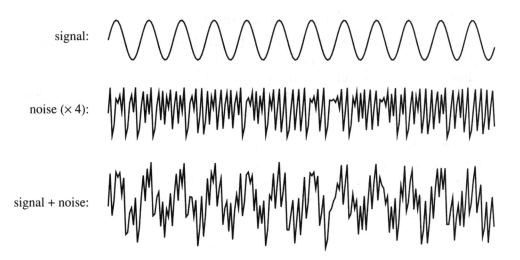

If we increase the original noise level by a factor of 10, we appear to lose the original signal altogether. The signal-to-noise ratio is now 1:2.5, and the signal + noise appears to be as random as the noise itself. In spite of appearances, the signal is still there, and it will be detected in the power spectrum!

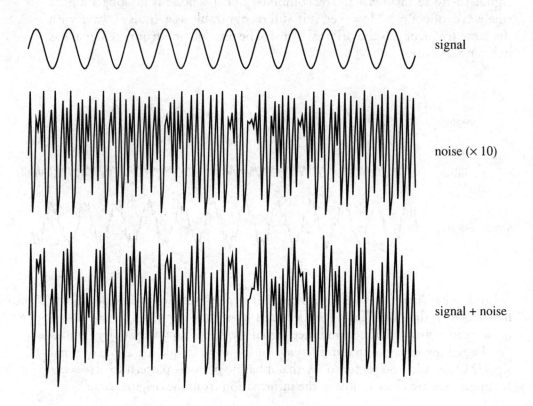

signal

noise (× 10)

signal + noise

Detecting the Frequency of a Signal

Compare the signal
to a probe whose
frequency can be
varied

Assume we have a signal that may be distorted by a lot of noise. We want to decide whether the signal has a periodic component; if it does, we want to determine its frequency. Our detector is based on this simple idea: *Compare the signal to a test probe of known frequency; vary the frequency of the probe until there is a positive response.* Of course, we still need to explain how the comparison is made, and what constitutes a positive response.

Although the detector will work on a very noisy signal, like the one above, we will understand it better if we first use it to analyze a signal whose periodic nature is evident. Let the signal $S(t)$ be a pure sine wave lying above the t-axis, and suppose that t is the time measured in seconds. Our **test probe** is the function

The test probe

$$P(t) = \sin(2\pi\omega t)$$

whose frequency is ω cycles per second. As its graph demonstrates, the values of P are equally likely to be positive or negative.

Is the same true for the product $P(t)S(t)$? Suppose first that $S(t)$ has the same frequency as $P(t)$ (below, left). As you can see, the positive values of $P(t)$ are always multiplied by the larger values of $S(t)$. By contrast, the negative values of $P(t)$ are always multiplied by the smaller values of $S(t)$. Consequently, the positive values of $P(t)S(t)$ outweigh the negative ones. On average, the value of the product is positive. In fact, the average value of the product is half the amplitude of the original signal. Later on we will see why this is so.

When the signal matches the test frequency

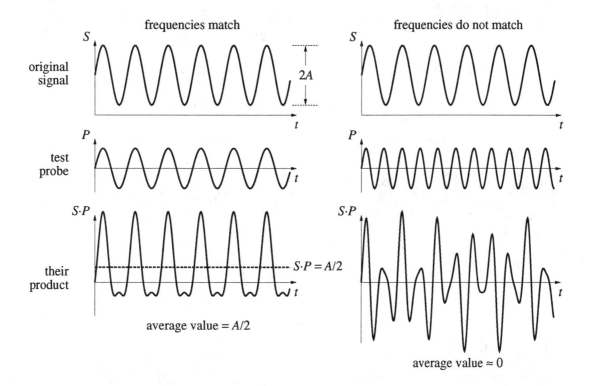

On the right we see what happens if $S(t)$ is not related to $P(t)$. In that case, a large value of $S(t)$ is just as likely to multiply a positive value of $P(t)$ as a negative one. Consequently, the product $P(t)S(t)$ will have both large positive and large negative values. On average, the value of the product will be about 0.

When the signal *doesn't* match the test frequency

Let's use the detector on the signals we constructed on page 717. In both we started with a pure sine wave and added some white noise. In the first, the signal-to-noise ratio was 4:1.

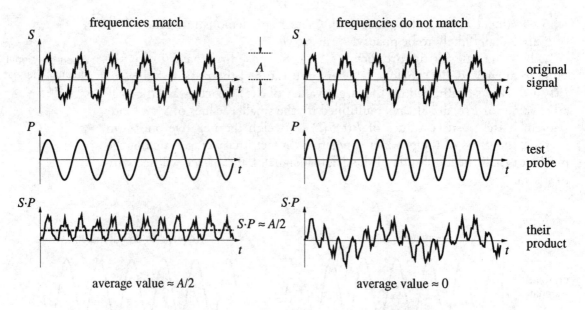

In the second the noise was stronger; the signal-to-noise ratio was 1:1.

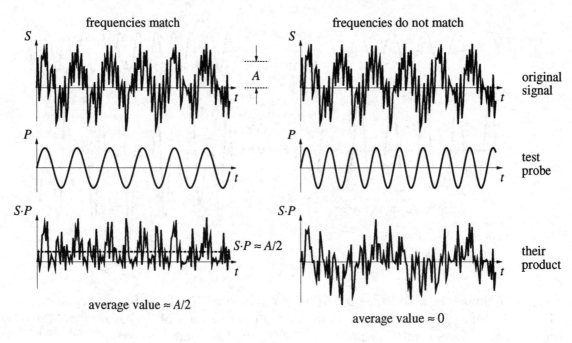

To use the detector yourself, you have to be able to calculate the average value of a function. This is discussed in section 6.3. The average value of $y = f(x)$ on the interval $a \leq x \leq b$ is

The average value of a function

$$\frac{1}{b - a} \int_a^b f(x)\, dx.$$

Our detector is the average value of the product of the signal $S(t)$ and the test probe $P(t) = \sin(2\pi\omega t)$.

> **Frequency detector:** $D(\omega) = \dfrac{1}{b-a} \displaystyle\int_a^b S(t)\sin(2\pi\omega t)\,dt.$

Clearly, the value of the detector depends on the frequency ω of the probe P. We have tried to reflect this in the notation: the detector is a function D whose input is the frequency ω. The output of the function is calculated as an integral in which the input ω plays the role of a parameter.

Integrals with parameters define functions

This is the first time we have defined a function as an integral with a parameter. Let's see how the detector works to analyze the signal $S(t) = 3\sin(5t)$ over the interval $0 \le t \le 10$. We have

$$D(\omega) = \frac{1}{10}\int_0^{10} 3\sin(5t)\sin(2\pi\omega t)\,dt.$$

In the exercises at the end of section 11.3 we obtained an explicit formula for the integral of the product of two sine functions. Find that formula and check that it yields the following:

$$D(\omega) = \frac{3}{10(4\pi^2\omega^2 - 25)}\left(5\cos(50)\sin(20\pi\omega) - 2\pi\omega\sin(50)\cos(20\pi\omega)\right).$$

Notice, in your own calculations, that ω emerges as the variable on which the whole expression depends.

The graph of $D(\omega)$ is shown below. You should plot it yourself, using a computer graphing utility. For most frequencies ω, the value of the detector D is close to 0. There is a single strong peak, which you can find at $\omega \approx .795$ cycle/sec. As it happens, the frequency of the signal $S = 3\sin(5t)$ is $5/2\pi = .79577\ldots$ cycle/sec! Moreover, the height of the peak is about 1.5, which is exactly half the amplitude of the signal.

$D(\omega)$ peaks when ω is the frequency of the signal

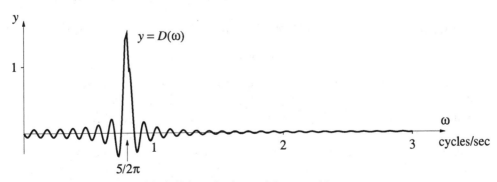

Detecting the frequency of $3\sin(5t)$ on the interval $0 \le t \le 10$

The graph on page 721 tests the signal $S(t)$ when the detector is integrated over a time interval that is 10 seconds long. That is, $0 \leq t \leq 10$ seconds. If we repeat the test by integrating over a much larger interval, the frequency detector gives us a sharper report on the frequency of the signal. In the graph below the function $D(\omega)$ was calculated by integrating over the interval $0 \leq t \leq 100$ seconds.

The peak in $D(\omega)$ is sharper if the signal is tested over a longer time interval

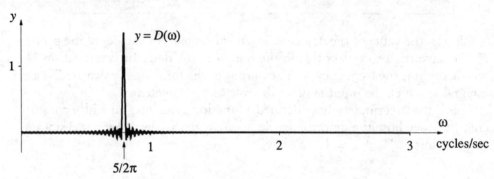

Detecting the frequency of $3 \sin(5t)$ on the interval $0 \leq t \leq 100$

Computation. Of course, it is rare to find a formula for $D(\omega)$ in terms of the frequency ω. For most signals $S(t)$, the best we can do is calculate the value of the integral numerically for a sequence of values of the parameter ω. The program DETECTOR, given below, does this. As it is written, it analyzes the function $3 \sin(5t)$ on the interval $0 \leq t \leq 10$, and it produces the graph $D(\omega)$ shown on page 723. The "outer loop"

The program DETECTOR

```
FOR j = 1 TO omegasteps  . . .  NEXT j
```

plots $D(\omega)$ over the interval $0 \leq \omega \leq 3$, using 2^{10} equally spaced values of ω. Each $D(\omega)$ is an integral whose value is first calculated as a midpoint Riemann sum with 2^7 steps. The calculation is carried out by the short "inner loop"

```
FOR k = 1 TO numberofsteps  . . .  NEXT k,
```

which you should recognize as an adaptation of the program RIEMANN from chapter 6.

If we modify the program DETECTOR so that it analyzes the function

$$S(t) = 3 \sin(5t) + \sin(8t),$$

we get the graph on the bottom of the next page. The scale on the ω-axis has also been modified to make it easier to read multiples of $1/2\pi$ cycles per second. Notice the strongest peak is at $\omega = 5/2\pi$ cycles/sec, and $D \approx 1.5$ there. But there is now a second peak at $\omega = 8/2\pi$ cycles/sec, where $D \approx .5$. Indeed, S consists of two periodic components, one with three times the amplitude of the other. The stronger component has frequency $5/2\pi$ cycles/sec, the weaker $8/2\pi$ cycles/sec.

Program: DETECTOR
To detect the frequency of a signal

```
Set up GRAPHICS
startomega = 0
endomega = 3
omegasteps = 2 ^ 10
deltaomega = (endomega - startomega) / omegasteps
twopi = 8 * ATN(1)
DEF fnf (t) = 3 * SIN(5 * t)
a = 0
b = 10
numberofsteps = 2 ^ 7
deltat = (b - a) / numberofsteps
omega = startomega
oldomega = omega
oldaccum = 0
FOR j = 1 TO omegasteps
    t = a + deltat / 2
    accum = 0
    FOR k = 1 TO numberofsteps
        deltaS = (fnf(t) * SIN(twopi * omega * t) * deltat) / (b - a)
        accum = accum + deltaS
        t = t + deltat
    NEXT k
    omega = omega + deltaomega
    Plot the line from (oldomega, oldaccum) to (omega, accum)
    oldomega = omega
    oldaccum = accum
NEXT j
```

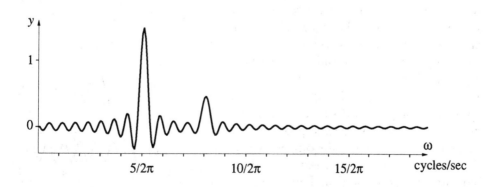

The following example first appeared in section 7.2. It is clear from the graph that it has a basic frequency of 5 Hz. The detector shows that it also has an equally strong component at 10 Hz and a much weaker component at 15 Hz. Can you guess a formula for $g(t)$?

A periodic signal ...

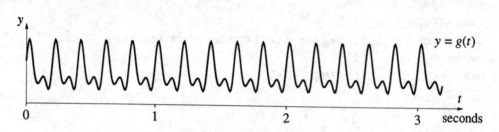

... and its frequency detector

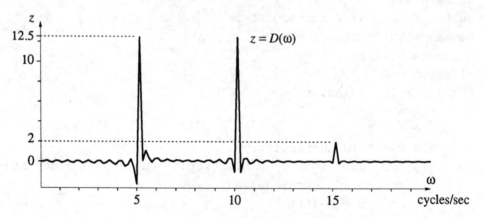

The graph of $z = D(\omega)$ was produced by DETECTOR. The integral was calculated for a $= 0$, b $= 10$, and numberofsteps $= 2 \, \hat{} \, 9$.

The Problem of Phase

Our detector is built on the premise that, if you take the product of two functions of the same frequency, its average value will be different from 0. This is illustrated by the three graphs on the right. The signal and the probe are both $\sin(t)$. Their product is a function that ranges between 0 and 1, and has average value 1/2. However, something quite different happens if we change the signal from $\sin(t)$ to $\cos(t)$. This doesn't change the period, but it does change the product, as you can see in the three graphs on the next page. The new product is centered around the t-axis; its average value is 0. Thus the detector

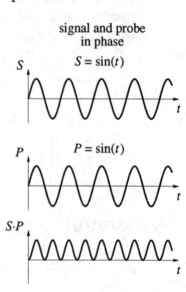

signal and probe
in phase

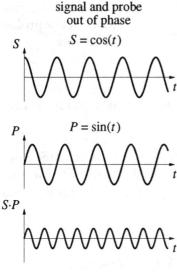

signal and probe
out of phase

$S = \cos(t)$

$P = \sin(t)$

$S \cdot P$

fails to reveal that the signal has the same frequency as the probe.

A closer look at the two sets of graphs will show what has happened. In the first case, when P is positive, so is S. When P is negative, so is S. Thus, the product $S \cdot P$ is never negative; on average, its value is positive. This is what we expect.

The second case is only a little more complicated. When P is positive, S is positive only half the time; the other half it is negative. Consequently, the product $S \cdot P$ takes both positive and negative values. The same thing happens when P is negative. On average, the value of the product is 0, *even though the frequencies of P and S match*.

The problem is that their *phases* don't match. The signal $S = \cos(t)$ hits its peak $\pi/2$ seconds before the probe $P = \sin(t)$. This kind of a difference is called a **phase shift**. In the exercises for section 7.2, you showed that if the phase of the sine function is shifted to the left by $\pi/2$, the result is the cosine function:

$$S = \sin(t + \pi/2) = \cos(t).$$

Since $\pi/2$ radians is the same as 90°, we sometimes express this equation by saying that "the sine and the cosine are 90° out of phase."

Of course the signal could involve a phase shift of any amount φ: $S = \sin(t - \varphi)$. All these signals have the same period as the probe $P = \sin(t)$. Exercise 20 of section 7.2 shows what happens if this signal is tested against the probe: the average value of the product $S \cdot P$ is $\cos(\varphi)/2$. Clearly, this depends on the size of the phase shift φ. In particular, if $\varphi = 0$ (so $S = \sin(t)$), the average value is 1/2. If $\varphi = -\pi/2$ (so $S = \cos(t)$), the average value is 0. The formula therefore agrees with what we already know for the two signals we considered as examples.

There is one more case worth glancing at: $\varphi = \pm\pi$. This is also called a phase shift of 180°. It doesn't matter whether you go forward 180° or backward; in either case $S = \sin(t \pm \pi) = -\sin(t)$. This time the average value of the product is $-1/2$.

The problem of phase is now clear: The probes $P = \sin(2\pi\omega t)$ have trouble detecting the frequency of a signal that is out of phase with them. However, any phase-shifted sine function can be expressed as a sum of pure sine and cosine functions:

$$\sin(bt - \varphi) = M\sin(bt) + N\cos(bt),$$

Arbitrary phase shifts

The average value varies with the phase

The problem of phase …

...and its solution

where $M = \cos(\varphi)$ and $N = -\sin(\varphi)$. (See the exercises.) Since the sine probes P will detect $M \sin(bt)$, we need only construct a second set of probes to detect $N \cos(bt)$. The test probes we add are the cosine functions

$$P_c = \cos(2\pi\omega t).$$

We use the subscript c to distinguish these from the sine probes, which henceforth will be denoted P_s.

Two new detectors

We must also construct a second detector, to handle the new cosine probes. Let's take this opportunity to make a technical adjustment: we redefine a detector to be *twice* the average value of the signal and the probe. In that way, the height of the detector at a peak equals the amplitude of the signal at that frequency—rather than half the amplitude.

> **Sine detector:** $D_s(\omega) = \dfrac{2}{b-a} \displaystyle\int_a^b S(t) \sin(2\pi\omega t)\, dt.$
>
> **Cosine detector:** $D_c(\omega) = \dfrac{2}{b-a} \displaystyle\int_a^b S(t) \cos(2\pi\omega t)\, dt.$

The graphs of D_s and D_c

You can modify the program DETECTOR to produce the graphs of $D_s(\omega)$ and $D_c(\omega)$. Here is how they analyze the signal $S = \cos(7t)$ over the interval $0 \le t \le 10$. The cosine detector D_c has a shape we've seen before. It has a single peak at $\omega = 7/2\pi$ cycles/sec, which is the frequency of the signal. The

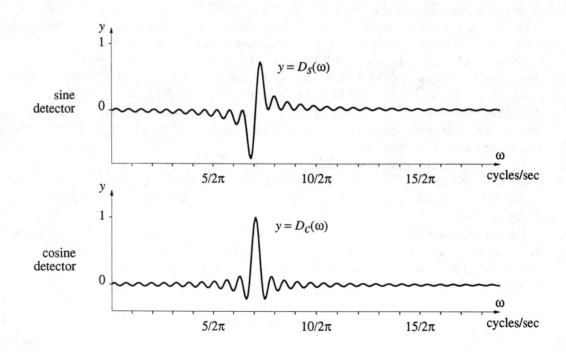

peak is 1 unit high, which is the amplitude of the signal. The sine detector has an unfamiliar shape. Notice first that $D_s(7/2\pi) = 0$. This confirms our earlier observation that the average value of the product of a sine and a cosine at the same frequency is 0. For values of ω slightly larger or smaller than $7/2\pi$, though, the sine detector swings relatively far from 0. This pattern is typical when a detector is analyzing a signal that is 90° out of phase with the probes.

Resonance.   Try this experiment. Sit at a piano and hold all the pedals down. Then sing a note. If you sing loudly enough, and hold the note long enough, one of the piano strings will start vibrating. If you stop abruptly and listen to the string, you will hear it sounding the same note you were singing. The piano has detected the frequency of your signal! It is the physical analog of our mathematical frequency detectors. The response of the string is called **resonance**. Had you sung a lower note, a larger string would have resonated.

The physical analog of a detector is a resonator

Resonance gives us a vivid language for describing how our detectors work. We can say a test probe "resonates" with a signal when their product is different from zero on average. The larger the average value, the stronger the resonance.

> Resonance occurs all around us. Sometimes it is a nuisance—for instance, when the windows in our house rattle while a heavy truck drives by, or an air conditioner runs. Sometimes we exploit it deliberately—for instance, when we use a radio tuner as an electronic resonator to detect and amplify certain electromagnetic waves.

Detector as Transform

We now have two distinct ways to describe a signal S. The function $S(t)$ is one way. It tells us how strong the signal is at each instant t. But we can also think of the signal as a mixture of sine and cosine waves of different frequencies. The detectors $D_s(\omega)$ and $D_c(\omega)$ tell us how strong the signal is at each frequency ω. That is the second way.

There is a direct connection between these two descriptions, of course. It is provided by the formulas

$$D_s(\omega) = \frac{2}{b-a} \int_a^b S(t) \sin(2\pi\omega t)\, dt \qquad D_c(\omega) = \frac{2}{b-a} \int_a^b S(t) \cos(2\pi\omega t)\, dt.$$

In effect, these formulas tell us how to *transform* the first description $S(t)$ into the second $D_s(\omega)$, $D_c(\omega)$. The transformation is so complete that even the input variable is changed—from t to ω. Look back at the formulas to see how the new variable ω is brought in.

Integrals transform S into D_s and D_c

Our detectors are essentially the same as the **Fourier sine transform** and the **Fourier cosine transform**. There is also an **inverse Fourier transform** that works in reverse: it produces $S(t)$ from the frequency data $D_s(\omega)$ and $D_c(\omega)$. The Fourier transforms are an important tool in mathematics and in science. For example, a hologram is the Fourier transform of an ordinary image. Fourier transforms and their inverses are used in photo restoration, in the enhancement of the digitized pictures sent back from cameras in space, and in filtering the signal in a stereo set.

> *The French mathematician Jean Baptiste Fourier (1768–1830) introduced what we call Fourier transforms and Fourier series to study the conduction of heat. Now his methods are used to study all sorts of periodic and nonperiodic phenomena. They are also the foundation for the part of pure mathematics called harmonic analysis.*

The Power Spectrum

A detector that ignores phase differences

The sine and cosine detectors provide enough information to reconstruct the original signal in complete detail—including phase. Often, though, they provide more detail than we want. We can use another tool—called the **power spectrum**—to determine only the strength of the different frequencies that occur in a signal, without regard to their phase. The power spectrum is constructed from the two detectors in the following way:

$$\text{Power spectrum:} \qquad P(\omega) = \sqrt{[D_s(\omega)]^2 + [D_c(\omega)]^2}.$$

To see how the power spectrum works, we'll consider the signal $S(t) = A\sin(7t - \varphi)$. This is a sine wave of frequency $\omega = 7/2\pi$ and amplitude A. Let's concentrate first on $\omega = 7/2\pi$. If there were no phase shift φ present, we would expect that

$$D_s(7/2\pi) = A \qquad D_c(7/2\pi) = 0.$$

However, because there is a phase shift, the actual values turn out to be

$$D_s(7/2\pi) = A\cos\varphi \qquad D_c(7/2\pi) = -A\sin\varphi.$$

(These calculations are given as exercises.) The values of the detectors clearly depend on the phase shift. By contrast,

$$
\begin{aligned}
P(7/2\pi) &= \sqrt{[D_s(7/2\pi)]^2 + [D_c(7/2\pi)]^2} \\
&= \sqrt{A^2\cos^2\varphi + A^2\sin^2\varphi} \\
&= A.
\end{aligned}
$$

We have used the fact that $\cos^2\varphi + \sin^2\varphi = 1$ for every φ. Thus, the power spectrum does *not* depend on the phase. It tells us only the amplitude of the signal at the frequency $\omega = 7/2\pi$.

The program POWER

If we calculate the power spectrum over all frequencies ω, we get the graph shown at the top of the next page. The program POWER generates this graph. It was derived from the program DETECTOR. Compare the two programs, particularly the terms `deltaS` and `deltaC`. In POWER, they have been multiplied by 2, to agree with our new definitions of D_s and D_c on page 726.

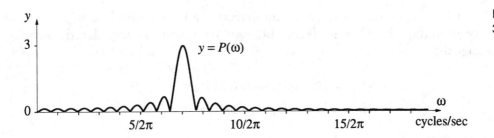

Power spectrum of
$3\sin(7t - \pi/3)$

Program: POWER
The power spectrum of a signal

```
Set up GRAPHICS
startomega = 0
endomega = 3
omegasteps = 2 ^ 9
deltaomega = (endomega - startomega) / omegasteps
pi = 4 * ATN(1)
twopi = 2 * pi
DEF fnf (t) = 3 * SIN(7 * t - pi / 3)
a = 0
b = 10
numberofsteps = 2 ^ 6
deltat = (b - a) / numberofsteps
omega = startomega
oldomega = omega
oldpower = 0
FOR j = 1 TO omegasteps
    t = a + deltat / 2
    accumS = 0
    accumC = 0
    power = 0
    FOR k = 1 TO numberofsteps
        deltaS = 2 * (fnf(t) * SIN(twopi * omega * t) * deltat) / (b - a)
        accumS = accumS + deltaS
        deltaC = 2 * (fnf(t) * COS(twopi * omega * t) * deltat) / (b - a)
        accumC = accumC + deltaC
        t = t + deltat
    NEXT k
    power = SQR(accumS ^ 2 + accumC ^ 2)
    omega = omega + deltaomega
    Plot the line from (oldomega, oldpower) to (omega, power)
    oldomega = omega
    oldpower = power
NEXT j
```

Two signals whose
components differ only
in phase

To see how the power spectrum detects the frequencies in a signal while overlooking the phases of the different components, consider these two signals:

$$g(t) = 10\sin(7t) + 7\cos(13t) + 5\cos(23t)$$
$$h(t) = 10\sin(7t) + 7\cos(13t) - 5\cos(23t)$$

They differ only in the sign of the last term. This is equivalent to a phase shift of 180° in that term. The graphs are drawn below (with constants added to separate them vertically). It is remarkable how different the graphs appear to be, considering how nearly alike their formulas are. You can find similarities if you look closely, though. For instance, the peaks of one graph tend to match the peaks of the other.

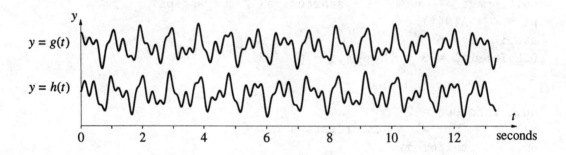

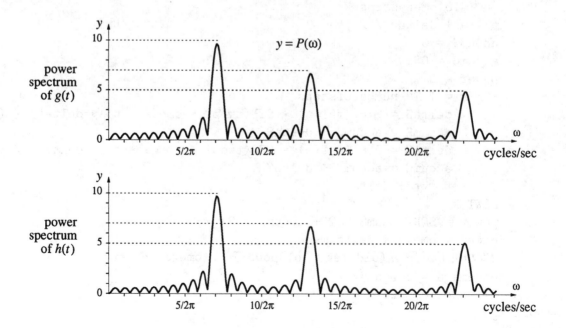

The power spectrum, however, has no trouble detecting the similarities between the two signals. As you can see, they indicate that the same dominant frequencies occur in g and h, and that corresponding frequencies occur with the same amplitude. We learn that the formula for g or h can be written as

$$10 \sin(7t - \varphi_1) + 7 \sin(13t - \varphi_2) + 5 \sin(23t - \varphi_3).$$

The only things we can't learn from the power spectrum are the three phase differences $\varphi_1, \varphi_2, \varphi_3$.

The graphs of the power spectra were drawn by POWER, using the following values:

```
endomega = 4
omegasteps = 2 ^ 8
numberofsteps = 2 ^ 7
```

These two graphs actually differ very slightly. You can see the difference most clearly near $\omega = 20/2\pi$.

For a final demonstration of the properties of the power spectrum, we return to the signal + noise problem that we raised at the beginning of this section. Let's see what happens to the power spectrum of a pure sine wave when we gradually add noise. For simplicity, we take the frequency of the pure signal to be 2 cycles/sec. The spectrum has a single strong spike at this frequency.

Detecting a periodic wave in a noisy signal

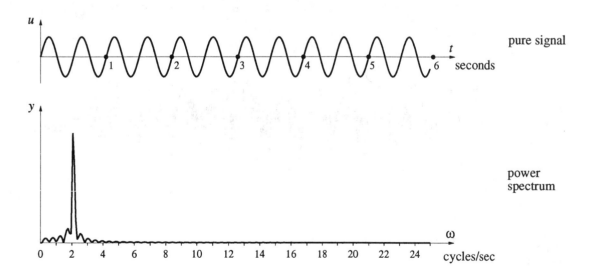

pure signal

power spectrum

On the following pages you can see what happens as the noise level is increased. The power spectrum, which was virtually zero for all $\omega > 3$ cycles/sec, is now nonzero for almost all frequencies in the range we have

In the power spectrum, noise and signal are separated

graphed. In other words, the noise is a mixture of many frequencies. Notice how the height of the power graph increases with the strength of the noise. This is most noticeable in the higher frequencies. Eventually, in the final graph, we lose sight of the signal; the noise has swamped it. The signal-to-noise ratio is 1:2.5, meaning that the noise is $2\frac{1}{2}$ times as strong as the signal. Nevertheless, the power spectrum still shows a strong spike at $\omega = 2$ cycles/sec. This corresponds to the signal. The power spectrum can still see the signal even when we can't!

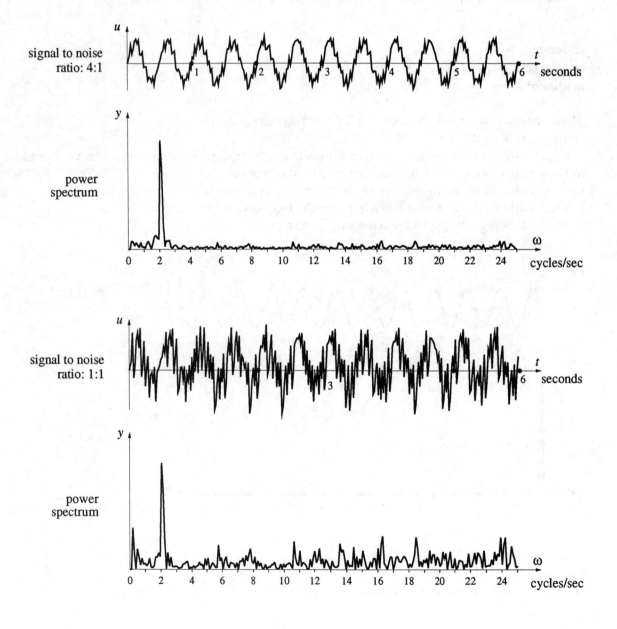

signal to noise ratio: 4:1

power spectrum

signal to noise ratio: 1:1

power spectrum

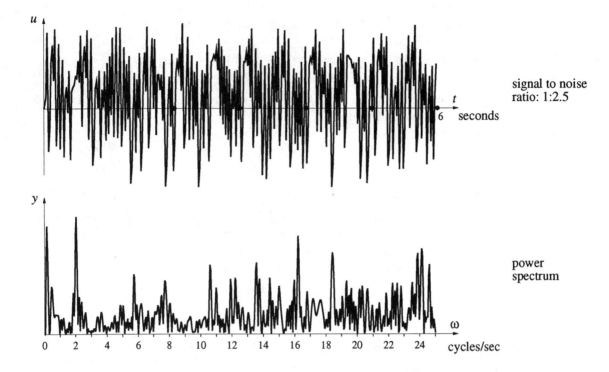

signal to noise
ratio: 1:2.5

power
spectrum

Exercises

The Problem of Phase

1. Use the "sum of two angles formula,"

$$\sin(A + B) = \sin(A)\cos(B) + \cos(A)\sin(B),$$

to show that the circular function $\sin(bt - \varphi)$ with period $2\pi/b$ and phase difference φ can be written as a combination of pure sine and cosine functions of the same period:

$$\sin(bt - \varphi) = M\sin(bt) + N\cos(bt).$$

Show that $M = \cos(\varphi)$ and $N = -\sin(\varphi)$. (Note that $M^2 + N^2 = 1$.)

2. a) Express $\sin(5t - \pi/3)$ as a sum of a pure sine function and a pure cosine function.

 b) Express $\sqrt{3}/2\sin(7t) + \frac{1}{2}\cos(7t)$ in the form $A\sin(bt - \varphi)$. To check your result, graph it together with the given function using a computer graphing utility.

 c) Express $f(t) = \sin(t) + 2\cos(t)$ in the form $A\sin(bt - \varphi)$. Notice that the formula in exercise 1 requires that $M^2 + N^2 = 1$, but in this example $M^2 + N^2 = 5$. Therefore, first write

$$f(t) = \sqrt{5}\left(\frac{1}{\sqrt{5}} \sin(t) + \frac{2}{\sqrt{5}} \cos(t)\right).$$

The expression in parentheses has the right form. Does your result check on a computer?

3. Suppose

$$A \sin(bt - \varphi) = M \sin(bt) + N \cos(bt).$$

How are A, M, and N related?

4. The functions $\sin(t) + 2\cos(t)$ and $2\sin(t) + \cos(t)$ have the same period but differ in phase. What is the phase difference? Determine this two ways: by graphing, and by writing each expression as a single function of the form $A \sin(bt - \varphi)$.

5. Choose values for A, b, and φ so that the function

$$3 \sin(2x) + 4 \cos(2x) + A \sin(bx - \varphi)$$

is *identically* zero—that is, equal to 0 for every value of x.

6. Choose values of A and φ so that the function

$$\sin(x) + \sin(x + 1) + \sin(x + 2) + A \sin(x - \varphi)$$

is identically zero.

The Programs DETECTOR and POWER

The purpose of these exercises is to give you experience interpreting the power spectrum of a known signal using the program POWER and the modifications of DETECTOR. The first exercise asks you to construct these modifications.

7. Modify DETECTOR to produce two new programs, SDETECTOR and CDETECTOR, which generate the sine detector and the cosine detector functions that appear on page 726.

8. a) Compare the outputs of $f(t) = \sin(t)$ and $g(t) = \cos(t)$ on POWER. Use the domain $0 \leq \omega \leq 1$. Does POWER distinguish between these functions? Would you expect it to?
   b) Compare $f(t)$ and $g(t)$ using SDETECTOR. Does SDETECTOR distinguish between these functions? Would you expect it to?
   c) Compare $f(t)$ and $g(t)$ using CDETECTOR. Does CDETECTOR distinguish between these functions? Why is the output of $g(t)$ on CDETECTOR the same as the output of $f(t)$ on SDETECTOR?

9. a) Describe the power spectrum of the signal $S = \sin(t) + \cos(t)$. How many peaks are there, and where are they?

b) How does the spectrum of S compare with the two generated in the last question?

c) Describe the output of SDETECTOR and CDETECTOR for the signal S. Compare these outputs to the corresponding outputs for f and g in the last exercise.

10. a) Graph the function

$$h(t) = 10\sin(7t) + 7\cos(13t) - 5\cos(23t)$$

over the domain $0 \le t \le 14$. Compare your result with the graph on page 730.

b) Graph the power spectrum of $h(t)$ over the frequency domain $0 \le \omega \le 4$. Compare your result with the text. How many peaks are there? Where are they? How high are they? Do these results agree with the amplitude and frequency information provided by the formula for $h(t)$?

11. (Continuation of the previous exercise.) Use SDETECTOR to analyze $h(t)$ over the same frequency domain. Compare the pattern near $\omega = 13/2\pi$ with the patterns generated by the sine and cosine detectors that appear on page 726. Compare the patterns near $\omega = 7/2\pi$ and near $\omega = 23/2\pi$ the same way. Would you expect the patterns near $\omega = 13/2\pi$ and $\omega = 23/2\pi$ to be similar? Are they? Are they similar to the pattern near $\omega = 7/2\pi$? Is this what you would expect?

12. (Continuation.) Use CDETECTOR to analyze $h(t)$. Follow the guidelines of the previous question.

A Grain of Salt

The purpose of the power spectrum is to make visible the periodic patterns contained within a given function. However, our *method of computing* the spectrum can introduce spurious information, too. It can tell us there are periods that are not really present in the function. So we must take the calculations with a grain of salt. The purpose of these exercises is to point out the spurious information, show why it arises, and how we can get rid of it.

∗13. Use the program POWER to graph the power spectrum of the function $\sin 2\pi x$ on the interval $0 \le x \le 10$. Let $0 \le \omega \le 3$. Set `numberofsteps = 100`, but let all the other parameters keep the values they have in the program.

14. Now increase the domain of integration to $0 \le x \le 30$, and set `numberofsteps = 300`, to adjust for the increase in the size of the domain. Use POWER again to graph the power spectrum. Compare this spectrum with the previous one.

*15. Leave $0 \le x \le 30$, but restore `numberofsteps = 100`. Use POWER once again to graph the power spectrum. Compare this spectrum with the previous two.

16. Let `numberofsteps = 50`, and calculate the power spectrum one more time. What happens?

   When we reduce the number of integration steps, new peaks appear in the power spectrum. These new peaks represent *spurious* information: the function $\sin 2\pi x$ has no components whose frequencies are 2/3, 7/3, or 8/3. Let's see why this happens. We'll concentrate on $\omega = 7/3$. First, you must decide whether the peak in the power spectrum at $\omega = 7/3$ comes from the sine or the cosine detector.

17. Use SDETECTOR and CDETECTOR to analyze $\sin(2\pi x)$. Take $0 \le x \le 30$, $0 \le \omega \le 3$, and set `numberofsteps = 100`. One of these detectors has the value 0 when $\omega = 7/3$. Which one?

18. According to the previous exercise, the peak in the power spectrum that is detected at $\omega \approx 7/3$ comes from the integral

$$\frac{2}{30} \int_0^{30} \sin(2\pi x)\sin(2\pi \tfrac{7}{3} x)\,dx,$$

   not from the cosine integral. By using one of the sine and cosine integrals from the exercises for section 11.3, determine the *exact* value of this integral. Is this the value you expected to get?

   The program POWER calculates the spectrum numerically. In particular, we used it to calculate

$$\int_0^{30} \sin(2\pi x)\sin(2\pi \tfrac{7}{3} x)\,dx,$$

   with 100 steps. The step size is therefore $\Delta x = .3$. In the following exercises duplicate this numerical work "by hand."

19. Make a sketch of the graph of the function

$$h(x) = \sin(2\pi x)\sin(2\pi \tfrac{7}{3} x)$$

   on an appropriate interval. What is the period of this function?

*20. Determine the value of $h(x)$ at $x = 0, .3, .6, .9, 1.2,$ and 1.5, and use these values to construct a Riemann sum for the integral

$$\int_0^{1.5} h(x)\,dx$$

   using left endpoints and a step size of $\Delta x = .3$. Mark these values of h on the sketch you made in the previous exercise.

*21. Evaluate the expression

$$\frac{1}{15}\int_0^{30} h(x)\,dx$$

using a left endpoint Riemann sum with a step size of $\Delta x = .3$. How can you use the previous exercise to answer this question?

22. Compare the values of the detector

$$\frac{2}{30}\int_0^{30} \sin(2\pi x)\sin(2\pi\tfrac{7}{3}x)\,dx$$

you have obtained by antidifferentiation and by numerical integration.

These exercises demonstrate that the exact and computed values of the power spectrum can be quite different, essentially because the steps in a Riemann sum can pick out very special values of the integrand.

One way to deal with the problem is to increase the number of steps. How will you know if you have gone far enough? Increase in stages until the graph of the power spectrum **stabilizes**—that is, until it no longer changes when you make a further increase in the number of steps.

The *true* spectrum is the limit of the computed graphs of the spectrum

Of course, increasing the number of steps increases computer time. This creates new problems. To deal with them, however, we can switch to more efficient numerical integration methods. Simpson's rule (section 11.6) is the most efficient method we have covered. You should try rewriting DETECTOR using Simpson's rule to see how it improves the performance.

▷ 12.4 FOURIER SERIES

In section 10.6 we obtained polynomials which were good approximations to a function over an interval, where "good" meant minimizing the *mean squared separation* between the function and the approximating polynomials.

While polynomials are the most obvious approximating functions to use due to the ease with which they can be evaluated, we have seen that finding good approximating polynomials leads to several serious technical complications. The first is that we have to solve systems of equations to determine the unknown coefficients, a procedure that is very time-consuming, even for a computer if we are trying to get a polynomial of, say, degree 30. Further, if we are trying to make the approximation over even a moderately sized interval, since we are evaluating expressions of the form x^n, we get large numbers very rapidly as x and n get large. This in turn leads to round-off problems in the computer routines.

Problems with polynomial approximations

Another aspect of these polynomial approximations that makes them complicated is that the values of the coefficients change as we change the degree of the approximating polynomial. Thus if we determine the least squares fourth-degree approximation and then decide we want the fifth-degree approximation instead, all the coefficients have to be recalculated. Knowing what the coefficient of x^3 was in the fourth-degree approximation is no help at all in knowing what the coefficient of x^3 will be in the fifth-degree approximation.

There is another class of approximating functions that avoids these problems. Moreover, this class is a natural one to use when we are trying to approximate *periodic* functions. In such cases it is reasonable to take the simplest periodic functions—the sine and cosine functions—and try to combine them to approximate more complicated periodic functions. This suggests that we want to look at functions of the form

Approximating periodic functions

$$\phi(x) = a_0 + a_1 \cos(x) + a_2 \cos(2x) + \cdots + a_n \cos(nx)$$
$$+ b_1 \sin(x) + b_2 \sin(2x) + \cdots + b_n \sin(nx)$$
$$= a_0 + \sum_{k=1}^{n} (a_k \cos(kx) + b_k \sin(kx)).$$

Such a combination is called a **trigonometric polynomial of degree n**. Note that any function of this form will in fact be periodic with period 2π. More generally, if we were interested in approximating a function of period T, we would want to look at trigonometric polynomials of the form

Trigonometric polynomial of degree n and period T

$$\phi(x) = a_0 + a_1 \cos \frac{2\pi x}{T} + a_2 \cos \frac{4\pi x}{T} + \cdots + a_n \cos \frac{2n\pi x}{T}$$
$$+ b_1 \sin \frac{2\pi x}{T} + b_2 \sin \frac{4\pi x}{T} + \cdots + b_n \sin \frac{2n\pi x}{T}$$
$$= a_0 + \sum_{k=1}^{n} \left(a_k \cos \frac{2k\pi x}{T} + b_k \sin \frac{2k\pi x}{T} \right).$$

You should verify that anything of this form will have period T.

Given a function $f(t)$, to find the coefficients a_k and b_k of the trigonometric polynomial that best fits f over the interval $[-\pi, \pi]$, we proceed exactly as we did in section 10.6, using the least squares criterion. That is, for a given degree n, we want to find coefficients $a_0, \ldots, a_n$ and $b_1, \ldots, b_n$ that minimize the integral

$$\int_0^{2\pi} (f(x) - \phi(x))^2 \, dx.$$

Coefficients are independent of n

The solution turns out to be remarkably compact and easy to state. Note one of the key features of the formulas for the coefficients: they are independent of each other and of the particular value of n being used. Thus if we find the value of a_3 in the seventh-degree approximation, we will have exactly the same value of a_3 in the 39th-degree approximation. This is a major advantage compared to the polynomial approximations we were working with in chapter 10.

Given a function $f(t)$, the least squares nth-degree trigonometric polynomial approximation to f over the interval $[-\pi, \pi]$ is

$$\phi_n(x) = a_0 + \sum_{k=1}^{n} (a_k \cos(kx) + b_k \sin(kx)),$$

where

$$a_0 = \frac{1}{2\pi} \int_{-\pi}^{\pi} f(x)\, dx$$

$$a_k = \frac{1}{\pi} \int_{-\pi}^{\pi} f(x) \cos(kx)\, dx \quad \text{for } k = 1, 2, \ldots, n$$

$$b_k = \frac{1}{\pi} \int_{-\pi}^{\pi} f(x) \sin(kx)\, dx \quad \text{for } k = 1, 2, \ldots, n.$$

The infinite series

$$a_0 + \sum_{k=1}^{\infty} (a_k \cos(kx) + b_k \sin(kx))$$

is called the **Fourier series** for f, and the coefficients a_k and b_k are called the **Fourier coefficients** for f. It turns out that any continuous function equals its Fourier series in the same sense we used earlier with Taylor series—for any x in the given interval, $f(x)$ is the limit as $n \to \infty$ of the nth-degree approximating trigonometric polynomials evaluated at x.

While the derivation is straightforward, we will leave it to the end of this section so we can look at some examples first.

Jean Baptiste Fourier (1768–1830) was active in both politics and in mathematics. He was an advocate of the French Revolution, worked as an engineer in Napoleon's army, and served as a prefect for a while. In mathematics he was interested in the mathematics of heat conduction and developed the series that now bear his name as a tool for investigating problems in this area. His ideas initially met with considerable resistance, but eventually became a central tool in mathematics.

While we have expressed everything here in terms of the interval $[-\pi, \pi]$, any other interval of width 2π would have done as well. In such cases we need only change the limits of integration to cover the interval we are interested in. In particular, we often want to use the interval $[0, 2\pi]$.

EXAMPLE 1.   Let's find the approximating trigonometric polynomials for

$$f(x) = \begin{cases} \pi + x & \text{if } -\pi \le x \le 0 \\ \pi - x & \text{if } 0 \le x \le \pi. \end{cases}$$

Then the graph of f simply consists of two line segments:

The graph of f is "triangular"

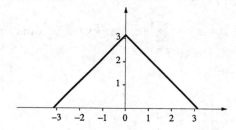

Finding the Fourier coefficients is straightforward. We can evaluate $a_0 = (1/2\pi)\pi^2 = \pi/2$ without any calculus at all—it's just the area of the triangle divided by 2π. The other coefficients can be evaluated with one use of integration by parts.

$$a_k = \frac{1}{\pi} \int_{-\pi}^{\pi} f(x) \cos(kx)\,dx$$

$$= \frac{1}{\pi} \int_{-\pi}^{0} (\pi + x) \cos(kx)\,dx + \frac{1}{\pi} \int_{0}^{\pi} (\pi - x) \cos(kx)\,dx.$$

The first of these integrals can be evaluated as

$$\int_{-\pi}^{0} (\pi + x) \cos(kx)\,dx = (\pi + x)\frac{\sin(kx)}{k}\bigg|_{-\pi}^{0} - \int_{-\pi}^{0} \frac{\sin(kx)}{k}\,dx$$

$$= 0 + \frac{\cos(kx)}{k^2}\bigg|_{-\pi}^{0}$$

$$= \begin{cases} \dfrac{2}{k^2} & \text{if } k \text{ is odd} \\ 0 & \text{if } k \text{ is even.} \end{cases}$$

Similarly we find

$$\int_{0}^{\pi} (\pi - x) \cos(kx)\,dx = (\pi - x)\frac{\sin(kx)}{k}\bigg|_{0}^{\pi} + \int_{0}^{\pi} \frac{\sin(kx)}{k}\,dx$$

$$= 0 - \frac{\cos(kx)}{k^2}\bigg|_{0}^{\pi}$$

$$= \begin{cases} \dfrac{2}{k^2} & \text{if } k \text{ is odd} \\ 0 & \text{if } k \text{ is even.} \end{cases}$$

Combining these two integrals we find

$$a_k = \begin{cases} \dfrac{4}{\pi k^2} & \text{if } k \text{ is odd} \\[2mm] 0 & \text{if } k \text{ is even.} \end{cases}$$

In the homework you will be asked to show that an analogous derivation shows that all the b_k are 0. We can thus write down the Fourier series for f:

$$f(x) = \frac{\pi}{2} + \frac{4}{\pi}\left(\frac{\cos(x)}{1} + \frac{\cos(3x)}{9} + \cdots + \frac{\cos((2n+1)x)}{(2n+1)^2} + \cdots\right)$$

The Fourier series for f

Let

$$\phi_n(x) = \frac{\pi}{2} + \frac{4}{\pi}\sum_{k=0}^{n}\frac{\cos((2k+1)x)}{(2k+1)^2}.$$

Here are the graphs of $\phi_1(x)$, $\phi_2(x)$, and $\phi_{10}(x)$:

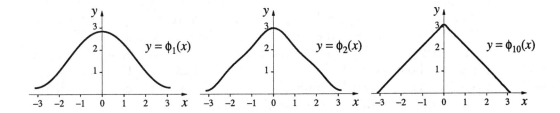

We see that $\phi_{10}(x)$ already appears to be a very good approximation to $f(x)$. If we look at the maximum separation between $f(x)$ and $\phi_n(x)$ over $[-\pi, \pi]$ for different values of n, we get the following:

n	1	2	10	50	100	1000
$\displaystyle\max_{-\pi \le x \le \pi} \lvert f(x) - \phi_n(x)\rvert$	.298	.156	.032	.0064	.0032	.00032

Since each $\phi_n(x)$ is periodic, if we graph it over a larger interval, we get an approximation to a **triangular waveform**. Here, for example, is the graph of $\phi_{20}(x)$ over the interval $[-\pi, 5\pi]$:

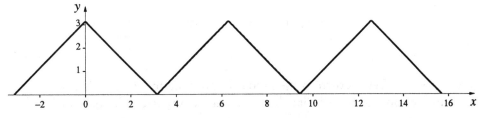

Approximating a triangular waveform

REMARK.   In addition to their use in approximating functions, Fourier series can lead to some interesting, and nonobvious, mathematical results. For instance, in the preceding example, we have $f(0) = \pi$. On the other hand, we should get the same value if we set $x = 0$ in the Fourier series for f. This leads to the identity

$$\pi = \frac{\pi}{2} + \frac{4}{\pi}\left(\frac{1}{1} + \frac{1}{9} + \frac{1}{25} + \frac{1}{49} + \frac{1}{81} + \cdots\right).$$

With a little rearranging, this can be rewritten as

$$\frac{\pi^2}{8} = 1 + \frac{1}{9} + \frac{1}{25} + \frac{1}{49} + \frac{1}{81} + \cdots$$

that is, if we add up the reciprocals of the squares of all the odd integers, we get $\pi^2/8$!

The formulas given on page 738 for approximating functions over the interval $[-\pi, \pi]$ extend readily to approximating periodic functions of any period T. For instance, if we wanted to approximate some function f over the interval $[-T/2, T/2]$, we have the following formulas. Check that when $T = 2\pi$, these equations reduce to the earlier ones. Again, there is nothing particularly important about the interval $[-T/2, T/2]$. If we had wanted to make the approximation over any other interval of length T, we simply change the limits of integration to be the endpoints of the interval.

The general rule for calculating Fourier series

Given a function $f(t)$, the least squares nth-degree trigonometric polynomial approximation to f over the interval $[-T/2, T/2]$ is

$$\phi_n(x) = a_0 + \sum_{k=1}^{n}\left(a_k \cos\frac{2k\pi x}{T} + b_k \sin\frac{2k\pi x}{T}\right),$$

where

$$a_0 = \frac{1}{T}\int_{-T/2}^{T/2} f(x)\,dx$$

$$a_k = \frac{2}{T}\int_{-T/2}^{T/2} f(x)\cos(kx)\,dx \quad \text{for } k = 1, 2, \ldots, n$$

$$b_k = \frac{2}{T}\int_{-T/2}^{T/2} f(x)\sin(kx)\,dx \quad \text{for } k = 1, 2, \ldots, n.$$

EXAMPLE 2.   Let's return to the predator–prey model of May that we examined in section 7.3. Recall that there we had two species, the predator y and the prey x. We used the following model:

Fourier series for the
periodic functions
produced by the
predator–prey model
of May

$$\text{prey:} \quad x' = .6x\left(1 - \frac{x}{10}\right) - \frac{.5xy}{x+1}$$

$$\text{predator:} \quad y' = .1y\left(1 - \frac{y}{2x}\right),$$

and discovered that the populations seemed to move toward the same periodic cycles, regardless of the initial conditions (although the phase of the cycles did depend on the starting values). In particular, if we begin with values on this cycle, our solution should be perfectly periodic with period, it turned out, $T = 38.6$ days. So let's start with $x = 7.75$ and $y = 2.38$, values that put us at the peak of the prey cycle. Now go to the differential equations and compute the solution numerically, storing the x-values as an array. We can then use these values to calculate all the integrals needed to find the Fourier coefficients to approximate the function $x(t)$. Here are the first 13 terms of the series:

$$
\begin{aligned}
x(t) = {}& 3.7951 + 3.8125\cos\frac{2\pi x}{T} + .1514\cos\frac{4\pi x}{T} + .0326\cos\frac{6\pi x}{T} \\
& - .0303\cos\frac{8\pi x}{T} - .0609\cos\frac{10\pi x}{T} + .0308\cos\frac{12\pi x}{T} + \cdots \\
& + 1.1724\sin\frac{2\pi x}{T} - .0867\sin\frac{4\pi x}{T} - .3954\sin\frac{6\pi x}{T} \\
& + .0639\sin\frac{8\pi x}{T} - .0142\sin\frac{10\pi x}{T} + .0129\sin\frac{12\pi x}{T} + \cdots.
\end{aligned}
$$

Let $\phi_3(t)$ be the seven-term trigonometric polynomial going out to the $\cos(6\pi t/T)$ and $\sin(6\pi t/T)$ terms in this expansion. If we graph $\phi_3(t)$ (dashed line) and $x(t)$ (solid line) together, they are almost indistinguishable. We let $\phi_3(t)$ run on a little beyond $x(t)$ so you can see it's there.

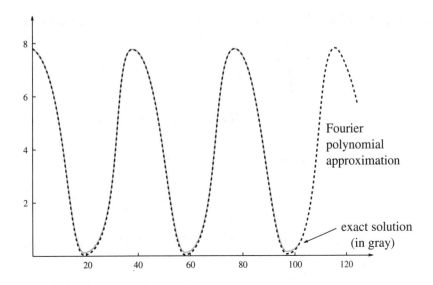

Fourier polynomial approximation

exact solution (in gray)

Derivation of the Formula for the Fourier Coefficients

The logic behind the derivation is the same as that used in section 10.6 to find the least squares polynomial approximations. Fix n and let

$$\phi_n(x) = a_0 + \sum_{k=1}^{n} \left(a_k \cos \frac{2k\pi x}{T} + b_k \sin \frac{2k\pi x}{T} \right),$$

where now we want to choose values of the a_k and b_k to minimize the integral

$$\int_{-\pi}^{\pi} (f(x) - \phi_n(x))^2 \, dx.$$

The value of this integral is thus a function of the undetermined coefficients $a_0, \ldots, a_n$ and $b_1, \ldots, b_n$. To find the coefficients that minimize this integral, we calculate the partial derivatives with respect to $a_0, a_1, \ldots$ as before and set them equal to 0. Note that

$$\frac{\partial}{\partial a_k} \phi(x) = \cos(kx) \quad \text{and} \quad \frac{\partial}{\partial b_k} \phi(x) = \sin(kx),$$

so that

$$\frac{\partial}{\partial a_k} \int_{-\pi}^{\pi} (f(x) - \phi(x))^2 \, dx = \int_{-\pi}^{\pi} 2 (f(x) - \phi(x)) (- \cos(kx)) \, dx$$

and

$$\frac{\partial}{\partial b_k} \int_{-\pi}^{\pi} (f(x) - \phi(x))^2 \, dx = \int_{-\pi}^{\pi} 2 (f(x) - \phi(x)) (- \sin(kx)) \, dx.$$

Setting the partial derivatives equal to zero gives the equations
The condition that all the partial derivatives must be 0 thus leads to the equations

$$\int_{-\pi}^{\pi} 2 (f(x) - \phi(x)) (-1) \, dx = 0$$

$$\int_{-\pi}^{\pi} 2 (f(x) - \phi(x)) (- \cos(x)) \, dx = 0$$

$$\int_{-\pi}^{\pi} 2 (f(x) - \phi(x)) (- \cos(2x)) \, dx = 0$$

$$\vdots$$

$$\int_{-\pi}^{\pi} 2 (f(x) - \phi(x)) (- \cos(nx)) \, dx = 0$$

and

$$\int_{-\pi}^{\pi} 2\left(f(x) - \phi(x)\right)\left(-\sin(x)\right) dx = 0$$

$$\int_{-\pi}^{\pi} 2\left(f(x) - \phi(x)\right)\left(-\sin(2x)\right) dx = 0$$

$$\vdots$$

$$\int_{-\pi}^{\pi} 2\left(f(x) - \phi(x)\right)\left(-\sin(nx)\right) dx = 0,$$

which can be rewritten as

$$\int_{-\pi}^{\pi} f(x)\, dx = \int_{-\pi}^{\pi} \phi(x)\, dx$$

$$\int_{-\pi}^{\pi} f(x) \cos(x)\, dx = \int_{-\pi}^{\pi} \phi(x) \cos(x)\, dx$$

$$\int_{-\pi}^{\pi} f(x) \cos(2x)\, dx = \int_{-\pi}^{\pi} \phi(x) \cos(2x)\, dx$$

$$\vdots$$

$$\int_{-\pi}^{\pi} f(x) \cos(nx)\, dx = \int_{-\pi}^{\pi} \phi(x) \cos(nx)\, dx,$$

and

$$\int_{-\pi}^{\pi} f(x) \sin(x)\, dx = \int_{-\pi}^{\pi} \phi(x) \sin(x)\, dx$$

$$\int_{-\pi}^{\pi} f(x) \sin(2x)\, dx = \int_{-\pi}^{\pi} \phi(x) \sin(2x)\, dx$$

$$\vdots$$

$$\int_{-\pi}^{\pi} f(x) \sin(nx)\, dx = \int_{-\pi}^{\pi} \phi(x) \sin(nx)\, dx.$$

To use this information to determine the coefficients, we need some facts you will derive in the exercises. They are based on integration formulas from the exercises at the end of section 11.3.

Integration formulas help

$$\int_{-\pi}^{\pi} \sin(kx) \cos(mx)\, dx = 0 \qquad \text{for all } k \text{ and } m.$$

$$\int_{-\pi}^{\pi} \sin(kx)\sin(mx)\,dx = \begin{cases} \pi & \text{if } k = m \\ 0 & \text{otherwise} \end{cases}$$

$$\int_{-\pi}^{\pi} \cos(kx)\cos(mx)\,dx = \begin{cases} \pi & \text{if } k = m \neq 0 \\ 2\pi & \text{if } k = m = 0 \\ 0 & \text{otherwise.} \end{cases}$$

Now we can evaluate the integrals we need. For instance, to determine a_k, where $k \geq 1$,

$$\int_{-\pi}^{\pi} \phi(x)\cos(kx)\,dx = \int_{-\pi}^{\pi} \Big[a_0 + \cdots + a_k\cos(kx) + \cdots + a_n\cos(nx)$$

$$+ b_1\sin(x) + \cdots + b_n\sin(nx) \Big]\cos(kx)\,dx$$

$$= a_0\int_{-\pi}^{\pi}\cos(kx)\,dx + \cdots + a_k\int_{-\pi}^{\pi}\cos^2(kx)\,dx$$

$$+ \cdots + a_n\int_{-\pi}^{\pi}\cos(nx)\cos(kx)\,dx$$

$$+ b_1\int_{-\pi}^{\pi}\sin(x)\cos(kx)\,dx + \cdots$$

$$+ b_n\int_{-\pi}^{\pi}\sin(nx)\cos(kx)\,dx$$

$$= a_0 \cdot 0 + a_1 \cdot 0 + \cdots + a_k\pi + \cdots + a_n \cdot 0$$

$$+ b_1 \cdot 0 + \cdots + b_n \cdot 0$$

$$= \pi a_k.$$

Thus

$$\int_{-\pi}^{\pi} f(x)\cos(kx)\,dx = \int_{-\pi}^{\pi} \phi(x)\cos(kx)\,dx = \pi a_k,$$

so

$$a_k = \frac{1}{\pi}\int_{-\pi}^{\pi} f(x)\cos(kx)\,dx \qquad \text{for } k = 1, 2, \ldots, n$$

as claimed.

In a similar way, the formulas for a_0 and the b_k can be verified.

Exercises

1. Use the formulas on page 640 in section 11.3 to derive the following equalities.

   a) $\int_{-\pi}^{\pi} \sin(kx)\cos(mx)\,dx = 0$ for all k and m.

   b)
   $$\int_{-\pi}^{\pi} \sin(kx)\sin(mx)\,dx = \begin{cases} \pi & \text{if } k = m \\ 0 & \text{otherwise} \end{cases}$$

   c)
   $$\int_{-\pi}^{\pi} \cos(kx)\cos(mx)\,dx = \begin{cases} \pi & \text{if } k = m \neq 0 \\ 2\pi & \text{if } k = m = 0 \\ 0 & \text{otherwise.} \end{cases}$$

2. Show that in the Fourier series for the triangular function discussed in the text, all the coefficients of the sine terms really are 0.

*3. Find the Fourier series for the following functions over the interval $[-\pi, \pi]$:

   a) $f(x) = x$

   b) $f(x) = \pi^2 - x^2$

   c) $f(x) = \begin{cases} 0 & \text{if } -\pi \leq x \leq 0 \\ x^2 & \text{if } 0 \leq x \leq \pi. \end{cases}$

4. In May's predator–prey model, find the first seven terms of the Fourier series for the predator species, $y(t)$.

APPENDIX A

GRAPHING CALCULATOR PROGRAMS

This appendix contains translations of the BASIC programs that appear in the text for various graphing calculators. Each calculator program has the same name as the corresponding computer program. Wherever possible, the calculator's code has been written to mirror the computer's, and we note points where the two differ. In particular, a variable name may reduce to just a single letter in the calculator. Comments in the text concerning aspects of a computer program generally apply as well to the corresponding calculator program.

Two of the computer programs, DETECTOR and POWER, make unreasonable demands on the current graphing calculators. For example, POWER takes more than 45 minutes to run on a TI-82. For this reason, we have not included translations of these programs (except for the TI-82). However, the programs themselves contain no new elements; you should be able to translate them for any of the graphing calculators if you wish.

We assume that you are already familiar with the general aspects of programming a calculator and that you know which keystrokes on your own calculator will produce these programs. This appendix covers five of the popular kinds of graphing calculators that are available today:

1. Texas Instruments TI-81 Graphics Calculator
2. Texas Instruments TI-82 Graphics Calculator
3. Texas Instruments TI-85 Advanced Scientific Calculator
4. Casio *fx*-7700G and *fx*-9700GE Power Graphic Calculators
5. Sharp EL-9200/9300 Graphing Scientific Calculators

➤ A.1 TEXAS INSTRUMENTS TI-81

SIR

The program pauses each time it displays t, S, I, and R. Press ENTER to continue.

```
PROGRAM:SIR
:0→T
:45400→S
:2100→I
:2500→R
:Disp "T,S,I,R"
:Disp T
:Disp S
:Disp I
:Disp R
:1→D
:1→K
:Lbl 1
:-.00001*S*I→A
:.00001*S*I-I/14→B
:I/14→C
:A*D→U
:B*D→V
:C*D→W
:T+D→T
:S+U→S
:I+V→I
:R+W→R
:Pause
:Disp T
:Disp S
:Disp I
:Disp R
:IS>(K,3)
:Goto 1
```

SIRVALUE

```
PROGRAM:SIRVALUE
:0→F
:1→L
:F→T
:45400→S
:2100→I
:2500→R
:10→N
:(L-F)/N→D
:1→K
:Lbl 1
:-.00001*S*I→A
:.00001*S*I-I/14→B
:I/14→C
:A*D→U
:B*D→V
:C*D→W
:T+D→T
:S+U→S
:I+V→I
:R+W→R
:IS>(K,N)
:Goto 1
:Disp T
:Disp S
:Disp I
:Disp R
```

Note: Be sure to distinguish between "0" (*zero*) and "O" (*Oh*).

SIRPLOT

```
PROGRAM:SIRPLOT
:0→F
:3→L
:F→T
:45400→S
:2100→I
:2500→R
:30→N
:All-Off
:ClrDraw
:F→Xmin
:L→Xmax
:1→Xscl
:0→Ymin
:S→Ymax
:10000→Yscl
:(L-F)/N→D
:1→K
:Lbl 1
:-.00001*S*I→A
:.00001*S*I-I/14→B
:I/14→C
:A*D→U
:B*D→V
:C*D→W
:Line(T,S,T+D,S+U)
:T+D→T
:S+U→S
:I+V→I
:R+W→R
:IS>(K,N)
:Goto 1
```

DOWHILE

```
PROGRAM:DOWHILE
:0→F
:F→T
:45400→S
:2100→I
:2500→R
:.00001*S*I-I/14→P
:All-Off
:ClrDraw
:F→Xmin
:20→Xmax
:5→Xscl
:0→Ymin
:S→Ymax
:10000→Yscl
:.01→D
:Lbl 1
:If P≤0
:Goto 2
:-.00001*S*I→A
:.00001*S*I-I/14→B
:I/14→C
:A*D→U
:B*D→V
:C*D→W
:Line(T,S,T+D,S+U)
:T+D→T
:S+U→S
:I+V→I
:R+W→R
:B→P
:Goto 1
:Lbl 2
:Disp T
```

SEQUENCE

Because of the speed and screen size of the calculator, this program has been limited to plotting the first 8 approximations to $y(t)$.

```
PROGRAM:SEQUENCE
:All-Off
:ClrDraw
:0→Xmin
:75→Xmax
:10→Xscl
:0→Ymin
:1500→Ymax
:500→Yscl
:1→J
:Lbl 1
:0→F
:75→L
:F→T
:100→V
:2^(J-1)→N
:(L-F)/N→D
:1→K
:Lbl 2
:.1*V*(1-V/1000)→P
:P*D→C
:Line(T,V,T+D,V+C)
:T+D→T
:V+C→V
:IS>(K,N)
:Goto 2
:IS>(J,8)
:Goto 1
```

LENGTH

The program pauses after displaying the length of each segment. Press ENTER to continue.

```
PROGRAM:LENGTH
:"X²"→Y₁
:0→F
:1→L
:2→N
:(L-F)/N→D
:0→T
:1→K
:Lbl 1
:F+(K-1)*D→A
:A→X
:Y₁→Y
:F+K*D→B
:B→X
:√((B-A)²+(Y₁-Y)²)→S
:T+S→T
:Disp K
:Disp S
:Pause
:IS>(K,N)
:Goto 1
:Disp "TOTAL"
:Disp N
:Disp T
```

TABLE

The program pauses after displaying the current value of Δy, the accumulated Δy, and the ending t. Press ENTER to continue.

```
PROGRAM:TABLE
:"cos X²"→Y₁
:Rad
:0→F
:4→L
:2^3→N
:(L-F)/N→D
:F→T
:0→A
:1→K
:Lbl 1
:Y₁*D→Y
:A+Y→A
:T+D→T
:Disp Y
:Disp A
:Disp T
:Pause
:IS>(K,N)
:Goto 1
:Disp "ACCUMULATION"
:Disp A
```

PLOT

```
PROGRAM:PLOT
:ClrDraw
:0→Xmin
:4→Xmax
:1→Xscl
:0→Ymin
:1→Ymax
:1→Yscl
:"cos X²"→Y₁
:Rad
:All-Off
:0→F
:4→L
:400→N
:(L-F)/N→D
:F→T
:0→A
:1→K
:Lbl 1
:T→X
:Y₁*D→V
:Line(T,A,T+D,A+V)
:A+V→A
:T+D→T
:IS>(K,N)
:Goto 1
```

NEWTON

The program pauses after displaying each successive approximation. Press ENTER to continue.

```
PROGRAM:NEWTON
:-1→S
:8→N
:S→X
:0→J
:Lbl 1
:Disp J
:Disp X
:4X^3+3X²+2X+1→G
:12X²+6X+2→P
:X-G/P→X
:Pause
:IS>(J,N)
:Goto 1
```

RIEMANN

The program pauses after displaying each successive approximation. Press ENTER to continue.

```
PROGRAM:RIEMANN
:"√(1+X^3)"→Y₁
:1→A
:3→B
:4→N
:(B-A)/N→D
:A→X
:0→C
:1→K
:Lbl 1
:Y₁*D→S
:C+S→C
:X+D→X
:Disp S
:Disp C
:Pause
:IS>(K,N)
:Goto 1
```

BABYLON

```
PROGRAM:BABYLON
:5→A
:2→X
:6→N
:1→K
:Lbl 1
:(X+A/X)/2→X
:Disp X
:IS>(K,N)
:Goto 1
```

HARMONIC

```
PROGRAM:HARMONIC
:100→N
:0→S
:1→I
:Lbl 1
:S+1/I→S
:IS>(I,N)
:Goto 1
:Disp N
:Disp S
```

TAYLOR

There are *two* programs here. The first, PTAYLOR, evaluates a 17th-degree Taylor polynomial at the current value of *x*. It is a subroutine called by the main program, TAYLOR, which refers to it as PrgmZ. Replace PrgmZ by the name your calculator uses to refer to PTAYLOR.

```
PROGRAM:PTAYLOR
:X→S
:-1→U
:3→K
:Lbl 2
:S+U*X^K/K!→S
:-1U→U
:K+2→K
:If K≤17
:Goto 2
:End
```

```
PROGRAM:TAYLOR
:All-Off
:ClrDraw
:0→Xmin
:3.14→Xmax
:1→Xscl
:-1.5→Ymin
:1.5→Ymax
:1→Yscl
:0→X
:PrgmZ
:S→T
:.01→J
:Lbl 1
:J→X
:PrgmZ
:Line(J-.01,T,J,S)
:S→T
:J+.01→J
:If J≤3.14
:Goto 1
:Stop
```

SIRSERIES

This program clears the statistical registers, so take care to back them up if they contain important data. The first 50 coefficients in the power series for $S(t)$ (that is, the coefficients of t^0, ..., t^{49}) are stored in {x} and the coefficients for $I(t)$ are stored in {y}. Note that the elements in these TI-81 lists are numbered from 1 to 50, so the coefficient of t^{49} in the power series for $S(t)$ would be {x}(50).

```
PROGRAM:SIRSERIE
:.00001→A
:1/14→B
:ClrStat
:45400→{x}(1)
:2100→{y}(1)
:1→K
:Lbl 1
:0→M
:0→J
:Lbl 2
:M+{x}(J+1)*{y}(K-J)→M
:IS>(J,K-1)
:Goto 2
:-A*M/K→{x}(K+1)
:-{x}(K+1)-B*{y}(K)/K→{y}(K+1)
:IS>(K,49)
:Goto 1
```

➤ A.2 TEXAS INSTRUMENTS TI-82

SIR	SIRVALUE

SIR

The program pauses each time it displays t, S, I, and R. Press **ENTER** to continue.

```
PROGRAM:SIR
:0→T
:45400→S
:2100→I
:2500→R
:Disp "T,S,I,R",T,S,I,R
:1→D
:For(K,1,3)
:-.00001*S*I→A
:.00001*S*I-I/14→B
:I/14→C
:A*D→U
:B*D→V
:C*D→W
:T+D→T
:S+U→S
:I+V→I
:R+W→R
:Pause
:Disp T,S,I,R
:End
```

SIRVALUE

```
PROGRAM:SIRVALUE
:0→F
:1→L
:F→T
:45400→S
:2100→I
:2500→R
:10→N
:(L-F)/N→D
:For(K,1,N)
:-.00001*S*I→A
:.00001*S*I-I/14→B
:I/14→C
:A*D→U
:B*D→V
:C*D→W
:T+D→T
:S+U→S
:I+V→I
:R+W→R
:End
:Disp T,S,I,R
```

Note: Be sure to distinguish between "0" (*zero*) and "O" (*Oh*).

SIRPLOT

```
PROGRAM:SIRPLOT
:0→F
:3→L
:F→T
:45400→S
:2100→I
:2500→R
:30→N
:FnOff
:ClrDraw
:F→Xmin
:L→Xmax
:1→Xscl
:0→Ymin
:S→Ymax
:10000→Yscl
:(L-F)/N→D
:For(K,1,N)
:-.00001*S*I→A
:.00001*S*I-I/14→B
:I/14→C
:A*D→U
:B*D→V
:C*D→W
:Line(T,S,T+D,S+U)
:T+D→T
:S+U→S
:I+V→I
:R+W→R
:End
```

DOWHILE

```
PROGRAM:DOWHILE
:0→F
:F→T
:45400→S
:2100→I
:2500→R
:.00001*S*I-I/14→P
:FnOff
:ClrDraw
:F→Xmin
:20→Xmax
:5→Xscl
:0→Ymin
:S→Ymax
:10000→Yscl
:.01→D
:While P>0
:-.00001*S*I→A
:.00001*S*I-I/14→B
:I/14→C
:A*D→U
:B*D→V
:C*D→W
:Line(T,S,T+D,S+U)
:T+D→T
:S+U→S
:I+V→I
:R+W→R
:B→P
:End
:Disp T
```

SEQUENCE

Because of the speed and screen size of the calculator, this program has been limited to plotting the first 8 approximations to $y(t)$.

```
PROGRAM:SEQUENCE
:FnOff
:ClrDraw
:0→Xmin
:75→Xmax
:10→Xscl
:0→Ymin
:1500→Ymax
:500→Yscl
:For(J,1,8)
:0→F
:75→L
:F→T
:100→V
:2^(J-1)→N
:(L-F)/N→D
:For(K,1,N)
:.1*V*(1-V/1000)→P
:P*D→C
:Line(T,V,T+D,V+C)
:T+D→T
:V+C→V
:End
:End
```

LENGTH

The program pauses after displaying the length of each segment. Press ENTER to continue.

```
PROGRAM:LENGTH
:"X²"→Y₁
:0→F
:1→L
:2→N
:(L-F)/N→D
:0→T
:For(K,1,N)
:F+(K-1)*D→A
:F+K*D→B
:√((B-A)²+(Y₁(B)-Y₁(A))²)→S
:T+S→T
:Disp K,S
:Pause
:End
:Disp "steps",N
:Disp "length",T
```

TABLE

This program deletes the lists L_1, L_2, and L_3, so take care to back them up if they contain important data. The current value of Δy is stored in L_1; the accumulated Δy is stored in L_2; and the ending t is stored in L_3.

```
PROGRAM:TABLE
:"cos X² "→Y₁
:Radian
:ClrList L₁ ,L₂ ,L₃
:0→F
:4→L
:2^3→N
:(L-F)/N→D
:F→T
:0→A
:For(K,1,N)
:Y₁ (T)*D→Y
:A+Y→A
:T+D→T
:Y→L₁ (K)
:A→L₂ (K)
:T→L₃ (K)
:End
:Disp "ACCUMULATION",L₂ (N)
```

PLOT

```
PROGRAM:PLOT
:ClrDraw
:0→Xmin
:4→Xmax
:1→Xscl
:0→Ymin
:1→Ymax
:1→Yscl
:"cos X² "→Y₁
:FnOff
:Radian
:0→F
:4→L
:400→N
:(L-F)/N→D
:F→T
:0→A
:For(K,1,N)
:Y₁ (T)*D→V
:Line(T,A,T+D,A+V)
:A+V→A
:T+D→T
:End
```

NEWTON

The program pauses after displaying each successive approximation. Press ENTER to continue.

```
PROGRAM:NEWTON
:-1→S
:8→N
:S→X
:For(J,0,N)
:Disp J,X
:4X^3+3X²+2X+1→G
:12X²+6X+2→P
:X-G/P→X
:Pause
:End
```

RIEMANN

This program deletes the lists L_1 and L_2, so take care to back them up if they contain important data. Incremental values are stored in L_1 and the accumulation is stored in L_2.

```
PROGRAM:RIEMANN
:"√(1+X^3)"→Y₁
:ClrList L₁,L₂
:1→A
:3→B
:4→N
:(B-A)/N→D
:A→X
:0→C
:For(K,1,N)
:Y₁(X)*D→S
:C+S→C
:X+D→X
:S→L₁(K)
:C→L₂(K)
:End
:Disp L₂(N)
```

BABYLON

```
PROGRAM:BABYLON
:5→A
:2→X
:6→N
:For(K,1,N)
:(X+A/X)/2→X
:Disp X
:End
```

HARMONIC

```
PROGRAM:HARMONIC
:100→N
:0→S
:For(I,1,N)
:S+1/I→S
:End
:Disp N,S
```

TAYLOR

There are *two* programs here. The first, called PTAYLOR, evaluates a 17th-degree Taylor polynomial for the current value of x. It is a subroutine called by TAYLOR, the main program, when it is needed.

```
PROGRAM:PTAYLOR
:X→S
:-1→U
:For(K,3,17,2)
:S+U*X^K/K!→S
:-1U→U
:End
```

```
PROGRAM:TAYLOR
:FnOff
:ClrDraw
:0→Xmin
:3.14→Xmax
:1→Xscl
:-1.5→Ymin
:1.5→Ymax
:1→Yscl
:0→X
:prgmPTAYLOR
:S→T
:For(J,.01,3.14,.01)
:J→X
:prgmPTAYLOR
:Line(J-.01,T,J,S)
:S→T
:End
```

SIRSERIES

This program deletes the lists L_1 and L_2, so take care to back them up if they contain important data. The first 51 coefficients in the power series for $S(t)$ (that is, the coefficients of t^0, ..., t^{50}) are stored in L_1 and the coefficients for $I(t)$ are stored in L_2. Note that the elements in these TI-82 lists are numbered from 1 to 51, so the coefficient of t^{50} in the power series for $S(t)$ would be $L_1(51)$.

```
PROGRAM:SIRSERIE
:ClrList L₁,L₂
:.00001→A
:1/14→B
:45400→L₁(1)
:2100→L₂(1)
:For(K,1,50)
:0→S
:For(J,0,K-1)
:S+L₁(J+1)*L₂(K-J)→S
:End
:-A*S/K→L₁(K+1)
:-L₁(K+1)-B*L₂(K)/K→L₂(K+1)
:End
```

DETECTOR

```
PROGRAM:DETECTOR
:ClrDraw
:0→F
:3→L
:F→Xmin
:L→Xmax
:1/(2π)→Xscl
:-.2→Ymin
:1.4→Ymax
:.5→Yscl
:2^6→N
:(L-F)/N→D
:"3sin 5X"→Y₁
:Radian
:FnOff
:0→A
:10→B
:2^5→S
:(B-A)/S→T
:F→M
:M→G
:0→U
:For(J,1,N)
:A+T/2→X
:0→C
:For(K,1,S)
:Y₁(X)*sin (2π*M*X)*D/(B-A)→E
:C+E→C
:X+T→X
:End
:M+D→M
:Line(G,U,M,C)
:M→G
:C→U
:End
```

POWER

```
PROGRAM:POWER
:ClrDraw
:0→F
:3→L
:F→Xmin
:L→Xmax
:1/(2π)→Xscl
:-.2→Ymin
:3.5→Ymax
:1→Yscl
:2^7→N
:(L-F)/N→D
:"3sin (7X-π/3)"→Y₁
:Radian
:FnOff
:0→A
:10→B
:2^6→S
:(B-A)/S→T
:F→M
:M→G
:0→U
:For(J,1,N)
:A+T/2→X
:0→Q
:0→R
:0→P
:For(K,1,S)
:2*Y₁(X)*sin (2π*M*X)*T/(B-A)→V
:Q+V→Q
:2*Y₁(X)*cos (2π*M*X)*T/(B-A)→W
:R+W→R
:X+T→X
:End
:√(Q²+R²)→P
:M+D→M
:Line(G,U,M,P)
:M→G
:P→U
:End
```

➤ A.3   TEXAS INSTRUMENTS TI-85

SIR

The program pauses each time it displays t, S, I, and R. Press **ENTER** to continue.

```
PROGRAM:SIR
:0→t
:45400→S
:2100→I
:2500→R
:1→deltat
:Disp t,S,I,R
:For(K,1,3)
:-.00001*S*I→Sprime
:.00001*S*I-I/14→Iprime
:I/14→Rprime
:Sprime*deltat→deltaS
:Iprime*deltat→deltaI
:Rprime*deltat→deltaR
:t+deltat→t
:S+deltaS→S
:I+deltaI→I
:R+deltaR→R
:Pause
:Disp t,S,I,R
:End
```

SIRVALUE

```
PROGRAM:SIRVALUE
:0→tinitial
:1→tfinal
:tinitial→t
:45400→S
:2100→I
:2500→R
:10→numsteps
:(tfinal-tinitial)/numsteps
      →deltat
:For(K,1,numsteps)
:-.00001*S*I→Sprime
:.00001*S*I-I/14→Iprime
:I/14→Rprime
:Sprime*deltat→deltaS
:Iprime*deltat→deltaI
:Rprime*deltat→deltaR
:t+deltat→t
:S+deltaS→S
:I+deltaI→I
:R+deltaR→R
:End
:Disp t,S,I,R
```

Note: Be sure to distinguish between "0" (*zero*) and "O" (*Oh*).

SIRPLOT

```
PROGRAM:SIRPLOT
:0→tinitial
:3→tfinal
:tinitial→t
:45400→S
:2100→I
:2500→R
:30→numsteps
:FnOff
:ClDrw
:tinitial→xMin
:tfinal→xMax
:1→xScl
:0→yMin
:S→yMax
:10000→yScl
:(tfinal-tinitial)/numsteps
    →deltat
:For(K,1,numsteps)
:-.00001*S*I→Sprime
:.00001*S*I-I/14→Iprime
:I/14→Rprime
:Sprime*deltat→deltaS
:Iprime*deltat→deltaI
:Rprime*deltat→deltaR
:Line(t,S,t+deltat,S+deltaS)
:t+deltat→t
:S+deltaS→S
:I+deltaI→I
:R+deltaR→R
:End
```

DOWHILE

```
PROGRAM:DOWHILE
:0→tinitial
:tinitial→t
:45400→S
:2100→I
:2500→R
:FnOff
:ClDrw
:tinitial→xMin
:20→xMax
:5→xScl
:0→yMin
:S→yMax
:10000→yScl
:.00001*S*I-I/14→Iprime
:.01→deltat
:While Iprime>0
:-.00001*S*I→Sprime
:.00001*S*I-I/14→Iprime
:I/14→Rprime
:Sprime*deltat→deltaS
:Iprime*deltat→deltaI
:Rprime*deltat→deltaR
:Line(t,S,t+deltat,S+deltaS)
:t+deltat→t
:S+deltaS→S
:I+deltaI→I
:R+deltaR→R
:End
:Disp t
```

SEQUENCE

Because of the speed and screen size of the calculator, this program has been limited to plotting the first 8 approximations to $y(t)$.

```
PROGRAM:SEQUENCE
:FnOff
:ClDrw
:0→xMin
:75→xMax
:10→xScl
:0→yMin
:1500→yMax
:500→yScl
:For(J,1,8)
:0→tinitial
:75→tfinal
:tinitial→t
:100→y
:2^(J-1)→numsteps
:(tfinal-tinitial)/numsteps
    →deltat
:For(K,1,numsteps)
:.1*y*(1-y/1000)→yprime
:yprime*deltat→deltay
:Line(t,y,t+deltat,y+deltay)
:t+deltat→t
:y+deltay→y
:End
:End
```

LENGTH

The program pauses after displaying the length of each segment. Press ENTER to continue.

```
PROGRAM:LENGTH
:y1=x²
:0→xinitial
:1→xfinal
:2→N
:(xfinal-xinitial)/N→deltax
:0→total
:For(K,1,N)
:xinitial+(K-1)*deltax→XL
:xinitial+K*deltax→XR
:XL→x
:y→YL
:XR→x
:y→YR
:√((XR-XL)²+(YR-YL)²)→segment
:total+segment→total
:Disp K,segment
:Pause
:End
:Disp "steps",N,"total",total
```

TABLE

The program pauses after displaying the current value of Δy, the accumulated Δy, and the ending t. Press ENTER to continue.

```
PROGRAM:TABLE
:y1=cos x²
:Radian
:0→tinitial
:4→tfinal
:2^3→numsteps
:(tfinal-tinitial)/numsteps
    →deltat
:tinitial→t
:0→accum
:For(K,1,numsteps)
:t→x
:y1*deltat→deltay
:accum+deltay→accum
:t+deltat→t
:Disp deltay,accum,t
:Pause
:End
:Disp "accum",accum
```

PLOT

```
PROGRAM:PLOT
:ClDrw
:0→xMin
:4→xMax
:1→xScl
:0→yMin
:1→yMax
:1→yScl
:y1=cos x²
:FnOff
:Radian
:0→tinitial
:4→tfinal
:400→numsteps
:(tfinal-tinitial)/numsteps
    →deltat
:tinitial→t
:0→accum
:For(K,1,numsteps)
:t→x
:y1*deltat→deltay
:Line(t,accum,t+deltat,
    accum+deltay)
:accum+deltay→accum
:t+deltat→t
:End
```

NEWTON

The program pauses after displaying each successive approximation. Press ENTER to continue.

```
PROGRAM:NEWTON
:-1→start
:8→numsteps
:start→x
:For(N,0,numsteps)
:Disp N,x
:G=4x^3+3x²+2x+1
:Gprime=12x²+6x+2
:x-G/Gprime→x
:Pause
:End
```

RIEMANN

The program pauses after displaying each successive accumulation. Press ENTER to continue.

```
PROGRAM:RIEMANN
:y1=√(1+x^3)
:1→A
:3→B
:4→numsteps
:(B-A)/numsteps→deltax
:A→x
:0→accum
:For(K,1,numsteps)
:y1*deltax→deltaS
:accum+deltaS→accum
:x+deltax→x
:Disp deltaS,accum
:Pause
:End
```

BABYLON

```
PROGRAM:BABYLON
:5→A
:2→x
:6→N
:For(K,1,N)
:(x+A/x)/2→x
:Disp x
:End
```

HARMONIC

```
PROGRAM:HARMONIC
:100→N
:0→sm
:For(I,1,N)
:sm+1/I→sm
:End
:Disp N,sm
```

TAYLOR

There are *two* programs here. The first, called PTAYLOR, evaluates a 17th-degree Taylor polynomial for the current value of x. It is a subroutine called by TAYLOR, the main program, when it is needed.

```
PROGRAM:PTAYLOR
:x→sm
:-1→sgn
:For(K,3,17,2)
:sm+sgn*x^K/K!→sm
:-1*sgn→sgn
:End
:sm→Taylor
```

```
PROGRAM:TAYLOR
:FnOff
:ClDrw
:0→xMin
:3.14→xMax
:1→xScl
:-1.5→yMin
:1.5→yMax
:1→yScl
:0→x
:PTAYLOR
:Taylor→T
:For(x,.01,3.14,.01)
:PTAYLOR
:Line(x-.01,T,x,Taylor)
:Taylor→T
:End
```

SIRSERIES

This program deletes the contents of lists S and I, so take care to back them up if they contain important data. Note that the elements in these TI-85 lists are numbered from 1 to 51, so the coefficient of t^{50} in the power series for $S(t)$ would be S(51).

```
PROGRAM:SIRSERIE
:51→dimL S
:51→dimL I
:.00001→A
:1/14→B
:45400→S(1)
:2100→I(1)
:For(K,1,50)
:0→sm
:For(J,0,K-1)
:sm+S(J+1)*I(K-J)→sm
:End
:-A*sm/K→S(K+1)
:-S(K+1)-B*I(K)/K→I(K+1)
:End
```

➢ A.4   CASIO *fx*-7700G/*fx*-9700GE

SIR	SIRVALUE

SIR

The program pauses each time it displays t, S, I, and R. Press ENTER to continue.

```
SIR
"T, S, I, R"
0→T:T⏌
45400→S:S⏌
2100→I:I⏌
2500→R:R⏌
1→D
1→K
Lbl 1
-.00001SI→A
.00001SI-I÷14→B
I÷14→C
AD→U
BD→V
CD→W
T+D→T:T⏌
S+U→S:S⏌
I+V→I:I⏌
R+W→R:R⏌
Isz K:K≤3⇒Goto 1
```

SIRVALUE

```
SIRVALUE
0→F
1→L
F→T
45400→S
2100→I
2500→R
10→N
(L-F)÷N→D
1→K
Lbl 1
-.00001SI→A
.00001SI-I÷14→B
I÷14→C
AD→U
BD→V
CD→W
T+D→T
S+U→S
I+V→I
R+W→R
Isz K:K≤N⇒Goto 1
T⏌
S⏌
I⏌
R
```

Note: Be sure to distinguish between "0" (*zero*) and "O" (*Oh*).

SIRPLOT

```
SIRPLOT
0→F
3→L
F→T
45400→S
2100→I
2500→R
30→N
Range F,L,1,0,S,10000
(L-F)÷N→D
Plot T,S
1→K
Lbl 1
-.00001SI→A
.00001SI-I÷14→B
I÷14→C
AD→U
BD→V
CD→W
T+D→T
S+U→S
I+V→I
R+W→R
Plot T,S
Line
Isz K:K≤N⇒Goto 1◢
```

DOWHILE

```
DOWHILE
0→F
F→T
45400→S
2100→I
2500→R
.00001SI-I÷14→P
Range F,20,5,0,S,10000
.01→D
Plot T,S
Lbl 1
P≤0⇒Goto 2
-.00001SI→A
.00001SI-I÷14→B
I÷14→C
AD→U
BD→V
CD→W
T+D→T
S+U→S
I+V→I
R+W→R
Plot T,S
Line
B→P
Goto 1
Lbl 2
T
```

SEQUENCE

Because of the speed and screen size of the calculator, this program has been limited to plotting the first 8 approximations to $y(t)$.

```
SEQUENCE
Range 0,75,10,0,1500,500
1→J
Lbl 1
0→F
75→L
F→T
100→V
2x^y(J-1)→N
(L-F)÷N→D
Plot T,V
1→K
Lbl 2
.1V(1-V÷1000)→P
PD→C
T+D→T
V+C→V
Plot T,V
Line
Isz K:K≤N⇒Goto 2
Isz J:J≤8⇒Goto 1
```

LENGTH

Store the expression x^2 into Function Memory as f_1 before running this program. If you store it into another function memory, change the following program to correspond.

The program pauses after displaying the length of each segment. Press ENTER to continue.

```
LENGTH
0→F
1→L
2→N
(L-F)÷N→D
0→T
1→K
Lbl 1
F+(K-1)D→A
A→X
f₁→Y
F+KD→B
B→X
√((B-A)²+(f₁-Y)²)→S
T+S→T
K⌐
S⌐
Isz K:K≤N⇒Goto 1
"TOTAL":N⌐
T
```

Note: "x^y" is a single character that represents *exponentiation* in Casio programs; it corresponds to "^" in the computer programs in the text.

TABLE

Store the expression $\cos x^2$ into Function Memory as f_1 before running this program. If you store it into another function memory, change the following program to correspond.

The program pauses after displaying the current value of Δy, the accumulated Δy, and the ending t. Press ENTER to continue.

```
TABLE
Rad
0→F
4→L
2xʸ3→N
(L-F)÷N→D
F→X
0→A
1→K
Lbl 1
f₁D→Y:Y⌐
A+Y→A:A⌐
X+D→X:X⌐
Isz K:K≤N⇒Goto 1
"ACCUMULATION":A
```

PLOT

Store the expression $\cos x^2$ into Function Memory as f_1 before running this program. If you store it into another function memory, change the following program to correspond.

```
PLOT
Rad
Range 0,4,1,0,1,1
0→F
4→L
400→N
(L-F)÷N→D
F→T
0→A
Plot T,A
1→K
Lbl 1
T→X
f₁D→V
A+V→A
T+D→T
Plot T,A
Line
Isz K:K≤N⇒Goto 1
```

NEWTON

The program pauses after displaying each successive approximation. Press ENTER to continue.

```
NEWTON
-1→S
8→N
S→X
0→J
Lbl 1
J◢
X◢
4Xx^y3+3X²+2X+1→G
12X²+6X+2→P
X-G÷P→X
Isz J:J≤N⇒Goto 1
```

RIEMANN

Store the expression $\sqrt{1+x^3}$ into Function Memory as f_1 before running this program. If you store it into another function memory, change the following program to correspond.

The program pauses after displaying each successive accumulation. Press ENTER to continue.

```
RIEMANN
1→A
3→B
4→N
(B-A)÷N→D
A→X
0→C
1→K
Lbl 1
f₁D→S:S◢
C+S→C:C◢
X+D→X
Isz K:K≤N⇒Goto 1
```

BABYLON

The program pauses after displaying each successive value of x. Press ENTER to continue.

```
BABYLON
5→A
2→X
6→N
1→K
Lbl 1
(X+A÷X)÷2→X:X◢
Isz K:K≤N⇒Goto 1
```

HARMONIC

```
HARMONIC
100→N
0→S
1→I
Lbl 1
S+1÷I→S
Isz I:I≤N⇒Goto 1
N◢
S
```

TAYLOR

Notice that there are *two* programs here. The first, PTAYLOR, evaluates a 17th-degree Taylor polynomial at the current value of x. It is a subroutine called by the main program, TAYLOR, which refers to it as Prog Z. In the listing below, you should replace Prog Z by the name your calculator uses to refer to PTAYLOR.

```
PTAYLOR
X→S
-1→U
3→K
Lbl 2
S+UXxʸK÷K!→S
-1U→U
K+2→K:K≤17⇒Goto 2
```

```
TAYLOR
Range 0,3.14,1,-1.5,1.5,1
0→X
Prog Z
Plot X,S
.01→J
Lbl 1
J→X
Prog Z
Plot J,S
Line
J+.01→J:J≤3.14⇒Goto 1
```

SIRSERIES

The coefficients in the power series for $S(t)$ are stored as S[0] to S[50] and the coefficients for $I(t)$ are stored as S[51] to S[101]. So, for example, the coefficient of t^{23} in the power series for $I(t)$ is stored as S[23+51] = S[74].

```
SIRSERIES
Defm 102
.00001→A
1÷14→B
45400→S[0]
2100→S[51]
1→K
Lbl 1
0→M
0→J
Lbl 2
M+S[J]×S[K-J+50]→M
Isz J:J≤ K-1⇒Goto 2
-AM÷K→S[K]
-S[K]-B×S[K+50]÷K→S[K+51]
Isz K:K≤50⇒Goto 1
```

Note: "×" represents *multiplication* in Casio programs; it corresponds to "*" in the computer programs in the text.

➤ A.5   SHARP EL-9200/9300

SIR

The program waits each time it displays t, S, I, and R. Press ENTER to continue.

```
t=0
S=45400
I=2100
R=2500
deltat=1
Print t
Print S
Print I
Print R
Wait
k=1
Label loop
sprime=-.00001*S*I
iprime=.00001*S*I-I/14
rprime=I/14
deltas=sprime*deltat
deltai=iprime*deltat
deltar=rprime*deltat
t=t+deltat
S=S+deltas
I=I+deltai
R=R+deltar
Print t
Print S
Print I
Print R
Wait
k=k+1
If k<=3 Goto loop
End
```

SIRVALUE

```
tinitial=0
tfinal=1
t=tinitial
S=45400
I=2100
R=2500
numsteps=10
deltat=(tfinal-
     tinitial)/numsteps
k=1
Label loop
sprime=-.00001*S*I
iprime=.00001*S*I-I/14
rprime=I/14
deltas=sprime*deltat
deltai=iprime*deltat
deltar=rprime*deltat
t=t+deltat
S=S+deltas
I=I+deltai
R=R+deltar
k=k+1
If k<=numsteps Goto loop
Print t
Print S
Print I
Print R
```

Note: Be sure to distinguish between "0" (*zero*) and "O" (*Oh*).

SIRPLOT

```
tinitial=0
tfinal=3
t=tinitial
S=45400
I=2100
R=2500
numsteps=30
Range tinitial,tfinal,1,0,
    S,10000
deltat=(tfinal-
    tinitial)/numsteps
k=1
Label loop
sprime=-.00001*S*I
iprime=.00001*S*I-I/14
rprime=I/14
deltas=sprime*deltat
deltai=iprime*deltat
deltar=rprime*deltat
Line t,S,t+deltat,S+deltas
t=t+deltat
S=S+deltas
I=I+deltai
R=R+deltar
k=k+1
If k<=numsteps Goto loop
DispG
```

DOWHILE

The program waits after it displays the final t. Press ENTER to look at the graph.

```
tinitial=0
t=tinitial
S=45400
I=2100
R=2500
iprime=.00001*S*I-I/14
Range t,20,5,0,S,10000
deltat=.01
Label loop
If iprime<=0 Goto stop
sprime=-.00001*S*I
iprime=.00001*S*I-I/14
rprime=I/14
deltas=sprime*deltat
deltai=iprime*deltat
deltar=rprime*deltat
Line t,S,t+deltat,S+deltas
t=t+deltat
S=S+deltas
I=I+deltai
R=R+deltar
Goto loop
Label stop
Print t
Wait
DispG
```

SEQUENCE

Because of the speed and screen size of the calculator, this program has been limited to plotting the first 8 approximations to $y(t)$.

The program waits after plotting each approximation. Press ENTER to continue.

```
Range 0,75,10,0,1500,500
j=1
Label loop1
tinitial=0
tfinal=75
t=tinitial
y=100
numsteps=2^(j-1)
deltat=(tfinal-
    tinitial)/numsteps
k=1
Label loop2
yprime=.1*y*(1-y/1000)
deltay=yprime*deltat
Line t,y,t+deltat,y+deltay
t=t+deltat
y=y+deltay
k=k+1
If k<=numsteps Goto loop2
j=j+1
Wait
If j<=8 Goto loop1
DispG
```

LENGTH

The program waits after displaying the length of each segment. Press ENTER to continue.

```
Goto start
Label evaluate
y=x²
Return
Label start
xinitial=0
xfinal=1
n=2
deltax=(xfinal-xinitial)/n
total=0
k=1
Label loop
xl=xinitial+(k-1)*deltax
xr=xinitial+k*deltax
x=xl
Gosub evaluate
yl=y
x=xr
Gosub evaluate
yr=y
segment=√((xr-xl)²+(yr-yl)²)
total=total+segment
Print k
Print segment
Wait
k=k+1
If k<=n Goto loop
Print n
Print total
```

TABLE

The program waits after displaying the current value of Δy, the accumulated Δy, and the ending t. Press ENTER to continue.

```
Goto start
Label evaluate
y=cos t²
Return
Label start
tinitial=0
tfinal=4
numsteps=2^3
deltat=(tfinal-
    tinitial)/numsteps
t=tinitial
accumulation=0
k=1
Label loop
Gosub evaluate
deltay=y*deltat
accumulation=accumulation+
    deltay
t=t+deltat
Print deltay
Print accumulation
Print t
Wait
k=k+1
If k<=numsteps Goto loop
Print accumulation
```

PLOT

```
Range 0,4,1,0,1,1
Goto start
Label evaluate
y=cos t²
Return
Label start
tinitial=0
tfinal=4
numsteps=400
deltat=(tfinal-
    tinitial)/numsteps
t=tinitial
accumulation=0
k=1
Label loop
Gosub evaluate
deltay=y*deltat
Line t,accumulation,
    t+deltat,accumulation+
    deltay
accumulation=accumulation+
    deltay
t=t+deltat
k=k+1
If k<=numsteps Goto loop
DispG
```

NEWTON

The program waits after displaying each successive approximation. Press ENTER to continue.

```
start=-1
numsteps=8
x=start
n=0
Label loop
Print n
Print x
Wait
g=4x^3+3x²+2x+1
gprime=12x²+6x+2
x=x-g/gprime
n=n+1
If n<=numsteps Goto loop
End
```

BABYLON

The program waits after displaying each successive value of x. Press ENTER to continue.

```
a=5
x=2
n=6
k=1
Label loop
x=(x+a/x)/2
Print x
Wait
k=k+1
If k<=n Goto loop
End
```

RIEMANN

The program waits after displaying each successive accumulation. Press ENTER to continue.

```
Goto start
Label evaluate
y=√(1+x^3)
Return
Label start
a=1
b=3
numsteps=4
deltax=(b-a)/numsteps
x=a
accumulation=0
k=1
Label loop
Gosub evaluate
deltas=y*deltax
accumulation=accumulation+deltas
x=x+deltax
Print deltas
Print accumulation
Wait
k=k+1
If k<=numsteps Goto loop
End
```

HARMONIC

```
HARMONIC
n=100
sum=0
i=1
Label loop
sum=sum+1/i
i=i+1
If i<=n Goto loop
Print n
Print sum
```

TAYLOR

```
Range 0,3.14,1,-1.5,1.5,1
Goto start
Label taylor
sum=x
sign=-1
k=3
Label loopk
sum=sum+sign*x^k/k!
sign=-1*sign
k=k+2
If k<=17 Goto loopk
t=sum
Return
Label start
x=0
xl=x
Gosub taylor
yl=t
x=.01
Label loopx
xr=x
Gosub taylor
yr=t
Line xl,yl,xr,yr
xl=xr
yl=yr
x=x+.01
If x<=3.14 Goto loopx
End
```

SIRSERIES

Enter this program in matrix mode. It deletes the contents of matrices S and I, so take care to back them up if they contain important data. The first 51 coefficients in the power series for $S(t)$ are stored in S[1,k] and the coefficients for $I(t)$ are stored in I[1,k]. Note that the elements in these matrices are numbered from 1 to 51, so the coefficient of t^{50} in the power series for $S(t)$ is S[1,51].

```
dim S[1,51]
dim I[1,51]
a=.00001
b=1/14
S[1,1]=45400
I[1,1]=2100
k=1
Label loopk
sum=0
j=0
Label loopj
sum=sum+S[1,j+1]*I[1,k-j]
j=j+1
If j<=k-1 Goto loopj
S[1,k+1]=-a*sum/k
I[1,k+1]=-S[1,k+1]-b*I[1,k]/k
k=k+1
If k<=50 Goto loopk
End
```

APPENDIX B

QUICK REFERENCE

> ### FORMULAS FROM GEOMETRY
> ### AND ALGEBRA

Notation. A = area, V = volume, b = length of base, h = height (perpendicular to base), r = radius, C = circumference.

$$\text{Rectangle:} \quad A = bh.$$
$$\text{Triangle:} \quad A = \tfrac{1}{2}bh.$$
$$\text{Trapezoid:} \quad A = \tfrac{1}{2}h(b_1 + b_2).$$
$$\text{Circle:} \quad A = \pi r^2, \quad C = 2\pi r.$$
$$\text{Cylinder:} \quad V = \pi r^2 h.$$
$$\text{Right Circular Cone:} \quad V = \tfrac{1}{3}\pi r^2 h.$$
$$\text{Sphere:} \quad V = \tfrac{4}{3}\pi r^3, \quad A = 4\pi r^2.$$

Pythagorean theorem. If a right triangle has hypotenuse of length c and legs of lengths a and b, then $c^2 = a^2 + b^2$.

Quadratic formula. The roots of the quadratic equation $ax^2 + bx + c = 0$ are

$$x = \frac{-b \pm \sqrt{b^2 - 4ac}}{2a}.$$

➤ FORMULAS FROM TRIGONOMETRY

DEFINITION.   Begin with a unit circle centered at the origin. Given the input number t, locate a point P on the circle by tracing an arc of length t along the circle from the point $(1, 0)$. If t is positive, trace the arc counterclockwise; if t is negative, trace it clockwise. Because the circle has radius 1, the arc of length t subtends a central angle of **radian** measure t. The trigonometric functions $\cos t$ and $\sin t$ are defined as the coordinates of the point P,

$$P = (\cos t, \sin t).$$

The other trigonometric functions are defined in terms of the sine and cosine functions:

$$\tan t = \frac{\sin t}{\cos t} \qquad \sec t = \frac{1}{\cos t}$$

$$\cot t = \frac{\cos t}{\sin t} \qquad \csc t = \frac{1}{\sin t}.$$

We also have the following identities:

$$\sin^2 t + \cos^2 t = 1$$
$$\sin(-t) = -\sin t$$
$$\cos(-t) = \cos t$$
$$\sin(A + B) = \sin A \cos B + \cos A \sin B$$
$$\cos(A + B) = \cos A \cos B - \sin A \sin B$$
$$\cos^2 t = \tfrac{1}{2}(1 + \cos 2t)$$
$$\sin^2 t = \tfrac{1}{2}(1 - \cos 2t)$$

➤ THE DERIVATIVE

- A **locally linear** function has a graph that looks approximately straight when magnified under a computer microscope.
- The **slope of the graph** at any point is the **limit** of the slopes seen under a microscope.
- The **rate of change** of a function at a point is the slope of its graph at that point, and thus is also a **limit.** Its dimensional units are (units of output)/(unit of input).

- The **derivative** of $f(x)$ at $x = a$ is the name given to both the rate of change of f at a and the slope of the graph of f at $(a, f(a))$.
- The derivative of $y = f(x)$ at $x = a$ is written $f'(a)$. The **Leibniz notation** for the derivative is dy/dx.

- $f'(a) = \lim_{\Delta x \to 0} \dfrac{\Delta y}{\Delta x} = \lim_{h \to 0} \dfrac{f(a+h) - f(a-h)}{2h} = \lim_{h \to 0} \dfrac{f(a+h) - f(a)}{h}.$

- The **microscope equation** $\Delta y \approx f'(a) \cdot \Delta x$ describes the relation between x and $y = f(x)$ as seen under a microscope.
- The microscope equation says the change in output is proportional to the change in the input. The derivative $f'(a)$ is the **multiplier,** or scaling factor.
- Functions that have more than one input variable have **partial derivatives.** A partial derivative is the rate at which the output changes with respect to one variable when we hold all the others constant.
- A function $z = F(x, y)$ of two variables also has a **microscope equation:**

$$\Delta z \approx F_x(a, b) \cdot \Delta x + F_y(a, b) \cdot \Delta y.$$

The partial derivatives are the **multipliers** in the microscope equation.

▷ DIFFERENTIAL EQUATIONS

- **Euler's method** is a procedure to approximate a function defined by a set of differential equations and initial conditions. The approximation is a piecewise linear function.
- The exact function defined by a set of differential equations and initial conditions can be expressed as a limit of a sequence of successive Euler approximations with smaller and smaller step sizes.
- The programs SIR, SIRVALUE, SIRPLOT, SEQUENCE, TABLE, and PLOT all compute Euler approximations.
- The function $y = Ce^{kt}$ is the solution to the initial value problem

$$\frac{dy}{dt} = ky \qquad y(0) = C.$$

- In some cases, the method of **separation of variables** can be used to find a *formula* for the solution of a differential equation.

➢ **DERIVATIVES OF BASIC FUNCTIONS**

Function	Derivative
$mx + b$	m
x^r	rx^{r-1}
$\sin x$	$\cos x$
$\cos x$	$-\sin x$
$\tan x$	$\sec^2 x$
$\sec x$	$(\tan x)(\sec x)$
$\csc x$	$-(\cot x)(\csc x)$
e^x	e^x
$\ln x$	$1/x$
b^x	$\ln b \cdot b^x$

➢ **DERIVATIVES OF COMBINATIONS OF FUNCTIONS**

Function	Derivative
$g(f(x))$	$g'(f(x)) \cdot f'(x)$
$f(x) + g(x)$	$f'(x) + g'(x)$
$f(x) - g(x)$	$f'(x) - g'(x)$
$cf(x)$	$cf'(x)$
$f(x) \cdot g(x)$	$f'(x) \cdot g(x) + f(x) \cdot g'(x)$
$\dfrac{f(x)}{g(x)}$	$\dfrac{g(x) \cdot f'(x) - f(x) \cdot g'(x)}{[g(x)]^2}$

➢ **DERIVATIVES OF INVERSES**

The derivative of a function and the derivative of its inverse are reciprocals. Thus, when $y = f(t)$ and $t = g(y)$ are inverses:

$$\frac{dt}{dy} = \frac{1}{dy/dt}.$$

Here are the derivatives of some frequently encountered inverse functions:

Function	Derivative
arctan(x)	$\dfrac{1}{1 + x^2}$
arcsin(x)	$\dfrac{1}{\sqrt{1 - x^2}}$
arccos(x)	$\dfrac{-1}{\sqrt{1 - x^2}}$

➤ THE INTEGRAL

- A **Riemann sum** for the function $f(x)$ on the interval $[a, b]$ is a sum of the form

$$f(x_1) \cdot \Delta x_1 + f(x_2) \cdot \Delta x_2 + \cdots + f(x_n) \cdot \Delta x_n,$$

where the interval $[a, b]$ has been subdivided into n subintervals whose lengths are $\Delta x_1, \Delta x_2, \ldots, \Delta x_n$, and each x_k is a sampling point in the kth subinterval (for each k from 1 to n).

- Riemann sums can be used to approximate a variety of quantities expressed as *products* where one factor varies.

- Riemann sums give more accurate approximations as the lengths Δx_1, $\Delta x_2, \ldots, \Delta x_n$ are made small.

- If all the Riemann sums for a function $f(x)$ on an interval $[a, b]$ get arbitrarily close to a *single number* when the lengths $\Delta x_1, \Delta x_2, \ldots, \Delta x_n$ are made small enough, then that number is called the **integral** of $f(x)$ on $[a, b]$, and it is denoted

$$\int_a^b f(x)\, dx.$$

- The *exact* value of a quantity approximated by a Riemann sum is given by the corresponding integral. If x and $f(x)$ have units, then $\int_a^b f(x)\, dx$ has units equal to the product of the units of $f(x)$ times the units of x.

- *The Fundamental Theorem of Calculus.* The solution $y = A(x)$ of the initial value problem

$$y' = f(x) \qquad y(a) = 0$$

is the **accumulation function**

$$A(X) = \int_a^X f(x)\,dx.$$

- If $F(x)$ is an antiderivative of $f(x)$, then

$$\int_a^b f(x)\,dx = F(b) - F(a).$$

- The integral $\int_a^b f(x)\,dx$ equals the **signed area** between the graph of $f(x)$ and the x-axis.
- A numerical method for estimating an integral is **efficient** if it gets accurate results with relatively few calculations. The **trapezoidal approximation** is the average of a *left* and a *right endpoint* Riemann sum and is more efficient than either. *Midpoint* Riemann sums are even more efficient than trapezoidal approximations.
- The most efficient method developed in this text is **Simpson's rule.** Simpson's rule approximates an integral by

$$\int_a^b f(x)\,dx \approx \frac{1}{6}\,(\text{left sum} + \text{right sum} + 4 \times \text{midpoint sum}).$$

- An **improper integral** is one that cannot be calculated directly. The problem may be that its "domain of integration" is infinite or that the integrand becomes infinite on the domain, even if the domain itself is finite. The value of the integral is obtained as a limit of ordinary integrals.

➢   TAYLOR POLYNOMIALS

The **Taylor polynomial of degree n centered at $x = a$** approximating the function $f(x)$ is given by the formula

$$P_n(x) = f(a) + f'(a)(x - a) + \frac{f''(a)}{2!}(x - a)^2 + \cdots + \frac{f^n(a)}{n!}(x - a)^n$$

$$= \sum_{k=0}^{n} \frac{f^{(k)}(a)}{k!}(x - a)^k.$$

In particular, the **Taylor polynomial of degree n centered at $x = 0$** approximating the function $f(x)$ is given by the formula

$$P_n(x) = \sum_{k=0}^{n} \frac{f^{(k)}(0)}{k!} x^k.$$

Some Taylor Polynomials

$f(x)$	$P_n(x)$
$\sin(x)$	$x - \dfrac{x^3}{3!} + \dfrac{x^5}{5!} - \dfrac{x^7}{7!} + \cdots \pm \dfrac{x^n}{n!} (n \text{ odd})$
$\cos(x)$	$1 - \dfrac{x^2}{2!} + \dfrac{x^4}{4!} - \dfrac{x^6}{6!} + \cdots \pm \dfrac{x^n}{n!} (n \text{ even})$
e^x	$1 + x + \dfrac{x^2}{2!} + \dfrac{x^3}{3!} + \dfrac{x^4}{4!} + \cdots + \dfrac{x^n}{n!}$
$\ln(1 - x)$	$-\left(x + \dfrac{x^2}{2} + \dfrac{x^3}{3} + \dfrac{x^4}{4} + \cdots + \dfrac{x^n}{n} \right)$
$\dfrac{1}{1 - x}$	$1 + x + x^2 + x^3 + \cdots + x^n$

➢   TAYLOR'S THEOREM

- We say that $\varphi(x)$ has **the same order of magnitude** as $q(x)$ as $x \to a$, and we write $\varphi(x) = O(q(x))$ as $x \to a$, if there is a constant C for which

$$\lim_{x \to a} \frac{\varphi(x)}{q(x)} = C.$$

- *Taylor's theorem.* If $f(x)$ has derivatives up to order n at $x = a$, then

$$f(x) = f(a) + \frac{f'(a)}{1!}(x - a) + \cdots + \frac{f^{(n)}(a)}{n!}(x - a)^n + R(x),$$

where $R(x) = O((x - a)^{n+1})$ as $x \to a$. The term $R(x)$ is called the **remainder**.

- Here is a formula for the remainder:

$$R(x) = \frac{f^{(n+1)}(c_x)}{(n+1)!}(x-a)^{n+1},$$

where c_x is some number between a and x. (It is regrettable that we cannot say exactly *how* c_x is related to x—only that it exists.)

➣   ANTIDERIVATIVES

$\int f(x)\,dx$ denotes an **antiderivative** of $f(x)$, that is, a function whose derivative is $f(x)$.

A note on antidifferentiation formulas.   Remember that you can always add a constant to any antiderivative. Moreover, if the integrand has a break in its domain, then on different parts of the domain you can add *different* constants to the antiderivative. (See section 11.1 and the example $\int dx/x$ shown below.) We have marked with an asterisk $*$ any integral whose integrand has a break in its domain. (In cases where parameters are involved, we have starred the integral if there is any parameter value for which the domain of the integrand has a break.)

Why the $*$?

Antiderivatives of Combinations of Functions

$$\int cf(x)\,dx = c\int f(x)\,dx \qquad (c \text{ any constant})$$

$$\int f(x) \pm g(x)\,dx = \int f(x)\,dx \pm \int g(x)\,dx$$

$$\int G(f(x))G'(x)\,dx = \int G(u)\,du \qquad (\text{for } u = f(x))$$

$$\int f(x)g'(x)\,dx = f(x)g(x) - \int g(x)f'(x)\,dx$$

Antiderivatives of Basic Functions

$$\int x^r\,dx = \frac{1}{r+1}x^{r+1} \qquad (r \text{ any constant} \geq 0)$$

$$*\int x^{-r}\,dx = \frac{1}{1-r}x^{-r+1} \qquad (r > 0 \text{ and not equal to } 1)$$

$$*\int \frac{1}{x}\,dx = \ln|x| \qquad \text{that is:} \qquad \int \frac{1}{x}\,dx = \begin{cases} \ln(-x) + C_1 & \text{if } x < 0 \\ \ln(x) + C_2 & \text{if } x > 0 \end{cases}$$

$$\int e^x\,dx = e^x$$

$$\int \sin x \, dx = -\cos x$$

$$\int \cos x \, dx = \sin x$$

$$* \int \sec^2 x \, dx = \tan x$$

$$* \int \tan x \, dx = -\ln|\cos x|$$

$$* \int \sec x \, dx = \ln|\tan x + \sec x|$$

$$* \int \cot x \, dx = \ln|\sin x|$$

$$* \int \csc x \, dx = \ln|\csc x - \cot x|$$

$$\int \frac{1}{1 + x^2} \, dx = \arctan x$$

$$\int \frac{dx}{\sqrt{1 - x^2}} = \arcsin x,$$

Other Antiderivatives

$$* \int \frac{dx}{ax + b} \, dx = \frac{1}{a} \ln|ax + b|$$

$$* \int (ax + b)^n \, dx = \frac{(ax + b)^{n+1}}{a(n + 1)}, \quad n \neq -1$$

$$* \int x(ax + b)^n \, dx = \frac{(ax + b)^{n+1}}{a^2} \left(\frac{ax + b}{n + 2} - \frac{b}{n + 1} \right), \quad n \neq -1, -2$$

$$* \int \frac{x}{ax + b} \, dx = \frac{1}{a^2} (ax - b \ln|ax + b|)$$

$$* \int \frac{x}{(ax + b)^2} \, dx = \frac{1}{a^2} \left(\ln|ax + b| + \frac{b}{ax + b} \right)$$

$$* \int \frac{dx}{(ax + b)(cx + d)} = \frac{1}{ad - bc} \ln \left| \frac{ax + b}{cx + d} \right|$$

$$* \int \frac{x}{(ax + b)(cx + d)} \, dx = \frac{1}{ad - bc} \left(-\frac{b}{a} \ln|ax + b| + \frac{d}{c} \ln|cx + d| \right)$$

$$* \int \frac{dx}{(ax + b)^2(cx + d)} = \frac{1}{ad - bc} \left(\frac{-1}{ax + b} - \frac{c}{ad - bc} \ln \left| \frac{ax + b}{cx + d} \right| \right)$$

$$* \int \frac{x}{(ax + b)^2(cx + d)} \, dx = \frac{1}{ad - bc} \left(\frac{b}{a(ax + b)} + \frac{d}{ad - bc} \ln \left| \frac{ax + b}{cx + d} \right| \right)$$

$$\int x \sqrt{ax + b} \, dx = \frac{-2}{15a^2} (2b - 3ax)(ax + b)^{3/2}$$

$$* \int \frac{\sqrt{ax + b}}{x} \, dx = 2 \sqrt{ax + b} + \sqrt{b} \ln \left| \frac{\sqrt{ax + b} - \sqrt{b}}{\sqrt{ax + b} + \sqrt{b}} \right|, \quad b > 0$$

$$* \int \frac{\sqrt{ax + b}}{x} \, dx = 2 \sqrt{ax + b} - 2 \sqrt{-b} \arctan \sqrt{\frac{ax + b}{-b}}, \quad b < 0$$

$$* \int \frac{dx}{x \sqrt{ax + b}} = \frac{1}{\sqrt{b}} \ln \left| \frac{\sqrt{ax + b} - \sqrt{b}}{\sqrt{ax + b} + \sqrt{b}} \right|, \quad b > 0$$

$$* \int \frac{dx}{x \sqrt{ax + b}} = \frac{2}{\sqrt{-b}} \arctan \sqrt{\frac{ax + b}{-b}}, \quad b < 0$$

$$* \int \frac{\sqrt{ax + b}}{x^2} \, dx = -\frac{\sqrt{ax + b}}{x} + \frac{a}{2} \int \frac{dx}{x \sqrt{ax + b}}$$

$$* \int \frac{dx}{x^n \sqrt{ax + b}} = -\frac{\sqrt{ax + b}}{(n - 1)bx^{n-1}} - \frac{(2n - 3)a}{(2n - 2)b} \int \frac{dx}{x^{n-1} \sqrt{ax + b}}$$

$$\int \frac{dx}{x^2 + a^2} = \frac{1}{a} \arctan \frac{x}{a}, \quad a \neq 0$$

$$* \int \frac{dx}{x^2 - a^2} = \frac{1}{2a} \ln \left| \frac{x - a}{x + a} \right|, \quad a \neq 0$$

$$* \int \sqrt{x^2 \pm a^2} \, dx = \tfrac{1}{2}(x \sqrt{x^2 \pm a^2} \pm a^2 \ln |x + \sqrt{x^2 \pm a^2}|)$$

$$\int \sqrt{a^2 - x^2} \, dx = \frac{1}{2} \left(x \sqrt{a^2 - x^2} + a^2 \arcsin \frac{x}{a} \right)$$

$$* \int \frac{dx}{\sqrt{x^2 \pm a^2}} x \, dx = \ln |x + \sqrt{x^2 \pm a^2}|$$

$$\int \frac{dx}{\sqrt{a^2 - x^2}} = \arcsin \frac{x}{a}, \quad a > 0$$

$$* \int \frac{\sqrt{a^2 \pm x^2}}{x} \, dx = \sqrt{a^2 \pm x^2} - a \ln \left| \frac{a + \sqrt{a^2 \pm x^2}}{x} \right|$$

$$* \int \frac{\sqrt{x^2 - a^2}}{x} \, dx = \sqrt{x^2 - a^2} - a \operatorname{arcsec} \frac{x}{a}$$

$$\int \sin^2 x \, dx = \tfrac{1}{2} (x - \sin x \cos x)$$

$* \displaystyle\int \tan^2 x \, dx = \tan x - x$

$\displaystyle\int \sin^3 x \, dx = -\tfrac{1}{3} \cos x \, (2 + \sin^2 x)$

$* \displaystyle\int \tan^3 x \, dx = \tfrac{1}{2} \tan^2 x + \ln|\cos x|$

$* \displaystyle\int \sec^3 x \, dx = \tfrac{1}{2} (\sec x \, \tan x + \ln|\sec x + \tan x|)$

$* \displaystyle\int \csc^3 x \, dx = -\tfrac{1}{2} (\csc x \, \cot x + \ln|\csc x - \cot x|)$

$\displaystyle\int \sin^n x \, dx = -\frac{1}{n} \sin^{n-1} x \, \cos x + \frac{n-1}{n} \int \sin^{n-2} x \, dx, \quad n > 0$

$* \displaystyle\int \sin^{-n} x \, dx = -\frac{1}{n-1} \sin^{-n+1} x \, \cos x + \frac{n-2}{n-1} \int \sin^{-n+2} x \, dx,$

$\quad n > 0, n \neq 1$

$* \displaystyle\int \cos^n x \, dx = \frac{1}{n} \cos^{n-1} x \, \sin x + \frac{n-1}{n} \int \cos^{n-2} x \, dx, \quad n > 0$

$\displaystyle\int \sin^2 x \, \cos^2 x \, dx = \tfrac{1}{8}(-2 \sin x \, \cos^3 x + \sin x \, \cos x + x)$

$* \displaystyle\int \sin^n x \, \cos^m x \, dx$

$$= \frac{1}{n+m} \sin^{n+1} x \, \cos^{m-1} x + \frac{m-1}{n+m} \int \sin^n x \, \cos^{m-2} x \, dx, \quad n \neq -m$$

$\displaystyle\int x \sin x \, dx = \sin x - x \cos x$

$\displaystyle\int x \cos x \, dx = \cos x + x \sin x$

$\displaystyle\int x^2 \sin x \, dx = 2x \sin x + (2 - x^2) \cos x$

$\displaystyle\int x^2 \cos x \, dx = 2x \cos x + (x^2 - 2) \sin x$

$* \displaystyle\int x^n \sin x \, dx = -x^n \cos x + nx^{n-1} \sin x - n(n-1) \int x^{n-2} \sin x \, dx$

$* \displaystyle\int x^n \cos x \, dx = x^n \sin x + nx^{n-1} \cos x - n(n-1) \int x^{n-2} \cos x \, dx$

$\displaystyle\int xe^x \, dx = e^x(x - 1)$

$$\int x^2 e^x \, dx = e^x(x^2 - 2x + 2)$$

$$*\int x^n e^x \, dx = x^n e^x - n \int x^{n-1} e^x \, dx$$

$$\int \sin^2 \alpha x \, dx = \frac{x}{2} - \frac{1}{4\alpha} \sin 2\alpha x$$

$$\int \cos^2 \alpha x \, dx = \frac{x}{2} + \frac{1}{4\alpha} \sin 2\alpha x$$

$$\int \sin \alpha x \cos \alpha x \, dx = -\frac{1}{4\alpha} \cos 2\alpha x$$

If $\alpha \neq \beta$, then:

$$\int \sin \alpha x \sin \beta x \, dx = \frac{1}{\beta^2 - \alpha^2} (\alpha \cos \alpha x \sin \beta x - \beta \sin \alpha x \cos \beta x)$$

$$\int \cos \alpha x \cos \beta x \, dx = \frac{1}{\beta^2 - \alpha^2} (\beta \cos \alpha x \sin \beta x - \alpha \sin \alpha x \cos \beta x)$$

$$\int \sin \alpha x \cos \beta x \, dx = \frac{1}{\beta^2 - \alpha^2} (\beta \sin \alpha x \sin \beta x + \alpha \cos \alpha x \cos \beta x)$$

$$\int \cos \alpha x \sin \beta x \, dx = \frac{1}{\beta^2 - \alpha^2} (-\alpha \sin \alpha x \sin \beta x - \beta \cos \alpha x \cos \beta x)$$

APPENDIX C

ANSWERS

In this appendix, we have provided partial answers for a few of the exercises, which are marked in the text with an asterisk. These answers are just intended to help you get your bearings—to let you check if you are on the right track. In most cases, your own solutions should be more detailed and provide fuller explanations.

Section 1.1

9. By 1990, the river would have become $141\frac{1}{3}$ miles shorter; therefore, it would be only $958\frac{2}{3}$ miles long.

12. $L = 991\frac{1}{12} - 1\frac{1}{3}t$ miles.

17. $S = 43493.2$, $I = 3706.8$, $R = 2800.0$.

19. **b)** The new threshold level is $200000/14 \approx 14286$.
    c) Quarantine does *not* eliminate the epidemic, because S is still above the new threshold level. In other words, I will increase, at least as long as S stays above the threshold.

20. **d)** The susceptible population is now decreasing at the rate of 30 persons per day. Thus $S' = -30$.
    e) Currently, $S = 15000$ persons.

21. **b)** When S drops below 500 the infection rate I' becomes negative, so the illness fades away.

26. There is a threshold at $S = 5000$.

27. **a)** R increases.
    b) R decreases.

28. a) For I not to change, S must be 5000.
   b) For R not to change, it must be true that $I = \frac{5}{42}R$.
   e) R must satisfy the equation $\frac{47}{42}R = 45000$; thus $R \approx 40213$.

Section 1.2

5. (c), (d), and (e) are false.

7. b) The natural domain of R consists of all real numbers except $-1/2$ and 2.
   c) t must satisfy either $-2 < t \le -1$ or $1 \le t < 2$.

9. a) The x-intercepts are at $\pm 1/\sqrt{2} \approx \pm.7071068$.
   b) $x = .7071067812. \ldots$

10. b) The maximum occurs at two places: at $x = 1.57 \ldots$ and also at $x = 7.85. \ldots$

12. b) $x = .392699. \ldots$

14. c) The three graphs look like straight lines with different positive slopes. For $f(x) = 2^x$, the slope is about .69.

20. c) $m = 9/5; \mu = 1/m$.

21. d) The two values for 1970 agree, but the actual value is 6 PPM larger than the calculated value in 1980.

22. b) The multiplier is $k = .00013$ inches per degree Fahrenheit.

26. a) If P is the size of the colony in grams, and P' is its growth rate in grams per hour, then $P' = .0247 P$ grams per hour.

Section 1.3

12. The output is:

1	1
2	3
3	6
4	10
5	15

Notice that each x value is the sum of the first k whole numbers.

17. a) To the nearest tenth, $S = 133.1$, $I = 10329.7$, $R = 39537.2$.
   b) I is largest when $t = 13$.

19. Our estimate for the largest number of new infections in one day is 4609.1; it occurs when $t = 8$. This is found by checking the rate of new infections, which is $-S'$. Be sure to print out the values of t and S' after line 10 (following the directions in exercise 18) in order to match the value of S' to the correct day.

21. I is growing most rapidly when $t = 7$ and declining most rapidly when $t = 18$. These really are the correct days; be sure you followed the directions in exercise 18 (and see the answer to exercise 19, above).

Section 1.4

3. The maximum yield is $75 \times 90^2/14$ pounds of apples, when there are 90 trees on the acre.

Section 2.1

4. $S_3(2.5) = 42295.9$

13. a) Here is part of what you should do:

Replace the line: `FOR k = 1 TO numberofsteps`

with the line: `DO WHILE I >= 1`

b) $t = 160.2$

Section 2.2

2. $y(37) = 817.98\ldots$ (With 2^{18} steps, the third decimal place has stabilized at $817.987.\ldots$)

11. b) The graph levels off as t increases; the right half becomes a straight line with slope 0.

Section 2.3

1. There are three roots, and the smallest is $-.7666\ldots$ to four decimal places accuracy.

6. $1.478\,756\,512.$

15. The estimate $2.236\,067\,977\,499\,79$ is obtained at the fourth step. Its square is $5.000\,000\,000\,000\,001.$

Section 3.1

3. The three velocity estimates are 41.83, 64.13, and 52.98.

Section 3.2

1. a) $f'(1) \approx -4.000$, using $(f(1.001) - f(.999))/.002$.

2. a) Here is a possible sequence of estimates:

Δx	1	.1	.01	.001
$\Delta x/\Delta y$	0	-3.96	-3.9996	-3.999996

Section 3.3

2. a) $f'(2) \approx -.250\,000\,00$.

3. c) Q_1 is much better than Q_2. The best Q_2 estimate is 3.012, when $h = 1/2^8 = .004$, but Q_1 gives 3.004 when $h = 1/2^4 = .063$, an h-value that is 16 times larger.

10. a) $\Delta y \approx -.25\,\Delta x$.

11. b) The true value is .4878048..., and the microscope estimate is .4875. The difference is about .0003.

12. d) There is a savings of 1.25 hours.
    f) The total travel times are not equal. The 40-mph microscope equation predicts the trip takes 8.75 hours at 45 mph, while the 50-mph microscope equation predicts it takes 8.8 hours. (The exact formula predicts 8.89 hours!)

13. c) $f(5.3) \approx 12.12$.

18. a) 2.8.   d) 2.375.

Section 3.4

2. e) Wasting 10 ft^2 means that $\Delta A = -10$ ft^2, so $\Delta V \approx \frac{5}{2}\Delta A = -25$ ft^3, a decrease in volume of 25 cubic feet.

4. c) $\sqrt{101} \approx 10.05$.

Section 3.5

6. b) $\dfrac{d}{dx}\left(\sqrt{3}\,\sqrt{x} + \dfrac{7}{x^5}\right) = \dfrac{d}{dx}\left(\sqrt{3}x^{1/2} + 7x^{-5}\right) = \dfrac{\sqrt{3}}{2\sqrt{x}} - \dfrac{35}{x^6}$.

8. b) $f(x) = \dfrac{5x^8}{8}$ plus any constant.

9. $y' = 1 - \dfrac{1}{2\sqrt{x}}$, so $y'(4) = 3/4$.

11. b) $x = \sqrt{2.5}$.

13. b) $\sin.3 \approx .3$, according to the microscope equation, and $\sin.3 = .2955202$, according to a calculator. Be sure you are entering x in radians! The error in the approximation is less than .005.

15. b) $\sqrt{3592} \approx 59.933333$ according to the microscope equation, while $\sqrt{3592} = 59.933296$ according to a calculator. The difference is less than .00005.

18. d) The ball rises to a height of 201.5625 feet.

20. b) The ball never gets farther than 3 units from the center of the track.

Section 3.6

1. b) $\dfrac{dy}{dx} = -7\cos(4 - 7x)$.

2. b) $4w\big/\sqrt{4w^2 + 1}$.

5. $f'(1) = 160$.

9. c) $f'(x) = 0$ only when $x = 0$.

10. b) $y' = -(6x - 5)(3x^2 - 5x + 7)^{-2}$, and $y' = 0$ when $x = 5/6$.

Section 3.7

1. j) $\dfrac{\partial(x\tan y)}{\partial x} = \tan y$, $\dfrac{\partial(x\tan y)}{\partial y} = x\sec^2 y$.

3. a) $V_T = \dfrac{R}{P}$, $V_P = -R\dfrac{T}{P^2}$.

6. a) $w = G(u, 2) = 2u/(3 + 2)$, so $\Delta w/\Delta u = 2/5$ always.

7. b) The uncertainty is about 1188 square feet in the total area of 103,734 square feet. The percentage error is just over 1%.

8. a) $f(4, 10) \approx 239$.

Section 4.1

1. c) 25.258 months.

4. c) Here are estimates for P in 1980 (using different step sizes Δt):

Δt	1	2^{-1}	2^{-2}	2^{-3}	2^{-4}	2^{-5}	2^{-6}
P	4.00	4.05	4.07	4.08	4.09	4.09	4.09

12. b) Here are estimates that show $Q(20) \approx 29.47\,°C$:

Δt	1	2^{-2}	2^{-4}	2^{-6}	2^{-8}	2^{-10}
Q	27.66	29.00	29.35	29.44	29.47	29.47

15. There are 112 susceptibles after 40 days. The largest number of people are infected when $t = 6.33$ days. At that time, there are 8000 susceptibles.

Section 4.2

5. a) $C = 286$.

9. a) The doubling time is $t = .69/k$.

14. b) $k \approx .3364$; you need to explain *why* this is the value.

Section 4.3

3. d) $3.0\,e^{2t}$.

4. a) $\dfrac{\partial(e^{xy})}{\partial x} = ye^{xy}$, $\quad \dfrac{\partial(e^{xy})}{\partial y} = xe^{xy}$.

8. c) The formula predicts that the population of Poland will be about 44.9 million people in 2005.

Section 4.4

1. e) 3. **f)** 8. **l)** 1/2.

4. a) $1/x$.

6. The intensity is cut in half when $s \approx 1.98$ feet.

22. a) The inverse is $f(x) = x^2$, and $f'(x) = 2x$. Since $g(100) = 10$, $g'(100) = 1/f'(10) = 1/20$.

Section 4.5

1. b) $F(t) = \frac{1}{7}t^7 - \frac{8}{6}t^6 + 22\pi^3 t$.

2. $G(5) = 3.3$.

8. a) The accumulated change Δy is 0 as t increases from 0 to 1; however, $\Delta y = 4$ as t increases from 1 to 2.

Section 5.1

1. b) $60x^{11}(\pi - \pi^2 x^4) - 4\pi^2 x^3(5x^{12} + 2)$.

f) $\dfrac{-4\pi^2 x^3(5x^{12} + 2) - 60x^{11}(\pi - \pi^2 x^4)}{(5x^{12} + 2)^2}$.

8. The current daily per capita energy consumption is 8×10^5 BTU/person. This is *falling* at the rate of 10^3 BTU/person per year.

22. a) $f''(x) = 9e^{3x-2}$.

Section 5.2

1. e) $\dfrac{\partial}{\partial x}\dfrac{x+y}{y+z} = \dfrac{1}{y+z}$, $\quad \dfrac{\partial}{\partial y}\dfrac{x+y}{y+z} = \dfrac{z-x}{(y+z)^2}$, $\quad \dfrac{\partial}{\partial z}\dfrac{x+y}{y+z} = \dfrac{-(x+y)}{(y+z)^2}$.

4. d) $\dfrac{\partial^2}{\partial x^2}\left(\dfrac{y}{x}\right) = \dfrac{2y}{x^3}$, $\quad \dfrac{\partial^2}{\partial x \partial y}\left(\dfrac{y}{x}\right) = \dfrac{-1}{x^2} = \dfrac{\partial^2}{\partial y \partial x}\left(\dfrac{y}{x}\right)$, $\quad \dfrac{\partial^2}{\partial y^2}\left(\dfrac{y}{x}\right) = 0$.

Section 5.3

7. a) $x = 0$. **b)** $x = 1, -2$.

9. $x = \sqrt{7}$.

Section 5.5

2. The root is 2.094 551 481 542 327.

6. b) $1/3.4567 = .289\,293\,249\,327\,118$.

Section 6.1

1. a) 25 staff-hours total

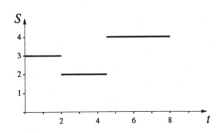

2. a) 6.9 hours.

3. In exercise 1, the average staffing is $25/8 = 3.125$ staff-hours per hour.

8. 60.29 megawatts is one estimate.

Section 6.2

1. 4.11 miles per hour.

3. About 183.5 grams of waste.

10. With 10000 equally spaced intervals, the left endpoint Riemann sum has the value .59485189.

16. b) $\sum_{k=1}^{5} k^2 = 55$.

Section 6.3

1. a) 39.

4. c) $\int_0^3 \frac{\cos x}{1 + x^2}\, dx = .6244\ldots$

9. b) error $< .00038$.   **d)** $\int_1^2 e^{-x^2}\, dx = .135\ldots$

13. b) The average value of $\sin x$ on $[0, 2\pi]$ is 0.

Section 6.4

4. a) $A'(X) = \cos(X)$.

5. There are five critical points in the interval $[0, 4]$. The first is a local maximum at $\sqrt{\pi/2}$.

10. a) $474\frac{2}{3}$.

11. b) $\dfrac{1 - \cos(\alpha\pi)}{\alpha}$.

12. a) $y = \frac{1}{3}x^3 - x^4$.

Section 7.2

1. b) $\omega = 1/5$.

7. a) $2/\pi$.

9. $\varphi = -\pi/2$.

12. b) $\pi/3$.

20. $F(\varphi) = \cos(\varphi)/2$.

24. The phase *difference* is $\varphi = 2\pi$. A phase *shift* of $\Delta x = 2\pi/3$ makes the two graphs coincide.

Section 7.3

2. b) The periods are the same, but the impulse solution has larger amplitude and is phase-shifted with respect to text solution.

7. b) $\frac{1}{2}p^2 + \frac{1}{2}b^2a^2$.

10. a) Solve the differential equations numerically. Then you should find that, in 10 cycles, $\Delta t \approx 13.7$ sec; thus the frequency is about $10/13.7 \approx .73$ cycles per second.

24. The threshhold impulse value is $p = \sqrt{2}$.

28. The period is 108.7 months (because the time unit in the original problem was 1 month).

Section 8.1

3. b) There are three equilibrium points, at $(0, 0)$, $(200, 0)$, and $(0, 200)$.

4. a) The three equilibrium points from question 3 reappear here, plus a fourth, at $(1000/7, 1000/35) \approx (142, 28)$.

    c) The species X grows to 200 while the species Y disappears. Nothing like this happens in exercise 3!

    d) X and Y coexist at the equilibrium $(1000/7, 1000/35)$, but this equilibrium is unstable. If x and y are changed even slightly from $1000/7$ and $1000/35$, they will gradually drift farther and farther away from these values. Eventually, either x or y becomes 0—one species disappears entirely.

    f) The equilibrium points at $(200, 0)$ and $(0, 200)$ are both stable; the other two are unstable.

8. c) The trajectory is a spiral, but it moves in very slowly.
   d) It will appear that, after an initial burst of infection, the disease has been effectively eliminated.

Section 8.2

2. b) To get the local linearization at $R = d/c$, $F = a/b$, introduce the variables $r = R - d/c$, $f = F - a/b$. Then the differential equations are

$$r' = -\frac{bd}{c}f \qquad f' = \frac{ac}{b}r.$$

Notice that the *nonlinear* parts of the differential equations have been removed. These are $-brf$ (in the equation for r') and crf (in the equation for f').

Section 8.3

4. a) $m = \dfrac{(d - a) \pm \sqrt{(a - d)^2 + 4bc}}{2b}.$

12. $a = -d.$

Section 8.5

2. b) The equilibrium points are $(x, y, z) = (0, 0, 0)$, $(1500, 250, 0)$, $(1000, 200, 4)$, $(1000, 0, 0)$, and $(0, 100, 0)$.
   c) For the case $(x, y, z) = (1000, 200, 4)$, let

$$x = u + 1000$$
$$y = v + 200$$
$$z = w + 4$$

and substitute these values into the differential equations to get the *localized* system

$$u' = -u - .001u^2 + 2v + .002uv - 100w - .1uw$$
$$v' = .2u - 2v + .001uv - .01v^2$$
$$w' = .004u + .001uw.$$

The *linearized* system derived from this is

$$u' = -u + 2v - 100w$$
$$v' = .2u - 2v$$
$$w' = .004u.$$

Section 9.1

1. a) Graph of $z = \sin x \sin y$. There are two minima, two maxima, and one saddle on the domain.

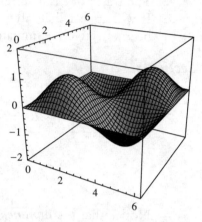

9. a)

b) The graph is a plane. Vertical slices in the y-direction are straight lines with slope .8, while those in the x-direction are straight lines with slope .3.

29. a) There is just one minimum; it is at the origin. Note: The range of values for v should be at least two or three times as large as the range for x. Otherwise the graph won't appear to have a minimum.

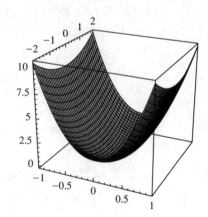

Section 9.2

7. a) $\Delta z = 122$.

13. $z = -4(x - 2) + 5(y + 9) = -4x + 5y + 53$.

15. a) $L_1(1, 0) = 15$.   **b)** $\Delta L_1 / \Delta x = 3$.

24. c) The difference is 10, independent of the formula you use.

26. a) y must increase by 125.

34. $\Delta z = 12\Delta x - 8\Delta y$.

40. a) $\dfrac{\partial Q}{\partial x}(27.6, 31.9) \approx -\dfrac{5}{6}$.

    b) $Q(27.6, 31.9) \approx 15.45$ and $Q(27, 31.9) \approx 15.95$.

44. b) At $(35.9, 25.564)$, H is about 2.796 units larger than it is at $(35, 26)$.

46. a) The error in the calculated volume will be $500\Delta x + 625\Delta y$ cubic inches.

47. A 1% error in the measurement of the base causes a 2% error in the volume, but a 1% error in the measurement of the height causes only a 1% error in the volume.

57. a) $z = 0$.

Section 9.3

4. The maximum is found at $(x, y) = (25/8, 15/8)$, as long as $a \geq 25/8$. When $a < 25/8$, the maximum is at $(x, y) = (a, 5 - a)$.

11. The maximum is $800/27$; it occurs where $(x, y) = (20/3, 2/3)$.

18. X should be at A. Does this surprise you?

20. Remarkable as it may seem, there is an entire line segment of solutions to this problem in the m, b plane. One end of the line is near $(m, b) = (.67, 2.3)$, the other is near $(m, b) = (1, 1)$.

Section 10.2

1. a) $\displaystyle\int \dfrac{\sin(x)}{x}\,dx \approx x - \dfrac{x^3}{3 \cdot 3!} + \dfrac{x^5}{5 \cdot 5!} - \dfrac{x^7}{7 \cdot 7!}$.

6. c) $\sin(3x) \approx -\dfrac{3(x - \pi)}{1!} + \dfrac{3^3(x - \pi)^3}{3!} - \dfrac{3^5(x - \pi)^5}{5!} - \dfrac{3^7(x - \pi)^7}{7!}$.

    This suggests that $\sin(3x) = -\sin(3(x - \pi))$ (do you see why?), and indeed, this is true.

12. $1 - \frac{1}{3}x^2 + \frac{2}{9}x^4 - \frac{14}{81}x^6$.

14. $\mathrm{erf}(.3) = .3286267\ldots$.

15. a) $\tan(x) \approx x + x^3/3 + 2x^5/15$.

23. a) 1.　　**b)** $1/2$.　　**d)** $1/6$.

29. c) As $x \to \infty$, $\dfrac{\sqrt{1 + x^2}}{x} = \sqrt{\dfrac{1 + x^2}{x^2}} = \sqrt{\dfrac{1}{x^2} + 1} \to 1$; hence

$\sqrt{1 + x^2} = O(x)$ as $x \to \infty$.

Section 10.3

2. a) The Taylor series converges for $-1 \le x < 1$.

3. a) $u - u^2/2 + u^3/3 - u^4/4 + u^5/5 + \cdots$.

5. a) You have to go out at least past the term of degree 4909; exactly how far may depend on the computer or calculator you use.

10. $3 + 4 + 4 + 4 + 4 + 1 = 20$.

11. The old way takes roughly 13 times as long.

Section 10.4

1. a) $y = a_0\left(1 + \dfrac{x^2}{1} + \dfrac{x^4}{2!} + \dfrac{x^6}{3!} + \cdots + \dfrac{x^{2n}}{n!} + \cdots\right)$.

5. a) $J_1(x) = \dfrac{x}{2} - \dfrac{1}{2!1!}\left(\dfrac{x}{2}\right)^3 + \dfrac{1}{3!2!}\left(\dfrac{x}{2}\right)^5 + \cdots + \dfrac{(-1)^n}{(n+1)!n!}\left(\dfrac{x}{2}\right)^{2n+1} + \cdots$.

Section 10.5

4. a) Converges.   **c)** Diverges.

5. a) We must have $\dfrac{n+1}{(n+1)^2+1} < .5 \times 10^{-6}$. This will happen, for example, if $n = 2 \times 10^6$. Finally,

$$\sum_{n=1}^{2\times10^6} (-1)^n \frac{n}{n^2+1} = -.269610\ldots$$

and this is the sum of the *infinite* series as well, to six decimal places accuracy.

8. c) $9.21044 < S_{10000} < 10.21035$.

d) It would take approximately 10^{87} cards—a number which is the same order of magnitude as the number of atoms in the universe!

10. c) $R = 1$.

11. a) $R = 1$.   **b)** $R = 2$.   **c)** $R = 1$.   **d)** $R = \infty$.   **e)** $R = 0$.

Section 10.6

1. a) $I_0 = 2$, and $I_1 = \pi$.

c) $I_2 = \pi^2 - 4$, $I_3 = \pi^3 - 6\pi$, and $I_4 = \pi^4 - 12\pi^2 + 48$.

d) $.00131 + .98260x + .05447x^2 - .23379x^3 + .03721x^4$, and the maximum difference occurs at the endpoints.

3. a) $I_0 = \arctan(2) = 1.10715$, and $I_1 = \frac{1}{2}\ln 5 = .804719$.

c) $I_2 = 2 - \arctan(2)$, $I_3 = 2 - \frac{1}{2}\ln 5$, $I_4 = \frac{2}{3} + \arctan(2)$.

Section 11.1

2. The inverse of $y = 1/x^3$ is $y = 1/\sqrt[3]{x}$.

7. $\displaystyle\int \frac{dx}{\sqrt{1-x^2}} = \arcsin x.$

10. $\displaystyle\int \left(7y + \frac{1}{y}\right) dy = \frac{7}{2}y^2 + \ln y.$

$\displaystyle\int (5\sin w - 2\cos w)\, dw = -5\cos w - 2\sin w.$

11. a) $F(x) = 7x + 12.$

16. b) $C = \ln 2.$

21. a) $F'(x) = x^2 + x^3.$

    d) $F'(x) = 2x\cos(x^2).$

22. b) $\displaystyle\int_0^5 \frac{t}{\sqrt{1+t^2}}\, dt = \sqrt{26} - 1.$

23. e) The *exact* value is $\sqrt{\pi/2}.$

24. $71\frac{1}{4}.$

27. c) The two areas are equal.

Section 11.2

1. c) $2e^{\sqrt{x}}.$ **g)** $-\ln|\cos x|.$

2. b) $\ln(e+1) - \ln(2).$ **d)** $-\frac{1}{2}\ln 5.$

4. b) $F(x) = \frac{1}{42}\left(1 + x^3\right)^{14} + 3\frac{41}{42}.$

Section 11.3

1. a) $\sin x - x\cos x.$

2. $\displaystyle\int (\ln x)^2\, dx = x\left[(\ln x)^2 - 2(\ln x) + 2\right].$

8. b) $1 - 5e^{-4}.$

21. b) $n\pi/2.$

Section 11.4

1. $y = Ce^t - 5.$

2. c) $y = \pm\sqrt{x^2 + C}.$

3. b) $Q(20) = 20 + 70e^{-.1\cdot 20} = 29.47\,°C.$

8. $Q = \left[A^{1-p} - (p-1)kt\right]^{1/(1-p)}.$ For supergrowth to occur, we must have $p > 1$, and then $\tau = A^{1-p}/k(p-1).$

9. **a)** $\alpha = \frac{1}{3}, \beta = -\frac{1}{3}.$

10. **a)** $\displaystyle\int \frac{3\,dx}{(x-1)(x+2)} = \ln\left|\frac{x-1}{x+2}\right|$

Section 11.5

1. **a)** $\frac{1}{2}\arcsin(2x).$

   h) $-\frac{1}{2}\ln\left|\dfrac{2 + \sqrt{4+x^2}}{x}\right|.$

4. **a)** $\ln\left|x - 1 + \sqrt{x^2 - 2x - 8}\right|.$

5. **d)** $\tan u = \dfrac{\sqrt{x^2 - a^2}}{a}.$

Section 11.6

3. **a)** When $n = 2^2$, the sum is $.\underline{746826}\ldots$, and the underlined digits have all stabilized.

Section 11.7

1. $\displaystyle\int_{-\infty}^{0} e^x\,dx = 1.$ $\displaystyle\int_0^1 \frac{dy}{y^2} = \infty.$

Section 12.1

5. The probability of exactly 50 heads in 100 tosses of a coin is

$$\frac{100!}{2^{100}(50!)^2} \sim \frac{\sqrt{2\pi} \cdot 100^{100} \cdot \sqrt{100} \cdot e^{-100}}{2^{100} \cdot 2\pi \cdot 50^{100} \cdot 50 \cdot e^{-100}}$$

$$= \frac{\sqrt{2\pi} \cdot 100^{100} \cdot 10 \cdot e^{-100}}{2\pi \cdot (2 \cdot 50)^{100} \cdot 50 \cdot e^{-100}} = \frac{1}{5\sqrt{2\pi}} \approx .0797885.$$

Incidentally, *Mathematica* gives the exact value:

$$\frac{12\,611\,418\,068\,195\,524\,166\,851\,562\,157}{158\,456\,325\,028\,528\,675\,187\,087\,900\,672} \approx 0.0795892.$$

The ratio

$$\frac{\text{Stirling's value}}{\text{Exact value}} \approx 1.0025$$

tells us Stirling's approximation is within $\frac{1}{4}\%$ of the exact value.

Section 12.2

1. The following table gives the probability of each outcome.

Outcome	H1	H2	H3	H4	H5	H6	TT	TH
Probability	1/12	1/12	1/12	1/12	1/12	1/12	1/4	1/4

7. Theoretical frequencies: 228, 212, 98, 30, 7, 1.

Section 12.3

13. The power spectrum has a single peak of height 1 at $\omega \approx 1$.

15. A new peak, of height 1, appears at $\omega \approx 7/3$.

20. The Riemann sum is $-.3(2\sin^2(2\pi/5) + 2\sin^2(\pi/5)) = -.75$.

21. -1. Since $h(x)$ is periodic with period $x = 1.5$, the interval $[0, 30]$ contains 20 periods of h. The integral of h over $[0, 30]$ is therefore 20 times its integral over $[0, 1.5]$.

Section 12.4

3. a) $2 \sum_{n=1}^{\infty} \dfrac{(-1)^{n-1} \sin nx}{n}$

Index